18 FE6425

Handbook of
Behavioral Neurobiology

Volume 13
Developmental Psychobiology

HANDBOOK OF BEHAVIORAL NEUROBIOLOGY

General Editor:
Norman T. Adler

University of Pennsylvania, Philadelphia, Pennsylvania

Handbook of
Behavioral Neurobiology

Volume 13
Developmental Psychobiology

Edited by

Elliott M. Blass

University of Massachusetts
Amherst, Massachusetts
and
Boston University School of Medicine
Boston, Massachusetts

KLUWER ACADEMIC / PLENUM PUBLISHERS
New York, Boston, Dordrecht, London, Moscow

Library of Congress Cataloging-in-Publication Data

Developmental psychobiology, developmental neurobiology, and behavioral ecology:
mechanisms and early principles/edited by Elliott M. Blass.
     p. cm. — (Handbook of behavioral neurobiology; v. 13)
  Includes bibliographical references and index.
  ISBN 0-306-46489-6
   1. Developmental neurobiology. 2. Developmental psychobiology. I. Blass, Elliott M.,
1940– II. Series.

QP363.5 .D495 2001
573.8—dc21

                  00-067113

ISBN 0-306-46489-6

©2001 Kluwer Academic / Plenum Publishers, New York
233 Spring Street, New York, N.Y. 10013

http://www.wkap.nl/

10 9 8 7 6 5 4 3 2 1

A C.I.P. record for this book is available from the Library of Congress

Printed in the United States of America

To Sarah, Edythe, Lizzie, Bridge and Joey
who have moved me in different ways at different times

Contributors

MARTIN BARTH, *Friedrich-Miescher-Laboratorium der Max-Planck Gesellschaft, Tübingen, Germany*

ELLIOTT M. BLASS, *Department of Psychology, University of Massachusetts, Amherst, Massachusetts, and Department of Pediatrics, Boston Medical Center, Boston University School of Medicine, Boston, Massachusetts*

MARK S. BLUMBERG, *Department of Psychology, University of Iowa, Iowa City, Iowa*

SUSAN A. BRUNELLI, *Developmental Psychobiology, New York State Psychiatric Institute, and College of Physicians and Surgeons, Columbia University, New York, New York*

GORDON M. BURGHARDT, *Departments of Psychology and Ecology & Evolutionary Biology, University of Tennessee, Knoxville, Tennessee*

JOHN C. FENTRESS, *Department of Psychology, Dalhousie University, Halifax, Nova Scotia, Canada and Department of Psychology, University of Oregon, Eugene, Oregon*

NANCY G. FORGER, *Department of Psychology and Center for Neuroendocrine Studies, University of Massachusetts, Amherst, Massachusetts*

SIMON GADBOIS, *Department of Psychology, Dalhousie University, Halifax, Nova Scotia, Canada*

HELEN GHIRADELLA, *Neurobiology Research Center and Department of Biological Sciences, The University at Albany, State University of New York, Albany, New York*

DAVID L. HILL, *Department of Psychology, Gilmer Hall, University of Virginia, Charlottesville, Virginia*

HELMUT V. B. HIRSCH, *Neurobiology Research Center and Department of Biological Sciences, The University at Albany, State University of New York, Albany, New York*

MYRON A. HOFER, *Developmental Psychobiology, New York State Psychiatric Institute, and College of Physicians and Surgeons, Columbia University, New York, New York*

WARREN G. HOLMES, *Psychology Department, University of Michigan, Ann Arbor, Michigan*

JERRY A. HOGAN, *Department of Psychology, University of Toronto, Toronto, Canada*

BRETT A. JOHNSON, *Department of Neurobiology and Behavior, School of Biological Sciences, University of California, Irvine, Irvine, California*

PRISCILLA KEHOE, *Department of Psychology, Trinity College, Hartford, Connecticut*

ANDREW P. KING, *Department of Psychology, Indiana University, Bloomington, Indiana*

CHRISTINE LAUAY, *Department of Psychology, Cornell University, Ithaca, New York*

MICHAEL LEON, *Department of Neurobiology and Behavior, School of Biological Sciences, University of California, Irvine, Irvine, California*

R. W. OPPENHEIM, *Department of Neurobiology and Anatomy, Wake Forest University School of Medicine, Winston-Salem, North Carolina*

WILLIAM SHOEMAKER, *Department of Psychiatry, University of Connecticut Health Center, Farmington, Connecticut*

SUZANNAH BLISS TIEMAN, *Neurobiology Research Center and Department of Biological Sciences, The University at Albany, State University of New York, Albany, New York*

TIMOTHY J. DEVOOGD, *Department of Psychology, Cornell University, Ithaca, New York*

ARON WELLER, *Departmental Psychobiology Laboratory, Department of Psychology, Bar Ilan University, Ramat-Gan, Israel*

MEREDITH J. WEST, *Department of Psychology, Indiana University, Bloomington, Indiana*

Contents

CHAPTER 5

The Developmental Context of Thermal Homeostasis

Mark S. Blumberg

CHAPTER 8

Play: Attributes and Neural Substrates

Gordon M. Burghardt

CHAPTER 9

Emerging Psychobiology of the Avian Song System

Timothy J. DeVoogd and Christine Lauay

CHAPTER 10

The Development of Action Sequences

John C. Fentress and Simon Gadbois

CHAPTER 11

Selective Breeding for an Infantile Phenotype (Isolation Calling):
    A Window on Developmental Forces

Susan A. Brunelli and Myron A. Hofer

CHAPTER 12

The Ontogeny of Motivation: Hedonic Preferences and Their Biolgoical
Basis in Developing Rats

Aron Weller

CHAPTER 13

Taste Development

David L. Hill

CHAPTER 14

Infant Stress, Neuroplasticity, and Behavior

Priscilla Kehoe and William Shoemaker

CHAPTER 15

Sciences Lies Its Way to the Truth ... Really

Meredith J. West and Andrew P. King

Introduction

ELLIOTT M. BLASS

Fifteen years have passed since the first volume on developmental psychobiology (Blass, 1986) appeared in this series and 13 since the publication of the second volume (Blass, 1988). These volumes documented the status of the broad domain of scientific inquiry called developmental psychobiology and were also written with an eye to the future. The future has been revolutionary in at least three ways.

First, there was the demise of a descriptive ethology as we had known it, to be replaced first by sociobiology and later by its more sophisticated versions based on quantitative predictions of social interactions that reflected relatedness and inclusive fitness. Second, there was the emergence of cognitive science, including cognitive development, as an enormously strong and interactive multidisciplinary effort. Making the "functional" brain more accessible made this revolution all the more relevant to our discipline. In the laboratory, immunocytochemical detection of immediate/early genes, such as *fos*, now allows us to trace neuronal circuits activated during complex behaviors. The "functional" brain of primates, especially humans, was also made very accessible through neuroimaging with which we can look at and into brains as they solve and attempt to solve particular tasks. Those of us who were trained in neurology as graduate students two or three decades ago recognize only the people in white coats and patients in beds or on gurneys when we visit neurological units today. The rest is essentially new.

Third, the explosion of neurobiology and molecular biology has had an impact on our scientific and occasionally our personal lives. We march full-steam ahead on the Human Genome Project, an enterprise that was hardly imaginable just a few years ago. Changes in the brain are starting to be understood on the basis of molecular alterations. Hope is expressed for uncovering a genetic basis of various forms of mental and physical illnesses and it is clear that future research and

ELLIOTT M. BLASS   Department of Psychology, University of Massachusetts, Amherst, Massachusetts 01003, and Department of Pediatrics, Boston Medical Center, Boston University School of Medicine, Boston, Massachusetts 02118.

Developmental Psychobiology, Volume 13 of *Handbook of Behavioral Neurobiology*, edited by Elliott Blass, Kluwer Academic / Plenum Publishers, New York, 2001.

treatment will incorporate a major molecular component. When reading and listening to these proclamations, one might even question if there is a place at the table for an organismic discipline such as developmental psychobiology.

Well, how have we fared during this epoch of revolution? Have we, as developmental psychobiologists, become more thoughtful and reflective? Have we chosen wisely from among the new paradigms and techniques to help advance some of the issues raised in previous volumes or have we become paradigm-bound and insular? Have we broadened our thinking to embrace new challenges? Or, in contrast, have we confined ourselves to older and dated paradigms that can no longer connect with the contemporary ideas and approaches that hold considerable promise toward unraveling some of the mysteries of the developing mind and brain?

The field has advanced both conceptually and empirically during the past epoch, and this is reflected in the chapters of this volume. We have incorporated the new waves into both our thinking and our actions. We have drawn from and have been enriched by the new biology. At the same time, we have maintained our balance. We have not lost to the siren of molecular biology our basic orientation of proximal analyses through the frameworks of behavioral ecology and organismic biology, despite regular pronouncements of genes that determine this or that behavior or pathological entity. Our interest in the epigenetic aspects of development has not been changed by the greater precision now available to the geneticists among us, although our understanding of these aspects has been exquisitely refined.

Our history of dealing with development in a principled manner has helped insulate us against simplistic notions of overdetermination. Thus, we have used the tools of contemporary neuroscience and molecular biology and appreciated their power without being overpowered by the accompanying hyperbole. With luck and resolve, we may be able to map particular behavioral events in terms of options made available by different genes and their interactions at different times.

As a group, developmental psychobiologists have not done as good a job in the domain of cognitive science. But that, too, will change. Many of us are becoming more steeped in this rich discipline and developmental psychologists are becoming more informed about the brain sciences and imaging opportunities, efforts that are being encouraged by the National Institute of Child Health and Human Development and the National Institute of Mental Health. Moreover, activity mapping techniques will soon allow us to help answer the sophisticated questions exploring the mind that have been posed in cognitive development so that regional activity in infant brains will become available "on-line" when infants engage in various cognitive tasks. The next generation of these devices is soon upon us. We should embrace this development, retrain ourselves (if need be), and certainly train our graduate students and postdoctoral fellows in both the neurology and physics necessary to help us understand both the capacities and limits of instrumentation that will help reveal the workings of the brain.

In this regard, we must seek collaborative efforts and develop research strategies with these experts. In my experience and that of other psychologists who have ventured into this tripartite marriage of cognitive, neurological, and physical scientists, the process and outcomes have been very successful. There is something in it for everyone and everyone has something to offer.

The current volume reflects the three mainstreams of our discipline of proximal and biographical influences and the connections among these disciplines: (1) neural mediation, (2) development and flexibility, and (3) functionally significant changes that occur in the habitats of many species. All of the chapters explicitly address at least two levels, and those that do not include the third have been influenced by that level nonetheless. A good start has been made on analyzing some underlying mechanisms and a number of chapters come to grips with the limits and the specifics of when and how restricted classes of experience can influence the developing nervous system and its behavioral output.

It is in the spirit of maintaining our balance while incorporating new procedures and techniques into our experimental armamentarium that we start by reprinting the chapter by Oppenheim and Havercamp from Volume 8 of the series, along with a contemporary introduction by Oppenheim. This chapter, written by a developmental neurobiologist, was chosen to lead the volume because Oppenheim's perspective, then and now, is that psychologists are struggling with problems that embryologists laid to rest many years ago. The message remains even more forceful today because of the much greater precision that can be brought to bear through molecular biology and genetics (witness the families of knockout mice that hold enormous potential as research tools for development). However, Oppenheim also cautions against misinterpreting what genomics and molecular biology can tell us about development. This area has been further advanced by identifying classes of experiences that contribute to neural growth and differentiation. Thus, neural activity is required for normal visual development to occur, but this activity is often intrinsic to the system and does not reflect an "experience" triggered by outside influences. We must attend to these distinctions, as Oppenheim correctly argues, and we must make smaller manipulations that are commensurate with the naturally occurring events that induce change. This is a time for subtlety.

I would add that because the nervous system appears to be organized, in large part, in a modular fashion and because natural selection chooses features, then we as developmental psychologists should look for the small matches of structure and selection that present opportunities for learning and change. For example, the coupling between activation prior to nursing in the presence of the mother's odor appears to define the mother to a variety of mammalian infants. Here the structure is both of the circumstances preceding suckling and of the preferential distribution of noradrenergic projections, in the case of rats into the olfactory bulb. The general learning theories of our predecessors are not very helpful here. It is the specificity of the match within a particular frame of reference that independently emerges and reemerges in this volume, and I, for one, welcome it (Blass, 1999).

Johnson and Leon's chapter on the effects of specific experiences on coding in the olfactory systems is revolutionary. It provides very strong evidence for specific regional coding at the glomerular level of olfactory processing. This is the first way station beyond the olfactory receptors themselves, which have been demonstrated through molecular techniques in the ground-breaking studies by Buck and others to code specifically along the olfactory receptor sheet. Johnson and Leon demonstrate that this specificity continues at the glomerular level and have now set the stage for looking more closely at the molecular effects that underlie these changes. At the same time, the next wave of studies can carry the coding issue into higher olfactory areas beyond the glomerular plane to determine storage and retrieval of this information and how the olfactory signals gain access over memories, motivational systems, and motor patterns. The history of this scientific enterprise exemplifies our field at its best. It started with the simple behavior of mother rats stimulating their infants prior to suckling by vigorous anogenital licking. This in turn caused a change in behavior; infants preferred the odors that had been in the surround during activation and in fact would only suckle nipples anointed with that odor. Then Leon and his colleagues brought the phenomenon under experimental control and with the aid of contemporary neurobiological and molecular techniques, launched the splendid analyses that they share in this progress report.

The next chapter, by Helmut Hirsch and his colleagues, on the development of vision in kittens and fruit flies was a revelation to me. It would only have occurred to a handful of people a short time ago that such an enterprise could be undertaken, let alone written up with the élan and power that these neurobiologists bring to the task. They demonstrate the homology between genes that determine visual structure between these orders and, remarkably, the parallel classes of events that the animals must experience in order for normal function to take place. In flies as in kittens, early visual experience affects structure and function. In addition, reproductive behavior was compromised through early, specific visual restrictions to which these flies were subjected. This chapter is a synthesis of a contemporary genetic analysis coupled with attention to an animal's life history and proximal changes that are born out of specific manipulations of those events. It expands the domain of biopsychological inquiry to invertebrates and provides a model (flies) in which both genetic and experiential factors can be manipulated and assessed with precision.

Nancy Forger has written an extraordinary chapter on sexual differentiation of the brain. This chapter is an important link between the molecular approaches of Hirsch and especially Oppenheim with the more ecologically oriented chapters of West and DeVoogd. They are of a mind in their perspectives on behavioral ecology providing the context for neurobiological analyses, but their emphases differ. The remarkable aspect of Forger's chapter is that she has written with an informed authority and lucid style on the history and characterization of sexual differentiation, on the putative evolutionary basis for this vertebrate feature, and on the neurobiological bases of gender differences. She goes beyond a simple listing of neural sex differences to examine what we know, and do not know, about the cellular and molecular mechanisms underlying sexual differentiation. Forger also raises several issues which collectively suggest that the dogma we learned as undergraduates may be on the cusp of overthrow. Genes may contribute directly to some brain sex differences, and within this framework, differential thresholds may determine levels of hormone necessary for short- and long-term changes to occur. When one factors in seasonal variations and effects of an organism's history on brain structure/function, the picture becomes yet more complex.

Forger does not back away from the hard issues. She raises the possibilities favored by DeVries, among others, that many brain sexual differences exist so that behavioral outcomes may be similar between genders, possibly even optimal. For example, elevated levels of neuropeptides in male rats may reduce differences in social memory otherwise caused by gender differences in the brain. It is an intriguing thought that has led me down some speculative functional paths. For example, can the elevated norepinephrine levels (see chapter by Johnson and Leon) in female rodents be understood within the context that their demands for social recognition may be greater than those of males (see chapter by Holmes)? There is an additional speculation on my part that arose from Forger's treatment of these complex issues; can we understand how the brain is put together from the perspective of an archeologist constructing a body from the finding of a single jaw, for example? The shape of the jaw leaves very few degrees of freedom for the type of skull to which it will articulate, and so on. So, perhaps the direct products of brain dimorphism were in the nuclei related to sexual function, and to the extent that these "stem" nuclei differed, changes had to occur in the neighboring nuclei and then again until so much of the brain—wherever the discipline has looked, in fact—has become different.

Mark Blumberg's is a most satisfying chapter. He chose to study the stability of temperature, the fundamental characteristic of all biological systems. As his point of departure he chose the salient of ontogenetic adaptations from the developmental psychobiology literature. In this instance, he focused on the different components of the infant rat thermoregulatory system, based on what was known, and speculated about the function of intrascapular brown adipose tissue (iBAT). He identified different means by which iBAT could be mobilized and within this framework made in-principle predictions about the consequences of different physiological and pharmacologic manipulations. Until his studies, the view was widely held that thermoregulation was not available to infants except through behavioral means.

The BAT system is strategically placed to warm the blood of the heart and the brain. It is under neural control and can be accessed through behavioral, ultrasonic vocalizations as well as through physiological alterations. The heat generated through this system does not reach the periphery, so that the standard measures of rectal temperature are misleading as indices of thermoregulatory capacity; a massive thermal gradient is found across a very short distance in the peritoneal cavity. BAT may be a unique thermoregulatory effector because it is not apparently derived from a process that originally evolved for some other purpose, cardiovascular circulation, for example. Remarkably, the different effectors have different thresholds for onset that reflect a calculus of ambient temperature, some correlate of pup temperature, other available effector systems, and rate of temperature change. In fact Blumberg demonstrates that huddling, which had been thought to be the only means of thermoregulation in rats, is dependent upon a functional physiological system for its efficacy in reducing heat loss. Rats with the effector systems discussed above are effective behavioral thermoregulators. Hamsters, which do not have iBAT deposits, are incompetent behavioral thermoregulators, even though they are motorically more competent than rats during the immediate postnatal days. None of this was known a decade ago and it is a great tribute to Blumberg and his colleagues that they have advanced this domain so far, so quickly, and so impressively.

The behavioral ecology section opens with Jerry Hogan's expansion of his influential chapter from the last edition. This is the most difficult of the chapters for me because it is really a book in itself, filled with information and insights that are revealed further with each successive reading. The oeuvre upon which Hogan bases his ideas is enormous, and accommodates a range from the ontogeny of feeding mechanisms in chickens to human speech development. Hogan's strokes are both broad and incredibly refined as he establishes the principles of developmental organization on the one hand and then tells us how it happens in the exquisite detail that is required both to evaluate his point of view and to understand the subtleties and the constraints through which different classes of experience may or may not cause behavioral change. Hogan emphasizes time and again how much of our behavioral systems are prefunctionaly organized (he provides precise definitions) and, by implication, how much of the central mechanisms underlying perception, motivation, and action are also constrained even before they are brought into play for their particular purpose for the first time. And yet, we are not locked into the final products. If we are to understand how the motor patterns come under control of the perceptual and central processing systems, then the details of causality must be examined because they are filled with subtle surprises. Thus, we learn that although the imprinting system in chicks is prefunctional, in order for a bird to follow its mother or a facsimile, it has to have had experience walking, even if in the dark, without visual information as to the consequences of that walk. As for infant ingestive systems, neither chicks nor suckling rats, at the beginning at least, are influenced by the nutritional or postingestive consequences of their actions. Thus, functional and causal analyses must be segregated conceptually, although they may overlap.

Hogan does not leave it there. He presents a wonderfully intelligent analysis at many levels of the concept of critical period, integrating the ideas of different authors, ranging from Waddington to Oppenheim, Lorenz, and Lehrman in discussing the process of timing and irreversibility. Out of this effort arises the idea that both prefunctional and experientially determined factors may at the end be working on a single system and possibly substrate. For those of us who earn a living in the brain, this poses an extraordinary challenge to understand the central mechanisms and their range of change and fixity during development. So in the final analysis Hogan is telling us about what we should be looking for in the brain and that we cannot possibly understand these mechanisms until we have a better understanding of the behaviors that they control, how they are packaged, and the changes that can and cannot occur at different points in time in different animals, including verbal ones. There is a lot of heavy lifting here; the chapter is difficult, but no more so than the questions that he addresses through his behavioral analyses.

I want to add a personal note of thanks to Jerry. When I was first becoming interested in developmental issues in 1973, I was blown away by his series of articles in the *Journal of Comparative and Physiological Psychology*. As a result, I called Jerry and invited myself to learn from him. He was a superb and responsible teacher who led a splendid developmental seminar for his young colleagues. They included Sara Shettleworth, Allison Flemming, Mertice Clark, David Sherry, and Derek van der Kooy. It was the best seminar that I had ever attended. I still refer to my notes.

Warren G. Holmes continues the behavioral ecology section with a superb example of how understanding the function of a behavior, nepotism in this case, in its natural ecology leads to in-principle analyses of its development. In a scholarly review, Holmes first takes us through the logic of nepotism, actions that are preferentially directed toward kin, anchoring the phenomenon quantitatively through Hamilton's notions of inclusive fitness. Holmes then documents nepotistic behavior in Belding's ground squirrel, *Spermophilus beldingi*, a species that has received considerable scrutiny, including influential studies by the author. Holmes continues his march toward understanding the underlying mechanisms. He identifies the circumstances underlying the formation of social preferences between littermates in the nest and provides a number of surprises that have to be understood proximately in order for the development of kin recognition and nepotism to be understood. Thus, even though *S. beldingi* young play preferentially with their littermates (nestmates) at the time of emergence from the nest into a common area inhabited by both littermates and unrelated juveniles, the preference is not sustained unless the mothers are available in the area. The crucial factor appears to be the fact that when mothers are present, they act as a "social magnet" for their offspring. When this occurs, the offspring follow the mothers; juveniles share their nest at night with their littermates rather than unrelated juveniles (although the mechanism has not been identified), and as a result, during the day, play more often occurs among littermates than among nonlittermates in the common area when above ground.

A large disruption occurs when *S. beldingi* are cross-fostered into different litters at 25 days of age, 2 days prior to natural nest emergence. The effects are astonishing. Frequency of play behavior drops markedly and the finely tuned discrimination in play that is seen between animals that are raised with littermates versus nonlittermates, and between littermates that have been isolated versus nonlittermates that were also isolated, disappears. It is reasonable to reflect whether these two disruptive phenomena of "last-minute introduction" and mother removal causing such low levels of play and the breakdown of kin preferences are two sides of the same coin of maternal actions within the nest during the eves of weaning. Holmes' chase will continue, and will, one hopes, evaluate interactions within a nest as emergence approaches.

Gordon Burghardt presents a chapter on the ontogeny of play from the perspective of one who has actively pursued this area both phylogeneticaly and ontogeneticaly for the past 30 or so years. He starts by providing a set of rigorous criteria for recognizing play and then takes us on a course that reviews various ideas as to the function of play. It is remarkable how the function of this familiar, accessible, and easily recognizable behavior has remained elusive, even though it has attracted a coterie of serious scientists. There have been a number of theories, of course, and Burghardt informs us of them. He comes down on the side of play having primary and secondary components, the latter helping refine skills and, especially, broaden the animals' social knowledge. This is certainly in keeping with the point of view expressed by Holmes. Burghardt then provides a series of testable hypotheses concerning the neural circuitry underlying play and its secondary processes. Although my own feeling about neural analyses is that they should await a better understanding of the behaviors in question, I agree with Burghardt's now championing this principled approach because it may yield important insights into the structure and function of play behavior, which have been elusive.

Tim DeVoogd has substantially advanced our understanding of bird song ontogeny and its underlying mechanisms through a judicious mixture of field and laboratory studies. His reviews have always revealed an appreciation of evolutionary, proximal, and neural mechanisms, and this broader, integrative perspective is also brought to the fore in the chapter in this volume with Christine Lauay.

A major focus of the chapter is an update of the functional neurology of the bird song circuit from the time of the chapter on this subject by Bottjer and Arnold (1986) to today. The progress has been impressive. The circuit has become far richer during the past 15 years. It is now tied to the rest of the brain, and functional connections have been established. This reflects the advances in behavioral studies that have shown the complex social nexus into which bird song is integrated. The decision to sing or to react to song depends upon a bird's own internal (hormonal) state, the time of year, the animal's recognition of the singer, and the singer's location, i.e., cognitive factors. This rich behavioral appreciation has allowed the anatomic and neurophysiologic studies to move forth. This includes studies in which early genetic markers were used to identify which aspects of the sensory and motor circuits are in play at a given time and in a given situation. The story starts to take on mammalian tones, as a number of laboratories have presented evidence for a sleep consolidation period of the bird's own song. The robust nucleus of the archistriatum is active in this process during sleep, a parallel to notions of mammalian hippocampal consolidation.

There are other interesting parallels. The song system appears to be primed even before song is heard. That is, there appears to be stimulus selectivity that arises from the isolated notes that the birds hear during the presong (parental) period. This is an interesting parallel to Pat Kuhl's important findings of language restrictions formed during the prebabbling period through the infant simply hearing the local tongue (see chapter by Hogan). DeVoogd and Lauay, like Holmes, tackle the major issue of what brings "early" learning (song, in this instance) or more generally, the critical period, to a close. They identify at least two factors, steroid levels and the fact that learning has already occurred. Whether these two factors act through a common mechanism is an open issue.

The ultimate question is, What is song's function? Is female choice on the basis of song going to enhance her inclusive fitness? Is song complexity, for example, actually a marker of a "better" feature that can be passed along to the next generation? There are small hints about differences in size of some song nuclei and a possible correlation with brain size and what that might entail. These are early times, however, and it is to the authors' credit that the question is put forward and that some possible directions for an answer are laid out. If the next decade is as productive as the last, the harvest will be incredibly bountiful, not only for the students of bird song, but for all of us interested in development and function.

The chapter by Fentress and Gadbois on the ontogeny of motor systems concludes the behavioral ecology section. It is the most thoughtful work that I have read on motor development. It seeks to unite a field that has often been fractious in debate as to the best way to approach the issue. Fentress and Gadbois provide an ineluctable argument for the necessity of both global and local approaches in the study of motor organization in general and its development in particular. They also strongly assert and bring to the fore the necessity of central nervous system analyses at all levels of study. It is hard for me to understand how this can be anything but so. The arguments for the integration of global and local approaches are compelling. Within a larger behavior pattern, say ingestion, there are postural adjustments and alternative strategies that can often be best appreciated through a dynamic approach. Selection within and among patterns, however, is a problem also faced on a more global level in reaching a particular endpoint or state. I was reminded of the influential studies of Alan Epstein on intragastric feeding by rats. These animals pressed a lever for food delivery that bypassed the oropharynx. They regulated their body weight beautifully, making adjustments to dietary dilutions and other challenges, although they were leaner than rats eating palatable diets.

Another major contribution of this chapter is its borrowing of the notion of prosody, a tool of linguistics, to help in understanding the development of other chains of behavior. The authors lay out a series of increasingly more complex sketches through which we might study and ultimately formally appreciate the characteristics of motor ontogeny, stability, and change. They then provide two very different behaviors that lend themselves to this form of analysis: caching by canids and grooming in mice. Attention to movement prosody within Fentress and Gadbois' paradigm made it possible for the authors to predict the length of a canid caching bout on the basis of its initial segments. What does this mean? Is there a catalog of sequences awaiting selection; does the initial segment drive the rest of the sequence; or is there feedback from the consequences of the first bout, as in the case of drinking in which the pattern becomes recalibrated, as W. G. Hall and I showed a number of years ago? Fentress and Gadbois discuss the implications of pattern length and its underlying motivational consequences for the ease and difficulty of pattern disruption, for example, at different points in both pattern and ontogenetic time.

The authors also discuss other aspects of development, for example, the prefunctional availability of complex patterns as discussed by Hogan (see also the chapter by Weller). It struck me, in reading Fentress and Gadbois' description of the complete, but misdirected "attacks" by very young wolf pups, of the similarity between these patterns and those described by Harlow in developmentally isolated adult rhesus who never got past this poorly directed stage in their sexual and maternal behaviors. A theme that we see here and in many of the other chapters is that particular experiences select the direction of available core behaviors (on a molar level) and, in parallel, that core processes are refined beyond their prefunctional status, although possibly through other experiences and mechanisms. Finally, the authors conclude with their approach to the developmental issue of motor expansion and restriction. The view is an interesting fusion of embryological and immunological approaches that awaits to be further developed and evaluated, but must be read in careful detail and evaluated empirically because it has a real chance of working.

The chapter by Brunelli and Hofer is the most heroic of this volume. They move into a new and complicated domain of behavioral genetics. After exploring a sometimes esoteric literature, they emerge ready to engage in the battle of determining whether an infantile trait can be selected and bred for. They also ask whether this trait can be used as a model for anxiety and its possible treatment. Brunelli and Hofer bring to this effort the strength and discipline of developmental psychobiology informing us about the larger possibilities of selecting for an ontogenetic adaptation, rather than a permanent adult trait, and identifying bases other than anxiety for differential calling rates (e.g., communication or thermoregulatory issues). They remind us, as others do throughout this volume (see especially the chapter by Burghardt), that the behaviors that we witness may be under differential proximal controls. They also demonstrate the complexity of the selective breeding enterprise. Rats bred for aggression when living individually revert to normal, indeed even low-aggressive profiles when either group-raised or raised by a "passive" mother, or after suffering a defeat in combat. From the beginning they emphasize the "considerable plasticity of neurological and functional organization during development." This is hard work.

Infant ultrasonic vocalization is an interesting choice for a model of anxiety. We know a lot about the behavior, its various controls, and its underlying neurology and neurochemistry. The behavior itself, especially when it occurs immediately at the time of maternal separation, has its own ontogenetic and proximal time courses and bases for modulation, and their laboratory has been a major player in this enterprise. Brunelli and Hofer then take us through the efforts of establishing and maintaining a breeding colony. They succeeded. Within one to three generations they produced high versus low responders to separation from the mother, and identified some of the characteristics that traveled with the phenomenon. They are now in the process of relating these differences to adult behaviors and provide us with an in-principle plan for evaluating adult differences, should they occur. I am reminded of the work of investigators who have selected a difference in naturally occurring early history phylogenetically and found substantial differences in adult behavior and neurochemistry. If Brunelli and Hofer meet equal success in their evaluations, they will have provided our discipline and biological psychiatry with a paradigm that will be extraordinarily powerful for teasing apart genetic–experiential determinants.

Aron Weller has written an illuminating chapter on the ontogeny of motivation. He first paints with very broad strokes by identifying core reward systems that appear to be characteristic of tactile, olfactory, and gustatory systems. He then discusses how these systems expand to recognize and prefer contact with objects and individuals in their environment that have caused either energy conservation or brain excitation (within limits). There are three noteworthy aspects. First, things change. Stimuli that had been rewarding and caused changes in behavior suddenly lose their efficacy. Each of the systems that Weller identifies appears to have its own epigenetic timetable that is not obviously linked to major changes that the infant is undergoing. There are many redundant paths toward forming preferences for mother or foods, for example. Thus, sweet tastes or fat flavors cause opioid release and induce a preference for the mother and for a particular diet. These preferences are reversible by the opioid antagonist naloxone. Fat also works through the cholecystokinin (CCK) pathway, which is not naloxone-reversible. Sweet tastes calm infant rats and humans and are analgesic. CCK only calms—it is not analgesic. In fact, later it appears to antagonize opioid analgesia systems. Contact calms and is analgesic, but uses completely different anatomic and neurochemical pathways. As Hill shows in the following chapter, this restriction of time and mechanisms holds for the development of sodium appetite. Holmes also sounds the restriction theme for the development of nepotism in Belding's ground squirrels.

Perhaps one way in which we may get better insights into these changes and restrictions is to explore the possibility that ontogenetic specializations (as suggested by Oppenheim and discussed by Hogan) determine the limits of aligning structure and functional changes during very brief ontogenetic epochs. This can be a dangerous gambit if not handled right, but it may make sense for us to look at restricted periods in greater depth than we have in the past. Thus, for example, we remain at a loss to explain some of the earlier findings concerning the circumstances under which regulatory systems are revealed at different ages. The restrictions for the younger rats are substantial. As Hall reveals, they must be deprived and excited in order to work for food. For older rats, as demonstrated by the Bruno and Stoloff studies, the mother must be present (even if anesthetized and largely covered up) in order for weanling-age rats to eat or drink or to respond to an acute drop in cellular water. These are important issues and Weller does not skirt them. In fact, he has started to take them head on by relating CCK to changes in vocalization in isolated rats, and how CCK treatment, diets that differently affect CCK release, and different CCK receptor blockers all affect the likelihood that the mother will retrieve her isolated pups. He has set the tone for the next generation of research in the ontogeny of motivation.

In a masterful review, David Hill focuses on one core property of the taste sense, the ontogeny of sodium taste. Hill places sodium gustatory development within a very broad organismic developmental framework. He then analyzes this system from the vantages of taste bud morphology and peripheral neural transmission of the chorda tympani nerve at various time points, as related to taste bud development. The animals of choice are rats and sheep, the animals most widely used in ontogenetic studies. The former has a slow postnatal period of orogustatory ontogeny; for sheep, most taste development is prenatal. Hill traces neural transmission as it moves through the first two central synapses in the n. tractus solitarius and the parabrachial nucleus. Having established for the reader the neurobiology of development, Hill proceeds to three experimental lines of manipulation that converge on experience dependence for the expression of the normal sodium gustation at cellular, neurophysiologic, and behavioral levels of structure and function.

The effects of early sodium restriction are unevenly distributed both in location and time. For rats that are in normal sodium balance but have not tasted fluids other than water (they are kept alive through intragastric infusions of liquid diet), there are specific impairments in chorda tympani transmission from tongue to brain. Yet, sodium information arising from palate and distal oropharynx is not compromised by this restriction. The same findings hold in rats made transiently agusic on the tongue through surgical intervention, but raised with a sodium-replete mother. Hill's studies on sodium restriction during gestation and the nursing period yield the same restricted set of findings. The gustatory system, as far as has been studied, is normal, except for the ability of the tongue to detect sodium taste.

Of considerable interest are the underlying mechanisms and the reversibility of the deficit. The proximal stimulus initiating the deficit has not been identified. Remarkably, it does not seem to be lowered levels of circulating sodium because placental sodium is protected (i.e., normal despite the mother's extreme hypo-natremia) and later, the mother sequesters sodium-adequate milk. Hill directs our attention to possible indirect factors including growth hormones, but it is not clear how these factors are triggered (or bypassed) in animals with this singular dietary history. Remarkably, the deficit can be reversed through a single ingestive bout of 30 ml of isotonic saline. The fluid must be absorbed, although the necessity of oral contributions within the absorptive context has not yet been established. These concerted approaches are a credit to the author and the field of chemical senses, one that has always been admirably respectful and integrative of different levels of analysis.

The chapter by Kehoe and Shoemaker is the final empirical chapter. Whereas the other empirical chapters look at particular systems (sexual differentiation, for example), their underlying mechanisms, and the effects of particular select experiences on the expressions of these systems, Kehoe and Shoemaker allow us to view a very different developmental enterprise. This one is concerned with the broader effects of what are apparently more general experiences, such as very brief separation from the mother. As editor, I must apologize for including only one chapter on this very important topic. But the chapter is superb in relating the influence of these early experiences and how far the field has progressed from its early days of extreme manipulations, such as separating infant rats from their mothers for almost 1 day at room temperature and even shaking them in a tomato juice can (what could the experimenters have been thinking?).

The contemporary appreciation of manipulations within the physiological range is widespread within the field and embraces studying separations from the mother for periods of time that occur under natural circumstances, to looking at the distribution of a mother–infant social trait throughout the population, e.g., grooming of pups by mother, and correlating this distribution with adult behaviors. In focusing on their own influential studies, Kehoe and Shoemaker reveal the enduring subtleties in both neurobiology and behavior that result from seemingly innocuous manipulations such as separating pups from their mothers for 1 hr daily (well within the normal range of mother absenteeism from the nest during the daily course of foraging for food and going to local watering spots). These separations are not sufficient to cause differences in body weight between experimental and control pups. Nor do they cause obvious behavioral changes. These occur when the animal undergoes low-grade behavioral or neurochemical challenges. Thus triggered, the gains in particular neurotransmitter and behavioral systems are substantial relative to handled controls and unhandled ones as well. In particular, dopamine levels are remarkably increased, as is hippocampal long-term potentiation and behavioral activity in a number of measures. A major challenge for future generations and for the editor of the next volume in this series will be to interface the particular systems that we have reported on in these volumes with the more general and profound effects reported here and by others in the field.

The study of early stress also has important clinical implications that has caught the eye of primatologists, behavioral neurochemists, psychiatrists, and social scientists with a biological bent. Individuals for whom chaos and stress have been defining features of their infancy and early childhood are prone to altered central nervous system function and to self-injurious behaviors including increased abuse of alcohol and other drugs. As one somewhat involved in bringing science to the community, it is interesting to witness the extreme asymmetry in press coverage and government actions between the rather modest effects of early drug history on later behavior versus the often disastrous effects of subtle and not so subtle deviations from stable and predictable experiences. It is always easier to pin the blame on "deviant mothers" than to correct deviant policies that only minimally alleviate the stress of poverty and overcrowding. Herein lies an opportunity for us to provide expert voices. The message of Kehoe and Shoemaker is clear on these points. In the spirit identified at the beginning of this introduction of collaborative investigations between psychobiologists and clinicians of various stripes, the domains laid out by Kehoe and Shoemaker very much deserve our attention and commitment.

I selected West and King for the final chapter because their message forces us to rethink our data and those in this volume. Their chapter is required reading especially at this time when many in the scientific community believe that the answer, through the genome, is near. Their metaphor that the "truth" seen in the early light soon changes to a mirage in later light is especially apt for today. The Information Sea is so vast that the early news becomes integrated into our thinking and the later news, without headlines, is buried in the back pages or does not surface at all.

West and King review their own scientific history and the various lights in which they have presented their own findings over the years. They justifiably lament that the scientific community's perception of their work has not really moved beyond their initial influential report. West and King take us beyond their initial findings, which were presented through a single lens that filtered out all lights except one, to show us that the female cowbird actually compiles a resume on the males in the neighborhood. In addition to their songs, the resumes also include behavioral considerations such as a male's attentiveness to females and his ability to defend himself and his territory against other males. What they have taught, and what has been shown repeatedly in this book, is the wonderful subtlety of behavior. The light is not so much different as it is filtered through a prism the refraction of which increases with our experimental analyses. This light is now able to reveal facets of behavior that had been previously hidden from us. So, singing is embedded within a visual context (DeVoogd and Lauay), ingestive behavior and body fluid control within the context provided by the anesthetized mother (Weller), and nepotism also demands the presence of the mother, although in a seemingly passive role (Holmes).

At the end, I think that this has been the main message of each of these chapters. Our field has grown in the appreciation of the subtleties and richness of the developmental material, and more is to come. This banner first raised by Oppenheim is carried forward in different ways by each of the authors and made very explicit and forceful by West and King. They all remind us of the texture of our enterprise and alert others that the path(s) from genome to behavior are at best torturous. I want to thank all of the authors both collectively and individually now that the process is over. My profound thanks to each of you for all of the time and good-spiritedness that you put into your chapters. I am enormously grateful to each of you for your seriousness, efforts, and good will.

And now, as before, I end on a personal note. This volume is dedicated to the four women who have contributed the most to my own development, each at different times and in different ways in my own journey, which does not seem to have critical periods so much as critical people. The women are my maternal grandmother, Sarah Horner; my mother, Edythe Horner Blass; my wife, Elizabeth Spelke; and our daughter, Bridget. This book is also dedicated to our youngest, our son, Joseph Alan Spelke Blass, named after his two grandfathers.

Sarah Horner lived with us for many years in a very small apartment in Brooklyn, where late at night she taught me Yiddish and I her English (or at least Brooklynese). The love and silent raucous laughter (we did not want to wake the sleeping household) of these lessons would have made the process remarkable even if not a word of those foreign tongues was ever learned.

My mother, Edythe Horner Blass, was the only sister to five brothers, raised two sons, had two grandsons, and only at the end had the special light of a granddaughter. Despite, or perhaps because of, this most unusual developmental history, hers was the clearest, most optimistic view of people that I have encountered and it remained so despite many hardships. She was a great fan of the Dodgers, and knew the statistics and the game itself. On that sour fall day in 1949 when Tommy Heinrich took the great Newcombe deep, her silence was most profound. It was my first metaphor of Sports and Life. She knew that it was more than a game. She felt a great injustice based on the conviction that this of all teams, the one that, with Robinson, had integrated the game and showed considerable individual and team commitment to something so profoundly important, should have won on moral grounds alone. Her joy was at least as great as ours was in 1955, although she did not go out in the streets.

As for Elizabeth Spelke, my wife, most of those reading this book know of her as a brilliant developmental cognitive scientist whose work has been enormously influential. This is all true, but it is the least of it. For me she has been both the fire and the light, a person whose zest for life and willingness to take it on and to savor it in all of its facets is unmatched. Generally, 47-year-olds (my age when we met) are pretty fixed in their ways and have finished expanding their horizons; they have not met Lizzie. It has been an exhilarating ride.

Bridget, our daughter, now 15 and going on 20, has been my guide to the new age. More than anyone I know, she is able to step back from the immediacy of a situation, see larger issues, and actually discuss them with us. That these lessons have been taught to me in Bridge's unique style and that we have been able to smile through most of them is a great tribute to her and has made them all the more worthwhile. I have learned much, although it does not always show, and am enormously grateful.

Finally, Joseph Alan Spelke Blass, the son of my later years. Those of you who have been blessed with another child when your peers, and possibly you too, are having grandchildren know of the joy of doing it all over again but at a time that you are already formed and have a sense of yourself rather than being formed and helping form the world about you. You realize that there is more of you to enjoy more of him (or her). And Joey, being an incredibly sweet person, with a droll teenage sense of humor accompanied, generally, by a gleam in his eye, has allowed that blessing to be savored all the more.

So to these remarkable people who have let me into their lives and helped shape my own development, even as I write, this book is dedicated—with love and gratitude.

REFERENCES

Blass, E. M. (Ed.). (1986). *Handbook of behavioral neurology*, Vol. 8: *Developmental psychobiology and developmental neurobiology.* New York: Plenum Press.

Blass, E. M. (1988). *Handbook of behavioral neurology*, Vol. 9: *Developmental psychobiology and behavioral ecology.* New York: Plenum Press.

Blass, E. M. (1999). The ontogeny of human face recognition: Orogustatory, visual and social influences. In P. Rochat (Ed.), *Early social cognition.* Hillsdale, NJ: Erlbaum.

Bottjer, S. W., & Arnold, A. P. (1986). The ontogeny of vocal learning in songbirds. In E. M. Blass (Ed.), *Handbook of behavioral neurology*, Vol. 8: *Developmental psychobiology and developmental neurobiology.* New York: Plenum Press.

1

Early Development of Behavior and the Nervous System, An Embryological Perspective

A Postscript from the End of the Millennium

R. W. OPPENHEIM

INTRODUCTION

A primary motivation for writing our original chapter 15 years ago (Oppenheim & Haverkamp, 1986) was to bring to the attention of developmental psychologists and psychobiologists a conceptual framework for studying neurobehavioral development that is derived principally from the field of embryology or developmental biology. It was our view that this perspective had been ignored and neglected in many conceptualizations of behavioral development. Although a casual perusal of textbooks and reviews in the areas of child psychology, developmental psychology, and developmental psychobiology that have since appeared indicates modest progress on this score, we are nonetheless discouraged that our efforts (Hall & Oppenheim, 1987) as well as that of others along these lines (Michel & Moore, 1995) have not had a greater impact on conceptualizations in those disciplines. For that reason, as well as because much of what we said in our previous review is as true now as it was then, I have agreed (at the suggestion of the editor) to republish the original chapter together with some brief thoughts on a few areas of major empirical progress in the field that have occurred since 1986. (See the original chapter beginning on page 23.)

At the time of publication of our chapter in 1986, the field of developmental

RONALD W. OPPENHEIM Department of Neurobiology and Anatomy, Wake Forest University School of Medicine, Winston-Salem, North Carolina 27157-1010.

Developmental Psychobiology, Volume 13 of *Handbook of Behavioral Neurobiology*, edited by Elliott Blass, Kluwer Academic / Plenum Publishers, New York, 2001.

biology was on the threshold of a major revolution unparalleled since the rise of experimental embryology earlier in the century (Horder, Witkowski, & Wylie, 1986). Since then, molecular biology has transformed the study of development such that many of the issues related to induction, determination, differentiation, and pattern formation have been answered or are within our grasp. Additionally, after a long hiatus, ontogeny and phylogeny (or development and evolution) are coalescing into a unified field in which all levels of analysis from cellular and molecular biology to ecology and ethology are making significant contributions (Raff, 1996). For anyone wishing to assess the progress made in all of these areas, the most recent edition of the standard textbook in the field, Gilbert's (2000) *Developmental Biology*, provides a readable, scholarly, and comprehensive analysis. For a recent review of progress in the specific field of developmental neurobiology, we recommend Cowan, Jessell, and Zipursky (1997), Zigmond, Bloom, Landis, Roberts, and Squire (1998), and Sanes, Reh, and Harris (2000).

As noted above, the recent use of molecular approaches for the study of developmental mechanisms has had a profound and lasting influence on the field of developmental biology. Too frequently, however, this has led to the mistaken impression that once we decipher the human (or any) genome and understand gene expression, development will be fully revealed. For instance, in a recent review on genomics (Lander & Weinberg, 2000), two prominent molecular biologists state that the "encapsulated instructions in the gametes are passed on to a fertilized egg and then they unfold spontaneously to give rise to offspring" (p. 1777) and these authors then go on to imply that it will soon be possible to "understand" an organism by knowing the DNA sequence of its genome.

If there is any single message to be learned from both our original chapter and the present comments—indeed, from this entire volume—it is that views such as those expressed by Lander and Weinberg are nonsense and have always been recognized as such by thoughtful developmental scientists (Gilbert, 2000). As three leading contemporary embryologists have recently put it, it is time "to move beyond the genomic analysis of protein and RNA components of the cell (which will soon become a thing of the past) and to turn to an investigation of the vitalistic properties of molecular, cellular and organismic function—it is obvious that there are many more possible outcomes than there are genes" (Kirschner, Geahart, & Mitchison, 2000, p. 87).*

NEURAL ACTIVITY AND NEUROBEHAVIORAL DEVELOPMENT

Although we began our 1986 chapter with a summary of some classic examples of how an embryologic framework can provide a fruitful means for understanding the role of intrinsic and extrinsic mechanisms in generating the nervous system and behavior, a major focus of the chapter was on the role of neural activity in its broadest sense (e.g., we included action potentials, synaptic transmission, sensory experience, and learning) as a key mechanism in controlling many aspects of neurobehavioral development. Since then, neural activity has become a prominent theme in developmental neurobiology and accordingly considerable progress has been

*The authors use the term vitalistic in the following way: "Although the units we consider, proteins, cells and embryos, are manifestly the products of genes, the mechanisms that promote their function are often far removed from sequence information. We might call research into this kind of light-hearted millenial cell and organismal physiology 'molecular vitalism' " (Kirschner *et al.*, p. 79).

made in understanding the cellular and molecular mechanisms involved. Ten years ago there was still lingering controversy over whether neural activity was a major feature in regulating neural development. Today the issue is not whether, but when, why, and how activity influences neural development.

However, we also pointed out in our review that not all aspects of neurobehavioral development are activity dependent, but rather, depending on the stage of development, the species, or the neuronal population involved, activity-independent, albeit epigenetic, mechanisms involving molecular cell–cell interactions are also crucial. Accordingly, it is now commonplace to consider the development of the nervous system and neuronal connectivity as being either *activity-dependent* or *activity-independent* (e.g., Crair, 1999; Goodman & Shatz, 1993; Katz & Shatz, 1996) and within the activity-dependent class, it is now possible to identify two subtypes, *spontaneous* versus *sensory-* or *experience-induced* activity-dependent processes. Some recent studies that confirm and extend our previous examples of activity-independent mechanisms include (1) evidence that the regulation of dendrite growth by afferent input in the developing amphibian Mauthner neuron is not mediated by neural activity (Goodman & Model, 1989), (2) activity also appears not to be required for the establishment of specific sensory–motor connections in the developing spinal cord (Frank & Jackson, 1986), (3) precise retinotopic maps and normal terminal arbors develop in the absence of activity in the visual system of zebrafish embryos (Stuermer, Rohrer, & Münz, 1990), and (4) normal precise motoneuron–muscle connections develop in zebrafish embryos in the absence of neuronal activity (Kimmel & Westerfield, 1990).

Furthermore, in accord with earlier data indicating an important role for spontaneous, non-experience-evoked neural activity in the regulation of neuronal survival (Pittman & Oppenheim, 1979), many other examples of the important role of spontaneous activity in survival as well as other aspects of neuronal development are also now available (Catsicas, Pequignot, & Clarke, 1992; Galli-Resta, Ensini, Fusco, Gravina, & Margheritti, 1993; Feller, Wellis, Stellwagen, Werblin, & Shatz, 1996; Katz & Shatz, 1996; Lippe, 1994; Meyer-Franke, Kaplan, Pfrieger, & Barnes, 1995; Oppenheim, 1991; Pequignot & Clarke, 1992; Primi & Clarke, 1997; Rubel *et al.*, 1990; Sanes & Takács, 1993). One aspect of neuronal development that has generally been considered exempt from activity-dependent signals is pathway formation and target selection by growing axons (Goodman & Shatz, 1993). These events were thought to be solely mediated by molecular cues along the pathway and at the target. However, even here important exceptions exist. In a recent report, it has been shown that blockade of endogenously generated action potentials in axons of lateral geniculate nucleus (LGN) neurons alters the targeting decisions of these projections in the cortex of fetal cats (Catalano & Shatz, 1998). In the absence of activity, significant numbers of LGN neurons send projections aberrantly to nonvisual cortical areas. In addition to endogenous or spontaneous neural activity, experience-dependent neural activity, including that associated with learning, has also been suggested to play a role in regulating such a fundamental event as the survival of developing and newly-generated adult neurons (Greenough, Cohen, & Juraska, 1999; Moore, Don, & Juraska, 1992; Najbauer & Leon, 1995; Truman & Schwartz, 1982).

Not surprisingly, the role of experience-dependent neuronal activity continues to be a primary focus of interest of developmental neurobiologists, and the mammalian visual system and neocortex remains a major arena in which such studies are being carried out. From the pioneering studies of Hubel and Wiesel in the 1960s to the molecular studies of today, visually driven neural activity has been the best-studied example of how experience can control and regulate neuronal development and connectivity (Katz & Shatz, 1996). Because this field has been the subject of many recent excellent reviews (e.g., Goodman & Shatz, 1993; Katz & Shatz, 1996), we do not wish to repeat that material here. However, one new and exciting development in this field with important implications for the whole issue of activity-dependent neuronal development and plasticity is the involvement of neurotrophic molecules as mediators of synaptic modifications.

Neurotrophic Molecules, Activity, Neural Development, and Plasticity

Since the discovery and characterization of the prototypical neurotrophic factor nerve growth factor (NGF) in the 1950s and 1960's (Hamburger, 1993; Levi-Montalcini, 1987; Oppenheim, 1996) many other neurotrophic molecules have been identified, including several additional members of the NGF family which collectively are known as neurotrophins (Lewin & Barde, 1996). Although several families of neurotrophic factors have been identified (e.g., ciliary neurotrophic factor (CNTF) the transforming growth factor beta, and insulin-like growth factor (IGF) family of proteins), the neurotrophins have been the most extensively studied agents in the context of activity-dependent neuronal development and plasticity. There are four major members of the neurotrophin family: NGF, brain-derived neurotrophic factor (BDNF), neurotrophin-3 (NT-3) and neurotrophin-4/5 (NT-4/5). Each of these factors acts on neurons via specific membrane-spanning tyrosine kinase receptors known as *trk* receptors (trkA, trkB, trkC). Until recently the classic biological actions of neurotrophic factors were thought to be restricted to neuronal growth and survival (Jacobson, 1991). Beginning in the 1990s, however, this classical perspective began to change as investigators realized that the originally defined action of these molecules on generalized nerve *growth* (hence the name nerve *growth* factor) could be extended to specific aspects of axonal and dendritic growth and the formation of synaptic connections (Purves, 1988). At about the same it time was postulated that activity-dependent competition between axons, dendrites, and synaptic connections may be mediated by neurotrophic factors (Purves, 1988; Snider and Lichtman, 1996). Accordingly, by the beginning of the 1990s the stage was set for exploring the interactions between neurotrophic factors and activity in development of the nervous system.

The first major step in this analysis was the demonstration that presynaptic excitatory neuronal activity can *increase* the synthesis and release of neurotrophic factors in postsynaptic neurons and that inhibitory synaptic input *decreases* neurotrophic factor synthesis and release (Thoenen, 1995). In this way neurotrophic factors provide a means by which presynaptic activity can regulate the development of postsynaptic cells as well as an avenue by which the activity-dependent release of neurotrophic factors (e.g., from dendrites or non-neuronal cells, such as skeletal muscle) can feedback on the presynaptic terminals that initiated the postsynaptic events in the first place. For example, the activation of muscle by motoneurons may result in the release by muscle and preferential uptake by axon terminals of a trophic agent that stabilizes or strengthens only those nerve terminals that initiated the muscle activation (Lichtman, Burden, Culican, & Wong, 1998). Accordingly, it was quite satisfying to investigators in the field when it was next discovered that neurotrophic factors could have both rapid (minutes) as well as longer lasting (hours, days) effects on the physiological, molecular, and structural aspect of synapses (Berninger & Poo, 1996; Cohen-Cory & Fraser, 1995; McAllister, Katz, & Lo, 1996, 1999; Prakash, Cohen-Cory, & Frostig, 1996; Schuman, 1999; Wang, Xie, & Lu, 1995; Wang, Berninger, & Poo, 1998). Recent studies indicate that neurotrophic factors can even directly depolarize neurons by binding to their own receptors (e.g., BDNF via trkB) on a time scale of milliseconds that is comparable to that of classic neurotransmitters (Kafitz, Rose, Thoenen, & Konnerth, 1999). This adds yet another level of regulation of synaptic events influenced by neurotrophic factors. Collectively, these data provided a means by which, in principle, activity-mediated changes in neurotrophic factor availability could regulate changes in the growth maintenance and regression of synaptic relationships among developing neurons, thereby providing a link between sensory input, neural plasticity, and behavioral development.

The next major step in the analysis of activity–neurotrophic interactions was the demonstration that these mechanisms actually do play a role in the normal development and plasticity of neurons, including neuronal connectivity. *In vivo* studies in the developing visual system indicate that neurotrophic factors mediate the classic activity-dependent segregation of visual system projections into ocular dominance columns in visual cortex of cat and rat during critical periods of development (Cabelli, Hohn, & Shatz, 1995; Maffei, Beradi, Domenici, Parisi, & Pizzorusso, 1992). One of the most exciting recent developments in this field is the demonstration that the activity-mediated synaptic changes that control learning-related events such as long-term potentiation in the hippocampus and neocortex in rodents also involve the action of neurotrophins (Akaneya, Tsuinoto, Kinoshita, & Hatanaka, 1997; Kang, Welcher, Shelton, & Schuman, 1997; Patterson *et al.*, 1996).

Although many important details regarding the specific role of neurotrophic factors in these events remain to be worked out, it is generally believed that neurotrophic molecules may act as coincidence detectors or retrograde signaling agents that mediate the physiological and structural changes occurring at synapses that act to strengthen or weaken them along the lines originally postulated by Hebb (1949)*. It is still too early, however, to know whether specific trophic factors regulate distinct activity-dependent events; whether different trophic factors are involved, for example, in developmental versus adult plasticity; and whether different animal species or types of neurons use particular trophic factors for controlling these activity-dependent processes.

In summary, neurotrophic factors, which were once only thought to regulate growth and survival of neurons, have now been implicated as one of the primary mediators of activity-dependent changes in the developing and adult nervous system. In this context, the statement made a decade ago regarding interactions between activity and trophic factors, that "the influence of activity on trophic effects

*It would be disingenuous of me if I failed to comment on the irony and apparent inconsistency in our criticism of Hebb in our previous review when contrasted with his now acknowledged pioneering role in conceptualizing the "Hebbian synapse," which is the foundation upon which all notions of activity-dependent neural plasticity are based (e.g., Stent, 1973; Lichtman & Purves, 1981; Fraser & Poo, 1982). In fact, we see no inconsistency in these two viewpoints. Despite his important insights on many matters, including his notion of a physiologic and anatomic basis for activity-dependent plasticity, we still argue that Hebb failed to appreciate the relevance of embryology to these issues. As we pointed out in our original chapter (Introduction), this failure on Hebb's part was not a reflection of a universal lack of appreciation of embryologic concepts by developmental psychologists, but rather appears to reflect a personal bias or lacuna in his own thinking.

may ultimately provide some insight into the way these two strategies interact" (Purves, 1998 p. 169), can be seen as prophetic and provides still another example of how embryology can inform our understanding of neurobehavioral development.

Conclusions

Despite significant progress in understanding basic neurobiological mechanisms involved in many aspects of behavioral development, the fields of developmental and child psychology have, in general, persisted in ignoring these successes in their pursuit of similar or related issues. Even when neurobiological mechanisms are invoked in discussions of behavioral development, they are often misinterpreted or erroneous (Bruer, 1999). This provides further evidence for the recent statements that "the behavioral sciences today are hardly more advanced than physics was in the seventeenth century, or biology was in the nineteenth" (Ramachandran & Smythies, 1997, p. 667) and that "the social and behavioral sciences have not enjoyed the dramatic theoretical and methodological advances that mark the last two decades in biology, chemistry, and astrophysics and, as a result, are not working well" (Kagen, 1998, p. 11). Ramachandran and Smythies argue that a major reason for psychology's lack of progress stems from a reliance on metaphorical explanations as substitutes for a real understanding of underlying neural mechanisms. This certainly seems to be true of developmental psychology. One means for attaining a more complete understanding of behavioral development is from integrated studies of behavior *and* the structure and function of the emerging nervous system in the embryo, neonate, and child. By stating this I do not mean to imply that all behavioral traits are easily (or will ever be) reducible to neurobiology, or that behavior cannot be fruitfully studied without reference to neural events. The fields of ethology and animal behavior are filled with elegant examples of the richness of purely behavioral analyses. However, I am convinced that a better understanding of the neurobiological basis of behavioral development will advance developmental psychology in the same way that knowledge about the structure of DNA has advanced our understanding of genetics, heredity, and embryology.

References

Akaneya, Y., Tsuinoto, T., Kinoshita, S., & Hatanaka, H. (1997). Brain-derived neurotrophic factor enhances long-term potentiation in rat visual cortex. *Journal of Neuroscience, 17,* 6707–6716.

Berninger, B., & Poo, M.-M. (1996). Fast actions of neurotrophic factors. *Current Opinion in Neurobiology, 6,* 324–330.

Bruer, J. T. (1999). *The myth of the first three years: A new understanding of early brain development and lifelong learning.* New York: Free Press.

Cabelli, R. J., Hohn, A., & Shatz, C. J. (1995). Inhibition of ocular dominance column formation by infusion of NT-4/5 or BDNF. *Science, 267,* 1662–1666.

Catalano, S. M., & Shatz, C. J. (1998) Activity-dependent cortical target selection by thalamic axons. *Science, 281,* 559–562.

Catsicas, M., Pequignot, Y., & Clarke, P. G. H. (1992). Rapid onset of neuronal death induced by blockade of either axoplasmic transport or action potentials in afferent fibers during brain development. *Journal of Neuroscience, 12,* 4642–4650.

Cohen-Cory, S., & Fraser, S. E. (1995). Effects of brain-derived neurotrophic factor on optic axon branching and remodeling *in vivo. Nature, 378,* 192–196.

Cowan, W. M., Jessell, T. M., & Zipursky, S. L. (1997). *Molecular and cellular approaches to neuronal development.* New York: Oxford University Press.

Crair, M. C. (1999). Neuronal activity during development: Permissive or instructive. *Current Opinion in Neurobiology, 9*, 88–93.

Feller, M. B., Wellis, D. P., Stellwagen, D., Werblin, F. S., & Shatz, C. J. (1996). Requirement for cholinergic synaptic transmission in the propagation of spontaneous retinal waves. *Science, 272*, 1182–1187.

Frank, E., & Jackson, P. C. (1986). Normal electrical activity is not required for the formation of specific sensory-motor synapses. *Brain Research, 378*, 147–151.

Fraser, S. E., & Poo, M.-M. (1982) Development, maintenance and modulation of patterned membrane topography. *Current Topics in Developmental Biology, 17*, 77–100.

Galli-Resta, L., Ensini, M., Fusco, E., Gravina, A., & Margheritti, B. (1993). Afferent spontaneous electrical activity promotes the survival of target cells in the developing retinotectal system of the rat. *Journal of Neuroscience, 13*, 243–250.

Gilbert, S. F. (2000). *Developmental Biology*, 6th ed. Sunderland, MA: Sinauer.

Goodman, L. A., & Model, P. G. (1989). Eliminating afferent impulse activity does not alter the dendritic branching of the amphibian Mauthner cell. *Journal of Neurobiology, 21*, 283–294.

Goodman, C. S., & Shatz, C. J. (1993). Developmental mechanisms that generate precise patterns of neuronal connectivity. *Cell, 72*, 77–98.

Greenough, W. T., Cohen, N. J., & Juraska, J. M. (1999). New neurons in old brains: Learning to survive. *Nature Neuroscience. 2*, 203–205.

Hall, W. G., & Oppenheim, R. W. (1987). Developmental psychobiology: Prenatal, perinatal and early postnatal aspects of behavioral development. *Annual Review of Psychology, 38*, 91–128.

Hamburger, V. (1993). The history of the discovery of the nerve growth factor. *Journal of Neurobiology, 24*, 893–897.

Hebb, D. O. (1949). *The organization of behavior*. New York: Wiley.

Horder, T. J., Witkowski, J. A., & Wylie, C. C. (1986). *A history of embryology*. Cambridge: Cambridge University Press.

Jacobson, M. (1991). *Developmental neurobiology*, 3rd ed. New York: Plenum Press.

Kafitz, K. W., Rose, C. R., Thoenen, H., & Konnerth, A. (1999). Neurotrophin-evoked rapid excitation through trkB receptors. *Nature, 401*, 918–926.

Kagen, J. (1998). *Three seductive ideas*. Cambridge, MA: Harvard University Press.

Kang, H., Welcher, R. A., Shelton, D., & Schuman, E. (1997). Neurotrophins and time: Different roles for trkβ signaling in hippocampal long-term potentiation. *Neuron, 19*, 653–664.

Katz, L. C., & Shatz, C. J. (1996). Synaptic activity and the construction of cortical circuits. *Science, 274*, 133–138.

Kimmel, C. K., & Westerfield, M. (1990). Primary neurons of the zebrafish. In G. Edelman & W. Gall (Eds.) *Signal and sense: Local and global order in perceptual maps* (pp. 175–196). New York: Wiley.

Kirschner, M., Geahart, J., & Mitchison, T. (2000). Molecular "vitalism." *Cell, 100*, 79–88.

Lander, E. S., & Weinberg, R. A. (2000). Genomics: Journey to the center of biology. *Science, 287*, 1777–1782.

Levi-Montalcini, R. (1987). The nerve growth factor: Thirty-five years later. *Science, 237*, 1154–1162.

Lewin, G. R., & Barde, Y.-A. (1996). Physiology of the neurotrophins. *Annual Review of Neuroscience, 19*, 289–317.

Lichtman, J. W., & Purves, D. (1981). Regulation of the number of axons that innervate target cells. In D. R. Garrod and J. D. Feldman (Eds.), *Development of the nervous system* (pp. 233–243). Cambridge: Cambridge University Press.

Lichtman, J. W., Burden, J. J., Culican, S. M., & Wong, R. O. L. (1998). Synapse formation and elimination. In M. J. Zigmond *et al.* (Eds.) *Fundamental neuroscience* (pp. 547–580). New York: Academic Press.

Lippe, W. R. (1994). Rhythmic spontaneous activity in the developing avian auditory system. *Journal of Neuroscience, 14*, 1486–1495.

Maffei, L., Beradi, N., Domenici, L., Parisi, V., & Pizzorusso, T. (1992). Nerve growth factor (NGF) prevents the shift in ocular dominance distribution of visual cortical neurons in monocularly deprived rats. *Journal of Neuroscience, 12*, 4651–4662.

McAllister, S. K., Katz, L. C., & Lo, D. C. (1996). Neurotrophin regulation of cortical dendritic growth requires activity. *Neuron, 17*, 1057–1064.

McAllister, S. K., Katz, L. C., & Lo, D. C. (1999). Neurotrophins and synaptic plasticity. *Annual Review of Neuroscience, 22*, 295–319.

Meyer-Franke, A., Kaplan, M. R., Pfrieger, F. W., & Barnes, B. S. (1995). Characterization of the signaling interaction that promote the survival and growth of developing retinal ganglion cells in culture. *Neuron, 15*, 805–819.

Michel, G. F., & Moore, C. L. (1995). *Developmental psychobiology: An interdisciplinary science*. Cambridge, MA: MIT Press.

Moore, C. L. Don, H., & Juraska, J. M. (1992). Maternal stimulation affects the number of motor neurons in a sexually dimorphic nucleus of the lumbar spinal cord. *Brain Research, 527*, 52–56.

Najbauer, J., & Leon, M. (1995). Olfactory experience modulates apoptosis in the developing olfactory bulb. *Brain Research, 674*, 245–251.

Oppenheim, R. W. (1991). Cell death during development of the nervous system. *Annual Review of Neuroscience, 14*, 453–502.

Oppenheim, R. W. (1996). The concept of uptake and retrograde transport of neurotrophic molecules during development: History and present status. *Neurochemical Research, 24*, 769–777

Oppenheim, R. W., & Haverkamp, L. J. (1986). Early development of behavior and the nervous system: An embryological perspective. In E. M. Blass (Ed.), *Handbook of behavioral neurobiology*, Vol. 8, *Developmental psychobiology and developmental neurobiology* (pp. 1–33). New York: Plenum Press.

Patterson, S. L., Abel, T., Deuel, T. A. S., Martin, K. C., Rose, J. C., & Kandel, E. R. (1996). Recombinant BDNF rescues deficits in basal synaptic transmission and hippocampal LTP in BDNF knockout mice. *Neuron, 16*, 1137–1145.

Pequignot, Y., & Clarke, P. G. H. (1992). Changes in lamination and neuronal survival in the isthmo-optic nucleus following the intraocular injection of TTX in chick embryos. *Journal of Comparative Neurology, 321*, 336–350.

Pittman, R., & Oppenheim, R. W. (1979). Cell death of motoneurons in the chick embryo spinal cord: IV. Evidence that a functional neuromuscular interaction is involved in the regulation of naturally occurring cell death and the stabilization of synapses. *Journal of Comprehensive Neurology, 187*, 425–466.

Prakash, N., Cohen-Cory, S., & Frostig, R. D. (1996). Rapid and opposite effects of BDNF and NGF on the functional organization of the adult cortex *in vivo*. *Nature, 381*, 702–706.

Primi, M. P., & Clarke, P. G. H. (1997). Pre-synaptic initiation by axon potentials of retrograde signals in developing neurons. *Journal of Neuroscience, 17*, 4253–4261.

Purves, D. (1988). *Body and brain, A trophic theory of neural connections*. Cambridge, MA: Harvard University Press.

Raff, R. A. (1996). *The shape of life: Genes, development and the evolution of animal form*. Chicago: University of Chicago Press.

Ramachandran, V. S., & Smythies, J. J. (1997). Shrinking minds and swollen heads. *Nature, 386*, 667–668

Sanes, D. H., & Takács, C. (1993). Activity-dependent refinement of inhibitory connections. *European Journal of Neuroscience, 5*, 570–574.

Sanes, D. H., Reh, T. A., & Harris, W. A. (2000). *Development of the nervous system*. New York: Academic Press.

Schuman, E. M. (1999). Neurotrophin regulation of synaptic transmission. *Current Opinion in Neurobiology, 9*, 105–109.

Snider, W. D., & Lichtman, J. W. (1996). Are neurotrophins synaptotrophins? *Molecular and Cellular Neurosciences, 7*, 433–442.

Stent, G. S. (1973). A physiological mechanism for Hebb's postulate of learning. *Proceedings of the National Academy of Sciences of the USA, 70*, 997–1001.

Stuermer, C. A. O., Rohrer, B., & Münz, H. (1990). Development of the retinotectal projection in zebrafish embryos under TTX-induced neural impulse blockade. *Journal of Neuroscience, 10*, 3615–3626.

Thoenen, H. (1995). Neurotrophins and neuronal plasticity. *Science, 270*, 593–598.

Truman, J. W., & Schwartz, L. M. (1982). Insect systems for the study of programmed neuronal death. *Neuroscience Comment, 1*, 66–72.

Wang, T., Xie, K., & Lu, B. (1995). Neurotrophins promote maturation of developing neuromuscular synapses. *Journal of Neuroscience, 15*, 4796–4805.

Wang, X.-H., Berninger, B., & Poo, M.-M. (1998). Localized synaptic action of NT-4. *Journal of Neuroscience, 18*, 4985–4992.

Zigmond, M. J., Bloom, F. E., Landis, S. C., Roberts, J. L., & Squire, L. R. (1998). *Fundamental neuroscience*. New York: Academic Press.

ADDENDUM
EARLY DEVELOPMENT OF BEHAVIOR AND THE
NERVOUS SYSTEM: AN EMBRYOLOGICAL PERSPECTIVE

R. W. OPPENHEIM AND L. HAVERKAMP

Dedicated to the memory of G. E. Coghill (1872–1941) whose writings on related issues have been an inspiration to our own thoughts on these matters.

I. INTRODUCTION

For much of the period between 1950 and 1970 the Canadian psychologist D. O. Hebb was a leading exponent of a developmental approach to neural and behavioral problems. Perhaps his best-known contribution in this regard was his now classic book *Organization of Behavior* (1949). In this work Hebb not only outlined a plausible and testable conceptual framework for understanding the role of sensory and perceptual experience in the development of the nervous system and behavior, but also provided empirical support for the model. By virtue of this, Hebb's book is considered to be an important milestone in the age-old debate over the role of nature versus nurture, heredity versus environment, or preformation versus epigenesis in the development of behavioral and psychological processes.

Although we have always been greatly impressed with Hebb's thoughtfulness on these issues, we have never understood why he failed to draw on the rich literature in embryology to strengthen his conceptual arguments. Hebb was not alone in this shortcoming, however, for most psychologists of his era either were unaware of or failed to appreciate the contributions that embryology could make to understanding principles of both neural and behavioral development. Yet there were important exceptions, Karl Lashley, for instance, was very much interested in the ways in which embryology might help elucidate neurobehavioral development (see, e.g., Lashley, 1938).* Other leading psychologists of this period with equally strong interests in embryology were L. Carmichael (1946), A. Gesell (1945), M. McGraw (1935), and J. Piaget (1969). Even if one attributes Hebb's failure on this score to the general lack of influence of embryology on psychological thought during this period, it does not explain why, in another book written many years later (Hebb, 1980), but which deals with similar conceptual issues, the same oversight is repeated.

In his more recent book a chapter devoted to the issue of heredity versus environment contains a tabular list of factors that Hebb claims cover all the various influences that control behavioral development (Table 1). However, only two of these categories deal at all with factors that might conceivably be studied by embryologists: number II, which includes "nutritive and toxic" influences, and number IV,

---

*The neuroembryologist Viktor Hamburger, who was at the University of Chicago with Lashley for several years in the 1930s, has related to me (R.W.O.) that he recalls having had many long conversations with Lashley about embryology, behavior, and neural development.

R. W. OPPENHEIM AND L. HAVERKAMP   Department of Anatomy, Wake Forest University, Bowman Gray School of Medicine, Winston-Salem, North Carolina 27103.   L. HAVERKAMP • Department of Neurology, Baylor College of Medicine, Houston, Texas 77004.

TABLE 1. CLASSES OF FACTORS IN BEHAVIORAL DEVELOPMENT[a]

No.	Class	Source, mode of action, etc.
I	Genetic	Physiological properties of the fertilized ovum.
II	Chemical, prenatal	Nutritive or toxic influence in the uterine environment
III	Chemical, postnatal	Nutritive or toxic influence; food, water, oxygen, drugs, etc.
IV	Sensory, constant	Pre- and postnatal experience normally inevitable for all members of the species
V	Sensory, variable	Experience that varies from one member of the species to another
VI	Traumatic	Physical events tending to destroy cells: an "abnormal" class of events to which an animal might conceivably never be exposed, unlike Factors I to V

[a]From Hebb, 1972, by permission.

which covers sensory experience. Although these are obviously important factors in neurobehavioral development, they do not begin to reflect the rich variety of normally occurring interactions that are the focus of most neuroembryological investigations. Indeed, to an embryologist, Hebb's categories ignore or exclude many of the most interesting and relevant phenomena involved in early development of the nervous system (and by implication in the earliest structural basis for behavior).

We have singled out Hebb in this context not because he is necessarily more vulnerable to criticism on this issue, but because it seems apparent that if someone as thoughtful and knowledgeable as Hebb could seemingly miss the relevance of embryology for helping to elucidate these issues, then there must certainly be many others in developmental psychology or psychobiology who also are not fully aware of the applicability of this field to their own work. We say this knowing, of course, that it is becoming increasingly fashionable to draw on certain embryological ideas for explicating neurobehavioral development. The writings of the late embryologist C. H. Waddington, for instance, have had some influence on a number of contemporary developmental psychobiologists (see, e.g., Scarr, 1976; Bateson, 1978), particularly his related concepts of the *epigenetic landscape* and *canalization*, which in principle at least can be easily utilized in conceptions of behavioral development.

Nonethelss, we are not convinced that such ideas have penetrated very deeply into the thinking of most behaviorists interested in ontogeny. Again, without wishing to draw undue attention to any particular individual, we note that in a recent chapter on "Genetics and Behavior," which is, in many ways, exemplary, the behavioral biologist G. Barlow has commented that, "most of us, with apologies to the embryologists, are concerned with the role of the environment that is *external to the animal.... experience received via sensory systems, since sensory competence develops late in the embryo's life ... we are mainly interested in experience after hatching or birth*" (Barlow, 1981, p. 237, italics added). This, of course, is basically the same position taken many years ago by the ethnologist K. Lorenz. In rebutting one of the points made by Lehrman (1953) in his classic critique of ethology, Lorenz argues that, "not being experimental embryologists but students of behavior, we begin our query not at the beginning of growth, but at the beginning of... innate mechanisms" (Lorenz, 1965, pp. 43–44). In addition to relegating embryology to a secondary position for the understanding of neurobehavioral development, such views also tend to assume that there are few, if any, significant contributions of experience (or of the external environment) to behavior prior to birth or hatching. In fact, there is now considerable evidence indicating that this is probably not a valid assumption (Gottlieb, 1976; Mistretta and Bradley, 1978; Oppenheim, 1982a). In apparent contrast to these views, we believe that it is impossible to formulate a valid theory of developmental psychobiology without including concepts, principles, and findings from neuroembryology and without considering the possible roles of experience, practice, or use during the embryonic and prenatal periods.

II. Some Embryological Findings

Although it seems entirely plausible to us to consider developmental psychology and psychobiology as branches of a single parent discipline, embryology (or developmental biology), we realize that this is not necessarily a widespread view. It is still somewhat surprising, however, that this is not the case. For it was experiments in embryology, starting even before the turn of the present century, that led to the establishment of the modern theory of epigenesis, the conceptual touchstone for all valid theories of ontogeny, whether behavioral, biological, or molecular.

In the late 1880s and early 1890s a major controversy was being debated that involved the proponents of two opposing, and seemingly irreconcilable, view of biological development: *neoepigenesis versus neopreformationism* (Oppenheim, 1982b). According to the supporters of neopreformationism, the full complement of hereditary material that was present in the zygote following conception was believed to be progressively reduced during development such that each cell type ended up with only a specific subset of genes. A nerve cell, for example, would retain only those genes necessary for carrying out neural functions (e.g., synaptic transmission) and those needed for other *general* metabolic functions common to all living cells. In other words, differentiation was thought to be mainly due to the actual loss of all specific genetic material not directly involved in the phenotypic expression of each cell type; thus each cell type in the mature animal was thought to retain only a small fraction of the genetic material (genes, chromosomes) originally present in the nucleus of the fertilized egg. In its most extreme form, neopreformationism held that by this means, differentiation occurred autonomously with little, if any, contribution from the environment.

By contrast, the advocates of neoepigenesis argued that all cells, at all stages of development, contained identical, complete complements of hereditary material. Thus differentiation was thought to be controlled not by the autonomous loss of specific genes or chromosomes from cells, but rather by a progressive, precisely timed sequence of intra- and extracellular events that selectively activated or suppressed the *expression* of particular genes. More recent experiments involving nuclear transplantation from differentiating or differentiated cells to unfertilized eggs, have shown conclusively that this viewpoint is substantially correct; even after differentiation is completed, sufficient genetic information exists in the nucleus of individual cells to support the development of a complete embryo (Gurdon, 1974) (Figure 1). Thus we now know that genes are suppressed or inhibited, and not actually lost, during ontogeny. In its most extreme form, however, the theory of neoepigenesis implied that all the genetic material in the fertilized egg was homogeneous and that the subsequent diversity of cells and functions was therefore due *entirely* to environmental influences.

In 1888 the pioneer embryologist W. Roux published the results of an experiment that appeared to refute the theory of neoepigenesis. Roux argued that if the neopreformation theory was correct, then when the fertilized egg divides for the first time, each cell (blastomere) should only contain the appropriate genetic material necessary to form one-half of the embryo. By contrast, if the neoepigenesis theory was true, then each blastomere should be capable of producing a complete, albeit quantitatively small, embryo. The outcome of Roux's experiment is illustrated in Figure 2. Since the destruction of one blastomere resulted in the development of a half embryo, the experiment seemed to settle unequivocally the dispute in favor of neopreformationism. A few years later, however, similar experiments, done independently by H. Driesch, T. H. Morgan, and H. Spemann, produced different results: a single blastomere was, in fact, shown to be capable of forming a complete, small

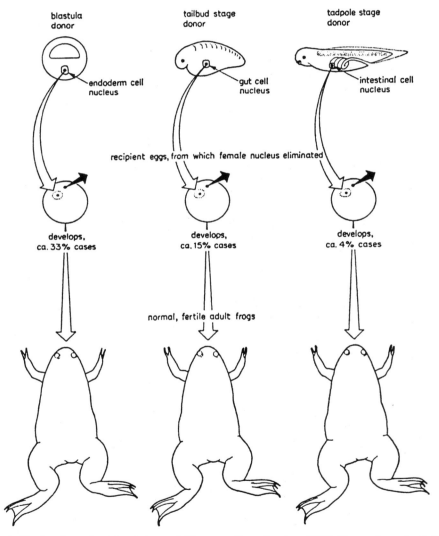

Figure 1. Nuclear transplantation into amphibian eggs (from Gurdon, 1974). The nucleus of cells from embryos and tadpoles at different stages of development when placed into egg cells support the development of normal adult frogs.

embryo (see Spemann, 1938). The discrepancy apparently resulted from an artifact of Roux's experimental procedures; by leaving the damaged or destroyed blastomere in contact with the intact cell, normal development of the intact blastomere was impaired. If one blastomere was completely isolated by a different procedure, then it was found that these cells formed a complete embryo (Figure 3).

Although these later experiments were interpreted as supporting neoepigenesis, additional experiments by Spemann and others showed that the ability of a single, isolated blastomere to form a complete embryo was restricted: after the four- to six-cell stage, only particle embryos were formed from single blastomeres. Collectively, such findings led to the formulation of the concepts of *determination* and *critical* or *sensitive periods*. Over the next 50 years, repeated demonstrations in a variety of tissues using diverse experimental designs showed that during development, cells gradually lose the potential to regulate or alter their fate when perturbed. In other words, the phenotypic expression of each cell becomes gradually restricted (determined) according to a rather rigidly controlled timetable (critical period). A particularly striking experiment (see Spemann, 1938) that clearly exemplified these phenomena is summarized in Figure 4.

Pieces of ectoderm were exchanged between two frog embryos that were in the beginning of the gastrula phase of early development. The dark tissue is derived from the future brain region of one frog species and the light piece from the future belly-skin region of another species. (In the absence of modern cell marking techniques, the embryologist ingeniously used the cells from two differently pigmented species as a naturally occurring marker for tracing the fate of the transplanted tissue). Following the transplantation of brain cells to the belly region or vice versa,

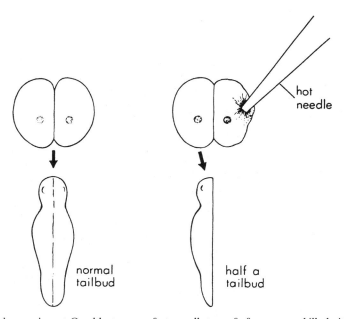

Figure 2. Roux's experiment. One blastomere of a two-cell stage of a frog egg was killed with a hot needle. The surviving cell gave rise to a half embryo. (From Hamburg, 1971). As explained in the text, this result was obtained because of an artifact of the technique employed to isolate the blastomere. See Figure 3 for a different result when the artifact is excluded.

the "foreign" tissue healed completely and continued to develop. The transplants adapted remarkably well to their new surroundings, and except for the different pigmentation, differentiated normally and fulfilled in every respect the demands of their new location. The transplanted skin cells became neurons and participated in the formation of the brain region to which they were transplanted. It would obviously be of considerable interest to know whether the transformed skin cells also *function* as nerve cells and participate in the mediation of behavior patterns appropriate to the brain region in which they are located. In any event, it can be concluded from this experiment that at the onset of the gastrula stage, brain and skin cells are not irrevocably determined. Rather, they are still able to alter their ultimate fate (phenotype) if provided with the appropriate environmental influences.

By contrast, if the very same experimental manipulations are done at the *end*, rather than the beginning, of the gastrula stage—a difference of only a few hours— the outcome is strikingly different. Although the tissues at this stage, when examined in histological sections, are virtually indistinguishable from those used in the early gastrula transplants, they respond quite differently to the same perturbation. Despite being placed in the same region of the future brain, the prospective skin now continues to develop into skin and persists in the brain as a disruptive foreign tissue. Similarly, the piece of transplanted brain tissue separates from the surrounding

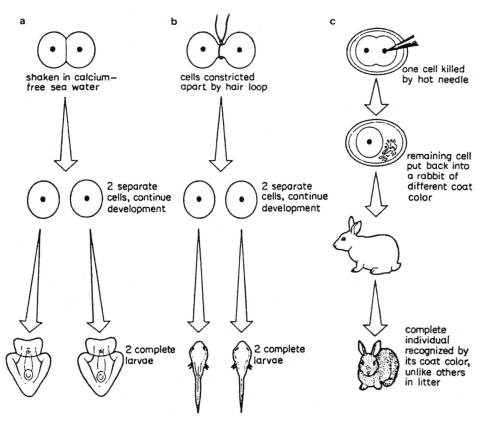

Figure 3. Examples of the adaptability of embryonic cells at the two-cell stage. Results of separating the first two blastomeras in (a) Echinoderms; (b) amphibians; (c) the rabbit. (From Deuchar, 1975.)

skin in the belly region and self-differentiates at this location into whatever part of the brain it would have formed in its original location; in the specific case show here, the prospective neural tissue develops into an eye and adjacent diencephalon.

Obviously, something decisive has happened to the ectoderm cells between the times of early versus late gastrulation. The *early* gastrula cells still have the potential to become either skin or brain (or perhaps a variety of other tissues, depending upon the site to which they are transplanted); that is, they have not yet been determined to form any specific cell type. By the *end* of gastrulation, however, those specific molecular and biochemical events that mediate the process of determination is these tissues (which are largely unknown) are completed. Even if the cells from a *late* gastrula are transplanted to a heterotopic position in an early gastrula embryo—where one might imagine conditions would be more favorable for transformation—they are still unable to alter their fate (e.g., "late" brain tissue in the belly region develops into brain). These experiments illustrate that whatever has transpired between the two stages involves changes inherent to the cells themselves and is not simply the result of a failure of the surrounding tissue environment of the

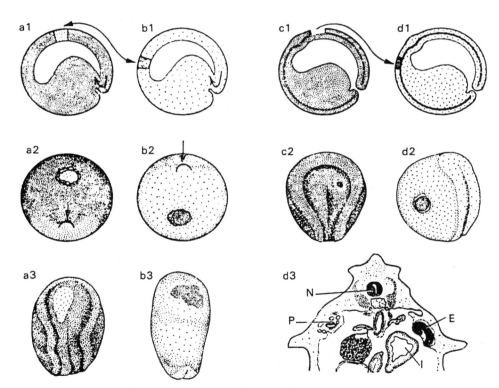

Figure 4. H. Spemann's classical exchange transplant experiments on newt embryos, performed either at the beginning (a and b) or after completion of gastrulation (c and d). (a1 and b1) Diagram of the operation whereby neural ectoderm from a1 (dark) is exchanged with skin ectoderm from b1 (light). The arrow within the dorsal blastopore lip indicates the direction of invagination. (a2 and b2) Location of the newly implanted pieces in relation to the blastopore (arrow). (a3) Implant is located in the brain part of the neural plate. (b3) The other implant is located in the ectoderm on the underside of the head. (c1 and c2) Donor of neural material (gap). (d1 and d2) Host neurula with implant taken from c in its flank. (d3) Further development of the larva (section). Implant develops into an eye (E). Host organs: (N) neural tube, (P) pronephros, (I) intestine. (From Hadorn, 1974.)

late gastrula to provide a signal. Once the cells have been determined, their subsequent course of development is virtually inevitable as well as irreversible. It is worth noting that experiments such as this provide an especially compelling illustration of the principle of sensitive or critical periods in development (see Erzurumlu and Killackey, 1982).

As already mentioned, the rationale for making transplants between differently pigmented species (heterospecific transplants) in these experiments was to provide a convenient cell marker. However, similar heterospecific transplants have often been used in embryology to reveal a number of important developmental principles. One example is show in Figure 5. This experiment capitalizes on the fact that salamander and frog tadpoles have different kinds of mouth parts that they use to stabilize themselves in the water; salamanders have balancers, whereas frogs have adhesive suckers.

It was known from previous experiments that when the oral ectoderm of a frog embryo is removed and then replaced by a piece of ectoderm from its own belly region, the belly ectoderm responds to signals from surrounding cells and develops into adhesive suckers rather than belly skin. The result is conceptually similar to the "brain-skin" transplants described earlier. What would happen, however, if the belly skin of a frog embryo were transplanted to the oral region of a salamander (newt) embryo or vice versa? The results of this experiment, which was first done by O. Schotte (see Spemann, 1938), are show in Figure 5.

As expected, the frog belly ectoderm when transplanted to the oral region of the newt embryo develops mouth parts rather than skin. Most interestingly, however, rather than developing mouth parts characteristic of the newt (i.e., balancers), the frog belly ectoderm now develops into typical frog suckers. The same thing occurs in the reciprocal newt → frog transplants (i.e., newt belly ectoderm placed in the oral region of a frog develops into balancers).

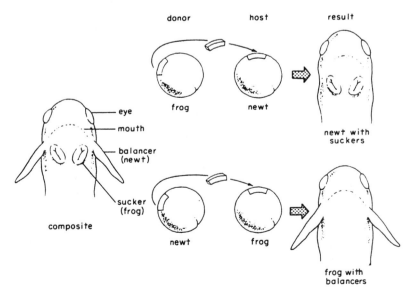

Figure 5. Schematic representation of Spemann and Schotte's experiment. Note prospective oral epithelium of a frog transplanted to a newt gave rise to oral suckers, typical of the mouth region of a frog. Conversely, oral epithelium from a newt transplanted to a host frog gave rise to balancers, typical mouth structures found in newts. (From Hamburg, 1971.)

The results of this rather remarkable experiment reveal an important developmental principle, namely, that the cells comprising a given piece of tissue (e.g., *frog* belly skin) can respond to signals from a heterospecific source (*newt* oral tissue) only within the limits of its own species-specific properties. Despite their ability to alter their fate when perturbed, the belly ectoderm cells can only express the specific properties programmed in their genome.

These kinds of experiments are entirely representative of a large literature in embryology that has repeatedly illustrated the inexorable interrelationship that exists between the genome and cytoplasm and between cells, tissues, and their internal and external environments. By the turn of the century, experiments such as these already had convinced most biologists that the extreme forms of both neoepigenesis and neopreformationism were invalid conceptualizations and that some sort of rapprochement was needed. E. B. Wilson, a leading biologist of the period, expressed what was rapidly to become a consensus on this issue. Writing in 1900, in the second edition of his popular book *The Cell in Development and Heredity,* Wilson stated:

> Every living organism at every stage of its existence reacts to its environment by physiological and morphological changes. The developing embryo is a moving equilibrium—a product of the response of the inherited organization to the external stimuli working upon it. If these stimuli be altered, development is altered. However, we cannot regard specific forms of development as directly caused by these external conditions, for the character of the response is determined not by the stimulus but by the inherited organization. [p. 428]

Writing at about the same time, the embryologist (and later animal behaviorist) C. O. Whitman (1894) expressed the same point even more succinctly: "Therefore, the intra and extra do not exclude each other but coexist and cooperate from the beginning to the end of development" (p. 222).

It is no coincidence that the views expressed here by both Wilson and Whitman have a remarkably modern ring to them. Despite their ignorance of most of the details of ontogeny, by the turn of the century embryologists were beginning to develop a conceptual framework (the modern theory of epigenesis) that has continued to guide the entire field right up to the present time. Although it does not take a great deal of imagination to recognize the significance of this framework for the study of neurobehavioral development, it has nonetheless taken a long time for most students of behavioral development to begin to appreciate this fact. For most of this century, behavioral scientists have often been involved in disputes over ideas and issues (e.g., nature versus nurture) that, conceptually at least, had been resolved by most biologists over three-quarters of a century ago. The reasons behind this constitute a fascinating episode in the history of science. However, since these matters have been recently discussed in detail by one of us (see Oppenheim, 1982b) and because it constitutes somewhat of an aside to the major focus of the present essay, we shall not attempt to repeat that material here.

III. Early Neural Development

The embryological studies reviewed previously help to reveal important principles of development. Although someone strictly concerned only with neurobehavioral development may find them interesting, they will probably also consider them rather remote from, or tangential to, behavioral problems. After all, it is still widely thought that those aspects of neural development most germane to behavior must involve interactions between the entire organism and the external (i.e., sensory) environment. While there is clearly some validity to this point, we would argue that the factors involved in behavioral development represent a continuum ranging from the kinds of cell and tissue interactions discussed previously all the way to the developmental role of complex social interactions. Consequently, what we wish to explore in this section are problems and issues that were motivated largely by embryological concerns but that also have important and, we believe, more clearly recognizable relevance to neurobehavioral development.

The first example in this section involves a population of transient embryonic cells known as the *neural crest*. The neural crest is composed of ectoderm cells that become segregated from the surrounding ectoderm during closure of the neural tube (Figure 6). Neural crest cells migrate throughout the head and trunk of the embryo and eventually give rise to a variety of cellular phenotypes, including peripheral neurons and glia, cartilage, endocine cells, and pigment cells. For the specific experiments we shall discuss, it is important to point out that in the trunk region, neural crest cells are the source of all the *adrenergic* neurons in the sympathetic paravertebral ganglia, as well as the source of all the *cholinergic* neurons in the parasympathetic ganglia. Andrenergic sympathetic neurons use norepineprine (NE) as a neurotransmitter, whereas the cholinergic parasympathetic neurons use acetylcholine (ACh).

On the basis of a large number of experiments too numerous to review here (see Le Douarin, 1980; Bunge *et al.*, 1978; Weston, 1982, for reviews), it now seems likely that most individual neural crest cells are not initially programmed to develop a specific phenotype. Rather, signals provided by other cells encountered during (or following) the migration of the neural crest are essential in determining, for in-

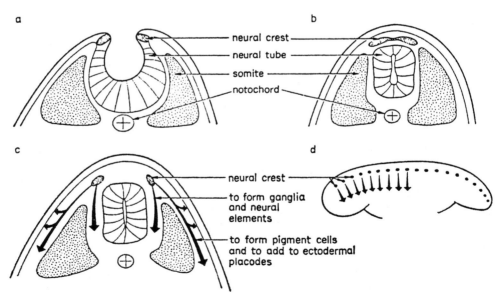

Figure 6. The origins and interactions of neural crest cells. (a–c) Diagrammatic transverse sections of the axial tissues of a vertebrate embryo, during and after the end of neurulation. Neural crest cells (shaded) become detached from the edges of the neural tube, then form dorsolateral clusters from which individual cells migrate in the directions shown by arrows in (c) and (d), to become pigment cells, nerve ganglion cells, or other cell types. In (d) a lateral view of the embryo is shown, with the neural crest clusters segmentally arranged. (From Deuchar, 1975).

stance, whether a crest cell will express adrenergic or cholinergic properties. As indicated in Figure 7, differentiated adrenergic and cholinergic ganglionic neurons can be reliably distinguished by cytological criteria, as well as by their biochemical, histochemical, and physiological properties.

Experiments by Paul Patterson and his colleagues (Walicke *et al.*, 1977; Patterson, 1978) have suggested that one of the extrinsic factors involved in the determination of phenotypic expression of neural crest cells is synaptic activity. They have shown that embryonic neural crest cells, placed in tissue culture and exposed to ionic changes that mimic electrical (synaptic) activity, develop primarily into adrenergic neurons, whereas the same cells when left untreated develop cholinergic

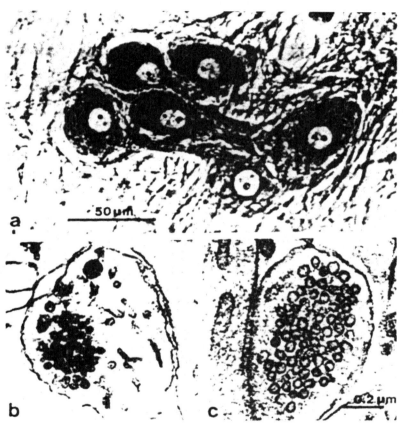

Figure 7. The "shift" from adrenergic to cholinergic function demonstrated in cultures of neonatal autonomic neurons. The pictures are of neurons from the superior cervical ganglion of newborn rats in dissociated cell cultures with supporting cells. (a) Light micrograph of a cluster of six neurons after 1 month in culture. The neurons are enmeshed in a network of neuronal processes, Schwann cells, and fibroblasts. (b) Synaptic vesicles after 1 week in culture [as in (a)]. This electron micrograph shows an axonal terminal on a neuronal soma after incubation in $10^{-5}\,M$ norepinephrine and fixation in potassium permanganate. As is characteristic of synaptic vesicle clusters in dissociated perinatal neurons after 1 week in culture, most vesicles contain dense cores, a cytochemical index of the norepinephrine content. Magnification as in (c). (c) Synaptic vesicles after 4 weeks in cultures. Preparation as in (b). Characteristically, the synaptic vesicles now show few dense cores. Correlating with this change in vesicle morphology, the cultures show increased levels of choline acetyltransferase and an increasing incidence of cholinergic synaptic interactions between the cultured neurons. (From Bunge *et al.*, 1978.)

properties. Although it is believed that the time period during which these same events occur *in vivo* coincides with the onset of innervation of the ganglion neurons by preganglionic cells in the spinal cord, it has not yet been shown that ganglionic blockade does, in fact, result in altered neural crest phenotypes. If this assumption is correct, however, then the results of the *in vitro* experiments imply that synaptic transmission (or its absence) may be important for regulating the initial expression of specific neurotransmitter properties of embryonic neural crest cells. A variety of other experiments have also implicated both pre- and postganglionic synaptic transmission in the regulation of other properties of adrenergic ganglion cells, albeit at somewhat later stages of development (Black, 1982).

Another example in which nerve activity plays an important role in development involves the relationship between motoneurons and the differentiation of different types of skeletal muscle. Most adult vertebrates have two basic types of skeletal muscle, fast-twitch and slow-tonic, which differ in their morphology, biochemistry, innervation pattern, and contraction properties (Goldspink, 1980). The fast-twitch or phasic muscles are those used primarily during short bursts of rapid movements, such as when a leopard is chasing prey, whereas the slow-tonic muscles are used primarily for the maintenance of posture, as occurs during amplexus of frogs or in the ability of birds to keep the wings from drooping when at rest.

There is considerable evidence that in both birds and mammals the nerves that innervate an embryonic muscle are involved in the developmental regulation of specific fast and slow muscle properties (Vrbová *et al.*, 1978). For instance, cross innervation of fast muscles with nerves that normally innervate slow muscles, when carried out early in development, alters subsequent differentiation, such that the prospective fast muscles take on many properties of slow muscles and vice versa. This influence appears to involve synaptic activity at the neuromuscular junction, and/or

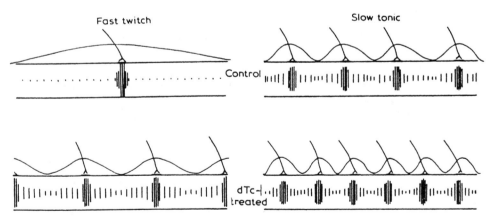

Figure 8. A schematic representation of the distribution of Ach sensitivity of skeletal muscle fibers in relation to their innervation is shown. The density and size of the vertical bars indicate the degree of chemosensitivity. The curve above each "muscle fiber" indicates the area along which the depolarization produced on nerve activity may spread and reduce chemosensitivty; on the fast twitch muscle fiber this spread would cover most of the muscle fiber surface, and on the slow tonic muscle fiber each nerve ending produces a depolarization that spreads only over a small distance. After curare, since the endplate potential is reduced, the area of spread of the depolarization is reduced, and chemosensitivity will decrease over a smaller area. The fast twitch muscle fibers will become multiply innervated and the distance between successive endplates on the slow tonic muscle will become reduced. (From Vrbová *et al.*, 1978.)

muscle contraction, since modifications of such activity, even in the absence of cross-innervation, result in similar changes in muscle properties (Figure 8).

Not only does neuromuscular activity regulate the differentiation of muscle, but recently it has been reported that the survival and maintenance of the motoneurons innervating skeletal muscle may also be regulated by such activity. During the embryonic development of vertebrates, many thousands of nerve cells begin to develop, only to die subsequently (Oppenheim, 1981; Hamburger and Oppenheim, 1982). This naturally occurring cell loss is especially clear in the case of chick spinal motoneurons, in which 40–60% of the cells in a given population die during the early stages of embryogenesis (Figure 9). The number of motoneurons that die can be altered by increasing or decreasing the size of the peripheral targets (skeletal muscle). Removal of a limb-bud early in development results in the death of virtually all the motoneurons in the limb-innervating population, whereas the addition of a supernumerary limb-bud rescues many of the motoneurons that normally would have died. These results have strongly implicated neuromuscular interactions in the regulation of motoneuron survival.

Recently, it has been shown that synaptic activity is also somehow involved in this phenomenon (Pittman and Oppenheim, 1978, 1979). Blocking neuromuscular activity with drugs such as curare during the entire period when motoneurons are normally dying results in the prevention of virtually all such cell loss (Figure 9). This result can be accomplished by either pre- or postsynaptic blockade of the neuromuscular junction. If the pharmacological blockade is stopped during embryonic development and neuromuscular activity is allowed to recover, the excess (rescued) motoneurons then degenerate. Furthermore, electrical stimulation of the nerves and limb muscles during the period of naturally occurring motoneuron death enhances or accelerates the normal cell loss (Oppenheim and Nunez, 1982). Al-

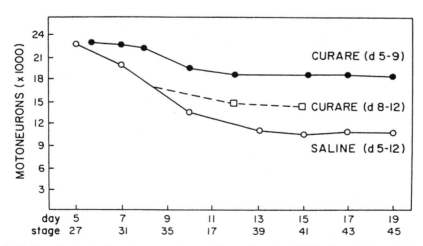

Figure 9. Cell number in the lateral motor column of the embryonic chick lumbar spinal cord. Two mg of curare in 0.2 ml saline were administered daily from either day 5 to day 9 or day 8 to day 12. Saline (0.2 ml) was used as a control. (Data from Pittman and Oppenheim, 1979; Oppenheim, unpublished data.). The E19 value for the curare d 5–9 group is retained for at least 4 days after hatching (Oppenheim, 1982c). (From Hamburger and Oppenheim, 1982.)

though it remains to be seen whether other kinds of neurons respond in the same way (but see Oppenheim *et al.*, 1982), collectively, these findings indicate that, at least for motoneurons, even such a basic aspect of neurogenesis as the control of final cell number in a population may involve neural activity.

The experiments cited previously illustrate some of the ways by which neural activity, use, or experience can influence the normal development of the nervous system during the early embryonic or prenatal period. We do not wish to give the impression, however, that we believe such factors to be involved in all stages or aspects of neurobehavioral development. White it is true that the roles of neural activity are receiving increasing attention in neuroembryology (Harris, 1981), there are also many interesting cases where such activity seems to play little, if any, role. Because we shall review some of this literature in detail, in the present context we only wish to mention the well-known studies of Sperry (1951), Weiss (1941), Detwiler (1936), Coghill (1929), and others in which embryological procedures were used to demonstrate the capacity of portions of the early nervous system to develop autonomously, independent of environmental influences, and despite the maladaptive nature of the resulting behavior.

An interesting example illustrating this latter point involves the morphological and functional development of spinal cord regions in the chick embryo. The histological and morphological appearance of the various levels of the adult spinal cord (cervical, thoracic, lumbar, etc.) exhibit characteristic differences. Limb-innervating regions, for instance, have large motor columns and lack autonomic preganglionic neuronal groups (Figure 10). By transplanting portions of prospective spinal cord early in development from limb to nonlimb regions (and vice versa), Wenger (1951) showed that by the time of neural tube closure, the regional-specific spinal cord differences were already determined. That is, even though the limb and nonlimb regions appear indistinguishable at the time of neural tube closure, they each exhibit characteristic developmental differences when placed in the same environment. For instance, the neural tube from the brachial region develops normally even when placed in the cervical or thoracic regions. In the reverse experiment, Wenger noted that wings innervated by transplanted cervical or thoracic cord showed spontaneous neuromuscular activity. However, his study did not attempt to analyze whether such movements were characteristic of, or different from, wings innervated by normal brachial spinal cord.

Two more recent experiments have, however, carried the functional analysis of these preparations a step further. Independently, Straznicky (1967) and Narayanan and Hamburger (1971) have shown that, in the chick, replacing brachial (wing-innervating) spinal cord with lumbo-sacral (leg) segments, or vice versa, leads to characteristic abnormalities in limb function. After hatching, wings innervated by lumbo-sacral cord behave like legs in that they perform alternating movements that are in synchrony with the alternating leg movements seen during normal walking. Similarly, legs innervated by brachial (wing) spinal cord do not exhibit alternating movements but rather move together in synchrony with the simultaneous movements (wings–flaps) of the wings.

Although there obviously are a great number of epigenetic events intervening between the time of neural tube closure and hatching, these findings demonstrate that the gradual structural and functional differentiation of these spinal cord segments are, nonetheless, irreversibly determined by 48 hr of incubation. While there appears to be little, if any, capacity for regulation of this basic structure–function pattern following translocation of gross spinal cord regions after this time, it remains to be seen whether more detailed analyses of the microanatomy, physiology, and biochemistry of the transplanted tissue will support the implications of the behavioral results. Even if this should prove to be the case, it will remain to be shown that such results are applicable to nervous tissue that subserves more complex behavioral processes (e.g., cerebral cortex) or whether the same phenomena can be demonstrated in mammals (but see Lund and Hauschka, 1976; Perlow, 1980; Roberts, 1983). Nonetheless, the results are encouraging and indicate that an embryological perspective on these issues may be a valuable addition to the range of experimental approaches used by developmental psychobiologists.

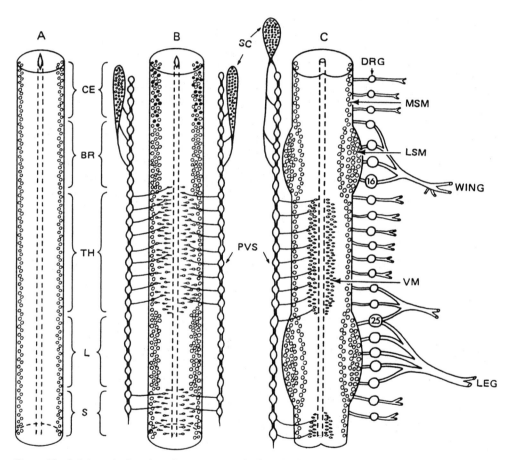

Figure 10. Origin and migration of motoneurons in the spinal cord of the chick embryo at 3 days (A), 4.5 days (B), and 8 days (C) of incubation. The motoneurons originate uniformly at all levels from the germinal zone surrounding the central canal. Visceromotor neurons (spindle-shaped) migrate toward the central canal to form the visceromotor column (of Terni) at thoracic levels. Degenerating motoneurons at cervical levels are shown in black. The somatic motoneurons form into a medial column that supplies axial muscles and into a lateral motor column that supplies limb muscles. In (C) the spinal nerves and ganglia have been omitted on one side and the sympathetic ganglia on the other. (BR, Brachial; CE, cervical; DRG, dorsal root ganglia; L, lumbar; LSM, lateral somatic motor column; MSM, medial somatic motor column; S, sacral; PVS, sympathetic ganglia; SC, superior cervical ganglion; TH, thoracic; VM, visceromotor.) (From Jacobson, 1978.)

IV. NEUROBEHAVIORAL DEVELOPMENT: THE ROLE OF NEURAL ACTIVITY

The examples presented in the previous section have demonstrated some of the means by which the environment external to the nervous system can influence neural development. They also indicate that functional activity *within* the nervous system may influence the ontogeny of that system. Growing neurons exhibit membrane depolarizations, propagated action potentials, synaptic transmitter release, and many other functional properties of mature neurons. One reflection of this activity within the developing nervous system is spontaneous embryonic motility—a phenomenon observed in every species in which it has been studied (Corner *et al.*, 1979; Bekoff, 1981; Hamburger, 1972; Oppenheim, 1974). In those systems that are amenable to manipulation, these embryonic movements have been shown to be, at least initially, entirely independent of sensory control (Hamburger *et al.*, 1966). It is not surprising, therefore, that this phylogenetically stable phenomenon has been conscripted frequently into the repertoire of interactions that guide normal development (see Jacob, 1977).

As noted in the previous section, however, functional activity within the developing nervous system is not involved in all aspects of neural and behavioral development. In the experiments discussed earlier in which a pharmacological blockade of neuromuscular transmission prevented naturally occurring motoneuron death, it was shown that recovery from the blocking agent during embryonic life leads to a gradual return of normal levels of motility (and belated degeneration of rescued cells). At least some of these embryos later hatch and appear to behave relatively normally after hatching (Oppenheim *et al.*, 1978). Indeed, it has been possible to bring a few animals through hatching (at 21 days of incubation) that have been exposed continually to curare since the normal time of cell death (5–10 days of incubation). For reasons not yet understood, the *posthatch* return of movement that occurs in these animals does not result in the delayed degeneration of excess motoneurons that had been rescued from naturally occurring cell death during embryonic life. Even in the presence of excessive numbers of motoneurons, however, observations during the first 3–4 days after hatching show these chicks to perform many normal-appearing behaviors vigorously (although detailed analyses could not be carried out because of the effects of immobilization on joint and muscle development; Oppenheim, 1982c). Thus the absence of motor activity during substantial portions of embryonic development does not appear to have any significant effects on at least some aspects of subsequent behavior.

These observations in the chick are reminiscent of the results of the classic deprivation experiments performed by Harrison (1904), Carmichael (1926, 1927), and Matthews and Detwiler (1927). In each of these studies, amphibian embryos were immersed in the immobilizing agent chloretone prior to the onset of embryonic movements. When they were removed from the drug solution to pond water a number of days later, the swimming of the experimental embryos soon appeared equivalent to that of normally reared animals. The conclusion thus drawn from these frequently cited works was that function or practice was not necessary for the normal development of amphibian swimming behavior. These classic studies have been influential in the formulation of a basic tenet of development neuroembryology (i.e., that during the *earliest* stages of neurobehavioral ontogeny, the nervous system develops in "forward reference" to, and without benefit from, functional activity). For a number of reasons, however (most notably the purely observational nature of behavioral scoring, with no application of quantitative measures, as well as the lack of direct studies of neuronal development), the adequacy of these works as bases for such a conclusion is questionable.

A replication of these early works has recently been performed in which the basic paradigm of the original studies was extended and many of the procedural and interpretational ambiguities of the classic studies addressed (Haverkamp, 1983). Embryos of the amphibian *Xenopus laevis* were immersed very early in development (late neural fold state) in solutions of either chloretone or lidocaine (which blocks conduction of neural impulses) or were injected with alpha-bungarotoxin (which blocks neuromuscular transmission). Approximately 36 hr later, when stage-matched, normally reared control embryos from the same spawning were swimming well, the experimental animals were removed, or had recovered, from the immobilizing drugs. Two hours after regaining motility, many of the previously immobilized embryos *appeared* to behave quite normally. However, detailed quantitative analyses performed at this time revealed significant deficits in their swimming. When behavioral quantifications were repeated approximately 24, 48, and 168 hr after recovery from the drug, however, the previously immobilized animals swimming was indistinguishable from the control embryos (Figure 11). Therefore, preventing all neural

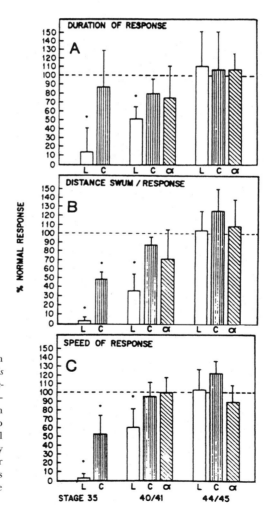

Figure 11. Quantitative measures of swimming in response to tactile stimulation by *Xenopus laevis* embryos, immobilized by immersion in chloretone (C) or lidocaine (L) solutions from st. 17–35, or through injection of alpha-bungarotoxin (α) at st. 23. The bars (±S.D.) represent group means of measures obtained 2 hr after removal from the drug solutions (st. 35), approximately 24 hr later (st. 40/41), and approximately 48 hr after recovery from the drug (st. 44/45). Stars denote values that significantly differ from those of normally reared embryos.

activity during that period when the motor system is forming had no *permanent* effects upon later behavioral performance. The primary conclusions of Harrison's (1904) and the later studies were thus supported in a different species, using a number of drugs to induce immobilization, and with the application of sensitive quantitative measures.

These observations that the initial development of motor capacities can proceed normally in the absence of neural activity imply that the growth, differentiation, and connectivity of the neurons underlying these behaviors also proceeds normally when impulse activity is blocked. Nonetheless, the many demonstrations of seemingly normal behavior in the presence of distinctly abnormal neuroanatomy (e.g., Brodal, 1981; Caviness and Rakic, 1978) preclude the ready acceptance of such an assumption. Also, in the experiment just described, there was a period of time, immediately after removal of experimental embryos from the drug solutions, during which deficits in quantitative measures of behavior could be demonstrated. While the source of these transient effects may simply have been the effects of a residual drug, it might also be argued that this period of behavioral deficit was a time during which developmental events dependent upon function belatedly occurred (i.e., a period of rapid behavioral "practice").

Further experiments upon *Xenopus* embryos treated under the immobilization paradigm just outlined indicate, however, that at least certain essential components of the motor system had differentiated normally in the absence of impulse activity. Physiological recordings of activity in the motor nerves of previously immobilized animals revealed no effects of the drug treatments (Figure 12). When these record-

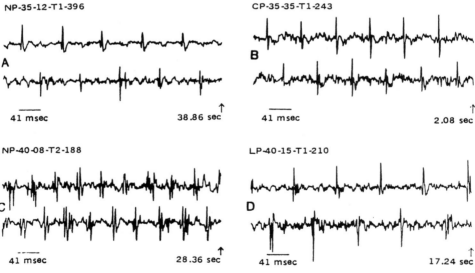

Figure 12. Representative traces of ventral root activities recorded from *Xenopus* embryos. Each pair of traces are those recorded simultaneously from the left and right sides of an immobilized embryo. Traces (A) and (B) are from st. 35 embryos (approx. 50 hr postfertilization), traces (C) and (D) from st. 40 embryos (approx. 65 hr postfertilization). Traces (A) and (C) were obtained from normally reared embryos, while (B) and (D) were obtained from embryos reared in chloretone and lidocaine, respectively, from st. 17–35. The animal from which the traces in (B) were obtained had been removed from the anesthetic solution just 2 hr prior to recording. In each of the tracings, note the regular, alternating bursting pattern of ventral root activity on either side.

ings were quantified, no significant differences between experimental and control animals were seen, either in the period of time between successive motor nerve bursts during fictive swimming episodes or in the phase relationship of bursts on either side of the animal. Experimental and control animals did not differ in these measures even during that time after drug release when behavioral effects of immobilization could be demonstrated. In addition, anatomical studies of the dendritic arborization of the primary motoneurons of *Xenopus* embryos that had *never* been allowed to recover from the anesthetizing agents, revealed no significant effects of neural blockade on this measure of differentiation (Figure 13).

Conceptually related studies done in tissue culture by Crain and his co-workers (Crain *et al.*, 1968; Model *et al.*, 1971) also support the premise that during the earliest stages of neural development, differentiation and connectivity take place normally in the absence of impulse activity. When small pieces of embryonic neural tissue are maintained in solutions composed of certain salts and sera, cells will grow and differentiate in an apparently normal manner. Synapses will form within such explants and, after a period of time, the tissue will evince complex electrical activities in response to local stimulation. When Crain's group maintained explants of embryonic rat cortex in solutions that continued the neural conduction blocker lidocaine, they observed that the treated tissue formed synapses that were ultrastructurally indistinguishable from those found in normally reared explants. Furthermore, these explanted tissues produced normal-appearing electrically evoked responses immediately upon withdrawal of the drug from the tissue environment.

This ability of neurons to develop structurally and functionally effective connec-

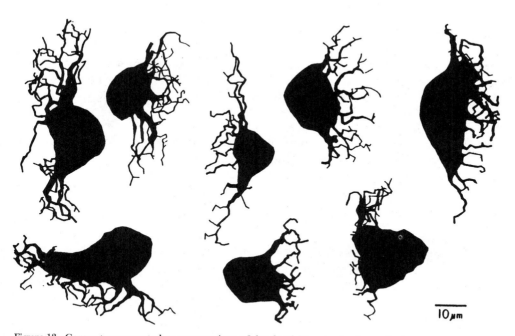

Figure 13. Computer-generated reconstructions of the dendritic arborizations of primary motoneurons of st. 35 *Xenopus* embryos. The cells were visualized through reaction of peripherally applied horseradish peroxidase, and morphometric quantifications made by computer-assisted digitization of their processes. Depicted here are representative neurons from both normally reared embryos, as well as from embryos continuously exposed to either chloretone or lidocaine from st. 17.

tivity in the absence of activity has been shown in a number of different systems, using a variety of means to block neural conduction or synaptic transmission (e.g., Bird, 1980; Obata, 1977; Romijn *et al.*, 1981). It should be noted, however, that a number of these works have demonstrated certain *quantitative* effects of such blockade (e.g., decreased numbers of synapses formed; Janka and Jones, 1982). In addition, one set of studies (Bergey *et al.*, 1981; Christian *et al.*, 1980) has shown that blocking neural activity in *dissociated* cell cultures with the potent neurotoxin, tetrodotoxin (TTX), results in the death of one of the interacting cell types contained in the culture.

A particularly elegant demonstration in the intact animal of normal differentiation and connectivity in the absence of neural activity has been performed by Harris (1980). It was first demonstrated by the embryologist Twitty (1937) that TTX, which blocks all neural impulse propagation, is a normal constituent of the body fluids of the California newt. Though this newt's tissues are, of course, insensitive to the effects of TTX, neural tissue from other salamander species, transplanted into the California newt, is incapable of generating action potentials. Harris exploited this system by transplanting the eye primordia of the Mexican axolotl into the head region of newt embryos (Figure 14). After the animals matured, Harris found that not only had the axolotl eyes correctly projected to the optic tectum, but the spatial distribution of these projections on the tectum was also normal and the synapses formed were indistinguishable from the tectal synapses formed by the host retina. Thus in the absence of not only light-induced activity but also spontaneous discharge, retinal ganglion cells had apparently differentiated normally and projected to form connections with that portion of their target appropriate to the position of the ganglion cell within the eye. This finding is all the more remarkable considering that the transplanted eyes were in direct competition at the target site with the normal, functionally intact, host eye.

Results of a recent study by Meyer (1982) are similar to those of Harris' work yet demonstrate that while functional activity may play no role in one aspect of a system's development, it may be integral to another step in the same system's ontogeny. When the two optic nerves of a goldfish are transected and forced to reinnervate a single opt tectum, the projections from either eye initially show extensive overlap at the target site. After a few weeks of development, however, the projections from the left and right eyes segregate, forming multiple bands or the tectum, each of which contain terminals from only one optic nerve. Meyer (1982) injected TTX into the eyes of goldfish treated in this manner and found that the eyes nevertheless projected to, and distributed over the surface of the tectum normally (as Harris found for the supernumerary axolotl eyes). The subsequent segregation of retinal afferent terminals on the tectal surface did not occur, however, with the TTX-induced block of neural activity in both eyes. Interestingly, when TTX was injected into only one eye, the segregation seen in control animals occurred normally.

This surgically induced column formation on the goldfish optic tectum is analogous to the formation of ocular dominance columns in the mammalian visual cortex. In this latter system there is a similar initial extensive overlap of afferent terminals driven by either eye. This overlap is later modified so that bands are formed in the cortex that show primary responses to stimulation of one or the other retina. This segregation of ocular dominance columns can be prevented, however, as in the goldfish, by bilateral intraocular injections of TTX into neonatal kittens (Stryker, 1981).

In addition to pointing out the selective effects of neural activity on neural

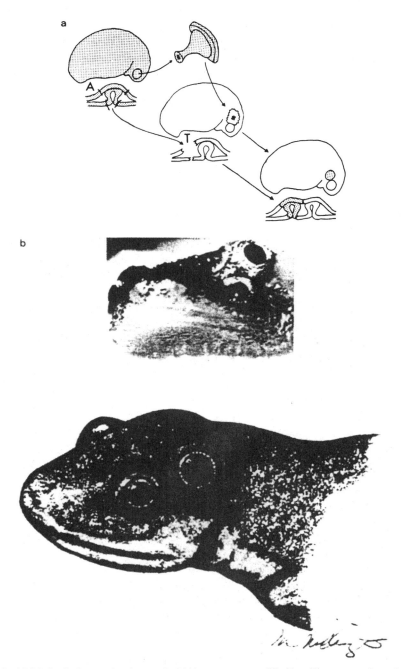

Figure 14. (a) Method of transplanting axolotl (A) eyes to newts (T). Top: The eye cup is excised along with the entire optic stalk. Middle: Wounds are made in the host's epidermis and neuropithelium, dorsal and posterior to the eye on the same side as the one being transplanted. Bottom: The transplant is put into place. Underneath each embryo is a schematic cross section of the operation. (b) Photograph and drawing of metamorphic newt with a third eye from a axolotl. (From Harris, 1980.)

development, these experiments on ocular dominance column formation further illustrate the nonequivalence of sensory deprivation and total blockade of neural activity. While binocular deprivation (lid suturing, dark rearing) may retard the development of the ocular dominance columns in the cortex, such treatment is not effective in totally preventing their formation in the manner that TTX injections are. To the same point, while monocular visual deprivation has distinct effects on retinal projections to the lateral geniculate nucleus, Archer *et al.* (1982) have shown that monocular TTX injections alter retino-geniculate projections in a manner totally different from that which results from sensory deprivation.

V. Behavioral Embryology: Practice and Perceptual Experience

In the preceding examples the overlap of concern of developmental psychology and of developmental biology is apparent. The role that environment plays in behavior development is readily seen as one aspect of the study of the modes and effects by which cellular and tissue interactions bring about development of the embryonic nervous system. Equally obvious is the fact that an understanding of behavioral and biological development cannot be gained by confining our investigations to a period after an often arbitrary (albeit significant) developmental landmark such as birth or hatching (or metamorphosis in invertebrates and amphibians).

Locomotor development in a number of species further illustrates this point. The ability of butterflies and moths to fly soon after emergence from their pupal cases is a striking event, so much so that it has been used as a paradigm case of the lack of necessity for practice in the performance of complex motor patterns (see, e.g., Hinde, 1970). Recordings from the motor nerves of moth pupae reveal, however, that a primitive form of the flight motor pattern is apparent long before emergence, and it is gradually refined to adult form while the animal is still encased in the pupal sheath (Kammer and Kinnamon, 1979). The formation and refinement of intra- and interlimb coordination are also apparent in the chick embryo during early embryonic development (Bekoff, 1981). In the frog, patterned activity in the ventral roots that innervate the hindlimbs is first apparent during those stages of development during which the limbs are merely buds of undifferentiated mesenchyme (Stehouwer and Farel, 1983). Thus even such a basic behavior as locomotion has an embryonic history that must be taken into account for any consideration of behavioral development. The question of what role this precocious function plays in development of mature movements is unanswered, however. The demonstrations of normal development of swimming behavior in immobilized tadpoles, discussed earlier, while compelling, do not necessarily imply that more complex forms of motor activity are similarly independent of function during ontogeny (but see Oppenheim, 1982a).

A particularly striking example of prenatal experience that clearly *does* affect later behavior has been elucidated by G. Gottlieb (1980a,b, 1982, 1983). Gottlieb has demonstrated that the embryo of the domestic mallard duck is specifically responsive to the mallard maternal call prior to any auditory experience. In normal ducklings this responsiveness is largely directed toward the repetition rate of the call, and the specificity of the response is maintained through hatching. If, however, duckling embryos are devocalized and incubated in isolation, this selective responsiveness is lost and the ducklings will respond equally well to calls of a number of other species (performed at different species-specific rates). The experiential variable missing from the environment of these experimental animals appears to be the perception of their own, or sib's, vocalizations while *in ovo*. If the devocalized embryos are exposed to recordings of their own "contact-contentment" calls, selective responsiveness to species-specific maternal vocalizations remains intact in the posthatch ducklings.

Surprisingly, when Gottlieb performed a finer analysis of the perceptual development of these embryos reared in auditory isolation, he found that a 48-hour "consolidation" period was necessary for the ducklings to exhibit normal responsiveness to recordings of maternal vocalizations. When tested 24 hr after exposure to contact calls, the experimental animals did not show the selectivity of response that could be demonstrated 24 hr later. Gottlieb noted that in the natural environment, the ducklings and embryos hear not only contact-contentment calls (normally produced at a repetition rate of 4.0 notes/sec), but a wide range of other calls from sibs and self as well, which vary in repetition rates from 2.1 to 5.8 notes/sec. A group of devocalized, isolated embryos was therefore exposed to the contact call replayed at three repetition rates: of 2.1, 4.0, and 5.8 notes/sec. Unlike experimental embryos exposed only to the contact call replayed at its normal rate, embryos exposed to all three rates of the call did not require a 48-hr consolidation period, but would respond to the correct call in choice tests given 24 hr after stimulation.

While exposure to a single type of duckling call is a sufficient stimulus, therefore, for the young birds to later exhibit a preference for the maternal call, exposure to the normally encountered range of repetition rates at which different duckling vocalizations are normally produced is necessary for this preference to be contiguously maintained. In further experiments, Gottlieb has shown that the frequency modulation of the contact call is also required for later preference for the repetition rate of the maternal call to be exhibited; exposure of the ducklings to white noise pulsed at the correct rate is an ineffective stimulus.

As Gottlieb (1982) points out, these often subtle and "nonobvious" requirements for normal perceptual development in embryonic ducklings are quite reminiscent of the requirements in cellular differentiation for interaction with the normally occurring *tissue* environment. In the same manner that the particular stimulus and specific response may differ between species in this type of cellular interaction with the environment, behavioral, and perceptual developmental interactions with the environment also show species variabilities.

A variation on this theme of prenatal auditory experience affecting postnatal perception as demonstrated in the mallard duck is seen in a related species, the wood duck. Embryos of this species also develop a selective responsiveness to a component of the maternal call, in this case a descending frequency modulation of the notes. As with the mallard duckling, specific auditory experience is required to maintain this preference. In the wood duck, however, this requirement is for exposure to the duckling's alarm call (rather than the contact call). A further difference between the two species is that while self-stimulation is sufficient to maintain perceptual preference in the mallard duckling, the young of the wood duck must be exposed to the calls of siblings to maintain a preference.

These demonstrations of the role of sensory experience in the development of ducklings bring to mind the many investigations performed on the development of adult song in various other avians. These works have shown song development in passarines to be dependent upon the complex interactions of genetic, hormonal, and experiential factors. The degrees to which vocal development can be regulated by the manipulation of each of these factors likely describes a continuum of control, but may be grouped into three broad categories (1) no auditory feedback is required for normal song development, (2) auditory feedback (self-stimulation) is necessary to normal song development but experience in hearing conspecifics is not, and (3) exposure to conspecific song is necessary to normal development of adult song (Nottebohm, 1968).

As the anatomical, physiological, and pharmacological bases of these differences are further revealed, this system may provide valuable insights into the neural mechanisms underlying a variety of types of behavioral polymorphism and development strategies (see Chapter 4). The recent extension of neurobiological methods to the study of these systems (see, e.g., Nottebohm, 1980) does not imply, of course, that further behavioral analyses will not continue to be of great value. While behavioral investigations will continue to suggest courses of study at the suborganismal level of analysis, so will these latter works undoubtedly contribute to defining routes for further behavioral studies.

This impetus to behavioral investigation and interpretation may be either indirect (i.e., by analogy to interactions at the cellular level) or direct (i.e., a testing or proposal of behavioral mechanisms from the demonstrated function of neurons and their assemblies). A number of examples of both types of crossover have been noted in the previous papers. A more recent example, relating to a system just discussed, is the epigenetic narrowing of the vocal repertoire of the male swamp sparrow. Marler and Peters (1982) have demonstrated that, prior to these birds' adult performance of a stabilized song, they exhibit a phase during which their vocalization contains up to five times as many song types as in the mature state. During the intervening period the components of adult song are derived from this enlarged repertoire by an actively selective process. Marler and Peters (1928) draw the apt analogy of this developmental restriction of song types to the phenomenon of neuronal cell death and related regressive phenomena during development, citing the initial overproliferation of units in each, followed by their subsequent attrition.

Another, more detailed, analysis that relates behavioral to physiological data has been published recently by Bischof (1983). Two of the best studied phenomena in behavioral development and developmental neurobiology are, respectively, imprinting and the development of the mammalian visual cortex. Although the similarities in the two systems are too extensive to relate here (see, e.g., Figure 15), it is for this very reason (i.e., extensive descriptive coincidence) that the two phenomena have often been linked, and it has even been suggested that the same sort of neural mechanisms are operating in each (Rose, 1981). While we do not believe that such linkage is necessarily indicated by the data from each field, this type of exercise in contrast and comparison is exactly the sort of insight into the methods and findings of separate fields that is too often lacking in many research approaches. Not only can the works of developmental psychologists and psychobiologists be extended through such comparisons of data, but that of neurobiologists and embryologists can profit as well. Through his comparative analysis of imprinting and visual cortical plasticity, Bischof (1983) concludes that a similarity of mechanisms underlying the two systems is unlikely but that each may reflect the expression of a common developmental process. He goes on to propose some characterizations of this process (i.e., its self-limiting nature and its basis in the alteration of neural development). Whether such characterizations are justified or will hold up to experimentation is irrelevant to this illustration of the new insight and predictive value that such comparative consideration of data from different levels of analysis imparts to each of the fields.

These last few examples have extended the range of discussion from the prenatal to the postnatal period, yet continue to emphasize, it is hoped, the value of relating the findings of different levels of analyses. Investigations of bird song development, mentioned earlier, are revealing a system that illustrates the interacting contributions of the genome (sexual genotype), hormones (presence of testosterone), and experience (the role of practice and learning) in neural and behavioral development.

Another fascinating example of the interplay of sexual genotype and hormonal influence is to be found in mammalian development. Hauser and Gandelman (1983) have each presented evidence that during intrauterine development, the sex of adjacent embryos markedly affects adult behavior of rats. To cite one portion of these data, male pups that develop between two female litter mates were significantly more sexually active and less aggressive as adults than were male pups that gestated between two other males. This effect is presumably due to increased levels of estrogens available to the female-contiguous male pups. In the same manner, prenatal androgen levels are held responsible for the altered avoidance responding of adult female rats that have developed between two males *in utero*.

As a final example of such prenatal influences on later behavior, we shall mention a number of studies performed on nipple attachment and nursing in newborn mammals. Two of the sensory systems that are most saliently associated with this behavior (olfaction and taste) become functional during prenatal life. A response to substances injected into the amniotic fluid that presumably stimulate receptors of both types can be perceived prior to birth (Bradley and Mistretta, 1973). When such substances are paired with negative stimuli *in utero*, postnatal exposure to that odor results in an avoidance of the odiferous source postnatally (Smotherman, 1982). When exposure to such substances is not paired with negative stimuli *in utero*, and exposure is extended to the period immediately following birth, the infant pup

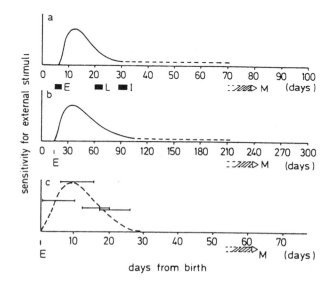

Figure 15. Time course of the efficiency of external stimulation in three different paradigms: (a) Sexual imprinting in the zebra finch (Immelman, 1972), (b) Ocular dominance in the cat (Blakemore, 1980), and (c) Sexual imprinting in the Japanese quail (Gallagher, 1967). Ordinate scale is arbitrary. (E, eye opening, L, leaving the nest, I, independent, M, sexual maturity.). (From Bischof, 1983.)

will preferentially attach to a nipple that is coated with the same substance (Figure 16). The significance in the natural state of this fetal experiential ability and its postnatal retention is hinted at by the observation that the mother rat, at birth, may have significant oral contact with amniotic fluid and her own nipples. When the nipple is washed, her pups fail to attach to the nipple and do not nurse (Pedersen and Blass, 1982).

These examples have clearly shown, then the fallacy in assuming that a prior behavior is innate and independent of experiential effect simply because initial perceptual responses or behavioral performances are performed soon after birth. It should be obvious that investigations of behavioral development cannot, therefore, be confined to the influences of obvious postnatal experience and easily recognizable performance. The examples given earlier have pointed out not only the necessity of extending manipulation and analysis of behavioral development to the prenatal period, but also the advantages of viewing neural and behavioral development from an embryological perspective.

Modifications of the nervous system, which are, of course, the bases of behavioral change, may be either transient or resistant to further alterations. Whatever the time course or persistence of these changes, they are always the result of activity within cells and due to the interactions of the neurons with both the intra- and extraorganismal environment. Decades of study in developmental biology have led to an increased understanding of many of the epigenetic interactions that influence ontogeny, including the one class of those interactions reviewed here, the roles of neural activity, use, and sensory input on behavioral development. We hope, therefore, that through reflection on the experimental examples outlined previously, the reader may come to share our belief that the development of behavior is an inextricable part of developmental biology and may be fully understood only when viewed within that framework. To attempt a complete understanding of a system's functions, it is indeed necessary at times to "begin our query at the beginning of growth."

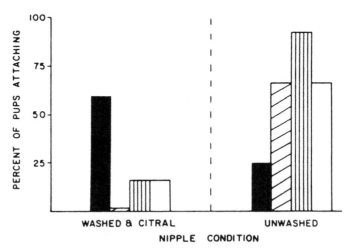

Figure 16. Percent of pups experiencing citral pre- and postnatally (dark bar), citral prenatally (hatched), citral postnatally (striped), or no citral (open bar) and attaching to nipples washed and scented with citral (left) or unwashed (right). (From Pedersen and Blass, 1982.)

REFERENCES

49

EARLY
DEVELOPMENT
OF BEHAVIOR
AND THE
NERVOUS SYSTEM

Archer, S. M., Dubin, M. W., and Stark, L. A. Abnormal development of kitten retino-geniculate connectivity in the absence of action potentials. *Science*, 1982, *217*, 743–745.

Bateson, P. P. G. How does behavior develop? In P. P. G. Bateson and P. Klopfer (Eds.), *Perspectives in ethology*. vol. 3. Cambridge: Cambridge University Press, 1978.

Barlow, G. Genetics and development of behavior, with special reference to patterned motor output. In K. Immelmann, G. W. Barlow, L. Petrinovich, and M. Main (Eds.), *Behavioral development*. Cambridge: University of Cambridge Press, 1981.

Bekoff, A. Embryonic development of the neural circuitry underlying motor coordination. In W. M. Cowan (Ed.), *Studies in developmental neurobiology: Essays in honor of Viktor Hamburger*. New York: Oxford University Press, 1981.

Bergey, G. K., Fitzgerald, S. C., Schrier, B. L, and Nelson, P. G. Neuronal maturation in mammalian cell culture is dependent on spontaneous electric activity. *Brain Research*, 1981, *207*, 49–58.

Bird, M. M. The morphology of synaptic profiles in explants of foetal and neonatal mouse cerebral cortex maintained in a magnesium-enriched environment. *Cell and Tissue Research*, 1980, *206*, 115–122.

Bischof, H.-J. Imprinting and cortical plasticity: A comparative review. *Neuroscience and Biobehavioral: Reviews*, 1983, *7*, 213–225.

Black, I. Stages of neurotransmitter development in autonomic neurons. *Science*, 1982, *215*, 1198–1204.

Bradley, R. M., and Mistretta, C. M. Fetal sensory receptors. *Physiological Review*, 1973, *55*, 352–381.

Brodal, A. *Neurological anatomy in relation to clinical medicine*, 3rd ed. New York: Oxford University Press, 1981.

Bunge, R., Johnson, M., and Ross, C. D. Nature and nurture in development of the autonomic neuron. *Science*, 1978, *199*, 1409–1416.

Carmichael, L. The development of behavior in vertebrates experimentally removed from the influence of external stimulation. *Psychological Review*, 1926, *33*, 51–58.

Carmichael, L. A further study of the development of behavior in vertebrates experimentally removed from the influence of external stimulation. *Psychological Review*, 1927, *34*, 34–47.

Carmichael, L. The onset and early development of behavior. In L. Carmichael (Ed.), *Manual of child psychology*. New York: Wiley, 1946.

Caviness, V. S., Jr., and Rakic, P. Mechanisms of cortica development: A view from mutations in mice. *Annual Review of Neuroscience*, 1978, *1*, 297–326.

Christian, C. N., Bergey, G. K., Daniels, M. P., and Nelson, P. G. Cell interactions in nerve and muscle cell cultures. *Journal of Experimental Biology*, 1980, *89*, 85–101.

Coghill, G. E. *Anatomy and the problem of behavior*. Cambridge: Cambridge University Press, 1929.

Corner, M. A., Bour, H. L., and Mirmiran, M. Development of spontaneous motility and its physiological interpretation in the rat, chick, and frog. In E. Meisami and M. A. B. Brazier (Eds.), *Neural growth and differentiation*. New York: Raven Press, 1979.

Crain, S. M., Bornstein, M. B., and Peterson, E. R. Maturation of cultured embryonic CNS tissues during chronic exposure to agents which prevent bioelectric activity. *Brain Research*, 1968, *8*, 363–372.

Detwiler, S. R. *Neuroembryology: An experimental study*. New York: Macmillan, 1936.

Deucher, E. *Cellular interactions in animal development*. London: Chapman and Hall, 1975.

Erzurumlu, R. S., and Killackey, H. P. Critical and sensitive periods in neurobiology. *Current Topics in Developmental Biology*, 1982, *17*, 207–240.

Gesell, A. *The embryology of behavior*. New York: Harper, 1945.

Goldspink, D. F. *Development and specialization of skeletal muscle*. Cambridge: Cambridge University Press, 1980.

Gottlieb, G. The role of experience in the development of behavior and the nervous system. In G. Gottlieb (Ed.), *Neural and behavioral specificity*. New York: Academic Press, 1976.

Gottlieb, G. Development of species identification in ducklings: VI. Specific embryonic experience required to maintain species-typical perception in peking ducklings. *Journal of Comparative and Physiological Psychology*, 1980a, *94*, 579–587.

Gottlieb, G. Development of species identification in ducklings: VII. Highly specific early experience fosters species-specific perception in wood ducklings. *Journal of Comparative Physiological Psychology*, 1980b, *94*, 1019–1027.

Gottlieb, G. Development of species identification in ducklings: IX. The necessity of experiencing normal variations in embryonic auditory stimulation. *Developmental Psychobiology*, 1982, *15*, 507–517.

Gottlieb, B. Development of species identification in ducklings: X. Perceptual specificity in the wood duck embryo requires sib stimulation for maintenance. *Developmental Psychobiology*, 1983, 16, 323–334.

Gurdon, J. B. *The control of gene expression in animal development.* Cambridge, MA: Harvard University Press, 1974.

Hadorn, E. *Experimental studies of amphibian development.* New York: Springer Verlag, 1974.

Hamburg, M. *Theories of Differentiation.* New York: American Elsevier, 1971.

Hamburger, V., and Oppenheim, R. W. Naturally-occurring neuronal death in vertebrates. *Neuroscience Commentaries*, 1982, *1*, 39–55.

Hamburger, V. Anatomical and physiological basis of embryonic motility in birds and mammals. In G. Gottlieb (Ed.), *Studies in the development of behavior and the nervous system.* Vol. 1. *Behavioral embryology*, New York: Academic Press, 1973.

Hamburger, V., Wenger, E., and Oppenheim, R. Motility in the chick embryo in the absence of sensory input. *Journal of Experimental Zoology*, 1966, *162*, 133–160.

Harris, W. A. Neural activity and development. *Annual Review of Physiology*, 1981, *43*, 689–710.

Harris, W. A. The effects of eliminating impulse activity on the development of retinotectal projections in salamanders. *Journal of Comparative Neurology*, 1980, *194*, 303–317.

Harrison, R. G. An experimental study of the relation of the nervous system to the developing musculature in the embryo of the frog. *American Journal of Anatomy*, 1904, *3*, 197–220.

Hauser, H., and Gandelman, R. Contiguity to males *in utero* affects avoidance responding in adult female mice. Science, 1983, *220*, 437–438.

Haverkamp, L. J. Neurobehavioral development with blockade of neural function in embryos of *Xenopus laevis*, Ph.D. dissertation, University of North Carolina, Chapel Hill, 1983.

Hebb, D. O. *Organization of behavior.* New York: Wiley, 1949.

Hebb, D. O. *Essay on mind.* Hillsdale, NJ: Erlbaum, 1980.

Hinde, R. A. *Animal behavior: A synthesis of ethology and comparative psychology.* New York: McGraw-Hill, 1970.

Jacob, F. Evolution and tinkering. *Science*, 1977, *196*, 1161–1166.

Jacobson, M. *Developmental neurobiology.* New York: Plenum Press, 1978.

Janka, Z., and Jones, D. G. Junctions in rat neocortical explants cultured in TTX-, GABA-, and MG^{++}-environments. *Brain Research Bulletin*, 1982, *8*, 273–278.

Kammer, A. E., and Kinnamon, S. C. Maturation of the flight motor pattern without movement in *Manduca sexta. Journal of Comparative Physiology*, 1979, *130*, 29–37.

Lashley, K. Experimental analysis of instinctive behavior. *Psychological Review*, 1938, *45*, 445–471.

Le Douarin, N. Migration and differentiation of neural crest cells. *Current Topics in Developmental Biology*, 1980, *16*, 32–85.

Lehrman, D. S. A critique of Konrad Lorenz's theory of instinctive behavior. *Quarterly Review of Biology*, 1953, *28*, 337–363.

Lorenz, K. *Evolution and modification of behavior.* Chicago: University of Chicago Press, 1965.

Lund, R. D., and Hauschka, S. D. Transplanted neural tissue develops connections with host strain. *Science*, 1976, *193*, 582–584.

Marler, P., and Peters, S. Developmental overproduction and selective attrition: New processes in the epigenesis of birdsong. *Developmental Psychobiology*, 1982, *15*, 369–378.

Matthews, S. A., and Detwiler, S. W. The reaction of *Amblystoma* embryos following prolonged treatment with chloretone. *Journal of Experimental Zoology*, 1926, *45*, 279–292.

McGraw, M. *Growth: A study of Johnny and Jimmy.* New York: Appleton, 1935.

Meyer, R. L. Tetrodotoxin blocks the formation of ocular dominance columns in goldfish. *Science*, 1982, *218*, 589–591.

Mistretta, C. M., and Bradley, R. M. Effects of early sensory experience on brain and behavioral development. In G. Gottlieb (Ed.), *Early influences.* New York: Academic Press, 1978.

Model, P. G., Bornstein, M. B., Crain, S. M., and Pappas, G. D. An electron microscopic study of the development of synapses in cultured fetal mouse cerebrum continuously exposed to xylocaine. *Journal of Cell Biology*, 1971, *49*, 362–371.

Narayanan, C. H., and Hamburger, V. Motility in chick embryos with substitution of lumbosacral by bracial and brachial by lumbosacral spinal cord segments. *Journal of Experimental Zoology*, 1971, *178*, 415–432.

Nottebohm, F. Auditory experience and song development in the chaffinch, *Fringilla coelebs. Ibis*, 1968, *110*, 549–568.

Nottebohm, F. Brain pathways for vocal learning in birds: A review of the first 10 years. *Progress in Psychobiology and Physiological Psychology*, 1980, *9*, 85–124.

Obata, K. Development of neuromuscular transmission in culture with a variety of neurons and in the presence of cholinergic substances and tetrodotoxin. *Brain Research*, 1977, *119*, 141–153.

Oppenheim, R. W. The ontogeny of behavior in the chick embryo. In D. S. Lehrman, R. A. Hinde, E. Shaw, and J. Rosenblatt (Eds.), *Advances in the study of behavior.* Vol. 5. New York: Academic Press, 1974.

Oppenheim, R. W. Neuronal cell death and some related regressive phenomena during neurogenesis: A selective historical review and progress report. In W. M. Cowan (Ed.), *Studies in developmental neurobiology: Essays in honor of Viktor Hamburger.* New York: Oxford University Press, 1981.

Oppenheim, R. W. The neuroembryology of behavior: Progress, problems, perspectives. *Current Topics in Developmental Biology*, 1982, *17*, 257–309.

Oppenheim, R. W. preformation and epigenesis in the origins of the nervous system and behavior. In P. P. G. Bateson and P. Klopfer. *Perspectives in ethology.* vol. 5. New York: Plenum Press, 1982b.

Oppenheim, R. W. Cell death of motoneurons in the chick embryo spinal cord: VIII. Motoneurons prevented from dying in the embryo persist after hatching. *Developmental Biology*, 1982c, *101*, 35–39.

Oppenheim, R. W., Maderdrut, J. L., and Wells, D. Reduction of naturally-occurring cell death in the thoraco-lumbar preganglionic cell column of the chick embryo by nerve growth factor and hemi-cholinium-3. *Developmental Brain Research*, 1982, *3*, 134–139.

Oppenheim, R. W., and Nunez, R. Electrical stimulation of hindlimb increases neuronal cell death in chick embryo. *Nature*, 1982, *295*, 57–59.

Oppenheim, R. W., Pittman, R., Gray, M., and Maderdrut, J. L. Embryonic behavior, hatching and neuromuscular development in the chick following a transient reduction of spontaneous motility and sensory input by neuromuscular blocking agents. *Journal of Comparative Neurology*, 1978, *179*, 619–640.

Patterson, P. H. Environmental determination of autonomic neurotransmitter functions. *Annual Review of Neuroscience*, 1978, *1*, 1–17.

Perlow, M. J. Functional brain transplants. *Peptides*, 1980, *1*, 101–110.

Pedersen, P. E., and Blass, E. M. Prenatal and postnatal determinants of the 1st suckling episode in albino rats. *Developmental Psychobiology*, 1982, *15*, 349–355.

Piaget, J. and Inhelder, B. *The psychology of the child.* London: Routledge and Kegan, 1969.

Pittman, R., and Oppenheim, R. W. Neuromuscular blockade increases motoneuron survival during normal cell death in the chick embryo. *Nature*, 1978, *271*, 364–366.

Pittman, R., and Oppenheim, R. W. Cell death of motoneurons in chick embryo spinal cord: IV. Evidence that a functional neuromuscular interaction is involved in the regulation of naturally-occurring cell death and the stabilization of synapses. *Journal of Comparative Neurology*, 1979, *187*, 425–446.

Roberts, L. Brain grafting: Surgery reduces neurological damage. *Bioscience*, 1983, *33*, 80–83.

Romijn, H. J., Mud, M. T., Habets, A. M. M. C., and Wolters, P. S. A quantitative electron microscopic study of synapse formation in dissociated fetal rat cerebral cortex *in vitro. Developmental Brain Research*, 1981, *1*, 591–605.

Rose, S. P. R. From causation to translations: What biochemists can contribute to the study of behaviour. In P. O. G. Bateson and P. H. Klopfer (Eds.), *Perspectives in ethology.* IV. *Advantages of diversity*, New York: Plenum Press, 1981.

Roux, W. Contributions to the developmental mechanics of the embryo (1888). In B. H. Willier and J. Oppenheimer (Eds.), *Foundations of experimental embryology.* Englewood Cliffs, NJ: Prentice-Hall, 1967.

Scarr-Salapatek, S. An evolutionary perspective on infant intelligence: Species patterns and individual variations. In M. Lewis (Ed.). *Origins of intelligence.* New York: Plenum Press, 1976.

Smotherman, W. P. Odor aversion learning by the rat fetus. *Physiology Behavior*, 1982, *29*, 769–771.

Spemann, H. *Embryonic development and induction.* New Haven: Yale University Press, 1938.

Sperry, R. W. Mechanisms of neural maturation. In S. S. Stevens (Ed.), *Handbook of experimental psychology.* New York: Wiley, 1951.

Stehouwer, D. J., and Farel, P. B. Development of hindlimb locomotor activity in the bullfrog (*Rana catesbeiana*) studied *in vitro. Science*, 1983, *219*, 516–518.

Straznicky, K. Function of heterotopic spinal cord segments investigated in the chick. *Acta Biologica Hungarium*, 1967, *14*, 145–155.

Stryker, M. P. Late segregation of geniculate afferents to the cat's visual cortex after recovery from binocular impulse blockade. *Society for Neuroscience Abstracts*, 1981, *7*, 842.

Twitty, V. C. Experiments on the phenomenon of paralysis produced by a toxin occurring in *Triturus* embryos. *Journal of Experimental Zoology*, 1937, *76*, 67–104.

von Saal, F. S., Grant, W. M., McMullen, C. W., and Laves, K. S. High fetal estrogen concentrations: Correlation with increased adult sexual activity and decreased aggression in male mice. *Science*, 1983, *220*, 1306–1309.

Vrbová G., Gordon, T., and Jones, R. *Nerve–muscle interaction.* Chapman and Hall: London, 1978.

Walicke, P. A., Campenot, R. B., and Patterson, P. H. Determination of neurotransmitter function by neuronal activity. *Proceedings National Academy of Sciences USA*, 1977, *74*, 5767–5771.

Weiss, P. Self-differentiation of the basic patterns of coordination. *Comparative Psychology Monographs*, 1941, *17*, 1–96.

Wenger, B. S. Determination of structural patterns in the spinal cord of the chick embryo studied by transplantation between brachial and adjacent levels. *Journals of Experimental Zoology*, 1951, *116*, 123–146.

Weston, J. A. Neural crest cell development. In M. Burger and R. Weber (Eds.), *Embryonic development. Part B, cellular aspects.* New York: Alan Liss, 1982.

Whitman, C. O. Evolution and epigenesis. *Woods Hole Biological Lectures*, 1894, No. 10, 203–224.

Wilson, E. B. *The cell in development and heredity.* 2nd ed. New York: Macmillan, 1900.

2

Spatial Coding in the Olfactory System

The Role of Early Experience

BRETT A. JOHNSON AND MICHAEL LEON

SPATIALLY SPECIFIC ALTERATIONS IN OLFACTORY BULB FUNCTION AND STRUCTURE FOLLOWING EARLY ODOR LEARNING

Infant rats are born with a functional olfactory system (Guthrie & Gall, 1999). Within the first days of their life they begin to approach the odor of their mother in preference to the odor of a virgin female (Leon & Moltz, 1971). These preferences can be seen when the pups are placed in an apparatus designed to allow them to approach one of two areas on the basis of odor cues alone. Such a preference also can be induced when the natural situation is mimicked experimentally by pairing a nonmaternal odor (such as peppermint extract) with tactile stimulation of the kind that a mother might impose on her pups (Coopersmith & Leon, 1984). These data indicate that pups acquire their preference for the mother's odor postnatally, rather than being born with that ability. In addition, the individuality of the odor of one mother compared to another is due to differences in their diet; mothers with identical diets are equally approached by their pups (Leon, 1975).

Learning under natural circumstances seemed likely to be the kind of behavioral change that should induce large changes in the relatively simple brain of developing rats, large enough to visualize through relatively straightforward analyses. Even small changes in the brain induced by the kind of experience described above would be magnified as the brain developed, allowing us to visualize changes that are difficult to see in the brains of experienced adults. We also wanted to

BRETT A. JOHNSON AND MICHAEL LEON   Department of Neurobiology and Behavior, School of Biological Sciences, University of California, Irvine, Irvine, California 92697-4550.

Developmental Psychobiology, Volume 13 of *Handbook of Behavioral Neurobiology,*
edited by Elliott Blass, Kluwer Academic / Plenum Publishers, New York, 2001.

BRETT A.
JOHNSON AND
MICHAEL LEON

capitalize on the fact that the responses of the olfactory system to specific odorants is highly localized at one level of their processing, thereby allowing us to focus our initial efforts on a very small area in the brain.

As can be seen in Figure 1, mammalian olfactory coding starts with the binding of airborne chemicals to receptors located on the olfactory receptor neurons deep within the nasal turbinates. This induces a cascade of events in the olfactory receptor neuron that eventually causes it to become active. Thousands of such neurons with identical chemical receptors converge in the olfactory bulb and the site of this convergence is the glomeruli that ring the outer lamina of the bulb (Ressler, Sullivan, & Buck, 1994; Mombaerts *et al.*, 1996; Vassar *et al.*, 1994). Within the glomeruli, the axons of homologous olfactory receptor neurons synapse with the dendrites of the second set of neurons in the olfactory pathway and it is within these glomeruli that localization of activity in response to different odorants occurs and can be observed through contemporary neuroscience processes. Specifically, when different odorants are presented to rats, different olfactory glomeruli are activated: activation patterns for individual odorants are consistent among animals (Coopersmith, Henderson, & Leon, 1986; Bell, Laing, & Panhuber, 1987; Guthrie, Anderson, Leon, & Gall, 1993; Royet, Sicard, Souchier, & Jourdan, 1987; Stewart, Kauer, & Shepard, 1979). These specific activation patterns were first seen with a technique in which rats were injected with tracer amounts of a radiolabeled glucose analog ([^{14}C]2-deoxyglucose) before exposure to an odorant. Unlike glucose, this analog is incompletely metabolized, leaving its radioactively tagged product in the cell. The assumption is that increased neural activity increases the use of glucose in brain cells and active cells will accumulate increased amount of the radiolabeled product. The density of the radiolabel in sections of the olfactory bulb reveals localized specific responses to different odorants, and refined image analysis with respect to radioactive standards allow differences in glomerular activity to be quantified.

In our initial investigation of the neurobiological correlates of olfactory learning, we wanted to determine whether the uptake of [^{14}C]2-deoxyglucose (2-DG) in the glomerular layer of the olfactory bulb occurred differentially in response to learned and control odorants (Coopersmith & Leon, 1984). The volatile components of peppermint extract evoked multiple foci of 2-DG uptake that were reliably associated with regional glomerular activity even in control rats. After early olfactory preference training, however, the focal glomerular response to peppermint odor was elevated relative to the response of pups that previously received either repeated presentations of odor or tactile stimulation alone, or unpaired presentations of odor, or tactile stimulation alone (Coopersmith & Leon, 1984; Sullivan & Leon, 1986; Sullivan, Wilson, & Leon, 1989). Only the pairing of odor and tactile stimulation induced both a behavioral preference and an enhanced uptake of radiolabeled 2-DG in response to subsequent presentations of the training odor. Moreover, a variety of stimuli that are associated with the mother, such as oral infusion of milk, can be paired with the odor to induce both a behavior preference and an increase in focal 2-DG uptake in the olfactory bulb glomerular layer (Do, Sullivan, & Leon, 1988; Sullivan & Leon, 1986; Sullivan, Wilson, Wong, Correa, & Leon, 1990). The enhanced 2-DG uptake induced by early learning persisted into adulthood (Coopersmith & Leon, 1986).

One possibility for the increase in the glomerular response is that the pups simply increase their respiration in the presence of an odor that they have come to like. The increase in the resultant amount of the training odor that reaches the nasal

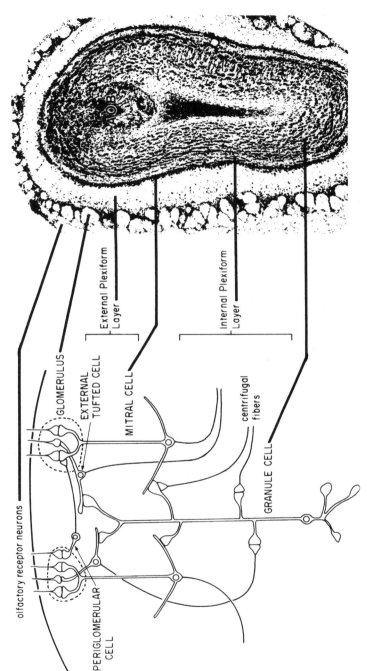

Figure 1. The olfactory receptor neurons transduce chemical stimulation and project into the glomeruli of the olfactory bulb. The axons of the olfactory receptor neurons synapse with mitral, tufted and periglomerular cells within the glomeruli. Mitral cells are the dominant output neurons of the bulb, tufted cells project both within the bulbs as well as to the olfactory cortex, and periglomerular cells appear to mediate interglomerular inhibition. The external plexiform layer contains secondary dendrites of mitral and tufted cells as well as tufted cell bodies and granule cell dendrites. The internal plexiform layer contains granule cell dendrites as well as mitral and tufted axons. The granule cells mediate self-inhibition of mitral cells, inhibition among neighboring mitral cells, and inhibition of mitral cells evoked by efferent neurons. The accessory olfactory bulb (AOB) receives input from a secondary chemical sensing system, which has its receptors in the vomeronasal organ. (Reprinted from Leon, 1987, with permission. Copyright 1987, Elsevier Science.)

epithelium would therefore be expected to increase, perhaps driving an increase in the glomerular response. However, when we monitored the respiration patterns of trained pups in the presence of the training odorant, we found no difference in such breathing patterns (Coopersmith & Leon, 1984; Coopersmith *et al.*, 1986), suggesting that there were no increases in the amount of odorant reaching the olfactory sensory neurons. It was possible that we did not identify a special sniffing pattern that could underlie the differential glomerular response in the absence of a simple increase in respiration. Therefore, we decided to determine whether the glomerular response would differ between trained and control pups after we imposed the same number of identical "sniffs" on them (Sullivan, Wilson, Kim, & Leon, 1988). Anesthetized pups were tested for their response to the trained odorant after they had been placed on a respirator and the odorant was pulled through their nares with another respirator. Only those pups that had been trained to prefer the odor had an increase in 2-DG following identical stimulus exposure.

Central changes caused by early preference training were not confined to the glomeruli. c-Fos expression also specifically increased in periglomerular cells surrounding the glomerular foci with enhanced 2-DG uptake (Johnson, Woo, Duong, Nguyen, & Leon, 1995). Mitral or tufted cells in the region of the bulb deep to the foci of high 2-DG uptake displayed an increased proportion of inhibitory responses, measured electrophysiologically, following early preference training (Wilson, Sullivan, & Leon, 1987). Furthermore, learning decreased the c-Fos response of inhibitory granule cell interneurons located deep to the glomerular 2-DG foci (Woo, Oshita, & Leon, 1996). Granule cells are responsible for lateral inhibition within the bulb by way of their reciprocal dendrodendritic connections with excitatory mitral and tufted cells. Together, these data suggest that neurobiological changes localized within the glomerular layer engage a circuit that modulates at least one output response of the olfactory bulb.

To investigate possible structural changes that could underlie such a long-lasting change in response, we measured the width of the glomerular layer in sections containing 2-DG foci. Trained animals had a wider glomerular layer in focal response regions than did control animals (Woo, Coopersmith, & Leon, 1987). This increased width included both an increase in the diameter of glomeruli and an increase in the number of periglomerular cells surrounding the active glomeruli (Woo & Leon, 1991). The density of glial cell processes within active glomeruli also increased in trained animals (Matsutani & Leon, 1993). In contrast, the density of mitral cells and granule cells in areas deep to the altered glomeruli did not change with learning (McCollum, Woo, & Leon, 1997).

Learning-induced changes in the olfactory bulb appear to be largely confined to the areas showing focal 2-DG uptake. Measurements of glomerular width in nonfocal regions taken systematically with respect to high-uptake foci revealed that experience-dependent increases in width were found only in the regions of high uptake (Woo *et al.*, 1987). The odorant cyclohexanone (with a powerful, solvent-like odor) evoked 2-DG uptake in posterior, medial glomeruli, and early preference training with cyclohexanone-enhanced 2-DG uptake in these glomeruli (Coopersmith *et al.*, 1986). Glomerular 2-DG uptake did not increase in these cyclohexanone-responsive regions following training with peppermint odor, and in animals trained with cyclohexanone, 2-DG uptake did not increase in foci evoked by peppermint extract (Coopersmith *et al.*, 1986). Similarly, changes in the activity of mitral cells following odor preference learning occurred in the lateral bulb, where foci of high 2-DG uptake are observed, but such changes were not found in a region of the bulb remote from these foci (Wilson & Leon, 1988). This specificity impresses us. It implies that anatomically and functionally distinct regions of the bulb have access to motivational and motoric systems that are engaged when animals make choices or engage in acts (Holmes, this volume, Chapter 7) that reflect their particular early olfactory experiences.

Within the portion of the lateral bulb where 2-DG uptake is stimulated by peppermint odor, learning changed the Fos-like response of periglomerular cells associated with midlateral 2-DG foci, but learning did not change the Fos-like response in ventrolateral 2-DG foci (Johnson *et al.*, 1995). These findings of regional heterogeneity in the modification of the bulb by olfactory experience prompted us to explore more intensively the spatial distribution of experience-dependent changes in 2-DG uptake throughout the glomerular layer. To the extent that the regions that we are exploring are representative of others in the bulb, the following analyses may serve as a model for future regional studies.

Important to an analysis of the glomerular code is to have a technique that is capable of resolving activity of individual glomeruli, such as the uptake of radio-labeled 2-DG. To characterize at this level of resolution the spatial distribution of activity evoked by peppermint extract, we generated maps of 2-DG uptake across the glomerular layer and averaged the maps obtained for individual trained and control pups. The maps of the entire glomerular layer throughout the bulb reveal the entire pattern of activity that each odorant evokes. Since the maps are essentially data matrices, this technique also allows us to average individual maps to obtain a group response that can be compared statistically to the responses of other group re-sponses. Naive animals increased activity to peppermint extract in four circum-scribed regions in the posterior half of the lateral glomerular layer (Johnson & Leon, 1996). Within this general region training did not cause a large change in the spatial pattern of 2-DG uptake (Johnson & Leon, 1996). There were circumscribed regional changes, however. Three *midlateral* fields of high uptake in the bulb were markedly enhanced. In contrast, the *ventrolateral* field of peppermint-evoked 2-DG uptake was unchanged after the training (Johnson & Leon, 1996). This pattern of midlateral enhancement and ventrolateral stability parallels what we had seen previ-ously for Fos-like responses (Johnson *et al.*, 1995). Differences between the learning-dependent changes in midlateral versus ventrolateral peppermint-activated glomeruli were not limited to these measures of neural activity. We also found that training decreased β-adrenergic receptor binding in the midlateral glomerular layer, while ventrolateral β-adrenergic receptors were unaffected (Woo & Leon, 1995b). It may be that the medial glomeruli convey different information regarding specific odor-ants than do lateral responses. The lateral representation may contain additional information regarding the odorant that includes affective or experiential informa-tion in addition to the normal information carried by the medial representation. While differential projections of medial and lateral bulb have not been well studied, there are separate lateral and medial olfactory tracts in lower vertebrates that project to distinct forebrain targets where the different kinds of information may be put to different uses (Eisthen, 1997).

Given that odor learning can cause spatially specific increases in focal glomeru-lar activity and changes in local glomerular structure, as well as lead to specific behavioral change, it is important to understand how individual glomeruli contrib-ute to odor processing. This knowledge would provide a framework in which to interpret the changes brought about by experience. It has long been known that different odorants activate different, but overlapping, areas within the glomerular layer of mammalian olfactory bulbs (Bell *et al.*, 1987; Guthrie *et al.*, 1993; Jourdan, Duveau, Astic, & Holley, 1980; Royet *et al.*, 1987; Shepherd, 1991; Stewart *et al.*, 1979). It now is appreciated that the activation of specific glomeruli by particular odorants is a phenomenon that generalizes across a number of species (Cinelli, Hamilton, & Kauer, 1995; Friedrich and Korsching, 1997; Galizia, Sachse, Rappert, & Menzel, 1999; Galizia, Menzel, & Holldobler, 1999; Joerges, Küttner, Galizia, & Menzel, 1997). Only recently, however, have we begun to understand how spatial distributions of glomerular activity might be involved in the processing of volatile chemical cues. We now review recent developments in the interpretation of glomerular activity patterns. Then we address how these developments may relate to the learning-dependent changes we have observed in the olfactory bulb.

Spatial Representations of Chemical Cues

Understanding the code through which different odors are represented in the nervous system is basic to understanding olfactory neurobiology. Since each of the glomeruli receives input from a set of olfactory receptor neurons with a single type of receptor protein, the differential activation of glomeruli would be an ideal place to use differential activity to try to understand what the system is coding. Yet the evidence that individual odorants generate characteristic spatial patterns of activity has generally been ignored. Most models of olfactory coding have favored a low-specificity, broadly tuned system with widely distributed neural connections. Indeed, there has been little emphasis on specificity at any anatomic/functional level of the olfactory brain that has been presented in models of coding. Unlike other sensory systems, spatial information was considered to contribute little to chemical coding. Rather, olfactory information has been thought to been extracted from a distributed excitation of the system. Such a system would likely have few receptor types, but those receptors would be expected to respond to a wide variety of chemicals that exist in nature. For example, electrophysiologic recordings of isolated salamander olfactory receptor neurons indicated that most of these neurons respond to most odorants (Firestein, Picco, & Menini, 1993). Subsequent processing of this broadly tuned response would then allow the signal representing the chemicals to be decoded. In addition, all of the "electronic noses" developed to recognize airborne chemicals have utilized electrochemical sensors that respond to a broad range of odorants. Subsequent computational processing of the information is used to allow the electronic systems to sense and discriminate odorants (Dickinson, White, Kauer, & Walt, 1998).

Olfactory receptor neurons synapse with olfactory bulb cells within the densely synaptic glomeruli. These connections have been reported to be imprecise in nature, with many of the connections demonstrating no obvious organization (Astic & Saucier, 1986; Kauer, 1987). The response of the glomerular layer to different odorants also seemed to be nonspecific because different odorants appeared to stimulate much of the glomerular layer in salamander olfactory bulbs (Cinelli *et al.*, 1995; Kauer & Cinelli, 1993).

Moreover, mitral cells, the dominant output neuron emanating from glomeruli and projecting to the olfactory cortex, were reported to have a nonspecific response to odorant stimulation. Motokizawa (1996), who recorded from mitral cells in rats, found very few differences in their response patterns to different odorants. Mitral cells were broadly tuned; they responded to a wide range of airborne chemicals.

Similar conclusions were reached using electroencephalographic recordings of the bulb to monitor responses to odorants in rabbits (Freeman & Skarda, 1985).

Finally, mitral cell projections to the olfactory cortex in rats appeared to be broadly distributed (Haberly & Price, 1977). Similarly, responses to odorants in the olfactory cortex were difficult to correlate with odorant stimulation, suggesting a random, distributed representation of information regarding olfactory cues in the olfactory cortex (Haberly & Bower, 1989). In all, the entire system appeared to be broadly tuned, with coding and decoding accomplished without using spatial arrangements of projections. Rather, complex computational methods were thought to accomplish odor coding and decoding (Alkasab et al., 1999).

The lack of spatial specificity of the system was further emphasized by complementary lesion studies in which large parts of the olfactory bulb could be removed without affecting the ability of an animal to detect and discriminate a variety of odorants (Lu & Slotnick, 1994, 1998; Slotnick, Graham, Laing, & Bell, 1987; Slotnick, Bell, Panhuber, & Laing, 1997). Even when these lesions targeted specific focal areas of 2-DG uptake evoked by propionic acid, lesioned animals detected the odorant with high sensitivity (Slotnick et al., 1997) and responded as if the perceived odor of the chemical had been unaltered by the lesion (Lu & Slotnick, 1994). Preservation of function following removal of identified 2-DG foci led to the suggestion that these focal responses might not be important for odor processing. Together, these multiple approaches to olfactory function pointed to broadly distributed networks of low specificity.

The discovery of a superfamily of putative olfactory receptor genes that are expressed by sensory neurons in the mammalian olfactory epithelium (Buck & Axel, 1991) began to change the perception of the olfactory system from a low-specificity to a high-specificity system. Mammals appear to have a single olfactory receptor type expressed in each olfactory sensory neuron, thereby conferring an unexpected, but exquisite specificity to the system at its outset (Chess, Simon, Cedar, & Axel, 1994; Malnic, Hirono, Sato, & Buck, 1999). Indeed, recent functional studies indicate that responses of individual receptor proteins to specific odorants are narrowly tuned with a high degree of ligand specificity (Krautwurst, Yu, & Reed, 1998; Zhao et al., 1998).

The response of individual rodent olfactory sensory neurons seems to reflect the specificity of the receptor proteins they express (Malnic et al., 1999). Sato, Hirono, Tonoike, and Takebayashi (1994) recorded the responses of a subpopulation of receptor neurons in deeply anesthetized rabbits, and found highly specific response patterns to low concentrations of chemically related odorants. Thus each olfactory receptor neuron seems to relay specific odorant information to the olfactory bulb.

If the olfactory receptor neurons distribute widely to various parts of the bulb as originally believed, then one would imagine that this specificity would be lost at that next level of processing. Figure 2 illustrates the findings of Mombaerts et al. (1996a), who identified individual olfactory receptor neurons expressing the same olfactory receptor gene in mice and found that all of their axons converged on as few as two glomeruli in each olfactory bulb, one lateral and one medial. The medial projection area was located more caudally and ventrally than the lateral projection (Sullivan & Dyer, 1996; Ressler et al., 1994; Mombaerts et al., 1996a; Vassar et al., 1994), and these identified glomeruli are consistent in their location across animals (Mombaerts et al. 1996a; Ressler et al., 1994; Vassar et al., 1994). Accordingly, there appears to be a high degree of specificity in the organization of information as it arrives in the olfactory

BRETT A.
JOHNSON AND
MICHAEL LEON

bulb with respect to both the tuning of individual receptors and the projections of receptor neurons. The convergence of olfactory information into a small number of glomeruli constitutes a system that is ideally suited to amplify the olfactory signal. In such a system, many thousands of receptor neurons carrying the same information stimulate a small number of mitral cells. The spatial specificity of the organization of the mammalian olfactory system is carried through to the level of the mitral cells because each mitral cell receives afferent input only from a primary dendrite that extends into a single glomerulus. Moreover, the response patterns of neighboring mitral cells resemble those of each other more closely than those of distant mitral cells (Buonviso & Chaput, 1990), thereby further enhancing the regional efferent signal toward the control of processes influenced by early olfactory experience.

Because the projections of homologous olfactory receptor neurons to the bulb appear to be specific to a small number of glomeruli, it seems possible to form a map of the representation of all possible odorants by assessing with high resolution the responses of glomeruli to different odorants. We then could see whether the functional response matched the anatomic specificity in the system. The vast number of odorants to which the system is responsive, however, complicates this possibility. Estimates of the number of possible odorants range from a low of 10,000 to hundreds

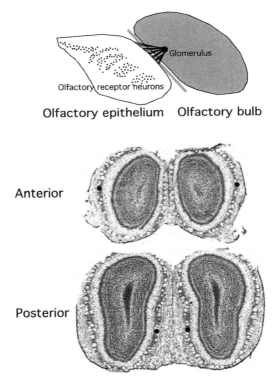

Figure 2. The projection patterns of olfactory sensory neurons that express the same putative olfactory receptor gene. The receptors of these homologous olfactory receptor neurons are distributed across the olfactory epithelium and these neurons project to a single glomerulus on the medial aspect and a single glomerulus on the lateral aspect of the bulb, albeit in different anterior–posterior planes shown in coronal sections.

of thousands. If the number of odorants matched the number of putative odorant receptors, then it would be possible to predict that each receptor could bind one odorant maximally and then communicate this information to the olfactory cortex. However, the number of chemicals with distinct odors is at least 10 times the number of odorant receptors. Therefore, it is unlikely that each odorant is bound to a single olfactory receptor that recognizes principally that chemical. It is far more likely that most olfactory receptor proteins function in the same way as do other receptors, by binding to and detecting specific molecular features (Figure 3) (Buck & Axel, 1991; Imamura, Mataga, & Mori, 1992; Katoh, Koshimoto, Tani, & Mori, 1993; Sato *et al.*, 1994). The idea that olfactory receptors are molecular feature detectors follows from studies of the binding of neurotransmitters and pharmacologic agents to their receptors, for which specific molecular features determine the binding affinity (Dean, 1987).

The change in the conception of how olfactory systems are organized, at least in mammals, has been dramatic. The dominant view of its organization has changed from a broadly tuned, randomly organized system with widely distributed activity to that of a narrowly tuned system with a high degree of spatial specificity of neural activity playing a key role in the coding of chemical cues. We will first show that the coding at the level of the glomeruli is quite specific and linked to specific aspects of the airborne molecules that are perceived as odors.

Airborne chemicals are probably represented by a combination of either shared

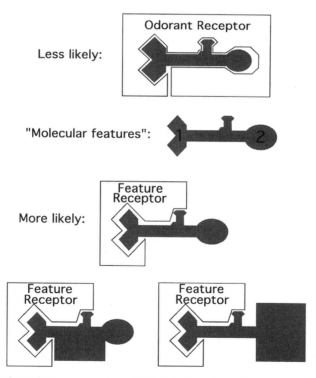

Figure 3. Two hypotheses for how odorants would likely be bound to olfactory receptor proteins. The less likely possibility is that each odorant binds to a single receptor in its entirety. A more likely hypothesis is that the molecular features that comprise each odorant bind separately to each type of receptor.

BRETT A.
JOHNSON AND
MICHAEL LEON

or unique features of odorant molecules, termed "primitives," "odotopes," or "epitopes" (Buck, 1996; Shepherd, 1991; 1994), that would allow discrimination among odorants (Figure 3). Since several odorants would be expected to activate some of the same receptors, each odor would be coded by the particular combination of receptors activated by the features of that odorant, as illustrated in Figure 4. This concept is the foundation for recent "combinatorial" proposals for the mechanism underlying olfactory coding (Buck, 1996; Friedrich and Korsching, 1997; Johnson *et al.*, 1998; Kauer and Cinelli, 1993; Malnic *et al.*, 1999; Ressler *et al.*, 1994; Shepherd, 1991, 1994; Vassar *et al.*, 1994; Vickers and Christensen, 1998).

How would a combinatorial code of odor quality be relayed from the olfactory receptor proteins to the central nervous system? Recall that axons from homologous olfactory receptor neurons converge onto a small number of glomeruli in the main olfactory bulb (Mombaerts *et al.*, 1996a; Ressler *et al.*, 1994; Vassar *et al.*, 1994). Given this information, the pattern of glomerular activation should be determined by the specific olfactory receptor proteins that recognize distinct features of an odorant molecule. Thus, each pure odorant should generate a characteristic spatial pattern of glomerular activation. Figures 3 and 4 shows that any two odorants sharing a chemical feature recognized by a single receptor should activate the same glomeruli. Odorants with few molecular features should generate simpler patterns of glomerular activation than odorants with many molecular features because the latter should bind to additional receptor proteins. Finally, if the specific receptor proteins encode

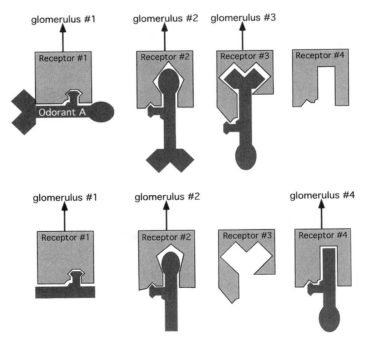

Figure 4. The combinatorial model of olfactory coding. The combination of molecular features that describe an odorant constitutes the representation for any odor in the olfactory system. One odor may be represented by activation of receptors 1, 2, and 3, while another odor may be represented by activation of receptors 1, 2, and 4. The code for each odor would be revealed by the pattern of glomerular activation because olfactory receptor neurons expressing the same receptors for a specific molecular feature project to individual glomeruli.

information in a highly tuned way, the glomerular layer activity evoked by an odorant feature should be represented both laterally and medially because olfactory sensory neurons expressing a given type of receptor protein project to both lateral and medial glomeruli in each bulb. On the other hand, if the olfactory information is widely distributed and the code is represented by very broad activity patterns in a broadly tuned system, we should not see a focal organization of activity in the glomerular layer of each bulb.

Our strategy has been to test the idea that the olfactory system functions as a detector of molecular features, and involves characterizing spatial responses to odorants that shared some molecular features, while not sharing others (Johnson *et al.*, 1998). We selected four chemicals for our initial study, all with fruit odors: ethyl butyrate, isoamyl butyrate, isoamyl acetate, and ethyl acetate (Figure 5). These chemicals were selected because they have a common structure and because pairs of the chemicals share particular potential molecular features. If the patterns of activity revealed a common representation among all odorants and the pairs that shared a particular molecular feature also shared a focal representation, then the data would support the notion of a system that was sensitive to combinations of molecular features. We then mapped [^{14}C] 2-deoxyglucose uptake across the entire glomerular layer of the rat olfactory bulb to examine activity patterns evoked by these odorants.

Each of the four distinctive odorants evoked a distinct spatial pattern of glomerular activation, despite the fact that they differed only slightly in their molecular features. More importantly, odorants that shared molecular features stimulated overlapping areas of activity in the glomerular layer, as can be seen in Figure 6. Both odorants that possessed an isoamyl group activated glomerular areas that were not activated by the two odorants that did not have an isoamyl group (Johnson *et al.*,

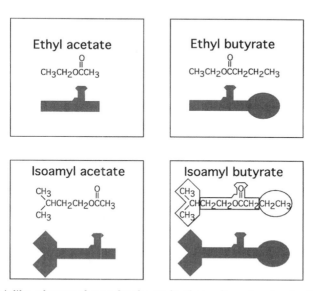

Figure 5. Four fruit-like odorants chosen for the study of encoding of molecular features by the rat olfactory system. All of the odorants share a common molecular structure, but only two odorants share the isoamyl feature. Another pair shares the butyrate feature. Common responses to shared features would indicate that the olfactory system encodes information based on the sensation of the discrete combination of the molecular features of each odorant.

1998). Similarly, both odorants that possessed an ethyl group stimulated areas that were not stimulated by the two odorants that did not have an ethyl group. Finally, all four molecules also activated overlapping parts of the bulb, possibly reflecting the common core structure that they shared. These data are consistent with a combinatorial mechanism of olfactory coding wherein unitary responses of olfactory receptors to particular features of a given odorant generate spatial patterns of bulbar activity that are characteristic of that odorant. This implies that the specific pattern of activity generated by a particular odorant could be predicted from the activity patterns of other odorants with which it shares nonoverlapping common elements. Another observation that was consistent with such a combinatorial mechanism was that the complexity of the spatial patterns increased with odorant molecular size (Johnson *et al.*, 1998). Increasing molecular size would likely increase the number of molecular features bound by olfactory receptor proteins. The increase in the number of molecular features that would be processed simultaneously would be expected to increase the number of focal regions activated in the glomerular layer.

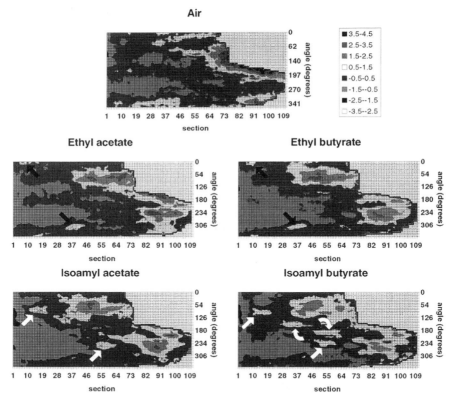

Figure 6. Maps of 2-DG activity comparing groups of rats exposed to odorants sharing or not sharing specific molecular features. Straight white arrows point to common unique areas of activation and can be seen for the two groups exposed to the isoamyl feature, and straight black arrows point to different foci that are shared by the two groups exposed to the ethyl feature. Curved arrows point to unique foci present only in isoamyl butyrate. All groups share common regions of activity. All foci are paired, with one medial and one lateral region of activity, mirroring the bilateral projections of homologous olfactory receptor neurons. (Reprinted from Johnson *et al.*, 1998, with permission. Copyright 1998 Wiley-Liss, Inc., a subsidiary of John Wiley & Sons, Inc.)

As predicted by the projections of homologous olfactory receptor neurons, all maps of activity revealed paired foci, with one focus on the lateral and one focus on the medial aspect of the bulb. Moreover, the lateral focus was always more anterior than the medial focus, again as predicted from the olfactory neuron projection patterns (Mombaerts *et al.*, 1996a; Sullivan and Dyer, 1996). Given the similar projection patterns of olfactory receptor neurons expressing individual genes, the spatial relationship between these foci of activity supports the functional relevance of the putative olfactory receptor genes.

While the ester odorants that we studied are related, other odorants are even more closely related. One possibility is that the olfactory system encodes similar odorants in widely separated parts of the bulb to allow them to be easily discriminated. Alternatively, the system may cluster the representations of very similar odorants to allow their identification as a group and to sharpen the discrimination among similar odorants by lateral inhibition. To distinguish between these two hypotheses, we tested the spatial patterns of activity evoked by odorants that differed by as little as one methylene group. We exposed rats to a series of straight-chained acids varying in length from two to eight carbons, all with odors characteristic of mammalian bodies. Then we mapped [^{14}C]2-DG uptake across the glomerular layer. We focused on the question of whether glomerular responses to these odorants were clustered or dispersed in the bulb (Johnson, Woo, Hingco, Pham, & Leon, 1999).

The acid odorants evoked activity in four discrete glomerular regions that involved two lateral/medial pairs of fields (Figure 7A) Johnson *et al.*, 1999). Odorants differing by a single step in carbon chain length stimulated overlapping, but distinct sets of glomeruli within each of these four fields. The locations of response within the fields were calculated as centroids of 2-DG uptake (Figure 7B). For each field, the centroid of activity shifted progressively toward more ventral and/or rostral positions with increasing carbon chain length of the straight-chain acid odorants (Johnson *et al.*, 1999). Thus, within each functional field, glomeruli appear to be chemotopically arranged such that the nearest neighbors have the most similar odorant specificity. Clustered responses for aldehydes also have been observed in the rat bulb (Rubin & Katz, 1999).

Single-unit recording studies of mitral/tufted cells within a region of the rabbit olfactory bulb that corresponds to a dorsomedial field that we identified in rats reveal responses to various straight-chain acids (Imamura *et al.*, 1992; Mori, Mataga, & Imamura, 1992). Furthermore, individual mitral/tufted cells in this region are tuned to odorants possessing a limited range of carbon chain lengths (Imamura *et al.*, 1992; Mori *et al.*, 1992). On the basis of their findings, Mori and coworkers proposed that neighboring glomeruli within the dorsomedial bulb may respond to carbonyl-containing odorants of different chain lengths, with a preference for a given chain length. Imamura *et al.* (1992) also suggested that lateral inhibitory interactions between glomeruli and/or mitral/tufted cells could sharpen the specificity of an individual mitral/tufted cell to a more limited range of stimuli by suppressing neighboring responses to closely related odorants. Consistent with this hypothesis, they found that a reduction of lateral inhibition among adjacent mitral cells produced a broader response specificity by these cells to different straight-chain acids (Yokoi, Mori, & Nakanishi, 1995). These suggestions are supported further by our data that reveal a chemotopic arrangement of glomerular responses within this area of the bulb. The use of chemotopic arrangements of glomeruli to achieve odorant tuning provides evidence that spatial distributions of responses may be used by the olfactory system to encode odor quality.

BRETT A.
JOHNSON AND
MICHAEL LEON

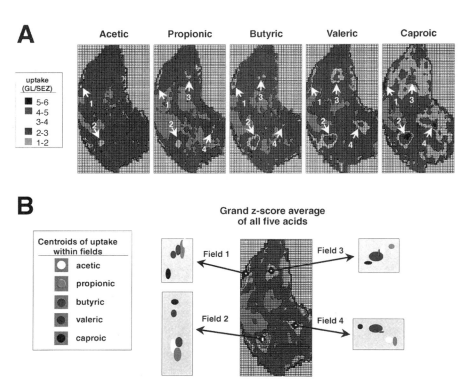

Figure 7. (A) Averaged 2-DG uptake in the glomerular layer of the olfactory bulb of groups of rats exposed to different volatile acid odorants. Contour charts show different average levels of 2-DG uptake principally in two paired fields (medial and lateral) of the glomerular layer of the olfactory bulb. (B) To compare the spatial distributions of glomerular activity produced by different odorants, we describe the entire pattern of activity in each animal using a single value. We compare centroids of activity as that single value; this calculation describes the location within an array of values in two coordinates in which the sums of values on all sides of that point are equal. In addition, the standard error around the centroids is represented in two dimensions as an ellipse. These centroids of activity are clustered, but distinct in each of the paired fields, suggesting a chemotopic organization of coding of closely related molecules. (Reprinted from Johnson *et al.*, 1999 with permission. Copyright 1999 Wiley-Liss, Inc., a subsidiary of John Wiley & Sons, Inc.)

In our study of ester odorants, we identified high-uptake foci that responded specifically to isoamyl butyrate (Johnson *et al.*, 1998). These foci were associated with very few glomeruli in any given coronal section. In some cases, the foci seemed to involve only a single glomerulus in a section. Such a finding would be the predicted consequence of the activation of a single class of olfactory receptor protein expressed by olfactory receptor neurons that all project to a single glomerulus. However, most other responses to the esters involved foci of 2-DG uptake that were associated with more than one glomerulus in any given coronal section. Our ability to find a close correspondence between individual glomeruli and high-uptake foci for the isoamyl butyrate-specific fields suggests that the large number of glomeruli associated with other foci may be due to the activation of closely related receptor neurons projecting to adjacent sites. The number of glomeruli associated with a given focus differed for the different esters evoking the focus (Johnson *et al.*, 1998), which also suggests that neighboring glomeruli might have related, but distinct, specificities.

Clustering of similarly specific glomeruli is further supported by data showing that increases in odorant concentration lead to increases in the area of either focal 2-DG uptake or c-Fos mRNA hybridization signal (Bell *et al.*, 1987; Guthrie and Gall, 1995; Johnson *et al.*, 1999; Stewart *et al.*, 1979). The increased areas at higher odorant concentrations could be interpreted as a recruitment of nearby glomeruli receiving projections from olfactory receptors with lower affinity for the odorant, suggesting that just as in all receptor–ligand relationships, a receptor will bind to a range of molecular features. Clustering of glomeruli with specificities for chemically related amino acid odorants also has been observed through Ca^{2+}-sensitive dye recordings of zebrafish olfactory bulbs, indicating that the use of spatial relationships to achieve finer odorant tuning might be a phenomenon that can be generalized across species (Friedrich & Korsching, 1997).

Johnson and Leon (2000) presented two types of odorants at different concentrations. One set of odorants, at least in humans, demonstrates what we may call odor constancy throughout the range of concentrations. The other set of odorants was selected because humans report that these change in quality with increasing concentration. A combinatorial code would predict that there would be distant glomeruli activated at these higher concentrations for those stimuli that change with concentration. For the odorants with presumed odor constancy, increased concentration correlated with increased 2-DG uptake in all glomerular areas of activation. Indeed, increased glomerular activity with increasing odorant concentration has been noted repeatedly with a variety of techniques (Cinelli *et al.*, 1995; Friedrich and Korsching, 1997; Guthrie and Gall, 1995; Joerges *et al.*, 1997; Johnson *et al.*, 1999; Rubin and Katz, 1999; Stewart *et al.*, 1979). At the same time, those odorants selected for their ability to change quality evoked activity in parts of the bulb that were not activated at lower concentrations.

Increasing odorant concentrations are thought to recruit additional types of odorant receptors that have lower affinity for the odorants (Malnic *et al.*, 1999). The activation of new glomeruli at higher odorant concentrations probably indicates that increasing odorant concentrations recruit new olfactory receptor proteins, since sensory neurons expressing the same receptor gene project to as few as one lateral and one medial glomerulus (Mombaerts *et al.*, 1996; Ressler *et al.*, 1994; Vassar *et al.*, 1994; Wang, Nemes, Mendelsohn, & Axel, 1998).

The recruitment of different kinds of olfactory sensory neurons at higher odor concentrations might not always result in a different perceived odor. Activation of distinct olfactory sensory neurons probably depends on the spatial arrangements of the glomeruli to which different types of sensory neurons project. Glomeruli responding maximally to aliphatic acids of a given carbon chain length appear to be located near glomeruli that respond maximally to acids of slightly different carbon chain lengths, suggesting that sensory neurons expressing receptors of similar selectivity may project to nearby glomeruli (Johnson *et al.*, 1999). Indeed, olfactory receptor neurons expressing similar receptors project to neighboring glomeruli (Tsuboi *et al.*, 1999). Similar specificities of nearby glomeruli also would explain why higher concentrations of odorants increase the area of focal glomerular responses detected using 2-DG uptake (Johnson *et al.*, 1999; Stewart *et al.*, 1979), *in situ* hybridization for c-Fos mRNA (Guthrie & Gall, 1995), voltage-dependent dye recording (Cinelli *et al.*, 1995; Friedrich and Korsching, 1997), or optical recording (Rubin

& Katz, 1999). The increased area of response may reflect the recruitment of neighboring glomeruli that are maximally responsive to odorants of slightly different chemistries, but that also can respond to higher concentrations of suboptimal stimuli (Johnson et al., 1999).

Spatial clustering of glomeruli with similar, but not identical response profiles has been proposed to stimulate a lateral inhibitory network that tunes projection neurons to specific odorant carbon chain lengths (Friedrich & Korsching, 1997; Johnson et al., 1999; Yokoi et al., 1995). Since neighboring mitral cells can inhibit one another via granule cell dendrites, the recruitment of a neighboring glomerulus at higher odorant concentrations may not override the greater inhibition caused by increasing activity of the originally active mitral cells. Thus, mitral cells associated with a neighboring glomerulus may not change their activity even if their associated glomeruli do. In that case, there would not be a change in odor quality. On the other hand, odorants that evoked new areas of activity far enough from the original glomerular activity to avoid lateral inhibition and their associated mitral cells could well evoke the additional activity that could underlie an altered perception with increasing concentration.

In mice, the olfactory receptor proteins expressed by sensory neurons appear to be involved in targeting the axons of the cells to the appropriate regions of the olfactory bulb (Mombaerts et al., 1996a; Wang et al., 1998). If receptors with similar specificity also contain a high degree of amino acid sequence homology, and if similar sequences cause guidance to nearby glomerular locations, then axonal guidance by the receptor proteins would be an efficient means to construct clusters of similarly specific glomeruli. The reliability in the locations of uptake across different animals exposed to the same odorant, which must have been present to obtain statistically significant results in our analyses, implies that there is a profoundly rigorous set of parameters used to establish the topography of the projections of homologous sensory neurons.

The 2-DG maps suggest that the olfactory system is encoding the information provided by molecular features of chemicals. If this mechanism is the means by which the system codes chemical information, then destruction of such foci should prevent the animals from normal recognition of the odorants. As discussed above, the results of olfactory bulb lesion studies questioned the importance of focal areas of 2-DG uptake. Prior to our work, there were numerous reports that propionic acid evoked only a single major focus of activity that was located in the dorsomedial bulb (Mori et al., 1992; Sallaz & Jourdan, 1993; Slotnick et al., 1987; Slotnick, Panhuber, Bell, & Laing, 1989). The removal of this focus did not affect the ability of an animal to detect propionic acid, to distinguish propionic acid from other chemicals, or to remember the odor of propionic acid learned prior to the lesion (Lu & Slotnick, 1994, 1998; Slotnick et al., 1987, 1997). However, by systematically mapping the response of the entire glomerular layer, we found four fields (i.e., two paired fields) that contained reliable, robust responses to propionic acid (Johnson et al., 1999). These four fields were distributed in anterior, posterior, medial, and lateral portions of the bulb, including regions not removed in the prior lesion studies (Johnson et al., 1999). Because important focal areas of 2-DG uptake were not removed in the lesion studies, it remains possible that these areas are critical for odor coding. It would seem possible to be able to remove all of the areas that are stimulated by a particular odorant and thereby prevent its normal perception.

Our data on glomerular activity patterns suggested that there was considerable localization of responses within the olfactory bulb, with some areas being stimulated far more than others. Because individual mitral cells in rats extend their apical dendrites into only a single glomerulus, one would predict that the mitral cells connected to active glomeruli would also be more active than mitral cells located elsewhere in the bulb. Such localization of active mitral cells was reported in studies performed on deeply anesthetized rabbits in which recording was deep in the glomerular layer (Imamura *et al.*, 1992; Mori *et al.*, 1992). However, other studies indicated that widely distributed mitral cells in lightly anesthetized rats responded similarly to many odorants (Motokizawa, 1996). These units may have been secondary or tertiary and not the primary ones recorded by Mori. To help resolve this issue, we mapped 2-DG uptake in the external plexiform layer and in the internal plexiform/superficial granule cell layers of rats exposed to straight-chain acid odorants (Johnson *et al.*, 1999). Uptake in these layers must reflect the activation of tufted or mitral cells because the projections from olfactory sensory neurons do not penetrate further than the glomerular layer. We found focal areas of response in these deeper layers that mirrored those detected in the glomerular layer (Johnson *et al.*, 1999). These distinctive responses were located in the same dorsomedial part of the bulb where Mori *et al.* found specific responses to straight-chain acids (Imamura *et al.*, 1992; Mori *et al.*, 1992). Furthermore, the uptake in the internal plexiform/superficial granule cell layer differed systematically in location across straight-chain acid odorants differing in carbon chain length, which likely indicates an orderly spatial arrangement of projection neurons exhibiting different optimal specificities (Johnson *et al.*, 1999). Our results arose from studies of rats that were not subjected to anesthesia, so that a picture not contaminated by anesthesia has been obtained in rats.

Since homologous olfactory sensory neurons converge into specific glomeruli within the bulb, glomerular activity should reflect the narrow tuning of particular olfactory receptor proteins to the specific molecular features of odorants. This raises the question of why so many salamander olfactory receptor neurons respond to a given odorant, and why a large number of odorants can stimulate any particular salamander olfactory receptor neuron *in vitro* (Firestein *et al.*, 1993). One simple possibility is that salamanders and rodents have evolved different kinds of tuning mechanisms for their olfactory systems (Sato *et al.*, 1994), a possibility that will be explored in more detail below. Another possible answer may be similar to that for the low-specificity mitral/tufted cell responses detected in the olfactory bulb by Motokizawa (1996). The study showing broad specificity of salamander sensory neurons used high odorant concentrations and did not preselect cells giving the largest responses. Indeed, in studies that did preselect for mouse olfactory receptor neurons located in a portion of the septum of the epithelium, Sato *et al.* (1994) found good tuning with respect to the carbon chain length of straight-chain acids and alcohols, especially when the odorants were presented at low concentrations. Similarly, in a study that preselected for mouse sensory neurons sending axons to glomeruli of the dorsomedial olfactory bulb, where responses to straight-chain acids are observed, Bozza and Kauer (1998) found that individual sensory neurons responded to straight-chain acids but not to alcohols of a similar carbon chain length. Selection for a particular rat olfactory receptor gene also resulted in sensory neuron responses that both were finely tuned and highly sensitive (Zhao *et al.*, 1998).

Despite the evidence that the olfactory system is selective for a narrow range of chemical structures, crude counts of glomeruli underlying high-uptake foci evoked by high concentrations of valeric acid suggest that up to 5% of all glomeruli are located in the four fields that are activated by this odorant. High concentrations of caproic acid can stimulate an even greater proportion of the glomeruli (Johnson *et al.*, 1999), as can high concentrations of isoamyl butyrate (Johnson *et al.*, 1998). Thus, at high odorant concentrations, the inherent specificity of the olfactory system may be obscured. Behavioral studies would help to resolve these issues.

In our study of straight-chain acid odorants, both the amount of 2-DG uptake and the number of responsive glomeruli increased with increasing size of the odorant molecule (Johnson *et al.*, 1999), a finding consistent with other observations demonstrating elevated responses with increases in odorant molecular size and/or hydrophobicity. For example, Sato *et al.* (1994) found that the number of responsive mouse olfactory receptor neurons increases with increasing carbon number of straight-chain acids and alcohols. Ottoson (1958) found that increasing odorant hydrophobicity correlated with increased electroolfactograph response amplitude in frog epithelium. Increased responses of sensory neurons and glomeruli with increasing odorant size may underlie the inverse correlation that exists between the carbon chain length of straight-chain acids or aldehydes and the concentration of these odorants required for threshold odor detection in humans (Cometto-Muñiz, Cain, & Abraham, 1998).

The spatial overlap in responses at the level of the glomerular layer even for odorants that differ greatly in certain molecular features suggests the possibility that even chemically dissimilar odorants might be confused on a perceptual level. Humans show surprisingly poor initial performance during odorant identification tasks, correctly identifying only about half of common, familiar odorants (Cain, 1979; Cain & Potts, 1996). While there are many misidentifications of similar odorants, there also are dramatic mistakes involving very different odorants (Cain, 1979; Cain & Potts, 1996) that appear to be related to perceptual confusion rather than to problems in verbal labeling (Cain & Potts, 1996). Odorant identification becomes virtually perfect, however, with feedback about odorant identification (Cain, 1979), suggesting that humans may learn to use information that allows them to identify odors. Following training, humans can discriminate between very closely related acids, although their difficulties increase when these compounds differ by only a single carbon or by changes in branch structure (Laska & Teubner, 1998). Rats can be trained to discriminate between propionic acid and acetic acid, which differ by a single step in carbon chain length and which to humans have very similar odors (Lu & Slotnick, 1994, 1998; Slotnick *et al.*, 1997). However, the possibility that rats would confuse these odorants (or even more dissimilar ones) without prior discrimination training has not been evaluated.

A SIMPLE MODEL OF ODORANT MOLECULAR FEATURE PROCESSING

The spatial patterns of odorant-evoked activity observed when 2-DG uptake is surveyed across the entire glomerular layer are consistent with a simple model of the processing of odorant molecular features in the rat olfactory bulb. The model presented below synthesizes aspects of several, previously proposed models of odor coding by the olfactory system (Axel, 1995; Buck, 1996; Imamura *et al.*, 1992; Kauer and Cinelli, 1993; Mori and Yoshihara, 1995; Shepherd, 1994).

In a sense, the odorant molecule is broken up into its component features at the level of the olfactory epithelium and the information regarding those features is then sent to the olfactory bulb for further processing. Each receptor protein expressed by neurons in the olfactory epithelium should exhibit a preferential response to a given molecular feature that is present in odorant molecules. The receptor also would respond with lower affinity to variants of that molecular feature. Each olfactory sensory neuron in rats appears to expresses only one receptor protein (Chess *et al.*, 1994; Malnic *et al.*, 1999; Mombaerts *et al.*, 1996b) and therefore each sensory neuron should respond preferentially to a single molecular feature.

Olfactory receptor neurons expressing the same kind of receptors converge on a limited number of glomeruli (Mombaerts *et al.*, 1996a). The convergence of many neurons processing the same information regarding a molecular feature should increase the sensitivity of the system because it maximizes the number of neurons that will stimulate the second-order neurons in the bulb. Mitral/tufted cells have a single apical dendrite that arborizes in a single glomerulus (Mori, 1987), thereby assuring that each mitral/tufted cell receives direct projections from sensory neurons in which activity is related to the binding status of a single receptor protein.

Glomeruli responding to the same molecular feature appear to be clustered together, creating a functional field that can be revealed by studying 2-DG uptake. Within a given cluster, glomeruli appear to be distributed in such a way that nearest neighbors exhibit the smallest differences in response specificity, perhaps related to odorant size or hydrophobicity. This chemotopy is the strongest evidence for spatial coding in the olfactory system. The rigorous spatial arrangement would insure that lateral inhibition by synaptic interactions with interneurons in the glomerular and external plexiform layers would sharpen mitral and tufted cells to the specific molecular attributes that would be difficult to distinguish without tuning.

The effects of tuning would be most pronounced during later responses after the lateral inhibition by neighboring mitral cells has been initiated. Initial mitral/tufted cell responses may transmit information about the general class of the odorant (floral), while later, sharpened responses may transmit more specific information (rose). Indeed, one computational model of coding in the olfactory system predicted a similar hierarchical organization of olfactory coding (Ambros-Ingerson, Granger, & Lynch, 1990).

In this model, the bulb would be envisioned as an array of functional fields, in which each field accomplishes tuning of odorant features by way of lateral inhibition. It is not yet clear that there will be any psychophysical meaning associated with the spatial relationships between distinct fields evoked by structurally different odorants. If the basal dendrites of mitral cells connected to glomeruli at the boundary of a functional field branch radially in all directions, they may induce inhibition of neighboring mitral cells connected to glomeruli within a separate functional field. Thus, there may be interactions between parts of adjacent fields that have different molecular feature specificities. These interactions would lead to the additional prediction that certain odorant features and thereby certain odorants that evoke activity at the boundaries of adjacent fields may mask each other more effectively (e.g., in odorant mixtures) than would others that are represented in more distant fields. On the other hand, it may be that mitral cells located at the edge of a given functional field have basal dendrites extending only in the direction of neighboring mitral cells within the same field, thereby creating an isolated anatomical unit for the processing of chemicals with similar specificities.

At some point in this processing system, the molecular features that had been separately processed in the olfactory bulb must be brought together to form the perception of a specific odor. The axon of a mitral cell leaves the olfactory bulb to synapse with cortical and other forebrain targets (Macrides & Davis, 1983). A single cortical pyramidal cell may receive input from various mitral cells, each of which transmits information about a single feature present in an odorant. Such an idealized convergence could provide coincidence detection to reassemble the features of a pure odorant. The previously reported lack of topography in mitral cell-to-cortex projections (Haberly & Price, 1977) could insure the convergence of most possible sets of molecular features that could be present in an odorant or odorant mixture. Recent studies, however, have shown that mitral cells associated with a single glomerulus may project to only a limited domain within the olfactory cortex (Puche, Aroniadou-Anderjaska, & Shipley, 1998). Activation of one glomerulus thus leads to a modular activation of cortical pyramidal cells (Puche *et al.*, 1998). Therefore, a hierarchical reassembly of the information describing distinct molecular features carried by mitral cell axons may occur in the olfactory cortex and the features of other odors may be rejoined at an even higher level of processing involving neuron assemblies and/or secondary projections of cortical pyramidal cells.

ALTERNATIVE MODELS OF ODORANT PROCESSING

The study of olfaction has employed an extremely wide range of animal species, and alternative models of odor quality coding have combined results from these numerous species (Kauer & Cinelli, 1993). However, there are both fundamental species-dependent differences in the anatomy of the olfactory system (Eisthen, 1997) and important ethological concerns that suggest caution in considering one species as a model for another. There also are considerable gaps in our knowledge concerning some of the species that make it difficult to judge the suitability of a given model.

Studies of locusts and honeybees have revealed odorant-specific temporal patterns of responses by individual projection neurons in the antennal lobe that are phase-linked to each other and to widespread odorant-induced oscillations (Laurent, Wehr, & Davidowitz, 1996; Stopfer, Bhagavan, Smith, & Laurent, 1997; Wehr & Laurent, 1996). Pharmacologic blockade of the oscillations disrupts the learning and memory of a certain pair of similar chemical odorants detected by honeybees, but not of a pair of more dissimilar chemicals (Stopfer *et al.*, 1997). Different odorants also activate distinct subsets of glomeruli in the honeybee (Galizia, Sachse, Rappert, & Menzel, 1999; Joerges *et al.*, 1997) and the persistence of discrimination between dissimilar odorants in the absence of oscillations suggests the possibility of spatial coding to distinguish some odorants (Stopfer *et al.*, 1997). Nevertheless, the rigorous correlation between temporal parameters of the projection neuron activity and the identity of the similar odorants raises the possibility that these species also may use a critical mechanism involving temporal coding to discriminate between similar odorants. On the other hand, there are alternative interpretations of these data that suggest that the locusts or bees function very much like rats and rabbits in using a highly specific spatial code. The loss of synchronized activity in the antennal lobes of these insects is accomplished by blocking γ-aminobutyric acid (GABA) receptors, and as noted above, when these receptors are blocked in rabbits, there is a loss of response specificity to odorants, probably due to the loss of lateral inhibition (Yokoi *et al.*, 1995). Since the spatial localization of the initial odorant response in bees and locusts was not identified before and after drug administration, it is possible that blocking GABA in these insects simply allows a less specific spatial localization activity, with consequent loss of perceptual specificity. Under these conditions, one would expect to observe difficulties in discriminating between similar odors and little difficulty discriminating between dissimilar odors.

It has been suggested that these data from insects also may apply to odor coding in mammals, including rats (Dorries, 1998; Laurent, 1997), despite remarkable anatomic differences between the species. A given locust projection neuron has direct connections to approximately 11 noncontiguous glomeruli (Laurent & Naraghi, 1994), in contrast to the one-to-one relationship between glomeruli and projection neurons in the rat. To our knowledge, olfactory receptor genes in these species have not yet been cloned, so the relationship between glomeruli and the projections of homologous sensory neurons is unknown. On the other hand, an olfactory receptor gene family has been identified in fruit flies, and that species seems to express one receptor per olfactory receptor neuron (Clyne *et al.*, 1999; Vosshall, Amrein, Morozov, Rzhetsky, & Axel, 1999). If there were a one-to-one relationship between homologous sensory neurons and glomeruli, then a given insect projection neuron could receive direct excitatory input relay from about 11 different olfactory receptors. Such an organization would be expected to increase sensitivity to a broad group of odorants (e.g., food odors). However, it could be that all of the glomeruli to which a given locust projection neuron is connected receive axons only from homologous sensory neurons, thereby producing the same type of convergence present in a rodent. Such an anatomic organization could be expected to increase sensitivity to a specific odorant. If this were the case, rules for the clustering of glomeruli with related specificities may be different in the locust, given that the multiple glomeruli influencing a single projection neuron are themselves separated by additional glomeruli.

There is direct evidence in other invertebrates of excitatory and inhibitory actions of different odorants on the same sensory neuron (Michel & Ache, 1994). This suggests both that a single sensory neuron may express multiple olfactory receptor proteins and that some degree of odor processing may be accomplished by the first neuron in the pathway. Direct evidence for such processing in mammals has not been reported, although peripheral interactions among distinct odorants remain a possibility (Bell *et al.*, 1987).

Our data certainly cannot rule out the possibility of temporal representations of odorant quality in the rat, and temporal factors may figure heavily in coincidence detection of multiple molecular features. Odorant-evoked oscillations also may be involved in the discrimination between similar odorants that evoke overlapping responses in the rat glomerular layer, especially when these odorants are presented as mixtures. Indeed, mixtures of propionic and caproic acids have been reported to induce oscillatory responses in the dorsomedial region of the rabbit olfactory bulb (Mori *et al.*, 1992).

Other studies with salamanders have described broad patterns of activation across bulbar layers, and the patterns evoked by odorants of different chemistries overlap extensively (Cinelli *et al.*, 1995). These data have been used to support diffusely distributed models of odor processing in mammals. The broad patterns of voltage-sensitive dye responses to odorants measured in salamanders (Cinelli *et al.*, 1995) contrast greatly with the focal patterns of 2-DG uptake observed in rats using the same odorant, ethyl butyrate (Johnson *et al.*, 1998). Although these findings may indicate a different sensitivity of the two imaging techniques, it also may be indicative of an important species difference. As in the insect, individual mitral cells of the salamander olfactory bulb have apical dendrites in multiple glomeruli (Eisthen, 1997; Nieuwenhuys, 1967), and the relationships between glomeruli and sensory neurons expressing single types of olfactory receptor proteins are not yet known. As discussed above, it also is possible that sensory neurons in salamander (Firestein *et al.*, 1993) are more broadly tuned than are sensory neurons in a rodent (Bozza and Kauer, 1998; Sato *et al.*, 1994; Zhao *et al.*, 1998). Because of the numerous anatomic differences in the olfactory systems of amphibians and rodents, odor-coding principles also may differ between them.

Fish and rodent olfactory systems are even more different in anatomy than are the olfactory systems of salamanders and rodents (Eisthen, 1997; Nieuwenhuys, 1967). Therefore, it is remarkable that clustering of glomeruli of related specificities is observed in fish (Friedrich & Korsching, 1997) as well as in rodents. This finding suggests that some tuning principles may be present across widely different species. Indeed, tuning of individual mitral cells to straight-chain compounds of a given chain length also has been found in the frog olfactory bulb (Dôving, 1966).

Obviously, the possible discovery of subtle differences in the specificities of neighboring glomeruli, as have been observed in fish and rats, requires the presentation of odorants differing systematically in chemical structure. It therefore would be interesting to determine whether insects and salamanders employ more spatially specific principles when tested with sets of chemically related odorants. The behavioral relevance of the odorants selected for study also should be considered. For example, the amino acid odorants that are represented by clusters of glomeruli with similar specificity in zebrafish (Friedrich & Korsching, 1997) are nonvolatile, water-soluble compounds. These compounds presumably would not reach the main olfactory epithelium of a land-dwelling mammal, but they represent naturally occurring, important olfactory stimuli related to food in fish. It is likely that each species is somewhat specialized in terms of which odorants it best discriminates, and it also is possible that rats and other omnivorous species may discriminate a broad range of odorants related to a great variety of food sources.

DEVELOPMENTAL PLASTICITY OF THE OLFACTORY BULB IN LIGHT OF THE SIMPLE MODEL OF ODORANT FEATURE PROCESSING

The simple model of odorant feature processing in the olfactory system that we described above provides several interesting predictions concerning changes in olfactory bulb structure and function caused by early learning about odors.

One specific prediction of the model is that increased glomerular activity, such as occurs following learning, may lead to increased behavioral sensitivity to odors. That the amount of glomerular activity can be correlated with increased sensitivity of odorant detection follows from our studies of 2-DG uptake in response to different odorants. For example, larger odorants are detected at lower concentrations than are smaller odorants of the same chemical class (Cometto-Muñiz *et al.*, 1998), and larger odorants stimulate more glomerular 2-DG uptake than do smaller odorants presented at the same vapor phase concentration (Johnson *et al.*, 1998; 1999). Increases in sensitivity of detection following prior exposure to certain odorants have been described for adult mice (Wang, Wysocki, & Gold, 1993) and humans (Stevens & O'Connell, 1995; Wysocki, Dorries, & Beauchamp, 1989). In mice, this increase in sensitivity is associated with increased activity that can be measured in the olfactory nerve, suggesting that the underlying neurobiological changes are likely to involve the olfactory sensory neurons themselves (Wang *et al.*, 1993). Following discrimination learning in adult rats, spatially restricted regions of the olfactory epithelium increase in activity (Youngentob & Kent, 1995). Changed activity through an increase either in the number of receptors expressed in each sensory neuron or in the number of sensory neurons expressing a given receptor is likely to cause changed activity in a limited number of glomeruli in the olfactory bulb.

In the simple model of odorant feature processing, glomeruli are envisioned as functional units that serve to segregate responses to individual molecular features of odorants. Reassembly of these features into odor perceptions is likely to occur at some higher level of processing. The hedonic value of the odor would likely emerge at still some later stage. In other words, the increased glomerular activity during and following learning would not reflect odor preference per se. Consistent with this prediction, increased glomerular uptake of 2-DG also occurs once young rats have learned an aversion to an odorant (Sullivan & Wilson, 1991).

Different odorants possessing a common molecular feature activate the same glomeruli. Therefore, after an animal has learned about the significance of one odorant and thereby has an enhanced response in certain glomeruli, a subsequent exposure to another odorant possessing a molecular feature in common with the learned odorant would stimulate some of the same glomeruli and these responses also would be enhanced. This other odorant also would activate additional glomeruli that do not overlap with those stimulated by the learned odorant. An interesting prediction of this partial overlap is that the learning of one odorant could change the spatial pattern of activity evoked by the other odorant by altering the relative contributions of different glomeruli to the pattern. Because spatial patterns of activity are likely to be related to odor perception, early learning of one odorant then would be predicted to change the perceived odor of other volatile chemicals. For odorants that are very closely related in chemistry to a previously learned odorant, the stimulation of a large number of overlapping, enhanced glomeruli may lead to a greater perceived similarity between these odorants and the learned one. A potential benefit of this predicted increase in odor generalization is a decreased significance of minor variations in the compositions of odorant mixtures. It also would make it possible for animals to generalize among odorants typical of food classes. For example, prior preference learning of particular ester odorants associated with one kind of fruit may become more readily generalized to esters associated with some other fruit.

In a related issue, recall that there are glomeruli in the posterior, ventrolateral bulb that respond to peppermint extract, but that do not increase their 2-DG uptake or Fos-like response following early learning (Johnson & Leon, 1996; Johnson et al., 1995). One possible explanation for this finding is that control animals previously also had experienced the odorant molecular features detected by these glomeruli. For example, the volatile components responsible for the odor of peppermint extract are primarily terpenes related to menthol, and many other plants produce the terpenes that occur in peppermint. Both control and trained animals would have been exposed to volatile compounds from the hardwood chips used to line their cages, and it is possible that these compounds had the same molecular feature responsible for activity in the ventrolateral glomeruli that is evoked by components of peppermint extract. An alternative explanation for the differential effects of learning on midlateral and ventrolateral glomeruli is that some glomeruli are more plastic than others (Johnson & Leon, 1996). Differential plasticity would be consistent with the heterogeneous glomerular distribution of proteins that may play a role in plasticity, including the beta-adrenergic receptor (Woo & Leon, 1995a), nitric oxide synthase, and certain growth factor receptors. It would be interesting to determine with a wide variety of odorants where in the bulb the glomerular responses are plastic in response to early olfactory learning.

SUMMARY

The olfactory system appears to encode odorants by initially breaking them up into their component molecular features. A large family of odorant receptors is present at this first level to bind these features with different levels of specificity. The olfactory neurons bearing a specific kind of receptor all project to a glomerulus either on the lateral or the medial aspect of olfactory bulbs, thereby allowing the signal generated by one kind of feature to be amplified. As the signal is transmitted into the bulb at the level of the glomerular layer, the signal for a specific feature also can be sharpened by local inhibition of the responses to similar features in the bulb. The signal representing each feature then is transmitted into the olfactory cortex, where it is combined with the information regarding other features to form the perception of the odor. While a representation for odorants is present in the bulb at birth, early learning regarding odors can change the subsequent response in the bulb.

In the coming years, a study of the perceptual consequences of early alterations in the coding of odorants may reveal the connection between what is represented and what is perceived in the olfactory system. Another realm of inquiry to be explored is the microcircuitry of the olfactory bulb and olfactory cortex; much should be learned with a careful respect for the spatially specific organization of the coding process. In addition, the differences among species may be appreciated in relation to the different olfactory tasks that each is asked to accomplish by the natural world. Finally, we have to render a more detailed map of molecular features across the bulbar surface and then we must correlate those spatial activation patterns with their psychophysical properties to gain a full understanding of odor coding.

Acknowledgment

Supported by grant DC03545 from the National Institute for Deafness and Communication Disorders.

REFERENCES

Alkasab, T. K., Bozza, T. C., Cleland, T. A., Dorries, K. M., Pearce, T. C., White, J., & Kauer, J. S. (1999). Characterizing complex chemosensors: information-theoretic analysis of olfactory systems. *Trends in Neurosciences, 22,* 102–108.

Astic, L., & Cattarelli, M. (1982). Metabolic mapping of functional activity in the rat olfactory system after a bilateral transection of the lateral olfactory tract. *Brain Research, 245,* 17–25.

Astic, L. & Sacier, D. (1986). Anatomical mapping of the neuroepithelial projection to the olfactory bulb in the rat. *Brain Research Bulletin, 16,* 445–454.

Axel, R. (1995). The molecular logic of smell. *Scientific American, 273,* 154–159.

Bell, G. A., Laing, D. G., & Panhuber, H. (1987). Odour mixture suppression: Evidence for a peripheral mechanism in human and rat. *Brain Research, 426,* 8–18.

Bozza, T. C., & Kauer, J. S. (1998). Odorant response properties of convergent olfactory receptor neurons. *Journal of Neuroscience, 18,* 4560–4569.

Buck, L. B. (1996). Information coding in the vertebrate olfactory system. *Annual Review of Neuroscience, 19*, 517–544.

Buck, L. B., & Axel, R. (1991) A novel multigene family may encode odorant receptors: A molecular basis for odor recognition. *Cell, 65*, 175–187.

Buonviso, N., & Chaput, M. A. (1990). Response similarity to odors in olfactory bulb output cells presumed to be connected to the same glomerulus: Electrophysiological study using simultaneous single-unit recordings. *Journal of Neurophysiology, 63*, 447–454.

Cain, W. S. (1979). To know with the nose: Keys to odor identification. *Science, 203*, 467–470.

Cain, W. S., & Potts, B. C. (1996). Switch and bait: Probing the discriminative basis of odor identification via recognition memory. *Chemical Senses, 21*, 35–44.

Chess, A., Simon, I., Cedar, H., & Axel, R. (1994). Allelic inactivation regulates olfactory receptor gene expression. *Cell, 78*, 823–834.

Cinelli, A. R., Hamilton, K. A., & Kauer, J. S. (1995). Salamander olfactory bulb neuronal activity observed by video rate, voltage-sensitive dye imaging. III. Spatial and temporal properties of responses evoked by odorant stimulation. *Journal of Neurophysiology, 73*, 2053–2071.

Clyne, P. J., Warr, C. G., Freeman, M. R., Lessing, D., Kim, J., & Carlson, J. R. (1999). A novel family of divergent seven-transmembrane proteins: Candidate odorant receptors in *Drosophila. Neuron, 22*, 327–338.

Cometto-Muñiz, J. E., Cain, W. S., & Abraham, M. H. (1998). Nasal pungency and odor of homologous aldehydes and carboxylic acids. *Experimental Brain Research, 118*, 180–188.

Coopersmith, R., & Leon, M. (1984). Enhanced neural response to familiar olfactory cues. *Science, 225*, 849–851.

Coopersmith, R., & Leon, M. (1986). Enhanced neural response by adult rats to odors experienced early in life. *Brain Research, 371*, 400–403.

Coopersmith, R., Henderson, S. R., & Leon, M. (1986). Odor specificity of the enhanced neural response following early odor experience in rats. *Developmental Brain Research, 27*, 191–197.

Dean, P. M. (1987). *Molecular foundations of drug–receptor interaction.* Cambridge: Cambridge University Press.

Dickinson, T. A., White, J., Kauer, J. S., & Walt, D. R. (1998). Current trends in 'artificial-nose' technology. *Trends in Biotechnology, 16*, 250–258.

Do, J. T., Sullivan, R. M., & Leon, M. (1988). Behavioral and neural correlates of postnatal olfactory conditioning: II. Respiration during conditioning. *Developmental Psychobiology, 21*, 591–600.

Dorries, K. M. (1998). Olfactory coding: Time in a model. *Neuron, 20*, 7–10.

Dôving, K. B. (1966). An electrophysiological study of odour similarities of homologous substances. *Journal of Psychology, 186*, 97–109.

Eisthen, H. L. (1997). Evolution of vertebrate olfactory systems. *Brain Behavior and Evolution, 50*, 222–233.

Firestein, S., Picco, C., & Menini, A. (1993). The relation between stimulus and response in olfactory receptor cells of the tiger salamander. *Journal of Physiology, 468*, 1–10.

Freeman W. J., & Skarda, C. A. (1985). Spatial EEG patterns, non-linear dynamics and perception: The neo-Sherringtonian view. *Brain Research, 357*, 147–175.

Friedrich, R. W., & Korsching, S. I. (1997). Combinatorial and chemotopic odorant coding in the zebrafish olfactory bulb visualized by optical imaging. *Neuron, 18*, 737–752.

Galizia, C. G., Menzel, R., & Holldobler, B. (1999). Optical imaging of odor-evoked glomerular activity patterns in the antennal lobes of the ant *Camponotus rufipes. Naturwissenschaften, 86*, 533–537.

Galizia, C. G., Sachse, S., Rappert, A., & Menzel, R. (1999). The glomerular code for odor representation is species specific in the honeybee *Apis mellifera. Nature Neuroscience, 2*, 473–478.

Guthrie, K. M., & Gall, C. (1995). Functional mapping of odor-activated neurons in the olfactory bulb. *Chemical Senses, 20*, 271–282.

Guthrie, K. M., & Gall, C. (1999). Functional mapping of the developing olfactory bulb. *21st Annual Meeting of the Association for Chemoreception Sciences (AChemS), Abstracts*, p. 17.

Guthrie, K. M., Anderson, A. J., Leon, M., & Gall, C. (1993). Odor-induced increases in c-*fos* mRNA expression reveal an anatomical unit for odor processing in olfactory bulb. *Proceedings of the National Academy of Sciences of the USA, 90*, 3329–3333.

Haberly L. B., & Bower, J. M. (1989). Olfactory cortex: Model circuit for study of associative memory? *Trends in Neuroscience, 12*, 258–264.

Haberly, L. B., & Price, J. L. (1977). The axonal projection patterns of the mitral and tufted cells of the olfactory bulb in the rat. *Brain Research, 129*, 152–157.

Imamura, K., Mataga, N., & Mori, K. (1992). Coding of odor molecules by mitral/tufted cells in rabbit olfactory bulb. I. Aliphatic compounds. *Journal of Neurophysiology, 68*, 1986–2002.

Joerges, J., Küttner, A., Galizia, C. G., & Menzel, R. (1997). Representations of odours and odour mixtures visualized in the honeybee brain. *Nature, 387,* 285–288.

Johnson, B. A., & Leon, M. (1996). Spatial distribution of [^{14}C] 2-deoxyglucose uptake in the glomerular layer of the rat olfactory bulb following early olfactory preference learning. *Journal of Comparative Neurology, 376,* 557–566.

Johnson, B. A., & Leon, M. (2000) Modular glomerular representations of odorants in the rat olfactory bulb: The effects of stimulus concentration. *Journal of Comparative Neurology, 426,* 496–509.

Johnson, B. A., Woo, C. C., Duong, H., Nguyen, V., & Leon, M. (1995). A learned odor evokes an enhanced Fos-like glomerular response in the olfactory bulb of young rats. *Brain Research, 699,* 192–200.

Johnson, B. A., Woo, C. C., & Leon, M. (1998). Spatial coding of odorant features in the glomerular layer of the rat olfactory bulb. *Journal of Comparative Neurology, 393,* 457–471.

Johnson, B. A., Woo, C. C., Hingco, E. E., Pham, K. L., & Leon, M. (1999). Multidimensional chemotopic responses to n-aliphatic acid odorants in the rat olfactory bulb. *Journal of Comparative Neurology, 409,* 529–548.

Jourdan, F., Duveau, A., Astic, L., & Holley, A. (1980). Spatial distribution of [^{14}C]2-deoxyglucose uptake in the olfactory bulbs of rats stimulated with two different odours. *Brain Research, 188,* 139–154.

Katoh, K., Koshimoto, H., Tani, A., & Mori, K. (1993). Coding of odor molecules by mitral/tufted cells in rabbit olfactory bulb. II. Aromatic compounds. *Journal of Neurophysiology, 70,* 2161–2175.

Kauer, J. S. (1987). Coding in the olfactory system. In T. E. Finger & W. S. Silder (Eds.), *Neurobiology of taste and smell.* (pp. 205–231). New York: Wiley.

Kauer, J. S., & Cinelli, A. R. (1993). Are there structural and functional modules in the vertebrate olfactory bulb? *Microscopy Research and Technique, 24,* 157–167.

Krautwurst D., Yau, K. W., & Reed, R. R. (1998). Identification of ligands for olfactory receptors by functional expression of a receptor library. *Cell, 95,* 917–926.

Laska, M., & Teubner, P. (1998). Odor structure–activity relationships of carboxylic acids correspond between squirrel monkeys and humans. *American Journal of Physiology, 274,* R1639–R1645.

Laurent, G. (1997). Olfactory processing: Maps, time and codes. *Current Opinion in Neurobiology, 7,* 547–553.

Laurent, G., & Naraghi, M. (1994). Odorant-induced oscillations in the mushroom bodies of the locust. *Journal of Neuroscience, 14,* 2993–3004.

Laurent, G., Wehr, M., & Davidowitz, H. (1996). Temporal representations of odors in an olfactory network. *Journal of Neuroscience, 16,* 3837–3847.

Leon, M. (1975). Dietary control of maternal pheromone in the lactating rat. *Physiology and Behavior, 14,* 311–319.

Leon, M. (1987). Plasticity of olfactory output circuits related to early olfactory learning. *Trends in Neurosciences, 10,* 434–438.

Leon, M., & Moltz, H. (1971). Maternal pheromone: Discrimination by preweanling albino rats. *Physiology and Behavior, 7,* 265–267.

Lu, X.-C. M., & Slotnick. B. M. (1994). Recognition of propionic acid vapor after removal of the olfactory bulb area associated with high 2-DG uptake. *Brain Research, 639,* 26–32.

Lu, X.-C. M., & Slotnick, B. M. (1998). Olfaction in rats with extensive lesions of the olfactory bulbs: Implications for odor coding. *Neuroscience, 84,* 849–866.

Macrides, F., & Davis, B. J. (1983). The olfactory bulb. In P. C. Emson (Ed.), *Chemical neuroanatomy.* (pp. 391–426). New York: Raven Press.

Malnic, B., Hirono, J., Sato, T., & Buck, L. (1999). Combinatorial receptor codes for odors. *Cell, 96,* 713–723.

Matsutani, S., & Leon, M. (1993). Elaboration of glial cell processes in the rat olfactory bulb associated with early learning. *Brain Research, 613,* 317–320.

McCollum, J. F., Woo, C. C., & Leon, M. (1997). Granule and mitral cell densities are unchanged following early olfactory preference training. *Developmental Brain Research, 99,* 118–120.

Michel, W. C., & Ache, B. W. (1994). Odor-evoked inhibition in primary olfactory receptor neurons. *Chemical Senses, 19,* 11–24.

Mombaerts, P., Wang, F., Dulac, C., Chao, S. K., Nemes, A., Mendelsohn, M., Edmonson, J., & Axel, R. (1996a). Visualizing an olfactory sensory map. *Cell, 87,* 675–686.

Mombaerts, P., Wang, F., Dulac, C., Chao, S. K., Nemes, A., Mendelsohn, M., Edmonson, J., & Axel, R. (1996b). The molecular biology of olfactory perception. *Cold Spring Harbor Symposium on Quantitative Biology, 61,* 135–145.

Mori, K. (1987). Membrane and synaptic properties of identified neurons in the olfactory bulb. *Progress in Neurobiology, 29,* 275–320.

Mori, K., & Yoshihara, Y. (1995). Molecular recognition and olfactory processing in the mammalian olfactory system. *Progress in Neurobiology, 45*, 585–619.

Mori, K., Mataga, N., & Imamura, K. (1992). Differential specificities of single mitral cells in rabbit olfactory bulb for a homologous series of fatty acid odor molecules. *Journal of Neurophysiology, 67*, 786–789.

Motokizawa, F. (1996). Odor representation and discrimination in mitral/tufted cells of the rat olfactory bulb. *Experimental Brain Research, 112*, 24–34.

Nieuwenhuys, R. (1967). Comparative anatomy of olfactory centres and tracts. In Y. Zotterman (Ed.), *Progress in brain research* (pp. 1–64). New York: Elsevier.

Ottoson, D. (1958). Studies on the relationship between olfactory stimulating effectiveness and physico-chemical properties of odorant compounds. *Acta Physiologica Scandanavica, 43*, 167–181.

Puche, A., Aroniadou-Anderjaska, V., & Shipley, M. (1998). Olfactory bulb–olfactory cortex slices in the study of central olfactory CNS circuits. *Society for Neuroscience Abstracts, 34*, 1885.

Ressler, K. J., Sullivan, S. L., & Buck, L. B. (1994). Information coding in the olfactory system: Evidence for a stereotyped and highly organized epitope map in the olfactory bulb. *Cell, 79*, 1245–1255.

Royet, J. P., Sicard, G., Souchier, C., & Jourdan, F. (1987). Specificity of spatial patterns of glomerular activation in the mouse olfactory bulb: Computer-assisted image analysis of 2-deoxyglucose auto-radiograms. *Brain Research, 417*, 1–11.

Rubin, B. D., & Katz, L. C. (1999). Optical imaging of odorant representations in the mammalian olfactory bulb. *Neuron, 23*, 499–511.

Sallaz, M., & Jourdan, F. (1993). C-fos expression and 2-deoxyglucose uptake in the olfactory bulb of odour-stimulated awake rats. *NeuroReport, 4*, 55–58.

Sato, T., Hirono, J., Tonoike, M., & Takebayashi, M. (1994). Tuning specificities to aliphatic odorants in mouse olfactory receptor neurons and their local distribution. *Journal of Neurophysiology, 72*, 2980–2989.

Shepherd, G. M. (1991). Computational structure of the olfactory system. In J. Davis and H. Eichenbaum (Eds.), *Olfaction as a model system for computational neuroscience* (pp. 3–41). Cambridge, MA: MIT Press.

Shepherd, G. M. (1994). Discrimination of molecular signals by the olfactory receptor neuron. *Neuron, 13*, 771–790.

Slotnick, B. M., Graham, S., Laing, D. G., & Bell, G. A. (1987). Detection of propionic acid vapor by rats with lesions of olfactory bulb areas associated with high 2-DG uptake. *Brain Research, 417*, 343–346.

Slotnick, B. M., Panhuber, H., Bell, G. A., & Laing, D. G. (1989). Odor-induced metabolic activity in the olfactory bulb of rats trained to detect propionic acid vapor. *Brain Research, 500*, 161–168.

Slotnick, B. M., Bell, G. A., Panhuber, H., & Laing, D. G. (1997). Detection and discrimination of pro-pionic acid after removal of its 2-DG identified major focus in the olfactory bulb: A psychophysical analysis. *Brain Research, 762*, 89–96.

Stevens, D. A., & O'Connell, R. J. (1995). Enhanced sensitivity to androstenone following regular exposure to pemenone. *Chemical Senses, 20*, 413–419.

Stewart, W. B., Kauer, J. S., & Shepherd, G. M. (1979). Functional organization of rat olfactory bulb analyzed by the 2-deoxyglucose method. *Journal of Comparative Neurology, 185*, 715–734.

Stopfer, M., Bhagavan, S., Smith, B. H., & Laurent, G. (1997) Impaired odour discrimination on de-synchronization of odour-encoding neural assemblies. *Nature, 390*, 70–74.

Sullivan, R. M., & Leon, M. (1986). Early olfactory learning induces an enhanced olfactory bulb response in young rats. *Developmental Brain Research, 27*, 278–282.

Sullivan, R. M., & Wilson, D. A. (1991). Neural correlates of conditioned odor avoidance in infant rats. *Behavioral Neuroscience, 103*, 307–312.

Sullivan, R. M., Wilson, D. A., Kim, M. H., & Leon, M. (1988). Behavioral and neural correlates of postnatal olfactory conditioning: I. Effect of respiration on conditioned neural responses. *Physiology and Behavior, 44*, 85–90.

Sullivan, R. M., Wilson, D. A., & Leon, M. (1989). Norepinephrine and learning-induced plasticity in infant rat olfactory system. *Journal of Neuroscience, 9*, 3998–4006.

Sullivan, R. M., Wilson, D. A., Wong, R., Correa, A., & Leon, M. (1990). Modified behavioral and olfactory bulb responses to maternal odors in preweanling rats. *Developmental Brain Research, 53*, 243–247.

Sullivan, S. L., & Dyer, L. (1996). Information processing in mammalian olfactory system. *Journal of Neurobiology, 30*, 20–36.

Tsuboi, A., Yoshihara, S., Yamazaki, N., Kasai, H., Asai-Tsuboi, H., Komatsu, M., Serizawa, S., Ishii, T., Matsuda, Y., Nagawa, F., & Sakano, H. (1999). Olfactory neurons expressing closely linked and homologous odorant receptor genes tend to project their axons to neighboring glomeruli on the olfactory bulb. *Journal of Neuroscience, 19*, 8409–8418.

Vassar, R., Chao, S. K., Sitcheran, R., Nuñez, J. M., Vosshall, L. B., & Axel, R. (1994). Topographic organization of sensory projections to the olfactory bulb. *Cell, 79*, 981–991.

Vickers, N. J., & Christensen, T. A. (1998). A combinatorial model of odor discrimination using a small array of contiguous, chemically defined glomeruli. *Annals of the New York Academy of Sciences, 855*, 514–516.

Vosshall, L. B., Amrein, H., Morozov, P. S., Rzhetsky, A., & Axel, R. (1999). A spatial map of olfactory receptor expression in the *Drosophila* antenna. *Cell, 96*, 725–736.

Wang, H.-W., Wysocki, C. J., & Gold, G. H. (1993). Induction of olfactory receptor sensitivity in mice. *Science, 260*, 998–1000.

Wang, F., Nemes, A., Mendelsohn, M., & Axel, R. (1998). Odorant receptors govern the formation of a precise topographic map. *Cell, 93*, 47–60.

Wehr, M., & Laurent, G. (1996). Odour encoding by temporal sequences of firing in oscillating neural assemblies. *Nature, 384*, 162–166.

Wilson, D. A., & Leon, M. (1988). Spatial patterns of olfactory bulb single-unit responses to learned olfactory cues in young rats. *Journal of Neurophysiology, 59*, 1770–1782.

Wilson, D. A., Sullivan, R. M., & Leon, M. (1987). Single-unit analysis of postnatal olfactory learning: Modified olfactory bulb output response patterns to learned attractive odors. *Journal of Neuroscience, 7*, 3154–3162.

Woo, C. C., & Leon, M. (1991). Increase in a focal population of juxtaglomerular cells in the olfactory bulb associated with early learning. *Journal of Comparative Neurology, 305*, 49–56.

Woo, C. C., & Leon, M. (1995a). Distribution and development of beta-adrenergic receptorsinthe rat olfactory bulb. *Journal of Comparative Neurology, 352*, 1–10.

Woo, C. C., & Leon, M. (1995b). Early olfactory enrichment and deprivation both decrease fl-adrenergic receptor density in the main olfactory bulb of the rat. *Journal of Comparative Neurology, 360*, 634–642.

Woo, C. C., Coopersmith, R., & Leon, M. (1987). Localized changes in olfactory bulb morphology associated with early olfactory learning. *Journal of Comparative Neurology, 263*, 113–125.

Woo, C. C., Oshita, M. H., & Leon, M. (1996). A learned odor decreases the number of Fos-immunopositive granule cells in the olfactory bulb of young rats. *Brain Research, 716*, 149–156.

Wysocki, C. J., Dorries, K. M., & Beauchamp, G. K. (1989). Ability to perceive androstenone can be acquired by ostensibly anosmic people. *Proceedings of the National Academy of Sciences of the USA, 86*, 7976–7978.

Yokoi, M., Mori, K., & Nakanishi, S. (1995). Refinement of odor molecule tuning by dendrodendritic synaptic inhibition in the olfactory bulb. *Proceedings of the National Academy of Sciences of the USA, 92*, 3371–3375.

Youngentob, S. L., & Kent, P. F. (1995). Enhancement of odorant-induced mucosal activity patterns in rats trained on an odorant identification task. *Brain Research, 670*, 82–88.

Zhao, H., Ivic, L, Otaki, J. M., Hashimoto, M., Mikoshiba, K., & Firestein, S. (1998). Functional expression of a mammalian odorant receptor. *Science, 279*, 237–242.

3

Tunable Seers

Activity-Dependent Development of Vision in Fly and Cat

Helmut V. B. Hirsch, Suzannah Bliss Tieman, Martin Barth, and Helen Ghiradella

Introduction

Programs guiding nervous system development must achieve two goals. First, they must generate the species-specific behaviors needed for survival and procreation. Second, they must incorporate into the nervous system information about the environment in order to "tune" the organism to current conditions. Such "experience-dependent assembly" of the nervous system has long been seen as a hallmark of development among the vertebrates (especially mammals), whereas "hardwiring" and inflexibility were seen as hallmarks of development among invertebrates.

We here question the "hard-wired" view of invertebrates with evidence from flies and conclude that experience-dependent development is characteristic of both vertebrate and invertebrate nervous systems and that those genetic and epigenetic programs that control nervous system assembly likely have fundamental commonalities across widely diverse species (Murphey, 1986b; Murphey & Hirsch, 1982). This view in turn "simplifies" the course of evolution, which appears to be conservative in its approach to the construction of systems for receiving, processing, interpreting, storing and acting upon information crucial for enhanced fitness. This is our perspective, and we will depend particularly on a comparison of elements in the visual development of two very different animals, cats and fruit flies, to make the case.

Helmut V. B. Hirsch, Suzannah Bliss Tieman, and Helen Ghiradela   Neurobiology Research Center and Department of Biological Sciences, The University at Albany, State University of New York, Albany, New York 12222.   Martin Barth   Friedrich-Miescher-Laboratorium der Max-Planck-Gesellschaft, 72076 Tübingen, Germany.

Developmental Psychobiology, Volume 13 of *Handbook of Behavioral Neurobiology,* edited by Elliott Blass, Kluwer Academic / Plenum Publishers, New York, 2001.

The fruit fly, *Drosophila melanogaster* has been a useful model system for the discovery and study of genes vital to the development of sensory, integrative, and behavioral systems. These genes are highly conserved and have common functions in diverse species. For example, genes critical to visual development in *Drosophila* have similar function in mice (Macdonald & Wilson, 1996), zebra fish, frogs, turtles, quail, rats (Zuker, 1994), squid (Tomarev *et al.*, 1997), and nematodes (Duncan *et al.*, 1997). However, the prevailing view that such small and short-lived invertebrates as *Drosophila* are hard-wired has not encouraged the use of these animals for the study of *plasticity*, experience-guided or epigenetic changes during the ontogeny of an individual (Murphey, 1986b). This view is no longer tenable: in this chapter we will summarize recent evidence that experience-dependent assembly is as important a force in shaping the ontogeny, physiology, and behavior in short-lived invertebrates as it is in long-lived vertebrates, and that epigenetic mechanisms for nervous system development may be as much a part of a "common heritage" as genetic ones (West, King, & Arberg, 1988; Wu, Renger, & Engel, 1998). To *Drosophila*'s many contributions to neurobiological research, we may now add that of modeling experience-dependent plasticity of brain and behavior.

We first briefly describe the organization of the cat's visual system and summarize recent findings (with a little help from studies of ferrets) concerning its development. This provides a context in which to evaluate the contribution of experience to the development of other visual systems. Next, we describe the organization and development of the *Drosophila* visual system. We then review evidence that genes important in visual system development of *Drosophila* are also major determinants of visual system development of vertebrates. Next we focus on particular experience-dependent influences on visual morphology in *Drosophila*. We also review the known commonality of developmental mechanisms in mammals and flies, with particular attention to the evidence that very similar molecules regulate the synaptic modifiability that presumably underlies much experience-dependent plasticity. Finally, we place these insights in a functional context by examining correlations between experience-dependent changes in brain and those in behavior. We emphasize the common themes that inform nervous system development in all animals and suggest that other invertebrates will have similar mechanisms.

THE CAT AS A MODEL SYSTEM

Much of what is known about developmental plasticity in the nervous system in mammals has come from studies in cats of the pathways linking the retina to the first cortical visual processing centers. The pathway from the retina through the lateral geniculate nucleus to the visual cortex (the retinogeniculocortical pathway) has been reviewed extensively (Hirsch, 1985; Sherman & Spear, 1982). Here we highlight those details needed to understand the recent work on activity-dependent development that we will describe later in the chapter.

DESCRIPTION OF THE RETINOGENICULOCORTICAL PATHWAY IN THE CAT

RETINAL ORGANIZATION. The organization of the retina has been elegantly reviewed by Dowling (1987) and Sterling, Freed, and Smith (1986). Briefly, within the retina, light is sensed by photoreceptors (rods and cones). The photoreceptors synapse onto bipolar cells, which in turn synapse onto retinal ganglion cells. The axons of the retinal ganglion cells synapse centrally with neurons in the lateral geniculate nucleus (LGN) and superior colliculus. Synapses between photoreceptors and bipolar cells are found in the outer plexiform layer of the retina, those between bipolar cells and ganglion cells in the inner plexiform layer. In addition to these "vertical" connections, there are lateral connections as well; these are made by horizontal cells in the outer plexiform layer and amacrine cells in the inner plexiform layer. Both bipolar cells and retinal ganglion cells have receptive fields that are concentrically organized into a center and a surround. Input for the center of the receptive field is provided by direct, vertical connections; input for the surround is provided by lateral connections.

ON and OFF Pathways. "ON-center" cells ("ON cells") respond both to "light-on" in the center of the field, and to "light-off" in the surround. "OFF-center" cells ("OFF cells") respond to light-off in the center of the field and to light-on in the surround. The connections of ON and OFF cells are segregated: OFF bipolar cells synapse with the dendrites of OFF-center ganglion cells (in the outer half of the inner plexiform layer), whereas ON bipolar cells synapse with ON-center ganglion cells (in the inner half of the inner plexiform layer) (Famiglietti & Kolb, 1976; Nelson, Famiglietti, & Kolb, 1978; Saito, 1983). The ON and OFF pathways remain segregated in the connections they make with cells in the LGN (Berman & Payne, 1989; Bowling & Caverhill, 1989; Bowling & Wieniawa-Narkiewicz, 1986).

X, Y, and W Pathways. Retinal ganglion cells can also be subdivided into three groups on the basis of their physiologic responses, retinal distribution, and central projections (for reviews see Garraghty, 1995; Rodieck, 1979; Sherman, 1985; Stone, Dreher, & Leventhal, 1979). The three classes are defined physiologically as X cells, Y cells, and W cells. Each of these classes is associated with a morphologically defined class of ganglion cell: beta cells, alpha cells, and gamma cells, respectively (Fukuda, Hsia, & Watanabe, 1985; Saito, 1983; Stanford & Sherman, 1984; Wässle, 1982). X cells tend to have (1) medium sized cell bodies and small dendritic fields, (2) small receptive fields, (3) slowly conducting and small diameter axons, (4) fine discrimination for spatial patterning, (5) sustained response to local contrast, (6) strong response to relatively slowly moving targets, and (7) linear summation within the receptive field; Y cells generally have the opposite properties. W cells form a heterogeneous group characterized by very slow conduction times. These W-cells tend to have small cell bodies, large dendritic and receptive fields, and axons of very small diameter.

LATERAL GENICULATE NUCLEUS. The distinctions among retinal X, Y, and W cells are preserved in the optic nerve and in the LGN, where each cell receives direct, excitatory monosynaptic inputs from only one type of retinal cell (Bullier & Norton, 1979; Cleland, Dubin, & Levick, 1971; Hoffmann, Stone, & Sherman, 1972; Wilson & Stone, 1975). The cells in the LGN display, with minor variations, response properties and distributions similar to those of the retinal ganglion cells, and each physiologically characterized LGN cell group is also associated with a distinct morphologically defined cell type (Friedlander, Lin, Stanford, & Sherman, 1981; Stanford, Friedlander, & Sherman, 1983).

LGN relay cells, in turn, send their axons to various cortical visual areas, each receiving a different selection of inputs from the various classes of retinal ganglion cells. Within each area, the excitatory connections maintain the segregation of the afferent streams; few cortical cells are excited by more than one type of afferent (Dreher, Leventhal, & Hale, 1980; Henry, Mustari, & Bullier, 1983; Singer, Tretter, & Cynader, 1975), and thus at least some response properties of cortical cells reflect those of the afferent stream that is the source of their input (Dreher *et al.*, 1980; Henry *et al.*, 1983). In summary, the afferent streams originating in the retina are largely segregated in the LGN and at the level of the first-order cells in cortical visual areas.

TOPOGRAPHY.   The two eyes view overlapping regions of the visual world. The area of overlap is called the binocular segment of the visual field. It is surrounded on either side by the monocular segments, areas seen by only one eye. The visual image is reversed by the eye so that the left half of each retina sees the right visual field, and vice versa. Fibers from the nasal hemiretina of each eye cross in the optic chiasm to innervate the top layer, layer A, of the contralateral LGN, whereas those from the temporal hemiretina remain uncrossed and innervate the second layer, layer A1, of the ipsilateral LGN (Hickey & Guillery, 1974). Thus, the left LGN views the right half of the visual world through both eyes, and similarly the right LGN views the left half of the visual world. The retinal projections to the LGN are topographic, and although the inputs from the two eyes terminate in separate layers, the topographic maps in these layers are in register (Sanderson, 1971). Each LGN projects directly to cortical areas 17, 18 and 19, where the inputs from the two eyes are combined. In each area the combined inputs form a topographic map (Hubel & Wiesel, 1962, 1965a; Tusa, Palmer, & Rosenquist, 1978; Tusa, Rosenquist, & Palmer, 1979).

VISUAL CORTEX.   The cortex is composed of horizontal layers, numbered 1–6 from superficial to deep. Afferents from the LGN to visual cortex terminate primarily in layer 4 with a minor projection to layer 6 (LeVay & Gilbert, 1976; Shatz, Lindström, & Wiesel, 1977). Cells in layer 4 relay visual information to cells in other layers above and below them, with a major projection into layers 2 and 3 (Martin & Whitteridge, 1984). Cells in layer 6 project back to the LGN (Gilbert & Kelly, 1975).

Ocular Dominance.   Cells from corresponding regions of geniculate layers A and A1 project to partially overlapping "ocular dominance" patches in layer 4 of areas 17 and 18 (LeVay & Gilbert, 1976; Shatz *et al.*, 1977). The inputs from the two eyes are thus combined in visual cortex, and most cortical cells can be driven by stimulation of either eye, although they tend to be dominated by one eye or the other (Hubel & Wiesel, 1962). Cells in the monocular segments of visual cortex, of course, can be driven by stimulation of only the contralateral eye.

Within the binocular regions of areas 17 and 18, but not area 19, cells driven by one eye are grouped together in ocular dominance columns that span the layers of cortex and extend above and below the patches of afferents representing that eye (Hubel & Wiesel, 1965b; S. B. Tieman & Tumosa, 1983). Columns of cells dominated by one eye alternate with columns of cells dominated by the other.

Orientation Selectivity.   Most cells in the visual cortex are also selective for the orientation of visual stimuli and have the same preferred orientation whether driven through the ipsilateral or the contralateral eye. Cells with the same orientation preference are clustered together in their own domains or columns, which again extend throughout the layers (Albus, 1979; Hubel & Wiesel, 1962, 1963b). Adjacent columns contain cells with similar preferences, so that an electrode traversing cortex parallel to the layers encounters cells with orientation preferences that generally shift in an orderly progression (Albus, 1979; Bonhoeffer & Grinvald, 1991; Hubel & Wiesel, 1963b; Schoppmann & Stryker, 1981).

In summary, within the retina, light activates photoreceptors, which synapse onto bipolar cells, which in turn synapse onto retinal ganglion cells. Retinal ganglion cells project to the LGN, the cells of which project to cortical areas 17, 18, and 19. Bipolar cells, retinal ganglion cells, and LGN cells all have concentric receptive fields and may be classified as ON or OFF. The ON and OFF pathways remain segregated, at least to the level of the LGN. In addition, there are at least three distinct types of retinal ganglion cells, X, Y, and W cells, which give rise to separate afferent streams that remain segregated in the LGN and in the first-order cells of cortical visual areas; response properties of the cortical cells reflect some of the properties of the retinal afferent streams. In visual cortex, cells respond to either eye, but tend to be dominated by one eye. Cortical cells are also orientation selective.

The next section presents data to show that neuronal activity, generated both intrinsically and in response to external stimulation, is important in the segregation of afferent streams and in the development of ocular dominance and orientation selectivity. Furthermore, experience-dependent neuronal activity affects the various afferent streams differently; the pathway arising from retinal Y cells is particularly affected by sensory stimulation during early life. (As we will see below, in flies, too, there is evidence for parallel afferent streams that differ in experience sensitivity).

ACTIVITY DEPENDENCE OF DEVELOPMENT OF CAT AND FERRET VISUAL SYSTEMS

Many recent studies have focused on three basic questions: First, how is segregation among the afferent streams that arise in the retina achieved during development? Second, what interactions among the streams occur during development? Third, how do cells develop complex, adult response properties such as orientation selectivity? Specifically, we will consider the separation of afferent streams (ON and OFF streams in the retina, afferents from the two eyes in the LGN, afferents from the two eyes in visual cortex), the competitive interactions between the afferents from the two eyes, and the development of orientation domains in visual cortex. In this discussion we will distinguish two types of neuronal activity: that generated intrinsically and that generated by sensory stimulation. We will refer to the latter as "experience."

Much of the data in this section were obtained from ferrets. These animals are of particular interest because the visual system of ferrets shares many traits with that of cats (Law, Zahs, & Stryker, 1988) . Moreover, ferrets are born about three weeks earlier in their development, so that much of the development that is prenatal in the kitten is postnatal in the ferret and therefore easier to study. For example, although the segregation of retinogeniculate axons by eye in cats occurs prenatally, in ferrets it occurs between birth and 22 days of age (Linden, Guillery, & Cucchiaro, 1981), which is still well before their eyes open, 30–34 days after birth (Chapman, Stryker, & Bonhoeffer, 1996).

DEVELOPMENT THAT DEPENDS ON INTRINSICALLY GENERATED NEURONAL ACTIVITY. *ON and OFF Channels within the Retina.* The dendrites of ON and OFF ganglion cells in the adult retina lie in different sublaminae of the inner plexiform layer (Famiglietti & Kolb, 1976; Nelson *et al.*, 1978; Saito, 1983). This stratification develops from an initially diffuse distribution, wherein the dendrites of the retinal ganglion

cells are found throughout the inner plexiform layer (Maslim & Stone, 1988). In cats, stratification begins about 2 weeks before birth, continues for about 2 weeks postnatally, and depends on the activity of the bipolar cells. Chronically blocking the response of the ON bipolar cells by applying the glutamate analog 2-amino-4-phosphonoamino butyrate (APB) during the first 2 weeks of postnatal life blocks dendritic stratification so that the dendrites retain their diffuse distribution (Bodnarenko & Chalupa, 1993; Bodnarenko, Jeyarasasingam, & Chalupa, 1995). The ganglion cells with unstratified dendrites receive inputs from both ON- and OFF bipolar cells, becoming in essence "mixed" ON/OFF cells (Bisti, Gargini, & Chalupa, 1998). Thus, activity of the bipolar cells is important in the development of the adult pattern of dendritic stratification in retinal ganglion cells.

It should be noted that although the process of dendritic stratification is activity dependent, it may not be experience dependent; that is, it may not require external stimulation. Some stratification occurs before birth, and most of it has occurred before the eyes open at 7–10 days after birth. It is essentially complete by about 2 weeks of age when kittens first start behaviorally to use vision (Hirsch, 1985).

ON and OFF Pathways to the LGN. Segregation of ON and OFF pathways to the LGN is known to require retinal activity, since it is prevented by eliminating that activity (Archer, Dubin, & Stark, 1982; Cramer & Sur, 1997; Dubin, Stark, & Archer, 1986). It also requires activation of N-methyl-D-aspartate (NMDA) receptors. These receptors, which are blocked at resting potential, are unblocked when the membrane is depolarized (Nowak, Bregestovski, Ascher, Herbert, & Prochiantz, 1984); thus they function as AND gates: they open only when they are bound by glutamate and the cell is also depolarized by other active receptors. In ferret LGN, ON and OFF pathways are segregated into sublaminae of laminae A and A1 (Stryker & Zahs, 1983), and this segregation can be prevented by administration of NMDA antagonists (Hahm, Langdon, & Sur, 1991). Thus, to influence segregation of ON and OFF pathways, neuronal activity must activate a specific class of receptors, the NMDA receptors, in the LGN.

Pathways from the Two Eyes to the LGN. The segregation of inputs from the two eyes to layers A and A1 of the LGN develops prenatally from an initially diffuse projection (Sretavan & Shatz, 1986). This segregation also depends on activity (Shatz, 1996). It can be prevented by using tetrodotoxin (TTX) to chemically block action potentials in the LGNs of fetal kittens (Shatz & Stryker, 1988). Because segregation occurs prenatally, it is independent of experience. The only postnatal manipulation known to affect segregation of retinogeniculate afferents in cats is perinatal enucleation of one eye, in which case a few sprouts from the other eye enter the denervated layer (Garraghty, Sur, Weller, & Sherman, 1986; Guillery, 1972b); these sprouts appear to be exclusively from Y-cell axons (Garraghty *et al.,* 1986). No such sprouting occurs in adult cats.

In contrast to the segregation of ON and OFF inputs, segregation of the inputs from the two eyes does not appear to depend on the NMDA receptor. Administration of an NMDA antagonist failed to disrupt laminar segregation in ferrets (Smetters, Hahm, & Sur, 1994), in which this segregation takes place postnatally (Linden *et al.,* 1981). Thus, although activity is required for eye-specific segregation in the LGN (Penn, Riquelme, Feller, & Shatz, 1998), that activity need not activate NMDA receptors.

Having shown the importance of intrinsically generated neuronal activity, we next consider how the retina is activated long before it is bombarded by visual stimuli.

All theories of the activity-dependent development of retinotopic maps and eye-specific segregation of afferents require correlated firing of neighboring retinal ganglion cells (Goodhill, 1993; Miller, Keller, & Stryker, 1989; Schmidt, 1985; Schmidt & Tieman, 1985; Stent, 1973; von der Malsburg & Willshaw, 1976). Recent work has shown that such locally correlated activity is present well before visual stimulation. Waves of spontaneous activity can be recorded from the retinas of rats prenatally (Maffei & Galli-Resta, 1990) and from the retinas of ferrets shortly after birth, before their eyes have opened (Wong, Meister, & Shatz, 1993). Feller, Welllis, Stellwagen, Werblin, and Shatz (1996) used calcium imaging combined with intracellular recording to examine the waves of spontaneous activity in developing ferret retina, and showed that the waves consisted of spatially restricted domains of activity that formed a mosaic pattern over the entire retinal ganglion cell layer. Furthermore, in the only animals in which it has been tested, mice and ferrets, these waves of spontaneous retinal activity are relayed to the cells in the LGN (Mooney, Penn, Gallego, & Shatz, 1996; Weliky & Katz, 1999), where they are modified by intrinsic geniculate properties and by feedback from visual cortex (Weliky & Katz, 1999).

How are local correlations in activity generated prior to visual experience? Many retinal cells are electrically coupled through gap junctions, which connect neighboring amacrine cells with each other, especially the A II, or cholinergic amacrine cells. Gap junctions also connect neighboring ganglion cells with each other and with neighboring amacrine cells (Famiglietti & Kolb, 1975; Penn, Wong, & Shatz, 1994; Vaney, 1991, 1994; Vardi & Smith, 1996). Electrical coupling increases the likelihood that neighboring cells will fire together, but cannot by itself explain the waves of spontaneous activity in immature retina because beta ganglion cells are not electrically coupled to other cells, yet participate in the waves of spontaneous activity (Penn et al., 1994; Vaney, 1991, 1994). Thus, other types of coupling, perhaps involving the release of neurotransmitters, must play a role.

In cat retina, and presumably also in ferret retina, lateral chemical synapses (amacrine-to-amacrine and amacrine-to-ganglion cell synapses) develop very early, well before the vertical synapses (photoreceptor-to-bipolar cell and bipolar-to-ganglion cell) do (Crooks & Morrison, 1989; Maslim & Stone, 1986), and thus lateral synapses can participate in generating spontaneous activity before the retinal circuits are mature. Amacrine cell synapses, particularly those of the A II amacrine cells, which release acetylcholine, appear to be critical (Feller et al., 1996; Penn et al., 1998). The waves of retinal activity can be initiated by local application of acetylcholine, and their propagation is blocked by the cholinergic antagonist curare (Feller et al., 1996). Note that although cholinergic transmission is required, it may not be sufficient; the effects of blocking the gap junctions have not been determined.

This "organized" intrinsic retinal activity plays a role in the segregation of afferents to the LGN. Blocking these waves with chronic binocular intraocular injections of epibatidine, which desensitizes the acetylcholine receptor (Marks, Robinson, & Collins, 1996), prevents the eye-specific segregation of the retinogeniculate axons in developing ferrets (Penn et al., 1998). Blocking activity in only one eye results in asymmetric projections from the two eyes. The projection from the active eye fails to retract; rather it remains in "inappropriate" regions of the LGN. In contrast, the projection from the inactive eye shrinks, so that it is virtually restricted to the monocular segment of the contralateral eye. These results suggest that afferents from the two eyes compete with each other during the process of segregation, and that the outcome of this competition is determined by the relative activity of the two eyes (Penn *et al.*, 1998). In summary, development of the segregation of afferents, both within the retina and between the retina and LGN, depends upon intrinsically generated neuronal activity. As we will see, the segregation of afferents from LGN to cortex, in contrast, is influenced by experience-dependent neuronal activity.

DEVELOPMENT THAT DEPENDS ON NEURONAL ACTIVITY GENERATED BY EXPERIENCE. *Distribution of Pathways from Layers A and A1 of the LGN to Visual Cortex.* Although the afferents representing the two eyes are segregated in the visual cortex of adult cats (and many other animals; see Schmidt & Tieman, 1985), the initial projection from the LGN is diffuse, with no evidence of segregation (LeVay, Stryker, & Shatz, 1978). Ocular dominance patches develop during the first 6 weeks of life. Anatomic segregation is first seen around 4 weeks of age (Antonini & Stryker, 1993b; LeVay *et al.*, 1978), but functional segregation may occur earlier. As early as 14 days, optical imaging shows patchiness in neural activity driven by the ipsilateral eye (Crair, Gillespie, & Stryker, 1998).

In contrast to the segregation of retinogeniculate afferents, that of geniculocortical afferents is experience dependent: The development of ocular dominance patches is delayed by dark-rearing (Mower, Caplan, Christen, & Duffy, 1985; Swindale, 1981) and prevented by intraocular injection of TTX (Stryker & Harris, 1986). Furthermore, this development reflects interactions between the afferents from the two eyes: In strabismic cats (in which one eye is surgically deviated by cutting either the medial or lateral rectus muscle) (Löwel & Singer, 1993; Shatz *et al.*, 1977) and in cats reared with alternating monocular exposure (in which the two eyes take turns in receiving exposure on alternate days) (Tumosa, Tieman, & Tieman, 1989), ocular dominance domains are much sharper (better segregated) than are those in normally reared cats. In monocularly deprived cats, in which the lids of one eye are sutured shut during the first week of life, the domains representing the deprived eye are smaller than normal, whereas those representing the experienced eye are larger than normal (Antonini & Stryker, 1996; Shatz & Stryker, 1978; Tumosa *et al.*, 1989). Similarly, in cats reared with *unequal* alternating monocular exposure (in which one eye consistently receives longer exposure than the other), the ocular dominance domains representing the less experienced eye are smaller than normal and those representing the more experienced eye are larger than normal (Hirsch, Tieman, Tieman, & Tumosa, 1987; Tumosa *et al.*, 1989).

Although the relative sizes of the ocular dominance domains are determined by the balance of activity in the two eyes (see above), the periodicity of the domains (that is, their center-to-center spacing) had been thought to be determined by factors intrinsic to the cortex or to the afferent fibers, factors such as local interactions among cortical cells or the spread of afferent arbors (Miller *et al.*, 1989; Swindale, 1980; von der Malsburg, 1979; von der Malsburg & Willshaw, 1976). It is now clear that experience can affect the periodicity of ocular dominance domains. Strabismus, which causes discordant inputs to the two eyes, increases the periodicity of ocular dominance columns by about 33% (Löwel, 1994). This had been predicted by Goodhill (1993), who modeled the effects of varying the degree of correlation between the patterns of activity in the two eyes. The results of this modeling suggested that the stronger the correlation between the two eyes, the narrower the columns will be. Further support for Goodhill's theory comes from the results of S. B. Tieman and Tumosa (1997), who found that alternating monocular exposure, which also decorrelates the activity of the two eyes, also increases the periodicity of the ocular dominance domains, although less than does strabismus. Thus, an animal's early experience determines all aspects of its ocular dominance domains: sharpness, relative size, and spacing.

Interactions between Afferents: Binocular Competition.   Because binocular deprivation (produced by suturing the lids of both eyes closed before eye opening) has a less devastating effect for a deprived eye than monocular deprivation (Wiesel & Hubel, 1963b, 1965), Wiesel and Hubel (1965) postulated that the geniculocortical afferents representing the two eyes compete with each other for access to cortical cells during development. Consistent with this hypothesis, the less-experienced eye of cats reared with unequal alternating monocular exposure was far more affected than an eye given the same amount of exposure in a cat reared with equal alternating monocular exposure (Hirsch *et al.*, 1987). That is, in both monocular deprivation and unequal alternating exposure, imbalanced exposure has greater effects than equivalent, but balanced exposure. This differential effect of an imbalance in stimulation provided the first evidence for binocular competition.

Differential effects within the binocular and monocular segments, first seen in the LGN, provided further support for binocular competition. Within the A layers of the LGN of monocularly deprived cats, the cell bodies of neurons in the deprived layers are shrunken in comparison to those of the experienced layer (Wiesel & Hubel, 1963a) or to those of normal cats (Hickey, Spear, & Kratz, 1977) (Figure 1). The fact that this shrinkage is largely restricted to the binocular segment (Guillery & Stelzner, 1970; Hickey *et al.*, 1977) suggests that the shrinkage is not due to deprivation per se, but is secondary to binocular competition. This suggestion has been

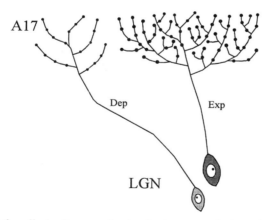

Figure 1. Summary of the effects of monocular deprivation on geniculocortical relay cells and their terminal arbors in area 17 (A17). Two relay cells of the lateral geniculate nucleus (LGN) are shown, a deprived one (Dep) and a nondeprived one (Exp). The deprived cell, located in layer A1, is smaller and terminates in a smaller arbor containing fewer branches and smaller boutons than does the nondeprived cell, which is located in layer A. In addition, the deprived cell contains less of the possible neurotransmitter *N*-acetylaspartylglutamate (indicated by the lighter shading). (Figure redrawn from Figure 10 of S. B. Tieman, 1991).

confirmed in later experiments which, by preventing competition or its consequences locally in small regions of the binocular segment of the geniculocortical pathway, also prevented cell shrinkage in those same regions (Bear & Colman, 1990; Guillery, 1972a; but see B. Gordon & Bremiller, 1992).

Neurobiological consequences of monocular deprivation are severe: synaptic input to visual cortex is virtually lost (reviewed by S. B. Tieman, 1991). First, in monocularly deprived cats, many cells in the LGN but not in the cortex can be driven by the deprived eye (Wiesel & Hubel, 1963a), suggesting that binocular competition disrupts the geniculocortical connections for the deprived pathway. Second, deprived geniculocortical cells project to smaller ocular dominance patches (Shatz & Stryker, 1978), where their axonal arbors branch less extensively (Antonini & Stryker, 1993a, 1996) (Figure 1). According to Antonini and Stryker (1998), this loss of arbor results from competition, as it does not occur after binocular deprivation. Furthermore, these smaller arbors establish fewer synapses that are also abnormal (S. B. Tieman, 1984, 1985). These deficits result in ocular dominance columns that are small, faint, and usually fail to extend into cortical layers above and below layer 4 (Tumosa *et al.*, 1989).

How does competition work? Active inputs win the competition when they drive the postsynaptic cell; they lose the competition when the postsynaptic cell remains silent. In an impressive series of experiments, Stryker and colleagues (Hata & Stryker, 1994; Hata, Tsumoto, & Stryker, 1999; Reiter & Stryker, 1988) infused the γ-aminobutyric acid (GABA) agonist muscimol to silence the visual cortex on one side during a period of monocular deprivation. In the control cortex, with no muscimol, the open eye came to dominate cortical cells and to project to a larger proportion of layer 4. In the region inactivated by the muscimol, however, the afferents from the open eye became extremely shrunken so that they were smaller than deprived arbors in control monocularly deprived cats, and the *deprived* eye dominated the cortical cells and projected to a greater part of layer 4. This relative dominance by the deprived eye in muscimol-treated cortex is not due to expansion of the deprived arbors (Hata *et al.*, 1999), which are no larger than those in normal kittens of the same age.

It seems likely that active postsynaptic cells normally give off a substance (perhaps a neurotrophin; see below) that influences presynaptic cells; in principle, active terminals may obtain more of this substance than inactive ones (Harris, Ermentrout, & Small, 1997). This differential uptake could occur for any number of reasons: because the greater recycling of synaptic vesicles in active terminals allows for greater uptake of extracellular substances (Schacher, Holtzman, & Hood, 1974; Singer, Holländer, & Vanegas, 1977), because release of the substance by the postsynaptic cell is activity dependent (Blochl & Thoenen, 1995, 1996; Harris *et al.*, 1997; Lledo, Zhang, Südhof, Malenka, & Nicoll, 1998; Maletic-Savatic & Malinow, 1998), or because active terminals have more receptors for the substance (Castrén, Zafra, Thoenen, & Lindholm, 1992). Since the resource is in limited supply, less active inputs do not get enough to sustain themselves. According to Stryker and colleagues (Hata & Stryker, 1994; Hata *et al.*, 1999; Reiter & Stryker, 1988), presynaptic activity is deleterious *unless* the synaptic terminal can receive this critical substance from the postsynaptic cell. In the absence of this substance, it is better to be inactive. In other words, it is likely that active terminals require more of this substance than do inactive ones.

According to the theory of binocular competition, effects of an imbalance in stimulation should never be observed in the monocular segment, where competition cannot take place. However, there are at least three instances in which the effects of an imbalance *are* observed within the monocular segment. First, both monocular deprivation and unequal alternating monocular exposure decrease behavioral responses to visual targets in the monocular segment of the less experienced eye. In cats reared with alternating monocular exposure, the extent of this decreased responsiveness is a function of the amount of exposure given the other eye (Tumosa, Tieman, & Hirsch, 1982). Responses are greater if the other eye received less exposure, intermediate if the other eye received the same exposure, and fewer if the other eye received more.

Further evidence for an effect of an imbalance on the monocular segment comes from studies on the pattern of staining for two different molecules in the LGN. In normal adult cats, enzyme histochemistry for NADPH-diaphorase (a marker for nitric oxide synthase) labels the axons and terminals arising from the parabrachial region of the brain stem; no cell bodies in the LGN are labeled in adults (Bickford, Günlük, Guido, & Sherman, 1993), although transient cellular labeling is seen in kittens (Guido, Scheiner, Mize, & Kratz, 1997). In monocularly deprived cats, however, a number of large cells in the experienced laminae are labeled (Günlük, Bickford, & Sherman, 1994). Surprisingly, these cells are found in the monocular segment as well as in the binocular segment. The staining is clearly due to the imbalance, since it does not occur in normal cats, but it is presumably not due to competition, since it occurs in the monocular segment. Similar results are seen with staining for *N*-acetylaspartylglutamate (NAAG), a possible neurotransmitter (Blakely & Coyle, 1988). In normally reared cats, NAAG is found in the relay cells, but not the interneurons, of the LGN (Xing & Tieman, 1993). Monocular deprivation decreases labeling for NAAG in the deprived layers of the LGN (S. B. Tieman, 1991) (Figure 1), whereas it increases labeling for NAAG in the largest cells of the experienced layers; these changes are seen in the monocular segment as well as in the binocular segment (S. B. Tieman, unpublished results). Thus, changes due to an imbalance can apparently occur in the monocular segment, where the effects of competition should not occur. What accounts for these changes is unclear.

Nevertheless, relay cells in the binocular segment of the A layers of the lateral geniculate nucleus do appear to compete with each other, and the cells of the experienced layer win this competition. Evidence presented below suggests that they compete for a neural trophic factor, or neurotrophin.

Development of Orientation Selectivity. Orientation-selective cells are present in kitten visual cortex almost as soon as the eyes open (Crair *et al.*, 1998; Hirsch & Tieman, 1987; Hubel & Wiesel, 1963a), although selectivity is less precise than in the adult (Crair *et al.*, 1998; Hirsch & Tieman, 1987). Orientation selectivity increases until 3–6 weeks of age (Crair *et al.*, 1998; Hirsch & Tieman, 1987). Until at least 3 weeks of age, i.e., during the first stage of development (Hirsch, 1985; Hirsch & Tieman, 1987), its development appears to be experience independent (Crair *et al.*, 1998). In contrast, during the second stage of development, from 3 to 6 weeks of age, experience dependence reaches a peak. Crair *et al.* (1998) used optical imaging to study the development of orientation-selective domains in normally reared and binocularly deprived (lid-sutured) kittens. Prior to 3 weeks of age, the organization of these domains was very similar in the two groups. In 2-week-old kittens from either group, optical imaging revealed that cortical activity driven through the contralateral eye was organized into orientation-selective domains; activity driven through the ipsilateral eye was less so. The difference in orientation preferences for the ipsilateral and contralateral eyes was greater than in the adult. For activity driven through the contralateral eye, the development of orientation-selective domains increased at the same rate in normal and binocularly deprived kittens, as did the similarity in orientation preferences for the two eyes. Prior to 3 weeks of age, the only difference between normal and binocularly deprived cats was that orientation-selective domains in the activity driven by the ipsilateral eye developed more slowly in the binocularly deprived cats. Beginning at 3 weeks of age, differences between normal and binocularly deprived kittens increased: visually driven activity was less clearly orientation selective in the binocularly deprived kittens than in the normal ones, and the difference in orientation preferences for the two eyes increased.

Although the initial development of orientation selectivity may be experience independent, it is clear that orientation preferences can be modified by selective visual exposure. Thus, exposure to lines of a single orientation ("stripe-rearing") during the second developmental stage constrains the distributions of both the physiological orientation preferences of cortical cells and the anatomic orientation of their dendritic fields (Hirsch, 1985; Hirsch & Tieman, 1987) (Figure 2).

In effect, early experience alters the response properties of the cells so that they match stimulus attributes of the environment experienced during rearing. Cats reared viewing vertical lines have more cells (and a greater percentage of visual cortex) devoted to responding to vertical lines, whereas cats reared viewing horizontal lines (or diagonal lines) have more cells that respond to horizontal (or diagonal) lines (Flood & Coleman, 1979; Hirsch & Spinelli, 1970; Rauschecker & Singer, 1981; Singer *et al.*, 1981).

Some, but not all, cells preferring horizontal and vertical stimuli are relatively

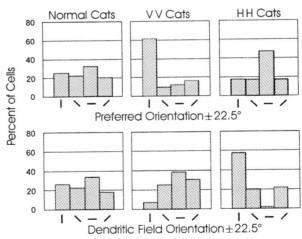

Figure 2. Stripe-rearing affects the distribution of preferred orientations and dendritic field orientations in the visual cortex of the cat. Top: Distribution of preferred orientations of orientation-selective cells recorded from Area 17 of 15 normal cats, 6 cats exposed to only vertical lines (VV cats), and 6 cats exposed to only horizontal lines (HH cats). The histograms plot the percentage of cells having preferred orientations within 22.5° of vertical, 135°, horizontal, and 45°. Note that for each group of stripe-reared cats the distribution peaks at the exposed orientation. Data from Hirsch (1985) and Leventhal and Hirsch (1980). Bottom: Distribution of the orientations of the basal dendritic fields of Golgi-impregnated layer III pyramidal cells from area 17 of 10 normal cats, 5 VV cats, and 3 HH cats. Note that most of the cells from stripe-reared cats have dendritic fields oriented orthogonal to the cortical representation of the exposed orientation. Data from Tieman and Hirsch (1982) and Zec and Tieman (1994).

insensitive to the effects of early experience. These experience-insensitive cells are among the first to develop orientation selectivity (Chapman & Bonhoeffer, 1998; Chapman *et al.*, 1996; Frégnac & Imbert, 1978), they are more likely to survive prolonged periods of dark-rearing (Frégnac & Imbert, 1978; Leventhal & Hirsch, 1977, 1980), they survive in cats reared viewing only diagonal lines (Leventhal & Hirsch, 1975), and they are overrepresented in the visual cortex of normal adults (Coppola, White, Fitzpatrick, & Purves, 1998; Mansfield, 1974; Mansfield & Ronner, 1978). Furthermore, as Hirsch, Leventhal, McCall, and Tieman (1983) showed, their physiological response properties (small receptive fields and poor response to rapid stimulus motion; Leventhal & Hirsch, 1980) indicate that they probably receive direct excitatory input from X-type cells in the LGN (Dreher *et al.*, 1980; Mustari, Bullier, & Henry, 1982). In contrast, cells preferring diagonally oriented stimuli are more sensitive to early experience; they are rarely observed in either dark-reared cats (Frégnac & Imbert, 1978; Leventhal & Hirsch, 1977, 1980) or cats reared viewing horizontal or vertical lines (Leventhal & Hirsch, 1975). Their response properties (large receptive fields and strong response to rapid stimulus motion; Leventhal & Hirsch, 1980) indicate that they probably receive direct excitatory input from Y-type cells in the LGN (Dreher *et al.*, 1980; Hirsch *et al.*, 1983; Mustari *et al.*, 1982). These results are consistent with earlier findings that Y cells in LGN require visual stimulation to develop normally, whereas X cells do not (Kratz, 1982; Kratz, Sherman, & Kalil, 1979).

In summary, orientation selectivity of cells in visual cortex develops relatively independently of visual experience during the first developmental stage. During the second developmental stage, it requires normal visual experience for its maintenance. Prolonged selective visual exposure at this time can have dramatic effects. Those cells preferring horizontal or vertical stimuli and having response properties indicating a direct input from X-type LGN cells are less susceptible to the effects of experience than are cells preferring diagonal lines, which have response properties indicating a direct input from Y-type LGN cells. As we will see below, there may be comparable differences in experience sensitivity of different afferent streams in the fly.

With the cat as introduction, we now address the use of *Drosophila* as a model system for studying experience-dependent development.

THE FLY AS A MODEL SYSTEM

Traditionally, the invertebrate nervous system has been viewed as developing independently of experience (Easter, Purves, Rakic, & Spitzer, 1985). Yet, considerable evidence stands contrary to that view: in invertebrates, as in mammals and other higher vertebrates, sensory experience and neuronal activity contribute selectively and significantly to the establishment of neural connectivity. This evidence has been gathered for many species including locust (Bloom & Atwood, 1980, 1981; Rogers & Simpson, 1997), cricket (Charii & Lambin, 1988; Deruntz, Palévody, & Lambin, 1994; Lambin, Charii, Meille, & Campan, 1990; Matsumoto & Murphey, 1977, 1978; Meille, Campan, & Lambin, 1994; Murphey & Chiba, 1990), bee (Hertel, 1982, 1983), crayfish (Lnenicka, 1991; Lnenicka & Atwood, 1985), and fly (Schuster, Davis, Fetter, & Goodman, 1996). For the sake of manageability, in this chapter we will limit our discussion to flies, especially *D. melanogaster*.

The fly nervous system consists of a relatively small number (perhaps a few hundred thousand) of neurons that are organized in a modular fashion, can often be identified anatomically from individual to individual, and have been described in exquisite detail. In addition, flies have a rich and varied repertoire of behaviors— many involving the use of vision—that are both reproducible and quantifiable and thus well suited for analysis (e.g., Heisenberg & Wolf, 1984; Hirsch *et al.*, 1990; Wehner, 1981). This combination of accessible neuronal morphology and accessible behavior makes flies ideal subjects for the study of experience-dependent development of brain and behavior.

Drosophila present scientists with the added advantage of a substantial number of nervous system mutants. Many of these affect the visual system (Fischbach & Heisenberg, 1984; Hall, 1982; Heisenberg, 1979) and its development (Cutforth & Gaul, 1997; Duncan *et al.*, 1997), whereas others affect learning (Belvin & Yin, 1997; for recent review, see Davis, 1996). This wide array of mutants can provide excellent tools for analyzing the experience-dependent development of the visual system.

STRUCTURE AND ORGANIZATION OF THE VISUAL SYSTEM IN THE ADULT FLY

The compound eye of *Drosophila* and other cycloraphe flies is the subject of dozens of articles and reviews; a concise overview of structure and function may be found in Young (1989), and the reader wishing more depth is referred to Cutforth and Gaul (1997), Laughlin (1981a), Meinertzhagen and Hanson (1993), Strausfeld and Lee (1991), Strausfeld and Nässel (1981), and Wolfe, Martin, Rubin, and Zipursky (1997). A brief overview forms the basis of our discussion on visual system development.

The compound eye has about 800 facets, or *ommatidia*, each a precise assembly of 20 cells: 8 photoreceptors, 4 cone cells, 6 pigment cells, and 2 bristle cells (survivors from an initial complex of 4). The schematic view of the eye in Figure 3 shows 11 such ommatidia. The lens system consists of an outer cornea and inner crystalline cone. These lens elements focus light on the photoreceptors, a cluster of eight long *retinula* cells, usually characterized as R1–R8. Six of these photoreceptors (R1–R6) are wedged together like segments of a pie and the centralmost border of each photoreceptor bears a *rhabdomere*, a dense fringe of microvilli. Rhabdomeres contain the photopigments that transduce the light signal.

The axon of each retinula cell passes through the basement membrane of the retina to make synaptic contact with second-order neurons in the optic lobes, which lie just below (Figure 3). The optic lobes consist of four ganglia: lamina, medulla (divided in Figure 3 into distal and proximal), lobula, and lobula plate. These ganglia represent progressive stages of visual processing. As in vertebrate visual systems, there are different types of photoreceptors that extract different kinds of information from the visual array and provide starting points for at least two parallel afferent streams that appear to serve different motor control systems.

Afferents from retinula cells R1–R6 terminate in the first ganglion, the lamina (Figure 3). These receptors are very sensitive and can be considered analogous to mammalian rods. Afferents from the remaining retinula cells, R7 and R8, the ultraviolet- and blue-sensitive photoreceptors, respectively (Strausfeld & Lee, 1991), project past the lamina to synapse in distal layers of the second ganglion, the medulla (Figure 3).

The lamina is divided into a series of discrete neural units, the *lamina cartridges*, which are separated from each other by glial cells. From a given ommatidium, retinula cells R1–R6 project to six different cartridges; conversely, a given cartridge receives input from single retinula cells (those whose microvilli are aligned along the same optical axis) of each of six neighboring ommatidia. This "neural superposition" ensures that neighboring photoreceptors that are viewing the same position in space project to the same cartridge. It follows that there is complex interweaving of axons to make these connections and, further, that the number of cartridges precisely matches the number of ommatidia.

Within each lamina cartridge, the photoreceptors make multiple synapses with large monopolar cells (Figure 3). R1–R6 synapse with L1 and L2; L2, in turn, synapses back onto the photoreceptors (Strausfeld & Campos-Ortega, 1977); these synapses are likely to be involved in feedback regulation of receptor output, introducing a phasic component to the response of the monopolar cells so that they can signal temporal contrast (Shaw, 1984). The axons of the monopolar cells join those of other lamina cells to project across an optic chiasm to the second ganglion, the medulla. The medulla is organized into columns, each of which receives axons from a single cartridge. Also projecting to the same column are the axons of R7 and R8 cells that share the same optical axis as that of the R1–R6 cells in the above-mentioned cartridge, so that now information from eight retinula cells looking at a single point in space converges in a single column; the pathways from R1–R6 and from R7 and R8, however, appear to remain distinct in the medulla and ultimately to project to different targets.

Working with *Calliphora*, Strausfeld and Lee (1991) demonstrated two separate pathways for information traveling from the medulla to the lobula and lobula plate

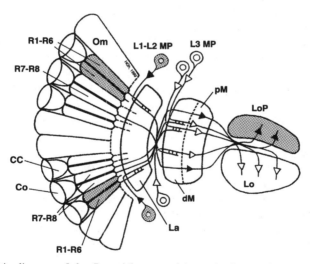

Figure 3. Schematic diagram of the *Drosophila* eye and its projection to the optic lobes. The eye is composed of about 800 ommatidia (Om; only 11 are shown here). The lens system, consisting of the cornea (Co) and the crystalline cone (CC), focuses the light onto a cluster of eight photoreceptors, or retinula cells (R1–R8). The flow of information from these retinula cells to the optic lobes is segregated. Shaded pathway: R1–R6 (motion sensitive but color insensitive) synapse in the lamina (La) with L1 and L2 monopolar cells, which in turn project through the medulla to the lobula plate (LoP), which ultimately projects largely to the neck muscles and flight motor. Unshaded pathway: R7–R8 (color sensitive and motion insensitive) project through the lamina to synapse in the distal medulla (dM; proximal medulla; pM) with the L3 monopolar cells, which in turn project to the lobula (Lo) and thence to the walking muscles. Thus the two information streams, which are strikingly similar to the Y and X pathways in the cat, remain essentially distinct. (Based in part on Young, 1989.)

(collectively referred to as lobular complex). The lobula plate is supplied by transmedullary neurons associated with the L1 and L2 monopolar cells, which receive their input from the R1–R6 retinula cells. These neurons appear to be color-insensitive and concerned rather with the parameters of motion detection. Many of the lobula plate cells onto which they synapse are large and have large receptive fields. The lobula, on the other hand, receives transmedullary neurons associated with the L3 monopolar cells as well as with the UV- and blue-sensitive retinula cells R7 and R8. Many of the postsynaptic cells in the lobula have smaller receptive fields and are not as responsive to motion as are the cells in the lobula plate. Strausfeld and Lee (1991) interpret these results in terms of possible separation of function: the motion-sensitive but spectrally insensitive output of the lobula plate ultimately projects to the neck muscles and the flight motor, while the motion-insensitive but color-sensitive output from the lobula projects to the body wall and leg muscles. Thus, as these authors point out, flying insects discriminate motion, whereas walking ones discriminate color.

The actual relationships between lobula and lobula plate are in fact considerably more complex than this capsule description allows (Douglass & Strausfeld, 1998). Strausfeld and Lee (1991) relate the characteristics of cells in the afferent stream passing through the lobula plate, on the one hand, and the color sensitive afferent stream passing through the lobula, on the other hand, to the magnocellular and parvicellular pathways, respectively, in primates (Hubel & Livingstone, 1987; Livingstone & Hubel , 1987); we see a resemblance to the cat's Y and X pathways.

The fly's motion-sensitive pathway to the lobula plate resembles the cat's Y pathway: Both have large neurons, large receptive fields, short latency, high contrast sensitivity, are better able to detect quickly moving stimuli, and are more sensitive to early stimulation (see below). Similarly, the color-sensitive path way to the lobula in many ways resembles the cat's X pathway. While color is not a salient behavioral cue for the cat (Clayton & Kamback, 1966; Loop, Bruce, & Petuchowski, 1979; Mello & Peterson, 1964; Pearlman & Daw, 1971), nor are color-opponent cells restricted to X-type retinal ganglion cells in cat (Ringo & Wolbarsht, 1986; Ringo, Wolbarsht, Wagner, Crocker, & Amthor, 1977), both the lobula plate pathway and the X pathway have other common attributes: smaller cells, smaller receptive fields, a greater latency, a role in acuity, and less sensitivity to modification as a result of early stimulation (see below).

The projections from the photoreceptors to the optic lobes are highly retinotopic. Their texture is made denser by interconnections between synaptic classes and by retrograde innervation from the brain to more proximal regions of the optic lobes; indeed, Meinertzhagen and Hanson (1993) remark that the insect optic lobes are the most precisely arranged neuropil known; and with each new study the complexity of that neuropil appears to increase.

Anatomical Development of the Visual System in Flies

Overview of Development.   The development of this system has been ably reviewed in many publications including Burke and Basler (1997), Cheyette *et al.* (1994), Cutforth and Gaul (1997), Fischbach (1983), Green, Hartenstein, and Hartenstein (1993), Lawrence (1992), Meinertzhagen and Hanson (1993), Mukhopadhyay and Campos (1995), Wolfe *et al.* (1997), and Zuker (1994). Briefly, both larval and adult eyes arise from a patch of dorsolateral head ectoderm. Part of this patch comprises the optic lobe primordium; during mid-embryogenesis, the posterior part of the primordium invaginates and fuses to the developing brain. The region immediately anterior to the optic lobe primordium invaginates to form a pocket of columnar epithelium, the imaginal disc, that will form the retina of the adult eye.

The *Drosophila* larval stages last about 4 days. During the first half of larval life, cells within the optic lobe primordium and the eye disc proliferate; during the second half, they start differentiating into their neuronal form. In the eye disc, morphogenesis of the ommatidia starts early in the third larval instar at the posterior edge and sweeps anteriorly as an organized wave whose leading edge is marked by a dorsoventrally oriented shallow groove, the *morphogenetic furrow*. The furrow is the scene of major change. As it moves across the eye from posterior to anterior, cells far ahead of it appear unpatterned and undifferentiated. Those cells immediately in front are starting to express such patterning genes as *Decapentaplegic* (*Dpp*), *eyeless* (*eye*), *Hedgehog* (*Hh*) and *sine oculis* (*so*) and are changing shape, constricting apically to produce the furrow morphology. Cells immediately behind the furrow are visibly organizing into clusters that will differentiate progressively into the ommatidia that characterize the adult structure. Cells within a normal ommatidium will differentiate in strict order, starting with R8, but the assignment of roles appears to be the result of cell–cell interactions rather than expression of cell lineages. These changes take place over a period of about 2 days.

During the first larval instar, the optic lobe primordium separates into two epithelial fields, or proliferation centers, that will generate the adult optic lobes. The outer (distal) proliferation center gives rise to the lamina and the outer medulla, while the inner (proximal) center gives rise to the inner medulla and the lobula complex. Considering the precise architecture and complex interweaving of projections between retina and lamina, there must be exquisite matching of pre- and postsynaptic cell populations during development and the appropriate fiber arborizations must be maintained. This development, particularly of the lamina, is critically dependent on proper retinal innervation. During normal development, the axons of the developing retina reach the developing lamina through the *optic stalk*, which also serves as a conduit for the axons of the larval eye. Projection is believed to occur in the order in which the photoreceptors differentiate in the ommatidium: R8, followed by R1–R6, and finally R7. R1–R6 from a given ommatidium first project to the same developing cartridge, but then they diverge to neighboring cartridges, thereby establishing the highly interwoven structure described above.

Retinal projections in the optic lobes appear to have at least two functions; inducing mitotic division and directing the postmitotic cells to differentiate into neurons. Failure of this projection, as in eyeless flies or retinal mutants, results in reduced and/or abnormal optic lobes. This effect is not reciprocal. Using genetic mosaics with the mutant genotype in the eye and the wild type in the lamina, or vice versa, Meyerowitz and Kankel (1978) showed that where mutant clones in the retina overlie wild-type laminae, the axon arrays in the laminae and medullae develop disorderly patches. The inverse situation, wild-type retinae overlying mutant optic lobes, generates wild-type neuropil, strongly suggesting that the genotype of the eye determines the phenotype of the optic lobes. Meinertzhagen and Hanson (1993) summarize apparent exceptions to this rule. It seems likely that retrograde influence of the optic lobes supports long-term maintenance of normal retinal structure rather than its development.

This dependence on proper retinal innervation for normal retinal structure is apparently shared by glial cells but may not hold true for all cells in medulla or lobula: sprouting and compensatory innervation may also direct development of these structures. Nevertheless, columnar elements in the optic lobes may be thought to be established by a wave of differentiation that starts in the retina and spreads centrally. As the lamina organizes, it in turn propagates a wave of differentiation that spreads to the medulla, although some cells there anticipate the event and begin differentiation beforehand.

In response to surgical or genetic lesions, optic lobe neurons may demonstrate plasticity of volume and connectivity (Meinertzhagen & Hanson, 1993) that, together with other aspects of development reviewed by these authors, resemble effects of growth factors in vertebrates. Even in unmanipulated systems, synaptogenesis apparently continues into the adult, and synaptic populations may reflect experience-dependent changes at the synaptic level.

COMMON GENES INVOLVED IN INVERTEBRATE AND VERTEBRATE VISUAL SYSTEM DEVELOPMENT.   The evidence that many of the genes involved in patterning in *Drosophila* are similarly involved in vertebrates is convincing (for reviews dealing with visual systems, see Barinaga, 1995; Cutforth & Gaul, 1997; Halder, Callaerts, & Gehring, 1995; Quiring, Walldorf, Kloter, & Gehring, 1994; Wolfe *et al.*, 1997; and Zuker, 1994). *Eyeless* in *Drosophila* is homologous to mouse and human homeobox genes (genes that contain specific conserved sequences that code for DNA-binding proteins, which regulate the transcription of other genes). In particular, *eyeless* is homologous to *pax-6*, a vertebrate gene important in differentiation of many organs, for example, neural tube primordium, developing brain and eye primordia, and later lens, retina, and corneal tissue in mice. A homologue has also been found to be active in the brain, olfactory organs, and eye of the squid (Tomarev *et al.*, 1997); others are being discovered in zebra fish, frogs, turtles, quail, rats (Zuker, 1994), and nematodes (Duncan *et al.*, 1997). Mouse *pax-6* introduced into *Drosophila* discs produces ectopic *Drosophila* eyes (Halder *et al.*, 1995). *Drosophila sine oculis* has a conserved mouse analog, *Six3*, that controls development in the anterior neural plate, including the eye and optic tract. These commonalities, together with shared photoreceptor biochemistry (all photoreceptors appear to have vitamin A-derived chromophores), suggest that animal eyes are fundamentally homologous and that genes like *pax-6* may have participated in specification of an ancestral photoreceptive cell, a root from which the various eyespots and image-forming eyes eventually evolved and diverged. Indeed, the presence of *vab-3*, the homologue of *pax-6*, in nematodes, which lack photoreceptors and where the gene regulates head morphogenesis, suggests an earlier origin for this gene, perhaps as a general regulator of anterior structures (Wolfe *et al.*, 1997).

EXPERIENCE-DEPENDENT CHANGES IN THE VOLUMES OF VISUAL STRUCTURES AND CENTRAL BRAIN.   *Changes in the Optic Lobes.*   Commonalities in the development of vertebrate and invertebrate visual systems include experience dependence. In flies kept in light–dark cycling illumination (LD) for 4 days after eclosion, "monocular occlusion" produced by coating one eye with opaque black paint within 1 hour of eclosion alters the volume of the optic lobes (Barth, Hirsch, Meinertzhagen, & Heisenberg, 1997). The lamina, medulla, and lobula plate are significantly smaller on the occluded side (a smaller difference in lobula volume is not significant; Figure 4A). To control for possible toxic effects of the paint, additional monocularly occluded flies were kept in constant darkness (DD) during the first 4 days of adult life. As Figure 4A shows, in these dark-reared flies, there were no significant differences in volume between the two sides of the brain; thus the left/right volume differences seen in the occluded flies kept in a lighted environment must depend upon visual stimulation.

In general, dark-rearing produces greater changes than monocular occlusion alone. The difference between the unoccluded side in LD flies and *either* side in DD flies is much larger than the left/right difference in monocularly occluded LD flies (Figures 4B, 5). As is summarized in the bottom panel in Figure 5, in monocularly occluded DD flies, both laminae and lobula plates are significantly smaller than their counterparts on the unoccluded side of LD flies. Thus, unlike the case for the LGN of the cat, volume changes in the laminae and lobula plates appear to be related to the amount of stimulation per se, rather than to an imbalance in stimulation.

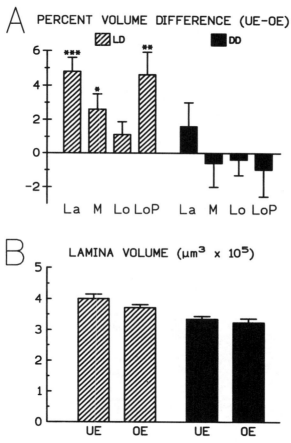

Figure 4. Effects of monocular occlusion and of rearing in darkness on volumes of optic lobes in *D. melanogaster.* (A) Effects of painting one eye on the left/right difference in volumes of the lamina, medulla, lobula, and lobula plate. For each fly the difference in volume (Unoccluded Eye − Occluded Eye) was computed and the differences were tested against zero; hatched bars represent data for LD flies ($n = 130$) and solid bars represent data for DD flies ($n = 80$). Since there were no obvious sex differences, data for males and females were pooled. Note that for LD flies there are significant left/right differences for all but the lobula. $*p < .005$; $**p < .0005$; $***p < .00001$. (B) Same animals as in panel A. Lamina volume for DD (solid bars; $n = 24$) and LD (hatched bars; $n = 37$) monocularly deprived flies. Note that lamina volume in DD flies is significantly smaller (25%) than the lamina receiving input from the open eye of the monocularly occluded flies. (Redrawn from Figure 2 in Barth, Hirsch, Meinertzhagen, & Heisenberg, 1997.)

Furthermore, the size of the lamina is correlated directly with the amount of visual stimulation: in nonoccluded flies, the lamina was smallest in DD flies, intermediate in LD flies, and largest in flies reared in constant light (LL) (Barth, Hirsch, Meinertzhagen, & Heisenberg, 1997). These results suggest that some development is facilitated by stimulation-induced neuronal activity. On the other hand, volume changes in the medulla occur only where there is an imbalance in stimulation (Figure 5), suggesting that some development is sensitive to competition between afferents from the two eyes.

Changes in optic lobe volume reflect changes in the axonal terminals of the

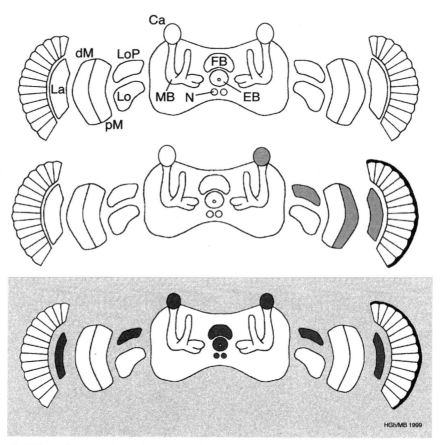

Figure 5. Effects of monocular occlusion and dark-rearing on optic lobes and brain in *Drosophila*. Top: Stylized section through the eyes, optic lobes, and brain. The areas of interest include the optic lobes: lamina (La), distal and proximal medullae (dM and pM), lobulae (Lo), and lobula plates (LoP). In the brain, they include the mushroom bodies (MB) with their attached calyces (Ca) and the central complex (which consists of four parts: the fan-shaped body, FB; the ellipsoid body, EB; two noduli, N; and the protocerebral bridge, not analyzed and not shown. Middle: In monocularly occluded flies, the experienced side (left) shows a relative increase in volume of the lamina, distal medulla, lobula plate, and calyx (the central complex was not analyzed in these flies). Bottom: Dark-rearing very effectively blocks experience-dependent volume increases on both sides of the laminae, lobula plates, calyces, and central complexes but does not appear to affect the medulla, suggesting that unilateral experience-dependent volume increase of the latter in the monocularly occluded flies may be due to imbalance of stimulation rather than deprivation per se.

photoreceptor cells. Figure 6 shows transverse sections through two lamina cartridges, one from an LL fly and one from a DD fly; the photoreceptor processes (R1–R6) are much smaller (28.5%) in the DD fly (Barth, Hirsch, Meinertzhagen, & Heisenberg, 1997). This difference, seen only in the distal lamina, was significant. In contrast, there was little difference (3%, not significant) in the proximal lamina. The measurements made at the distal and proximal boundaries of the lamina are consistent with an average change of 15.7%, very close to the overall measured volume change for this lobe. The photoreceptor processes make up about 70% of the cartridge cross-sectional area (Barth, Hirsch, Meinertzhagen, & Heisenberg, 1997; but see also Hauser-Holschuh, 1975) and are responsible for most of the change in volume. Monopolar cells undergo similar size changes but make up only about 5% of the cross-sectional area (Barth, Hirsch, Meinertzhagen, & Heisenberg, 1997) and therefore contribute less to the total change.

In the medulla, experience-dependent changes are seen only in monocularly occluded flies and only in the distal part, where the processes of the R7 and R8 photoreceptors terminate (Barth, Hirsch, Meinertzhagen, & Heisenberg, 1997). The fact that changes are restricted to distal medulla suggests that in the medulla, as in the lamina, photoreceptor terminals are responsible for most of the change. In summary (see Figure 5), degree of stimulation (lamina) or balance in stimulation between the two eyes (medulla) affects the cross-sectional area of photoreceptor processes and, at least in the lamina, of the interneurons with which they make synaptic contacts. The aggregate of these changes results in overall volume changes in those optic ganglia, lamina, and medulla, in which the photoreceptor processes terminate. Which cells might account for volume changes in lobula plate is not yet known.

Experience-dependent changes also occur in more central brain regions, such as the calyces of the mushroom bodies and the central complex (Barth, 1997), which is composed of the fan-shaped body, the ellipsoid body, and two noduli (Fig. 5). In DD flies, the volume of these structures is smaller than in flies given visual exposure (Fig. 5). The calyces are also affected by monocular occlusion (the central complex

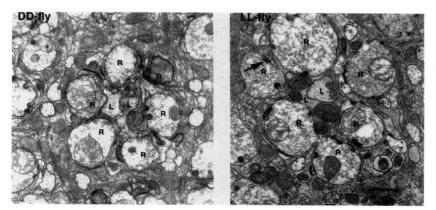

Figure 6. Cross sections of two lamina cartridges with six photoreceptor terminals (R) surrounding two monopolar cells (L) in the middle. The cartridge in turn is surrounded by epithelial glial cells. The left micrograph is from a fly reared in constant darkness (DD) and that on the right is from a fly reared in constant light (LL). The arrow points to a synaptic profile. Note how much larger the photoreceptor terminals are in the LL fly. Scale bar, 1.0 μm.

was not analyzed in the monocularly occluded flies). The finding that the calyces and the central complex are affected by visual stimulation during development provides indirect evidence for a link between visual structures and more central brain regions, such as the mushroom bodies (Barth, 1997). (In bees, there is direct evidence for such a link; Fahrbach & Robinson, 1995; Mobbs, 1982.)

Time Course and Critical Period. The optic lobes increase in volume during the first 4 days after eclosion. This increase is greater in flies reared with both eyes open in continuous light (LL) than in DD flies (Figure 7). These effects of visual experience are evident in the lobula plate by 9 hr of age and in the lamina by 12 hr (Barth, Hirsch, Meinertzhagen, & Heisenberg, 1997). Thus, continuous exposure to light can significantly enhance age-related increases in the volumes of both lamina and lobula plate relative to those occurring in deprived flies. Continuous exposure to

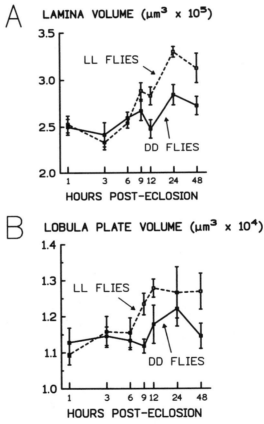

Figure 7. Semilogarithmic plot of volume of the lamina and lobula plate during the first 48 hr after eclosion. (A) The volume of the lamina increases in both LL and DD flies especially after the first 6 hr, and the change is larger in the LL group. Differences between LL and DD flies become significant starting at 12 hr after eclosion ($n_{LL} = 20_{(1\,hr)}, 10_{(3\,hr)}, 10_{(6\,hr)}, 11_{(9\,hr)}, 8_{(12\,hr)}, 6_{(24\,hr)}, 9_{(48\,hr)}; n_{DD} = 20_{(1\,hr)}, 10_{(3\,hr)}, 9_{(6\,hr)}, 10_{(9\,hr)}, 8_{(12\,hr)}, 10_{(24\,hr)}, 11_{(48\,hr)}$. (B) Same animals as in panel A. The volume of the lobula plate increases in both LL and DD flies after the first 6 hr after eclosion, and the change is again larger in the LL group. Differences between the LL and DD flies become significant at 9 hr after eclosion. (Redrawn from Figure 9, Barth, 1997.)

light (LL) also has similar effects on volume of the calyces and central complex (Barth, 1997).

There is a critical period during which light exposure must occur if it is to increase lamina volume. Deprivation during larval or pupal stages has no effect (M. Barth, unpublished observation). During the first 12 hr after eclosion, as little as 6 hr of darkness in an LL regime that extends until day 4 results in a significantly smaller lamina; 12 hr of darkness (followed by light exposure until day 4) leads to a lamina comparable in volume to that found in flies kept in darkness for 4 days (Barth, 1997; Barth, Hirsch, Meinertzhagen, & Heisenberg, 1997). Thus, light exposure is required during the first 12 hr after eclosion to enhance lamina volume; maintenance of light exposure during the first 4–5 days after eclosion is needed to preserve this experience-dependent increase in lamina volume. Unpublished results by M. Barth suggest that light deprivation during subjective night has less effect than does light deprivation during subjective day. Thus in flies, as in cats, there is a critical period during which at least part of the visual system is sensitive to visual experience. How experience sensitivity of the medulla or lobula plate might change over the first days of posteclosion life is not known.

Genetic Dissection of Possible Sites of Experience-Dependent Changes in the Optic Lobes. As expected, anatomic consequences of visual experience require photo-transduction. In the blind mutant *norpA^P24^*, phototransduction is blocked because of a defect in phospholipase C, an enzyme essential for the phototransduction process (Bloomquist *et al.*, 1988; McKay *et al.*, 1995). In these mutants, the volume of the lamina is similar in LL and DD flies and in both cases is comparable to the wild-type DD group (Barth, Hirsch, Meinertzhagen, & Heisenberg, 1997). This mutation also abolishes experience-dependent changes in the calyx and central complex (Barth, 1997). Thus, at least in the lamina, calyx, and central complex, experience-dependent changes require transduction by the photoreceptors.

Histamine is very likely to be the neurotransmitter between the photoreceptors and the monopolar cells (Hardie, 1987). Blocking it by use of the presumed histamine-null mutant *histidine decarboxylase^jk910^* (*hdc^jk910^*; Melzig *et al.*, 1996), prevents experience-dependent growth of the lobular plate and calyx, but not of the lamina (Barth, Hirsch, Meinertzhagen, & Heisenberg, 1997): despite the probable block in synaptic transmission, there is a significant difference in lamina volume between LL and DD flies. This is consistent with the evidence given above that much of the overall volume change in the lamina reflects experience-dependent changes in cross-sectional area of the photoreceptor processes themselves, and suggests that these changes do not depend upon synaptic activation of the second-order cells. Furthermore, since the photoreceptors should be unable to drive the monopolar cells in these mutants, visually driven feedback from the latter onto the photoreceptors (Meinertzhagen & O'Neil, 1991) is not likely to be necessary for experience-dependent enlargement of the photoreceptor processes. Thus, stimulation-dependent activation of the photoreceptors alone is apparently sufficient to increase the cross-sectional area of their processes and thereby increase lamina volume. Whether the size of the monopolar cells is affected by light in the *hdc^jk910^* mutants is an open question, although in the *hdc^jk910^* flies the volume of the lobula plate is significantly smaller than that of the wild-type LL and even DD flies, suggesting that these mutants may lack some intrinsically generated activity important in development (Barth, Hirsch, Meinertzhagen, & Heisenberg, 1997). Thus it appears that, in flies as in mammals, spontaneous activity may play an important role in development.

In summary, the photoreceptor processes account for most of the stimulation-dependent growth of the lamina. Photoreceptor processes are also important in experience-dependent changes in the medulla: in the medulla of monocularly occluded flies, changes are restricted to the distal region, where the axons from R7 and R8 terminate. Results obtained with the mutant hdc^{jk910} demonstrate the importance of synaptic transmission between the photoreceptors and the monopolar cells in the production of experience-dependent changes in the lobula plate and in central brain. Finally, the R1–R6 pathway (motion sensitive) appears to be more sensitive to the effects of visual experience than does the R7–R8 pathway (color sensitive). Both lamina and lobula plate (components of the R1–R6 pathway) are affected by both monocular occlusion and dark-rearing. In contrast, the lobula (part of the R7–R8 pathway) is unaffected by either exposure regime, and the distal medulla (also part of the R7–R8 pathway) is affected only by monocular occlusion (Figure 5).

Experience-Dependent Synaptic Changes in the Lamina. In *Musca domestica*, visual experience alters the number of synapses between the photoreceptors and the monopolar cells. Briefly exposing DD flies to light for periods lasting from 20 sec to several hours (maximum effect obtained at 15 min) increases the frequency and decreases the mean size of the synapses from the receptor processes onto the monopolar cells (Rybak & Meinertzhagen, 1997).

In contrast, visual experience *decreases* the frequency of the feedback synapses from the L2 monopolar cell back onto the photoreceptor processes. This effect is superimposed on an already complex pattern of developmental changes in the frequency of these feedback synapses in normal flies, which increases by about 25% during the first day posteclosion, then declines by about 47% until the eighth day, when it seems to reach an asymptote (Kral & Meinertzhagen, 1989). Monocularly occluded flies given visual stimulation show a left/right difference in frequencies of feedback synapses after 1 or 2 days, with a consistently higher frequency on the occluded side (Kral & Meinertzhagen, 1989). A similar increase in synaptic frequencies, but no left/right difference, occurs after only 1 day in monocularly occluded DD flies (Kral & Meinertzhagen, 1989). Thus, the higher frequency of synaptic contacts on the occluded side probably results from stimulus deprivation rather than from an imbalance in stimulation of the two eyes. The effectiveness of 1 or 2 days of monocular occlusion increases between days 1 and 4 and then falls to zero on days 6–8 (Kral & Meinertzhagen, 1989). This pattern provides additional support for the existence of a critical period during the first days of adult life, in this case for experience-dependent changes in the frequency of feedback synapses.

In summary, in *Musca domestica*, the frequency of the first-order afferent synapses from photoreceptors onto monopolar cells increases with light exposure, whereas that of the feedback synapses from the L2 monopolar cells back onto photoreceptors R1–R6 decreases. Developmental changes in these synaptic connections can be rapid, and some, but clearly not all, appear to be the result of experience-dependent activity of the visual pathway during an early critical period.

Is the Cyclic AMP Cascade Important for Experience-Dependent Changes in Optic Lobe and Central Brain? Mutations affecting cyclic AMP signaling have been shown in *Drosophila* to interfere with behavioral plasticity such as learning (Davis, 1996) and with structural plasticity such as that involved in the regulation of fiber number in the peduncle of the mushroom bodies (Balling, Technau, & Heisenberg, 1987;

Technau, 1984), the principal centers mediating olfactory learning (Davis, 1996; de

Technau, 1984), the principal centers mediating olfactory learning (Davis, 1996; de Belle, 1995; de Belle & Heisenberg, 1994; Heisenberg, Borst, Wagner, & Byers, 1985; Mobbs, 1982; Tully, Preat, Boynton, & Del Vecchio, 1994). To determine whether the cyclic AMP pathway is also important for experience-dependent growth of the optic lobe, LL and DD groups were compared for three mutants that affect the cyclic AMP pathway. Two of these (*dunce¹* and *rutabaga¹*), are defective in enzymes that regulate levels of cyclic AMP:phosphodiesterase and adenylate cyclase, respectively (Davis, Cherry, Dauwalder, Han, & Skoulakis, 1995); the third (*amnesiac¹*) encodes a neuropeptide involved in regulation of the cyclic AMP pathway (Feany & Quinn, 1995). In all three mutants, the volumes of both lamina and lobula plate were significantly larger in LL than in DD flies, just as in wild-type flies (Barth, Hirsch, Meinertzhagen, & Heisenberg). The experience-dependent changes in volume of structures in the visual pathway thus do not appear to be vulnerable to defects in the cyclic AMP signaling system; consistent with this, the expression in the optic lobes of the *dunce¹* and *rutabaga¹* mutations is low (Han, Levin, Reed, & Davis, 1992; Levin *et al.*, 1992; Nighorn, Healy, & Davis, 1991).

In more central brain regions, two of the cyclic AMP mutants, *dunce¹* and *amnesiac¹*, abolished experience-dependent changes; specifically, volumes of the calyx and the central complex were not significantly different in LL and DD flies (Barth, 1997), suggesting that the cyclic AMP cascade is involved in the effects of visual experience on central brain, and that common molecular mechanisms may be involved in developmental and other forms of neural plasticity. In contrast, in *rutabaga¹* differences between LL and DD flies were comparable to those occurring in wild-type flies (Barth, 1997). The differences among the effects of the three cAMP mutants on experience sensitivity of central brain may reflect differences in the sites at which their gene products are expressed. In *dunce¹* and *amnesiac¹*, gene products are expressed in the calyces, which are concerned with inputs to the mushroom bodies, whereas in *rutabaga¹* they are expressed in the peduncles of the mushroom bodies, which are concerned with outputs from the mushroom bodies (Davis *et al.*, 1995). This suggests that expression of experience-dependent volume changes depends upon intact inputs, but is not blocked when the outputs are disturbed. In this way, changes in central brain are comparable to those in the photoreceptor terminals in the lamina, which are affected by blockage of the input (the blind mutant *norpA^{P24}*), but not by blockage of the output (histamine-null mutant *hdc^{jk910}*). These results are summarized in Table 1.

TABLE 1. ARE THERE EXPERIENCE-DEPENDENT CHANGES IN BRAIN VOLUME?[a]

Strain	Lamina	Medulla	Lobula	Lobula plate	Central brain
Wild type	+	+[b]	−	+	+
Rutabaga	+	?	?	+	+
Dunce	+	?	?	+	−
Amnesiac	+	?	?	+	−

[a]Strain differences in susceptibility to experience-dependent changes in brain volume, compared in LL and DD flies. +, significant difference between LL and DD flies; −, no different; ?, not tested.

[b]Changes in brain volume may depend upon competition between two eyes. The lobula pathway, which is involved in walking in flies, is analogous to the X-cell pathway in cats, while the lobula plate pathway, which is involved in flying, resembles the Y-cell pathway in cats.

SUMMARY OF EXPERIENCE-DEPENDENT ANATOMIC CHANGES. Visual stimuli enter the system via the photoreceptors of the compound eyes and from there the information proceeds to the neuropil of the optic lobes (Figure 3). Developmental effects of visual experience are greatest in the lamina, where changes in photoreceptor processes account for most of these effects (Barth, Hirsch, Meinertzhagen, & Heisenberg, 1997). Volume changes in the medulla are smaller and may reflect changes in the processes of the R7 and R8 photoreceptors. Because these changes occur only in flies in which the two eyes receive different stimulation, they parallel the competitive changes seen in the cat visual system. Visual experience also regulates volumes of lobula plate, calyx and central complex, but apparently not of lobula. The calyx, which receives input from other sensory modalities, has also been shown to be affected by "social" experience, defined as exposure to other conspecifics in a complex, highly structured environment (Balling *et al.*, 1987; Barth & Heisenberg, 1997; Technau, 1984), an issue to which we return below.

Two distinct processes that differ in their dependence on cyclic AMP underlie structural plasticity in the adult fly brain. Cyclic AMP signaling does not contribute to stimulation-induced changes in the volume of the lamina and lobula plate, suggesting that the underlying mechanisms for these changes may differ from those involved in other forms of neuronal plasticity such as associative learning, which does appear to involve cyclic AMP. In contrast, cyclic AMP signaling does influence the volume of the calyx and central complex, suggesting possible commonalities with other forms of neuronal plasticity.

POSSIBLE MECHANISMS OF ACTIVITY-DEPENDENT CHANGES

Much of the evidence concerning possible mechanisms comes from studies of cats, but there is increasing evidence that the mechanisms of plasticity are conserved across species. We now address four main issues: (1) changes in synaptic morphology observed in both mammals and invertebrates, (2) the possibility that afferents from the two eyes compete for neurotrophins, (3) the relevance of long-term potentiation (LTP) as a model for experience-dependent synaptic changes, and (4) experience-induced changes in gene expression.

MORPHOLOGY OF SYNAPTIC CHANGE

Learning and experience-dependent development both involve morphological changes in chemical synapses, which are the site of chemical transmission between neurons and their targets, and can be identified only through electron microscopy. However, because synapses usually occur at the axon terminal or in swellings called boutons or varicosities that can be seen through light microscopy, the number of varicosities along the arbor is often taken as being indicative of synaptic number. In the discussion below, the terms bouton, varicosity, or terminal each refer to a structure that can be seen with the light microscope; the term synapse is reserved for the specialized contacts that can be demonstrated only with the electron microscope.

Experience affects the number of synaptic boutons or varicosities in both invertebrates and vertebrates. Long-term facilitation increases the number of synaptic terminals at a sensorimotor synapse in *Aplysia* (Bailey & Chen, 1988, 1991). In monocularly deprived cats, the geniculocortical axons of the experienced pathway have more synaptic boutons than do those of the deprived pathway (Antonini &

Stryker, 1993a, 1996; S. B. Tieman, 1984, 1991). Similarly, the synaptic arbors of tonically active motoneurons in crayfish contain more varicosities than do those of phasic motoneurons, and repeated stimulation transforms phasic arbors into ones resembling tonic arbors (Lnenicka, 1991; Lnenicka, Atwood, & Marin, 1986). In other words, the experienced pathway of monocularly deprived cats, the facilitated pathway in *Aplysia*, and the tonic pathway in crayfish all have more synaptic boutons or varicosities than their less active counterparts. In the house fly, *Musca domestica*, visual experience increases the number of synapses made by photoreceptors, but not the number of photoreceptor terminals themselves (Rybak & Meinertzhagen, 1997).

The experienced pathway of monocularly deprived cats and the tonic pathway of crayfish resemble each other in two other ways: both have larger terminal boutons and both have higher mitochondrial content (Lnenicka, 1991; Lnenicka *et al.*, 1986; S. B. Tieman, 1984, 1991). Increased terminal size is associated with decreased susceptibility to fatigue (Lnenicka, 1991; S. B. Tieman, 1991), with maturation (Dyson & Jones, 1980; Mason, 1982), and with kindling in rat cortical synapses (Racine & Zaide, 1978). In *Drosophila*, experience is associated with larger photoreceptor terminals in visually experienced flies (Barth, Hirsch, Meinertzhagen, & Heisenberg, 1997). Similarly, the boutons at the neuromuscular junction are larger (G. A. Lnenicka, personal communication) and more numerous (Budnik, Zhong, & Wu, 1990) in mutant flies with high synaptic activity than in those with low synaptic activity.

Finally, synapses of the experienced pathway in monocularly deprived cats are more likely to have a gap in the post synaptic density (S. B. Tieman, 1985, 1991). Synapses with gaps ("perforated" synapses) may be more effective because they contain a higher density of neurotransmitter receptors (Desmond & Weinberg, 1998). Perforated synapses are associated with synaptic maturation in the visual cortex of the rat and rabbit (Dyson & Jones, 1984; Greenough, West, & DeVoogd, 1978; Müller, Pattiselanno, & Vrensen, 1981), with environmental enrichment in the visual cortex of rats (Greenough *et al.*, 1978), with both kindling (Geinisman, Morrell, & De Toledo-Morrell, 1990) and LTP (Geinisman, De Toledo-Morrell, & Morrell, 1991) in the hippocampus, and with visual discrimination learning in the visual cortex of the rabbit (Vrensen & Nunes Cardozo, 1981). The proportion of perforated synapses decreases with aging, and it has been suggested that this decrease may underlie the memory deficit seen in aged rats (Geinisman, De Toledo-Morrell, & Morell, 1986). In summary, plasticity is associated with changes in synaptic number and morphology in a wide variety of systems, and in animals that range phylogenetically from mollusk to mammal.

COMPETITION FOR NEUROTROPHINS

Competition implies limited resources. Recent experiments suggest that for the geniculocortical relay cells of mammals, the limited resource is a neurotrophin, or neural trophic factor. Trophic factors are one way in which targets can affect the status of cells that project to them. For a neuron to be affected by a trophic factor in its target, it must have receptors for that trophic factor on its surface, especially on the surface of its axon terminals. An appropriate trophic factor secreted by the target will be taken up by the terminals of the projecting neurons and transported back to the parent cell bodies, where it can affect gene expression.

If geniculocortical afferents are competing for neurotrophins during segregation, then supplying exogenous neurotrophins should supply *all* afferents with enough neurotrophin to maintain themselves, thereby preventing segregation and the effects of monocular deprivation. Maffei and others have studied the effects of supplying nerve growth factor (NGF) on the consequences of monocular deprivation. Intraventricular administration of NGF prevents both physiological (Berardi *et al.*, 1993; Maffei, Berardi, Domenici, Paris, & Pizzorusso, 1992; Yan, Mazow, & Dafny, 1996) and anatomic (Domenici, Cellerino, & Maffei, 1993) deterioration in monocularly deprived rats, as well as behavioral changes in monocularly deprived kittens (Fiorentini, Berardi, & Maffei, 1995). Further, intraventricular injection of hybridoma cells secreting monoclonal antibodies to NGF caused shrinkage of LGN cells in normally reared rats (Berardi *et al.*, 1994), as might be expected if these antibodies deprived the LGN cells of needed trophic support. However, substances injected intraventricularly are likely to spread and to affect many parts of the brain, and thus such substances may not be acting directly on the visual pathway. Furthermore, NGF acts through the tyrosine kinase receptor trkA. Although trkB and trkC receptors have both been found in LGN (Allendoerfer *et al.*, 1994; Merlio, Ernfors, Jaber, & Persson, 1992), trkA receptors have not. It is therefore unlikely that NGF could act directly on the LGN cells or their terminals in visual cortex. More likely, the NGF affects the geniculocortical projection indirectly, through effects on some other system such as the cholinergic projection to visual cortex from the basal forebrain, which is sensitive to NGF (Chen *et al.*, 1997) and participates in cortical plasticity (Bear & Singer, 1986). In support of this interpretation, Domenici, Fontanesi, Cattaneo, Bagnoli, and Maffei (1994) found labeled cells in basal forebrain but not in the LGN after injections of ^{3}H-NGF into the visual cortex of rats.

Although geniculocortical axons are not likely to be competing for NGF (since they do not have the appropriate receptor), they may be competing for either brain-derived neurotrophic factor (BDNF) or neurotrophin 4/5 (NT-4/5). Both BDNF and NT-4/5 act at the trkB receptor, which *is* found in the geniculocortical pathway (Cabelli *et al.*, 1996). If geniculocortical axons are competing for a neurotrophin, and the outcome of that competition can be affected by visual stimulation, then the expression of that neurotrophin should be regulated by visual stimulation. Light regulates messenger RNA (mRNA) for both BDNF and trkB, but not for NGF, in rat visual cortex (Castrén *et al.*, 1992; Schoups, Elliott, Friedman, & Black, 1995).

Supplying the relevant neurotrophin during development should stabilize the initial, widespread connections, and ocular dominance patches should not form (Harris *et al.*, 1997). This prediction has been supported: administration of exogenous BDNF or NT-4/5 prevents the segregation of afferents into ocular dominance patches. Cabelli, Hohn, and Shatz (1995) used osmotic minipumps to infuse neurotrophins into the visual cortex of young kittens between 4 and 6 weeks of age, the time when the geniculocortical afferents normally segregate into ocular dominance patches. Infusion of either BDNF or NT-4/5 prevented the formation of ocular dominance patches in the site around the cannula, whereas two other neurotrophins, NGF and neurotrophin-3 (NT-3), did not. As noted above, BDNF and NT-4/5 both act through trkB, whereas NGF acts through trkA; NT-3 acts through trkC.

Cabelli, Shelton, Segal, and Shatz (1997) next examined the effects of blocking endogenous neurotrophins. They used antagonists for BDNF and NT-4/5 that were generated by making "pseudo-receptors" composed of the extracellular (neurotrophin-binding) domain of trkB fused to the complement-binding portion of the human immunoglobulin G (IgG) molecule. Intracortical infusion of this pseudo-receptor, trkB–IgG, for 2–3 weeks during the period 21–46 days after birth prevented segregation of geniculocortical axons, whereas infusion of trkA–IgG and trkC–IgG, which bind NGF and NT-3, respectively, did not. Furthermore, labeling of LGN terminals in layer 4 of visual cortex was decreased by infusion of trkB–IgG, almost as if the geniculocortical axons, failing to find any unbound BDNF or NT-4/5, were unable to maintain healthy terminals in the area.

If geniculocortical axons compete for a neurotrophin, and the cells in the deprived layers of the LGN in a monocularly deprived animal shrink because they lose this competition, then it should be possible to prevent cell shrinkage by supplying exogenous neurotrophin. Riddle, Lo, and Katz (1995) deprived young ferrets of vision in one eye for brief periods, and supplied small, localized areas of the cortex with neurotrophins. To identify neurotrophin-exposed LGN cells, a fluorescent retrograde tracer was attached to the neurotrophin. Supplying NT-4/5 (but not BDNF, NT-3, or NGF) prevented the geniculate cell shrinkage normally produced by monocular deprivation, but only in areas identified by the tracer as NT-4/5-exposed.

Intracortical infusion of the relevant neurotrophin should also prevent the physiologic consequences of monocular deprivation. Galuske, Kim, Castrén, Thoenen, and Singer (1996) found that BDNF, but not NGF, alters the consequences of monocular deprivation in kitten visual cortex. In control monocularly deprived cats, most cells are driven only by the initially open eye, and the deprived eye loses the ability to drive normally responsive cells, that is, cells that are selective for the orientation of the visual stimulus. In BDNF-infused, monocularly deprived cats, the deprived eye drives most of the cells near the infusion site, but these cells are not orientation selective. Slightly farther away, both eyes drive orientation-selective cells. Even further from the infusion site, the experienced eye drives most of the cells, and the deprived eye loses the ability to drive orientation-selective cells; that is, the situation is the same as in an otherwise untreated monocularly deprived cat.

In summary, during development, inputs from the two eyes compete with each other for some limited resource in visual cortex, probably a neurotrophin that acts at the trkB receptor. The patterns of neuronal activity in the two eyes determine the outcome of this competition, that is, whether the geniculocortical afferents segregate into ocularity domains, the sharpness of these domains, their relative sizes, and their spacing. The degree to which competition participates in *Drosophila* neural development is unknown, although there appears to be a competitive effect in the medulla following monocular occlusion (Figure 5). Furthermore, given that competition affects development in another insect, the cricket (Murphey, 1986a; Murphey & Lemere, 1984; Shepherd & Murphey, 1986), the possibility of competitive effects in *Drosophila* should be further examined.

There is one molecule with structural homology to NGF in *Drosophila* (DeLotto & DeLotto, 1998), a number of neural-specific trk receptors in both *Drosophila* (Haller, Cote, Bronner, & Jackle, 1987; Oishi *et al.*, 1997; Pulido, Campuzano, Koda, Modolell, & Barbacid, 1992) and mollusk (van Kesteren *et al.*, 1998), and a neurotrophic factor in mollusk (Fainzilber *et al.*, 1996). In addition, Meinertzhagen and Hanson (1993) discuss neuron–target interactions during *Drosophila* development that resemble those produced by growth factors in vertebrates. Thus neurotrophins are likely to exist in *Drosophila*, but their relation to development and learning has not yet been explored.

SIMILARITIES BETWEEN DEVELOPMENTAL PLASTICITY AND LONG-TERM POTENTIATION

Although connections in the nervous system are most readily modified during development, they retain a certain degree of plasticity during adult life, and adult and developmental plasticity are likely to share mechanisms (e.g., Kandel & O'Dell, 1992; Shatz, 1990; Wolpaw & Schmidt, 1991). Indeed, several authors have argued that the long-term facilitation that follows repeated activation of a synapse (long-term potentiation, or LTP), commonly used to model learning, may underlie some of the changes that occur in development (e.g., Artola & Singer, 1987; Kirkwood, Silva, & Bear, 1997; Schmidt, 1990). Certainly, LTP is more readily elicited in young than in adult visual cortex (Kato, Artola, & Singer, 1991; Kirkwood, Lees, & Bear, 1995). Furthermore, both LTP (Artola & Singer, 1987; Collingridge & Bliss, 1987; Frégnac, Shulz, Thorpe, & Bienensock, 1988; Sakimura *et al.*, 1995) and developmental plasticity of ocular dominance (Bear & Colman, 1990; Bear, Kleinschmidt, & Singer, 1990; Bear, & Singer 1990, Gu, Bear, & Singer, 1989; Kleinschmidt, Bear, & Singer, 1987; Roberts *et al.*, 1998) involve NMDA receptors.

Both LTP and developmental plasticity involve neurotrophins. The role of neurotrophins in ocular dominance plasticity was reviewed above. Here we will review the role of neurotrophins in LTP. First, neuronal activity itself increases the expression of neurotrophins, especially BDNF (Castrén, Zafra, Thoenen, & Lindholm, 1992; Castrén *et al.*, 1993; Dragunow *et al.*, 1993; Lindholm, Castrén, Berzaghi, Blöchl, & Thoenen, 1994; Patterson, Glover, Schwartzkroin, & Bothwell, 1992; Schoups *et al.*, 1995; Zafra, Castrén, Thoenen, Lindolm, 1991; Zafra, Lindholm, Castrén, Hartikka, & Thoenen, 1992). Second, BDNF enhances synaptic transmission (Akaneya, Tsumoto, Kinoshita, & Hatanaka, 1997; Carmignoto, Pizzorusso, Tia, & Vicini, 1997; Kang & Schuman, 1995, 1996; Kang, Jia, Su, Tang, & Shuman, 1996; Levine, Dreyfus, Black, & Plummer, 1995). Third, BDNF enhances LTP in young rats (Figurov, Pozzo-Miller, Olafsson, Wang, & Lu, 1996). Fourth, BDNF enhances transmission at NMDA receptors (Jarvis *et al.*, 1997). Fifth, LTP is deficient in mice that are deficient in BDNF (Korte *et al.*, 1995; Patterson *et al.*, 1996) and is restored by providing BDNF, either directly (Patterson *et al.*, 1996), or by transfecting hippocampal cells so that they are able to make their own BDNF (Korte, Staiger, Griesbeck, Thoenen, & Bonhoeffer, 1996). Finally, as noted above, neurotrophins act through tyrosine kinase receptors, and inhibitors of tyrosine kinase block LTP in hippocampus (O'Dell, Kandel, & Grant, 1991). Thus, neurotrophins are very much involved in the changes associated with LTP, as they are with experience-dependent changes in ocular dominance.

Much has been learned about the molecular basis of LTP by using ''knockout'' mice (Abeliovich *et al.*, 1993; Bourtchuladze *et al.*, 1994; Chen & Tonegawa, 1997; Korte *et al.*, 1995; Mansuy, Mayford, Jacob, Kandel, & Back, 1998; Qi *et al.*, 1996; Rotenberg, Mayford, Hawkins, Kandel, & Muller, 1996; Son *et al.*, 1996). The same approach could be used to determine whether LTP and developmental plasticity share molecular mechanisms: one could examine developmental plasticity in those knockout mice that show deficits in LTP. Stryker and colleagues have developed a model of ocular dominance plasticity in mice (J. A. Gordon & Stryker, 1996) and have begun to examine this plasticity in various strains of knockout mice in which LTP is known to be abnormal (J. A. Gordon, Cioffi, Silva, & Stryker, 1996; Hensch *et al.*, 1998). They found decreased ocular dominance plasticity in mice deficient in calcium-calmodulin protein kinase II (CaM kinase II) (J. A. Gordon *et al.*, 1996), but not in mice deficient in the RIß subunit of protein kinase A (Hensch *et al.*, 1998). Both strains of mice show deficits in LTP (Hensch *et al.*, 1998; Huang *et al.*, 1995; Silva, Stevens, Tonegawa, & Wang, 1992). Apparently LTP and developmental plasticity share some, but not all, molecular bases.

In summary, LTP and ocular dominance plasticity share many attributes. Both are more readily elicited in young animals and both involve NMDA receptors. Neurotrophins play an important role in both, as does CaM kinase II. In contrast, protein kinase A, which is involved in LTP, may not be involved in ocular dominance plasticity. This result in cat is consistent with the results in flies described earlier, in which defects in the cyclic AMP pathway (which activates protein kinase A) failed to disrupt experience-dependent changes in the optic pathway.

LEARNING INVOLVES ACTIVITY-DEPENDENT CHANGES IN GENE EXPRESSION

The old dichotomy of nature versus nurture is an oversimplification. Neuronal activity regulates gene expression and thus can control changes in neural function during development and in the adult (learning). Since gene expression is controlled by transcription factors, recent research has focused on how they may be involved in learning. One such factor is the cyclic AMP response element-binding protein (CREB; Silva, Kogan, Frankland, & Kida, 1998). CREB normally binds to a particular DNA sequence, called the cyclic AMP response element (CRE). Genes that can be activated by cyclic AMP tend to have CRE sequences in the upstream region that controls their expression. CREB is inactive until it is phosphorylated at its Ser133 residue. Once phosphorylated, it binds CREB-binding protein (CBP). The CREB–CBP complex is then capable of recruiting appropriate basal transcription factors and RNA polymerase II to initiate transcription of the gene (Figure 8). Neuronal activity can induce CREB phosphorylation (Moore, Waxham, & Dash, 1996), and CREB phosphorylation has been implicated in learning in flies, mollusks, and mammals (Frank & Greenberg, 1994; Silva *et al.*, 1998).

There are two major ways of activating CREB (Figure 8). First, cyclic AMP can activate protein kinase A , releasing the catalytic subunits of the kinase so that they are then free to move into the nucleus, where they phosphorylate CREB (Silva *et al.*, 1998). A second major way in which CREB can be activated is through increases in intracellular calcium, which activate calcium–calmodulin-dependent (CaM) kinases (Dash, Karl, Colicos, Prywes, & Kandel, 1991). In particular, CaM kinase IV, a nuclear kinase, appears to be involved in phosphorylating CREB (Deisseroth, Heist, & Tsien, 1998). Suzuki *et al.* (1998) showed that CREB is present in the postsynaptic densities of rat forebrain. This postsynaptic CREB could be phosphorylated by either protein kinase A or CaM kinase II, another component of the postsynaptic density (Gardoni *et al.*, 1998; Kennedy, Bennett, & Erondu, 1983; Klauck & Scott, 1995). Once phosphorylated, CREB dissociates from the postsynaptic density and is free to move to the nucleus, where it can bind CREs (Figure 8). In short, all the necessary elements are present in the postsynaptic region, and experience-dependent neuronal activity could activate CREB and thus change gene expression. In support of this, Pham, Impey, Storm, and Stryker (1999) demonstrated that monocular deprivation alters CREB/CRE-mediated transcription in mouse visual cortex.

Both activators of CREB phosphorylation, the cyclic AMP–protein kinase A cascade and the CAM kinases, have been implicated in learning. Two of the learning mutants (*dunce¹*, *rutabaga¹*) that have been described in *Drosophila* involve mutations in key enzymes that regulate levels of intracellular cyclic AMP: adenylate cyclase (Dudai, Uzzan, & Zvi, 1983; Levin *et al.*, 1992) and phosphodiesterase (Byers, Davis, & Kiger, 1981; Dauwalder & Davis, 1995), respectively. Another learning mutant, *amnesiac¹*, involves a defect in an adenylate cyclase-activating peptide (Davis, 1996). An additional learning mutant has been produced in *Drosophila* by creating a defect in the regulatory subunit of protein kinase A (Goodwin *et al.*, 1997). As noted above, *dunce¹* and *amnesiac¹*, but not *rutabaga¹*, affect experience-dependent changes in the calyx and central complex of *Drosophila*, but none of them affects plasticity in the

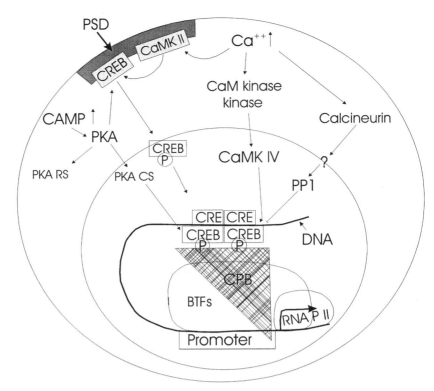

Figure 8. Schematic diagram of activity-dependent transcriptional regulation by the cyclic AMP respon-sive element-binding protein (CREB). Activity may increase calcium and/or cyclic AMP (cAMP), which in turn affect the phosphorylation and dephosphorylation of CREB. Cyclic AMP affects CREB phosphoryla-tion through protein kinase A (PKA), whereas calcium affects CREB phosphorylation through calcium–calmodulin-dependent kinases or calcineurin. Only phosphorylated CREB can interact with CREB-binding protein (CBP). The CREB–CBP complex can recruit basal transcription factors (BTF) and RNA polymerase II on the promoter and thus initiate transcription of the relevant gene, perhaps a neuro-trophin. PKA, Protein kinase A; PKA RS, PKA regulatory subunit; PKA CS, PKA catalytic subunit; PP1, protein phosphatase 1; PSD, postsynaptic density; CRE, cyclic AMP-responsive element; CaMK II, calcium–calmodulin-dependent protein kinase II; CaMK IV, calcium–calmodulin dependent protein kinase IV.

optic lobes. Cyclic AMP and protein kinase A are also involved in hippocampal LTP in rodents (Abel *et al.*, 1997; Bolshakov, Golan, Kandel, & Siegelbaum, 1997; Huang, Li, & Kandel, 1994; Huang *et al.*, 1995; Matthies & Reymann, 1993; Nguyen & Kandel, 1996; Qi *et al.*, 1996; Roberson & Sweatt, 1996; Slack & Walsh, 1995; Weisskopf, Castillo, Zalutsky, & Nicoll, 1994) and in long-term facilitation in *Aplysia* (Kaang, Kandel, & Grant, 1993; Schacher, Castellucci, & Kandel, 1988; Sweatt & Kandel, 1989).

CaM kinases have also been implicated in learning in both mammals and flies. Inhibition of postsynaptic CaM kinase II blocks the induction of LTP in rats (Mali-now, Schulman, & Tsien, 1989; Otmakhov, Griffith, & Lisman, 1997), whereas injecting constitutively active CaM kinase II postsynaptically enhances synaptic transmission and mimics LTP (Lledo *et al.*, 1995). Further, CaM kinase II-induced enhancement and LTP occlude each other, suggesting that they act by the same mechanism. CaM kinase II knockout mice are deficient in both LTP (Silva *et al.*, 1992; Stevens, Tonegawa, & Wang, 1994) and developmental plasticity of ocular dominance (J. A. Gordon *et al.*, 1996). LTP also increases the expression of CaM kinase II (Roberts *et al.*, 1996, 1998). In flies, CaM kinase II participates in both learning and memory for associative courtship conditioning (Joiner & Griffith, 1997). Transformed strains of *Drosophila* that express a specific inhibitor of CaM kinase II are impaired in both acoustic sensitization and courtship conditioning (Griffith *et al.*, 1993). Finally, CaM kinase IV, the nuclear CaM kinase, has been implicated in LTP in rats (Tokuda et al., 1997).

CREB's role in learning was first shown in Aplysia, where blocking CREB selectively blocked long-term, but not short-term, facilitation (Dash, Hochner, & Kandel, 1990). In *Drosophila*, blocking CREB prevented long-term memory for olfactory learning (Yin *et al.*, 1994), whereas overexpressing an activated form of CREB enhanced learning (Yin, Del Vecchio, Zhou, & Tully, 1995). In rodents, high-frequency stimulation that induces LTP also induces biphasic phosphorylation of CREB, with the second phase beginning 2 hr after stimulation and lasting at least 24 hr (Schulz, Siemer, Krug, & Höllt, 1999). Mutant mice in which the alpha and delta forms of CREB were knocked out showed learning deficits in a variety of contexts (Bourtchuladze *et al.*, 1994; Kogan et al., 1997). Finally, blocking CREB mRNA impaired long-term spatial memory in rats (Guzowski & McGaugh, 1997). There is thus direct evidence for CREB's role in learning in a wide range of species.

CREB phosphorylation is involved in the effects of neurotrophins, which may relate to its role in experience-dependent changes in ocular dominance (Pham *et al.*, 1999). First, neurotrophins appear to activate CREB phosphorylation (Finkbeiner *et al.*, 1997; Riccio, Pierchala, Ciarallo, & Ginty, 1997), apparently by increasing intracellular calcium and activating CaM kinase IV (Finkbeiner *et al.*, 1997). In addition, the expression of neurotrophins is controlled by CREB, probably in conjunction with other factors (Shieh, Hu, Bobb, Timmusk, & Ghosh, 1998; Tao, Finkbeiner, Arnold, Shaywitz, & Greenberg, 1998). Specifically, the activity-inducible transcript of BDNF has a CRE sequence in its regulatory domain (Shieh, Hu, Bobb, Timmusk, & Ghosh, 1998; Tao *et al.*, 1998). This participation of CREB in both the regulation and the effects of neurotrophins could provide a positive feedback loop whereby neurotrophins could increase their own expression. A similar positive feedback loop has been observed by Canossa *et al.* (1997), who found that neurotrophins can stimulate their own release. CREB's role in various forms of neuronal plasticity are summarized in Figure 9.

In summary, the transcription factor CREB has been implicated in learning in flies, mollusks, and mammals. CREB is active only when phosphorylated, and its phosphorylation is controlled by agents that have also been implicated in learning: protein kinase A and CaM kinases. Finally, CREB is involved in both the regulation and effects of neurotrophins, which, as we have seen, have been implicated in both learning and developmental plasticity.

FUNCTIONAL SIGNIFICANCE OF EXPERIENCE-DEPENDENT DEVELOPMENT

We started this chapter by suggesting that experience-dependent development of the visual system helps to ensure that the organism will be better tuned to the

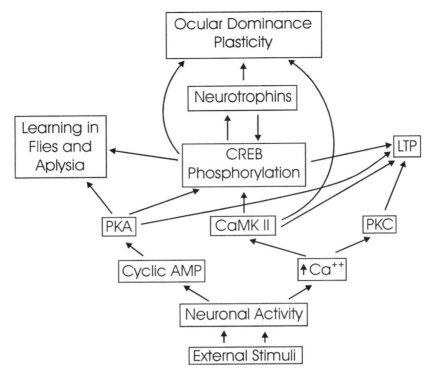

Figure 9. Schematic diagram illustrating CREB's central role in plasticity and summarizing the interrelationships described in this section. The arrows indicate that the process in the first box is known to affect the process in the secondbox. This effect need not be direct; rather, it is likely to involve several inter mediate steps, some of which may also be shown. Abbreviations as in Figure 8.

environment within which it will have to live and reproduce. We now use this hypothesis as a perspective from which to view the behavioral effects of rearing cats and flies in different visual environments.

DEVELOPMENT OF BEHAVIOR IN THE CAT

Kittens' use of vision changes dramatically during the first three months of postnatal life; these developmental changes were summarized by Hirsch (1985) and Hirsch and Tieman (1987). In brief, during the first three postnatal weeks (stage 1), kittens do not make much use of vision. Between the third and sixth postnatal weeks (stage 2), visually guided behaviors increase dramatically. It is during this period that the system is most sensitive to the effects of early experience.

Depriving a developing kitten of normal visual stimulation affects its later behavior, and these effects are summarized here. Specifically, an imbalance in the stimulation of the two eyes imposes or aggravates visual field deficits (Sherman, 1973; D. G. Tieman, Tumosa, & Tieman, 1983; Tumosa *et al.*, 1982; Van Hof-van Duin, 1977) and exacerbates difficulties in learning to discriminate among patterns (Ganz, Hirsch, & Tieman, 1972). Exposure to blurred or otherwise degraded visual images causes later deficits in visual acuity; for example, acuity deficits are severe and permanent after binocular deprivation, but not after dark-rearing (Mitchell, 1988;

Mower, Caplan, & Letsou, 1982; Smith, Lorber, Stanford, & Loop, 1980; Timney, Mitchell, & Griffin, 1978). In the absence of normal binocular visual experience, cats fail to develop the ability to use binocular cues to discriminate depth (Blake & Hirsch, 1975; Distler & Hoffmann, 1991; Kalil, 1978; Packwood & Gordon, 1975). When animals are exposed to only a limited range of stimulus orientations, there are orientation-specific deficits in acuity and in the ability to use vision to guide behavior (Blakemore & Cooper, 1970; Blasdel, Mitchell, Muir, & Pettigrew, 1977; Fiorentini & Maffei, 1978; Hirsch, 1972; Muir & Mitchell, 1973, 1975; Thibos & Levick, 1982; Wark & Peck, 1982). Finally, early experience alters attention; it seems to tune the visual system to attend selectively to certain stimuli, those present in the animal's early visual world (Hirsch, 1972; Tumosa, Nunberg, Hirsch, & Tieman, 1983). We see below that there is some evidence for similar behavioral effects of early experience in flies (Hirsch & Tompkins, 1994; Barth, 1997; Barth, Hirsch, & Heisenberg, 1997).

DEVELOPMENT OF BEHAVIOR IN FLIES

The behavior of adult flies changes dramatically over the course of the first few days after eclosion (Ford, Napolitano, McRobert, & Tompkins, 1989; Mimura, 1986). There are increases in responsiveness to gravity, light, odorants, and gustatory stimuli. In flies as in cats, visually guided behaviors change throughout the period during which sensory stimulation has been shown to affect visual system development. Also during this period, the courtship behavior of male *Drosophila* develops gradually to adult levels.

DEVELOPMENT OF SIMPLE, VISUALLY GUIDED BEHAVIORS *Experience-Dependent Development of Phototaxis and of Stimulus Preferences in the Flesh Fly,* Boettcherisca peregrina.   Visual exposure during the first days after eclosion affects behavioral development in the flesh fly. Phototaxis (approaching a lighted pattern in a dark arena) develops in these flies during the first day after eclosion, whether or not they receive visual exposure (Mimura, 1986). By the second day, LD flies begin to approach some lighted patterns (e.g., star-shaped patterns) in preference to others. These preferences become stronger over the next 2–4 days of exposure to a lighted environment (Mimura, 1986). Dark-rearing (DD) blocks the development of differential preferences for patterns, but if the deprivation does not exceed 4 days, then 4–6 days of subsequent light exposure will result in normal visual preferences. Flies dark-reared for more than 4 days before being placed in a lighted environment develop phototaxis but not differential preferences for patterns (Mimura, 1986). Finally, if flesh flies are exposed to a restricted visual environment (e.g., one consisting only of 45° lines) during the first few days post-eclosion, they preferentially approach the "familiar" stimuli (e.g. 45° rather than 135° lines) (Mimura, 1986). These last results bear striking similarity to those for cats exposed to lines of one orientation.

Changes in Responses to Moving Vertical Stripes.   One common behavioral test of the visual responsiveness of flies is the *optomotor response*, a compensatory turning of the animal in response to a movement of the entire visual scene (Wehner, 1981). Barth (1997) found that the optomotor response to wider stripes (18° spatial wavelength) was weaker in LL than in DD flies, and that in both groups, responses decreased as illumination decreased. The peak optomotor response for DD flies occurred at lower levels of illumination than that for LL flies (Figure 10). Flies that experienced 2 days of constant darkness followed by 2 days of constant light

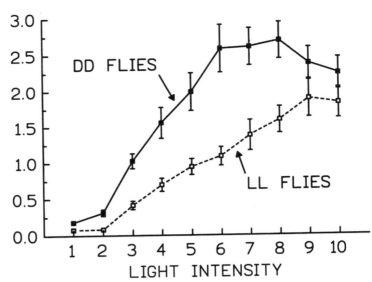

Figure 10. Optomotor responses of LL and DD flies, in arbitrary units. A vertically striped ($\delta = 18°$) cylinder, at whose center a fly was tethered, was rotated, 1 revolution/sec, and turning responses of the fly were measured. Light intensity was varied from a high level of 350 lux (10) to a low level that was below the range of the instrument (1). DD flies (26 runs from $n = 8$ flies) responded more than LL flies (28 runs from $n = 7$ flies) over the range of illumination tested, and the peak response of the DD flies was at a lower intensity than that of the LL flies. (Redrawn from Barth, 1997, Figure 26A.)

(DD_2LL_2) performed similarly to flies raised for 4 days in darkness, suggesting both that short-term adaptation effects were not a factor and that visual exposure during the first 2 days was critical. DD_2LL_2 rearing caused DD-like decreases in both lamina volume and optomotor performance (M. Barth, unpublished observation). When narrower (12°) stripes were used to test optomotor responses at moderate light levels (at or below 350 lux), DD flies responded whereas LL flies did not; responses of LD flies were intermediate (Barth, 1997). This suggests that rearing in constant darkness may adapt the system—perhaps by altering connectivity in the lamina—to function better under moderately low light conditions than does rearing in constant, bright light. At the lowest light levels, none of the flies were responsive. How well these flies might perform at intensities (ca. 10,000 lux) matching those to which the LD and LL flies were exposed is an open question.

Experience-Dependent Changes in Responses to Stationary Vertical Stripes. Flies walking about in an environment with stationary vertical stripes can use these stripes in two ways: (1) as a fiduciary point, to help them maintain a constant heading (Wehner & Wehner-von Segesser, 1973) and (2) as a target, the darkest region in their visual field being particularly attractive (Osorio, Srinivassan, & Pinter, 1990). Preferences for stripes of different widths can be studied in a large, cylindrical arena on whose inside wall are several groups of black vertical stripes, the groups differing in the width of their stripes. A "flightless" fly (either a flightless mutant or one in which the wings have been amputated) that is released at the center of such an arena usually walks rapidly toward one of the sets of stripes (Hirsch *et al.*, 1990; Wehner, 1972; Wehner & Horn, 1975; Wehner, Gartenmann, & Jungi, 1969); early experience affects which set it will choose (Hirsch *et al.*, 1990).

Responses from flies that were dark-reared (DD) from the egg stage through days 1–8 of the adult stage and then exposed to light 1–3 hours before testing were compared to those of LD controls (Hirsch *et al.*, 1990). The vast majority of both LD and DD flies walked to the three patterns composed of the wider stripes, avoiding both the narrow-striped pattern and the blank wall. At the earliest age tested (day 1), preferences of DD flies for different stripe widths were indistinguishable from those of LD controls. However, as the duration of light deprivation and age of testing increased, responses of the two groups (DD and LD) began to diverge. Starting at day 4, the DD flies were significantly more likely than the LD flies to approach the two sets of wide stripes (Figure 11).

The timing of the deprivation was as important as its duration. Light exposure during larval and pupal stages had little or no effect on later visual preferences. Darkness for several days during later adulthood was also ineffective; flies reared in normal illumination until day 4 of adult life and then kept in total darkness for the next 6 days did not differ in their preferences from 10-day-old LD controls. This suggests that there is a visual "critical period" during the days immediately after eclosion (Hirsch *et al.*, 1990). This coincides well with the critical period for plasticity of lamina volume.

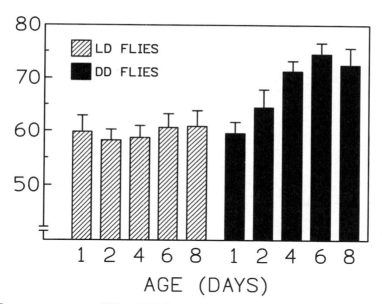

Figure 11. Percentage response of LD and DD flies to wide stripes. Flies were released one at a time at the center of a cylindrical arena with four sets of black vertical lines mounted against the white background of the cylinder wall. The line sets were as follows: eight lines (2.6 cm wide), four lines (5.2 cm wide), two lines (10.4 cm wide), or one line (22.8 cm wide). If flies approached one of the sets of lines or the blank wall within 2 min, this was scored as a response. The flightless *Drosophila* mutant *raised* (*rsd*) was used in these tests, and responses to the one- and two-line patterns were combined to give a response to wide stripes. The duration of the rearing and the age at testing were varied. Since there were no significant differences between males and females, their data were combined. At day 1 ($n = 200$) or day 2 ($n = 340$) posteclosion, there were no significant differences between LD and DD groups; at day 4 ($n = 840$), day 6 ($n = 460$) and day 8 ($n = 300$), the DD flies were significantly more likely than the LD flies to approach the two sets of wide stripes. Note that the percentage of LD flies that approached the two sets of wide stripes did not change with age, while for DD flies it did. Comparable results were obtained with Canton S wild-type flies that had their wings surgically removed prior to testing. (Redrawn from Hirsch *et al.*, 1990, Figure 5.)

If flies approach the stationary black stripes in part because they are attracted by "dark targets," then the DD flies, which select the wider stripes and thus the locally darker regions of the wall, show a stronger preference for these targets than do the LD flies. Again, a smaller lamina may reflect an adaptation to and thus a preference for a "darker" area.

EXPERIENCE-DEPENDENT CHANGES IN COURTSHIP.   Courtship is a complex but particularly well-studied behavior that normally involves considerable use of visual information. Courtship itself is experience dependent: young males are often courted by mature males, and as a result of this "homosexual experience," when they mature, they are more successful in courting females (McRobert & Tompkins, 1983, 1988; Tompkins, Hall, & Hall, 1980). While the advantage and mechanism of this "social stimulation" are unknown, as we will see below, early *visual* experience also influences the development of courtship patterns in *Drosophila.*

An Introduction to Courtship Behavior.   In most species, including *D. melanogaster,* courtship involves the exchange between male and female of information in the form of reciprocal types of sensory input as the sexes respond to each other (Averhoff & Richardson, 1974; Markow, 1981, 1987; Tompkins, Gross, Hall, Gailey, & Siegel, 1982). By observing the courtship behavior of wild-type females paired with a variety of mutant males who suffered from various sensory losses, Markow (1987) showed that visual information, and to a lesser extent chemical information, is important for the male to orient to the female, helps the male avoid directing courtship at inappropriate targets, and affects both the "transition probability" that he will switch from one courtship behavior to another and the likelihood that he will copulate successfully (Tompkins, 1984). Schäffel and Willmund 1985) showed that tracking, the most important visually guided behavior of the male, requires intact photoreceptors R1–R6. Lack of input from R7 appears not to affect tracking (Willmund & Ewing, 1982). Female receptivity is also influenced by visual cues provided by the male, especially the light reflection from and the shape of his eyes (Schäffel & Willmund, 1985).

Courtship can easily be studied under a dissecting microscope by placing two or more flies into a small (0.2 cm^3) chamber (Hirsch *et al.,* 1995). The behavior consists of a complex sequence of component responses. Within a few seconds of introduction into the chamber, a male approaches a female, chases her if she moves about, and orients his body so that he is facing her abdomen; while standing beside her, he may briefly vibrate his own abdomen. He then taps her abdomen with one of his foretarsi and vibrates the wing closest to her to produce a courtship song ("singing"). Should the female run away, as she often does, the male maintains his focus on her, pivoting to maintain his orientation relative to her abdomen ("orient back") or just simply follows her from behind ("chasing"). Once the male has been courting for a few minutes, he may lick the female's genitalia ("licking") and then attempt to copulate ("attempted copulation"). A rather striking response of the male is one in which he moves in a semicircle so that he winds up in front of the female "eye-to-eye" ("orient front"); this could provide an opportunity for the female to see clearly the shape of the male's eyes and the light reflected from them (Schäffel & Willmund, 1985). Should the female be stimulated by the male's courtship to become receptive to him, she will typically start to move more slowly and to extend her ovipositor more frequently (Barth, 1997; Barth, Hirsch, & Heisenberg, 1997); eventually she will respond to one of the copulation attempts by opening her vaginal plates to permit copulation (for review see Hirsch & Tompkins, 1994).

Effects of Visual Experience on Mate Choice, Copulation Latency, and Success. Hirsch *et al.* (1995) tested pairs of flies (all 4 days old and virgin) for copulation latency/ success. Pairs that had received the same early visual exposure (both LD or both DD) copulated more quickly than did those whose early experience differed (one LD, the other DD) (Hirsch *et al.*, 1995). This remained true when additional exposure groups (LL males and LL females) were added to the mix (Barth, Hirsch, & Heisenberg, 1997). Nine groups were tested: LL, LD, and DD females, each paired with an LL, LD, or DD male. Most of these pairs mated (average 92%), and the proportion of successful pairs did not vary among groups; unsuccessful pairs were thus excluded. In all cases, pairs with matching exposure history had shorter copulation latencies than did pairs with nonmatching exposure history (Figure 12). LD females in particular preferred LD males over DD males. They showed the same preferences even when tested under dim red light (Hirsch *et al.*, 1995) or in total darkness (Barth, Hirsch, & Heisenberg, 1997), although latency increased (Hirsch *et al.*, 1995) and copulation success decreased (Barth, Hirsch, & Heisenberg, 1997). These results raise the possibility that for the LD females, changes in mate choice may involve more than the use of vision.

Female preferences for males with exposure history matching their own was not present at 2 days of age but was observed at 4, 6, and 8 days of age (Barth, 1997). For females, but not for males, there was an overall effect of exposure history on copulation latency: regardless of the male's exposure history, DD females copulated faster than LD females, which in turn were faster than LL females (Barth, Hirsch, & Heisenberg, 1997).

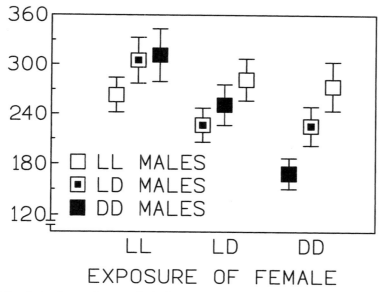

Figure 12. Copulation latency in seconds. A virgin male and a virgin female were placed into a small courtship chamber and observed for up to 15 min to determine when they mated (*n* = 558 pairs). Nine groups were tested: LL, LD, and DD females, each paired with an LL, LD, or DD male. There were 515 successful pairs (range of percentage of successful pairs in the nine groups 86–100%) and their distribution was not related to the rearing history of either sex. Data for unsuccessful pairs were thus excluded. Mean copulation latencies for LL, LD, and DD females, each tested with an LL, LD, or DD male, are shown. Note that copulation latencies are shortest when the male and female were exposed to the same light conditions during the first 4 days after eclosion, and that on average, DD females mated faster than LD females, which in turn were faster than LL females. (Redrawn from Barth, 1997, Figure 38.)

To look for critical periods in the effects of visual exposure on courtship preferences, male and female LL flies were placed into the dark for 2 days at different times during the first 4 days of adult life. For males, 2 days of darkness at any time during the first 4 days was sufficient for DD females to prefer them to LL males (Barth, 1997). For females, darkness for days 3–4, but not days 1–2, caused them to mate sooner with DD than with LL males, suggesting that for them, the critical period is days 3–4 of adult life. This timing is consistent with the above-mentioned changes in female preferences arising between day 2 and day 4.

When two males were forced to compete for one female, if the female had any visual exposure, the winner was usually the male with the same exposure history as the female. LD females tended to chose LD males over DD or LL males, and LL females tended to chose LL males over DD or LD males (Barth, Hirsch, & Heisenberg, 1997; Hirsch *et al.*, 1995). In contrast, DD females showed little or no preference for DD males. The results of competition were thus predictable from the latency data, but only for females that had some visual exposure. For DD females, which copulate soonest in pairwise testing, the mere presence of a DD male may be sufficient to stimulate her, but may not guarantee his success in competition with LL or LD males. Interactions between the males may help to determine the outcome.

Effects on Particular Components of Courtship Behavior.   Male flies with different rearing histories were videotaped during courtship (Barth, 1997; Barth, Hirsch, & Heisenberg, 1997) and the videotapes analyzed with regard to (1) the frequency and duration of component behaviors, (2) the "transition probabilities," the likelihood that two particular component behaviors will occur in sequence, (3) a *courtship* index, reflecting the proportion of time that the male was actually engaged in courtship, and (4) the locomotor activity of the female, since her slowing is prerequisite to successful copulation (Tompkins *et al.*, 1982; Barth, Hirsch, & Heisenberg, 1997). In addition, to eliminate as much as possible the effects on the male of female responsiveness, courtship was also recorded between mature males and immature females (who were unresponsive and unable to mate). There are at least two "distinguishing characteristics" between DD and LL males (at least when tested with an immature female): First, although the rapidity and intensity of courtship (courtship index) always increased during the session, it did so at a higher rate for DD than for LL males. Second, LL but not DD males increased the relative frequency of the component "orient-front-and-sing" over time. Since female receptivity appears to be positively influenced by the sight of the males eyes (Schäffel & Willmund, 1985), the male's "orient-front-and-sing" response may be an important source of sensory information for her. The lower frequency of this behavior in male DD flies could affect the outcome of their courtship.

Interactions between the male and the female affected the courtship behavior of the female. Both LL and DD females slowed down with time when courted by males whose exposure history matched theirs (Figure 13), but not significantly when the males' exposure history differed from theirs ("mismatched pairs"). Thus both LL and DD females change their locomotor activity during the first minutes of courtship, but only in response to the appropriate males. We suggest that experience affects the female's readiness to respond to the male's signals: she responds preferentially when, as a result of a shared early history, there is a match between both members of the pair.

Interactions between the male and the female also affected courtship behavior of the male. Specifically, when a male was paired with a female whose exposure history matched his own, he was more likely to engage in "licking" and "attempted copulation" during the first few minutes of courtship. These behaviors, which are associated with shorter copulation latencies, may help stimulate the female to copulate (Barth, Hirsch, & Heisenberg, 1997).

Thus experience has both complex and subtle effects on courtship behavior of *Drosophila melanogaster*. Male behaviors are affected in a rather direct way by early visual experience, and these behaviors affect how females respond, but again, the precise nature of the female's response depends upon her own history. Our findings indicate that flies are more likely to mate with individuals whose exposure history matches their own. We speculate (Hirsch & Tompkins, 1994) that these "cultural preferences" in mate choice may affect fitness by facilitating the female's ability to recognize and mate with males who have successfully grown up in the same local environment.

We may summarize the known effects of visual experience on courtship as follows (Barth, 1997; Barth, Hirsch, & Heisenberg, 1997; see Figure 14): (1) exposure to light during the first 4 days of adult life results in specific "labeling" of male and female flies in an as-yet-unknown manner; (2) flies prefer mates that have a label similar to their own; (3) the discrimination is achieved by way of a communications loop in which each fly signals his or her label to the prospective mate; (4) the male

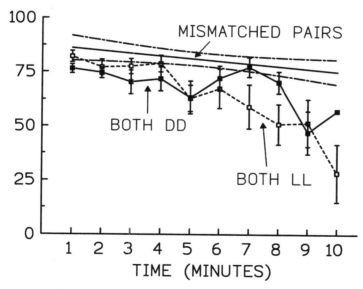

Figure 13. Percentage of time a virgin female, placed in a small courtship chamber with a virgin male, is in motion. Means and error bars are shown for pairs in which both flies had the same exposure (BOTH LL or BOTH DD). Once a female started to mate, her movement was no longer scored. Thus, the number of pairs gradually decreased ($n_{\text{BOTH LL}}$ = 32,31,24,20,18,12,10,8,7,5; $n_{\text{BOTH DD}}$ = 32,29,22,18,16,12,9,9,7,3). Note that both LL and DD females slowed down during the first minutes of courtship. The differences between the LL and DD groups were not judged significant. Data for pairs in which early exposure of the male and female was not matched (MISMATCHED PAIRS) were similar and therefore were pooled and a line fitted to the data points; the 95 % confidence envelope is shown. Note that in these mismatched pairs, the females did not slow down very much during the courtship. (Redrawn from Barth, 1997, Figure 49.)

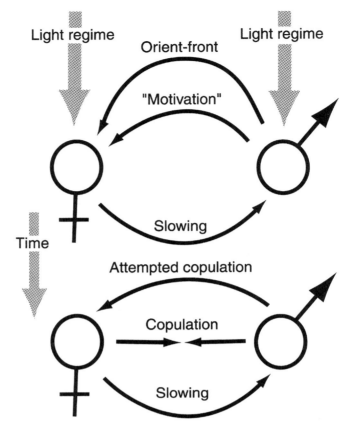

Figure 14. Summary diagram of part of the communication web leading to positive assortative mating. The light regine labels both male and female (upper level of the diagram). These labels are probably expressed in many ways. For males, light-rearing increases the frequency of "orient-front" behavior nd the level of general excitability ("motivation"). Females slow down in response to courtship activity by the appropriately labeled male. Her responsiveness in turn influences the male's frequency of attempted copulation during the later stages of courtship (bottom) and ultimately the likelihood and latency of copulation. (Redrawn from Barth, 1997, Figure 51.)

transmits his label using at least two signals (one visual and one nonvisual); (5) the female's response to the male's label depends upon her own visual exposure, and she indicates a willingness to copulate by reducing her locomotor activity and probably also with pheromones; and (6) in response to these signals, the male increases licking and copulation attempts, thereby affecting copulation latency and success.

CONCLUSIONS

Remarkably many characteristics of visual development are shared by fly and cat. This allows us to seek common developmental principles. First, in both species, some of the same genes control development. Second, both visual systems are experience dependent for the completion of their anatomic, physiologic, and behavioral development. Third, although the time scales differ, there is in both a critical period during which experience-dependent development takes place. Fourth, experience-dependent development involves changes in synaptic number and structure; normally, synaptic numbers increase with experience, but they can also decrease. Fifth, experience-dependent development involves changes in physiology and morphology that are correlated with properties of the animal's rearing conditions. Sixth, both flies and cats demonstrate perceptual changes that reflect the physiologic and anatomic changes, and we suggest that these changes are adaptations to the environmental conditions that were present during their early development and that, in a more natural setting, would be predictive of habitat conditions throughout their lives.

Commonalities between experience-dependent development in fly and cat also provide insights into the organization of sensory systems. In both cat and fly, there is evidence that two or more parallel afferent streams arise in the retina; these remain segregated at later stages of visual processing and differ in experience sensitivity. In the fly, the R1–R6 (motion-sensitive) afferent stream passing through the lobula plate is experience sensitive. This stream is analogous to the Y pathway in the cat, which is also experience sensitive. In contrast, in the fly, the R7–R8 (motion insensitive) afferent stream passing through the lobula shows relatively little evidence of experience sensitivity. We have compared this stream to the X pathway in the cat, which is also relatively insensitive to the effects of early visual experience.

Why should two such very different animals maintain at least two separated guidance systems? Perhaps a small brain cannot simultaneously process two very different classes of input, so that the animal must switch from one to the other as its locomotion changes. The example of the foraging bees who fly by day and walk by night (Chittka, Williams, Rasmussen, & Thomson, 1999) (presumably switching from one afferent stream to the other) supports this possibility. It may also be that a need for rapid, even simultaneous, processing may also encourage the development of parallel processing systems and their maintenance even in animals with larger brains. It might also be interesting to reexamine the older notion that there are two (or perhaps more) "visual systems" that divide the functions associated with foveal vision (i.e., high-resolution identification of images) and with peripheral vision (i.e., rapid detection of stimulus movement to facilitate both orientation and rapid response to change) (Schneider, 1969; Trevarthen, 1968). Whatever the reasons, the existence of dual processing systems seems to be widespread.

Spontaneous activity, competition, and neurotrophins all play a role in the visual developmental plasticity of cats, but there is limited evidence for their contribution in developing flies: First, the significantly smaller brain and optic ganglion volumes in hdc^{jk910} flies suggest that this mutation is blocking spontaneous activity that is important for growth of the lobula plate and central brain regions. Second, we saw some evidence for competition in that monocular occlusion, but not dark-rearing, affected the volume of the distal medulla. Finally, molecules resembling both neurotrophins and trk receptors have been observed in *Drosophila* (DeLotto & DeLotto, 1998; Haller *et al.*, 1987; Oishi *et al.*, 1997; Pulido *et al.*, 1992). Given the impressive parallels between cat and fly, these avenues should be explored further. In addition, while dark-rearing has been shown in flies to affect courtship, comparable experiments have not been done in cats (or in any other mammals). Although both social experience (Carter & Getz, 1993; Fillion & Blass, 1986) and maternal interactions (Moore, 1992) affect adult mammalian sexual behavior, it is not known whether visual deprivation will affect later courtship and reproductive success.

Experience-dependent development and plasticity provide the organism with a selective advantage by allowing it to match its nervous system to unpredictable features of its environment. Developmental programs are under the guidance of both genetic information, which reflects adaptation to past conditions, and epigenetic information, which should be more immediately predictive of future conditions. In the case of the cat, initial development of orientation selectivity, for example, is under genetic control, but later development requires visual experience, which tunes the system to respond to lines of the same orientation as those viewed during development. In the case of the fly, initial development of the photoreceptors and the lamina is likely to be under control of intrinsic developmental programs, whereas experience alters the frequency of both the synapses from the receptors onto the monopolar cells and the feedback synapses to the photoreceptors. Thus there is a flexible link between the "sensor" and the nervous system, a link that can be optimized on the basis of information about the prevailing environment.

Laughlin (1981b) suggested a strategy by which sensory systems might optimize themselves. If the range of a particular stimulus parameter (such as light intensity) in the environment is known along with the likelihood that any given value of that parameter will be encountered, then the organism can be set to respond optimally to the most common value. To see whether visual systems are indeed tuned to the range of stimulus values in the environment, Laughlin (1981b) analyzed the distribution of light levels in photographed scenes and determined the range, the most common levels, and the frequency with which light levels above and below these occurred. He next measured the response of the second-order neurons in the blow fly to changes in light levels and found a very close match between the distribution function for environmental light levels and the response function of the second-order cells. Meinertzhagen (1989) suggested that this matching could be achieved by using the developing fly's visual exposure to adjust the frequency of feedback synapses, and he and Kral showed that the frequency of these feedback synapses is indeed affected by visual stimulation during development (Kral & Meinertzhagen, 1989).

As Laughlin would predict, dark-rearing increases visual sensitivity in flies. Deimel and Kral (1992) compared visually driven responses (electroretinograms ERGs) in LD- and DD-reared *Musca domestica*. Specifically, they measured both the light intensity needed to obtain 50% of the maximal on-transient ERG and the change in intensity needed to increase the on-transient from 40% to 60% of its maximal level (the change being defined as "contrast sensitivity"). The 50% criterion response was obtained at a lower stimulus intensity in DD flies than in LD controls. Furthermore, in DD flies it took less of an increase in light intensity to achieve the same increase in response, and the point of saturation was reached at a lower light intensity. Therefore, flies deprived of light exposure (and of the neuronal activity it generates) were more sensitive to light, and the operating range of their visual systems was compressed and shifted toward lower light intensity levels. These effects were apparently long term, being present 3 weeks later. These findings support the hypothesis that during early stages of adult life, the response properties of the visual system are "adjusted" to take into account the conditions prevailing in the animal's visual world.

Further support for the idea of tuning comes from the results, presented earlier, that DD flies show peak optomotor responses at lower light intensities than do LL flies (Figure 10). Similarly, visual experience tunes the visual system of cats. Stripe-rearing, for example, alters the distribution of orientation preferences and of dendritic field orientations in visual cortex (Figure 2).It also alters the cats' behavioral responses to lines of different orientations.

The effects of light exposure on the development of courtship in *Drosophila* can similarly be interpreted as tuning the organism to its environment. While little is known about dispersal of *Drosophila melanogaster* (McDonald & Parsons, 1973), species of the *Drosophila obscura* group are known to disperse over considerable distances, the males farther than the females (Jones, Bryant, Lewontin, Moore, & Prout, 1981; Powell, Dobzhansky, Hook, & Wistrand, 1976; Taylor, Powell, Kebic, Andjelkovic, & Burla, 1984). *Drosophila melanogaster* males might be expected to migrate in similar fashion. If females remain closer to "home," and if, as Hirsch and Tompkins (1994) point out, a female lays her eggs in the food sources upon which she mated (Spieth & Ringo, 1983), her offspring will emerge and develop in the same environment that she did. The female's ability to recognize and mate preferentially with males who have matured in the same environment and thus are well adapted to it (having survived to adulthood and presumably having been fine tuned by the rearing conditions) should thus enhance her fitness. In effect, individuals would be recognizing others whose genotypic range includes survival in that environment, and their progeny would inherit from both parents alleles that facilitate survival under those same conditions. In choosing a mate, flies would be selecting for individuals whose sensory systems function well in the prevailing visual world. Plasticity of courtship behavior helps insure that "action is directed not only to what is sensed but to what is likely to happen, and it is this that matters" (Gregory, 1978, p. 222).

The last 40 years, which have seen great changes in our understanding of how experience might affect the developing nervous system, have also witnessed the discovery of many different model systems for the study of such development. The pioneering work of Hubel and Wiesel (Hubel, 1982; Wiesel, 1982) established the retinogeniculocortical pathway in the cat, and to a lesser degree in the monkey, as the model system for studying activity-dependent development (this viewpoint was reflected in the earlier version of the present chapter; Hirsch, 1985). In the early 1980s, it became clear that lower vertebrates, such as fish and frogs, also constituted good model systems for studying activity-dependent development (e.g., Schmidt, 1985), and now it is clear that invertebrates can also serve this role. Experience-dependent structural plasticity, once assumed to be a curiosity playing at most a minor role in the development of insect brain and behavior, has been shown to exist in many regions of the adult brain in *Drosophila* and in other invertebrates. It seems to be the rule rather than the exception that experience is needed to complete development of adult brain structures. A corollary of this view is that the ability to undergo experience-dependent development is heritable—certainly the genes controlling both development and learning are conserved—and that it should be possible to select strains of flies or mice that are more or less modifiable by early experience, thereby generating additional models for studying developmental plasticity.

One major task will now be to unravel the molecular mechanisms underlying the developmental "adjustments" in model systems such as *Drosophila*, and thereby to learn more about the often less accessible mechanisms underlying plastic processes in vertebrate brains. On a molecular level, there are likely to be several systems responsible for structural plasticity (for example, the cyclic AMP cascade seems to be involved in the central brain, but not in the optic lobes). There is a rich future for research in unraveling the genes involved in the plasticity of the visual system, and

Drosophila is likely to continue to be a central figure in attempts to identify and characterize these genes. Among the many powerful techniques available with *Drosophila* is the capability of selectively expressing any cloned gene in specific regions of the fly, providing very fine "molecular scalpels" (Calleja, Moreno, Pelaz, & Morata , 1996; Gustafson & Boulianne, 1996; Phelps & Brand, 1998; Rorth *et al.*, 1998). The usefulness of these "lessons from the world of insects" will underscore the basic unity of the mechanisms that appear to govern the link between brain and behavior in all organisms.

Another important lesson that we can learn as we come to understand more about brains and about what is needed to ensure their development is that "higher" and "lower" animals have much more in common than had been thought possible 20 years ago. This commonality in our origins, and especially the realization that we are affected by our environment during development, means that we have much to learn not only from our closest relatives, but from those who appear on the surface much more distant, and even alien. We hope it means also that our respect for these life forms, so like us in their vulnerability during development, will be enhanced. Finally, it means that we must add all of these life forms, and for the "longer haul" of development, to the eloquent statement of Gregory (1978), talking then only about those of us with large brains:

> The large brains of mammals, and particularly humans, allow past experience and anticipation of the future to play a large part in augmenting sensory information, so that we do not perceive the world merely from sensory information available at any given time, but rather use this information to test hypotheses of what lies before us. Perception becomes a matter of suggesting and testing hypotheses ... action is directed not only to what is sensed but to what is likely to happen, and it is this that matters. (Gregory, 1978, pp. 221–222)

Acknowledgments

Several colleagues read earlier versions of this chapter and provided helpful criticisms and comments: N. Daw, G. Lnenicka, K. Kral, J. Schmidt, B. Szaro, L. Tompkins, D. Tieman and N. Tumosa. D. Tieman also participated in preparing some of the figures. L. Benowitz made valuable suggestions about topics to include, and E. Blass encouraged us to be both critical and provocative. Support for much of the work on *Drosophila* was provided by a grant to H.V.B.H. from the Whitehall Foundation.

REFERENCES

Abel, T., Nguyen, P. V., Barad, M., Deuel, T. A., Kandel, E. R., & Bourtchouladze, R. (1997). Genetic demonstration of a role for PKA in the late phase of LTP and in hippocampus-based long-term memory. *Cell, 88,* 615–626.

Abeliovich, A., Chen, C., Goda, Y., Silva, A .J., Stevens, C. F., & Tonegawa, S. (1993). Modified hippocampal long-term potentiation in PKC gamma-mutant mice. *Cell, 75,* 1253–1262.

Akaneya, Y., Tsumoto, T., Kinoshita, S., & Hatanaka, H. (1997). Brain-derived neurotrophic factor enhances long-term potentiation in rat visual cortex. *Journal of Neuroscience, 17,* 6707–6716.

Albus, K. (1979). ^{14}C-deoxyglucose mapping of orientation subunits in the cat's visual cortical areas. *Experimental Brain Research, 37,* 609–613.

Allendoerfer, K. L., Cabelli, R. J., Escandón, E., Kaplan, D. R., Nikolics, K., & Shatz, C. J. (1994). Regulation of neurotrophin receptors during the maturation of the mammalian visual system. *Journal of Neuroscience, 14,* 1795–1811.

Antonini, A., & Stryker, M. P. (1993a). Rapid remodeling of axonal arbors in the visual cortex. *Science, 260,* 1819–1821.

Antonini, A., & Stryker, M. P. (1993b). Development of individual geniculocortical arbors in cat striate cortex and effects of binocular impulse blockade. *Journal of Neuroscience, 13,* 3549–3573.

Antonini, A., & Stryker, M. P. (1996). Plasticity of geniculocortical afferents following brief or prolonged monocular occlusion in the cat. *Journal of Comparative Neurology, 369,* 64–82.

Antonini, A., & Stryker, M. P. (1998). Effect of sensory disuse on geniculate afferents to cat visual cortex. *Visual Neuroscience, 15,* 401–409.

Archer, S. M., Dubin, M. W., & Stark, L. A. (1982). Abnormal development of kitten retinogeniculate connectivity in the absence of action potentials. *Science, 217,* 743–745.

Artola, A., & Singer, W. (1987). Long-term potentiation and NMDA receptors in rat visual cortex. *Nature, 330,* 649–652.

Averhoff, W. W., & Richardson, R. H. (1974). Pheromonal control of mating patterns in *Drosophila melanogaster. Behavior Genetics, 4,* 207–225.

Bailey, C. H., & Chen, M. (1988). Long-term memory in *Aplysia* modulates the total number of varicosities of single identified sensory neurons. *Proceedings of the National Academy of Sciences of the USA, 85,* 2373–2377.

Bailey, C. H., & Chen, M. (1991). Morphological aspects of synaptic plasticity in *Aplysia:* An anatomical substrate for long-term memory. *Annals of the New York Academy of Sciences, 627,* 181–196.

Balling, A., Technau, G. M., & Heisenberg, M. (1987). Are the structural changes in adult *Drosophila* mushroom bodies memory traces? Studies on biochemical learning mutants. *Journal of Neurogenetics, 4,* 65–73.

Barinaga, M. (1995). Focusing on the *eyeless* gene. *Science, 267,* 1766–1767.

Barth, M. (1997). The flexible fly: Experience-dependent plasticity of brain and behavior in the laboratory fruitfly *Drosophila melanogaster.* 1–112. Unpublished doctoral dissertation. Bayerischen Julius–Maximilians–Universität, Würzburg, Germany.

Barth, M., & Heisenberg, M. (1997). Vision affects mushroom bodies and central complex in *Drosophila melanogaster. Learning and Memory, 4,* 219–229.

Barth, M., Hirsch, H. V. B., & Heisenberg, M. (1997). Rearing in different light regimes affects courtship in *Drosophila melanogaster. Animal Behavior, 53,* 25–38.

Barth, M., Hirsch, H. V. B., Meinertzhagen, I. A., & Heisenberg, M. (1997). Experience-dependent developmental plasticity in the optic lobe of *Drosophila melanogaster. Journal of Neuroscience, 17,* 1493–1504.

Bear, M. F., & Colman, H. (1990). Binocular competition in the control of geniculate cell size depends upon visual cortical N-methyl-D-aspartate receptor activation. *Proceedings of the National Academy of Sciences of the USA, 87,* 9246–9249.

Bear, M. F., & Singer, W. (1986). Modulation of visual cortical plasticity by acetylcholine and noradrenaline. *Nature, 320,* 172–176.

Bear, M. F., Kleinschmidt, A., Gu, Q., & Singer, W. (1990). Disruption of experience-dependent synaptic modifications in striate cortex by infusion of an NMDA receptor antagonist. *Journal of Neuroscience, 10,* 909–925.

Belvin, M. P., & Yin, J. C. (1997). *Drosophila* learning and memory: Recent progress and new approaches. *BioEssays, 19,* 1083–1089.

Berardi, N., Domenici, L., Parisi, V., Pizzorusso, T., Cellerino, A., & Maffei, L. (1993). Monocular deprivation effects in the rat visual cortex and lateral geniculate nucleus are prevented by nerve growth factor (NGF). I. Visual cortex. *Proceedings of the Royal Society of London [Biology], 251,* 17–23.

Berardi, N., Cellerino, A., Domenici, L., Fagiolini, M., Pizzorusso, T., Cattaneo, A., & Maffei, L. (1994). Monoclonal antibodies to nerve growth factor affect the postnatal development of the visual system. *Proceedings of the National Academy of Sciences of the USA, 91,* 684–688.

Berman, N. E. J., & Payne, B. R. (1989). Modular organization of on and off responses in the cat lateral geniculate nucleus. *Neuroscience, 32,* 721–737.

Bickford, M. E., Günlük, A. E., Guido, W., & Sherman, S. M. (1993). Evidence that cholinergic axons from the parabrachial region of the brainstem are the exclusive source of nitric oxide in the lateral geniculate nucleus of the cat. *Journal of Comparative Neurology, 334,* 410–430.

Bisti, S., Gargini, C., & Chalupa, L. M. (1998). Blockade of glutamate-mediated activity in the developing retina perturbs the functional segregation of ON and OFF pathways. *Journal of Neuroscience, 18,* 5019–5025.

Blake, R., & Hirsch, H. V. B. (1975). Deficits in binocular depth perception in cats after alternating monocular deprivation. *Science, 192,* 1114–1116.

Blakely, R. D., & Coyle, J. T. (1988). The neurobiology of N-acetylaspartylglutamate. *International Review of Neurobiology, 30,* 39–100.

Blakemore, C., & Cooper, G. F. (1970). Development of the brain depends on the visual environment. *Nature, 228,* 477–478.

Blasdel, G. G., Mitchell, D. E., Muir, D. W., & Pettigrew, J. D. (1977). A physiological and behavioural study in cats of the effect of early visual experience with contours of a single orientation. *Journal of Physiology (London), 265,* 615–636.

Blochl, A., & Thoenen, H. (1995). Characterization of nerve growth factor (NGF) release from hippocampal neurons: Evidence for a constitutive and an unconventional sodium-dependent regulated pathway. *European Journal of Neuroscience, 7,* 1220–1228.

Blochl, A., & Thoenen, H. (1996). Localization of cellular storage compartments and sites of constitutive and activity-dependent release of nerve growth factor (NGF) in primary cultures of hippocampal neurons. *Molecular and Cellular Neuroscience, 7,* 173–190.

Bloom, J. W., & Atwood, H. L. (1980). Effects of altered sensory experience on the responsiveness of the locust descending contralateral movement detector neuron. *Journal of Comparative Physiology A, 135,* 191–199.

Bloom, J. W., & Atwood, H. L. (1981). Reversible ultrastructural changes in the rhabdom of the locust eye are induced by long term light deprivation. *Journal of Comparative Physiology, 144,* 357–365.

Bloomquist, B. T., Shortridge, R. D., Schneuwly, S., Perdew, M., Montell, C., Steller, H., Rubin, G., & Pak, W. L. (1988). Isolation of a putative phospholipase C gene of *Drosophila, norpA,* and its role in phototransduction. *Cell, 54,* 723–733.

Bodnarenko, S. R., & Chalupa, L. M. (1993). Stratification of ON and OFF ganglion cell dendrites depends on glutamate-mediated afferent activity in the developing retina. *Nature, 364,* 144–146.

Bodnarenko, S. R., Jeyarasasingam, G., & Chalupa, L. M. (1995). Development and regulation of dendritic stratification in retinal ganglion cells by glutamate-mediated afferent activity. *Journal of Neuroscience, 15,* 7037–7045.

Bolshakov, V. Y., Golan, H., Kandel, E. R., & Siegelbaum, S. A. (1997). Recruitment of new sites of synaptic transmission during the cAMP-dependent late phase of LTP at CA3–CA1 synapses in the hippocampus. *Neuron, 19,* 635–651.

Bonhoeffer, T., & Grinvald, A. (1991). Iso-orientation domains in cat visual cortex are arranged in pinwheel-like patterns. *Nature, 353,* 429–431.

Bourtchuladze, R., Frenguelli, B., Blendy, J., Cioffi, D., Schutz, G., & Silva, A. J. (1994). Deficient long-term memory in mice with a targeted mutation of the cAMP-responsive element-binding protein. *Cell, 79,* 59–68.

Bowling, D. B., & Caverhill, J. I. (1989). ON/OFF organization in the cat lateral geniculate nucleus: Sublaminae vs. columns. *Journal of Comparative Neurology, 283,* 161–168.

Bowling, D. B., & Wieniawa-Narkiewicz, E. (1986). The distribution of on- and off-centre X- and Y-like cells in the A layers of the cat's lateral geniculate nucleus. *Journal of Physiology (London), 375,* 561–572.

Budnik, V., Zhong, Y., & Wu, C. F. (1990). Morphological plasticity of motor axons in *Drosophila* mutants with altered excitability. *Journal of Neuroscience, 10,* 3754–3768.

Bullier, J., & Norton, T. T. (1979). Comparison of receptive-field properties of X and Y ganglion cells with X and Y lateral geniculate cells in the cat. *Journal of Neurophysiology, 42,* 274–291.

Burke, R., & Basler, K. (1997). Hedgehog signaling in *Drosophila* eye and limb development—Conserved machinery, divergent roles? *Current Opinions in Neurobiology, 7,* 55–61.

Byers, D., Davis, R. L., & Kiger, J. A., Jr. (1981). Defect in cyclic AMP phosphodiesterase due to the *dunce* mutation of learning in *Drosophila melanogaster. Nature, 289,* 79–81.

Cabelli, R. J., Hohn, A., & Shatz, C. J. (1995). Inhibition of ocular dominance column formation by infusion of NT-4/5 or BDNF. *Science, 267,* 1662–1666.

Cabelli, R. J., Allendoerfer, K. L., Radeke, M .J., Welcher, A. A., Feinstein, S. C., & Shatz, C. J. (1996). Changing patterns of expression and subcellular localization of trkB in the developing visual system. *Journal of Neuroscience, 16,* 7965–7980.

Cabelli, R. J., Shelton, D. L., Segal, R. A., & Shatz, C .J. (1997). Blockade of endogenous ligands of trkB inhibits formation of ocular dominance columns. *Neuron, 19,* 63–76.

Calleja, M., Moreno, E., Pelaz, S., & Morata, G. (1996). Visualization of gene expression in living adult *Drosophila. Science, 274,* 252–255.

Canossa, M., Griesbeck, O., Berninger, B., Campana, G., Kolbeck, R., & Thoenen, H. (1997). Neurotrophin release by neurotrophins: Implications for activity-dependent neuronal plasticity. *Proceedings of the National Academy of Sciences of the USA, 94,* 13279–13286.

Carmignoto, G., Pizzorusso, T., Tia, S., & Vicini, S. (1997). Brain-derived neurotrophic factor and nerve growth factor potentiate excitatory synaptic transmission in the rat visual cortex. *Journal of Physiology (London), 498,* 153–164.

Carter, C. S., & Getz, L. L. (1993). Monogamy and the prairie vole. *Scientific American, 268,* 100–106.

Castrén, E., Zafra, F., Thoenen, H., & Lindholm, D. (1992). Light regulates expression of brain-derived neurotrophic factor mRNA in rat visual cortex. *Proceedings of the National Academy of Sciences of the USA, 89,* 9444–9448.

Castrén, E., Pitkanen, M., Sirvio, J., Parsadanian, A., Lindholm, D., Thoenen, H., & Riekkinen, P. J. (1993). The induction of LTP increases BDNF and NGF mRNA but decreases NT-3 mRNA in the dentate gyrus. *NeuroReport, 4,* 895–898.

Chapman, B., & Bonhoeffer, T. (1998). Overrepresentation of horizontal and vertical orientation preferences in developing ferret area 17. *Proceedings of the National Academy of Sciences of the USA, 95,* 2609–2614.

Chapman, B., Stryker, M. P., & Bonhoeffer, T. (1996). Development of orientation preference maps in ferret primary visual cortex. *Journal of Neuroscience, 16,* 6443–6453.

Charii, F., & Lambin, M. (1988). Ontogenèse du contrôle visuel de la marche orientée chez le grillon *Gryllus bimaculatus. Biology of Behavior, 13,* 49–58.

Chen, C., & Tonegawa, S. (1997). Molecular genetic analysis of synaptic plasticity, activity-dependent neural development, learning, and memory in the mammalian brain. *Annual Review of Neuroscience, 20,* 157–184.

Chen, K. S., Nishimura, M. C., Armanini, M. P., Crowley, C., Spencer, S. D., & Phillips, H. S. (1997). Disruption of a single allele of the nerve growth factor gene results in atrophy of basal forebrain cholinergic neurons and memory deficits. *Journal of Neuroscience, 17,* 7288–7296.

Cheyette, B. N., Green, P. J., Martin, K., Garren, H., Hartenstein, V., & Zipursky, S. L. (1994). The *Drosophila* sine oculis locus encodes a homeodomain-containing protein required for the development of the entire visual system. *Neuron, 12,* 977–996.

Chittka, L., Williams, N. M., Rasmussen, H., & Thomson, J. D. (1999). Navigation without vision: Bumblebee orientation in complete darkness. *Proceedings of the Royal Society of London, B. Biological Science, 266,* 45–50.

Clayton, K. N., & Kamback, N. (1966). Successful performance by cats on several colour discrimination problems. *Canadian Journal of Psychology, 20,* 173–182.

Cleland, B. G., Dubin, M. W., & Levick, W. R. (1971). Sustained and transient neurones in the cat's retina and lateral geniculate nucleus. *Journal of Physiology (London), 217,* 473–496.

Collingridge, G. L,. & Bliss, T. V. P. (1987). NMDA receptors—Their role in long-term potentiation. *Trends in Neurosciences, 10,* 288–293.

Coppola, D. M., White, L. E., Fitzpatrick, D., & Purves, D. (1998). Unequal representation of cardinal and oblique contours in ferret visual cortex. *Proceedings of the National Academy of Sciences of the USA, 95,* 2621–2623.

Crair, M. C., Gillespie, D. C., & Stryker, M. P. (1998). The role of visual experience in the development of columns in cat visual cortex. *Science, 279,* 566–570.

Cramer, K. S., & Sur, M. (1997). Blockade of afferent impulse activity disrupts on/off sublamination in the ferret lateral geniculate nucleus. *Developmental Brain Research, 98,* 287–290.

Crooks, J., & Morrison, J. D. (1989). Synapses of the inner plexiform layer of the area centralis of kitten retina during postnatal development: A quantitative study. *Journal of Anatomy, 163,* 33–47.

Cutforth, T., & Gaul, U. (1997). The genetics of visual system development in *Drosophila*: Specification, connectivity and asymmetry. *Current Opinions in Neurobiology, 7,* 48–54.

Dash, P. K., Hochner, B., & Kandel, E. R. (1990). Injection of the cAMP-responsive element into the nucleus of *Aplysia* sensory neurons blocks long-term facilitation. *Nature, 345,* 718–721.

Dash, P. K., Karl, K. A., Colicos, M. A., Prywes, R., & Kandel, E. R. (1991). cAMP response element-binding protein is activated by Ca^{2+}/calmodulin- as well as cAMP-dependent protein kinase. *Proceedings of the National Academy of Sciences of the USA, 88,* 5061–5065.

Dauwalder, B., & Davis, R. L. (1995). Conditional rescue of the *dunce* learning/memory and female fertility defects with *Drosophila* or rat transgenes. *Journal of Neuroscience, 15,* 3490–3499.

Davis, R. L. (1996). Physiology and biochemistry of *Drosophila* learning mutants. *Physiological Reviews, 76,* 299–317.

Davis, R. L., Cherry, J., Dauwalder, B., Han, P. L., & Skoulakis, E. (1995). The cyclic AMP system and *Drosophila* learning. *Molecular and Cellular Biochemistry, 149,* 271–278.

de Belle, J. S. (1995). *Drosophila* mushroom body subdomains: Innate or learned representations of odor preference and sexual orientation? *Neuron, 15,* 245–247.

de Belle, J. S., & Heisenberg, M. (1994). Associative odor learning in *Drosophila* abolished by chemical ablation of mushroom bodies. *Science, 263,* 692–695.

Deimel, E., & Kral, K. (1992). Long-term sensitivity adjustment of the compound eyes of the housefly *Musca domestica* during early adult life. *Journal of Insect Physiology, 38,* 425–430.

Deisseroth, K., Heist, E. K., & Tsien, R. ,W. (1998). Translocation of calmodulin to the nucleus supports CREB phosphorylation in hippocampal neurons. *Nature, 392,* 198–202.

DeLotto, Y., & DeLotto, R. (1998). Proteolytic processing of the *Drosophila* Spatzle protein by easter generates a dimeric NGF-like molecule with ventralising activity. *Mechanisms of Development, 72,* 141–148.

Deruntz, P., Palévody, C., & Lambin, M. (1994). Effect of dark rearing on the eye of *Gryllus bimaculatus* crickets. *Journal of Experimental Zoology, 268*, 421–427.

Desmond, N. L., & Weinberg, R. J. (1998). Enhanced expression of AMPA receptor protein at perforated axospinous synapses. *NeuroReport, 9*, 857–860.

Distler, C., & Hoffmann, K.-P. (1991). Depth perception and cortical physiology in normal and innate microstrabismic cats. *Visual Neuroscience, 6*, 25–41.

Domenici, L., Cellerino, A., & Maffei, L. (1993). Monocular deprivation effects in the rat visual cortex and lateral geniculate nucleus are prevented by nerve growth factor (NGF). II. Lateral geniculate nucleus. *Proceedings of the Royal Society of London [Biology], 251*, 25–31.

Domenici, L., Fontanesi, G., Cattaneo, A., Bagnoli, P., & Maffei, L. (1994). Nerve growth factor (NGF) uptake and transport following injection in the developing rat visual cortex. *Visual Neuroscience, 11*, 1093–1102.

Douglass, J. K., & Strausfeld, N. J. (1998). Functionally and anatomically segregated visual pathways in the lobula complex of a calliphorid fly. *Journal of Comparative Neurology, 396*, 84–104.

Dowling, J. E. (1987). *The retina: An approachable part of the brain.* Cambridge MA: Belknap/Harvard University Press

Dragunow, M., Beilharz, E., Mason, B., Lawlor, P., Abraham, W., & Gluckman, P. (1993). Brain-derived neurotrophic factor expression after long-term potentiation. *Neuroscience Letters, 160*, 232–236.

Dreher, B., Leventhal, A. G., & Hale, P. T. (1980). Geniculate input to cat visual cortex: A comparison of area 19 with areas 17 and 18. *Journal of Neurophysiology, 44*, 804–826.

Dubin, M. W., Stark, L. A., & Archer, S. M. (1986). A role for action-potential activity in the development of neuronal connections in the kitten retinogeniculate pathway. *Journal of Neuroscience, 6*, 1021–1036.

Dudai, Y., Uzzan, A., & Zvi, S. (1983). Abnormal activity of adenylate cyclase in the *Drosophila* memory mutant rutabaga. *Neuroscience Letters, 42*, 207–212.

Duncan, M. K., Kos, L., Jenkins, N. A., Gilbert, D. J., Copeland, N. G., & Tomarev, S. I. (1997). Eyes absent: A gene family found in several metazoan phyla. *Mammalian Genome, 8*, 479–485.

Dyson, S. E., & Jones, D. G. (1980). Quantitation of terminal parameters and their interrelationships in maturing central synapses: A perspective for experimental studies. *Brain Research, 183*, 43–59.

Dyson, S. E., & Jones, D. G. (1984). Synaptic remodelling during development and maturation: Junction differentiation and splitting as a mechanism for modifying connectivity. *Developmental Brain Research, 13*, 125–137.

Easter, S. S., Jr., Purves, D., Rakic, P., & Spitzer, N. C. (1985). The changing view of neural specificity. *Science, 230*, 507–511.

Fahrbach, S. E., & Robinson, G. E. (1995). Behavioral development in the honey bee: Toward the study of learning under natural conditions. *Learning and Memory, 2*, 199–224.

Fainzilber, M., Smit, A. B., Syed, N. I., Wildering, W. C., Hermann, P. M., Van der Schors, R. C., Jiménez, C., Li, K. W., Van Minnen, J., Bulloch, A. G. M., Ibáñez, C. F., & Geraerts, W. P. M. (1996). CRNF, a molluscan neurotrophic factor that interacts with the p75 neurotrophin receptor. *Science, 274*, 1540–1543.

Famiglietti, E. V., Jr., & Kolb, H. (1975). A bistratified amacrine cell and synaptic circuitry in the inner plexiform layer of the retina. *Brain Research, 84*, 293–300.

Famiglietti, E. V., Jr., & Kolb, H. (1976). Structural basis for ON- and OFF-center responses in retinal ganglion cells. *Science, 194*, 193–195.

Feany, M. B., & Quinn, W. G. (1995). A neuropeptide gene defined by the *Drosophila* memory mutant amnesiac. *Science, 268*, 869–873.

Feller, M. B., Wellis, D. P., Stellwagen, D., Werblin, F. S., & Shatz, C. J. (1996). Requirement for cholinergic synaptic transmission in the propagation of spontaneous retinal waves. *Science, 272*, 1182–1187.

Figurov, A., Pozzo-Miller, L. D., Olafsson, P., Wang, T., & Lu, B. (1996). Regulation of synaptic responses to high-frequency stimulation and LTP by neurotrophins in the hippocampus. *Nature, 381*, 706–709.

Fillion, T. J., & Blass, E. M. (1986). Infantile experience with suckling odors determines adult sexual behavior in male rats. *Science, 231*, 729–731.

Finkbeiner, S., Tavazoie, S. F., Maloratsky, A., Jacobs, K. M., Harris, K. M., & Greenberg, M. E. (1997). CREB: A major mediator of neuronal neurotrophin responses. *Neuron, 19*, 1031–1047.

Fiorentini, A., & Maffei, L. (1978). Selective impairment of contrast sensitivity in kittens exposed to periodic gratings. *Journal of Physiology (London), 277*, 455–466.

Fiorentini, A., Berardi, N., & Maffei, L. (1995). Nerve growth factor preserves behavioral visual acuity in monocularly deprived kittens. *Visual Neuroscience, 12*, 51–55.

Fischbach, K. F. (1983). Neural cell types surviving congenital sensory deprivation in the optic lobes of *Drosophila melanogaster. Developmental Biology, 95*, 1–18.

Fischbach, K. F., & Heisenberg, M. (1984). Neurogenetics and behaviour in insects. *Journal of Experimental Biology, 112*, 63–95.

Flood, D. G., & Coleman, P. D. (1979). Demonstration of orientation columns with [^{14}C]2-deoxyglucose in a cat reared in a striped environment. *Brain Research, 173,* 538–542.

Ford, S. C., Napolitano, L. M., McRobert, S. P., & Tompkins, L. (1989). Development of behavioral competence in young *Drosophila melanogaster* adults. *Journal of Insect Behavior, 2,* 575–588.

Frank, D. A., & Greenberg, M. E. (1994). CREB: A mediator of long-term memory from mollusks to mammals. *Cell, 79,* 5–8.

Frégnac, Y., & Imbert, M. (1978). Early development of visual cortical cells in normal and dark-reared kittens: Relationship between orientation selectivity and ocular dominance. *Journal of Physiology (London), 278,* 27–44.

Frégnac, Y., Shulz, D., Thorpe, S., & Bienenstock, E. (1988). A cellular analogue of visual cortical plasticity. *Nature, 333,* 367–370.

Friedlander, M. J., Lin, C.-S., Stanford, L. R., & Sherman, S. M. (1981). Morphology of functionally identified neurons in the lateral geniculate nucleus of the cat. *Journal of Neurophysiology, 46,* 80–129.

Fukuda, Y., Hsia, A. Y., & Watanabe, M. (1985). Morphological correlates of Y, X and W type ganglion cells in the cat's retina. *Vision Research, 25,* 319–327.

Galuske, R. A. W., Kim, D. S., Castrñ, E., Thoenen, H., & Singer, W. (1996). Brain-derived neurotrophic factor reverses experience-dependent synaptic modifications in kitten visual cortex. *European Journal of Neuroscience, 8,* 1554–1559.

Ganz, L., Hirsch, H. V. B., & Tieman, S. B. (1972). The nature of perceptual deficits in visually deprived cats. *Brain Research, 44,* 547–568.

Gardoni, F., Caputi, A., Cimino, M., Pastorino, L., Cattabeni, F., & Di Luca, M. (1998). Calcium/calmodulin-dependent protein kinase II is associated with NR2A/B subunits of NMDA receptor in postsynaptic densities. *Journal of Neurochemistry, 71,* 1733–1741.

Garraghty, P. E. (1995). Connectional specificity in the cat's retinogeniculate system. *International Journal of Neuroscience, 80,* 31–40.

Garraghty, P. E., Sur, M., Weller, R. E., & Sherman, S. M. (1986). Morphology of retinogeniculate x and y axon arbors in monocularly enucleated cats. *Journal of Comparative Neurology, 251,* 198–215.

Geinisman, Y., De Toledo-Morrell, L., & Morrell, F. (1986). Loss of perforated synapses in the dentate gyrus: Morphological substrate of memory deficit in aged rats. *Proceedings of the National Academy of Sciences of the USA, 83,* 3027–3031.

Geinisman, Y., Morrell, F., & De Toledo-Morrell, L. (1990). Increase in the relative proportion of perforated axospinous synapses following hippocampal kindling is specific for the synaptic field of stimulated axons. *Brain Research, 507,* 325–331.

Geinisman, Y., DeToledo-Morrell, L., & Morrell, F. (1991). Induction of long-term potentiation is associated with an increase in the number of axospinous synapses with segmented postsynaptic densities. *Brain Research, 566,* 77–88.

Gilbert, C. D., & Kelly, J. P. (1975). The projections of cells in the different layers of the cat's visual cortex. *Journal of Comparative Neurology, 163,* 81–105.

Goodhill, G. J. (1993). Topography and ocular dominance: A model exploring positive correlations. *Biological Cybernetics, 69,* 109–118.

Goodwin, S. F., Del Vecchio, M., Velinzon, K., Hogel, C., Russell, S. R. H., Tully, T., & Kaiser, K. (1997). Defective learning in mutants of the *Drosophila* gene for a regulatory subunit of cAMP-dependent protein kinase. *Journal of Neuroscience, 17,* 8817–8827.

Gordon, B., & Bremiller, R. (1992). Decreasing the cortical response to monocular deprivation need not decrease cell shrinkage in cat lateral geniculate nucleus. *Experimental Brain Research, 92,* 79–84.

Gordon, J. A., & Stryker, M. P. (1996). Experience-dependent plasticity of binocular responses in the primary visual cortex of the mouse. *Journal of Neuroscience, 16,* 3274–3286.

Gordon, J. A., Cioffi, D., Silva, A. J., & Stryker, M. P. (1996). Deficient plasticity in the primary visual cortex of α-calcium/calmodulin-dependent protein kinase II mutant mice. *Neuron, 17,* 491–499.

Green, P., Hartenstein, A. Y., & Hartenstein, V. (1993). The embryonic development of the *Drosophila* visual system. *Cell and Tissue Research, 273,* 583–598.

Greenough, W. T., West, R. W., & DeVoogd, T. J. (1978). Subsynaptic plate perforations: Changes with age and experience in the rat. *Science, 202,* 1096–1098.

Gregory, R. L. (1978). *Eye and brain: The psychology of seeing.* New York: McGraw-Hill.

Griffith, L. C., Verselis, L. M., Aitken, K. M., Kyriacou, C. P., Danho, W., & Greenspan, R. J. (1993). Inhibition of calcium/calmodulin-dependent protein kinase in *Drosophila* disrupts behavioral plasticity. *Neuron, 10,* 501–509.

Gu, Q., Bear, M. F., & Singer, W. (1989). Blockade of NMDA-receptors prevents ocularity changes in kitten visual cortex after reversed monocular deprivation. *Developmental Brain Research, 47,* 281–288.

Guido, W., Scheiner, C. A., Mize, R. R., & Kratz, K. E. (1997). Developmental changes in the pattern of NADPH-diaphorase staining in the cat's lateral geniculate nucleus. *Visual Neuroscience, 14,* 1167–1173.

Guillery, R. W. (1972a). Binocular competition in the control of geniculate cell growth. *Journal of Comparative Neurology, 144,* 117–130.

Guillery, R. W. (1972b). Experiments to determine whether retinogeniculate axons can form translaminar collateral sprouts in the dorsal lateral geniculate nucleus of the cat. *Journal of Comparative Neurology, 146,* 407–420.

Guillery, R. W., & Stelzner, D. J. (1970). The differential effects of unilateral lid closure upon the monocular and binocular segments of the dorsal lateral geniculate nucleus in the cat. *Journal of Comparative Neurology, 139,* 413–422.

Günlük, A. E., Bickford, M. E., & Sherman, S. M. (1994). Rearing with monocular lid suture induces abnormal NADPH-diaphorase staining in the lateral geniculate nucleus of cats. *Journal of Comparative Neurology, 350,* 215–228.

Gustafson, K., & Boulianne, G. L. (1996). Distinct expression patterns detected within individual tissues by the GAL4 enhancer trap technique. *Genome, 39,* 174–182.

Guzowski, J. F., & McGaugh, J. L. (1997). Antisense oligodeoxynucleotide-mediated disruption of hippocampal cAMP response element binding protein levels impairs consolidation of memory for water maze training. *Proceedings of the National Academy of Sciences of the USA, 94,* 2693–2698.

Hahm, J.-O., Langdon, R. B., & Sur, M. (1991). Disruption of retinogeniculate afferent segregation by antagonists to NMDA receptors. *Nature, 351,* 568–570.

Halder, G., Callaerts, P., & Gehring, W. J. (1995). Induction of ectopic eyes by targeted expression of the *eyeless* gene in *Drosophila. Science, 267,* 1788–1792.

Hall, J. C. (1982). Genetics of the nervous system in *Drosophila. Quarterly Review of Biophysics, 15,* 223–479.

Haller, J., Cote, S., Bronner, G., & Jackle, H. (1987). Dorsal and neural expression of a tyrosine kinase-related *Drosophila* gene during embryonic development. *Genes and Development, 1,* 862–867.

Han, P. L., Levin, L. ., Reed, R. R., & Davis, R. L. (1992). Preferential expression of the *Drosophila rutabaga* gene in mushroom bodies, neural centers for learning in insects. *Neuron, 9,* 619–627.

Hardie, R. C. (1987). Is histamine a neurotransmitter in insect photoreceptors? *Journal of Comparative Physiology [A]. 161,* 201–213.

Harris, A. E., Ermentrout, G. B., & Small, S. L. (1997). A model of ocular dominance column development by competition for trophic factor. *Proceedings of the National Academy of Sciences of the USA, 94,* 9944–9949.

Hata, Y., & Stryker, M. P. (1994). Control of thalamocortical afferent rearrangement by postsynaptic activity in developing visual cortex. *Science, 265,* 1732–1735.

Hata, Y., Tsumoto, T., & Stryker, M. P. (1999). Selective pruning of more active afferents when cat visual cortex is pharmacologically inhibited. *Neuron, 22,* 375–381.

Hauser-Holschuh, H. (1975). Vergleichende quantitative Untersuchungen an den Sehganglien der Fliege *Musca domestica* und *Drosophila melanogaster.* Unpublished doctoral dissertation. Eberhard–Karls–Universität, Tübingen, Germany

Heisenberg, M. (1979). Genetic approach to a visual system. In H. Autrum (Ed.), *Handbook of Sensory Physiology* (pp. 665–679). New York: Springer-Verlag.

Heisenberg, M., & Wolf, R. (1984). *Vision in* Drosophila: *Genetics of microbehavior, Vol. 12.* In V. Braitenberg (Series Ed.), *Studies of brain function.* Berlin: Springer-Verlag.

Heisenberg, M., Borst, A., Wagner, S., & Byers, D. (1985). *Drosophila* mushroom body mutants are deficient in olfactory learning. *Journal of Neurogenetics, 2,* 1–30.

Henry, G. H., Mustari, M. J., & Bullier, J. (1983). Different geniculate inputs to B and C cells of cat striate cortex. *Experimental Brain Research, 52,* 179–189.

Hensch, T. K., Gordon, J. A., Brandon, E. P., McKnight, G. S., Idzerda, R. L., & Stryker, M. P. (1998). Comparison of plasticity *in vivo* and *in vitro* in the developing visual cortex of normal and protein kinase A RIβ-deficient mice. *Journal of Neuroscience, 18,* 2108–2117.

Hertel, H. (1982). The effect of spectral light deprivation on the spectral sensitivity of the honey bee. *Journal of Comparative Physiology, 147,* 365–369.

Hertel, H. (1983). Change of synapse frequency in certain photoreceptors of the honeybee after chromatic deprivation. *Journal of Comparative Physiology, 151,* 477–482.

Hickey, T. L., & Guillery, R. W. (1974). An autoradiographic study of retinogeniculate pathways in the cat and the fox. *Journal of Comparative Neurology, 156,* 239–254.

Hickey, T. L., Spear, P. D., & Kratz, K. E. (1977). Quantitative studies of cell size in the cat's dorsal lateral geniculate nucleus following visual deprivation. *Journal of Comparative Neurology, 172,* 265–282.

Hirsch, H. V. B. (1972). Visual perception in cats after environmental surgery. *Experimental Brain Research, 15,* 405–423.

Hirsch, H. V. B. (1985). The tunable seer. Activity-dependent development of vision. In E. M. Blass (Ed.), *Handbook of behavioral neurobiology* (pp. 237–295). New York: Plenum Press.

Hirsch, H. V. B., & Spinelli, D. N. (1970). Visual experience modifies distribution of horizontally and vertically oriented receptive fields in cats. *Science, 168,* 869–871.

Hirsch, H. V. B., & Tieman, S. B. (1987). Perceptual development and experience-dependent changes in cat visual cortex. In M. Bornstein (Ed.), *Sensitive periods in development: Interdisciplinary perspectives* (pp. 39–79). Hillsdale, NJ: Erlbaum.

Hirsch, H. V. B., & Tompkins, L. (1994). The flexible fly: Experience-dependent development of complex behaviors in *Drosophila melanogaster. Journal of Experimental Biology, 195,* 1–18.

Hirsch, H. V. B., Leventhal, A. G., McCall, M. A., & Tieman, D. G. (1983). Effects of exposure to lines of one or two orientations on different cell types in striate cortex of cat. *Journal of Physiology (London), 337,* 241–255.

Hirsch, H. V. B., Tieman, D. G., Tieman, S. B., & Tumosa, N. (1987). Unequal alternating exposure: Effects during and after the classical critical period. In J. P. Rauschecker, & P. Marler (Eds.), *Imprinting and cortical plasticity* (pp. 286–320). New York: Wiley.

Hirsch, H. V. B., Potter, D., Zawierucha, D., Choudhri, T., Glasser, A., Murphey, R. K., & Byers, D. (1990). Rearing in darkness changes visually-guided choice behavior in *Drosophila. Visual Neuroscience, 5,* 281–289.

Hirsch, H. V. B., Barth, M., Luo, S., Sambaziotis, H., Huber, M., Possidente, D., Ghiradella, H., & Tompkins, L. (1995). Early visual experience affects mate choice of *Drosophila melanogaster. Animal Behavior, 50,* 1211–1217.

Hoffmann, K.-P., Stone, J., & Sherman, S. M. (1972). Relay of receptive-field properties in dorsal lateral geniculate nucleus of the cat. *Journal of Neurophysiology, 35,* 518–531.

Huang, Y. Y., Li, X. C., & Kandel, E. R. (1994). cAMP contributes to mossy fiber LTP by initiating both a covalently mediated early phase and macromolecular synthesis-dependent late phase. *Cell, 79,* 69–79.

Huang, Y. Y., Kandel, E. R., Varshavsky, L., Brandon, E. P., Qi, M., Idzerda, R. L., McKnight, G. S., & Bourtchouladze, R. (1995). A genetic test of the effects of mutations in PKA on mossy fiber LTP and its relation to spatial and contextual learning. *Cell, 83,* 1211–1222.

Hubel, D. H. (1982). Exploration of the primary visual cortex, 1955–1978. *Nature, 299,* 515–524.

Hubel, D. H., & Livingstone, M. S. (1987). Segregation of form color, and stereopsis in primate area 18. *Journal of Neuroscience, 7,* 3378–3415.

Hubel, D. H., & Wiesel, T. N. (1962). Receptive fields, binocular interaction and functional architecture in the cat's visual cortex. *Journal of Physiology (London), 160,* 106–154.

Hubel, D. H., & Wiesel, T. N. (1963a). Receptive fields of cells in striate cortex of very young, visually inexperienced kittens. *Journal of Neurophysiology, 26,* 994–1002.

Hubel, D. H., & Wiesel, T. N. (1963b). Shape and arrangement of columns in cat's striate cortex. *Journal of Physiology (London), 165,* 559–568.

Hubel, D. H., & Wiesel, T. N. (1965a). Receptive fields and functional architecture in two nonstriate visual areas (18 and 19) of the cat. *Journal of Neurophysiology, 28,* 229–289.

Hubel, D. H., & Wiesel, T. N. (1965b). Binocular interaction in striate cortex of kittens reared with artificial squint. *Journal of Neurophysiology, 28,* 1041–1059.

Jarvis, C. R., Xiong, Z. G., Plant, J. R., Churchill, D., Lu, W. Y., MacVicar, B. A., & MacDonald, J. F. (1997). Neurotrophin modulation of NMDA receptors in cultured murine and isolated rat neurons. *Journal of Neurophysiology, 78,* 2363–2371.

Joiner, M. L. A., & Griffith, L. C. (1997). CaM kinase II and visual input modulate memory formation in the neuronal circuit controlling courtship conditioning. *Journal of Neuroscience, 17,* 9384–9391.

Jones, J. S., Bryant, S. H., Lewontin, R. C., Moore, J. A., & Prout, T. (1981). Gene flow and the geographical distribution of a molecular polymorphism in *Drosophila pseudoobscura. Genetics, 98,* 157–178.

Kaang, B.-K., Kandel, E. R., & Grant, S. G. (1993). Activation of cAMP-responsive genes by stimuli that produce long-term facilitation in *Aplysia* sensory neurons. *Neuron, 10,* 427–435.

Kalil, R. (1978). Dark rearing in the cat: Effects on visuomotor behavior and cell growth in the dorsal lateral geniculate nucleus. *Journal of Comparative Neurology, 178,* 451–467.

Kandel, E. R., & O'Dell, T. J. (1992). Are adult learning mechanisms also used for development? *Science, 258,* 243–245.

Kang, H., & Schuman, E. M. (1995). Long-lasting neurotrophin-induced enhancement of synaptic transmission in the adult hippocampus. *Science, 267,* 1658–1662.

Kang, H. J., & Schuman, E. M. (1996). A requirement for local protein synthesis in neurotrophin-induced hippocampal synaptic plasticity. *Science, 273,* 1402–1406.

Kang, H., Jia, L. Z., Su, K. Y., Tang, L., & Schuman, E. M. (1996). Determinants of BDNF-induced hippocampal synaptic plasticity: Role of the trk B receptor and the kinetics of neurotrophin delivery. *Learning and Memory, 3,* 188–196.

Kato, N., Artola, A., & Singer, W. (1991). Developmental changes in the susceptibility to long-term potentiation of neurones in rat visual cortex slices. *Developmental Brain Research, 60,* 43–50.

Kennedy, M. B., Bennett, M. K., & Erondu, N. E. (1983). Biochemical and immunochemical evidence that the "major postsynaptic density protein" is a subunit of a calmodulin-dependent protein kinase. *Proceedings of the National Academy of Sciences of the USA, 80,* 7357–7361.

Kirkwood, A., Lee, H.-K., & Bear, M. F. (1995). Co-regulation of long-term potentiation and experience-dependent synaptic plasticity in visual cortex by age and experience. *Nature, 375,* 328–331.

Kirkwood, A., Silva, A., & Bear, M. F. (1997). Age-dependent decrease of synaptic plasticity in the neocortex of αCaMKII mutant mice. *Proceedings of the National Academy of Sciences of the USA, 94,* 3380–3383.

Klauck, T. M., & Scott, J. D. (1995). The postsynaptic density: A subcellular anchor for signal transduction enzymes. *Cellular Signaling, 7,* 747–757.

Kleinschmidt, A., Bear, M. F., & Singer, W. (1987). Blockade of "NMDA" receptors disrupts experience-dependent plasticity of kitten visual cortex. *Science, 238,* 355–358.

Kogan, J. H., Frankland, P. W., Blendy, J. A., Coblentz, J., Marowitz, Z., Schutz, G., & Silva, A. J. (1997). Spaced training induces normal long-term memory in CREB mutant mice. *Current Biology, 7,* 1–11.

Korte, M., Carroll, P., Wolf, E., Brem, G., Thoenen, H., & Bonhoeffer, T. (1995). Hippocampal long-term potentiation is impaired in mice lacking brain-derived neurotrophic factor. *Proceedings of the National Academy of Sciences of the USA, 92,* 8856–8860.

Korte, M., Staiger, V., Griesbeck, O., Thoenen, H., & Bonhoeffer, T. (1996). The involvement of brain-derived neurotrophic factor in hippocampal long-term potentiation revealed by gene targeting experiments. *Journal de Physiologie, 90,* 157–164.

Kral, K., & Meinertzhagen, I. A. (1989). Anatomical plasticity of synapses in the lamina of the optic lobe of the fly. *Philosophical Transactions of the Royal Society of London, Series B [Biological], 323,* 155–183.

Kratz, K. E. (1982). Spatial and temporal sensitivity of lateral geniculate cells in dark-reared cats. *Brain Research, 251,* 55–63.

Kratz, K. E., Sherman, S. M., & Kalil, R. (1979). Lateral geniculate nucleus of dark reared cats: Loss of Y cells without changes in cell size. *Science, 203,* 1353–1355.

Lambin, M., Charii, F., Meille, O., & Campan, R. (1990). Effets de la privation précoce de lumière sur l'orientation viuso-guidée chez un grillon *Gryllus bimaculatus. Behavioral Processes, 22,* 165–176.

Laughlin, S. (1981a). Neural principles in the peripheral visual systems of invertebrates. In H. Autrum (Ed.), *Handbook of sensory physiology* (pp. 133–280). New York: Springer-Verlag.

Laughlin, S. (1981b). A simple coding procedure enhances a neuron's information capacity. *Zeitschrift für Naturforschung, 36c,* 910–912.

Law, M. I., Zahs, K. R., & Stryker, M. P. (1988). Organization of primary visual cortex (area 17) in the ferret. *Journal of Comparative Neurology, 278,* 157–180.

Lawrence, P. A. (1992). *The making of a fly.* Boston: Blackwell.

LeVay, S., & Gilbert, C. D. (1976). Laminar patterns of geniculocortical projection in the cat. *Brain Research, 113,* 1–19.

LeVay, S., Stryker, M. P., & Shatz, C. J. (1978). Ocular dominance columns and their development in layer IV of the cat's visual cortex: A quantitative study. *Journal of Comparative Neurology, 179,* 223–244.

Leventhal, A. G., & Hirsch, H. V. B. (1975). Cortical effect of early selective exposure to diagonal lines. *Science, 190,* 902–904.

Leventhal, A. G., & Hirsch, H. V. B. (1977). Effects of early experience upon orientation sensitivity and binocularity of neurons in visual cortex of cats. *Proceedings of the National Academy of Sciences of the USA, 74,* 1272–1276.

Leventhal, A. G., & Hirsch, H. V. B. (1980). Receptive field properties of different classes of neurons in the visual cortex of normal and dark-reared cats. *Journal of Neurophysiology, 43,* 1111–1132.

Levin, L. R., Han, P. L., Hwang, P. M., Feinstein, P. G., Davis, R. L., & Reed, R. R. (1992). The *Drosophila* learning and memory gene *rutabaga* encodes a Ca2+/calmodulin-responsive adenylyl cyclase. *Cell, 68,* 479–489.

Levine, E. S., Dreyfus, C. F., Black, I. B., & Plummer, M. R. (1995). Brain-derived neurotrophic factor rapidly enhances synaptic transmission in hippocampal neurons via postsynaptic tyrosine kinase receptors. *Proceedings of the National Academy of Sciences of the USA, 92,* 8074–8077.

Linden, D. C., Guillery, R. W., & Cucchiaro, J. (1981). The dorsal lateral geniculate nucleus of the normal ferret and its postnatal development. *Journal of Comparative Neurology, 203,* 189–211.

Lindholm, D., Castrén, E., Berzaghi, M., Blöchl, A., & Thoenen, H. (1994). Activity-dependent and hormonal regulation of neurotrophin mRNA levels in the brain—Implications for neuronal plasticity. *Journal of Neurobiology, 25,* 1362–1372.

Livingstone, M. S., & Hubel, D. H. (1987). Psychophysical evidence for separate channels for the perception of form, color, movement, and depth. *Journal of Neuroscience, 7,* 3416–3468.

Lledo, P. M., Hjelmstad, G. O., Mukherji, S., Soderling, T. R., Malenka, R. C., & Nicoll, R. A. (1995). Calcium calmodulin-dependent kinase II and long-term potentiation enhance synaptic transmission by the same mechanism. *Proceedings of the National Academy of Sciences of the USA, 92,* 11175–11179.

Lledo, P. M., Zhang, X. Y., Südhof, T. C., Malenka, R. C., & Nicoll, R. A. (1998). Postsynaptic membrane fusion and long-term potentiation. *Science, 279,* 399–403.

Lnenicka, G. A. (1991). Activity-dependent development of synaptic varicosities at terminals of an identified crayfish motoneuron. *Annals of the New York Academy of Sciences, 627,* 197–211.

Lnenicka, G. A., & Atwood, H. L. (1985). Age-dependent long-term adaptation of crayfish phasic motor axon synapses to altered activity. *Journal of Neuroscience, 5,* 459–467.

Lnenicka, G. A., Atwood, H. L., & Marin, L. (1986). Morphological transformation of synaptic terminals of a phasic motoneuron by long-term tonic stimulation. *Journal of Neuroscience, 6,* 2252–2258.

Loop, M. S., Bruce, L. L., & Petuchowski, S. (1979). Cat color vision: The effect of stimulus size, shape and viewing distance. *Vision Research, 19,* 507–513.

Löwel, S. (1994). Ocular dominance column development: Strabismus changes the spacing of adjacent columns in cat visual cortex. *Journal of Neuroscience, 14,* 7451–7468.

Löwel, S., & Singer, W. (1993). Monocularly induced 2-deoxyglucose patterns in the visual cortex and lateral geniculate nucleus of the cat: II. Awake animals and strabismic animals. *European Journal of Neuroscience, 5,* 857–869.

Macdonald, R., & Wilson, S. W. (1996). Pax proteins and eye development. *Current Opinions in Neurobiology, 6,* 49–56.

Maffei, L., & Galli-Resta, L. (1990). Correlation in the discharges of neighboring rat retinal ganglion cells during prenatal life. *Proceedings of the National Academy of Sciences of the USA, 87,* 2861–2864.

Maffei, L., Berardi, N., Domenici, L., Parisi, V., & Pizzorusso, T. (1992). Nerve growth factor (NGF) prevents the shift in ocular dominance distribution of visual cortical neurons in monocularly deprived rats. *Journal of Neuroscience, 12,* 4651–4662.

Maletic-Savatic, M., & Malinow, R. (1998). Calcium-evoked dendritic exocytosis in cultured hippocampal neurons. Part I: Trans-Golgi network-derived organelles undergo regulated exocytosis. *Journal of Neuroscience, 18,* 6803–6813.

Malinow, R., Schulman, H., & Tsien, R. W. (1989). Inhibition of postsynaptic PKC or CaMKII blocks induction but not expression of LTP. *Science, 245,* 862–866.

Mansfield, R. J. W. (1974). Neural basis of orientation perception in primate vision. *Science, 186,* 1133–1135.

Mansfield, R. J. W., & Ronner, S. F. (1978). Orientation anistropy in monkey visual cortex. *Brain Research, 149,* 229–234.

Mansuy, I. M., Mayford, M., Jacob, B., Kandel, E. R., & Bach, M. E. (1998). Restricted and regulated overexpression reveals calcineurin as a key component in the transition from short-term to long-term memory. *Cell, 92,* 39–49.

Markow, T. A. (1981). Genetic and sensory aspects of mating success of phototactic strains of *Drosophila melanogaster*. *Behavior Genetics, 11,* 273–279.

Markow, T. A. (1987). Behavioral and sensory basis of courtship success in *Drosophila melanogaster*. *Proceedings of the National Academy of Sciences of the USA, 84,* 6200–6204.

Marks, M. J., Robinson, S. F., & Collins, A. C. (1996). Nicotinic agonists differ in activation and desensitization of 86Rb+ efflux from mouse thalamic synaptosomes. *Journal of Pharmacology and Experimental Therapeutics, 277,* 1383–1396.

Martin, K. A. C., & Whitteridge, D. (1984). Form, function, and intracortical projections of spiny neurones in the striate cortex of the cat. *Journal of Physiology (London), 353,* 463–504.

Maslim, J., & Stone, J. (1986). Synaptogenesis in the retina of the cat. *Brain Research, 373,* 35–48.

Maslim, J., & Stone, J. (1988). Time course of stratification of the dendritic fields of ganglion cells in the retina of the cat. *Developmental Brain Research, 44,* 87–93.

Mason, C. A. (1982). Development of terminal arbors of retino-geniculate axons in the kitten. I. Light microscopical observations. *Neuroscience, 7,* 541–559.

Matsumoto, S. G., & Murphey, R. K. (1977). Sensory deprivation during development decreases the responsiveness of cricket giant interneurones. *Journal of Physiology (London), 268,* 533–548.

Matsumoto, S. G., & Murphey, R. K. (1978). Sensory deprivation in the cricket nervous system: Evidence for a critical period. *Journal of Physiology (London), 285,* 159–170.

Matthies, H., & Reymann, K. G. (1993). Protein kinase A inhibitors prevent the maintenance of hippocampal long-term potentiation. *NeuroReport, 4,* 712–714.

McDonald, J., & Parsons, P. A. (1973). Dispersal activities of the sibling species *Drosophila melanogaster* and *Drosophila simulans*. *Behavior Genetics, 3,* 293–301.

McKay, R. R., Chen, D. M., Miller, K., Kim, S., Stark, W. S., & Shortridge, R. D. (1995). Phospholipase C rescues visual defect in *norpA* mutant of *Drosophila melanogaster*. *Journal of Biological Chemistry, 270,* 13271–13276.

McRobert, S. P., & Tompkins, L. (1983). Courtship of young males is ubiquitous in *Drosophila melanogaster. Behavior Genetics, 13*, 517–523.

McRobert, S. P., & Tompkins, L. (1988). Two consequences of homosexual courtship performed by *Drosophila melanogaster* and *Drosophila affinis* males. *Evolution, 42*, 1093–1097.

Meille, O., Campan, R., & Lambin, M. (1994). Effects of light deprivation on visually guided behavior early in the life of *Gryllus bimaculatus* (Orthoptera: Gryllidae). *Annals of the Entomological Society of America, 87*, 133–142.

Meinertzhagen, I. A. (1989). Fly photoreceptor synapses: Their development, evolution, and plasticity. *Journal of Neurobiology, 20*, 276–294.

Meinertzhagen, I. A., & Hanson, T. E. (1993). The development of the optic lobe. In *The development of Drosophila melangaster* (pp. 1363–1491). Cold Spring Harbor, NY: Cold Spring Harbor Laboratory Press.

Meinertzhagen, I. A., & O'Neil, S. D. (1991). Synaptic organization of columnar elements in the lamina of the wild type in *Drosophila melanogaster. Journal of Comparative Neurology, 305*, 232–263.

Mello, N. K., & Peterson, N. J. (1964). Behavioral evidence for color discrimination in the cat. *Journal of Neurophysiology, 27*, 323–323.

Melzig, J., Buchner, S., Wiebel, F., Wolf, R., Burg, M., Pak, W. L., & Buchner, E. (1996). Genetic depletion of histamine from the nervous system of *Drosophila* eliminates specific visual and mechanosensory behavior. *Journal of Comparative Physiology, 179*, 763–773.

Merlio, J. P., Ernfors, P., Jaber, M., & Persson, H. (1992). Molecular cloning of rat trkC and distribution of cells expressing messenger RNAs for members of the trk family in the rat central nervous system. *Neuroscience, 51*, 513–532.

Meyerowitz, E. M., & Kankel, D. R. (1978). A genetic analysis of visual system development in *Drosophilia melanogaster. Developmental Biology, 62*, 112–142.

Miller, K. D., Keller, J. B., & Stryker, M. P. (1989). Ocular dominance column development: Analysis and simulation. *Science, 245*, 605–615.

Mimura, K. (1986). Development of visual pattern discrimination in the fly depends on light experience. *Science, 232*, 83–85.

Mitchell, D. E. (1988). The extent of visual recovery from early monocular or binocular visual deprivation in kittens. *Journal of Physiology (London), 395*, 639–660.

Mobbs, P. G. (1982). The brain of the honey bee *Apis mellifera.* The connections and spatial organization of the mushroom bodies. *Philosophical Transactions of the Royal Society of London, Series B [Biological], 298*, 309–345.

Mooney, R., Penn, A. A., Gallego, R. & Shatz, C. J. (1996). Thalamic relay of spontaneous retinal activity prior to vision. *Neuron, 17*, 863–874.

Moore, C. L. (1992). The role of maternal stimulation in the development of sexual behavior and its neural basis. *Annals of the New York Academy of Sciences, 662*, 160–177.

Moore, A. N., Waxham, M. N., & Dash, P. K. (1996). Neuronal activity increases the phosphorylation of the transcription factor cAMP response element-binding protein (CREB) in rat hippocampus and cortex. *Journal of Biological Chemistry, 271*, 14214–14220.

Mower, G. D., Caplan, C. J., & Letsou, G. (1982). Behavioral recovery from binocular deprivation in the cat. *Behavioural Brain Research, 4*, 209–215.

Mower, G. D., Caplan, C. J., Christen, W. G., & Duffy, F. H. (1985). Dark rearing prolongs physiological but not anatomical plasticity of the cat visual cortex. *Journal of Comparative Neurology, 235*, 448–466.

Muir, D. W., & Mitchell, D. E. (1973). Visual resolution and experience: Acuity deficits in cats following early selective visual deprivation. *Science, 180*, 420–422.

Muir, D. W. & Mitchell, D. E. (1975). Behavioral deficits in cats following early selected visual exposure to contours of a single orientation. *Brain Research, 85*, 459–477.

Mukhopadhyay, M., & Campos, A. R. (1995). The larval optic nerve is required for the development of an identified serotonergic arborization in *Drosophila melanogaster. Developmental Biology, 169*, 629–643.

Murphey, R. K. (1986a). Competition and the dynamics of axon arbor growth in the cricket. *Journal of Comparative Neurology, 251*, 100–110.

Murphey, R. K. (1986b). The myth of the inflexible invertebrate: Competition and synaptic remodelling in the development of invertebrate nervous systems. *Journal of Neurobiology, 17*, 585–591.

Murphey, R. K., & Chiba, A. (1990). Assembly of the cricket cercal sensory system: Genetic and epigenetic control. *Journal of Neurobiology, 21*, 120–137.

Murphey, R. K., & Hirsch, H. V. B. (1982). From cat to cricket: The genesis of response selectivity of interneurons. *Current Topics in Developmental Biology, 17*, 241–256.

Murphey, R. K., & Lemere, C. A. (1984). Competition controls the growth of an identified axonal arborization. *Science, 224*, 1352–1355.

Müller, L., Pattiselanno, A., & Vrensen, G. (1981). The postnatal development of the presynaptic grid in the visual cortex of rabbits and the effect of dark-rearing. *Brain Research, 205,* 39–48.

Mustari, M. J., Bullier, J., & Henry, G. H. (1982). Comparison of response properties of three types of monosynaptic S-cell in cat striate cortex. *Journal of Neurophysiology, 47,* 439–454.

Nelson, R., Famiglietti, E. V., Jr., & Kolb, H. (1978). Intracellular staining reveals different levels of stratification for on- and off-center ganglion cells in cat retina. *Journal of Neurophysiology, 41,* 472–483.

Nguyen, P. V., & Kandel, E. R. (1996). A macromolecular synthesis-dependent late phase of long-term potentiation requiring cAMP in the medial perforant pathway of rat hippocampal slices. *Journal of Neuroscience, 16,* 3189–3198.

Nighorn, A., Healy, M. J., & Davis, R. L. (1991). The cyclic AMP phosphodiesterase encoded by the *Drosophila dunce* gene is concentrated in the mushroom body neuropil. *Neuron, 6,* 455–467.

Nowak, L., Bregestovski, P., Ascher, P., Herbert, A., & Prochiantz, A. (1984). Magnesium gates glutamate-activated channels in mouse central neurones. *Nature, 307,* 462–465.

O'Dell, T. J., Kandel, E. R., & Grant, S. G. N. (1991). Long-term potentiation in the hippocampus is blocked by tyrosine kinase inhibitors. *Nature, 353,* 558–560.

Oishi, I., Sugiyama, S., Liu, Z. J., Yamamura, H., Nishida, Y., & Minami, Y. (1997). A novel *Drosophila* receptor tyrosine kinase expressed specifically in the nervous system. Unique structural features and implication in developmental signaling. *Journal of Biological Chemistry, 272,* 11916–11923.

Osorio, D., Srinivasan, M. V., & Pinter, R. B. (1990). What causes edge fixation in walking flies? *Journal of Experimental Biology, 149,* 281–292.

Otmakhov, N., Griffith, L. C., & Lisman, J. E. (1997). Postsynaptic inhibitors of calcium/calmodulin-dependent protein kinase type II block induction but not maintenance of pairing-induced long-term potentiation. *Journal of Neuroscience, 17,* 5357–5365.

Packwood, J., & Gordon, B. (1975). Stereopsis in normal domestic cat, Siamese cat, and cat raised with alternating monocular occlusion. *Journal of Neurophysiology, 38,* 1485–1499.

Patterson, S. L., Grover, L. M., Schwartzkroin, P. A., & Bothwell, M. (1992). Neurotrophin expression in rat hippocampal slices: A stimulus paradigm inducing LTP in CA1 evokes increases in BDNF and NT-3 mRNAs. *Neuron, 9,* 1081–1088.

Patterson, S. L., Abel, T., Deuel, T. A. S., Martin, K. C., Rose, J. C., & Kandel, E. R. (1996). Recombinant BDNF rescues deficits in basal synaptic transmission and hippocampal LTP in BDNF knockout mice. *Neuron, 16,* 1137–1145.

Pearlman, A. L., & Daw, N. W. (1971). Behavioral and neurophysiological studies on cat color vision. *International Journal of Neuroscience, 1,* 357–360.

Penn, A. A., Wong, R. O. L., & Shatz, C. J. (1994). Neuronal coupling in the developing mammalian retina. *Journal of Neuroscience, 14,* 3805–3815.

Penn, A. A., Riquelme, P. A., Feller, M. B., & Shatz, C. J. (1998). Competition in retinogeniculate patterning driven by spontaneous activity. *Science, 279,* 2108–2112.

Pham, T. A., Impey, S., Storm, D. R., & Stryker, M. P. (1999). CRE-mediated gene transcription in neocortical neuronal plasticity during the developmental critical period. *Neuron, 22,* 63–72.

Phelps, C. B., & Brand, A. H. (1998). Ectopic gene expression in *Drosophila* using GAL4 system. *Methods, 14,* 367–379.

Powell, J. R., Dobzhansky, T., Hook, J. E., & Wistrand, H. (1976). Genetics of natural populations. XLIII. Further studies on rates of dispersal of *Drosophila pseudoobscura* and its relatives. *Genetics, 82,* 493–506.

Pulido, D., Campuzano, S., Koda, T., Modolell, J., & Barbacid, M. (1992). *Dtrk,* a *Drosophila* gene related to the trk family of neurotrophin receptors, encodes a novel class of neural cell adhesion molecule. *EMBO Journal, 11,* 391–404.

Qi, M., Zhuo, M., Skalhegg, B. S., Brandon, E. P., Kandel, E. R., McKnight, G. S., & Idzerda, R. L. (1996). Impaired hippocampal plasticity in mice lacking the Cβ catalytic subunit of cAMP-dependent protein kinase. *Proceedings of the National Academy of Sciences of the USA, 93,* 1571–1576.

Quiring, R., Walldorf, U., Kloter, U., & Gehring, W. J. (1994). Homology of the *eyeless* gene of *Drosophila* to the *Small eye* gene in mice and *Aniridia* in humans. *Science, 265,* 785–789.

Racine, R., & Zaide, J. (1978). A further investigation into the mechanisms underlying the kindling phenomenon. In K. E. Livingston & O. Hornykiewicz (Eds.). *Limbic mechanisms: The continuing evolution of the limbic system concept* (pp. 457–493). New York: Plenum Press.

Rauschecker, J. P., & Singer, W. (1981). The effects of early visual experience on the cat's visual cortex and their possible explanation by Hebb synapses. *Journal of Physiology (London), 310,* 215–239.

Reiter, H. O., & Stryker, M. P. (1988). Neural plasticity without postsynaptic action potentials: Less-active inputs become dominant when kitten visual cortical cells are pharmacologically inhibited. *Proceedings of the National Academy of Sciences of the USA, 85,* 3623–3627.

Riccio, A., Pierchala, B. A., Ciarallo, C. L., & Ginty, D. D. (1997). An NGF-TrkA-mediated retrograde signal to transcription factor CREB in sympathetic neurons. *Science, 277*, 1097–1100.

Riddle, D. R., Lo, D. C., & Katz, L. C. (1995). NT-4-mediated rescue of lateral geniculate neurons from effects of monocular deprivation. *Nature, 378*, 189–191.

Ringo, J. L., & Wolbarsht, M. L. (1986). Spectral coding in cat retinal ganglion cell receptive fields. *Journal of Neurophysiology, 55*, 320–330.

Ringo, J., Wolbarsht, M. L., Wagner, H. G., Crocker, R., & Amthor, F. (1977). Trichromatic vision in the cat. *Science, 198*, 753–755.

Roberson, E. D., & Sweatt, J. D. (1996). Transient activation of cyclic AMP-dependent protein kinase during hippocampal long-term potentiation. *Journal of Biological Chemistry, 271*, 30436–30441.

Roberts, E. B., Meredith, M. A., & Ramoa, A. S. (1998). Suppression of NMDA receptor function using antisense DNA blocks ocular dominance plasticity while preserving visual responses. *Journal of Neurophysiology, 80*, 1021–1032.

Roberts, L. A., Higgins, M. J., O'Shaughnessy, C. T., Stone, T. W., & Morris, B. J. (1996). Changes in hippocampal gene expression associated with the induction of long-term potentiation. *Molecular Brain Research, 42*, 123–127.

Roberts, L. A., Large, C. H., Higgins, M. J., Stone, T. W., O'Shaughnessy, C. T., & Morris, B. J. (1998). Increased expression of dendritic mRNA following the induction of long-term potentiation. *Molecular Brain Research, 56*, 38–44.

Rodieck, R. W. (1979). Visual pathways. *Annual Review of Neuroscience, 2*, 193–225.

Rogers, S. M., & Simpson, S. J. (1997). Experience-dependent changes in the number of chemosensory sensilla on the mouthparts and antenna of *Locusta migratoria*. *Journal of Experimental Biology, 200*, 2313–2321.

Rorth, P., Szabo, K., Bailey, A., Laverty, T., Rehm, J., Rubin, G. M., Weigmann, K., Milan, M., Benes, V., Ansorge, W., & Cohen, S. M. (1998). Systematic gain-of-function genetics in *Drosophila*. *Development, 125*, 1049–1057.

Rotenberg, A., Mayford, M., Hawkins, R. D., Kandel, E. R., & Muller, R. U. (1996). Mice expressing activated CaMKII lack low frequency LTP and do not form stable place cells in the CA1 region of the hippocampus. *Cell, 87*, 1351–1361.

Rybak, J., & Meinertzhagen, I. A. (1997). The effects of light reversals on photoreceptor synaptogenesis in the fly *Musca domestica*. *European Journal of Neuroscience, 9*, 319–333.

Saito, H.-A. (1983). Morphology of physiologically identified X-, Y-, and W-type retinal ganglion cells of the cat. *Journal of Comparative Neurology, 221*, 279–288.

Sakimura, K., Kutsuwada, T., Ito, I., Manabe, T., Takayama, C., Kushiya, E., Yagi, T., Aizawa, S., Inoue, Y., Sugiyama, H., & Mishina, M. (1995). Reduced hippocampal LTP and spatial learning in mice lacking NMDA receptor ε1 subunit. *Nature, 373*, 151–155.

Sanderson, K. J. (1971). The projection of the visual field to the lateral geniculate and medial interlaminar nuclei in the cat. *Journal of Comparative Neurology, 143*, 101–118.

Schacher, S. M., Holtzman, E., & Hood, D. C. (1974). Uptake of horseradish peroxidase by frog photoreceptor synapses in the dark and in the light. *Nature, 249*, 261–263.

Schacher, S., Castellucci, V. F., & Kandel, E. R. (1988). cAMP evokes long-term facilitation in *Aplysia* sensory neurons that requires new protein synthesis. *Science, 240*, 1667–1669.

Schäffel, F., & Willmund, R. (1985). Visual signals in the courtship of *Drosophila melanogaster*: Mutant analysis. *Journal of Insect Physiology, 31*, 899–907.

Schmidt, J. T. (1985). Formation of retinotopic connections: Selective stabilization by an activity-dependent mechanism. *Cellular and Molecular Neurobiology, 5*, 65–83.

Schmidt, J. T. (1990). Long-term potentiation and activity-dependent retinotopic sharpening in the regenerating retinotectal projection of goldfish: Common sensitive period and sensitivity to NMDA blockers. *Journal of Neuroscience, 10*, 233–246.

Schmidt, J. T., & Tieman, S. B. (1985). Eye-specific segregation of optic afferents in mammals, fish, and frogs: The role of activity. *Cellular and Molecular Neurobiology, 5*, 5–34.

Schneider, G. E. (1969). Two visual systems. *Science, 163*, 895–902.

Schoppmann, A., & Stryker, M. P. (1981). Physiological evidence that the 2-deoxyglucose method reveals orientation columns in cat visual cortex. *Nature, 293*, 574–576.

Schoups, A. A., Elliott, R. C., Friedman, W. J., & Black, I. B. (1995). NGF and BDNF are differentially modulated by visual experience in the developing geniculocortical pathway. *Developmental Brain Research, 86*, 326–334.

Schulz, S., Siemer, H., Krug, M., & Hölt, V. (1999). Direct evidence for biphasic cAMP responsive element-binding protein phosphorylation during long-term potentiation in the rat dentate gyrus *in vivo*. *Journal of Neuroscience, 19*, 5683–5692.

Schuster, C. M., Davis, G. W., Fetter, R. D., & Goodman, C. S. (1996). Genetic dissection of structural and functional components of synaptic plasticity. I. Fasciclin II controls synaptic stabilization and growth. *Neuron, 17,* 641–654.

Shatz, C. J. (1990). Impulse activity and the patterning of connections during CNS development. *Neuron, 5,* 745–756.

Shatz, C. J. (1996). Emergence of order in visual system development. *Proceedings of the National Academy of Sciences of the USA, 93,* 602–608.

Shatz, C. J., & Stryker, M. P. (1978). Ocular dominance in layer IV of the cat's visual cortex and the effects of monocular deprivation. *Journal of Physiology (London), 281,* 267–283.

Shatz, C. J., & Stryker, M. P. (1988). Prenatal tetrodotoxin infusion blocks segregation of retinogeniculate afferents. *Science, 242,* 87–89.

Shatz, C. J., Lindström, S., & Wiesel, T. N. (1977). The distribution of afferents representing the right and left eyes in the cat's visual cortex. *Brain Research, 131,* 103–116.

Shaw, S. R. (1984). Early visual processing in insects. *Journal of Experimental Biology, 112,* 225–251.

Shepherd, D., & Murphey, R. K. (1986). Competition regulates the efficacy of an identified synapse in crickets. *Journal of Neuroscience, 6,* 3152–3160.

Sherman, S. M. (1973). Visual field defects in monocularly and binocularly deprived cats. *Brain Research, 49,* 25–45.

Sherman, S. M. (1985). Functional organization of the W-, X-, and Y-cell pathways in the cat: A review and hypothesis. In J. M. Sprague & A. N. Epstein (Eds.) *Progress in psychobiology and physiological psychology* (Vol. 11, pp. 233–314). New York: Academic Press.

Sherman, S. M., & Spear, P. D. (1982). Organization of visual pathways in normal and visually deprived cats. *Physiological Reviews, 62,* 738–855.

Shieh, P. B., Hu, S. C., Bobb, K., Timmusk, T., & Ghosh, A. (1998). Identification of a signaling pathway involved in calcium regulation of *BDNF* expression. *Neuron, 20,* 727–740.

Silva, A. J., Stevens, C. F., Tonegawa, S., & Wang, Y. (1992). Deficient hippocampal long-term potentiation in α–calcium–calmodulin kinase II mutant mice. *Science, 257,* 201–206.

Silva, A. J., Kogan, J. H., Frankland, P. W., & Kida, S. (1998). CREB and memory. *Annual Review of Neuroscience, 21,* 127–148.

Singer, W., Tretter, F., & Cynader, M. (1975). Organization of cat striate cortex: A correlation of receptive-field properties with afferent and efferent connections. *Journal of Neurophysiology, 38,* 1080–1098.

Singer, W., Holländer, H., & Vanegas, H. (1977). Decreased peroxidase labeling of lateral geniculate neurons following deafferentation. *Brain Research, 120,* 133–137.

Singer, W., Freeman, B., & Rauschecker, J. (1981). Restriction of visual experience to a single orientation affects the organization of orientation columns in cat visual cortex. A study with deoxyglucose. *Experimental Brain Research, 41,* 199–215.

Slack, J. R., & Walsh, C. (1995). Effects of a cAMP analogue simulate the distinct components of long-term potentiation in CA1 region of rat hippocampus. *Neuroscience Letters, 201,* 25–28.

Smetters, D. K., Hahm, J., & Sur, M. (1994). An N-methyl-D-aspartate receptor antagonist does not prevent eye-specific segregation in the ferret retinogeniculate pathway. *Brain Research, 658,* 168–178.

Smith, D. C., Lorber, R., Stanford, L. R., & Loop, M. S. (1980). Visual acuity following binocular deprivation in the cat. *Brain Research, 183,* 1–11.

Son, H., Hawkins, R. D., Martin, K., Kiebler, M., Huang, P. L., Fishman, M. C., & Kandel, E. R. (1996). Long-term potentiation is reduced in mice that are doubly mutant in endothelial and neuronal nitric oxide synthase. *Cell, 87,* 1015–1023.

Spieth, H. T., & Ringo, J. M. (1983). Mating behavior and sexual isolation in *Drosophila*. In M. Ashburner, H. L. Carlson, & J. J. Thompson, Jr. (Eds.) *The genetics and biology of Drosophila* (pp. 223–284). New York: Academic Press.

Sretavan, D. W., & Shatz, C. J. (1986). Prenatal development of retinal ganglion cell axons: Segregation into eye-specific layers within the cat's lateral geniculate nucleus. *Journal of Neuroscience, 6,* 234–251.

Stanford, L. R., & Sherman, S. M. (1984). Structure/function relationships of retinal ganglion cells in the cat. *Brain Research, 297,* 381–386.

Stanford, L. R., Friedlander, M. J., & Sherman, S. M. (1983). Morphological and physiological properties of geniculate W-cells of the cat: A comparison with X- and Y-cells. *Journal of Neurophysiology, 50,* 582–608.

Stent, G. S. (1973). A physiological mechanism for Hebb's postulate of learning. *Proceedings of the National Academy of Sciences of the USA, 70,* 997–1001.

Sterling, P., Freed, M., & Smith, R. G. (1986). Microcircuitry and functional architecture of the cat retina. *Trends in Neurosciences, 9,* 186–192.

Stevens, C. F., Tonegawa, S., & Wang, Y. (1994). The role of calcium-calmodulin kinase II in three forms of synaptic plasticity. *Current Biology, 4,* 687–693.

Stone, J., Dreher, B., & Leventhal, A. (1979). Hierarchical and parallel mechanisms in the organization of visual cortex. *Brain Research Reviews, 1,* 345–394.

Strausfeld, N. J., & Campos-Ortega, J. A. (1977). Vision in insects: Pathways possibly underlying neural adaptation and lateral inhibition. *Science, 195,* 894–897.

Strausfeld, N. J., & Lee, J. K. (1991). Neuronal basis for parallel visual processing in the fly. *Visual Neuroscience, 7,* 13–33.

Strausfeld, N. J., & Nässel, D. R. (1981). Neuroarchitectures serving compound eyes of crustacea and insects. In H. Autrum (Ed.) *Handbook of sensory physiology* (pp. 1–132). New York: Springer-Verlag.

Stryker, M. P., & Harris, W. A. (1986). Binocular impulse blockade prevents the formation of ocular dominance columns in cat visual cortex. *Journal of Neuroscience, 60,* 2117–2133.

Stryker, M. P., & Zahs, K. R. (1983). On and off sublaminae in the lateral geniculate nucleus of the ferret. *Journal of Neuroscience, 3,* 1943–1951.

Suzuki, T., Usuda, N., Ishiguro, H., Mitake, S., Nagatsu, T., & Okumura-Noji, K. (1998). Occurrence of a transcription factor, cAMP response element-binding protein (CREB), in the postsynaptic sites of the brain. *Molecular Brain Research, 61,* 69–77.

Sweatt, J. D., & Kandel, E. R. (1989). Persistent and transcriptionally-dependent increase in protein phosphorylation in long-term facilitation of *Aplysia* sensory neurons. *Nature, 339,* 51–54.

Swindale, N. V. (1980). A model for the formation of ocular dominance stripes. *Proceedings of the Royal Society of London [Biology], 208,* 243–264.

Swindale, N. V. (1981). Absence of ocular dominance patches in dark reared cats. *Nature, 290,* 332–333.

Tao, X., Finkbeiner, S., Arnold, D. B., Shaywitz, A. J., & Greenberg, M. E. (1998). Ca^{2+} influx regulates *BDNF* transcription by a CREB family transcription factor-dependent mechanism. *Neuron, 20,* 709–726.

Taylor, C. E., Powell, J. R., Kekic, V., Andjelkovic, & Burla, H. (1984). Dispersal rates of species of the *Drosophila obscura* group: Implications for population structure. *Evolution, 38,* 1397–1401.

Technau, G. M. (1984). Fiber number in the mushroom bodies of adult *Drosophila melanogaster* depends on age, sex and experience. *Journal of Neurogenetics, 1,* 113–126.

Thibos, L. N., & Levick, W. R. (1982). Astigmatic visual deprivation in cat: Behavioral, optical and retinophysiological consequences. *Vision Research, 22,* 43–53.

Tieman, D. G., Tumosa, N., & Tieman, S. B. (1983). Behavioral and physiological effects of monocular deprivation: A comparison of rearing with occlusion and diffusion. *Brain Research, 280,* 41–50.

Tieman, S. B. (1984). Effects of monocular deprivation on geniculocortical synapses in the cat. *Journal of Comparative Neurology, 222,* 166–176.

Tieman, S. B. (1985). The anatomy of geniculocortical connections in monocularly deprived cats. *Cellular and Molecular Neurobiology, 5,* 35–45.

Tieman, S. B. (1991). Morphological changes in the geniculocortical pathway associated with monocular deprivation. *Annals of the New York Academy of Sciences, 627,* 212–230.

Tieman, S. B., & Hirsch, H. V. B. (1982). Exposure to lines of only one orientation modifies dendrite morphology of cells in the visual cortex of the cat. *Journal of Comparative Neurology, 211,* 353–362.

Tieman, S. B., & Tumosa, N. (1983). [^{14}C]-2-Deoxyglucose demonstration of the organization of ocular dominance in areas 17 and 18 of the normal cat. *Brain Research, 267,* 35–46.

Tieman, S. B., & Tumosa, N. (1997). Alternating monocular exposure alters the spacing of ocularity domains in area 17 of cat. *Visual Neuroscience, 14,* 929–938.

Timney, B., Mitchell, D. E., & Giffin, F. (1978). The development of vision in cats after extended periods of dark-rearing. *Experimental Brain Research, 31,* 547–560.

Tokuda, M., Ahmed, B. Y., Lu, Y. F., Matsui, H., Miyamoto, O., Yamaguchi, F., Konishi, R., & Hatase, O. (1997). Involvement of calmodulin-dependent protein kinases-I and -IV in long-term potentiation. *Brain Research, 755,* 162–166.

Tomarev, S. I., Callaerts, P., Kos, L., Zinovieva, R., Halder, G., Gehring, W., & Piatigorsky, J. (1997). Squid Pax-6 and eye development. *Proceedings of the National Academy of Sciences of the USA, 94,* 2421–2426.

Tompkins, L. (1984). Genetic analysis of sex appeal in *Drosophila*. *Behavior Genetics, 14,* 411–440.

Tompkins, L., Hall, J. C., & Hall, L. M. (1980). Courtship-stimulating volatile compounds from normal and mutant *Drosophila*. *Journal of Insect Physiology, 26,* 689–697.

Tompkins, L., Gross, A. C., Hall, J. C., Gailey, D. A., & Siegel, R. W. (1982). The role of female movement in the sexual behavior of *Drosophila melanogaster*. *Behavior Genetics, 12,* 295–307.

Trevarthen, C. B. (1968). Two mechanisms of vision in primates. *Psychologische Forschung, 31,* 299–337.

Tully, T., Preat, T., Boynton, S. C., & Del Vecchio, M. (1994). Genetic dissection of consolidated memory in *Drosophila*. *Cell, 79,* 35–47.

Tumosa, N., Tieman, S. B., & Hirsch, H. V. B. (1982). Visual field deficits in cats reared with unequal alternating monocular exposure. *Experimental Brain Research, 47,* 119–129.

Tumosa, N., Nunberg, S., Hirsch, H. V. B., & Tieman, S. B. (1983). Binocular exposure causes suppression of the less experienced eye in cats previously reared with unequal alternating monocular exposure. *Investigative Ophthalmology and Visual Science, 24,* 496–506.

Tumosa, N., Tieman, S. B., & Tieman, D. G. (1989). Binocular competition affects the pattern and intensity of ocular activation columns in the visual cortex of cats. *Visual Neuroscience, 2,* 391–407.

Tusa, R. J., Palmer, L. A., & Rosenquist, A. C. (1978). The retinotopic organization of area 17 (striate cortex) in the cat. *Journal of Comparative Neurology, 177,* 213–236.

Tusa, R. J., Rosenquist, A. C., & Palmer, L. A. (1979). Retinotopic organization of areas 18 and 19 in the cat. *Journal of Comparative Neurology, 185,* 657–678.

Van Hof-van Duin, J. (1977). Visual field measurements in monocularly deprived and normal cats. *Experimental Brain Research, 30,* 353–368.

van Kesteren, R. E., Fainzilber, M., Hauser, G., Van Minnen, J., Vreugdenhil, E., Smit, A. B., Ibanez, C. F., Geraerts, W. P. M., & Bulloch, A. G. M. (1998). Early evolutionary origin of the neurotrophin receptor family. *EMBO Journal, 17,* 2534–2542.

Vaney, D. I. (1991). Many diverse types of retinal neurons show tracer coupling when injected with biocytin or neurobiotin. *Neuroscience Letters, 125,* 187–190.

Vaney, D. I. (1994). Patterns of neuronal coupling in the retina. *Progress in Retinal and Eye Research, 13,* 301–355.

Vardi, N., & Smith, R. G. (1996). The aII amacrine network: Coupling can increase correlated activity. *Vision Research, 36,* 3743–3757.

von der Malsburg, C. (1979). Development of ocularity domains and growth behavior of axon terminals. *Biological Cybernetics, 32,* 49–62.

von der Malsburg, C., & Willshaw, D. J. (1976). A mechanism for producing continuous neural mappings: Ocularity dominance stripes and ordered retinotectal projections. *Experimental Brain Research, 1(Supplement),* 463–469.

Vrensen, G., & Nunes Cardozo, J. (1981). Changes in size and shape of synaptic connections after visual training: An ultrastructural approach of synaptic plasticity. *Brain Research, 218,* 79–97.

Wark, R. C., & Peck, C. K. (1982). Behavioral consequences of early visual exposure to contours of a single orientation. *Developmental Brain Research, 5,* 218–221.

Wässle, H. (1982). Morphological types and central projections of ganglion cells in the cat retina. In N. Osborne & G. Chader (Eds.). *Progress in retinal research* (pp. 125–152). New York: Pergamon Press.

Wehner, R. (1972). Spontaneous pattern preferences of *Drosophila melanogaster* to black areas in various parts of the visual field. *Journal of Insect Physiology, 18,* 1531–1543.

Wehner, R. (1981). Spatial vision in arthropods. In H. Autrum (Ed.). *Handbook of sensory physiology* (pp. 287–616). New York: Springer-Verlag.

Wehner, R., & Horn, E. (1975). The effect of object distance on pattern preferences in the walking fly (*Drosophila melanogaster*). *Experientia, 31,* 641-643.

Wehner, R., Gartenmann, G., & Jungi, T. (1969). Contrast perception in eye color mutants of *Drosophila melanogaster* and *Drosophila subobscura*. *Journal of Insect Physiology, 15,* 815-823.

Wehner, R., & Wehner-von Segesser, S. (1973). Calculation of visual receptor spacing in *Drosophila melanogaster* by pattern recognition experiments. *Journal of Comparative Physiology, 82,* 165–177.

Weisskopf, M. G., Castillo, P. E., Zalutsky, R. A., & Nicoll, R. A. (1994). Mediation of hippocampal mossy fiber long-term potentiation by cyclic AMP. *Science, 265,* 1878–1882.

Weliky, M., & Katz, L. C. (1999). Correlational structure of spontaneous neuronal activity in the developing lateral geniculate nucleus *in vivo. Science, 285,* 599–604.

West, M. J., King, A. P., & Arberg, A. A. (1988). An inheritance of niches: The role of ecological legacies in ontogeny. In E. M. Blass (Ed.), *Developmental psychobiology and behavioral ecology* (pp. 41–62). New York: Plenum Press.

Wiesel, T. N. (1982). Postnatal development of the visual cortex and the influence of environment. *Nature, 299,* 583–591.

Wiesel, T. N., & Hubel, D. H. (1963a). Effects of visual deprivation on morphology and physiology of cells in the cat's lateral geniculate body. *Journal of Neurophysiology, 26,* 978–993.

Wiesel, T. N., & Hubel, D. H. (1963b). Single-cell responses in striate cortex of kittens deprived of vision in one eye. *Journal of Neurophysiology, 26,* 1003–1017.

Wiesel, T. N., & Hubel, D. H. (1965). Comparison of the effects of unilateral and bilateral eye closure on cortical unit responses in kittens. *Journal of Neurophysiology, 28,* 1029–1040.

Willmund, R., & Ewing, A. W. (1982). Visual signals in the courtship of *Drosophila melanogaster. Animal Behavior, 30,* 209–215.

Wilson, P. D., & Stone, J. (1975). Evidence of W-cell input to the cat's visual cortex via the C laminae of the lateral geniculate nucleus. *Brain Research, 92,* 472–478.

Wolfe, T., Martin, K. A., Rubin, G. M., & Zipursky, S. L. (1997). The development of the Drosophila visual system. In W. M. Cowan, T. M. Jessel, & S. L. Zipursky (Eds.). *Molecular and cellular approaches to neural development* (pp. 474–508). New York: Oxford University Press.

Wolpaw, J. R., & Schmidt, J. T. (1991). Preface. *Annals of the New York Academy of Sciences, 627,* ix–xi

Wong, R. O. L., Meister, M., & Shatz, C. J. (1993). Transient period of correlated bursting activity during development of the mammalian retina. *Neuron, 11,* 923–938.

Wu, C.-F., Renger, J. J., & Engel, J. E. (1998). Activity-dependent functional and developmental plasticity of *Drosophila neurons.* In P. D. Evans (Ed.). *Advances in insect physiology* (pp. 385–440). San Diego, CA: Academic Press.

Xing, L.-C. S., & Tieman, S. B. (1993). Relay cells, not interneurons, of cat's lateral geniculate nucleus contain N-acetylaspartylglutamate. *Journal of Comparative Neurology, 330,* 272–285.

Yan, H. Q., Mazow, M. L., & Dafny, N. (1996). NGF prevents the changes induced by monocular deprivation during the critical period in rats. *Brain Research, 706,* 318–322.

Yin, J. C., Wallach, J. S., Del Vecchio, M., Wilder, E. L., Zhou, H., Quinn, W. G., & Tully, T. (1994). Induction of a dominant negative CREB transgene specifically blocks long-term memory in *Drosophila. Cell, 79,* 49–58.

Yin, J. C., Del Vecchio, M., Zhou, H., & Tully, T. (1995). CREB as a memory modulator: Induced expression of a dCREB2 activator isoform enhances long-term memory in *Drosophila. Cell, 81,* 107–115.

Young, D. (1989). *Nerve cells and animal behavior.* Cambridge: Cambridge University Press.

Zafra, F., Castrén, E., Thoenen, H., & Lindholm, D. (1991). Interplay between glutamate and gamma-aminobutyric acid transmitter systems in the physiological regulation of brain-derived neurotrophic factor and nerve growth factor synthesis in hippocampal neurons. *Proceedings of the National Academy of Sciences of the USA, 88,* 10037–10041.

Zafra, F., Lindholm, D., Castrén, E., Hartikka, J., & Thoenen, H. (1992). Regulation of brain-derived neurotrophic factor and nerve growth factor mRNA in primary cultures of hippocampal neurons and astrocytes. *Journal of Neuroscience, 12,* 4793–4799.

Zec, N., & Tieman, S. B. (1994). Development of the dendritic fields of layer 3 pyramidal cells in the kitten's visual cortex. *Journal of Comparative Neurology, 339,* 288–300.

Zuker, C. (1994). On the evolution of eyes: Would you like it simple or compound? *Science, 265,* 742–743.

4

Development of Sex Differences in the Nervous System

Nancy G. Forger

Where There Is Sex There Are Sex Differences

Before launching into a review of sexual differentiation, it seems appropriate to consider for a moment why there are two sexes. Sexual reproduction pervades the world of living organisms and yet sex itself remains a mystery. When an organism reproduces sexually only half of the population produces any offspring—a consequence that Maynard-Smith (1978) has referred to as "the cost of producing males" The sacrifice that sex entails can be appreciated by considering that a female could have twice as many grandchildren were she and her offspring to reproduce asexually. For the selfish gene, this would seem to be an enormous cost! So what maintains sex despite this major advantage of asexual reproduction?

The textbook explanation is that sexual reproduction allows for genetic variety, which enables a species to better adapt to variability in the environment. Put another way, sexual reproduction increases the rate of evolution and the chance that a given population will survive in a changing world. In fact, however, there is surprisingly little direct evidence that biparental sex is selected for in populations because it generates genetic variation, and there is some evidence to the contrary (Bell, 1982). One leading theory actually postulates that sexual reproduction is advantageous not because it allows for increased genetic variability, but because it provides a mechanism for *minimizing* the number of mutations within a population, most of which will be neutral or mildly deleterious (Kondrashov, 1988). It is not possible to do justice to any of the evolution-of-sex theories here, and several thoughtful reviews have recently been published on the subject (e.g., Hurst & Peck, 1996). For our

Nancy G. Forger   Department of Psychology and Center for Neuroendocrine Studies, University of Massachusetts, Amherst Massachusetts 01003.

Developmental Psychobiology, Volume 13 of *Handbook of Behavioral Neurobiology*, edited by Elliott Blass, Kluwer Academic / Plenum Publishers, New York, 2001.

purposes it is enough to marvel over the fact that despite the ubiquity of sex we still do not have a good explanation for it. As Margulis, Sagan, and Olendzenski. (1985) so colorfully put it in dismissing the adequacy of current theories to explain the sacrifice in reproductive potential that sex entails, "sexual reproduction is still a waste of time and energy."

Regardless of why so many species have opted for obligate sexual reproduction, the fact is that almost all vertebrates come in two sexes, and the two sexes are discernible on the basis of anatomic, physiological, and behavioral characteristics, i.e., they are sexually differentiated. Given that biparental sex appears to be here to stay, differences in anatomy, physiology, and behavior will be selected for whenever they enhance the chances of mating successfully. This, presumably, is the evolutionary motor that drives sexual differentiation. In this context, sex differences of the nervous system can be seen as evolutionary adaptations that ensure sex-appropriate behaviors.

In contrast to our fuzzy ideas about why sex (and, hence, sex differences) exist at all, we now know quite a lot about the ways in which the behavior and nervous systems of males and females differ. In this chapter I will review the process of mammalian sexual differentiation, with a focus on the current ideas and future directions regarding the cellular and molecular mechanisms underlying sex differences in the nervous system.

SEXUAL DIFFERENTIATION OF SOMATIC TISSUES

Our understanding of how neural sex differences develop grew out of observations on the sexual differentiation of somatic tissues. The fundamental mechanism of mammalian sexual differentiation was first described by Alfred Jost in the 1940s and 1950s (reviewed in Jost, 1972). Jost surgically removed the genital ridge (the tissue from which testes and ovaries are formed) in rabbit embryos and noted that the subsequent differentiation of all embryos, in terms of internal reproductive tracts and external genitalia, was female. From this observation he made two important deductions: (1) the specialization of the developing gonad into testis or ovary determines the subsequent sexual differentiation of the embryo and (2) the testes must produce substances that inhibit female development and that stimulate male development. Because the normal female phenotype resulted in the absence of any gonad, Jost also concluded that the fetal ovary is dispensable for feminine development. Thus, Jost focused attention on hormones from the testes as playing a crucial role in mammalian sexual differentiation. We now know that soon after the first signs of cellular differentiation of the fetal testes the Sertoli cells begin producing anti-Müllerian hormone (AMH; also called Müllerian inhibitory substance, MIS), and the Leydig cells make the androgenic steroid, testosterone. These two hormones then determine the subsequent differentiation of the internal reproductive tract and the external genitalia.

Prior to sexual differentiation all mammalian embryos possess two sets of ducts which are the precursors to *both* the male and female urogenital systems (Figure 1, top). The Müllerian duct has the capacity to develop into the female urogenital tract and the Wolffian duct is the precursor to male structures. In individuals with testes (i.e., most genetic males) the Müllerian ducts first regress due to the action of AMH and, subsequently, testosterone causes the Wolffian ducts to develop into the epididymis, vas deferens, and seminal vesicles. In females the Wolffian ducts degenerate due to the simple absence of testosterone, and the Müllerian ducts give rise to the fallopian tubes, uterus, and upper vagina (Wilson, George, & Griffin, 1981).

Similarly, differentiation of the external genitalia depends on hormones secreted by the fetal testes. The primordia of the external genitals are indistinguishable in the two sexes, consisting of a genital tubercle, genital folds, and genital swellings surrounding a single urogenital opening (Figure 1, bottom). In the absence of hormonal stimulation these structures form the clitoris and labia of females, whereas androgens direct these structures to form the penis and scrotum in males. In this case it is the conversion of testosterone to dihydrotestosterone (DHT), a more potent androgen, that is required for the formation of normal male genitalia (Imperato-McGinley, Guerrero, Gautier, & Peterson, 1974; Siiteri & Wilson, 1974). Although the basic process of sexual differentiation is essentially identical for all eutherian mammals, the timing of various events differs by species. For example, differentiation of the genitals and ducts occurs during the latter part of the first trimester in humans, whereas in rats these events occur during the last week or so of a 23-day gestation.

14!

DEVELOPMENT
OF SEX
DIFFERENCES IN
THE NERVOUS
SYSTEM

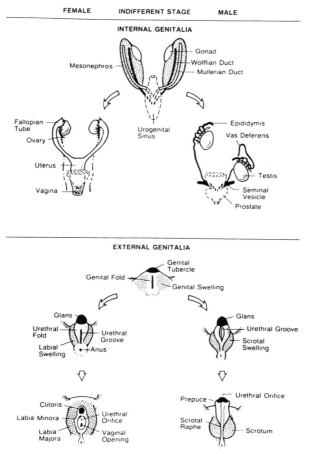

Figure 1. Differentiation of the internal and external genitalia in humans. (Reprinted from George & Wilson, 1994, with permission. Copyright 1994 Raven Press.)

THE TESTIS-DETERMINING FACTOR.   The essential role of gonadal hormones in the process of somatic sexual differentiation has been appreciated for about 50 years. But what determines whether the genital ridge will form a testis or ovary? Our understanding of the molecular mechanism underlying this process began in 1959 when, with advances in the methods of chromosomal analysis, it became clear that the Y chromosome carries genetic information required for the development of testes (Ford, Jones, Polani, de Almeida, & Briggs, 1959; Jacobs & Strong, 1959; Welshons & Russel, 1959). Prior to this realization it was assumed that, as in fruit flies, sex determination in mammals depended on the number of X chromosomes. The region of the Y-chromosome responsible for testis determination was progressively pinpointed by studying so-called XX males, mice and humans that appear to be genetically female, but are phenotypically male. In almost every case, careful scrutiny of the X chromosomes of XX males reveals a small piece of the Y chromosome that presumably has been inherited due to an abnormal crossover event during meiosis. Since XX males have testes, this Y-chromosome fragment must contain the gene for the testis determining factor. After several false starts (see Graves, 1990, for a lively discussion) a gene on the Y chromosome required for testis formation was finally identified in 1990 and was named SRY (sex-determining region of the Y chromosome; Gubbay et al., 1990; Sinclair et al., 1990). Although the prior false leads may have resulted in an unusually wary audience, the evidence now seems overwhelming that the SRY gene encodes the testis determining factor (Goodfellow & Lovell-Badge, 1993). Exactly how SRY causes testicular development is not known, but the protein product of this gene is capable of binding DNA and of directly or indirectly regulating the expression of anti-Müllerian hormone and several enzymes required for steroidogenesis (Haqq et al., 1994). Of course, the normal development of testes and ovaries will depend on the coordinated expression of many genes that are located on the autosomes and sex chromosomes. SRY may be thought of simply as the switch that turns on a genetic cascade of many essential secondary genes.

Despite the near universality of sex, it is interesting to note that the molecular mechanisms of sex determination and sexual differentiation are quite diverse. As mentioned above, sex determination in the fruit fly is controlled by the number of X chromosomes, or more specifically the ratio of X chromosomes to the non-sex chromosomes, or autosomes. The Y chromosome is present in male fruit flies but does not determine sex (Belote, 1989). Nematodes use a similar X-chromosome counting method for determining sex, but the specific genes involved are completely different (N. Williams, 1995), and yet another set of genes is invoked in mammalian sex determination. Given the strict conservation of genes that control many other aspects of development, this diversity of mechanisms is remarkable. Even among vertebrates the very fundamental process of sex determination is not conserved. Several reptilian species, including all crocodilians and some turtles, do not even have sex chromosomes, and the sex ratio of a given clutch depends on the temperature at which the eggs are incubated (Bull & Vogt, 1979; Crews et al., 1994). In this case, temperature determines the activity of genes encoding steroidogenic enzymes and steroid hormone receptors which, in turn, define the local hormonal environment and direct gonadal differentiation. The role of the SRY gene in determining testicular development apparently is a mammalian invention. Thus, the remarkable similarity of outcome in all these species (two sexes, one male and one female) is apparently the result of evolutionary convergence (Hodgkin, 1990).

The testicular hormones anti-Müllerian hormone (AMH) and testosterone have wide-ranging effects on the development of the body and brain. How do these hormones exert their effects on cells?

Anti-Müllerian Hormone

AMH is a glycoprotein and a member of the transforming growth factor-beta (TGFβ) family of growth and differentiation factors. Much of what we know about the mode of AMH action actually comes from observations on other members of the TGFβ family. TGFβ-like growth factors signal through two cell-surface receptors (type I and type II) that are serine/threonine kinases. The type II receptors bind hormone, but require the presence of the type I receptor for signaling (Massagué, 1996). Hormone binding then initiates a cascade of phosphorylations of membrane and cytoplasmic proteins that indirectly leads to changes in gene expression. Type II receptors for AMH have recently been identified and cloned (Baarends et al., 1994; di Clemente et al., 1994). Surprisingly, the Müllerian duct epithelium does not express AMH receptors during embryogenesis. Rather, AMH receptor expression is found in the mesenchyme adjacent to the Müllerian epithelium, indicating that regression of the Müllerian ducts occurs indirectly (Baarends et al., 1994). As might be expected, males that lack AMH or AMH receptors develop as pseudohermaphrodites (Behringer, Finegold, & Cate, 1994; Mishina et al., 1996). They have testes and fully differentiated Wolffian duct derivatives, but also have a uterus and oviducts. Conversely, female transgenic mice that chronically express AMH during embryogenesis are born without a uterus or oviducts (Behringer, Cate, Froelick, Palmiter, & Brinster, 1990). Although regression of the Müllerian ducts is the best-characterized action of AMH, recent reports have led to the realization that the actions of this hormone may be more widespread than previously thought (see below).

Gonadal Steroids and Their Intracellular Receptors

Steroid hormones such as testosterone (T) and estradiol (E) are lipophilic, which means that they can enter cells by passively diffusing through the lipid bilayer of the cell membrane. Once inside a cell the steroid may be converted to another hormone metabolite, depending on what enzymes are present. Testosterone, for example, can be converted into DHT by the action of the enzyme 5-α reductase, or can be metabolized to estradiol by the aromatase enzyme. The "active" steroid metabolite then binds a receptor protein (an androgen receptor for T and DHT; estrogen receptor for E) found in the cytoplasm or nucleus. Steroid hormone receptors are ligand-dependent transcription factors which, upon binding hormone, interact with specific DNA sequences to directly alter gene transcription (Figure 2) (Beato, 1989).* The expression of intracellular androgen or estrogen

---

*Although I focus here primarily on the classical mode of steroid hormone action, in which hormone binding to an intracellular receptor protein is required for gene activation, recent findings indicate that steroid receptors can sometimes be activated independent of their traditional hormonal ligands. Such ligand-independent activation of steroid receptors has been best documented for cells in culture, but recently has been demonstrated in the brain (see O'Malley et al., 1995, for review).

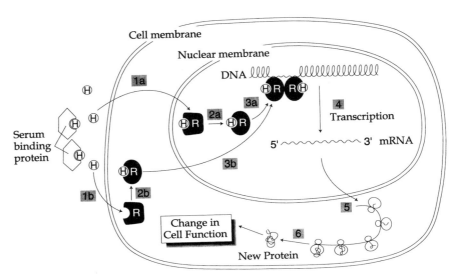

Figure 2. The "classical" mechanism of steroid hormone action. (1) The steroid hormone, Ⓗ diffuses across the plasma membrane and binds to a receptor protein located in the nucleus or (less often) in the cytoplasm. (2) Hormone binding results in receptor "activation" (this may involve a conformational change in the receptor, the shedding of associated heat shock proteins, and increased receptor phosphorylation). (3) The hormone–receptor unit dimerizes and binds to specific sites on DNA to (4) direct gene transcription. Messenger RNA is processed, exported to the cytoplasm (5), and translated into protein (6). The new protein then influences cell function.

receptors therefore identifies potential direct targets of gonadal steroid hormone action. At this point we know of only a single androgen receptor gene in mammals, whereas it has recently been discovered that there are at least two estrogen receptors which exhibit sequence homology but are encoded by separate genes (Kuiper, Enmark, Pelto-Huikko, Nilsson, & Gustafsson, 1996; Mosselman, Polman, & Dijkema, 1996).

MEMBRANE RECEPTORS FOR STEROID HORMONES

In the scenario described above, steroid hormones have effects on cells by binding to intracellular receptors and regulating gene transcription and protein synthesis. This is the "classical" mode of steroid hormone action. However, changes in protein synthesis take time, and a number of effects of steroid hormones have been reported which occur too rapidly to involve a genomic mechanism. For example, estrogens can influence the firing rates of neurons within minutes or seconds (Wong & Moss, 1991) and do so even in the presence of protein synthesis inhibitors (Nabekura, Oomura, Minami, Mizuno, & Fukuda, 1986). Such rapid effects are apparently mediated through specific membrane receptors (e.g., Pietras & Szego, 1977; Towle & Sze, 1983; for review see Ramirez & Zheng, 1996). Hormone binding to membrane receptors can change neuronal activity by altering the ionic conductance across the neuronal membrane (Nabekura et al., 1986), but rapid effects of steroid hormones need not be limited to changes in neural activity. Estradiol induces the filopodial growth of cultured hippocampal neurons within minutes of exposure

(Brinton, 1993), indicating that rapid effects (presumably mediated by membrane receptors) may also underlie some morphological actions of steroids.

149

DEVELOPMENT
OF SEX
DIFFERENCES IN
THE NERVOUS
SYSTEM

In the overwhelming majority of the discussion that follows, the tacit assumption is that effects of steroids are mediated by classical intracellular receptors. It is probably worth keeping in mind that in most cases this assumption is unproven. One way to determine whether a classical intracellular receptor is required for a given hormonal response is to examine animals with genetic mutations that interfere with the production or function of intracellular steroid hormone receptors. For example, we know that the classic androgen receptor is required for differentiation of the external genitalia because genetic males (whether mice or men) lacking functional intracellular androgen receptors develop as phenotypic females (Charest *et al.*, 1991; Griffin & Wilson, 1989). Such "androgen-insensitive" individuals possess testes and secrete testosterone, but in the absence of normal intracellular androgen receptors the tissues can not respond to the hormonal signal and female genital development results. Another way to test whether intracellular steroid receptors are involved in a given hormonal response is to administer pharmacologic agents that block receptor binding or activation. For example, flutamide is a nonsteroidal "anti-androgen" that binds the androgen receptor and interferes with androgen action. Genetically male rats that have been exposed to flutamide prenatally exhibit feminization of the genitalia and of spinal cord centers involved in copulation (Breedlove & Arnold, 1983; Nari, Florance, Koziol, & Van Cleave, 1972). For many of the neural sex differences discussed below, however, the role of classical, intracellular steroid receptors is commonly assumed but has not been directly examined.

Sex Differences in Behavior Point to the Existence of Sex Differences in the Nervous System

The Organizational Hypothesis

An important step forward in understanding brain sexual differentiation came about 40 years ago, and stemmed from observations that the very same hormones responsible for differentiating somatic tissues concurrently had permanent effects on *behavior*. For example, when a normal female guinea pig is primed with estrogens and progestins in adulthood, she will exhibit a posture indicative of sexual receptivity called lordosis. Female guinea pigs that have been exposed to testosterone prenatally, however, are much less likely than normal females to adopt the lordosis posture, and are more likely to exhibit male-typical sexual behaviors such as mounting (Phoenix, Goy, Gerall, & Young, 1959). Since all behaviors are generated by the nervous system, and since early exposure to testosterone exerts a permanent effect on behavior, one reasonable interpretation of these results is that testosterone must have a permanent, "organizing" effect on developing neural circuits. Specifically, the *organizational hypothesis* first elaborated by Phoenix *et al.* (1959) might be summarized as follows: *Testosterone, produced by the testes during a critical fetal/neonatal period, influences neural development such that the brain and behavior are permanently masculinized.* Although the initial form of this hypothesis emerged from work on guinea pigs, similar observations in other species soon followed (reviewed in Feder, 1981; Ward, 1992), and the organizational hypothesis took hold, with the laboratory rat as the favored subject of sexual differentiation studies.

The organizational hypothesis highlights a distinction that is often drawn between "organizational" and "activational" effects of gonadal steroids. Organizing effects are associated with relatively permanent changes in physiology, morphology, or behavior, and are generally thought to result from hormone action during early development. Activational effects of steroids are more transient, require the presence of the hormone, and typically occur in adulthood. The distinction between organizational and activational effects has had considerable heuristic value, although, as might be imagined, the boundaries between these two forms of steroid hormone action are in fact blurry. For example, activational (i.e., transient) effects of steroid hormone have been reported in developing animals, and gonadal steroid hormones can sometimes have enduring consequences even when exposure occurs in adulthood.

THE BRAIN AS A SITE OF STEROID HORMONE ACTION

The proposal that exposure to sex steroid hormones could have effects on the nervous system got a boost from reports in the 1960s demonstrating selective uptake of estrogens and androgens by the brain (Eisenfeld & Axelrod, 1965; Resko, Goy, & Phoenix, 1967). Although the earliest studies simply measured steroid retention in brain homogenates, the subsequent development of steroid autoradiography—and later, antibody and nucleic acid probes to steroid hormone receptors—allowed further refinements in anatomic localization (Anderson & Greenwald, 1969; Michael, 1965; Pfaff & Keiner, 1973; Sar & Parikh, 1986; Sar & Stumpf, 1975; Simerly, Chang, Muramatsu, & Swanson, 1990). We now know that androgen and estrogen receptors are widely expressed throughout the brain of adult mammals, but are found in highest concentrations within the hypothalamus, and in telencephalic regions that provide inputs to the hypothalamus including the lateral septum, amygdala, and bed nucleus of the stria terminalis (Simerly *et al.*, 1990). Importantly, neural receptors for gonadal steroids are also expressed in early development, during the period of sexual differentiation (Lieberburg, MacLusky, & McEwen, 1980; MacLusky, Lieberburg, & McEwen, 1979; Vito & Fox, 1979; Vito, Wieland, & Fox, 1979). A final crucial piece of the puzzle fell into place when investigators demonstrated that, as predicted, perinatal male mammals have higher circulating levels of androgens than do their female siblings. In the rat, for example, fetal males exhibit a surge in blood levels of testosterone on days 18 and 19 of a 23-day gestation, and a second surge on the day of birth (Baum, Woutersen, & Slob, 1991; Weisz & Ward, 1980). Thus, a mechanism is in place for differential exposure of the brain to testosterone in developing males and females.

A SEX DIFFERENCE IN BRAIN FUNCTION

The first reported "functional" sex difference in the brain was a difference in the ability of male and female rodents to generate a cyclic surge in luteinizing hormone (LH). In response to elevated estrogen levels, normal female rodents will produce a surge of LH from the pituitary, which leads to ovulation. Male rats normally do not produce an LH surge in response to estrogen, and cannot support ovulation (transplanted ovaries are supplied to the males for these studies). However, this basic sex difference can be reversed with early hormone treatments. Males castrated neonatally do produce LH surges and support ovulation in transplanted ovaries (Barraclough, 1967), whereas the positive feedback effect of estrogen is

151

DEVELOPMENT
OF SEX
DIFFERENCES IN
THE NERVOUS
SYSTEM

eliminated in females treated with testosterone during the early postnatal period (Barraclough, 1961; Barraclough & Gorski, 1961; Pfeiffer, 1936). Because LH release is controlled by the brain (e.g., Harris & Jacobsohn, 1952), these observations implied that there is a testosterone-dependent sex difference in brain connectivity and/or function that underlies the sex difference in the ability to generate an LH surge and support ovulation.

THE ROLE OF ESTROGENS

The work on sexual differentiation of the ovulatory LH surge uncovered an apparent paradox: dose for dose, estradiol benzoate is even more effective than testosterone in producing an anovulatory syndrome in rats (Gorski & Wagner, 1965). That is, when females are treated with estrogens neonatally they are masculinized and do not ovulate as adults. This is puzzling since estrogen is not a major product of the testes. The issue was partially resolved when it was demonstrated that testosterone can be aromatized to estradiol within the brain (Naftolin *et al.*, 1975), and it has since been well established that many effects of testosterone on the rat brain require the local conversion to estrogens. However, the resolution of one dilemma immediately posed another: since mammalian fetuses of *both* sexes are presumably exposed to estrogens from the placenta and the maternal circulation, why aren't female rats masculinized by fetal exposure to maternal estrogens? In rats and mice, the developing brain is apparently protected from circulating estradiol by α-fetoprotein, an estrogen-binding protein that circulates at high concentrations during the latter part of gestation (reviewed in MacLusky & Naftolin, 1981). Much of the estradiol present in fetal circulation is effectively sequestered by α-fetoprotein and therefore cannot interact with intracellular estrogen receptors. Because testosterone is not bound by α-fetoprotein, it is free to enter brain cells, where is can be locally converted to estrogens and activate estrogen receptors. In some of the experiments described below, the protective action of α-fetoprotein was bypassed by injecting enough estradiol to "swamp" the system, by injecting hormone directly into relevant brain sites, or by systemic injections of estrogen analogues, such as RU-2858 or diethylstilbestrol, that are not avidly bound by α-fetoprotein (e.g., Whalen & Etgen, 1978). The variant of α-fetoprotein found in humans does not bind much estradiol (Uriel, Bouillon, & Dupiers, 1975). It is not clear whether some other estrogen-binding protein(s) exists to "protect" human female fetuses from estrogen exposure, or whether such a mechanism is even necessary in primates. Whereas estradiol seems to be the most important hormonal metabolite for sexual differentiation of the rat brain, androgens play a more prominent role in other species such as guinea pigs and monkeys (Goy & McEwen, 1980).

EARLY REPORTS OF NEUROANATOMIC SEX DIFFERENCES

Although the existence of neural sex differences was predicted in the early 1960s on the basis of behavioral and "functional" differences between males and females, the first demonstrations of sex differences in neural morphology did not come until several years later (Pfaff, 1966; Raisman & Field, 1971, 1973). Initially, many investigators believed that if sex differences did exist in the brain, they would likely be relatively subtle and difficult to measure. Indeed, in one of the first demonstrations of a neural sex difference of any kind, Raisman and Field used electron microscopy to heroically reconstruct more than 25,000 synapses in the preoptic areas of male and female rats. Synapses were classified as being of amygdaloid or nonamygdaloid origin and as being on either dendritic spines or shafts (Raisman & Field, 1971). Although Raisman and Field were not looking for a sex difference when they undertook this study, they found that the proportion of synapses of nonamygdaloid origin making contacts directly upon dendritic shafts was significantly higher in males than in females. Suffice it to say that if all neural sex differences required the skill and patience evident in this early report, the field of neural sex differences would not have moved forward with great speed. In the late 1970s and early 1980s, however, several gross sex differences in the overall volumes of specific brain regions were identified, including differences that could be detected in sectioned and stained brain sections without any microscope at all (Gorski, Gordon, Shryne, & Southam, 1978; Nottebohm & Arnold, 1976). These reports opened the floodgates and were followed by a torrent of papers over the next 25 years reporting neural sex differences, large and small.

A Catalog of Neural Sex Differences

Sex differences have been described for a bewildering array of neural characteristics. One is left with the impression that if a given measure has not been reported as being sexually dimorphic in some neural area, then perhaps that is because no one has looked. In this section I will elaborate on some of the ways in which the brains of males and females have been reported to differ (Table 1). This is

TABLE 1. A Catalog of Sex Differences in the Nervous System

Nature of the difference	Examples[a]
Overall volume of neural areas	SDN-POA (rat); male nucleus (ferret); SDA (gerbil); HVC, RA (song bird); AVPv (rat); CA3 region of hippocampus (rate)
Neuron number	SNB (rat); HVC, RA (song bird); binocular area visual cortex (rat)
Neuronal soma size	SNB (rat); HVC, RA (song bird)
Dendritic arborization	RA (zebra finch); dentate granule cells and CA3 of hippocampus (rat)
Neural connectivity	
Number of synapsess	Binocular area of visual cortex (rat)
Types of synapses	mPOA, arcuate nucleus and SCN (rat)
Synaptic strength	Laryngeal neuromuscular junction (frog)
Neurochemistry	
Neuropeptides	Vasopressin (rat)
Neurotransmitters	Catecholamines, serotonin, dopamine (rat)
Hormone sensitivity and metabolism	
Hormone receptor expression	Estrogen (rat, vole, guinea pig), androgen (rat, guinea pig), and progesterone (rat) receptors
Steroidogenic enzymes	Aromatase (rat, quail, dove)
Glia	
Glial cell number/gliogenesis	SCN (gerbil); RA (zebra finch)
Glial morphology	Arcuate nucleus (rat)

[a]Only examples specifically discussed in the text are listed here. For most categories, many more examples exist.
Abbreviations: AVPv, anteroventral periventricular nucleus; HVC, high vocal center; mPOA, medial preoptic area; RA, robust nucleus of the archistratum; SCN, suprachiasmatic nucleus; SDA, sexually dimorphic area; SDN-POA, sexually dimorphic area of the preoptic area; SNB, spinal nucleus of the bulbocavernosus.

of necessity a selective review of what is now an enormous literature, and rather than an encyclopedic listing of neural sex differences I offer here only a few examples illustrating each point. The purpose is to come away with a sense of the types of neuroanatomic sex differences that have been found, and also to gain some familiarity with the best-studied model systems. These model systems will resurface repeatedly in the sections that follow, where cellular and molecular mechanisms underlying sex differences will be examined and where data that do not fit neatly into the accepted doctrine of neural sexual differentiation will be considered.

15?

DEVELOPMENT
OF SEX
DIFFERENCES IN
THE NERVOUS
SYSTEM

SEX DIFFERENCES IN OVERALL VOLUME, NEURONAL SIZE, AND NEURON NUMBER

As mentioned above, several of the earliest reported and most dramatic neural sex differences were differences in the volumes of specific brain areas in males and females. Not surprisingly, sex differences in overall volume generally are accounted for by differences in neuronal size, number, and/or packing density.

THE SEXUALLY DIMORPHIC NUCLEUS OF THE PREOPTIC AREA.

Roger Gorski and colleagues found that a region of the preoptic area of the hypothalamus is approximately five times larger in volume in adult male rats than in females (Gorski et al., 1978). Called the sexually dimorphic nucleus of the preoptic area (SDN-POA), this cell group stands out in Nissl-stained sections as more intensely stained and more cell-dense than surrounding areas. Adult hormone manipulations have little or no effect on SDN-POA volume, but neonatal castration reduces the size of the SDN-POA in males, and neonatal testosterone or estrogen treatments masculinize the nucleus in females (Gorski et al., 1978). In an elegant study demonstrating the role of classical estrogen receptors in this response, McCarthy, Schlenker, and Pfaff (1993) blocked the masculinizing effects of testosterone on SDN-POA volume by injecting antisense oligonucleotides to the estrogen receptor gene into the hypothalamus of neonatal rats.* Similar results are obtained by treating neonatal males with the antiestrogen tamoxifen (Döhler et al., 1984).

Sexual dimorphisms in the medial POA and adjacent anterior hypothalamus (AH) have also been described in other mammalian and nonmammalian species including gerbils, guinea pigs, ferrets, humans, lizards, and quail (Allen, Hines, Shryne, & Gorski, 1989; Byne & Bleier, 1987; Commins & Yahr, 1984a; Crews, Wade, & Wilczynski, 1980; Panzica et al., 1987; Swaab & Fliers, 1985; Tobet, Zahniser, & Baum, 1986). In each case, the nucleus is significantly larger in males than in females. Although in most cases the sex difference is a matter of degree, ferrets exhibit a true dimorphism: adult males have a densely staining "male nucleus" in the preoptic area of the hypothalamus which is completely undetectable in normal females (Tobet et al., 1986). Similarly, a small, dense cluster of cells in the POA/AH of gerbils, known as the sexually dimorphic area pars compacta (SDApc), is present in males and virtually absent in females (Figure 3) (Commins & Yahr, 1984a). As in the rat, the sex difference in the POA/AH of ferrets results from the perinatal action of estrogen formed in the brain from circulating testosterone (Tobet et al., 1986), whereas both perinatal and adult gonadal steroids apparently contribute to the sex difference in the gerbil SDA (Ulibarri & Yahr, 1988).

---

*Antisense oligonucleotides are short, single-stranded pieces of DNA which can selectively hybridize to target messenger RNA. Injection of antisense oligonucleotides to the estrogen receptor presumably prevents synthesis of new estrogen receptor protein within hypothalamic cells, and therefore blocks estrogen receptor-dependent responses.

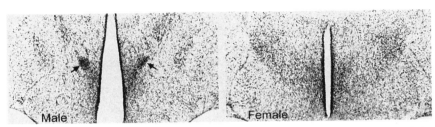

Figure 3. Photomicrographs of sections through the SDA of a male (left) and female (right) gerbil. Arrows point to the SDA pars compacta (SDApc), which is present only in males. (Adapted from Figure 1G, H in Ulibarri & Yahr, 1993 with permission. Copyright 1993 Elsevier Science.]

SONG CONTROL NUCLEI IN BIRDS.   Although this review focuses primarily on neural sex differences of mammals, sexual differentiation of the brain of song birds has contributed significantly to our understanding of the cellular mechanisms underlying neural sex differences.* Song plays an important role in courtship behavior in many species of birds and, commonly, males sing more than do females. Bird song is controlled by an identified network of brain nuclei that exhibit striking morphological sex differences. In zebra finches the behavioral dimorphism is extreme (normal females never sing), and several song control nuclei, including HVC (high vocal center) and RA (robust nucleus of the archistriatum), are much larger in volume in males than in females (Nottebohm & Arnold, 1976). These differences in overall volume are due to the fact that females have fewer, smaller, more densely packed neurons in song control nuclei than do males (Gurney, 1981; Konishi & Akutagawa, 1985; Nordeen & Nordeen, 1988). In addition, the dendritic trees of neurons within RA are more extensive in males than in females (DeVoogd & Nottebohm, 1981a). Castration of adult males or treatment of female zebra finches with gonadal steroids in adulthood results in few changes in song control nuclei or singing behavior (Arnold, 1975; Simpson & Vicario, 1991a). By contrast, treatment of hatchling female zebra finches with testosterone or estradiol masculinizes song control nuclei and, as adults, these estrogenized females can sing (Gurney, 1982; Simpson & Vicario, 1991a, b).

Canaries also exhibit large sex differences in the volumes of HVC and RA in adulthood. Unlike the song system of zebra finches, however, the size of song control nuclei in canaries is responsive to hormones throughout life. The hormone does not merely activate existing neural circuits, rather adult testosterone treatment of female canaries increases the volume of song control nuclei, induces dendritic growth, and results in a 50% increase in the number of synapses within RA (DeVoogd & Nottebohm, 1981b; DeVoogd, Nixdorf, & Nottebohm, 1985). Conversely, castration of adult male canaries results in the shrinkage of song control nuclei and the deterioration of song (Heid, Guttinger, & Prove, 1985).

THE SPINAL NUCLEUS OF THE BULBOCAVERNOSUS.   Sex differences in neuronal size and number are also seen in spinal cord centers involved in copulation. Moto-

*Those of us who study mammals often (regrettably) ignore the work on nonmammalian species. We do make exceptions when it suits our purposes, however, and within the sexual differentiation literature it is humorous to note that songbirds seem to have achieved the curious status of "honorary mammals." The details of sex determination and sexual differentiation are not identical in birds and mammals, however. As stated in the excellent review by Arnold and Schlinger (1993), "the zebra finch is not just a flying rat."

15[5]

DEVELOPMENT
OF SEX
DIFFERENCES IN
THE NERVOUS
SYSTEM

neurons of the spinal nucleus of the bulbocavernosus (SNB) are located in the lower lumbar spinal cord of rodents and innervate sexually dimorphic muscles that attach to the base of the penis. Adult male rats have three to four times as many SNB motoneurons as do females, and the soma size of the few SNB cells present in females is only about one-half as large as in males (Breedlove & Arnold, 1980). The sex difference in SNB cell number can be completely eliminated by treating females with testosterone perinatally (Nordeen, Nordeen, Sengelaub, & Arnold, 1985). In addition, SNB motoneuron number is completely feminine in androgen-insensitive (*Tfm*) male rats lacking functional androgen receptors, and in males prenatally exposed to the androgen-receptor blocker flutamide (Breedlove & Arnold, 1981, 1983). As the last two observations imply, it is androgens acting through classical androgen receptors that are crucial for the sexual dimorphism in SNB motoneuron number. A sex difference in the number of motoneurons innervating the striated penile muscles is also seen in mice, gerbils, dogs, hyenas, monkeys, and humans (Forger & Breedlove, 1986; Forger, Frank, Breedlove, & Glickman, 1996; Ueyama *et al.*, 1985; Ulibarri, Popper, & Micevych, 1995; Wagner & Clemens, 1989).

IT'S NOT JUST SEX.   Although most work on neural sex differences has focused on the hypothalamus or other neural regions directly concerned with reproduction, sex differences have also been reported in the volumes of "cognitive" regions not obviously related to sexual function. For example, overall cortical thickness is larger in male rats than in females in numerous areas (Juraska, 1991). In the binocular area of the rat visual cortex, larger cortical volume is associated with more neurons and consequently more synaptic junctions in males (Reid & Juraska, 1995).

Sex differences in spatial ability favoring males have been well documented in a number of species. At least in rodents, this sex difference is dependent on perinatal exposure to testosterone (C. L. Williams & Meck, 1991; Roof & Havens, 1992). Since the hippocampus is the brain region most clearly linked to spatial learning, investigators have looked to this structure for the neuroanatomic basis of the sex difference in performance on spatial tasks. Within the granule cell layer of the hippocampus, males of some strains of mice have a higher density of neurons than do females (Wimer & Wimer, 1985). There is also a sex difference in overall volume of the CA3 region of the rat hippocampus (male > female) which correlates with superior spatial memory performance in males. Early testosterone treatment increases total dendritic length and the total number of spines per pyramidal neuron in CA3 of rats and concomitantly enhances female spatial ability (Isgor & Sengelaub, 1996). Interestingly, there is also a sex difference in the complexity of dendritic trees of dentate granule cells of the hippocampus which depends on rearing condition. When raised under standard laboratory conditions, male rats have more extensive dendritic trees than do females, but the sex difference actually reverses when rats are reared in a "complex" environment, complete with playmates and toys (Juraska, Fitch, Henderson, & Rivers, 1985). Male rats castrated at birth show the female-like pattern of dendritic extent in each rearing condition (i.e., reduced dendritic trees when reared in standard conditions and enhanced dendritic trees in complex environments; Juraska, Kopcik, Washburne, & Perry, 1988).

Why should sex differences in "cognitive" areas of the brain exist? Although the cortex and hippocampus are not obviously related to sexual behavior per se, in order for such sex differences to have evolved they must contribute directly or indirectly to reproductive success. Given the role of the hippocampus in spatial learning, it has been proposed that a larger hippocampus in males may be selected for in species in which males must defend a large home range or must wander considerable distances from home in order to mate with multiple females (Sherry, Jacobs, & Gaulin, 1992).

Do All Sex Differences in Volume "Favor" Males? In most of the best-studied models of neural sexual differentiation males exhibit larger volumes of cell groups, bigger neurons, and/or more neurons. However, testosterone does not always mean bigger/better/more, and there are several neural regions that show the opposite pattern. The parastrial nucleus, which is found in the dorsal preoptic region just below the anterior commissure, is larger in volume in female rats than in males (del Abril, Segovia, & Guillamon, 1990), as is the anteroventral periventricular nucleus (AVPv; Bleier, Byne, & Siggelkow, 1982), which appears to control the cyclic release of gonadotrophins. In both cases, neonatal treatment of female rats with testosterone propionate (TP) causes a significant *reduction* in the volume of the nuclei (del Abril *et al.*, 1990; Sumida, Nishizuka, Kano, & Arai, 1993). Thus, regardless of the direction of the sex difference, testosterone exposure often can account for sex differences in neural morphology.

Neural Connectivity: Numbers and Types of Synapses

As mentioned above, one of the first neural sex differences reported was a difference in the types of synapses made onto neurons of the rat medial preoptic area (mPOA) (Raisman & Field, 1971). As with so many other neural sex differences, this difference in neural connectivity within the mPOA could be eliminated by treating neonatal females with testosterone (Raisman & Field, 1973). Subsequently, sex differences in synaptic input have been somewhat neglected as attention has instead been lavished on the easily measured sex differences in the volumes of various brain regions. This is unfortunate since the functional significance of volume differences is not always clear.

Although the literature in this area is relatively sparse, other examples of sex differences in synapse number and/or type include a difference in the proportion of axosomatic and axodendritic synapses in the arcuate nucleus of adult rats. Males have more axosomatic and fewer axodendritic spine synapses in the arcuate than do females, and this sexually dimorphic wiring pattern of the adult brain is dependent upon gonadal steroid exposure during the critical period of sexual differentiation (Matsumoto & Arai, 1980). There is also a sex difference in the number of asymmetric (presumably inhibitory) synapses versus symmetric (presumably excitatory) synapses in the suprachiasmatic nucleus of the adult rat (Güldner, 1982). Since the wiring pattern in a given neural area will determine electrical input, it seems reasonable to conclude that there are sex differences in neural processing within the arcuate, suprachiasmatic nucleus, and other brain regions.

Although the studies cited above emphasize the role of early exposure to gonadal hormones on the adult pattern of brain wiring, steroid hormones circulating in adulthood can also have profound and surprisingly rapid effects on synapse number or type. For example, the number of synapses on SNB motoneurons is increased in castrated rats by only 48 hr of testosterone treatment (Leedy, Beattie, & Bresnahan, 1987), and synaptic rearrangement occurs in the hippocampus over the course of each estrous cycle (Woolley, Gould, Frankfurt, & McEwen, 1990). This remarkable plasticity of the adult brain serves as a reminder that not all sex differences in neural wiring are necessarily due to developmental effects of steroid hormones.

NEUROCHEMISTRY

15'

DEVELOPMENT
OF SE>
DIFFERENCES IN
THE NERVOU!
SYSTEN

Sex differences in neuron number, dendritic extent, and synaptic connectivity are striking morphological effects of gonadal steroid hormones on the nervous systems. Equally important, however, are effects of hormones on neuronal signaling. The purpose of morphological structures such as dendrites and synapses is, after all, intercellular communication. At least one sex difference in synaptic strength has been reported, as have many sex differences in neurotransmitter and neuropeptide systems that are likely to affect the flow of information through the brain.

SYNAPTIC STRENGTH.   Synapses can vary in strength, with a "strong synapse" reliably causing large changes in postsynaptic potential and a "weak synapse" having only small, sub-threshold effects on the membrane potential of its synaptic partner. Although I am not aware of any reports of sex differences in synaptic strength in mammals, Darcy Kelley and colleagues have described a neuromuscular system in the African clawed frog, *Xenopus laevis*, which exhibits sex differences in synaptic strength. Male frogs generate courtship songs that are distinct from the vocalizations of females, and this sex difference is due to differences in the vocal control organ, the larynx, as well as in neural input to the larynx (see Kelley, 1988, for review). Although males have a much larger laryngeal muscle and more laryngeal moto-neurons innervating the muscle, the strength of laryngeal motoneuron synapses is greater in females. Quantal analyses of vocal synapses indicate that the sex difference in synaptic efficiency is presynaptic: laryngeal motoneurons of males release less neurotransmitter than do the motoneurons of females (Tobias, Kelley, & Ellisman, 1995). "Weak" synapses in the male apparently allow for amplitude-modulated trills, a sexually differentiated trait (Ruel, Kelley, & Tobias, 1998). The cause of the sex difference is not completely clear, but the female synapse increases in synaptic strength at reproductive maturity, probably in response to estrogen secretion (Tobias & Kelley, 1995).

NEUROTRANSMITTER SYSTEMS.   Sex differences have been observed in catecholaminergic, serotonergic, cholinergic, and GABAergic pathways. For example, volume and neuron number of the locus coeruleus (the major source of norepinephrine input for the central nervous system) is greater in adult female rats than in males, and testosterone treatment of females on postnatal day 1 eliminates the sex difference (Guillamon, de Blas, & Segovia, 1988). Serotonin content of various regions of the postnatal rat brain is reported to be higher in females than in males, and gonadal steroid treatments eliminate these sex differences as well (Giulian, Pohorecky, & McEwen, 1973; Ladosky & Gaziri, 1970). Throughout late prenatal development (E16–E21) female rats have higher densities of tyrosine hydroxylase (TH; the rate-limiting enzyme in dopamine production) immunoreactive axons in the striatum than do males (Ovtscharoff, Eusterschulte, Zienecker, Reisert, & Pilgrim, 1992), and dopaminergic innervation of the cortex develops earlier in postnatal females than in males (Stewart, Kuhnemann, & Rajabi, 1991). On the other hand, males have more dopamine in the median eminence and arcuate nucleus than do females (Crowley, O'Donohue, & Jacobowitz, 1978). Although the functional significance of most of these differences is not clear, a marked sex difference in the release of dopamine into the pituitary portal blood may underlie the sex difference in the control of prolactin secretion by pituicytes (Gudelsky & Porter, 1981).

NEUROPEPTIDES. Sex differences have also been reported in the expression of oxytocin, vasopressin, cholycystokinin, substance P, galanin, endogenous opiates, calcitonin gene-related peptide, and other neuropeptides (DeVries, 1990; Herbert, 1993). One of the best-studied examples of a sex difference in neuropeptide expression comes from the work of DeVries and colleagues on the vasopressinergic innervation of the rodent brain. Apart from its role as a posterior pituitary hormone controlling water balance, vasopressin made by neurons of the bed nucleus of the stria terminalis (BNST), medial amygdala (MA), and suprachiasmatic nucleus serves as a neuromodulator. Vasopressin projections from the BNST and MA are markedly sexually dimorphic (Figure 4). Males have two to three times as many vasopressinergic cell bodies in the BNST and MA as do females, and consequently, the projection of these neurons to the lateral septum and habenula is also much denser in males (DeVries, Crenshaw, & Al-Shamma, 1992; Van Leeuwen, Caffé, & De Vries,

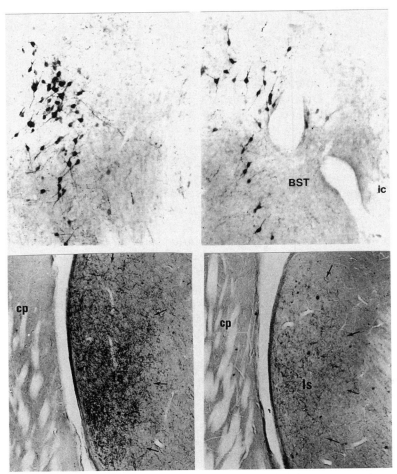

Figure 4. Immunohistochemical detection of vasopressin cell bodies in the BST (top), and projection fibers in the lateral septum (bottom) of adult rats. The vasopressin innervation in males (left) is about twice as dense as that in females (right). Abbreviations: BST, bed nucleus of the stria terminalis; ic, internal capsule; cp, caudate/putamen; ls, lateral septum. (Photomicrographs courtesy of Geert De Vries.)

1985). Castration of adult males leads to a gradual reduction in the vasopressin projection to the lateral septum (De Vries, Buijs, & Sluiter, 1984). In addition, neonatally castrated males have fewer vasopressin-immunoreactive cells in the BNST and MA than do sham-operated males, even when adult hormone levels are made equal (Wang, Bullock, & De Vries, 1993). Thus, the sex difference in vasopressin innervation of the brain is influenced by both developmental and adult exposure to gonadal steroids.

159

DEVELOPMENT
OF SEX
DIFFERENCES IN
THE NERVOUS
SYSTEM

HORMONE RECEPTORS AND STEROIDOGENESIS

Sex differences in behavior are often detected as a sex difference in adult responsiveness to gonadal steroids. Therefore, it is not surprising that sex differences in steroid binding have received considerable attention. In addition, a given dose of hormone may have different effects in males and females because the two sexes metabolize the hormone differently. For this reason sex differences in the enzymes that convert one hormone metabolite to another are of special interest.

ESTROGEN AND ANDROGEN RECEPTORS.   Male and female brains are quite similar with respect to the overall distribution of steroid hormone receptors. In rodents and primates of both sexes, the highest levels of intracellular androgen and estrogen receptors (AR and ER) are found in the lateral septum, bed nucleus of the stria terminalis, medial preoptic area, arcuate nucleus, ventromedial nucleus of the hypothalamus, and the amygdala (Handa, Reid, & Resko, 1986; Krey & McEwen, 1983; McGinnis & Katz, 1996; Sar & Parikh, 1986; Simerly *et al.*, 1990). However, using techniques as diverse as steroid autoradiography, nuclear exchange assay, immunohistochemistry and *in situ* hybridization, a number of reports do indicate regional sex differences in ER and AR levels in the rodent brain.

Female rats, voles, and guinea pigs exhibit higher levels of estrogen receptors or estrogen binding than do males, particularly within the preoptic area and ventromedial nucleus of the hypothalamus (T. J. Brown, Hochberg, Zielinski, & MacLusky, 1988; T. J. Brown, Naftolin, & MacLusky, 1992; T. J. Brown, Yu, Gagnon, Sharma, & MacLusky, 1996; Hnatczuk, Lisciotto, DonCarlos, Carter, & Morrell, 1994; Lauber, Mobbs, Muramatsu, & Pfaff, 1991; MacLusky, Bowlby, Brown, Peterson, & Hochberg, 1997). Estrogen receptor mRNA is also more abundant in the hypothalami of female rats perinatally (DonCarlos & Handa, 1994). Estradiol , derived from the local aromatization of testosterone, is apparently responsible for the downregulation of estrogen receptor expression in males: castration of male rats on the day of birth increases the level of estrogen receptor mRNA (DonCarlos, McAbee, Ramer-Quinn, & Stancik, 1995), and neonatal treatment with aromatase inhibitors similarly increases ER protein levels in males (Bakker, Pool, Sonnemans, van Leeuwen, & Slob, 1997). Because it was discovered only recently that there are (at least) two estrogen receptors (Kuiper *et al.*, 1996), most published papers do not distinguish between expression of ERα and ERβ. Future studies will no doubt address the functional significance of each estrogen receptor in nervous tissue, and will determine whether the sex differences in ER expression are due to sex differences in the abundance of ERα, ERβ, or both.

Reports concerning sex differences in brain AR levels have been somewhat conflicting. Early studies found no differences in AR binding in the brains of male and female rodents (Handa et al, 1986; Lieberburg & McEwen, 1977). However, several more recent papers have reported sex differences in androgen binding within discrete nuclei of the brains of rats and guinea pigs (Ahdieh & Feder, 1988; McGinnis & Katz, 1996; Roselli, 1991). In each case in which a sex difference is seen, androgen binding is greater in adult males than in females.

In summary, although the basic pattern of AR and ER expression is similar in male and female brains, a number of regional sex differences do exist. Females exhibit higher levels of ER and lower levels of AR in specific brain regions, and these sex differences may contribute to observed sex differences in the behavioral sensitivity to exogenously administered steroid. It should be noted, however, that receptor expression is only part of the story and cannot be equated with hormone responsiveness. For example, intracellular accessory proteins modulate the interactions of a given steroid with its receptor and of steroid–receptor complexes with DNA (Halachmi *et al.*, 1994; Lindzey, Kumar, Grossman, Young, & Tindall, 1994). Sex differences in the expression of such proteins could therefore result in marked sex differences in hormone-induced gene expression, *even in the absence of any difference in hormone or hormone receptor levels.* Steroid receptor accessory proteins, coactivators, and corepressors have been identified relatively recently, and our knowledge of how they work comes almost exclusively from studies of nonneural cells in culture. A significant challenge for the next decade will be to extend to neural cells *in vivo* the investigation of proteins that modulate the activity of steroid hormone receptors.

PROGESTERONE RECEPTORS. Christine Wagner and coworkers have recently reported a sex difference in steroid hormone receptor expression that may shift the way we think about hormone effects on brain differentiation. On the day before birth, male rats have more than 10 times as much progesterone receptor immunoreactivity in the preoptic area as do females (Figure 5) (Wagner, Nakayama, & De Vries, 1998). The magnitude of this sex difference is striking and far exceeds the more subtle sex differences in AR and ER expression described above. During late gestation, rat fetuses are exposed to high levels of progesterone, which is produced by the maternal ovary and crosses the placenta to enter fetal circulation. Because prenatal testosterone treatment of females completely eliminates the sex difference in progesterone receptor expression (C. Wagner, unpublished), one of the actions of testosterone appears to be altering the sensitivity of the developing brain to maternally derived progesterone. Exactly what progesterone does to POA neurons is not yet clear, but this result encourages a true paradigm shift in a field that has focused virtually exclusively on hormones from the *fetal gonad* as the source of sexually differentiating signals. Perhaps, as we all suspected, it is the mother's fault after all.

STEROIDOGENIC ENZYMES. The aromatase enzyme catalyzes the intracellular conversion of testosterone to estradiol. Within the brain, the highest concentrations of aromatase are found in the hypothalamus and amygdala (e.g., Roselli, Ellinwood, & Resko, 1984). Aromatase activity can be detected as early as E16 in the rat hypothalamus (George & Ojeda, 1982), an age which corresponds to the time that the fetal testes first become steroidogenic. Soon thereafter (E18) and continuing into neonatal life, a sex difference in aromatase activity emerges, with enzyme activities higher in the hypothalamus and amygdala of males than of females (MacLusky, Philip, Hurlburt, & Naftolin, 1985; Tobet, Baum, Tang, Shim, & Canick, 1985). In adulthood, males continue to have significantly higher aromatase activity than females in several brain regions. Exogenous testosterone increases aromatase activity in adult rats of both sexes, apparently by acting primarily through androgen recep-

161

DEVELOPMENT
OF SEX
DIFFERENCES IN
THE NERVOUS
SYSTEM

tors (Roselli & Resko, 1993). Thus, much of the sex difference in aromatase activity in the brains of adults may be "activational" and due to the sex difference in circulating levels of androgenic hormones. However, even when androgen levels are equalized in adult males and females, a sex difference in aromatase activity persists (Roselli & Resko, 1993). Observations in rats, quail, and doves all indicate that developmental steroid hormone exposure may also contribute to the sex differences in brain aromatase activity (Balthazart, 1989; Steimer & Hutchinson, 1990). The importance of these findings is that they imply that the brain "sees" different levels of estrogen in response to the same dose of testosterone. The sex differences in estrogen synthetic capacity of the brain may help to explain some of the many observed sex differences in hormone sensitivity of adults.

SEX DIFFERENCES IN GLIA

GLIAL CELL NUMBER AND GLIOGENESIS. Sex differences in the total number or density of glial cells have been reported in several brain areas. For example, the number of astroglia in the suprachiasmatic nucleus is significantly higher in neonatal male gerbils than in females (Collado, Beyer, Hutchison, & Holman, 1995). In one especially intriguing observation, a neural sex difference is actually preceded by a sex difference in neighboring glial cells. As mentioned above, adult male zebra finches have significantly more neurons in nucleus RA than do females, and this sex difference emerges between days 25 and 30 of life (Kim & DeVoogd, 1989; Konishi &

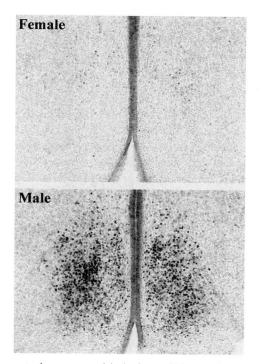

Figure 5. Progesterone receptor immunoreactivity in the medial preoptic area of male and female rats on the day of birth. (Adapted from Wagner *et al.*, 1998, with permission. Copyright 1998 The Endocrine Society.)

Akutagawa, 1985). However, before any difference in RA neuron number is evident, the rate of glial proliferation in and around RA is higher in males than in females (Nordeen & Nordeen, 1996). Astroglia are a rich source of trophic factor production (Lin, Doherty, Lile, Bektesh, & Collins, 1993; Müller, Junghans, & Kappler, 1995). Thus, the fact that immature males have more gliogenesis in the RA than females suggests the possibility that an increased availability of trophic factors is what rescues RA neurons of males (Nordeen & Nordeen, 1996). That is, a sex difference in glia may actually trigger the sexual dimorphism of neurons. Of course, this raises the question of how glia come to be sexually dimorphic in the first place. Steroid hormone receptors were identified in neurons in the 1960s. Although most of the evidence comes from *in vitro* work, several recent reports indicate that glia also express androgen and estrogen receptors and therefore may be direct sites of steroid hormone action (see Garcia-Segura, Chowen, & Naftolin, 1996 for review).

GLIAL MORPHOLOGY.   As discussed above, the ability to produce a luteinizing hormone (LH) surge is sexually dimorphic in the rat; females release a surge of LH in response to elevated estradiol levels, whereas males and androgenized females do not. The LH surge is triggered by the release of gonadotrophin-releasing hormone from nerve terminals in the arcuate nucleus and median eminence of the hypothalamus, and the plasticity of glia in the arcuate may underlie the sex difference in the ability to generate an LH surge. Prominent changes in the surface area covered by astroglia have been observed in the arcuate nucleus over the course of the estrous cycle. When estrogen is high there is an extension of astroglial processes in the arcuate nucleus of females, and these glial processes appear physically to block synapses by ensheathing arcuate nerve terminals (Olmos, Naftolin, Perez, Tranque, & Garcia-Segura, 1989). Since most of the affected synapses are inhibitory, the net effect is to reduce inhibition and this may explain the estrogen-dependent increase in the rate of arcuate neuron firing that accompanies the LH surge. These effects of estrogen on glial morphology appear to be indirect and mediated by neurons (Parducz, Szilagyi, Hoyk, Naftolin, & García-Segura, 1996; Torres-Alemán, Rejas, Pons, & Garcia-Segura, 1992). The phenomenon of cyclical changes in glial morphology has been called phasic synaptic remodeling and is seen only in adult female brains. The arcuate glia of males and of females masculinized by androgen treatments during the critical perinatal period do not respond to estrogen treatments in adulthood with similar changes in morphology (García-Segura, Chowen, Duenas, Torres-Aleman, & Naftolin, 1994). Mong, Kurzweil, Davis, Rocca, and McCarthy (1996) demonstrated that the morphology of glia in the arcuate nucleus of rats is sexually differentiated as early as postnatal day 1.

CELLULAR MECHANISMS UNDERLYING NEURAL SEX DIFFERENCES

A remarkably diverse array of differences has been described in the nervous systems of male and female vertebrates. However, a simple listing of neural sex differences cannot tell us *how* the nervous systems of males and females become different. If gonadal steroid hormones cause a particular sex difference in the brain, what are the cellular mechanisms affected by the hormones? What are the molecular mechanisms that underlie these changes at the cellular level?

To begin to address these questions, it is useful to consider the life history of a young neuron: a neuron is born (neurogenesis), moves to the proper location within the nervous system (migration), assumes a particular soma size and pattern of dendritic arborization (morphological differentiation), makes connections with other cell groups (synaptogenesis), and begins the production of characteristic neuropeptides and neurotransmitters (chemical differentiation). In addition, many neuronal populations undergo a period of cell death that usually takes place soon after contacts with synaptic targets have been established. In theory, sex differences in the nervous system could result from sex-specific modulation of any or all of these normal developmental processes. Surprisingly, as we will see below, there is firm evidence only for gonadal hormone regulation of neuronal cell death and morphological differentiation, although future work may very well reveal that other processes are involved as well.

163

DEVELOPMENT
OF SEX
DIFFERENCES IN
THE NERVOUS
SYSTEM

NEUROGENESIS

The large majority of neurons are born in proliferative zones surrounding the ventricles of the brain and the central canal of the spinal cord. These newly generated neurons then migrate out to populate all areas of the central nervous system. Regional sex differences in neuron number would result if there were sex differences in the rate of neurogenesis during specific developmental periods and in specific regions of the dividing neuroepithelium. A clear-cut example of a sex-specific program for proliferation of neuronal precursors is found in the abdominal ganglia of fruit flies, where a particular set of neuroblasts continues to divide in the male after proliferation has terminated in the female (Taylor & Truman, 1992). *At present, however, there is no unequivocal demonstration that sex differences in neurogenesis contribute to neural sex differences in vertebrates.**

As we have seen, gonadal steroid hormones can account for virtually all known neural sex differences. In order for neurogenesis to even be a logical candidate mechanism by which gonadal hormones cause a neural sex difference, there must be some overlap between the period of neurogenesis and the period of sexually dimorphic gonadal hormone secretion. The birth date of neurons in a given area can be determined by taking advantage of the fact that cells must replicate their DNA before dividing. When labeled DNA analogs such as ^{3}H-thymidine are injected in a discrete pulse during development, actively dividing cells become labeled and remain labeled throughout life. Based on such birth dating techniques, neurogenesis has been ruled out as a possible differentiating mechanism in some sexually dimorphic regions. For example, motoneuron number in the spinal nucleus of the bulbocavernosus (SNB) is androgen dependent, but the birth of SNB motoneurons is complete well before the embryonic testes begin to secrete testosterone (Breedlove, Jordan, & Arnold, 1983). In other cases (e.g., the SDN-POA of rats) the birth of neurons and gonadal hormone secretion overlap. An early study found a sex difference in the birth date of neurons comprising the mature SDN-POA. A higher proportion of SDN-POA neurons present at postnatal day 30 were born on E14 in females than in males, and a higher proportion were born on E17 in males (Jacobson & Gorski, 1981). This observation might suggest sex differences in neurogenesis, but could

*Note added: Tanapat, Hastings, Reeves, and Gould (1999) recently provided strong evidence that estradiol increases the rate of neurogenesis in the granule cell layer of the dentate gyrus of adult female rats. The significance of this finding for the establishment of neural sex differences is not clear, since the same study did not find any sex difference in either the volume of the granule cell layer or total number of granule neurons. Nonetheless, this is the best evidence to date for gonadal steroid hormone modulation of neurogenesis.

also be accounted for by differential cell death of cohorts of neurons born on different days. A subsequent study did not find evidence for a sex difference in neurogenesis among neurons contributing to the SDN-POA (Jacobson, Davis & Gorski, 1985).

Telencephalic neurons that populate vocal control regions are generated throughout life in the canary (Alvarez-Buylla & Nottebohm, 1988). The rate of incorporation of newly generated neurons into vocal control regions varies seasonally, contributing to seasonal variations in the size of brain regions controlling song (Nottebohm, 1981). Because testosterone levels also vary seasonally, it is tempting to conclude that testosterone regulates neurogenesis in the canary telencephalon. However, the number of ^{3}H-thymidine-labeled cells in the proliferative ventricular zone is not affected by treating adult female canaries with estrogens, testosterone, or inhibitors of androgen or estrogen synthesis (S. D. Brown, Johnson, & Bottjer, 1993; Rasika, Nottebohm, & Alvarez-Buylla, 1994), indicating that gonadal hormones do not alter rates of neurogenesis. Rather, seasonal changes in neuronal life span or in the recruitment of new neurons into song control circuits likely contribute to the seasonal variation in the size of song control nuclei.

Although there is no proof that gonadal hormones affect the number of neurons that are born in males and females, the timing of neurogenesis may still contribute to sex differences in the brain. Investigators have noted that neurons within several sexually dimorphic brain regions are born earlier or later than are the bulk of adjacent, nondimorphic neurons (Al-Shamma & De Vries, 1996; Jacobson & Gorski, 1981). The time of the last cellular division may play an important role in the expression of neuronal phenotype, and sexually dimorphic neurons might differ from other neurons because they are exposed to different environmental signals during a critical point in their development.

Migration

There is no direct evidence that any neural sex difference results from an effect of gonadal steroids upon neuronal migration. Having said that, however, let us step back to consider what kind of evidence would be required definitively to prove that migration contributes to a known sex difference. First, and most difficult, it would be necessary specifically to label newly generated neurons destined to populate a given sexually dimorphic nucleus. A method of tracking labeled cells over time, from their sites of origin to their destinations, would also be necessary. Because markers are not available that specifically label only presumptive sexually dimorphic neurons, the absence of evidence for steroid hormone-regulated migration may be more a reflection of technical limitations than of the underlying biology.

Stuart Tobet and colleagues have recently focused attention on the possibility that sex differences in neuronal migration contribute to the adult morphology of the preoptic area/anterior hypothalamus (POA/AH). Within the POA/AH some regions contain more neurons in males (e.g., the SDN-POA) whereas other nearby areas contain more neurons in females (e.g., AVPv). One possible explanation for these regional sex differences is that there is simply a different distribution of the pool of neurons contributing to the POA/AH in males and females. For example, neuroblasts that migrate to the SDN-POA in males might instead join the AVPv in females (Tobet & Hanna, 1997).

Because markers which identify cells destined for specific hypothalamic nuclei are not available, investigators have had to rely on less precise labeling techniques. For example, Jacobson *et al.* (1985) determined that a substantial proportion of cells contributing to the SDN-POA of the adult rat are born on embryonic day 18 (E18). By injecting ^{3}H-thymidine on E18 and sacrificing groups of rats at successively later time points after the injection, they were able to follow the migration of cells from the proliferative zone of the third ventricle into the region of the SDN-POA. No sex differences in migratory pattern or in the number of cells labeled on E18 were found (Jacobson *et al.*, 1985). One limitation of this approach, however, is that presumptive SDN-POA neurons are only a fraction of all cells dividing on E18, and many of the labeled cells will contribute to nuclei outside of the SDN-POA. Another issue is that assumptions must be made about likely migratory routes so that the relevant cells can be followed. Although the proliferative zone around the third ventricle has previously been thought to be the only birth site of POA/AH neurons, Tobet and colleagues find that neurons comprising the POA/AH actually arise from the more dorsal neuroepithelium surrounding the lateral ventricles as well as from the third ventricle proliferative zone (Tobet, Paredes, Cickering, & Baum, 1995).

A more direct approach to study hormonal influences on neuronal migration is the use of time-lapse video microscopy to follow the movements of labeled neuroblasts in an *in vitro* slice preparation (Tobet, Cickering, Hanna, Crandall, & Schwarting, 1994). Preliminary evidence using this technique indicates that there may in fact be differences in the mass movements of neurons in the brains of embryonic male and female mice. Over 2-hr sampling periods, more fluorescently labeled cells in the dorsal preoptic area moved in a medial-lateral direction in females than in males (Tobet, Henderson, Directo, & Dyer, 1997). Whether such differences contribute to sex differences in the organization of specific cell groups is not known.

How could steroid hormones influence migrating neuroblasts? At least some neurons appear to express estrogen receptors en route to the POA/AH, indicating that gonadal steroids could have a direct influence on migrating neuroblasts (Tobet, Basham, & Baum, 1993). Hormone receptors need not be expressed by the migrating neuroblasts themselves, however, in order for steroids to alter migration. For example, effects on migration could be achieved if hormones cause the release of tropic factors from resident glia or neurons, which then increase recruitment to a given area. Hormones might also affect migratory scaffolds. Migrating neuronal precursors in many regions of the brain move along the elongated processes of radial glia (Rakic, 1990). Within the mammalian POA/AH, radial glia are transiently found during development and are oriented along predicted pathways for cell migration. Radial glia express an unidentified antigen recognized by antibody AB-2, and expression of this antigen is both sexually dimorphic and altered by prenatal testosterone exposure (Tobet & Fox, 1989). Since the identity of the antigen is not known, the significance of this observation for neuronal migration is not clear, but one could imagine that sex differences in the cell surface composition of migratory guidance fibers could promote differences in the migration of neurons.

In summary, there has not been a direct demonstration that any neuroanatomic sex difference results from a sex difference in neural migration. Techniques are available for tracking migrating neuroblasts over time, but the definitive test of the migration hypothesis awaits the discovery of cellular markers that specifically identify cells destined for sexually dimorphic neural regions.

MORPHOLOGICAL DIFFERENTIATION

Morphological differentiation contributes to several known neural sex differences. For example, the overall volume of the birdsong nucleus, RA, is larger in adult male

165

DEVELOPMENT
OF SEX
DIFFERENCES IN
THE NERVOUS
SYSTEM

zebra finches and canaries than in females. This difference is due in part to the fact that individual RA neurons have larger somata and more extensive dendritic trees in males (Gurney, 1981; Nordeen & Nordeen, 1988; DeVoogd & Nottebohm, 1981b). Similarly, those few SNB motoneurons present in the spinal cords of female rats are about half the size of those in males (Breedlove & Arnold, 1980). Soma size of SNB cells is determined by circulating androgens in adulthood (Breedlove & Arnold, 1981) and also by perinatal androgen exposure (Breedlove, 1997; Breedlove & Arnold, 1983). The dendritic extent of both SNB motoneurons in rodents and RA neurons in canaries increases in response to testosterone in adulthood (DeVoogd & Nottebohm, 1981a; Forger & Breedlove, 1987; Kurz, Sengelaub, & Arnold, 1986). Thus, some sex differences in neuronal morphology are due to organizational effects of sex steroids, whereas others are activated by hormones in adulthood.

In other cases, effects on morphological differentiation seem likely simply because other mechanisms have been ruled out. For example, Area X is one of the brain areas controlling singing in birds. Area X is present in adult male zebra finches and not identifiable in Nissl-stained sections of females, yet rates of cell death in developing Area X are extremely low, and neither cell death nor cell birth seems adequate to account for the extreme sex difference observed. Kirn and DeVoogd (1989) therefore suggested that sex-specific patterns of differentiation may result in Area X being apparent in mature males, but not females. This differentiation would presumably involve some combination of morphological and/or chemical changes that make Area X stand out in Nissl-stained sections of males. Similarly, no sex differences in birth, death, or migration are observed in the developing "male nucleus" of the ferret POA/AH, leading to the proposal that differentiation (both morphological and chemical) may cause the marked sex difference in the apparent volume of this nucleus (Park, Baum, Paredes, & Tobet, 1996). Obviously, negative findings for alternative mechanisms do not provide resounding evidence in favor of sex differences in "neuronal differentiation," but this explanation seems reasonable given the data.

SYNAPTOGENESIS. On a more fine-grained level, sex differences in neural connectivity may also be considered "morphological differentiation" As discussed above, sex differences exist in the numbers and/or types of synapses in several neural regions, including the preoptic area, arcuate, suprachiasmatic nucleus, medial amygdaloid nucleus, and cortex of rats (Güldner, 1982; Matsumoto & Arai, 1976; Nishizuka & Arai, 1981; Raisman & Field, 1971; Reid & Juraska, 1995). Whether these findings are due to sex differences in synaptogenesis per se, however, is not clear. For example, if hormones influence cell survival in areas providing afferents to the POA, then a sex difference in the types and numbers of synapses in the POA could be expected, without an effect on synaptogenesis except in a trivial sense. In adulthood, steroid hormones can stimulate synaptogenesis, and can do so rapidly and without alterations in neuronal survival (e.g., Leedy, Micerych, & Arnold, 1987; Matsumoto, 1988; Woolley *et al.*, 1990).

Different numbers of synapses in a given neural area could also result from sex differences in axon outgrowth. For example, male rats have 20-fold more fibers extending from the bed nucleus of the stria terminalis (BNST) to the anteroventral periventricular nucleus of the hypothalamus (Hutton, Gu, & Simerly, 1998). This sex difference apparently results from greater axon outgrowth from BNST in males during the second postnatal week.

CHEMICAL DIFFERENTIATION

167

DEVELOPMENT
OF SEX
DIFFERENCES IN
THE NERVOUS
SYSTEM

Certainly, many hormone-dependent sex differences in the neurochemistry of adults have been described. However, *it is often difficult to determine whether the observed difference is due to an effect of gonadal hormones on the chemical differentiation of cells or on other processes.* For example, about twice as many neurons that produce the peptide vasopressin project from the bed nucleus of the stria terminalis to the lateral septum in male rats as in females (De Vries *et al.*, 1992), and this sex difference depends at least in part on exposure to perinatal testosterone in males (Wang *et al.*, 1993). One interpretation of these observations is that testosterone specifies the phenotype of vasopressinergic neurons (i.e., testosterone instructs the cell to make vasopressin). However, it is also possible that a sex difference in the number of vasopressinergic neurons is due to differential cell death (i.e., more of these neurons may die in females than in males). To decide between these alternative explanations, one would have to keep track of the number of presumptive vasopressin cells that are initially produced and that die in each sex. This would be nearly impossible unless the neurons conveniently express vasopressin prior to the time of cell death (they do not), and if they could still be identified as vasopressinergic neurons even when undergoing the morphological changes associated with cell death.

Similarly, there are more dopaminergic cell bodies in the AVPv region of the POA of female rats than of males (Simerly, Swanson, Handa, & Gorski, 1985). Perinatal testosterone treatment of females eliminates the sex difference, but, again, it is not clear whether testosterone acts by influencing cell survival or cellular differentiation. What would be required in order definitively to prove that a given sex difference found in adults is due to "chemical differentiation"? Ideally, prior to neuronal differentiation one would like to be able specifically to label all neurons or neuroblasts that will contribute to a given sexually dimorphic brain region. If the sex difference is in fact due only to chemical differentiation, then there should be no sex difference in the number of labeled neurons in developing animals or in adults. Only the neuropeptide/neurotransmitter expression of the labeled neurons would differ in adult males and females. The catch, as mentioned above, is that such specific cell markers are generally not available.

CELL DEATH

In contrast to the rather equivocal evidence in support of gonadal steroid hormone control of neurogenesis, migration, and differentiation, *a number of studies have documented hormonal regulation of cell death in sexually dimorphic neural regions.* Neuronal cell death can be directly measured by counting the number of degenerating cells based on morphological criteria (dying cells have a characteristic condensed, or "pyknotic," appearance) or by using molecular techniques to visualize the nuclear fragmentation that occurs during apoptosis. In the spinal nucleus of the bulbocavernosus, differential, hormone-regulated cell death appears to be responsible for the large sex difference in motoneuron number seen in adults. Prenatally, male and female rats have similar numbers of SNB motoneurons. Around the time of birth, however, the number declines precipitously in females, and more pyknotic cells are present in the SNB region of females than of males (Nordeen *et al.*, 1985). Testosterone treatment of perinatal females reduces the number of pyknotic cells in the SNB and results in a significant increase in the number of SNB motoneurons maintained into adulthood (Nordeen *et al.*, 1985).

Differential cell death also appears to be a major factor contributing to sex differences in several birdsong nuclei. For example, female zebra finches have about as many RA neurons as do males after the period of neurogenesis is complete (Kirn & DeVoogd, 1989; Konishi & Akutagawa, 1985). By adulthood, neuron number in the female RA has decreased by over 50% and elevated numbers of pyknotic cells can be seen in the RA of females between 4 and 6 weeks of age (Figure 6) (Kirn & DeVoogd, 1989).

The sex difference in the number of neurons of the SDN-POA of rats also appears to be due, at least in part, to hormonally regulated cell death. There are more pyknotic cells in the SDN-POA of neonatal females and castrated males than in intact males or in females treated with testosterone (Davis, Popper, & Gorski, 1996). Similarly, the number of cells undergoing nuclear fragmentation in the SDN-POA of neonatal females is reduced 24-hr after hormone treatments known to masculinize SDN-POA volume (Arai, Sekine, & Murakami, 1996). The anteroventral nucleus of the preoptic area (AVPv) is larger and more cell dense in female rats than in males

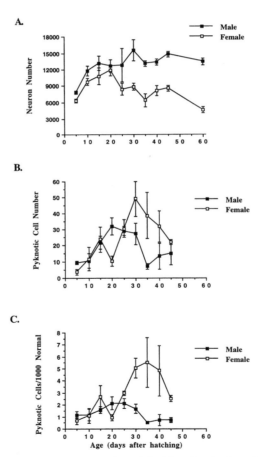

Figure 6. Developmental changes in RA of male and female zebra finches. (A) The total number of neurons. (B) The total number of pyknotic (degenerating) neurons. (C) The ratio of pyknotic cells per 1000 normal cells. (Reprinted from Kirn & DeVoogd, 1989, with permission. Copyright 1989 *The Journal of Neuroscience.*)

169

DEVELOPMENT
OF SEX
DIFFERENCES IN
THE NERVOUS
SYSTEM

(Bleier *et al.*, 1982). Prenatal or neonatal testosterone treatment of females decreases the nuclear volume of AVPv and increases the number of pyknotic cells (Murakami & Arai, 1989; Sumida *et al.*, 1993). Thus it appears that testosterone *increases* cell death in AVPv while at the same time decreasing cell death in the nearby SDN-POA and elsewhere. How could the same signal prevent cell death in one group of neurons while favoring death in another? If testosterone acts directly on AVPv and SDN-POA neurons to control their survival, then one must conclude that very different molecular cascades are turned on by testosterone in each case. Alternatively, testosterone may affect cell survival in AVPv and/or SDN-POA indirectly, for example by acting on afferents to these cell groups, on their targets, or on nearby glia. In this case, the very different effects of testosterone on cell survival may be due to intercellular signaling by trophic molecules (see below).

LOOKING WHERE THE LIGHT IS GOOD?

In summary, a critical look at the evidence favors cell death as the best-established mechanism for sculpting sex differences in the nervous system. However, the emphasis on cell death in the sex differences literature may be more an indication of what we are good at measuring than a true reflection of how the nervous system is put together. It is relatively easy to count cells in a given area and to distinguish degenerating neurons from healthy ones, and much more challenging to track sex differences in migration or chemical differentiation. Given the many different types of sex differences that have been reported, multiple underlying mechanisms seem likely. If the technical limitations discussed above can be overcome, we may find evidence for a greater diversity of cellular mechanisms contributing to the differentiation of sexually dimorphic neural areas.

MOLECULAR MECHANISMS UNDERLYING NEURAL SEX DIFFERENCES

Despite the substantial progress that has been made in identifying cellular processes, little is known about the molecular events underlying sexual differentiation of the nervous system. For starters, we rarely know the site of steroid hormone action. Investigators frequently assume that hormones act directly on the cells they are studying to change their survival, differentiation, or morphology, and this conclusion is especially tempting when those neurons express classical, intracellular steroid hormone receptors. In fact, the assumption is rarely tested and may often be incorrect. Many effects of gonadal steroids are probably mediated indirectly, and some may not require intracellular steroid receptors at all. Regardless of the site of action, it seems inevitable that changes in neurochemistry, neuronal morphology, and neuronal survival must involve changes in gene expression. Identifying hormone-regulated changes in gene expression in the developing nervous system is thus a valid way to begin dissecting the molecular cascades leading to neural sex differences.

CYTOSKELETAL PROTEINS

Several of the morphological sex differences described above require differential growth of dendrites, neurite outgrowth, and/or synaptogenesis in males and females. Axon elongation, dendritic growth, and synaptic remodeling all require alterations in the neuronal cytoskeleton. The activation of genes encoding cytoskeletal proteins is therefore one logical mechanism by which gonadal hormones may engender structural differences in the brain. Microtubules, neurofilaments, and actin are the major structural components of the neuronal cytoskeleton. Microtubules are formed by polymerization of tubulin subunits, and class II β-tubulin in particular is thought to be involved in neuronal plasticity because it is expressed in the brain and is upregulated during development and regeneration (Hoffman & Cleveland, 1988). The expression of actin and tubulin mRNAs in the POA/AH of female rats is increased by neonatal treatment with androgen (Stanley & Fink, 1986). Similarly, neonatal male rats have higher class II β-tubulin mRNA expression in the medial basal hypothalamus and preoptic area than do females (Rogers, Junier, Farmer & Ojeda, 1991). This sex difference in β-tubulin expression is site-specific and due to early hormone exposure. There is no sex difference in β-tubulin expression in the cerebral cortex or cerebellum, and the sex difference seen in the preoptic area is reduced by castrating males on the day of birth (Rogers *et al.*, 1991).

In adulthood, castration of male rats leads to a significant decrease in soma size and dendritic length of SNB motoneurons (Breedlove & Arnold, 1981; Kurz, Sengelaub, & Arnold, 1986). Androgen treatment of castrates reverses these morphological changes and concomitantly increases the expression of β-actin and β-tubulin mRNAs within SNB cells (Matsumoto, Arai, Urano, & Hyodo, 1992; Matsumoto, Arai, & Hyodo,1993). In contrast, androgen does not induce changes in the expression of β-tubulin mRNA in the neighboring retrodorsolateral nucleus, a motor pool consisting of less hormone-sensitive motoneurons which do not undergo such marked morphological changes after castration.

Taken together, these findings confirm that sex steroid-induced morphological changes in neurons are accompanied by increases in the expression of genes coding for cytoskeletal proteins. It is not clear, however, whether gonadal steroids directly or indirectly alter the transcription of these genes. Since the coordinated expression of a large number of genes is no doubt required for cell growth, steroid hormones may increase the expression of regulatory genes, which, in turn, initiate increased expression of individual cytoskeletal protein mRNAs and of other genes required for growth.

GAP-43

As noted above, there are sex differences in neuronal connectivity in several brain regions. GAP-43 (growth-associated protein 43, also known as neuromodulin) is a protein concentrated in the membranes of axonal growth cones and implicated in axonal elongation and synaptogenesis (Karns, Shi-Chung, Freeman, & Fishman, 1987). Overexpression of GAP-43 in transgenic mice leads to the formation of supernumerary synapses (Aigner *et al.*, 1995), and antibodies to GAP-43 can inhibit neurite outgrowth in neuronal cells grown in culture (Shea, Perrone-Bizzezero, Beermann, & Benowitz, 1991). Shughrue and Dorsa (1993a) demonstrated that GAP-43 levels are sexually dimorphic and regulated by testosterone in the mPOA, bed nucleus of the stria terminalis, and cerebral cortex of rats. On postnatal day 6, GAP-43 mRNA levels are high in intact males and testosterone-treated females, and low in untreated females and males castrated on day 1 (Figure 7). Although the exact function of GAP-43 is not known, testosterone regulation of GAP-43 expression may be a mechanism underlying regional sex differences in synaptic density. In addition, estrogen treatment rapidly augments GAP-43 mRNA in the arcuate and ventromedial nuclei of the basal hypothalamus of adult ovariectomized rats (Lustig, Sudol, Pfaff, & Federoff, 1991; Shughrue & Dorsa, 1993b), suggesting that changes in GAP-43 may underlie neural plasticity over the estrous cycle.

171

DEVELOPMENT
OF SEX
DIFFERENCES IN
THE NERVOUS
SYSTEM

CELL DEATH PROTEINS

In the previous section I argued that hormone regulation of neuronal cell death is the best-established cellular mechanism for generating sex differences in the nervous system. Over the past several years enormous strides have been made in

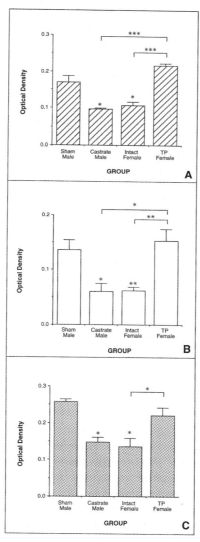

Figure 7. The level of GAP-43 mRNA on postnatal day 6 in the medial preoptic area (A), bed nucleus of the stria terminalis (B), and cerebral cortex (C) of intact male rats, males castrated on postnatal day 1, intact females, and females treated with testosterone on postnatal days 1–5. Asterisks: significantly different from intact males or groups indicated (*$p < .05$, **$p < .01$, ***$p < .001$). (Reprinted from Shughrue & Dorsa, 1993, with permission. Copyright 1993 Elsevier Science.)

understanding the molecular mechanisms of developmental cell death, and it is now clear that this process is an active one, requiring new gene expression (Oppenheim, Prevette, Tytell, & Homma, 1990). Once a cell is committed to die, an intracellular cascade of gene expression is triggered which leads to death. This cell-suicide program seems to be remarkably similar in different cell types and in very different organisms, ranging from worms to humans (e.g., Hengartner & Horvitz, 1994). Since gonadal steroid hormones regulate cell death in several regions of the nervous system, one might predict that steroids regulate the expression of cell death regulatory proteins. As far as I know, however, there are no reports directly addressing this issue in developing animals.

What are some candidate cell death regulatory proteins that might be targets for steroid hormone activation or suppression? Bcl-2 is the founding member of a family of proteins that regulates cell death in many cell types, including neurons (Davies, 1995). Bcl-2 itself inhibits cell death, although the expression of other family members, such as Bax, *increases* the rate of cell death (Merry & Korsmeyer, 1997). One leading hypothesis is that the ratio of pro-death and pro-life members of the Bcl-2 family determines whether a given cell will live or die (Oltvai, Milliman, & Korsmeyer, 1993). Consistent with this view, overexpression of Bcl-2 prevents the death of developing neurons that would otherwise die (Dubois-Dauphin, Frankowski, Tsujimoto, & Huarte, 1994; Martinou *et al.*, 1994). The "caspases" (cysteine-containing aspartate-specific proteases) are another family of proteins involved in programmed cell death. Family members cleave proteins, which eventually leads to cellular degradation (Schmitt, Sane, Steyaert, Cimoli, & Bertrand, 1997). Caspases appear to act downstream of Bcl-2 family members in the cell death cascade, and identification of the signals that activate the caspases is now an area of intense study.

It seems most parsimonious to assume that the molecular cascades leading to neuronal death in sexually dimorphic neurons is similar to that in other neurons, but that some initial step in the pathway has come under hormone control. Thus, any of the Bcl-2 or caspase family members are candidate genes for steroid regulation in sexually dimorphic neural areas. Recently, García-Segura, Cardona-Gomez, Naftolin, and Chowen (1998) reported that immunoreactivity for Bcl-2 in the arcuate nucleus of adult female rats is regulated by estradiol. Hormonal regulation of cell death proteins during development is a logical future area of research for investigators interested in understanding the molecular bases of neural sexual differentiation.

NEUROTROPHIC FACTORS

Many effects of sex steroid hormones on peripheral tissues are indirect, in the sense that hormone effects are mediated by secreted growth or trophic factors. For example, estrogens stimulate the growth of uterus and vagina, at least in part, by increasing the secretion of epidermal growth factor (Nelson, Takahashi, Bossert, Walmer, & McLachlan, 1991). An analogous mechanism operates in the androgenic regulation of Wolffian duct development (Gupta & Jaumotte, 1993). Evidence is mounting that within the nervous system many growth- or survival-promoting effects of androgens and estrogens are similarly mediated by trophic factors.

Neurotrophic factors are endogenous proteins that regulate the differentiation, survival, and phenotype of developing neurons. During the period of naturally occurring cell death, neurons are thought to compete for limited quantities of trophic factors produced by target cells, afferents, or neighboring glia (for reviews, see Davies, 1996; Korsching, 1993). By controlling the production of neurotrophic factors (or expression of trophic factor receptors), steroid hormones could indirectly influence neuronal cell death, neuronal morphology, axon outgrowth, or synaptogenesis.

173

DEVELOPMENT
OF SEX
DIFFERENCES IN
THE NERVOUS
SYSTEM

The best-studied class of neurotrophic factors is the neurotrophins, consisting of nerve growth factor (NGF), brain-derived neurotrophic factor (BDNF), neurotrophin 3 (NT-3), and NT-4/5. Toran-Allerand and colleagues have demonstrated colocalization of estrogen receptors and neurotrophin receptors in single neurons of the developing basal forebrain (Miranda, Sohrabji, & Toran-Allerand, 1993; Toran-Allerand et al., 1992). Moreover, reciprocal regulation of estrogen and neurotrophin receptors has been demonstrated in vitro: estradiol increases the expression of NGF receptors and NGF treatment increases estrogen binding (Sohrabji, Greene, Miranda, & Toran-Allerand, 1994). In vivo, ovariectomy of adult rats results in a significant decrease in basal forebrain expression of the high-affinity NGF receptor (trkA), and short-term estrogen replacement restores NGF receptor expression (Gibbs, 1998; McMillan, Singer, & Dorsa, 1996). Thus, estrogens appear to increase neurotrophin responsiveness in the basal forebrain, and such increased responsiveness to trophic support could underlie estrogenic effects on neuronal growth and survival.

More direct evidence for a role of neurotrophic factors in sexually dimorphic development comes from studies on motoneuron cell death in the rodent spinal nucleus of the bulbocavernosus. As discussed above, androgens rescue SNB motoneurons from cell death during perinatal development. However, SNB motoneurons themselves do not express androgen receptors until after the cell death period (Jordan, Padgett, Hershey, Prins, & Arnold, 1997). Rather, androgens apparently act at the bulbocavernosus/levator ani target muscles to indirectly control SNB motoneuron survival (Fishman & Breedlove, 1985; Freeman, Watson, & Breedlove, 1996). How is hormone action at the target muscles translated into a live-or-die decision by the motoneurons? We propose that androgens increase the availability of a neurotrophic molecule, which is picked up by SNB motor nerve terminals and regulates SNB cell survival. In support of this hypothesis, we have found that one neurotrophic molecule—ciliary neurotrophic factor (CNTF)—can mimic some effects of androgen in this system. When CNTF is injected into the perineal region of newborn female rats, SNB cell death is markedly reduced (Figure 8), and BC/LA muscle size is masculinized (Forger, Roberts, Wong, & Breedlove, 1993). CNTF rescues SNB motoneurons from cell death even in male Tfm rats lacking functional androgen receptors, indicating that the trophic factor does not work by merely increasing androgen production (Forger, Wong, & Breedlove, 1995). Receptors for CNTF are expressed by both the SNB motoneurons and BC/LA muscles of perinatal rats (Xu & Forger, 1998; J. Xu, unpublished). Finally, the normal sex difference in the SNB is absent in knockout mice lacking CNTF receptors: CNTF receptor knockout males have fewer than half as many SNB motoneurons as do wild-type males, and no more than do females (Figure 9) (Forger et al., 1997). Thus, androgens apparently cannot rescue SNB motoneurons in the absence of CNTF receptors. Taken together, these results are consistent with the proposal that testosterone normally engenders a sex difference in SNB motoneuron number by increasing the availability of a CNTF-like trophic factor produced by the SNB target muscles.

Direct versus Indirect Actions of Gonadal Steroids

Where do gonadal steroid hormones act to cause the many cellular and molecular changes described in the previous sections? For almost every sex difference

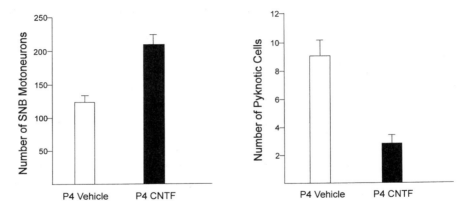

Figure 8. Left: The number of SNB motoneurons on postnatal day 4 (P4) in females treated with daily injections of ciliary neurotrophic factor (CNTF) or vehicle alone from embryonic day 22 through P3. CNTF significantly increased the number of SNB motoneurons surviving to P4 ($p < .0001$). Right: The number of pyknotic cells in vehicle- and CNTF-treated females on P4. Pyknotic profiles were 3.2 times more abundant in control than in CNTF-treated animals. (Modified from Figures 1 and 2C in Forger *et al.*, 1993, with permission. Copyright 1993 *The Journal of Neuroscience.*)

described in this chapter the answer is: we really don't know. Suppose, for example, that early exposure to testosterone causes a morphological change in the neurons of cell group A. If these neurons make estrogen or androgen receptors, then the effect may be direct. That is, the hormone may act directly on the neurons in A to alter their morphology. But maybe not. Hormones secreted from the testes or injected systemically may act at steroid receptors anywhere in the body. Steroid hormone binding at cell group B might change the activity or gene expression of B neurons, which then communicate with neurons in A to indirectly affect them.

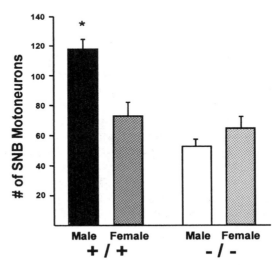

Figure 9. The number of SNB motoneurons on postnatal day 1 in wild type mice ($+/+$) and in mice lacking the gene for the CNTF α-receptor (CNTFRα $-/-$). SNB motoneuron number is sexually dimorphic in wild-type mice (*$p < .001$), but the normal sex difference is absent in CNTFRα $-/-$ mice. (Reprinted from Forger *et al.*, 1997, with permission. Copyright 1997 *The Journal of Neuroscience.*)

17!

DEVELOPMENT
OF SEX
DIFFERENCES IN
THE NERVOUS
SYSTEM

In some cases we are forced to consider indirect effects of steroid hormones because the neurons altered by hormone exposure do not themselves make the relevant hormone receptors. Effects of testosterone on SNB motoneuron number, discussed in the previous section, is one example. Similarly, estradiol treatment of hatchling zebra finches leads to a large increase in the volume of the song control nucleus, RA, even though very few neurons in RA express estrogen receptors at any time in the bird's life (Gahr & Konishi, 1988; Nordeen, Nordeen, & Arnold, 1987). This consideration has led investigators to propose that RA neurons are spared by an indirect mechanism, perhaps involving estradiol action on the afferent input to RA or on the glia surrounding RA (Nordeen & Nordeen, 1996; see section, A Catalog of Neural Sex Differences).

In other instances steroid hormone effects on neurons may be indirect, even when those neurons express classical steroid receptors. Androgen effects on the dendritic trees of adult SNB motoneurons fall into this category. As described above, the dendritic arbors of SNB motoneurons shrink after castration in adulthood, and this shrinkage can be completely prevented by treating castrates with testosterone (Kurz *et al.*, 1986). Although SNB motoneurons do not express androgen receptors during neonatal life, they do abundantly express androgen receptors in adulthood. Thus, it is tempting to conclude that androgens act directly on SNB motoneurons to alter gene expression required for dendritic growth. However, Rand and Breedlove (1995) have shown that, in fact, androgen can alter the dendritic processes of SNB motoneurons by acting at their target muscles (Figure 10). The mechanism mediating this indirect hormone effect is not known, but might involve hormone regulation of trophic factor expression from the target.

Are there instances where we know for sure that an effect of hormone on a given neural population is direct? One approach to this question is to study neurons in culture, where relatively pure populations can be observed. It was demonstrated 25 years ago that estradiol has growth-promoting effects on embryonic rat hypothalamic neurons grown in culture (Figure 11) (Matsumoto & Arai, 1976; Toran-Allerand, 1976). Estradiol also promotes the *in vitro* growth and survival of embryonic rat cortical neurons (Brinton, Tran, Proffitt, & Montoya, 1997). Since limited numbers of cell types are present in these cultures and normal synaptic connections are eliminated, these findings support a direct effect of steroid on neuronal morphology and survival. However, it is still possible that this "direct" effect is mediated by neurotrophic factors which act in autocrine or paracrine loops.

Demonstrating direct effects of steroids on neurons *in vivo* is much more difficult. The application of steroids into discrete brain areas can help to localize the site of hormone action, but even the smallest hormone implant will impinge on many cells and multiple cell types. Watson, Freeman, and Breedlove (2001) recently used genetically mosaic animals in an ingenious approach to get around this problem. The size of SNB somata and dendrites is androgen dependent in adulthood and, as just discussed, the SNB target muscles are one site of hormone action controlling dendritic extent. However, SNB *soma size* is not influenced by the local application of hormones at the target muscle (Figure 10) (Rand & Breedlove, 1995). Where, then, does androgen act to alter SNB soma size? Watson *et al.* (2001) took advantage of *Tfm* rats (androgen receptor mutants) to address this question. Because the androgen receptor gene is located on the X chromosome, random X inactivation during development renders female carriers of the *Tfm* mutation mosaics for androgen sensitivity (see Freeman *et al.*, 1996, for a further discussion of the mosaicism). Some SNB motoneurons of these females express the wild-type an-

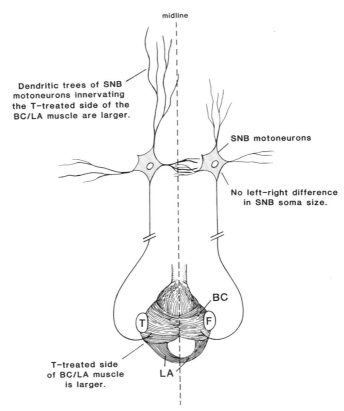

Figure 10. Schematic illustration of the experiment by Rand and Breedlove (1995). Small capsules were sutured onto the lateral aspect of the BC muscle of castrated adult male rats. One of these capsules contained crystalline testosterone (T), the other contained hydroxyflutamide (F; an antiandrogen). After 30 days, the BC muscles and SNB dendritic trees on the T side were larger than those on the F side. In contrast, SNB cell body size was not affected by the lateralized treatment. (Reprinted from Forger & Breedlove, 1991, with permission. Copyright 1991 Academic Press.)

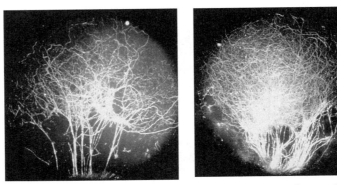

Figure 11. Effect of estradiol on neurite outgrowth in explants of the neonatal mouse hypothalamus and POA. Homologous explant pairs were cultured for 13 days under control conditions (left; normal horse serum only), or with added estradiol (right; horse serum plus 50 ng/ml estradiol). Estradiol caused a significant increase in neurite density. (Reprinted from Toran-Allerand, 1980, with permission. Copyright 1980 Elsevier Science.)

drogen receptor, whereas other SNB motoneurons sitting side by side in the same spinal cord express the mutant, nonfunctional receptor. When *Tfm* carrier females are treated with testosterone in adulthood, soma size is increased *only in those SNB motoneurons expressing functional androgen receptors*. This means that the response to testosterone is a cell-autonomous decision, and strongly supports the idea that androgen acts directly on SNB motoneurons to control soma size.

Given how difficult it is to pinpoint the direct site of hormone action *in vivo*, is the information really worth the effort? In other words, why is knowing the site of direct hormone action so important? If we are serious about understanding the molecular bases of sexual differentiation, then this knowledge is essential. We cannot hope to dissect the molecular cascades leading from steroid hormone binding to some change in the cellular or molecular characteristics of neurons unless we know *where* to start looking.

Sex Differences in the Nervous System: A Mechanism for *Generating* or for *Preventing* Sex Differences in Behavior?

The assumption of almost everyone working in the field of neural sex differences is that sex differences in the brain cause sex differences in behavior. Indeed, in the first section of this chapter I went one step further in implying that the very *purpose* of neural sex differences is to ensure sexually dimorphic behaviors that increase reproductive fitness. It is sobering to realize that despite the fact that we now have an extremely large literature documenting sex differences in the nervous system, there is little unequivocal proof relating any given neural sex difference to any particular behavioral sex difference.

One reason for this distressing lack of evidence may be the overwhelming complexity of the brain. We cannot accurately predict how a sex difference within a neural circuit will affect behavior unless we fully understand the neural underpinnings of the behavior in question. The difficulty in assigning a functional role to any neural sex difference can be appreciated by considering the well-studied sexual dimorphisms in the medial preoptic area (mPOA) discussed earlier. The discovery of the SDN-POA of rats was particularly exciting because the mPOA was an area that had previously been clearly linked to male sexual behavior. Large lesions of the mPOA abolish male sex behavior even in sexually experienced males of a wide range of vertebrate species including snakes, lizards, quail, rodents, carnivores, ungulates, and primates (reviewed in Kelley & Pfaff, 1978; Sachs & Meisel, 1988). As we saw above, rats, guinea pigs, ferrets, and gerbils all have cell groups within the mPOA that are either present in males and absent in females or much larger in males than in females. Furthermore, early hormone manipulations that increase the size of each of these sexually dimorphic areas correspondingly increase the likelihood that an animal will exhibit male sexual behavior in adulthood. One inviting conclusion from these observations is that the sexually dimorphic cell groups in the mPOA control male sexual behavior and that the size of these cell groups explains the sex difference in copulatory behaviors shown by males and females.

However, discrete lesions of the preoptic area limited specifically to the SDN-POA have no effect on copulatory behavior in experienced male rats (Arendash & Gorski, 1983), and lesions of the "male nucleus" of ferrets also have little or no effect on sexual performance measures (Cherry & Baum, 1990). Moreover, effects of hormones on behavior are not always reflected in effects on cytoarchitecture. For example, testosterone treatments of prenatal guinea pigs initiated after day 37 of a 60-day gestation masculinize mounting behavior, suppress lordosis, and inhibit the positive feedback effects of estrogen and progesterone on LH release, but have no effect on the size of the sexually dimorphic region of the medial preoptic nucleus (MPNc) (Byne & Bleier, 1987).

There are a number of important lessons here. One is that our knowledge of the behaviors controlled by a given neural area may not be complete enough to make precise predictions about the consequences of neural manipulations. For example, although lesions of the ferret "male nucleus" do not affect copulatory behavior, they do alter sexual partner preference (Paredes & Baum, 1995). SDNpc lesions in gerbils do disrupt male copulation, but also have marked effects on scent marking behavior (Commins & Yahr, 1984b). Changes in partner preference and scent marking behavior probably would not be detected by an experimenter focused only on measuring male copulatory performance. In other cases, we may be looking at the "right" behaviors in the wrong way. For example, when placed with maximally receptive females in a small, enclosed cage, male rats with SDN-POA lesions successfully mate (Arendash & Gorski, 1983). However, such a situation is unlikely to reflect the real world, where males must compete for females and where marginally receptive females have the option to run away. Indeed, when tested with females displaying low levels of receptivity or when given only limited access to estrous females, males with SDN-POA lesions do exhibit modest mating deficits compared to intact controls (De Jonge *et al.*, 1989).

A related consideration is that the mPOA, like most brain areas, is multimodal. For example, in addition to its involvement in copulation, the mPOA has been implicated in the regulation of sleep, thermoregulation, and maternal behavior (Imeri, Bianchi, Angeli, & Marcia, 1995; Lonstein, Simmons, Swann, & Stern, 1997; Numan, Corodimas, Numan, Facotr, & Piers, 1988; Thornhill & Halvorson, 1994). In general, behavioral functions have not been localized sufficiently for us to state whether there are separate, nonoverlapping populations of neurons within the mPOA dedicated to each of these functions or whether a single neural area may participate in more than one functional circuit. Therefore, a sex difference within the mPOA could contribute to any of these behaviors, and probably some that we do not even know about.

So where does this leave the theory that sexually dimorphic brains produce sexually dimorphic behaviors? I count myself among the faithful who believe that our inability to demonstrate the behavioral consequences of many of the neural sex differences we study is not so much a failure of the theory as it is a demonstration that our behavioral tests lack sophistication and that our knowledge of the neurocircuitry underlying most behaviors is still quite superficial. However, it may also be the case that not all sex differences in the nervous system cause differences in behavior. In fact, De Vries and coworkers recently argued for the view that neural sex differences sometimes *prevent* sex differences in behavior (De Vries & Boyle, 1998; De Vries & Villalba, 1997). In particular, neuroanatomic sex differences in the vasopressin innervation of the brain may allow males and females to behave similarly despite sex differences in adult levels of gonadal steroid hormones. For example, rats do not exhibit any obvious sex differences in social recognition memory (defined as the measure of an animal's ability to recognize a previously investigated conspecific). However, there appears to be a marked difference in the role that septal vasopressin has on social memory function in males and females. As mentioned above, male rats have a much denser vasopressinergic innervation of the lateral septum than do females (De Vries *et al.*, 1984). Social recognition memory in adult male rats is facilitated when vasopressin is injected directly into the lateral septum and is blocked by septal injections of a vasopressin antagonist (Dantzer, Koob, Bluthé, & Le Moal, 1988). In contrast, similar manipulations have no effect on social recognition memory of female rats (Bluthe & Dantzer, 1990). In other words, males have more vasopressin in the septum than do females, and endogenous vasopressin is essential for social recognition memory in males, but not in females. It is not clear why this should be true, but the explanation may be related to the fact that the lateral septum expresses very high levels of androgen receptor (Zhou, Blaustein, & De Vries, 1994). It is conceivable that the sexually dimorphic vasopressin innervation counteracts effects of androgen on septal neurons in males, allowing male and female rats to perform similarly on social recognition tasks (De Vries & Boyle, 1998). An analogous argument has been made for the neural control of parental behavior in male and female voles (De Vries & Villalba, 1997).

A final example of how sex differences in neural function may produce similar behavioral outcomes comes from studies of language processing in our own species. Shaywitz *et al.* (1995) used functional magnetic resonance imaging to map the brain areas activated during a language task requiring subjects to decide whether two nonsense words rhymed. Men and women performed equally well on the task, but the neural activity in the brains of men was lateralized predominantly to a language area in the left cerebral hemisphere, whereas women used both hemispheres more or less equally when performing the same task (Shaywitz *et al.*, 1995). Whether this sex difference in neural activation compensates for a sex difference in underlying neurocircuitry or simply reflects a sex difference in learned strategies is not clear. Nonetheless, as Kirkpatrick and Bryant (1995) put it, "it's worse than you thought"—sex differences in neuroanatomy may abound which go unsuspected and undetected because the behavioral outcome in males and females is similar.

Picking Away at the Dogma: Alternative Modes of Neural Differentiation?

In its original form, the organizational theory of sexual differentiation states that testosterone, produced by the fetal or neonatal testes, acts during a critical developmental period to permanently masculinize the brain. As we have seen, this idea has been remarkably influential and enjoys empirical support from many quarters. Not surprisingly, however, the dogma has had to be modified over the years to account for empirical observations. An early addendum to the organizational theory, discussed above, was that in many cases it is not testosterone per se that alters neural development. Rather, estradiol produced from local conversion of testosterone is the hormone crucial for sexual differentiation of many neural sites. Several other examples exist which indicate that effects of testosterone on neurons are not always direct. For example, the action of testosterone on SNB motoneuron number appears to be mediated by protein neurotrophic factors (Forger *et al.*, 1993, 1997). Nonetheless these are refinements, not rebuttals, of the organizational theory. A sex difference in gonadally produced testosterone is still the crucial factor, even if cellular effects are mediated by other hormone metabolites or trophic factors. Are there examples that genuinely challenge the dogma?

179

DEVELOPMENT
OF SEX
DIFFERENCES IN
THE NERVOUS
SYSTEM

SPOTTED HYENAS. The remarkable virilization of the female spotted hyena may be evidence of how nongonadal sources of steroids can contribute to sexual differentiation. The clitoris of the female spotted hyena is enlarged to essentially the size and shape of the male's penis, and the labia are fused to form a pseudo-scrotum (Figure 12) (see Glickman *et al.*, 1992, for review). There is no external vagina and female spotted hyenas urinate, copulate, and give birth through a single urogenital canal that traverses the length of the clitoris. Penile genital development has been clearly linked to developmental androgens in other mammalian species and, in fact, spotted hyena females *are* exposed to high levels of androgen *in utero*. The androgen, however, does not appear to come from the fetal gonad. Pregnant spotted hyenas produce elevated levels of the prohormone androstendione, which is converted to testosterone by enzymes present in the placenta (Yalcinkaya *et al.*, 1993). This testosterone of placental origin is then passed on to the fetus via the umbilical vein. In contrast, the human placenta normally acts as a barrier, converting maternal androgens to estrogens, which are then bound to large proteins and do not pass to the fetus (Glickman *et al.*, 1992; Yalcinkaya *et al.*, 1993).

The story is not quite so simple. Drea *et al.* (1998) treated pregnant spotted hyenas with antiandrogens throughout gestation in order to examine the effect of such treatments on the genital development of the offspring. If the unusual genital morphology of female spotted hyenas is due only to prenatal androgen exposure, then female offspring of antiandrogen-treated mothers should have "ordinary" female genitalia, with an external vagina and unfused labia. In fact, the genitalia of females exposed to antiandrogens *in utero* are feminized, in the sense that the clitoris is somewhat shorter and thicker than the clitoris of females from untreated pregnancies. However, the basic penile form of the genitalia remains even in the face of prenatal androgen blockade. This observation raises the possibility that the pseudo-penis may be genetically determined in female hyenas, with hormones playing only a modulatory role. Nonetheless, work on the spotted hyena highlights the fact that mammalian fetuses are exposed not just to steroid hormones produced by their own gonads, but to hormones of maternal/placental origin as well.

Figure 12. Erect clitoris in an adult female spotted hyena. (Reprinted from Forger *et al.*, 1996, with permission. Copyright 1996 John Wiley and Sons, Inc.)

THE BRAIN AS A SOURCE OF MASCULINIZING HORMONES. As we saw above, estrogens can masculinize the size of brain nuclei controlling birdsong. In an elegant series of experiments Schlinger and Arnold have shown that, at least for adult zebra finches, the brain itself may be the primary site of estrogen synthesis. First, high levels of the estrogen synthetic enzyme aromatase are present in cultures of zebra finch telencephalon (Schlinger & Arnold, 1992a). In addition, the zebra finch brain seems to contribute substantial quantities of estrogen to the plasma. Radiolabeled (^{3}H) androgen was injected into the systemic circulation of male birds and the amount of ^{3}H-estrogen going into the brain (carotid artery samples) and coming out of the brain (jugular vein samples) was compared. Jugular concentrations of ^{3}H-estradiol were consistently higher than in the carotid, indicating that the brain synthesizes estrogens from circulating androgens, which then enter the general circulation (Schlinger & Arnold, 1992b, 1993).

The work on placental testosterone production in hyenas and estrogen synthesis in the brains of songbirds is important in that these findings shift our attention away from the fetal gonad. However, in order for alternate sources of steroids to cause sex differences in the brain, there must be a way to differentially expose the two sexes. Aromatase levels in male and female zebra finch brain are quite similar (Wade, Schlinger, & Arnold, 1995) and sex differences in circulating androgens which might serve as a substrate for brain aromatase are not consistently found in hatchling birds. Thus, it is not clear that brain-produced estrogens can account for the sex differences in the songbird brain. Similarly, steroids of maternal origin will not cause sex differences unless there is a mechanism for delivering different levels of those hormones to male and female fetuses or unless there is differential sensitivity to maternal hormones. The work of Wagner and colleagues, mentioned above, demonstrates that progesterone receptors are expressed in a sexually dimorphic pattern in the perinatal rat brain (Figure 5) (Wagner *et al.*, 1998). This provides a mechanism for differential effects of maternal hormones in males and females, and indicates that maternal contributions to fetal sexual development may be more widespread than is currently appreciated. However, since the sex difference in progesterone responsiveness is itself dependent on testosterone (presumably secreted by the fetal testes; C. Wagner, unpublished), we find ourselves right back where we started. The dogma still stands.

A ROLE FOR GONADAL HORMONES OTHER THAN TESTOSTERONE?

One hint that gonadal substances other than testosterone may contribute to neural sex differences is that exogenous testosterone administration often only partially masculinizes neural centers of females. In South African clawed frogs, for example, males possess larger laryngeal muscles and more laryngeal motoneurons than do females, and only males produce courtship song (Kelley, 1997). Testosterone treatment of female frogs masculinizes some aspects of the song system, but laryngeal muscle fiber number, laryngeal nerve axon number, and singing behavior are not male-like in androgenized females. On the other hand, females implanted with testes *at any age* do produce the male vocal pattern and are masculinized with respect to laryngeal motor axon and muscle fiber number (Kelley, 1997; Watson, Robertson, Sachdev, & Kelley, 1993). Of course, it is always possible that exogenous hormone treatments do not perfectly mimic the amount or timing of testosterone production by the testes, and in general when investigators obtain only partial masculinization in hormone treated females it is assumed that the testosterone regimen was in some way suboptimal. It is also possible, however, that testosterone is not the only hormone secreted by the testes that can influence sexual differentiation.

In addition to testosterone, the fetal testes also produce the protein anti-Müllerian hormone. Regression of the Müllerian ducts is by far the best-characterized action of AMH, but this hormone may have additional developmental effects. For example, AMH is measurable in the blood of near-term human fetuses and AMH slows the maturation of lung tissue *in vitro* and *in vivo* (Catlin *et al.*, 1990). These findings may explain why newborn boys suffer from respiratory distress syndrome more frequently than do newborn girls. AMH might also play an autocrine/paracrine role in gonadal development. At high doses AMH directly inhibits testosterone production by Leydig cells (Rouiller-Fabre *et al.*, 1998), and male mice with overexpression of AMH have depressed circulating levels of testosterone and poorly differentiated Wolffian ducts (Lyet *et al.*, 1995). If lower, physiological, levels of AMH also influence Leydig cell activity, then AMH could influence sexual differentiation indirectly, by affecting testosterone production. In addition, because AMH is detectable in the general circulation, it is feasible that this hormone enters the brain to directly alter neural development. One way to test this hypothesis might be to examine the brains and behavior of transgenic mice lacking or overexpressing AMH and/or its receptors. To my knowledge this has not yet been done, and no effect of AMH on the nervous system has been demonstrated. The identification of a neural role for AMH or for other testicular products would not detract from the role played by the fetal testes in neural sexual differentiation, but would force testosterone to share the limelight with other hormones.

Hormone-Independent Differentiation of Nonneural Tissues

The importance of androgens in the differentiation of the penis and scrotum of mammals was stressed in the section on sexual differentiation of somatic tissues. Although this rule applies to eutherian mammals, other mechanisms operate among some marsupial species. In particular, development of the scrotum, pouch, and mammary glands seems to be under direct genetic control in the tamar wallaby (Renfree & Short, 1988). The first observation in support of this idea is that the scrotal anlagen develop in fetal males, and pouch anlagen are seen in fetal females, before the gonadal ridge has begun to differentiate into testis or ovary. Moreover, administration of testosterone during the period of genital differentiation does not induce scrotal formation in genetic female wallabies (Shaw, Renfree, Short, & O, 1988), as would be expected in eutherian mammals. Finally, observations of intersex tamar wallabies suggest that scrotal, pouch, and mammary gland formation are under the control of genes on the X chromosome (Renfree & Short, 1988; Sharman, Robinson, Walton, & Berger, 1970). A tamar wallaby that is XXY has testes and a normal male reproductive tract internally, but externally possesses a pouch, mammary glands, and no scrotum. An XO individual has ovaries, but no pouch or mammary glands.

There are also other, less well understood, examples indicating hormone-independent sexual dimorphism in mammals. For example, it was reported nearly 25 years ago that male rat embryos are heavier than female rat embryos on embryonic day 12 (Scott & Holson, 1977). A similar effect is seen in mice on day 9 of a 19-day gestation (Seller & Perkins-Cole, 1987). Since the gonads have not yet differentiated in E12 rat or E9 mouse embryos, it is hard to imagine how sex differences in

183

DEVELOPMENT
OF SEX
DIFFERENCES IN
THE NERVOUS
SYSTEM

gonadal hormone secretion could contribute to these sex differences in growth rate. More recently, these findings were confirmed and extended by two independent reports that male mouse embryos grow faster than female embryos during the preimplantation stage (Valdivia, Kuneida, Azuma, & Toyoda, 1993; Zwingman, Erickson, Boyer, & Ao, 1993). That is, sex differences in growth are observed within 24 hr of fertilization! Since this is clearly before there is a gonadal ridge, much less a gonad, gonadal hormones cannot be invoked to explain these findings. Zwingman *et al.* (1993) conclude that a growth factor is transcribed from the Y chromosome from very early on in development, and furthermore demonstrate the expression of both SRY (the testis-determining gene) and a second Y-chromosome-specific gene, Zfy, as early as the two-cell stage of mouse embryogenesis. Whether SRY or Zfy has anything to do with the sex difference in growth rate has not been demonstrated, but the fact that these genes are expressed at all supports the idea that Y-chromosome specific genes may contribute to sexually dimorphic development beginning at very early stages.

HORMONE-INDEPENDENT DIFFERENTIATION OF THE NERVOUS SYSTEM?

The previous section presents evidence for direct genetic control of sex differences in embryonic body growth. Do nonhormonal mechanisms of brain sexual differentiation also exist? Several investigators have recently argued for direct genetic control of some neural sex differences (Arnold, 1996; Diamond, Binstock, & Kohl, 1996; Reisert & Pilgrim, 1991), and there is a growing sense in the field that something new may be afoot.

SRY AND BRAIN DEVELOPMENT.   In principle, any gene on the Y chromosome that contributes to neural development could contribute to neural sex differences. Although it might seem that every chromosome must code for at least one gene important in neural development, in fact the case is questionable for the Y. The Y chromosome is the smallest chromosome, consists largely of heterochromatic DNA with highly repeated sequences, and is almost devoid of genes recognizable by inheritance markers. Thus, the pool of potential Y-linked genes involved in brain differentiation is limited. One Y-specific gene with a known function in sexual differentiation is SRY, the testis-determining gene. SRY is expressed in the developing genital ridge of the undifferentiated gonad and this expression directs testis formation. In addition, by sensitive polymerase chain reaction, mRNA transcripts for SRY recently have been detected in all tissues assayed from the fetal human male, including brain, adrenal, liver, pancreas, intestine, spleen, thymus, and heart (Cépet *et al.*, 1993), although expression in the brain was lower than that in any of the other tissues. Lahr *et al.* (1995) report the transcription of SRY in the brains of adult male mice, raising the possibility that ongoing expression of this Y-specific gene could play a differentiative role throughout life. It has not yet been demonstrated that any of these mRNA transcripts are made into functional protein, however, and a role for SRY in sexual differentiation of the nervous system remains speculative.

DEVELOPMENT OF DOPAMINERGIC NEURONS.   The strongest case for a nonhormonal contribution to brain sexual differentiation has been made by Pilgrim, Reisert, and collaborators, who have shown that sex differences in dopamine neurons may develop in the absence of sex steroids. In dissociated cell cultures from embryonic day 14 rat brain, region-specific morphological and functional sex differences are observed. Cultures from female midbrains contain more dopaminergic cells than do cultures from males, as measured by tyrosine hydroxylase immunohisto-chemistry (Reisert & Pilgrim, 1991). Regional sex differences were also found in the size of dopaminergic neurons and in the specific uptake of ^{3}H-dopamine, and were not affected by daily treatment of the cell cultures with testosterone or estradiol. The significance of these findings is that the age at which the cells were removed for culture (E14) is *before* the fetal rat testis is thought to be steroidogenic and well before the testosterone surge that occurs on E18/19. Thus, the authors argue that their findings challenge the hormonal theory of sexual differentiation. The gonads have begun to differentiate by E14, however, and in fact these investigators use the appearance of the gonad to sex their embryos. Might low levels of gonadal steroids, undetected in previous assays, contribute to the sexual differentiation of dopamine neurons before E14? This concern was addressed by treating pregnant dams with antiestrogens or antiandrogens to block activation of hormone receptors prior to removal of the embryos. Such antihormone treatments did not abolish the sex difference in tyrosine hydroxylase activity of cultured dopaminergic neurons (Beyer, Eusterschulte, Pilgrim, & Reisert, 1992), lending weight to the argument that the sex difference is independent of gonadal steroids. Of course, other gonadal products produced before E14 may be responsible for the sex differences in dopamine neurons, but this, too, would be a departure from the currently accepted dogma.

The implications of these *in vitro* findings for *in vivo* brain development and function are not clear. Reisert, Schuster, Zienecker, and Pilgrim (1990) report that there is a subtle and transient sex difference in dopaminergic cell number in the rat midbrain on E17, favoring females. The testes are steroidogenic by E17, however, which detracts somewhat from the ability of this finding to corroborate the *in vitro* work. Evidence for a sex difference *in vivo* before the onset of gonadal steroido-genesis is really needed before firm conclusions can be drawn. Nonetheless, the findings of Reisert and Pilgrim pose an interesting challenge to the view that sex differentiation of the brain is a purely epigenetic phenomenon due to the action of gonadal sex steroids, and support the alternative conclusion that some sex-linked differentiative events take place independent of the steroid environment.

DIFFERENTIATION OF SONG NUCLEI IN ZEBRA FINCHES.   One area of study that may soon produce evidence for direct genetic control of neural sexual differentia-tion is the study of differentiation of song control nuclei in birds. As we have seen, telencephalic nuclei controlling song are larger in male zebra finches than in females (Nottebohm & Arnold, 1976), and high doses of estrogen given to hatchling females can masculinize the size of song control nuclei and singing behavior (Gur-ney & Konishi, 1980; Simpson & Vicario, 1991a, b). On the surface, then, one might conclude that the scenario is quite similar to that in rodents, where testicular testosterone is converted to estrogens within the brain to masculinize brain structure and behavior. However, several findings cast serious doubt on this interpretation. For one thing, consistent sex differences in circulating steroid levels have not been demonstrated during the period of sexual differentiation of the zebra finch song system (Adkins-Regan, Abdelnabi, Mobarek, & Ottinger, 1990; Hutchinson, Wing-field, & Hutchinson, 1984; Schlinger & Arnold, 1992a). Castration of hatchling male zebra finches results in little or no change in adult singing behavior (Adkins-Regan & Ascenzi, 1990; Arnold, 1975), and posthatch treatment with antiestrogens or with aromatase inhibitors does not prevent masculine development of the song system

(Balthazart, Absil, Fiasse, & Ball, 1995; Matthews & Arnold, 1990; Wade & Arnold, 1994). One must conclude that posthatching exposure to gonadal steroid hormones is not critical for the development of song in genetic males.

To examine whether estrogens influence the development of song nuclei in males *prior to hatching*, Wade and colleagues treated embryonic zebra finches with fadrozole, a potent aromatase inhibitor (Wade & Arnold, 1996; Wade, Springer, Wingfield, & Arnold, 1996). Fadrozole treatment had little or no effect on the size of the song control nuclei in males and, if anything, fadrozole-treated males sang more than control males (Springer & Wade, 1997; Wade *et al.*, 1996). An even more puzzling result was seen in females treated with fadrozole. Because estrogen is required for normal ovarian development in birds, the gonads of fadrozole-treated female embryos developed into ovotestes. In fact, in two fadrozole-treated females only testicular tissue was present. Despite evidence that the testicular tissue was steroidogenically functional, the development of song control nuclei was not masculinized, and no fadrozole-treated females produced audible song as adults (Springer & Wade, 1997). Thus, even the presence of testes could not ensure male development in genetic females, raising the possibility that genes, not hormones, control differentiation of the song system.

If genes control sexual differentiation of song control nuclei, what are we to make of the many demonstrations that exogenous estrogen treatments can masculinize the song control nuclei of females? One possible conclusion is that estrogens do not normally play any role in the development of song control nuclei, and that it is only a coincidence that high doses of estrogen mimic normal masculinization. A more satisfying proposal that attempts to account for all the data is that there is a direct genetic effect that in some way biases the *response* to hormone. In this scenario, genetic males may require only low doses of estrogens for complete masculinization of the song control system, whereas genetic females require a very high level of estrogen stimulation for masculine development (Arnold, 1996).

Conclusions and Future Directions

Sex differences in the nervous system are remarkably numerous and diverse. It sometimes seems that almost any parameter one can think of has been found to be sexually dimorphic in one neural area or another, in one species or another. From even a cursory look at the literature, however, it is evident that although neural sex differences likely exist in all vertebrates, most of our information comes from the intense study of a relatively few model systems. In short, we know quite a lot about sexual differentiation of the rat brain and relatively little about any other species. Even in the rat, our understanding of the cellular and molecular mechanisms underlying most neural sex differences is still primitive. Cell death has enjoyed a preeminent position in the literature on the cellular mechanisms of neural sexual differentiation but, as discussed above, the emphasis may turn out to be misplaced. We simply have not yet come up with the paradigms or the tools properly to investigate other candidate mechanisms and this is an area clearly warranting further study. Our knowledge of the molecular cascades that lead to morphological changes in sexually dimorphic neural areas is extremely spotty. In most cases we do not even know the cellular site of steroid action, whether effects are mediated directly or via intercellular messengers. We know that the expression of cytoskeletal proteins and of some synapse-associated proteins is upregulated by gonadal steroids,

18[5

DEVELOPMENT
OF SEX
DIFFERENCES IN
THE NERVOUS
SYSTEM

but it is not known what other genes may be involved. It seems reasonable to expect that the next few years will see a significant expansion in this area as the techniques of molecular biology are increasingly adopted by investigators interested in neural sex differences.

Finally, as I have tried to convey, there is a growing sense among workers in the field that *in vivo* proof of direct genetic contributions to neural sex differences may be on the horizon. In general, I think this proof is awaited with excited anticipation. The overriding importance of testosterone and its metabolites in sexual differentiation of the nervous system can hardly be overstated, and there are literally hundreds of examples of neural sex differences that are dependent on gonadally produced testosterone. Only 30 years ago the theory that gonadal hormones control sexually dimorphic development of the brain and behavior was an innovative and somewhat radical idea, but today the paradigm is so well accepted that it is in danger of becoming passé. In writing the last half of this review I became painfully aware of the monotony of this single explanation for the existence of neural sex differences, as I struggled to find interesting ways to write the same basic sentence ("there is a sex difference in X, and testosterone treatment of females eliminates the difference") over and over again. We are ripe for a new paradigm. On the other hand, it is also the case that several recent observations are very difficult to square with a strictly hormonal explanation for neural sex differences. Although at this writing there is no direct *in vivo* evidence that a single sex difference anywhere in the nervous system is due to nonhormonal factors, the discovery in the near future of genetically determined sex differences or sex differences otherwise not dependent on gonadally produced testosterone seems likely. The challenge for investigators in this field will be to maintain a cool head and in a balanced way to incorporate new findings into the mountain of evidence demonstrating gonadal hormonal control of neural sexual differentiation.

Acknowledgments

I am indebted to Elliott Blass, Geert De Vries, Steve Glickman, Joe Lonstein, Juli Wade, and Christine Wagner, who offered many helpful comments on earlier versions of this manuscript. I am also grateful to Ernie Nordeen and Stuart Tobet for sharing unpublished results and updates of their research programs. This review was made possible by an Independent Scientist Award (#HD01188) from the NIH. Work reported from my own laboratory was supported by NIH grant HD33044 and a grant from the Whitehall Foundation.

REFERENCES

Adkins-Regan, E., & Ascenzi, M. (1990). Sexual differentiation of behavior in the zebra finch: Effect of early gonadectomy or androgen treatment. *Hormones and Behavior, 24,* 114–127.

Adkins-Regan, E., Abdelnabi, M., Mobarek, M., & Ottinger, M. A. (1990). Sex steroid levels in developing and adult male and female zebra finches (*Poephila guttata*). *General and Comparative Endocrinology, 78,* 93–109.

Ahdieh, H. B., & Feder, H. H. (1988). Sex differences in nucular androgen receptors in guinea pig brain and the effects of an α_2 nor adrenergic blocker on androgen receptors. *Brain Research, 456,* 275–279.

Aigner, L., Arber, S., Kapfhammer, J. P., Laux, T., Schneider, C., Botteri, F., Brenner, H.-R., & Caroni, P. (1995). Overexpression of the neural growth-associated protein GAP-43 induces nerve sprouting in the adult nervous system of transgenic mice. *Cell, 83,* 269–278.

187

DEVELOPMENT
OF SEX
DIFFERENCES IN
THE NERVOUS
SYSTEM

Allen, L. S., Hines, M., Shryne, J. E., & Gorski, R. A. (1989). Two sexually dimorphic cell groups in the human brain. *Journal of Neuroscience, 9*, 497–506.

Al-Shamma, H., & De Vries, G. J. (1996). Neurogenesis of the sexually dimorphic vasopressin cells of the bed nucleus of the stria terminalis and amygdala of rats. *Journal of Neurobiology, 29*, 91–98.

Alvarez-Buylla, A., & Nottebohm, F. (1988). Migration of young neurons in adult avian brain. *Nature, 335*, 353–354.

Anderson, C. H., & Greenwald, G. S. (1969). Autoradiographic analysis of estradiol uptake in the brain and pituitary of the female rat. *Endocrinology, 85*, 1160–1165.

Arai, Y., Sekine, Y., & Murakami, S. (1996). Estrogen and apoptosis in the developing sexually dimorphic preoptic area in female rats. *Neuroscience Research, 25*, 403–407.

Arendash, G. W., & Gorski, R. A. (1983). Effects of discrete lesions of the sexually dimorphic nucleus of the preoptic area or other medial preoptic regions on the sexual behavior of male rats. *Brain Research Bulletin, 10*, 147–154.

Arnold, A. P. (1975). The effects of castration on song development in zebra finches (*Poephila guttata*). *Journal of Experimental Zoology, 191*, 261–277.

Arnold, A. P. (1996). Genetically triggered sexual differentiation of brain and behavior. *Hormones and Behavior, 30*, 495–505.

Arnold, A. P., & Schlinger, B. A. (1993). Sexual differentiation of brain and behavior: The zebra finch is not just a flying rat. *Brain Behavior and Evolution, 42*, 231–241.

Baarends, W. M., van Helmond, M. J. L., Post, M., van der Schoot, J. C. M., Hoogerbrugghe, J. W., de Winter, J. P., Uilenbroek, J. T. J., Karels, B., Wilming, L. G., Meijers, J. H. C., Themmen, A. P. N., & Grootegoed, J. A. (1994). A novel member of the transmembrane serine/threonine kinase receptor family is specifically expressed in the gonads and in mesenchymal cells adjacent to the Müllerian duct. *Development, 120*, 189–197.

Bakker, J., Pool, C. W., Sonnemans, M., van Leeuwen, F. W., & Slob, A. K. (1997). Quantitative estimation of estrogen and androgen receptor-immunoreactive cells in the forebrain of neonatally estrogen-deprived male rats. *Neuroscience, 77*, 911–919.

Balthazart, J. (1989). Correlation between the sexually dimorphic aromatase of the preoptic area and sexual behavior in quail: Effects of neonatal manipulations of the hormonal milieu. *Archives Internationales de Physiologie de Biochimie, 97*, 465–481.

Balthazart, J., Absil, P., Fiasse, V., & Ball, G. F. (1995). Effects of the aromatase inhibitor R76713 on sexual differentiation of brain and behavior in zebra finches. *Behavior, 131*, 225–255.

Barraclough, C. A. (1961). Production of anovulatory sterile rats by single injections of testosterone propionate. *Endocrinology, 68*, 62–67.

Barraclough, C. A. (1967). Modifications in reproductive function after exposure to hormones during the prenatal and early postnatal period. In L. Martini, & W. F. Ganong (Eds.), *Neuroendocrinology* (pp. 61–99). New York: Academic Press.

Barraclough, C. A., & Gorski, R. A. (1961). Evidence that the hypothalamus is responsible for androgen-induced sterility of the female rat. *Endocrinology, 68*, 68–79.

Baum, M. J., Woutersen, P. J. A., & Slob, A. K. (1991). Sex difference in whole-body androgen content in rats on fetal days 18 and 19 without evidence that androgen passes from males to females. *Biology of Reproduction, 44*, 747–751.

Beato, M. (1989). Gene regulation by steroid hormones. *Cell, 56*, 335–344.

Behringer, R. R., Cate, R. L., Froelick, G. J., Palmiter, R. D., & Brinster, R. L. (1990). Abnormal sexual development in transgenic mice chronically expressing Müllerian inhibiting substance. *Nature, 345*, 167–170.

Behringer, R. R., Finegold, M. J., & Cate, R. L. (1994). Müllerian inhibiting substance function during mammalian sexual development. *Cell, 79*, 415–425.

Bell, G. (1982). *The masterpiece of nature.* Berkeley, CA: University of California Press.

Belote, J. M. (1989). The control of sexual development in *Drosophilia melanogaster*: Genetic and molecular analysis of a genetic regulatory hierarchy—A minireview. *Gene, 82*, 161–167.

Beyer, C., Eusterschulte, B., Pilgrim, C., & Reisert, I. (1992). Sex steroids do not alter sex differences in tyrosine hydroxylase activity of dopaminergic neurons in vitro. *Cell Tissue Research, 270*, 547–552.

Bleier, R., Byne, W., & Siggelkow, I. (1982). Cytoarchitectonic sexual dimorphisms of the medial preoptic and anterior hypothalamic areas in guinea pig, rat, hamster, and mouse. *Journal of Comparative Neurology, 212*, 118–130.

Bluthe, R., & Dantzer, R. (1990). Social recognition does not involve vasopressinergic neurotransmission in female rats. *Brain Research, 535*, 301–304.

Breedlove, S. M. (1997). Neonatal androgen and estrogen treatments masculinize the size of motoneurons in the rat spinal nucleus of the bulbocavernosus. *Cellular and Molecular Neurobiology, 17*, 687–697.

Breedlove, S. M., & Arnold, A. P. (1980). Hormone accumulation in a sexually dimorphic motor nucleus in the rat spinal cord. *Science, 210,* 564–566.

Breedlove, S. M., & Arnold, A. P. (1981). Sexually dimorphic motor nucleus in the rat lumbar spinal cord: Response to adult hormone manipulation, absence in androgen-insensitive rats. *Brain Research, 225,* 297–307.

Breedlove, S. M., & Arnold, A. P. (1983). Hormonal control of a developing neuromuscular system. I. Complete demasculinization of the male rat spinal nucleus of the bulbocavernosus using the anti-androgen flutamide. *Journal of Neuroscience, 3,* 417–423.

Breedlove, S. M., Jordan, C. L., & Arnold, A. P. (1983). Neurogenesis of motoneurons in the sexually dimorphic spinal nucleus of the bulbocavernosus in rats. *Developmental Brain Research, 9,* 39–43.

Brinton, R. D. (1993). 17β-Estradiol induction of filopodial growth in cultured hippocampal neurons within minutes of exposure. *Molecular and Cellular Neurosciences, 4,* 36–46.

Brinton, R. D., Tran, J., Proffitt, P., & Montoya, M. (1997). 17β-Estradiol enhances the outgrowth and survival of neocortical neurons in culture. *Neurochemical Research, 22,* 1339–1351.

Brown, S. D., Johnson, F., & Bottjer, S. W. (1993). Neurogenesis in adult canary telencephalon is independent of gonadal hormone levels. *Journal of Neuroscience, 13,* 2024–2032.

Brown, T. J., Hochberg, R. B., Zielinski, J. E., & MacLusky, N. J. (1988). Regional sex differences in cell nuclear estrogen-binding capacity in the rat hypothalamus and preoptic area. *Endocrinology, 123,* 1761–1770.

Brown, T. J., Naftolin, F., & MacLusky, N. J. (1992). Sex differences in estrogen receptor binding in the rat hypothalamus: Effects of subsaturating pulses of estradiol. *Brain Research, 578,* 129–134.

Brown, T. J., Yu, J., Gagnon, M., Sharma, M., & MacLusky, N. J. (1996). Sex differences in estrogen receptor and progestin receptor induction in the guinea pig hypothalamus and preoptic area. *Brain Research, 725,* 37–48.

Bull, J. J., & Vogt, R. C. (1979). Temperature-dependent sex determination in turtles. *Science, 206,* 1186–1188.

Byne, W., & Bleier, R. (1987). Medial preoptic sexual dimorphisms in the guinea pig. I. An investigation of their hormonal dependence. *Journal of Neuroscience, 7,* 2688–2696.

Catlin, E. A., Powell, S. M., Manganaro, T. F., Hudson, P. L., Ragin, R. C., Epstein, J., & Donahoe, P. K. (1990). Sex-specific fetal lung development and Müllerian inhibiting substance. *American Review of Respiratory Disease, 141,* 466–470.

Charest, N. J., Zhou, Z. X., Lubahn, D. B., Olsen, K. L., French, F. S., & Wilson, E. M. (1991). A frameshift mutation destabilizes androgen receptor messenger RNA in the TFM mouse. *Molecular Endocrinology, 5,* 573–581.

Cherry, J. A., & Baum, M. J. (1990). Effects of lesions of a sexually dimorphic nucleus in the preoptic/anterior hypothalamic area on the expression of androgen- and estrogen-dependent sexual behaviors in male ferrets. *Brain Research, 522,* 191–203.

Clépet, C., Schafer, A. J., Sinclair, A. H., Palmer, M. S., Lovell-Badge, R., & Goodfellow, P. N. (1993). The human SRY transcript. *Human Molecular Genetics, 2,* 2007–2012.

Collado, P., Beyer, C., Hutchison, J. B., & Holman, S. D. (1995). Hypothalamic distribution of astrocytes is gender-related in Mongolian gerbils. *Neuroscience Letters, 184,* 86–89.

Commins, D., & Yahr, P. (1984a). Adult testosterone levels influence the morphology of a sexually dimorphic area in the Mongolian gerbil brain. *Journal of Comparative Neurology, 224,* 132–140.

Commins, D., & Yahr, P. (1984b). Lesions of the sexually dimorphic area disrupt mating and marking in male gerbils. *Brain Research Bulletin, 13,* 185–193.

Crews, D., Wade, J., & Wilczynski, W. (1990). Sexually dimorphic areas in the brain of whiptail lizards. *Brain Behavior and Evolution, 36,* 262–270.

Crews, D., Bergeron, J. M., Bull, J. J., Flores, D., Tousignant, A., Skipper, J. K., & Wibbels, T. (1994). Temperature-dependent sex determination in reptiles: Proximate mechanisms, ultimate outcomes, and practical applications. *Developmental Genetics, 15,* 297–312.

Crowley, W. R., O'Donohue, T. L., & Jacobowitz, D. M. (1978). Sex differences in catecholamine content in discrete brain nuclei of the rat: Effects of neonatal castration or testosterone treatment. *Acta Endocrinologica, 89,* 20–28.

Dantzer, R., Koob, G. F., Bluthé, R., & Le Moal, M. (1988). Septal vasopressin modulates social memory in male rats. *Brain Research, 457,* 143–147.

Davies, A. M. (1995). The bcl-2 family of proteins, and the regulation of neuronal survival. *Trends in Neurosciences, 18,* 355–358.

Davies, A. M. (1996). Paracrine and autocrine actions of neurotrophic factors. *Neurochemical Research, 21,* 749–753.

189

DEVELOPMENT
OF SEX
DIFFERENCES IN
THE NERVOUS
SYSTEM

Davis, E. C., Popper, P., & Gorski, R. A. (1996). The role of apoptosis in sexual differentiation of the rat sexually dimorphic nucleus of the preoptic area. *Brain Research, 734,* 10–18.

De Jonge, F. H., Louwerse, A. L., Ooms, M. P., Evers, P., Endert, E., & Van de Poll, N. E. (1989). Lesions of the SDN-POA inhibit sexual behavior of male wistar rats. *Brain Research Bulletin, 23,* 483–492.

del Abril, A., Segovia, S., & Guillamon, A. (1990). Sexual dimorphism in the parastrial nucleus of the rat preoptic area. *Developmental Brain Research, 52,* 11–15.

DeVoogd, T. J., & Nottebohm, F. (1981a). Gonadal hormones induce dendritic growth in the adult avian brain. *Science, 214,* 202–204.

DeVoogd, T. J., & Nottebohm, F. (1981b). Sex differences in dendritic morphology of a song control nucleus in the canary: A quantitative golgi study. *Journal of Comparative Neurology, 196,* 309–316.

DeVoogd, T. J., Nixdorf, B., & Nottebohm, F. (1985). Synaptogenesis and changes in synaptic morphology related to aquisition of a new behavior. *Brain Research, 329,* 304–308.

De Vries, G. J. (1990). Sex differences in neurotransmitter systems. *Journal of Neuroendocrinology, 2,* 1–13.

De Vries, G. J., & Boyle, P. A. (1998). Double duty for sex differences in the brain. *Behavior in Brain Research, 92,* 205–213.

De Vries, G. J., & Villalba, C. (1997). Brain sexual dimorphism and sex differences in parental and other social behaviors. *Annals of the New York Academy of Sciences, 807,* 273–286.

DeVries, G. J., Buijs, R. M., & Sluiter, A. A. (1984). Gonadal hormone actions on the morphology of the vasopressinergic innervation of the adult rat brain. *Brain Research, 298,* 141–145.

De Vries, G. J., Crenshaw, B. J., & Al-Shamma, H. A.. (1992). Gonadal steroid modulation of vasopressin pathways. *Annals of the New York Academy of Sciences, 652,* 387–396.

Diamond, M., Binstock, T., & Kohl, J. V. (1996). From fertilization to adult sexual behavior. *Hormones and Behavior, 30,* 333–353.

di Clemente, N., Wilson, C., Faure, E., Boussin, L., Carmillo, P., Tizard, R., Picard, J.-Y., Vigier, B., Josso, N., & Cate, R. (1994). Cloning, expression, and alternative splicing of the receptor for anti-Müllerian hormone. *Molecular Endocrinology, 8,* 1006–1020.

Döhler, K. D., Srivastava, S. S., Shryne, J. E., Jarzab, B., Sipos, A., & Gorski, R. A. (1984). Differentiation of the sexually dimorphic nucleus in the preoptic area of the rat brain is inhibited by postnatal treatment with an estrogen antagonist. *Neuroendocrinology, 38,* 297–301.

DonCarlos, L. L., & Handa, R. J. (1994). Developmental profile of estrogen receptor mRNA in the preoptic area of male and female neonatal rats. *Developmental Brain Research, 79,* 283–289.

DonCarlos, L. L., McAbee, M., Ramer-Quinn, D. S., & Stancik, D. M. (1995). Estrogen receptor mRNA levels in the preoptic area of neonatal rats are responsive to hormone manipulation. *Developmental Brain Research, 84,* 253–260.

Drea, C. M., Weldele, M. L., Forger, N. G., Coscia, E. M., Frank, L. G., Licht, P., & Glickman, S. E. (1998). Androgens and masculinization of genitalia in the spotted hyaena (*Crocuta crocuta*). 2. Effects of prenatal anti-androgens. *Journal of Reproduction and Fertility, 113,* 117–127.

Dubois-Dauphin, M., Frankowski, H., Tsujimoto, Y., & Huarte, J. (1994). Neonatal motoneurons over-expressing the *bcl-2* protooncogene in transgenic mice are protected from axotomy-induced cell death. *Proceedings of the National Academy of Sciences of the USA, 91,* 3309–3313.

Eisenfeld, A. J., & Axelrod, J. (1965). Selectivity of estrogen distribution in tissues. *Journal of Pharmacology and Experimental Therapeutics, 150,* 469–475 .

Feder, H. H. (1981). Perinatal hormones and their role in the development of sexually dimorphic behaviors. In N. T. Adler (Ed.), *Neuroendocrinology of Reproduction* (pp. 158–228). New York: Plenum Press.

Fishman, R. B., & Breedlove, S. M. (1985). The androgenic induction of spinal sexual dimorphism is independent of supraspinal afferents. *Developmental Brain Research, 23,* 255–258.

Ford, C. E., Jones, K. W., Polani, P. E., de Almeida, J. C., & Briggs, J. H. (1959). A sex-chromosome anomaly in a case of gonadal dysgenesis (Turner's Syndrome). *Lancet, i,* 711–713.

Forger, N. G., & Breedlove, S. M. (1986). Sexual dimorphism in human and canine spinal cord: Role of early androgen. *Proceedings of the National Academy of Sciences of the USA, 83,* 7527–7531.

Forger, N. G., & Breedlove, S. M. (1987). Seasonal variation in mammalian striated muscle mass and motoneuron morphology. *Journal of Neurobiology, 18,* 155–165.

Forger, N. G., & Breedlove, S. M. (1991) Steroid influences on a mammalian neuromuscular system. *Seminars in the Neurosciences, 3,* 459–468.

Forger, N. G., Roberts, S. L., Wong, V., & Breedlove, S. M. (1993). Ciliary neurotrophic factor maintains motoneurons and their target muscles in developing rats. *Journal of Neuroscience, 13,* 4720–4726.

Forger, N. G., Wong, V., & Breedlove, S. M. (1995). Ciliary neurotrophic factor arrests muscle and motoneuron degeneration in androgen-insensitive rats. *Journal of Neurobiology, 28,* 354–362.

Forger, N. G., Frank, L. G., Breedlove, S. M., & Glickman, S. E. (1996). Sexual dimorphism of perineal muscles and motoneurons in spotted hyenas. *Journal of Comparative Neurology, 375*, 333–343.

Forger, N. G., Howell, M. L., Bengston, L., MacKenzie, L., DeChiara, T. M., & Yancopoulos, G. D. (1997). Sexual dimorphism in the spinal cord is absent in mice lacking the ciliary neurotrophic factor receptor. *Journal of Neuroscience, 17*, 9605–9612.

Freeman, L. M., Watson, N. V., & Breedlove, S. M. (1996). Androgen spares androgen-insensitive motoneurons from apoptosis in the spinal nucleus of the bulbocavernosus in rats. *Hormones and Behavior, 30*, 424–433.

Gahr, M., & Konishi, M. (1988). Developmental changes in estrogen-sensitive neurons in the forebrain of the zebra finch. *Proceedings of the National Academy of Sciences of the USA, 85*, 7380–7383.

Garcia-Segura, L. M., Chowen, J. A., Duenas, M., Torres-Aleman, I., & Naftolin, F. (1994). Gonadal steroids as promoters of neuroglial plasticity. *Psychoneuroendocrinology, 19*, 445–453.

Garcia-Segura, L. M., Chowen, J. A., & Naftolin, F. (1996). Endocrine glia: roles of glial cells in the brain actions of steroid and thyroid hormones and in the regulation of hormone secretion. *Frontiers in Neuroendocrinology, 17*, 180–211.

Garcia-Segura, L. M., Cardona-Gomez, P., Naftolin, F., & Chowen, J. A. (1998). Estradiol upregulates Bcl-2 expression in adult brain neurons. *NeuroReport, 9*, 593–597.

George, F. W., & Ojeda, S. R. (1982). Changes in aromatase activity in the rat brain during embryonic, neonatal, and infantile development. *Endocrinology, 111*, 522–529.

George, F. W., & Wilson, J. D. (1994). Sex determination and differentiation. In E. Knobil, & J. D. Neill (Eds.), *The physiology of reproduction* (2nd ed., pp. 3–28). New York: Raven Press.

Gibbs, R. B. (1998). Levels of trkA and BDNF mRNA, but not NGF mRNA, fluctuate across the estrous cycle and increase in response to acute hormone replacement. *Brain Research, 787*, 259–268.

Giulian, D., Pohorecky, L. A., & McEwen, B. S. (1973). Effects of gonadal steroids upon brain 5-hydroxytryptamine levels in the neonatal rat. *Endocrinology, 93*, 1329–1335.

Glickman, S. E., Frank, L. G., Licht, P., Yalcinkaya, T., Siiteri, P., & Davidson, J. (1992). Sexual differentiation of the female spotted hyena. *Annals of the New York Academy of Sciences, 662*, 135–159.

Goodfellow, P. N., & Lovell-Badge, R. (1993). Sry and sex determination in mammals. *Annual Review of Genetics, 27*, 71–92.

Gorski, R. A., & Wagner, J. W. (1965). Gonadal activity and sexual differentiation of the hypothalamus. *Endocrinology, 76*, 226–239.

Gorski, R. A., Gordon, J. H., Shryne, J. E., & Southam, A. M. (1978). Evidence for a morphological sex difference within the medial preoptic area of the rat brain. *Brain Research, 148*, 333–346.

Goy, R. W., & McEwen, B. S. (1980). *Sexual differentiation of the brain*. Cambridge, MA: MIT Press.

Graves, J. A. M. (1990). The search for the mammalian testis-determining factor is on again. *Reproduction Fertility and Development, 2*, 199–204.

Griffin, J. E., & Wilson, J. D. (1989). The androgen-resistance syndromes: 5 alpha reductase deficiency, testicular feminization, and related disorders. In C. R. Scriver, A. L. Beaudet, W. S. Sly, and D. Valle (Eds.), *The metabolic and molecular basis of inherited disease* (6th ed., pp. 1919–1944). New York: McGraw-Hill.

Gubbay, J., Collignon, J., Koopman, P., Capel, B., Economou, A., Münsterberg, A., Vivian, N., Goodfellow, P., & Lovell-Badge, R. (1990). A gene mapping to the sex-determining region of the mouse Y chromosome is a member of a novel family of embryonically expressed genes. *Nature, 346*, 245–250.

Gudelsky, G. A., & Porter, J. C. (1981). Sex-related difference in the release of dopamine into hypophysial portal blood. *Endocrinology, 109*, 1394–1398.

Guillamon, A., de Blas, M. R., & Segovia, S. (1988). Effects of sex steroids on the development of the locus coeruleus in the rat. *Developmental Brain Research, 40*, 306–310.

Güldner, F. H. (1982). Sexual dimorphisms of axo-spine synapses and postsynaptic density material in the suprachiasmatic nucleus of the rat. *Neuroscience Letters, 28*, 145–150.

Gupta, C., & Jaumotte J. (1993). Epidermal growth factor binding in the developing male reproductive duct and its regulation by testosterone. *Endocrinology, 133*, 1778–1782.

Gurney, M. (1981). Hormonal control of cell form and number in the zebra finch song system. *Journal of Neuroscience, 1*, 658–673.

Gurney, M. (1982). Behavioral correlates of sexual differentiation in the zebra finch song system. *Brain Research, 231*, 153–172.

Gurney, M., & Konishi, M. (1980). Hormone-induced sexual differentiation of brain and behavior in zebra finches. *Science, 208*, 1380–1383.

Halachmi, S., Marden, E., Martin, G., MacKay, H., Abbondanza, C., & Brown, M. (1994). Estrogen receptor-associated proteins: Possible mediators of hormone-induced transcription. *Science, 264*, 1455–1458.

Handa, R. J., Reid, D. L., & Resko, J. A. (1986). Androgen receptors in brain and pituitary of female rats: Cyclic changes and comparisons with the male. *Biology of Reproduction, 34*, 293–303.

Haqq, C. M., King, C., Ukiyama, E., Falsafi, S., Haqq, T. N., Donahoe, P. K., & Weiss, M. A. (1994). Molecular basis of mammalian sexual differentiation: Activation of Müllerian inhibiting substance gene expression by SRY. *Science, 266*, 1494–1500.

Harris, G. W., & Jacobsohn, D. (1952). Functional grafts of the anterior pituitary gland. *Proceedings of the Royal Society of London, Series B. Biological Sciences, 130*, 263–276.

Heid, P., Guttinger, H. R., & Prove, E. (1985). The influence of castration and testosterone replacement on the song architecture of canaries (*Serinus canaria*). *Zeitschrift für Tierpsychologie, 69*, 224–236.

Hengartner, M. O., & Horvitz, H. R. (1994). *C. elegans* cell survival gene *ced-9* encodes a functional homolog of the mammalian proto-oncogene *bcl-2*. *Cell, 76*, 665–676.

Herbert, J. (1993). Peptides in the limbic system. Neurochemical codes for coordinated adaptive responses to behavioral and physiological demand. *Progress in Neurobiology, 41*, 723–791.

Hnatczuk, O. C., Lisciotto, C. A., DonCarlos, L. L., Carter, S., & Morrell, J. I. (1994). Estrogen receptor immunoreactivity in specific brain areas of the prairie vole (*Microtus ochrogaster*) is altered by sexual receptivity and genetic sex. *Journal of Endrocrinology, 6*, 89–100.

Hodgkin, J. (1990). Sex determination compared in *Drosophila* and *Caenorhabditis*. *Nature, 344*, 721–728.

Hoffman, P. N., & Cleveland, D. W. (1988). Neurofilament and tubulin expression recapitulates the developmental program during axonal regeneration: Induction of a specific β-tubulin isotype. *Proceedings of the National Academy of Sciences of the USA, 85*, 4530–4533.

Hurst, L. D., & Peck, J. R. (1996). Recent advances in understanding of the evolution and maintenance of sex. *TREE, 11*, 46–52.

Hutchison, J. B., Wingfield, J. C., & Hutchison, R. E. (1984). Sex differences in plasma concentrations of steroids during the sensitive period for brain differentiation in the zebra finch. *Journal of Endocrinology, 103*, 363–369.

Hutton, L. A., Gu, G., & Simerly, R. B. (1998) Development of a sexually dimorphic projection from the bed nucleus of the stria terminalis to the anteroventral periventricular nucleus in the rat. *Journal of Neuroscience, 18*, 3003–3013.

Imeri, L., Bianchi, S., Angeli, P., & Mancia, M. (1995). Stimulation of cholinergic receptors in the medial preoptic area affects sleep and cortical temperature. *American Journal of Physiology, 269*, R294–R299.

Imperato-McGinley, J., Guerrero, L., Gautier, T., & Peterson, R. E. (1974). Steroid 5 alpha-reductase deficiency in man: An inherited form of male pseudohermaphroditism. *Science, 186*, 1213–1215.

Isgor, C., & Sengelaub, D. R. (1996). Neonatal androgens affect adult spatial behavior and CA3 pyramidal cell morphology in rats: A Golgi study. *Society for Neuroscience Abstracts, 22*, 756.

Jacobs, P. A., & Strong, J. A. (1959). A case of human intersexuality having a possible XXY sex-determining mechanism. *Nature, 183*, 302–303.

Jacobson, C. D., & Gorski, R. A. (1981). Neurogenesis of the sexually dimorphic nucleus of the preoptic area in the rat. *Journal of Comparative Neurology, 196*, 519–529.

Jacobson, C. D., Davis, F. C., & Gorski, R. A. (1985). Formation of the sexually dimorphic nucleus of the preoptic area: Neuronal growth, migration and changes in cell number. *Developmental Brain Research, 21*, 7–18.

Jordan, C. L., Padgett, B., Hershey, J., Prins, G., & Arnold, A. (1997). Ontogeny of androgen receptor immunoreactivity in lumbar motoneurons and in the sexually dimorphic levator ani muscle of male rats. *Journal of Comparative Neurology, 379*, 88–98.

Jost, A. (1972). A new look at the mechanisms controlling sex differentiation in mammals. *Johns Hopkins Medical Journal, 130*, 38–52.

Juraska, J. M. (1991). Sex differences in "cognitive" regions of the rat brain. *Psychoneuroendocrinology, 16*, 105–119.

Juraska, J. M., Fitch, J., Henderson, C., & Rivers, N. (1985). Sex differences in the dendritic branching of dentate granule cells following differential experience. *Brain Research, 333*, 73–80.

Juraska, J. M., Kopcik, J. R., Washburne, D. L., & Perry, D. L. (1988). Neonatal castration of male rats affects the dendritic response to differential environments in hippocampal dentate granule neurons. *Psychobiology, 16*, 406–410.

Karns, L. R., Shi-Chung, N., Freeman, J. A., & Fishman, M. C. (1987). Cloning of complementary DNA for GAP-43, a neuronal growth-related protein. *Science, 236*, 597–600.

Kelley, D. B. (1988). Sexually dimorphic behaviors. *Annual Review of Neuroscience, 11*, 225–251.

Kelley, D. B. (1997). Generating sexually differentiated songs. *Current Opinion in Neurobiology, 7*, 839–843.

Kelley, D. B., & Pfaff, D. W. (1978). Generalizations from comparative studies on neuroanatomical and endocrine mechanisms of sexual behavior. In J. B. Hutchison (Ed.), *Biological determinants of sexual behavior* (pp. 225–254). Chichester, UK: Wiley.

Kirkpatrick, B., & Bryant, N. L. (1995). Sexual dimorphism in the brain: It's worse than you thought. *Biological Psychiatry, 38*, 347–348

Kirn, J. R., & DeVoogd, T. J. (1989). Genesis and death of vocal control neurons during sexual differentiation in the zebra finch. *Journal of Neuroscience, 9*, 3176–3187.

Kondrashov, A. S. (1988). Deleterious mutations and the evolution of sexual reproduction. *Nature, 336*, 435–440.

Konishi, M., & Akutagawa, E. (1985). Neuronal growth, atrophy and death in a sexually dimorphic song nucleus in the zebra finch brain. *Nature, 315*, 145–147.

Korsching, S. (1993). The neurotrophic factor concept: A reexamination. *Journal of Neuroscience, 13*, 2739–2748.

Krey, L. C., & McEwen, B. S. (1983). Steroid hormone processing in the brains and pituitary glands of nonhuman primates: Mechanisms and physiological significance. In R. L. Norman (Ed.), *Neuroendocrine aspects of reproduction* (pp. 47–67). New York: Academic Press.

Kuiper, G. G. L. N., Enmark, E., Pelto-Huikko, M., Nilsson, S., & Gustafsson, J. A. (1996). Cloning of a novel estrogen receptor expressed in rat prostate and ovary. *Proceedings of the National Academy of Sciences of the USA, 93*, 5925–5930.

Kurz, E. M., Sengelaub, D. R., & Arnold, A. P. (1986). Androgens regulate the dendritic length of mammalian motoneurons in adulthood. *Science, 232*, 395–398.

Ladosky, W., & Gaziri, L. C. J. (1970). Brain serotonin and sexual differentiation of the nervous system. *Neuroendocrinology, 6*, 168–174.

Lahr, G., Maxson, S. C., Mayer, A., Just, W., Pilgrim, C., & Reisert, I. (1995). Transcription of the Y chromosome gene, *Sry*, in adult mouse brain. *Molecular Brain Research, 33*, 179–182.

Lauber, A. H., Mobbs, C. V., Muramatsu, M., & Pfaff, D. W. (1991). Estrogen receptor messenger RNA expression in rat hypothalamus as a function of genetic sex and estrogen dose. *Endocrinology, 129*, 3180–3186.

Leedy, M. G., Beattie, M. S., & Bresnahan, J. C. (1987). Testosterone-induced plasticity of synaptic inputs to adult mammalian motoneurons. *Brain Research, 424*, 386–390.

Lieberburg, I., & McEwen, B. S. (1977). Brain cell nuclear retention of testosterone metabolites, 5 α-dihydrotestosterone and estradiol-17β, in adult rats. *Endocrinology, 100*, 588–597.

Lieberburg, I., MacLusky, N., & McEwen, B. S. (1980). Androgen receptors in the perinatal rat brain. *Brain Research, 196*, 125–138.

Lin, L.-F. H., Doherty, D. H., Lile, J. D., Bektesh, S., & Collins, F. (1993). GDNF: A glial cell line-derived neurotrophic factor for midbrain dopaminergic neurons. *Science, 260*, 1130–1132.

Lindzey, J., Kumar, M. V., Grossman, M., Young, C., & Tindall, D. J. (1994). Molecular mechanisms of androgen action. *Vitamins and Hormones, 49*, 383–432.

Lonstein, J. S., Simmons, D. A., Swann, J. M., & Stern, J. M. (1997). Forebrain expression of *c-fos* due to active maternal behavior in lactating rats. *Neuroscience, 82*, 267–281.

Lustig, R. H., Sudol, M., Pfaff, D. W., & Federoff, H. J. (1991). Estrogenic regulation and sex dimorphism of growth-associated protein 43 kDa (GAP-43) messenger RNA in the rat. *Molecular Brain Research, 11*, 125–132.

Lyet, L., Louis, F., Forest, M. G., Josso, N., Behringer, R. R., & Vigier, B. (1995). Ontogeny of reproductive abnormalities induced by deregulation of anti-Müllerian hormone expression in transgenic mice. *Biology of Reproduction, 52*, 444–454.

MacLusky, N. J., & Naftolin, F. (1981). Sexual differentiation of the central nervous system. *Science, 211*, 1294–1303.

MacLusky, N. J., Lieberburg, I., & McEwen, B. S. (1979). The development of estrogen receptor systems in the rat brain: Perinatal development. *Brain Research, 178*, 129–142.

MacLusky, N. J., Philip, A., Hurlburt, C., & Naftolin, F. (1985). Estrogen formation in the developing rat brain: Sex differences in aromatase activity during early post-natal life. *Psychoneuroendocrinology, 10*, 355–361.

MacLusky, N. J., Bowlby, D. A., Brown, T. J., Peterson, R. E., & Hochberg, R. B. (1997). Sex and the developing brain: Suppression of neuronal estrogen sensitivity by developmental androgen exposure. *Neurochemical Research, 22*, 1395–1414

Margulis, L., Sagan, D., & Olendzenski, L. (1985). What is sex? In H. O. Halvorson, & A. Monroy, (Eds.), *The origin and evolution of sex* (Vol. 7, pp. 67–86). New York: Liss.

Martinou, J., Dubois-Dauphin, M., Staole, J. K., Rodriguez, I., Frankowski, H., Missotten, M., Albertini, P., Talabot, D., Catsicas, S., Pietra, C., & Huarte, J. (1994). Overexpression of BCL-2 in transgenic mice protects neurons from naturally occurring cell death and experimental ischemia. *Neuron, 13*, 1017–1030.

Massagué, J. (1996). TGF signaling: Receptors, transducers, and Mad proteins. *Cell, 85*, 947–950.

193

DEVELOPMENT
OF SEX
DIFFERENCES IN
THE NERVOUS
SYSTEM

Matsumoto, A., & Arai, Y. (1976). Effect of estrogen on early postnatal development of synaptic formation in the hypothalamic arcuate nucleus of female rats. *Neuroscience Letters, 2*, 79–82.

Matsumoto, A., & Arai, Y. (1980). Sexual dimorphism in "wiring pattern" in the hypothalamic arcuate nucleus and its modification by neonatal hormonal environment. *Brain Research, 190*, 238–242.

Matsumoto, A., Micevych, P. E., & Arnold, A. P. (1988). Androgen regulates synaptic input to motoneurons of the adult rat spinal cord. *Journal of Neuroscience, 8*, 4168–4176.

Matsumoto, A., Arai, Y., Urano, A., & Hyodo, S. (1992). Effect of androgen on the expression of gap junction and β-actin mRNAs in adult rat motoneurons. *Neuroscience Research, 14*, 133–144.

Matsumoto, A., Arai, Y., & Hyodo, S. (1993). Androgenic regulation of expression of β-tubulin messenger ribonucleic acid in motoneurons of the spinal nucleus of the bulbocavernosus. *Journal of Neuroendocrinology, 5*, 357–363.

Matthews, G., & Arnold, A. P. (1990). Antiestrogens fail to prevent the masculine ontogeny of the zebra finch song system. *General and Comparative Endocrinology, 80*, 48–58.

Maynard-Smith, J. (1978). *The evolution of sex.* Cambridge: Cambridge University Press.

McCarthy, M. M., Schlenker, E. H., & Pfaff, D. W. (1993). Enduring consequences of neonatal treatment with antisense oligodeoxynucleotides to estrogen receptor messenger ribonucleic acid on sexual differentiation of rat brain. *Endocrinology, 133*, 433–439.

McGinnis, M. Y., & Katz, S. E. (1996). Sex differences in cytosolic androgen receptors in gonadectomized male and female rats. *Journal of Neuroendocrinology, 8*, 193–197.

McMillan, P. J., Singer, C., & Dorsa, D. M. (1996). The effects of ovariectomy and estrogen replacement on trkA and choline acetyltransferase mRNA expression in the basal forebrain of the adult female Sprague-Dawley rat. *Journal of Neuroscience, 16*, 1860–1865.

Merry, D. E., & Korsmeyer, S. J. (1997). BCL-2 gene family in the nervous system. *Annual Review of Neuroscience, 20*, 245–267.

Michael, R. P. (1965). Oestrogens in the central nervous system. *British Medical Bulletin, 21*, 87–90.

Miranda, R. C., Sohrabji, F., & Toran-Allerand, C. D. (1993). Presumptive estrogen target neurons express mRNAs for both the neurotrophins and neurotrophin receptors: A basis for potential developmental interactions of estrogen with the neurotrophins. *Molecular and Cellular Neurosciences, 4*, 510–525.

Mishina, Y., Rey, R., Finegold, M. J., Matzuk, M. M., Josso, N., Cate, R. L., & Behringer, R. R. (1996). Genetic analysis of the Müllerian-inhibiting substance signal transduction pathway in mammalian sexual differentiation. *Genes and Development, 10*, 2577–2587.

Mong, J. A., Kurzweil, R. L., Davis, A. M., Rocca, M. S., & McCarthy, M. M. (1996). Evidence for sexual differentiation of glia in rat brain. *Hormones and Behavior, 30*, 553–562.

Mosselman, S., Polman, J., & Dijkema, R. (1996). ER beta: Identification and characterization of a novel human estrogen receptor. *FEBS Letters, 392*, 49–53.

Müller, H. W., Junghans, U., & Kappler, J. (1995). Astroglial neutrophic and neurite-promoting factors. *Pharmacology and Therapeutics, 65*, 1–18.

Murakami, S., & Arai, Y. (1989). Neuronal death in the developing sexually dimorphic periventricular nucleus of the preoptic area in the female rat: Effect of neonatal androgen treatment. *Neuroscience Letters, 102*, 185–190.

Nabekura, J., Oomura, Y., Minami, T., Mizuno, Y., & Fukuda, A. (1986). Mechanism of the rapid effect of 17β-estradiol on medial amygdala neurons. *Science, 233*, 226–228.

Naftolin, F., Ryan, K. J., Davies, I. J., Reddy, V. V., Flores, F., Petor, Z., Kuhn, M., White, R. J., Takaoka, Y., & Wolin, L. (1975). The formation of estrogens by central neuroendocrine tissues. *Recent Progress in Hormone Research, 31*, 295–319.

Nelson K. G., Takahashi, T., Bossert, N. L., Walmer, D. K., & McLachlan, J. A. (1991) Epidermal growth factor replaces estrogen in the stimulation of female genital-tract growth and differentiation. *Proceedings of the National Academy of Sciences of the USA, 88*, 21–25.

Neri, R., Florance, K., Koziol, P., & Van Cleave, S. (1972). A biological profile of a nonsteroidal antiandrogen, SCH 13521 (4′-nitro-3′-trifluoromethylisobutyranilide). *Endocrinology, 91*, 427–437.

Nishizuka, M., & Arai, Y. (1981). Sexual dimorphism in synaptic organization in the amygdala and its dependence on neonatal hormone environment. *Brain Research, 212*, 31–38.

Nordeen, E. J., & Nordeen, K. W. (1988). Sex and regional differences in the incorporation of neurons born during song learning in zebra finches. *Journal of Neuroscience, 8*, 2869–2874.

Nordeen, E. J., & Nordeen, K. W. (1996). Sex difference among nonneuronal cells precedes sexually dimorphic neuron growth and survival in an avian song control nucleus. *Journal of Neurobiology, 30*, 531–542.

Nordeen, E. J., Nordeen, K, W., Sengelaub, D. R., & Arnold, A. P. (1985). Androgens prevent normally occurring cell death in a sexually dimorphic spinal nucleus. *Science, 229*, 671–673.

Nordeen, K. W., Nordeen, E. J., & Arnold, A. P. (1987). Estrogen accumulation in zebra finch song control nuclei: Implications for sexual differentiation and adult activation of song behavior. *Journal of Neurobiology, 18,* 569–582.

Nottebohm, F. (1981). A brain for all seasons: Cyclic anatomical changes in song control nuclei of the canary brain. *Science, 214,* 1368–1370.

Nottebohm, F., & Arnold, A. P. (1976). Sexual dimorphism in vocal control areas of the song bird brain. *Science, 194,* 211–213.

Numan, M., Corodimas, K. P., Numan, M. J., Facotr, E. M., & Piers, W. D. (1988). Axon-sparing lesions of the preoptic region and substantia innominata disrupt maternal behavior in rats. *Behavioral Neuroscience, 102,* 381–396.

Olmos, G., Naftolin, F., Perez, J., Tranque, P. A., & García-Segura, L. M. (1989). Synaptic remodeling in the rat arcuate nucleus during the estrous cycle. *Neuroscience, 32,* 663–667.

Oltvai, Z. N., Milliman, C. L., & Korsmeyer, S. J. (1993). Bcl-2 heterodimerizes *in vivo* with a conserved homolog, Bax, that accelerates programmed cell death. *Cell, 74,* 609–619.

O'Malley, B. W., Schrader, W. T., Mani, S., Smith, C., Weigel, N. L., Conneely, O. M., & Clark, J. H. (1995). An alternative ligand-independent pathway for activation of steroid receptors. *Recent Progress in Hormone Research, 50,* 333–347.

Oppenheim, R. W., Prevette, D., Tytell, M., & Homma, S. (1990). Naturally occurring and induced neuronal death in the chick embryo *in vivo* requires protein and RNA synthesis: Evidence for the role of cell death genes. *Developmental Biology, 138,* 104–113.

Ovtscharoff, W., Eusterschulte, B., Zienecker, R., Reisert, I., & Pilgrim, C. (1992). Sex differences in densities of dopaminergic fibers and GABAergic neurons in the prenatal rat striatum. *Journal of Comparative Neurology, 323,* 299–304.

Panzica, G. C., Viglietti-Panzica, C., Calacagni, M., Anselmetti, G. C., Schumacher, M., & Balthazart, J. (1987). Sexual differentiation and hormonal control of the sexually dimorphic medial preoptic nucleus of the quail. *Brain Research, 416,* 59–68.

Parducz, A., Szilagyi, T., Hoyk, S., Naftolin, F., & García-Segura, L.-M. (1996). Neuroplastic changes in the hypothalamic arcuate nucleus: The estradiol effect is accompanied by increased exoendocytotic activity of neuronal membranes. *Cellular and Molecular Neurobiology, 16,* 259–269.

Paredes, R. G., & Baum, M. J. (1995). Altered sexual partner preference in male ferrets given excitotoxic lesions of the preoptic area/anterior hypothalamus. *Journal of Neuroscience, 15,* 6619–6630.

Park, J., Baum, M. J., Paredes, R. G., & Tobet, S. A. (1996). Neurogenesis and cell migration into the sexually dimorphic preoptic area/anterior hypothalamus of the fetal ferret. *Journal of Neurobiology, 30,* 315–328.

Pfaff, D. W. (1966). Morphological changes in the brains of adult male rats after neonatal castration. *Journal of Endocrinology, 36,* 415–416.

Pfaff, D. W., & Keiner, M. (1973). Atlas of estradiol-concentrating cells in the central nervous system of the female rat. *Journal of Endocrinology, 151,* 121–158.

Pfeiffer, C. A. (1936). Sexual differences of the hypophyses and their determination by the gonads. *American Journal of Comparative Neurology, 58,* 195–225

Phoenix, C. H., Goy, R. W., Gerall, A. A., & Young, W. C. (1959). Organizing action of prenatally administered testosterone propionate on the tissues mediating mating behavior in the female guinea pig. *Endocrinology, 65,* 369–382.

Pietras, R. J., & Szego, C. M. (1977). Specific binding sites for oestrogen at the outer surfaces of isolated endometrial cells. *Nature, 205,* 69–72.

Raisman, G., & Field, P. M. (1971). Sexual dimorphism in the preoptic area of the rat. *Science, 173,* 731–733.

Raisman, G., & Field, P. M. (1973). Sexual dimorphism in the neuropil of the preoptic area and its dependence on neonatal androgen. *Brain Research, 54,* 1–29.

Rakic, P. (1990). Principles of neural cell migration. *Experientia, 46,* 882–891.

Ramirez, V. D., & Zheng, J. (1996) Membrane sex-steroid receptors in the brain. *Frontiers in Neuroendocrinology, 17,* 402–439.

Rand, M. N., & Breedlove, S. M. (1995). Androgen alters the dendritic arbors of SNB motoneurons by acting upon their target muscles. *Journal of Neuroscience, 15,* 4408–4416.

Rasika, S., Nottebohm, F., & Alvarez-Buylla, A. (1994). Testosterone increases the recruitment and/or survival of the new high vocal center neurons in adult female canaries. *Proceedings of the National Academy of Sciences of the USA, 91,* 7854–7858.

Reid, S. N. M., & Juraska, J. M. (1995). Sex differences in the number of synaptic junctions in the binocular area of the rat visual cortex. *Journal of Comparative Neurology, 352,* 560–566.

Reisert, I., & Pilgrim, C. (1991). Sexual differentiation of monoaminergic neurons—Genetic or epigenetic? *Trends in Neurosciences, 14,* 468–473.

195

DEVELOPMENT
OF SEX
DIFFERENCES IN
THE NERVOUS
SYSTEM

Reisert, I., Schuster, R., Zienecker, R., & Pilgrim, C. (1990). Prenatal development of mesencephalic and diencephalic dopaminergic systems in the male and female rat. *Developmental Brain Research, 53,* 222–229.

Renfree, M. B., & Short, R. V. (1988). Sex determination in marsupials: Evidence for a marsupial–eutherian dichotomy. *Philosophical Transactions of the Royal Society of London, B322,* 41–53.

Resko, J. A., Goy, R. W., & Phoenix, C. H. (1967). Uptake and distribution of exogenous testosterone[1,2,3]H in neural and genital tissues of the castrate guinea pig. *Endocrinology, 80,* 490–498.

Rogers, L. C., Junier, M., Farmer, S. R., & Ojeda, S. R. (1991). A sex-related difference in the developmental expression of class II β-tubulin messenger RNA in rat hypothalamus. *Molecular and Cellular Neurosciences, 2,* 130–138.

Roof, R. L., & Havens, M. D. (1992). Testosterone improves maze performance and induces development of a male hippocampus in females. *Brain Research, 572,* 310–313.

Roselli, C. E. (1991). Sex differences in androgen receptors and aromatase activity in microdissected regions of the rat brain. *Endocrinology, 128,* 1310–1316.

Roselli, C. E., & Resko, J. A. (1993). Aromatase activity in the rat brain: Hormonal regulation and sex differences. *Journal of Steroid Biochemistry and Molecular Biology, 44,* 499–508.

Roselli, C. E., Ellinwood, W. E., & Resko, J. A. (1984). Regulation of brain aromatase activity in rats. *Endocrinology, 114,* 192–200.

Rouiller-Fabre, V., Carmona, S., Merhi, R. A., Cate, R., Habert, R., & Vigier, B. (1998) Effect of anti-Mullerian hormone on Sertoli and Leydig cell functions in fetal and immature rats. *Endocrinology, 139,* 1213–1220.

Ruel, T. D., Kelley, D. B., & Tobias, M. L. (1998). Facilitation at the sexually differentiated laryngeal synapse of *Xenopus laevis. Journal of Comparative Physiology A, 182,* 35–42.

Sachs, B. D., & Meisel, R. L. (1988). The physiology of male sexual behavior. In E. Knobil, & J. Neill (Eds.), *The physiology of reproduction* (pp. 1393–1485). New York: Raven Press.

Sar, M., & Parikh, I. (1986). Immunohistochemical localization of estrogen receptor in rat brain, pituitary and uterus with monoclonal antibodies. *Journal of Steroid Biochemistry, 24,* 409–442.

Sar, M., & Stumpf, W. E. (1975). Distribution of androgen-concentrating neurons in rat brain. In W. E. Stumpf, & L. D. Grant (Eds.), *Anatomical neuroendocrinology* (pp. 120–133). Basel, Switzerland: Karger.

Schlinger, B. A., & Arnold, A. P. (1992a). Plasma sex steroids and tissue aromatization in hatchling zebra finches: Implications for the sexual differentiation of singing behavior. *Endocrinology, 130,* 289–299.

Schlinger, B. A., & Arnold, A. P. (1992b). Circulating estrogens in a male songbird originate in the brain. *Proceedings of the National Academy of Sciences of the USA, 89,* 7650–7653.

Schlinger, B. A., & Arnold, A. P. (1993). Estrogen synthesis *in vivo* in the adult zebra finch—Additional evidence that circulating estrogens can originate in brain. *Endocrinology, 133,* 2610–2616.

Schmitt, E., Sane, A. T., Steyaert, A., Cimoli, G., & Bertrand, R. (1997). The Bcl-xl and Bax-alpha control points: Modulation of apoptosis induced by cancer chemotherapy and relation to TPCK-protease and caspase activation. *Biochemistry and Cell Biology, 75,* 301–314.

Scott, W. J., & Holson, J. F. (1977). Weight differences in rat embryos prior to sexual differentiation. *Journal of Embryology and Experimental Morphology, 40,* 259–263.

Sellar, M. J., & Perkins-Cole, K. J. (1987). Sex difference in the mouse embryonic development at neurulation. *Journal of Reproduction and Fertility, 79,* 159–161.

Sharman, G. B., Robinson, E. S., Walton, S. M., & Berger, P. J. (1970). Sex chromosomes and reproductive anatomy of some intersex marsupials. *Journal of Reproduction and Fertility, 21,* 57–68.

Shaw, G., Renfree, M. B., Short, R. V., & O, W.-S. (1988). Experimental manipulation of sexual differentiation in wallaby pouch young treated with exogenous steroids. *Development, 104,* 689–701.

Shaywitz, B. A., Shaywitz, S. E., Pugh, K. R., Constable, R. T., Skudlarski, P., Fulbright, R. K., Bronen, R. A., Fletcher, J. M., Shankweller, D. P., Katz, L., & Gore, J. C. (1995). Sex differences in the functional organization of the brain for language. *Nature, 373,* 607–609.

Shea, T. B., Perrone-Bizzozero, N. I., Beermann, M. L., & Benowitz, L. I. (1991) Phospholipid-mediated delivery of anti-GAP-43 antibodies into neuroblastoma cells prevents neuriteogenesis. *Journal of Neuroscience, 11,* 1685–1690.

Sherry, D. F., Jacobs, L. F., & Gaulin, S. J. C. (1992). Spatial memory and adaptive specialization of the hippocampus. *Trends in Neurosciences, 15,* 298–303.

Shughrue, P. J., & Dorsa, D. M. (1993a). Gonadal steroids modulate the growth-associated protein GAP-43 (neuromodulin) mRNA in postnatal rat brain. *Developmental Brain Research, 73,* 123–132.

Shughrue, P. J., & Dorsa, D. M. (1993b). Estrogen modulates the growth-associated protein GAP-43 (neuromodulin) mRNA in the rat preoptic area and basal hypothalamus. *Neuroendocrinology, 57,* 439–447.

Siiteri, P. K., & Wilson, J. D. (1974). Testosterone formation and metabolism during male sexual differentiation in the human embryo. *Journal of Clinical Endocrinology and Metabolism, 38,* 113–125.

Simerly, R. B., Swanson, L. W., Handa, R. J., & Gorski, R. A. (1985). Influence of perinatal androgen on the sexually dimorphic distribution of tyrosine hydroxylase-immunoreactive cells and fibers in the anteroventral periventricular nucleus of the rat. *Neuroendocrinology, 40,* 501–510.

Simerly, R. B., Chang, C., Muramatsu, M., & Swanson, L. W. (1990). Distribution of androgen and estrogen receptor mRNA-containing cells in the rat brain: An *in situ* hybridization study. *Journal of Comparative Neurology, 294,* 76–95.

Simpson, H. B., & Vicario, D. S. (1991a). Early estrogen treatment alone causes female zebra finches to produce learned, male-like vocalizations. *Journal of Neurobiology, 22,* 755–776.

Simpson, H. B., & Vicario, D. S. (1991b). Early estrogen treatment of female zebra finches masculinizes the brain pathway for learned vocalizations. *Journal of Neurobiology, 22,* 777–793.

Sinclair, A. H., Berta, P., Palmer, M. S., Hawkins, J. R., Griffiths, B. L., Smith, M. J., Foster, J. W., Frischauf, A., Lovell-Badge, R., & Goodfellow, P. N. (1990). A gene from the human sex-determining region encodes a protein with homology to a conserved DNA-binding motif. *Nature, 346,* 240–244.

Sohrabji, F., Greene, L. A., Miranda, R. C., & Toran-Allerand, C. D. (1994). Reciprocal regulation of estrogen and NGF receptors by their ligands in PC12 cells. *Journal of Neurobiology, 25,* 974–988.

Springer, M. L., & Wade, J. (1997). The effects of testicular tissue and prehatching inhibition of estrogen synthesis on the development of courtship and copulatory behavior in zebra finches. *Hormones and Behavior, 32,* 46–59.

Stanley, H. F., & Fink, G. (1986). Synthesis of specific brain proteins is influenced by testosterone at mRNA level in the neonatal rat. *Brain Research, 370,* 223–231.

Steimer, T., & Hutchison, J. B. (1990). Is androgen-dependent aromatase activity sexually differentiated in the rat and dove preoptic area? *Journal of Neurobiology, 21,* 787–795.

Stewart, J., Kuhnemann, S., & Rajabi, H. (1991). Neonatal exposure to gonadal hormones affects the development of monoamine systems in rat cortex. *Journal of Endocrinology, 3,* 85–93.

Sumida, H., Nishizuka, M., Kano, Y., & Arai, Y. (1993). Sex differences in the anteroventral periventricular nucleus of the preoptic area and in the related effects of androgen in prenatal rats. *Neuroscience Letters, 151,* 41–44.

Swaab, D. F., & Fliers, E. (1985). A sexually dimorphic nucleus in the human brain. *Science, 228,* 1112–1115.

Tanapat, P, Hastings, N. B., Reeves, A. J., & Gould, E. (1999). Estrogen stimulates a transient increase in the number of new neurons in the dentate gyrus of the adult female rat. *Journal of Neuroscience, 19,* 5792–5801.

Taylor, B. J., & Truman, J. W. (1992). Commitment of abdominal neuroblasts in *Drosophila* to a male or female fate is dependent on genes of the sex-determining hierarchy. *Development, 114,* 625–642.

Thornhill, J., & Halvorson, I. (1994). Activation of shivering and non-shivering thermogenesis by electrical stimulation of the lateral and medial preoptic areas. *Brain Research, 656,* 367–374.

Tobet, S. A., & Fox, T. O. (1989). Sex- and hormone-dependent antigen immunoreactivity in developing rat hypothalamus. *Proceedings of the National Academy of Sciences of the USA, 86,* 382–386.

Tobet, S. A., & Hanna, I. K. (1997). Ontogeny of sex differences in the mammalian hypothalamus and preoptic area. *Cellular and Molecular Neurobiology, 17,* 565–601.

Tobet, S. A., Baum, M. J., Tang, H. B., Shim, J. H., & Canick, J. A. (1985). Aromatase activity in the perinatal rat forebrain: Effects of age, sex, and intrauterine position. *Developmental Brain Research, 23,* 171–178.

Tobet, S. A., Zahniser, D. J., & Baum, M. J. (1986). Sexual dimorphism in the preoptic/anterior hypothalamic area of ferrets: Effects of adult exposure to sex steroids. *Brain Research, 364,* 249–257.

Tobet, S. A., Basham, M. E., & Baum, M. J. (1993). Estrogen receptor immunoreactive neurons in the fetal ferret forebrain. *Developmental Brain Research, 72,* 167–180.

Tobet, S. A., Chickering, T. W., Hanna, I., Crandall, J. E., & Schwarting, G. A. (1994). Can gonadal steroids influence cell position in the developing brain? *Hormones and Behavior, 28,* 320–327.

Tobet, S. A., Paredes, R. G., Chickering, T. W., & Baum, M. J. (1995). Telencephalic and diencephalic origin of radial glial processes in the developing preoptic area/anterior hypothalamus. *Journal of Neurobiology, 26,* 75–86.

Tobet, S. A., Henderson, R. G., Directo, C. O., & Dyer, B. V. (1997). Regional difference in cellular development in the preoptic area and anterior hypothalamus. *Society for Neuroscience Abstracts, 22,* 344.

Tobias, M. L., Kelley, D. B. (1995). Sexual differentiation and hormonal regulation of the laryngeal synapse in *Xenopus laevis. Journal of Neurobiology, 28,* 515–526.

Tobias, M. L., & Kelley, D. B., & Ellisman, M. (1995). A sex difference in synaptic efficacy at the laryngeal neuromuscular junction of *Xenopus laevis. Journal of Neuroscience, 15,* 1660–1668.

Toran-Allerand, C. D. (1976). Sex steroids and the development of the newborn mouse hypothalamus and preoptic area *in vitro*: Implications for sexual differentiation. *Brain Research, 106,* 407–412.

Toran-Allerand, C. D. (1980). Sex steroids and the development of the newborn mouse hypothalamus

197

DEVELOPMENT
OF SEX
DIFFERENCES IN
THE NERVOUS
SYSTEM

and preoptic area *in vitro.* II. Morphological correlates and hormonal specificity. *Brain Research, 189,* 413–427.

Toran-Allerand, C. D., Gerlach, J. L., & McEwen B. S. (1980) Autoradiographic localization of [³H]estradiol related to steroid responsiveness in cultures of the newborn mouse hypothalamus and preoptic area. *Brain Research, 184,* 517–522.

Toran-Allerand, C. D., Miranda, R. C., Bentham, W. D. L., Sohrabji, F., Brown, T. J., Hochberg, R. B., & MacLusky, N. J. (1992). Estrogen receptors colocalize with low-affinity nerve growth factor receptors in cholinergic neurons of the basal forebrain. *Proceedings of the National Academy of Sciences of the USA, 89,* 4668–4672.

Torres-Alemán, I., Rejas, M. T., Pons, S., & García-Segura, L. M. (1992). Estradiol promotes cell shape changes and glial fibrillary acidic protein redistribution in hypothalamic astrocytes *in vitro*. A neuronal-mediated effect. *Glia, 6,* 180–187.

Towle, A. C., & Sze, P. Y. (1983). Steroid binding to synaptic plasma membrane: Differential binding of glucocorticoids and gonadal steroids. *Journal of Steroid Biochemistry, 18,* 135–143.

Ueyama, T., Mizuno, N., Takahashi, O., Nomura, S., Arakawa, H., & Matsushima, R. (1985). Central distribution of efferent and afferent components of the pudendal nerve in macaque monkeys. *Journal of Comparative Neurology, 232,* 548–556.

Ulibarri, C., & Yahr, P. (1988). Role of neonatal androgens in sexual differentiation of brain structure, scent marking and gonadotrophin secretion in gerbils. *Behavioral and Neural Biology, 49,* 27–44.

Ulibarri, C. M., & Yahr, P. (1993). Ontogeny of the sexually dimorphic area of the gerbil hypothalamus. *Developmental Brain Research, 74,* 14–24.

Ulibarri, C. P., Popper, P., & Micevych, P. E. (1995). Motoneurons dorsolateral to the central canal innervate perineal muscles in the mongolian gerbil. *Journal of Comparative Neurology, 356,* 225–237.

Uriel, J., Bouillon, D., & Dupiers, M. (1975). Affinity chromatography of human, rat and mouse alpha-fetoprotein on estradiol-Sepharose adsorbents. *FEBS Letters, 53,* 305–308.

Valdivia, R. P. A., Kuneida, T., Azuma, S., & Toyoda, Y. (1993). PCR sexing and development rate differences in preimplantation mouse embryos fertilized and cultured *in vitro. Molecular Reproduction and Development, 35,* 121–126.

Van Leeuwen, F. W., Caffé, A. R., & De Vries, G. J. (1985). Vasopressin cells in the bed nucleus of the stria terminalis of the rat: Sex differences and the influence of androgens. *Brain Research, 325,* 391–394.

Vito, C. C., & Fox, T. O. (1979). Embryonic rodent brain contains estrogen receptors. *Science, 204,* 517–519.

Vito, C. C., Wieland, S. J., & Fox, T. O. (1979). Androgen receptors exist throughout the "critical period" of brain sexual differentiation. *Nature, 282,* 308–310.

Wade, J., & Arnold, A. P. (1994). Post-hatching inhibition of aromatase activity does not alter sexual differentiation of the zebra finch song system. *Brain Research, 693,* 347–350.

Wade, J., & Arnold, A. P. (1996). Functional testicular tissue does not masculinize development of the zebra finch song system. *Proceedings of the National Academy of Sciences of the USA, 93,* 5264–5268.

Wade, J., Schlinger, B. A., & Arnold, A. P. (1995). Aromatase and 5β-reductase activity in cultures of developing zebra finch brain: An investigation of sex and regional differences. *Journal of Neurobiology, 27,* 240–251.

Wade, J., Springer, M. L., Wingfield, J. C., & Arnold, A. P. (1996). Neither testicular androgens nor embryonic aromatase activity alter morphology of the neural song system in zebra finches. *Biology of Reproduction, 55,* 1126–1132.

Wagner, C. K., & Clemens, L. G. (1989). Anatomical organization of the sexually dimorphic perineal neuromuscular system in the house mouse. *Brain Research, 499,* 93–100.

Wagner, C. K., Nakayama, A. Y., & De Vries, G. J. (1998). Potential role of maternal progesterone in the sexual differentiation of the brain. *Endocrinology, 139,* 3658–3661.

Wang, Z., Bullock, N. A., & De Vries, G. J. (1993). Sexual differentiation of vasopressin projections of the bed nucleus of the stria terminalis and medial amygdaloid nucleus in rats. *Endocrinology, 132,* 2299–2306.

Ward, I. (1992). Sexual behavior: The product of perinatal hormonal and prepubertal social factors. In A. Gerall, H. Moltz, & L. L. Ward. (Eds.), *Handbook of Behavioral Neurobiology* (pp. 157–180). New York: Plenum.

Watson, J. T., Robertson, J., Sachdev, U., & Kelley, D. B. (1993). Laryngeal muscle and motor neuron plasticity in *Xenopus laevis*: Testicular masculinization of a developing neuromuscular system. *Journal of Neurobiology, 24,* 1615–1625.

Watson, N., Freeman, L., & Breedlove, S. M. (2001). Neuronal size in the spinal nucleus of the bulbocavernosus: Direct modulation by androgen in rats with mosaic androgen sensitivity. *Journal of Neuroscience, 21,* 1062–1066.

Weisz, J., & Ward, I. L. (1980). Plasma testosterone and progesterone titers of pregnant rats, their male and female fetuses, and neonatal offspring. *Endocrinology, 106*, 306–316.

Welshons, W. J., & Russell, L. B. (1959). The Y-chromosome as the bearer of male determining factors in the mouse. *Proceedings of the National Academy of Sciences of the USA, 45*, 560–566.

Whalen, R. E., & Etgen, A. M. (1978). Masculinization and defeminization induced in female hamsters by neonatal treatment with estradiol benzoate and RU-2858. *Hormones and Behavior, 10*, 170–177.

Williams, C. L., & Meck, W. H. (1991). The organizational effects of gonadal steroids on sexually dimorphic spatial ability. *Psychoneuroendocrinology, 16*, 155–176.

Williams, N. (1995). Tracing how the sexes develop. *Science, 269*, 1822–1827.

Wilson, J. D., George, F. W., & Griffin, J. E. (1981). The hormonal control of sexual development. *Science, 211*, 1278–1284.

Wimer, R. E., & Wimer, C. (1985). Three sex dimorphisms in the granule cell layer of the hippocampus in house mice. *Brain Research, 328*, 105–109.

Wong, M., & Moss, R. L. (1991). Electrophysiological evidence for a rapid membrane action of the gonadal steroid, 17β-estradiol, on CA1 pyramidal neurons of the rat hippocampus. *Brain Research, 543*, 148–152.

Woolley, C. S., Gould, E., Frankfurt, M., & McEwen, B. S. (1990). Naturally occurring fluctuation in dendritic spine density on adult hippocampal pyramidal neurons. *Journal of Neuroscience, 10*, 4035–4039.

Xu, J., & Forger, N. G. (1998) Expression and androgen regulation of the ciliary neurotrophic factor receptor (CNTFRα) in muscles and spinal cord. *Journal of Neurobiology, 35*, 217–225,

Yalcinkaya, T. M., Siiteri, P., Vigne, J., Licht, P., Pavgi, S., Frank, L. G., & Glickman, S. E. (1993). A mechanism for virilization of female spotted hyenas *in utero*. *Science, 260*, 1929–1931.

Zhou, L., Blaustein, J. D., & De Vries, G. J. (1994). Distribution of androgen receptor immunoreactivity in vasopressin- and oxytocin-immunoreactive neurons in the male rat brain. *Endocrinology, 134*, 2622–2627.

Zwingman, T., Erickson, R. P., Boyer, T., & Ao, A. (1993). Transcription of the sex-determining region genes *Sry* and *Zfy* in the mouse preimplantation embryo. *Proceedings of the National Academy of Sciences of the USA, 90*, 814–817.

5

The Developmental Context
of Thermal Homeostasis

MARK S. BLUMBERG

INTRODUCTION

Developmentalists are concerned with origins. For some, the search for origins in developing animals provides little more than an opportunity to clarify the factors that contribute to adult behavior and physiology. Although this is an understandable justification for developmental research, it is also important to understand that developing organisms are not simply small adults or adults-in-waiting. On the contrary, infant animals face many problems that are unique to their physical, physiological, social, and ecological circumstances (Alberts & Cramer, 1988; Hall & Oppenheim, 1987; West, King, & Arberg, 1988). These problems cannot be put off; rather, to survive, infants must solve each problem as it is encountered during ontogeny. Therefore, the "dual infant" must meet the needs of the moment as well as prepare for later life, a vital combination of adaptation and anticipation (Alberts & Cramer, 1988).

Because some of the problems faced by infants are transient, some of the solutions are transient as well, and these solutions are referred to as ontogenetic adaptations (Alberts, 1987; Hall & Oppenheim, 1987). Although a central theme of developmental psychobiology, the concept of an ontogenetic adaptation can be easily misused, as can any adaptationist concept (Williams, 1966). Nonetheless, there is no other concept that so powerfully captures the temporary utility of the umbilical cord or of suckling behavior. There is also no other concept that so simply reminds us of the dangers of judging the physiological and behavioral capabilities of infants against an adult standard. When holding infants to adult standards, we lose sight of

MARK S. BLUMBERG   Department of Psychology, University of Iowa, Iowa City, Iowa 52242.

Developmental Psychobiology, Volume 13 of *Handbook of Behavioral Neurobiology*, edited by Elliott Blass, Kluwer Academic / Plenum Publishers, New York, 2001.

the context of development and the necessity of using appropriate experimental tools and methods to reveal the organizational complexity of our infant subjects.

This chapter examines a recent transformation in our understanding of the homeostatic control of body temperature in developing infants and the physiological and behavioral implications of endothermy. This chapter especially focuses on a specialized thermogenic organ called brown adipose tissue, the source of endothermy in the infants of many mammalian species. Contrary to a long-standing perspective, it will become clear that thermogenesis by brown adipose tissue during cold exposure protects and modulates diverse aspects of the infant rat's physiology and behavior. The picture of infant thermoregulation that is now emerging reveals previously unappreciated competence and complexity and points the way to a richer understanding of the developmental and evolutionary significance of endothermy. Despite these broad implications, the focus of this chapter is necessarily limited. Therefore, for broader perspectives on physiological and behavioral thermoregulation in infants and the influence of temperature on morphology and function during development, the reviews of Leon (1986), Brück and Hinckel (1996), Satinoff (1991, 1996), and Blumberg and Sokoloff (1998) are of interest.

HOMEOSTASIS, HOMEOTHERMY, AND DEVELOPMENT

EARLY EFFORTS

Thermoregulation in adult mammals is a classic example of a homeostatic process and was addressed at length in W. B. Cannon's (1932) treatise, *The Wisdom of the Body*. Cannon's ideas emerged from the foundation provided by Claude Bernard, who first described the compensatory physiological processes that establish constancy in the internal environment. In coining the term *homeostasis*, Cannon aimed to call attention to those "coordinated physiological processes which maintain most of the steady states in the organism [and which] are so complex and so peculiar to living beings" (p. 24). Furthermore, although he titled his chapter on thermoregulation "The constancy of body temperature," he was careful to stress that homeostasis denotes a condition of *relative constancy*, not stagnation, and that this relative constancy arises from "the various physiological arrangements which serve to restore the normal state when it has been disturbed" (p. 25). The early focus on physiological mechanisms was offset somewhat when, beginning with the work of Curt Richter, the contributions of behavior to homeostasis also came to be appreciated (Richter, 1943; see also Bartholomew, 1964, for a forceful endorsement of an integrative approach to homeostasis).

The publication of Wiener's (1961) book on cybernetics and the explanatory power of control systems theory (Houk, 1988) consolidated a connection between homeostasis and a few specific mechanisms derived from control theory. W. B. Cannon (1932) himself contributed to this connection when, in his chapter on body temperature, he stated that the "delicate control of body temperature indicates that somewhere in the organism a *sensitive thermostat* exists" and that this mechanism "is located in the base of the brain, in the diencephalon" (p. 199, italics added). The promise of control systems theory seemed to be realized when investigators began providing explicit mathematical descriptions of the set point mechanism (e.g., Hammel, Jackson, Stolwijk, Hardy, & Strömme, 1963). Over the years, however, numerous authors have cautioned against the reflexive invocation and reification of such control-theoretic concepts as set point and negative feedback, and have emphasized instead the hierarchical, dynamic control of interacting homeostatic mechanisms (e.g., Booth, 1980; Hogan, 1980; Mrosovsky, 1990; Satinoff, 1978). Thus, given that stability in some physiological systems can be achieved without comparators or feedback loops, it is argued that the concept of homeostasis provides a functional description of a regulatory system but does not proscribe the mechanisms involved (Hogan, 1980).

Developmental studies of homeostasis have been relatively rare. Nonetheless, in the field of thermoregulation, over 50 years of research has contributed to a perspective according to which infant mammals, especially those of altricial species such as the Norway rat, are homeostatically immature. This perspective was the logical outcome of the use of a particular methodology that, in turn, derived from a theoretical stance that was fundamentally nondevelopmental. That methodology entailed the isolation of an individual infant rat in a cold environment (e.g., 15°C: Brody, 1943; 10°C: Hahn, 1956) and the measurement of rectal temperature throughout the period of isolation. Pups of different ages were tested using identical procedures and in this way the researcher tracked the "development of homeothermy," that is, the development of a high and stable body temperature. For example, using this experimental approach, Hahn (1956) concluded that "physical thermoregulatory mechanisms develop between the 14th and 18th day" postpartum (p. 430). In a similar vein, Brody (1943) stated that it "is generally known that human infants, and young mammals of many species, do not possess the ability to maintain a constant body temperature; their homeothermic mechanisms are not well developed" (p. 230).

With the benefit of hindsight, we can identify the fundamental flaws in these early studies. First, the exposure of pups of different ages to the same *arbitrary* air temperature structured the outcome of the experiments. For example, if Hahn (1956) had exposed all pups to 0°C rather than 10°C, homeothermy would have appeared to develop at a later age; conversely, if he had exposed all pups to 30°C, homeothermy would have appeared to develop at an earlier age. Although this criticism may now seem obvious, it was not obvious to earlier researchers because their interest was not in the regulatory capabilities of infant rats but in the attainment of a particular *adult characteristic*. The notion that infant capabilities reflect "a balance between meeting the needs of the moment and preparing for later life" (Hall & Oppenheim, 1987, p. 95) was not yet appreciated.

A related, and perhaps more fundamental flaw in the approach of Brody (1943) and Hahn (1956) is the confusion of process (i.e., homeostatic regulation) and product (i.e., homeothermy). Exposing pups of different ages to the same cold environment provides a measure of the development of insulation (i.e., fur and subcutaneous fat) or, in other words, the developing control of heat loss. The rate of heat loss, however, is largely (but not entirely) a passive feature of an organism; for example, a dead elephant may have more thermal inertia than a dead mouse and thus may cool more slowly, but we do not conclude that the dead elephant, even in the short term, is exhibiting homeostasis. On the contrary, we reserve the concept of homeostasis for the correction of a system variable in response to a disturbance and, in the case of isolated infant rats, such mechanisms primarily involve the activation and deactivation of heat production, not the regulation of heat loss. Therefore, early investigations of the development of homeothermy provided little information about the development or control of homeostatic mechanisms.

The confusion between homeostasis and homeothermy was exacerbated further by the method of exposing pups to, what is for younger infants, an extremely cold environment. As was eventually learned, exposure to extreme cold and the rapid heat loss that results overwhelms the infant's ability to produce heat endogenously. It was only as investigators began testing pups in moderately cold environments that it was found that a newborn rat can increase metabolic heat production (Taylor, 1960). This point was brought home most clearly by Conklin and Heggeness (1971) when they stated that their infant subjects were tested at air temperatures "that would evoke, but not overwhelm, regulatory functions" (p. 333).

Brown Adipose Tissue

Although Taylor (1960) demonstrated that newborn rats can increase metabolic heat production in the cold, the source of this heat production was still not known. At that time, physiologists recognized two forms of heat production—shivering and nonshivering thermogenesis. Although a chemical, or nonshivering, form of thermogenesis was well established by the 1950s (Hsieh, Carlson, & Gray, 1957), the source of this heat production remained a mystery until Smith's (1961) suggestion that it was brown adipose tissue (BAT). In short order, investigators were using a variety of pharmacologic and physiological techniques to demonstrate that, during cold exposure, BAT is an important and sometimes exclusive source of heat production in the newborns of a number of species including rabbits, guinea pigs, and humans (e.g., Brück & Wünnenberg, 1970; Dawkins & Scopes, 1965; Hull & Segall, 1965). Over the next three decades, the "hibernating gland" first described by Gesner in the 16th century (Smith & Horwitz, 1969) was to become a focus of considerable scientific interest.

DISTRIBUTION AND ANATOMY.   Brown fat has been reported in a wide variety of mammalian species but not in birds (Johnston, 1971; Smith & Horwitz, 1969). Among mammals, BAT is found most notably in newborns, hibernators, and cold-acclimated adults. In rodents, BAT is distributed throughout the body but is primarily concentrated in large lobes surrounding the thoracic cavity (Smith & Horwitz, 1969). The largest deposits are the interscapular pad (iBAT), which is easily identified in newborn rats as a butterfly-shaped mass lying just under the skin in the interscapular region, and the superior cervical pad, which overlies the cervical spinal cord and is separated from the interscapular pad by layers of muscle.

It has been noted that the various deposits of BAT appear ideally suited for the warming of venous blood returning from the periphery (Smith, 1964). Furthermore, iBAT appears to exhibit a unique vascular anatomy; specifically, the bilateral arterial vessels supplying iBAT lie next to corresponding veins, thus allowing countercurrent heat exchange between the arteries and veins (for a lucid introduction to countercurrent exchange mechanisms, see Schmidt-Nielsen, 1981). This countercurrent system can be bypassed by diverting venous blood flowing from BAT to the Sulzer vein, a large, unpaired vessel that returns warmed blood directly to the heart (Smith & Roberts, 1964). There is little information that bears directly on the neural control of venous outflow from iBAT.

The location of iBAT in the week-old rat and the thermal gradient established by BAT thermogenesis are shown in Figure 1. This figure presents an infrared thermograph of the dorsal surface of a week-old pup that has been isolated at room temperature (~22°C). The white zone at the center of the thermograph overlies iBAT and indicates the region of highest temperature. Furthermore, the thermal gradient from the interscapular region to the base of the tail is substantial, approximately 2.5°C over a distance of approximately 3.5 cm. Therefore, this thermograph makes clear that heat production, at least as viewed from the skin surface, is localized to the thoracic region.

BIOCHEMISTRY AND PHARMACOLOGY.   Brown fat gets its name from its high concentration of mitochondria filled with cytochrome C. When a lipid-filled brown adipocyte is stimulated by norepinephrine, a chain of events ensues in which free fatty acids are used by the mitochondria as substrate for heat production. Moreover, the free fatty acids may also participate in activating or modulating the activity of thermogenin, the uncoupling protein that is necessary for mitochondrial heat production (B. Cannon, Jacobsson, Rehnmark, & Nedergaard, 1996). Thermogenin is believed to be the rate-limiting factor for BAT thermogenesis (Nedergaard, Connolly, & Cannon, 1986).

Although it has been known for many years that norepinephrine stimulates BAT thermogenesis, the receptor subtype responsible for this effect was unknown until recently. It now appears that activation of the β_3 adrenoceptor, a subtype found both on white and brown adipocytes, triggers the liberation of free fatty acids in the brown adipocyte, which in turn initiates the cascade of events culminating in heat production (B. Cannon *et al.*, 1996; Zhao, Unelius, Bengtsson, Cannon, & Nedergaard, 1994). The contributions of other adrenergic receptor subtypes to BAT thermogenesis are still less well understood, although it appears that α_1 and β_1 receptors participate in brown fat cell proliferation (B. Cannon *et al.*, 1996). In addition, blood flow to iBAT, which increases appreciably during heat production to sustain the 10- to 40-fold increase in cellular respiration, may be modulated via α-adrenergic receptors on the vascular smooth muscle of BAT (Girardier & Seydoux, 1986).

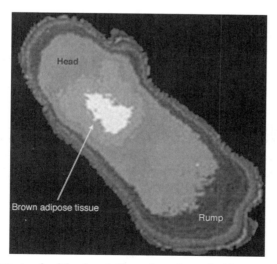

Figure 1. Infrared thermograph of the dorsal skin surface of a week-old rat during cold exposure. The region of highest temperature is coded in white, with decreasing skin temperatures coded by successively darker shades. The white region overlies the interscapular deposit of brown adipose tissue. A thermal gradient of approximately 2.5°C extends from the interscapular region to the base of the tail.

NEURAL CONTROL. It has been long recognized that BAT thermogenesis is under sympathetic nervous system control, although only the interscapular pad has been studied extensively (Girardier & Seydoux, 1986). In iBAT, each brown adipocyte is surrounded by fine fibers whose varicosities are filled with norepinephrine. These fibers arise from five pairs of nerves that enter the tissue after emerging from the intercostal muscles; these nerves originate in the first five thoracic ganglia. A final pair of perivascular nerves terminate along the arterial supply to iBAT. As would be expected from physiological studies of BAT thermogenesis, functional innervation of iBAT is present in newborn rats (Schneider-Picard & Girardier, 1982).

The central nervous system circuitry governing iBAT activation is largely unknown. Interest in brown fat as a regulator of energy balance has led some to investigate hypothalamic structures implicated in feeding. Specifically, it has been shown that electrical (Freeman & Wellman, 1987) or pharmacologic (Amir, 1990) stimulation of the periventricular nucleus in adult rats induces thermogenesis by iBAT. Investigations of the neural control of BAT at other levels of the neuraxis, however, suggest hierarchical control with multiple excitatory and inhibitory outputs. For example, decerebration experiments in anesthetized adult rats indicate a locus of inhibitory control in the region of the lower midbrain and the upper pons (Rothwell, Stock, & Thexton, 1983; Shibata, Benzi, Seydoux, & Girardier, 1987). This inhibitory influence is already present in infant rats (Bignall, Heggeness, & Palmer, 1975; Blumberg, Schalk, & Sokoloff, 1995). In addition, even in 1-day-old rats, a second locus of inhibitory control appears to exist in the basal forebrain (Blumberg *et al.*, 1995). Unfortunately, beyond this relatively crude outline, we have little understanding of the input/output relations of this hierarchically organized system, and we also know little about the neural control of the multitude of other BAT deposits throughout the body.

BROWN FAT THERMOGENESIS AND HOMEOTHERMY

An explosion of interest in brown adipose tissue over the last decade derives from its hypothesized role in the regulation of energy balance and feeding behavior and the onset of obesity (Lowell *et al.*, 1993). Our main concern here, however, is with the behavioral and physiological significance of BAT thermogenesis for infant animals; ironically, our increasing understanding of BAT as a thermogenic organ over the last 30 years has not substantially modified the prevailing view that cold-exposed altricial infants exhibit many of the characteristics of poikilothermy (e.g., Johanson, 1979; Satinoff, 1996).

The perspective that altricial infant mammals are poikilothermic reflects a number of common assumptions, some implicit and some explicit. First, Alberts' (1978) demonstration that rat pups within a litter actively huddle and, by doing so, reap substantial metabolic savings during cold exposure placed a new emphasis on the thermoregulatory capacities of the group while seemingly clarifying and justifying the deficiencies of the individual. Second, the conclusion that BAT thermogenesis is nearly useless for the individual is based upon the cold-exposed pup's inability to maintain *rectal* temperature at *adult* levels. Implicit in this conclusion is the notion that heat, whatever its source of production, is dispersed uniformly throughout the body core. Finally, when the concepts of homeothermy and homeostasis are commingled, it is not difficult to understand how a falling rectal temperature in the cold can be interpreted as evidence that homeostatic mechanisms are also absent.

Some investigators have departed somewhat from the above perspective. Brück and Hinckel (1996), for example, are particularly clear:

> Temperature regulation is frequently referred to as immature at the time of birth. However, one should be very cautious in asserting immaturity of the thermoregulatory mechanisms, even though the neonate shows more fluctuations in body temperature than does the adult. Greater fluctuations of body temperature are to be expected in smaller organisms because of their large surface area/volume ratio, the relatively small insulating body shell, and the smaller body mass that acts as a heat buffer in large organisms. Because of these peculiarities in body size and shape, a *reduced range of regulation* may be expected in the neonate. (p. 602, italics added)

It appears, however, that Brück and Hinckel (1996) are specifically directing the above comments toward the precocial infant mammals (e.g., guinea pig, lamb) that are typically larger in size when born, with fur, open eyes, significant behavioral mobility, and the capability of producing heat using shivering and nonshivering thermogenesis. In contrast, when discussing altricial species (e.g., rat, rabbit) they state that "either the capacity of the effector systems or the threshold temperatures or both are not sufficiently adjusted to the smaller body size. *Body temperature drops on exposure to environmental temperatures slightly less than thermal neutrality*" (p. 603, italics added). Although this statement is not inaccurate, we will see that the physiological and behavioral responses of even altricial infants within their "reduced range of regulation" are more interesting and complex than was previously thought.

A NEW VIEW OF THERMOREGULATORY COMPETENCE IN INFANT RATS

THERMOREGULATORY EFFECTORS

Temperature is a fundamental physical property that modulates nearly all chemical and biological activity. Perhaps the most obvious expression of temperature's central biological importance is the lawful, exponential relationship between temperature and cellular metabolism. This relationship generally expresses itself as a doubling or tripling of metabolic rate with each 10°C increase in ambient temperature (Schmidt-Nielsen, 1991). Although the vast majority of the earth's biomass is constrained by this lawful relationship, some animals have responded to this constraint by evolving diverse physiological and behavioral mechanisms for controlling heat exchange between the internal and external environments.

Because of their relatively large surface-to-volume ratio, infant mammals are primarily concerned with problems of heat loss. In the case of newborn rats, lack of fur and subcutaneous fat add to the problem of small size to pose a continual threat of heat loss and hypothermia. This problem is countered in rats and other altricial species through the production of large litters, which provides each individual pup with huddling littermates (Alberts, 1978). Because the surface area of a huddle of pups is less than the cumulative surface areas of the individual pups, huddling provides a behavioral means of reducing the individual's radiative heat loss at any given air temperature. Heat loss is further reduced by the construction and maintenance of an insulating nest by the mother (Jans & Leon, 1983; Kinder, 1927).

In addition to behavior, infant mammals possess a variety of physiological mechanisms that contribute to survival in the cold. Chief among these mechanisms is nonshivering thermogenesis by BAT, which we discussed at length above. In addition to nonshivering thermogenesis, precocial infants and older altricial infants also exhibit shivering thermogenesis, a prominent form of heat production in adult mammals (Brück & Wünnenberg, 1970; Taylor, 1960). For reasons that are not clear, however, nonshivering thermogenesis appears to be the favored form of heat production in young mammals. For example, in newborn guinea pigs, heat production by BAT suppresses the expression of shivering by inhibiting neural elements in the cervical spinal cord (Brück & Wünnenberg, 1970).

While most thermoregulatory effectors "evolved out of systems that were originally used for other purposes" (Satinoff, 1978, p. 21), BAT thermogenesis may be the rare exception. Thus, while shivering entails a unique, desynchronized form of thermogenic muscle activity and peripheral vasomotor activity entails the use of the circulatory system for modulating heat flow to the external environment, BAT likely evolved primarily as a thermogenic organ. In infant rats, at least until the age of 10 days, BAT is the sole source of heat production (Taylor, 1960) and is therefore the primary focus of the research described below.

Responses of Isolated Infants to Thermal Challenge

Conceptual and Methodological Considerations.   Adult mammals use a diversity of heat gain and heat loss mechanisms, both physiological and behavioral, to maintain a stable core temperature under a wide range of environmental conditions. This range is restricted, however, in infant mammals, especially those of altricial species, by small body size, poor insulation, limited locomotor abilities, and inadequate physiological mechanisms for heat loss and gain. With age, and as body size, insulation, and thermal stability increase, the range of air temperatures tolerated by the infant expands. These thermal factors contribute to the process by which pups engage in repeated egressions from the nest and eventually are weaned (Gerrish & Alberts, 1996).

Although much has been made of the thermoregulatory limitations of infant rats, these deficits should be interpreted in context. That context includes the habitat in which pups are reared, consisting of external sources of heat (i.e., littermates and mother) and thermal buffering from the ambient environment (i.e., huddle, nest, and burrow). When those external sources of heat and insulation are removed, the physiological capabilities of the infant can be probed; in doing so, however, we cannot use the same thermal challenges that we would use to probe the thermoregulatory capabilities of a huddle. Rather, we must scale the thermal challenges to the size and insulation of the individual. Again, as Conklin and Heggeness (1971) stressed, if our aim is to understand regulation, then an animal should be tested in conditions that evoke regulatory functions without overwhelming them. Thus, is it possible that the common conclusion that infant rats "quickly become hypothermic at ambient temperatures below their thermoneutral zone" (Satinoff, 1991, p. 171) is derived from testing conditions that overwhelm their regulatory capabilities?

The answer to this last question is "no": even when infant rats are gradually exposed to cold environments and heat production is observed to increase gradually, rectal temperature nonetheless decreases (e.g., Conklin & Heggeness, 1971). Thus, although it may be somewhat of an exaggeration to say that an "individual newborn rat's rectal temperature closely approximates the surrounding temperature" (Satinoff, 1991, p. 172), it is clearly the case that even moderate decreases in air temperature elicit what is, by common definition, a state of hypothermia.

Common definitions and concepts, however, may not be adequate here. Specifically, it is not necessarily the case that rectal temperature is the variable of interest for assessing the thermoregulatory success of infant rats. First, even though rectal temperature can sometimes provide valuable information about the balancing of heat gain and heat loss, its value is limited when trying to assess the dynamics of the processes involved (similarly, it would be difficult to understand the operating principles of Hoover Dam by simply monitoring the water level downstream). Second, even in adults, rectal temperature is not always the most relevant measure for assessing thermoregulatory mechanisms (Lovegrove, Heldmaier, & Ruf, 1991). For example, the singular importance of rectal temperature is contradicted by the fact that many avian and mammalian species are capable of selectively regulating brain temperature (Baker, 1979; Caputa, 1984). In other words, there is no single body temperature that is likely to provide a comprehensive description of an animal's thermal state.

In our attempt to reassess the thermoregulatory capabilities of infant rats, we evaluated the impact of a series of air temperatures on individual pups. Our approach identifies three ranges of air temperature, the exact specifications of which depend on the species, strain, age, and body size of the infant, in addition to other factors. The *thermoneutral zone*, or the zone of least thermoregulatory effort, includes those air temperatures at which an animal does not exhibit increased metabolic heat production (Satinoff, 1996); typically, oxygen consumption is minimal within this zone. As air temperature decreases, a point is reached where heat production and oxygen consumption begin to increase; this point is called the lower critical temperature and, for a week-old rat, is approximately 34°C (Spiers & Adair, 1986). As air temperature decreases further, oxygen consumption increases progressively until an air temperature is reached where it no longer increases; for a week-old rat, this occurs at an air temperature of approximately 25°C. The range of air temperatures defined from the lower critical temperature to the point of maximal oxygen consumption has been designated as *moderate*. Finally, air temperatures below the moderate zone have been designated as *extreme*.

In addition to clearly operationalizing the definitions of moderate and extreme cold exposure, it was also important to reevaluate the logic of designating rectal temperature as the variable by which to judge thermoregulatory success or failure. Given that compartmentalization of heat is now known to be a common thermoregulatory strategy, and keeping in mind that infant rats generate heat primarily in the BAT deposits in the interscapular–cervical region, it could be argued that measuring temperature in that region of the body, at the source of heat production, might provide greater insight into the regulatory dynamics of the system. Furthermore, just as moths selectively warm their thorax before taking flight (Heinrich, 1993), perhaps infant rats selectively warm their thorax during cold exposure. If so, then rectal temperature might prove to be an inadequate measure of the thermoregulatory concerns of infant rats and we would once again see that concepts derived from research on adults are sometimes inappropriately applied to infants. Of course, the flip side of this latter notion is also important: that the new concepts derived from developmental research can in turn inform our understanding of the more complex and interwoven systems of adults.

PHYSIOLOGICAL AND BEHAVIORAL RESPONSES TO COLD EXPOSURE.   A guiding theme of the research described below is that BAT thermogenesis modulates the physiological and behavioral responses of infant rats during cold exposure. In developing this theme, it has been useful to perform experiments on infants that lack the ability to produce heat endothermically. One obvious and effective strategy, discussed below, is to block BAT thermogenesis pharmacologically in infant rats; specifically, we have used chlorisondamine, a ganglionic blocker, to prevent neural activation of BAT during cold exposure (e.g., Sokoloff & Blumberg, 1998). Of course, pharmacological interventions are never as clean as one would like; specificity is always an issue and a specific antagonist for the β_3 adrenoceptor is not yet available. In addition, techniques for denervating BAT, either chemically or surgically, suffer from limitations as well (Girardier & Seydoux, 1986).

Pharmacologic or surgical interventions are not the only experimental options available for assessing the importance of a given structure or system for a physiological or behavioral process. The comparative method also provides a useful framework for such investigations by specifying a procedure for comparing different species that differ along a dimension of interest to the investigator. Syrian golden hamsters (*Mesocricetus auratus*) were chosen by us as a comparison species because they do not develop the capacity for BAT thermogenesis until they are 2 weeks of age (Hissa, 1968; Sundin, Herron, & Cannon, 1981). In other words, unmanipulated infant hamsters exhibit a natural "blockade" of BAT thermogenesis, thus providing a second avenue for assessing the contributions of BAT thermogenesis to the infant's physiological and behavioral responses to cold.

Of course, infant golden hamsters differ from rats on dimensions other than the ability to activate BAT thermogenesis. Their gestation length of just 16 days is 6 days shorter than that of rats; nonetheless, they are relatively precocial with respect to rats in that they locomote more effectively at birth and become independent feeders, develop fur, and open their eyes at an earlier postnatal age (Daly, 1976; Schoenfeld & Leonard, 1985). Their designation by some as an "immature" species (e.g., Nedergaard *et al.*, 1986) reflects the fact that they are not easily designated either as a precocial or altricial species. Thus, one challenge for our understanding of species differences in infant thermoregulation is to assess which developmental features are related to the presence or absence of endothermy.

Brown Fat Thermogenesis and the Significance of Interscapular Temperature. In rats, BAT depots are located in a number of places throughout the body, but the largest are the cervical and interscapular BAT pads (Smith & Horwitz, 1969). Under most circumstances, activation of heat production by BAT can be inferred from (1) a significant difference between skin temperatures measured in the interscapular and lumbar regions and (2) increased oxygen consumption (Blumberg & Stolba, 1996; Heim & Hull, 1966). The usefulness of these measures is illustrated in Figure 2, which presents the real-time thermoregulatory responses of two individual week-old rats acclimated at a thermoneutral air temperature (i.e., 35°C) and then challenged at either a moderate (30°C) or extreme (21°C) air temperature. These plots demonstrate that thermal challenges evoke pronounced increases in BAT thermogenesis, as indicated by the increasing difference between interscapular temperature (T_{is}) and a neutral skin temperature measured in the lumbar region (T_{back}), coupled with an increase in oxygen consumption. It can also be seen that while moderate cold exposure results in a fall in interscapular temperature, this fall is less than that which occurs during extreme cold exposure. Moreover, the moderately cooled pup stabilizes its interscapular temperature without having to maximize heat production.

We can compress much of the information contained in Figure 2 into a two-dimensional state space in which oxygen consumption is plotted against interscapular temperature. Figure 3 presents five such plots for individual week-old rats that were exposed to varying degrees of cold challenge. Relative to Figure 2, these state space plots do not convey some relevant information (e.g., time, air temperature); they do, however, readily capture the orderly relationships between the metabolic and thermal dimensions of BAT thermogenesis. Most importantly for this discussion, these plots reveal at a glance the fundamental distinction between moderate and extreme cold exposure. In addition, these state space plots help us to focus upon and appreciate the regulatory processes that underlie BAT thermogenesis.

The two upper plots in Figure 3 present the data from two week-old rats. In both cases, the pup was acclimated at an air temperature of 35°C; as indicated on the state space, each pup settled at a high interscapular temperature (~38°C) and a low rate of oxygen consumption (~4 ml O_2/100 g/min). After the acclimation period, air temperature was decreased to 30°C or 27°C and the two pups exhibit similar trajectories; in both cases, interscapular temperature falls passively before oxygen consumption increases (the horizontal portion of the plots). In contrast, when oxygen consumption begins to increase at an interscapular temperature of approximately 36.5°C, the trajectories now follow vertical paths. The verticality of the trajectories indicates that the pups are modulating heat production such that the temperature of iBAT remains relatively constant.

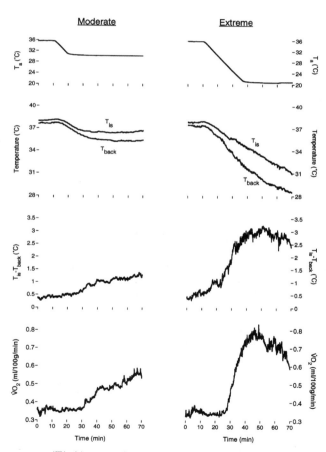

Figure 2. Air temperature (T_a), interscapular temperature (T_{is}), lower back temperature (T_{back}), $T_{is} - T_{back}$, and oxygen consumption ($\dot{V}O_2$) for two individual week-old rats challenged in a moderate (30°C) or extreme (21°C) environment. Data are from Blumberg and Stolba (1996).

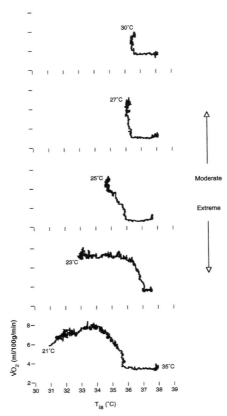

Figure 3. Space-state diagrams for individual week-old rats in which oxygen consumption ($\dot{V}O_2$) is plotted against interscapular temperature (T_{is}). Air temperature at the beginning of each test was 35°C. Temperatures at the left of each plot indicate the final air temperature for that test. Data from the top and bottom plots correspond to the T_{is} and $\dot{V}O_2$ values in Figure 2. (Reprinted from Blumberg & Sokoloff, 1998, with permission. Copyright 1998 John Wiley & Sons, Inc.)

The lower three plots in Figure 3 present the trajectories of three week-old rats cooled to 25°C, 23°C, and 21°C. We see that the vertical portions of the trajectories become progressively skewed toward lower interscapular temperatures as BAT thermogenesis is maximized. Furthermore, as cooling becomes even more extreme in the lowest plot, oxygen consumption for this pup decreases due to the suppression of cellular metabolism by cold.

As Figure 3 illustrates, pups arrive at settling points regardless of whether they are exposed to moderate or extreme air temperatures. To determine whether pups adjust metabolic heat production to defend these settling points, and to determine whether these adjustments are different at moderate and extreme air temperatures, perturbation experiments were performed (Blumberg & Sokoloff, 1997). Pups were cooled either to a moderate air temperature (30.5°C) or to an extreme air temperature (23°C). After they had settled at one of these two air temperatures, pups were exposed to a brief 3.7°C air temperature perturbation in the positive or negative direction. It was found that the pups in the moderate condition were better able to defend interscapular temperature against the perturbation than were the pups in the extreme condition. Moreover, the moderately cooled pups rapidly and effectively defended interscapular temperature to negative and positive perturbations by increasing or decreasing BAT thermogenesis, respectively. In other words, these pups responded to air temperature perturbations in either direction by exhibiting symmetric homeostatic responses that are similar to those that would be expected from any successful thermoregulator.

Respiratory Adjustments. BAT thermogenesis is an aerobic process and thus one would expect to see respiratory adjustments to cold challenge that mirror changes in heat production. In one study, strain gauges were used to measure respiratory rate during moderate and extreme cold exposure (Sokoloff & Blumberg, 1997). It was found that week-old rats increase respiratory rate approximately 80% over baseline values during cold exposure (i.e., from ~2.5 Hz at thermoneutral to ~4.5 Hz at the boundary between moderate and extreme air temperatures). It was also found that at extreme air temperatures and as oxygen consumption decreases, respiratory rate decreases. The correspondence between the responses of oxygen consumption and respiratory rate during cold exposure reflects the supportive role of respiration in providing oxygen to BAT. Furthermore, it is likely that maximum levels of BAT thermogenesis are determined in part by limitations in the delivery of oxygen to BAT (Dotta & Mortola, 1992), which in turn are partly due to biomechanical and energetic constraints on the respiratory system of the infant (Mortola, 1983, 1987).

Cardiovascular Adjustments: Cardiac Rate. Smith and Roberts (1964), in noting the venous outflow from BAT, suggested that the "direct venous connection between interscapular brown fat and the azygous vein could be readily adduced as indicative of a direct convective heat transfer to the heart" (p. 146). This observation, in conjunction with studies demonstrating the direct effects of cooling on heart rate *in vitro* (Lyman & Blinks, 1959; see also Fairfield, 1948; Tazawa & Nakagawa, 1985), suggested a possible role for BAT thermogenesis in the maintenance of cardiac function during cold challenge in infants.

Given that the largest deposits of BAT are located in the thoracic region, and in light of Smith and Roberts' (1964) hypothesis that BAT thermogenesis is primarily directed toward maintenance of cardiac function, it is reasonable to suspect that heat produced by BAT is compartmentalized in the thoracic cavity (Blumberg & Sokoloff, 1998; Blumberg, Sokoloff, & Kirby, 1997). If so, then it is certainly justified to question the customary reliance on rectal temperature as an indicator of thermoregulatory success in infant rats. Instead, interscapular temperature may provide more direct and useful information.

As suggested above, there is now substantial evidence that compartmentalization is a general theme in thermoregulatory biology. Just as oxygenated blood is directed toward metabolically active organs, so, too, may thermal resources be directed toward selected bodily compartments during cold exposure. In addition, numerous thermoregulatory mechanisms have been studied that protect particular organ systems from excessive heating or cooling. These processes and mechanisms are ubiquitous, and are found in a wide variety of insects (Heinrich, 1993), fish (Block, 1986), reptiles (Crawford, 1972), birds (Caputa, 1984), and mammals (Baker, 1979). Perhaps most striking is the finding that some moths, weighing less than 250 mg, can exhibit thermal gradients between thorax and abdomen of as much as 25°C (Heinrich, 1987); keeping the thorax warm is important because the temperature-sensitive flight muscles are located in that region. Similarly, in adult golden hamsters, the thoracic cavity is substantially warmer than the abdominal cavity during cold exposure as well as during rewarming from hibernation (Lyman & Chatfield, 1950; Pohl, 1965). Therefore, in assessing the thermoregulatory capabilities of any animal at any age, it is essential that the thermoregulatory needs of the animal be considered. As we will now see, for infant rats those needs may center on the cardiovascular system.

To assess the contribution of BAT thermogenesis to cardiac rate regulation, the electrocardiogram of week-old rats was monitored as pups were exposed to a series of moderate and extreme air temperatures (Blumberg *et al.*, 1997). When pups were cooled from thermoneutral to moderate air temperatures, cardiac rate did not change; in contrast, when pups were cooled further to extreme air temperatures, and when interscapular temperature fell, cardiac rate fell significantly (in fact, over the small air temperature drop from 26°C to 21°C, cardiac rate fell from 388 to 227 beats/min, a decrease of over 40%). In a separate experiment, when pups were pretreated with chlorisondamine to block control of the heart as well as BAT, heart rate fell in lock step with air temperature and interscapular temperature. Overall, in both untreated and ganglionically blocked pups, over 95% of the variance in cardiac rate could be accounted for by interscapular temperature.

In order to explore further the relations between cardiac rate and BAT thermogenesis, a β_3 agonist was used to selectively stimulate BAT thermogenesis in pups acclimated in a thermoneutral environment (Sokoloff, Kirby, & Blumberg, 1998). As expected, it was found that cardiac rate increased when BAT thermogenesis was stimulated and as interscapular temperature increased; this tachycardia occurred even when pups were pretreated with the ganglionic blocker, thus providing further support for the notion that cardiac rate was affected directly by warming of the heart. Finally, using a water-cooled thermode to manipulate interscapular temperature, we cooled the interscapular region of pups that had been pretreated with the β_3 agonist and found that the increase in cardiac rate could be reversed. Therefore, the results of these studies provide the strongest support for Smith and Roberts' (1964) original hypothesis that BAT thermogenesis is ideally suited to the maintenance of cardiac function during cold exposure. It is not known, however, whether rat pups enhance the effectiveness of BAT thermogenesis by selectively distributing blood flow to the thoracic cavity, as has been shown to occur in adult hamsters rewarming from hibernation (Lyman & Chatfield, 1950).

Based on the above findings relating BAT thermogenesis and cardiac rate in infant rats, we hypothesized that hamsters can only maintain cardiac rate in the cold when thermogenic mechanisms begin maturing at the end of the second week postpartum (Hissa, 1968; Sundin *et al.*, 1981). When this hypothesis was tested, it was found that cardiac rate fell in lock step with interscapular temperature even during mild cooling in hamsters 12 days of age and younger (Blumberg, 1997). In contrast, by 13 days of age when pups began exhibiting BAT thermogenesis, interscapular temperature and cardiac rate were maintained. Therefore, as in infant rats, young hamsters exhibit a striking relation between endogenous heat production and cardiac rate regulation.

It should be noted that the absence of metabolic heat production in infant hamsters precludes the use of the term "moderate" to describe the air temperatures to which they are exposed. That term is defined above on the basis of thermogenic responses to cold exposure. Therefore, given the absence of thermogenic responding by infant hamsters younger than 13 days of age, any air temperature below thermoneutral constitutes an extreme challenge for these animals.

Cardiovascular Adjustments: Arterial Pressure. Cardiac rate is just one of two variables that determine cardiac output (i.e., the amount of blood pumped by the heart per unit time). The second component of cardiac output is stroke volume, that is, the volume of blood forced out of the left ventricle during each individual contraction. In those infant mammals studied thus far, cardiac output is primarily determined by cardiac rate due to apparent limitations in the infant's ability to increase stroke volume above resting levels (e.g., Shaddy, Tyndall, Teitel, Li, & Rudolph, 1988; Teitel *et al.*, 1985). Therefore, the falling cardiac rate of infant rats during extreme cooling suggested that cardiac output was falling as well, which in turn suggested that such pups were facing related hemodynamic difficulties such as decreased arterial pressure.

To address this issue, the blood pressure responses of unanesthetized week-old rats were monitored during moderate and extreme cooling (Kirby & Blumberg, 1998). To do this, it was necessary to fashion a hair-thin catheter, insert this catheter into the femoral artery, and guide the tip to the junction of the descending abdominal aorta. When this catheter was used to monitor blood pressure during cooling to air temperatures as low as 17°C, we were surprised to find that pups were able to maintain blood pressure even in the face of substantial decreases in cardiac rate. Therefore, blood pressure is the one variable measured thus far that does not exhibit a change across the transition from moderate to extreme cold exposure. Pups maintain arterial pressure by increasing peripheral resistance (Blumberg, Knoot, & Kirby, submitted). In addition, as is discussed below, pups may engage an additional mechanism to help them maintain cardiovascular function during extreme cold exposure.

Ultrasonic Vocalizations. When isolated from the mother and littermates, infant rats emit a high-frequency, 40-kHz vocalization that is perceived by the mother and that can elicit retrieval behaviors (Allin & Banks, 1972). For many years, it has been recognized that an important stimulus for ultrasound production is the decrease in ambient temperature that normally accompanies isolation from the nest (Allin & Banks, 1971; Blumberg, Efimova, & Alberts, 1992a, b; Okon, 1971). Although the importance of cold exposure as a stimulus for ultrasound production is acknowledged, it appears that olfactory and tactile stimuli can have modulatory effects on ultrasound production (e.g., Hofer, Brunelli, & Shair, 1994).

While some investigators value the vocalizing rat pup as a potential model for the study of distress or anxiety (e.g., Miczek, Weerts, Vivian, & Barros, 1995; Winslow & Insel, 1991), we have chosen instead to focus on the environmental stimuli that elicit ultrasound production and the physiological changes that accompany it (Blumberg & Sokoloff, 2001). In one study, Blumberg and Alberts (1990) showed that rat pups emit ultrasonic vocalizations as heat production commences during cold exposure; at that time, however, we were not distinguishing between moderate and extreme air temperatures. Subsequently, when we examined ultrasound production in response to moderate and extreme cold exposure (Blumberg & Stolba, 1996), we found to our surprise that moderately cooled rat pups remain quiet even as BAT thermogenesis and oxygen consumption increase. In contrast, exposure to an extreme air temperature evoked high rates of ultrasound production. Furthermore, emission of the vocalization was found to be sensitive to air temperature differences of only 2°C across the bondary from moderate to extreme exposure (Sokoloff & Blumberg, 1997).

Although ultrasound production is described here as a behavioral response to cold, the underlying mechanisms that give rise to this behavior may actually fall within the purview of traditional physiology (for our purposes here we will leave aside the issue as to whether such distinctions between behavior and physiology are useful). Based on our earlier work on this vocalization and the physiological responses that accompany it, we hypothesized that ultrasound production is an acoustic by-product of laryngeal braking, a respiratory maneuver that is used, for example, by premature human infants as a mechanism for improving gas exchange in the lungs (Blumberg & Alberts, 1990); significantly, human infants emit an audible grunt during laryngeal braking. As we refined our procedures and measured more variables, however, an additional explanation for the rat pup's vocalization presented itself.

As described above, our investigation of arterial pressure in week-old rats indicated maintenance of pressure even during extreme cold exposure when cardiac rate (and presumably cardiac output) fell substantially (Kirby & Blumberg, 1998). It was surmised that maintenance of arterial pressure under these conditions required increased peripheral resistance, which turned out to be the case (Blumberg et al., submitted). In addition, the cooccurrence of decreased cardiac rate and increased ultrasound production raised the possibility of a causal connection between them. Specifically, a decreasing cardiac output creates a backward force that impedes venous return (Guyton & Hall, 1996), and one little-known maneuver for increasing venous return is the abdominal compression reaction (ACR). The ACR entails contraction of the abdominal muscles during or after expiration, thus increasing intraabdominal pressure and thereby propelling blood back toward the heart (Youmans et al., 1963; Youmans, Tjioe, & Tong, 1974). We hypothesized that pups employ the ACR during extreme cold exposure and that the ultrasonic vocalization is the acoustic by-product of the forceful expulsion of air during the ACR. We examined this possibility by monitoring intraabdominal pressure during cold exposure (Kirby & Blumberg, 1998). As expected, we detected sizable increases in intraabdominal pressure during each ultrasonic pulse, suggesting that ultrasound production is attended by forceful abdominal compressions that could, in theory, propel blood back to the heart during periods of decreased cardiac output. In addition, detection of arterial pressure pulses during ultrasound production provided additional support for this hypothesis.

Although it is clear that ultrasound production in infant rats during extreme cold exposure is associated with pulsatile increases in intraabdominal and arterial pressure, direct measurement of venous pressure during ultrasound production would provide even stronger evidence that venous return is increasing. Therefore, we conducted a critical test of the ACR hypothesis by measuring venous pressure in 15-day-old rats after injection with the α_2-adrenoceptor agonist clonidine (Blumberg, Sokoloff, Kirby, & Kent, 2000). Clonidine, like extreme cold exposure, evokes high rates of ultrasound production and simultaneously produces substantial bradycardia (Blumberg, Sokoloff, & Kent, 2000). We chose 15-day-olds for this study because they exhibit maximal ultrasonic responses to clonidine administration (Hård, Engel, & Lindh, 1988; Kehoe & Harris, 1989) and because their body size is sufficiently large to permit catheterization for measurement of venous pressure. As required by the ACR hypothesis, we found that ultrasonic vocalizations are associated with substantial pulsatile increases in venous pressure, indicative of increased venous return. Thus, these results suggested a functional linkage between the physiological conditions that evoke ultrasound production (i.e., decreased venous return) and the physiological consequences of the maneuver that produces ultrasound (i.e., increased venous return).

One possible depiction of the physiological and behavioral consequences of extreme cold exposure is presented in Figure 4. The cascade of events begins with the direct effects of cooling on the functional properties of heart muscle and blood. First, cardiac rate falls and results in decreases in cardiac output and venous return; second, the infant rat's blood becomes more viscous as it cools (Blumberg, Sokoloff, & Kent, 1999), thus resulting in an additional impediment to venous return (Goslinga, 1984). Recruitment of the ACR improves venous return and, in conjunction with increased peripheral resistance, contributes to the maintenance of arterial pressure. Initiation of the ACR also results in the production of ultrasound as a by-product. If, as a consequence of ultrasound production, the mother retrieves the pup to the nest, the pup's exposure to extreme cold terminates and the stresses on the cardiovascular system are alleviated.

Expression of Active Sleep Behaviors during Cold Exposure and the Role of BAT Thermogenesis. In addition to physiological mechanisms of heat production, animals use behavior to manipulate their thermal environment and to locomote to a more hospitable thermal environment (Satinoff, 1996). In addition to its role in thermoregulation, behavior can also provide important information to the investigator regarding thermoregulatory success. Specifically, as is shown below, changes in behavioral state during cold exposure provide an index by which to judge an infant rat's thermoregulatory capabilities.

For our studies of behavioral state changes during cold exposure, we focused on

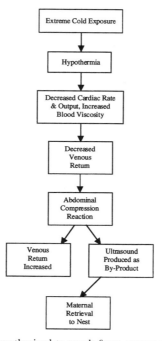

Figure 4. Cascade of events hypothesized to result from extreme cold exposure in infant rats.

myoclonic twitching, a behavior closely associated with active sleep in infants (Jouvet-Mounier, Astic, & Lacote, 1970). Myoclonic twitching consists of rapid, independent, and phasic movements of the distal limbs and tail of infant rats (Gramsbergen, Schwartze, & Prechtl, 1970). Moreover, epochs of twitching occupy as much as 70–80% of a rat pup's daily activity (Jouvet-Mounier *et al.*, 1970). Thus, although active (or rapid eye movement, REM) sleep in adults is defined on the basis of many components including a desynchronized electroencephalogram, rapid eye movements, and muscle atonia, the situation is more complicated in infants because these components are not expressed or are difficult to detect (Blumberg & Lucas, 1996). In the work described below, we have avoided the difficulties inherent in defining active sleep and other behavioral states in infants by focusing primarily on the presence or absence of myoclonic twitching.

In our first study of sleep behavior during cold exposure, isolated week-old rats initially were acclimated to a metabolic chamber at a thermoneutral air temperature (35.5°C). After acclimation, the test itself consisted of a 10-min baseline period followed by a 60-min period during which air temperature was decreased to either a moderate (30°C) or extreme (21°C) level (Blumberg & Stolba, 1996). The observer scored the incidence of myoclonic twitching continuously throughout the 70-min test. As expected, extreme cooling substantially reduced myoclonic twitching over the course of the test, but moderate cooling did not. Thus, during moderate cold exposure, it appeared that BAT thermogenesis maintains a permissive thermal environment for active sleep maintenance, as indexed by the expression of myoclonic twitching.

This hypothesis that BAT thermogenesis "protects" the expression of sleep during moderate cold exposure was tested further in a series of experiments (Sokoloff & Blumberg, 1998). The results of these experiments are presented in Figure 5. First, the findings of Blumberg and Stolba (1996) were replicated and extended in week-old rats by testing them at two moderate and two extreme air temperatures. As expected, twitching rates were high at the moderate, but not extreme air temperatures. Furthermore, rates of twitching were more closely associated with changes in interscapular temperature than oxygen consumption (Figure 5). Awake behaviors (e.g., kicking, stretching, locomotion) were also monitored in this experiment and it was found that, in general, the incidence of awake behaviors increased as rates of twitching decreased.

In the second experiment in this series (Sokoloff & Blumberg, 1998), pups were pretreated with a ganglionic blocker (i.e., chlorisondamine) and exposed to cold; now, with BAT thermogenesis inhibited, rates of twitching decreased in lock step with decreases in air temperature and interscapular temperature. Then, in the final experiment, week-old golden hamsters were tested as well and it was found that their twitching profiles during cold exposure were nearly identical to the ganglionically blocked rats. Overall, approximately 75% of the variance in levels of twitching was accounted for by interscapular temperature for the three experimental groups.

The contribution of endothermy to the maintenance of sleep behaviors in infant rats was explored further in two additional experiments (Sokoloff & Blumberg, 1998). For one experiment, a small thermode was constructed through which temperature-controlled water could flow. With this thermode attached to the interscapular region, and with a pup exposed to an extreme air temperature, supplemental heat could be added to that provided by iBAT. When this experiment was performed, pups given supplemental heat exhibited higher levels of myoclonic twitching than control animals. In a second experiment, pups were pretreated with the ganglionic blocker and were then exposed to an air temperature of 30°C (as shown in Figure 5, for ganglionically blocked pups, an air temperature of 30°C was sufficient to suppress twitching). Then, when the pups were injected with a β_3 agonist to stimulate BAT thermogenesis, both interscapular temperature and levels of myoclonic twitching increased significantly.

In their totality, these experiments on infants rats and hamsters provide strong evidence that myoclonic twitching is sensitive to the prevailing air temperature and the activation of BAT thermogenesis. Furthermore, the studies of myoclonic twitching provide additional behavioral evidence of thermoregulatory competence in untreated infant rats in that these sleep behaviors are maintained at baseline levels during moderate cold exposure but decrease significantly during extreme cold exposure.

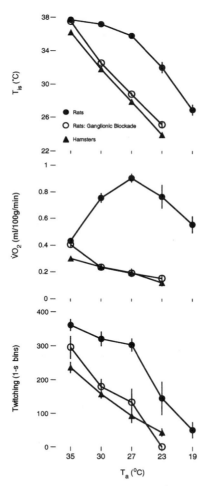

Figure 5. Interscapular temperature (T_{is}), oxygen consumption ($\dot{V}O_2$), and myoclonic twitching (per 15-min observation period) for week-old rats and hamsters during cold exposure. Rat pups were either untreated or pretreated with the ganglionic blocker chlorisondamine. The data demonstrate a strong relationship between BAT thermogenesis in the cold and the maintenance of myoclonic twitching, an index of active sleep. (Adapted from Sokoloff & Blumberg, 1998.)

Individual Differences and Developmental Changes. The research described above has concentrated on week-old rats in which the moderate air temperature range extends from approximately 25°C to 34°C. The lower boundary of the range (i.e., 25°C) is an approximation because, as stated above, the exact value can differ depending on the strain, age, size, insulation, and physiological condition of the pup, among other factors. Indeed, any experiential factor (e.g., rearing temperature, litter size) or experimental manipulation (e.g., food deprivation, surgery) that influences a pup's ability to produce heat can expand or contract the size of the moderate "zone." From a developmental standpoint, it is particularly important to recognize that the range of moderate air temperatures expands rapidly during the first 2 weeks postpartum, in part due to increases in body size (and therefore decreases in relative surface area). For example, some of the studies described above were also conducted on 2-day-old rats whose range of moderate air temperatures extends from thermoneutrality down to only 30–31°C. Nonetheless, within this narrow moderate range, 2-day-olds maintain cardiac rate (Blumberg *et al.*, 1997) and remain asleep (Blumberg & Stolba, 1996). Therefore, the concepts of moderate and extreme cold exposure are defined in relation to the individual pup being studied and its prevailing physiological state.

Summary. Figure 6 schematically summarizes the findings described above, based primarily on week-old rats. The figure should be scanned from left to right as if a pup were being exposed sequentially to decreasing air temperatures, from thermoneutral to moderate to extreme. For each variable, thicker lines represent higher values. Thus, we see that air temperature, the independent variable, decreases steadily throughout the test while the other variables exhibit distinct patterns during cooling. First, interscapular temperature, cardiac rate, and myoclonic twitching all retain high levels throughout moderate cooling and only decrease during extreme

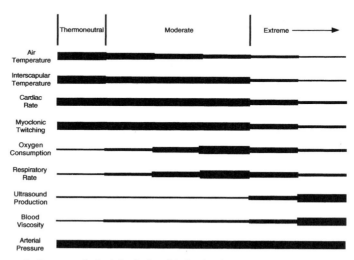

Figure 6. Schematic diagram of physiological and behavioral responses of infant rats to moderate and extreme cold exposure. Lines of greater thickness denote higher values for the given variable. The diagram is based on data collected from week-old rats, for whom the moderate range of air temperature is from approximately 25°C to 34°C. Extreme air temperatures below 17°C have not been systematically investigated.

cooling. Second, oxygen consumption and respiratory rate are similar in that they increase progressively during moderate cooling and then decrease during extreme cooling. Third, ultrasound production is rarely detected during moderate cooling but increases substantially during extreme cooling, as does blood viscosity. Finally, arterial pressure is maintained throughout the range of air temperatures tested.

BEHAVIORAL THERMOREGULATION AND GROUP REGULATORY PROCESSES IN INFANTS. With a new-found appreciation for the thermoregulatory capabilities of infant rats, we must now reconsider the individual in its natural environment, that is, as a member of a huddle of littermates. Huddling is a ubiquitous feature in mammals that give birth to multiple young. Its thermoregulatory properties were brought into sharp focus by Alberts' (1978) classic examination of huddling in rats. Huddling has also been investigated in other species including rabbits (Hull & Hull, 1982), Syrian golden hamsters (Leonard, 1982), Djungarian hamsters (Newkirk, Silverman, & Wynne-Edwards, 1995), and mice (Ogilvie & Stinson, 1966). Huddles are not static entities: individual pups within a huddle actively regulate their own temperature by diving into the middle of the huddle when cold and emerging when warm (Alberts, 1978). Therefore, individual behavioral thermoregulation is an essential component of group regulatory behavior. What has been less clear over the last two decades is how endogenous heat production contributes to the efficacy of the huddle. Stated simply, do rat pups use BAT thermogenesis while huddling, or is BAT thermogenesis activated only as an "emergency mechanism" when huddling littermates are not available?

Alberts (1978) showed that each individual pup, by huddling with its littermates, saves metabolic energy at any given air temperature. Looked at another way, the huddle can be seen as a means of expanding the range of air temperatures within which the littermates can thermoregulate. Thus, one should be able to define moderate and extreme air temperatures in the context of a huddle in the same way that this has been done for the individual. We have conducted such an experiment using infant rats and hamsters in huddles composed of two, four, or six pups (Sokoloff, Blumberg, & Adams, 2000). Not surprisingly, we have found that, for infant rats, the opportunity to huddle expands the range of air temperatures that can be defined as moderate. In contrast to rats, however, infant hamsters gain little advantage from huddling; even in huddles of six pups, heat loss cannot be significantly retarded. In other words, it appears that huddling benefits the individual only if endogenous heat production is also available. Thus, these data suggest that BAT thermogenesis is not an emergency mechanism in huddling infant rats but is an integral contributor to the huddle's effectiveness during cold exposure.

The ineffectiveness of hamster huddling does not imply that they are poor behavioral thermoregulators. On the contrary, newborn hamsters are generally considered superior to rats in their ability to move rapidly toward warmth (rats: Johanson, 1979; Kleitman & Satinoff, 1982; hamsters: Leonard, 1974, 1982). Although infants of these two species need to be tested using identical methodologies before a firm conclusion can be drawn, it is tempting to speculate with Leonard (1982) that the infant rat's larger size and endothermic capability retards cooling during cold exposure and thus suppresses its behavioral response relative to the hamster. Furthermore, it is possible that the infant hamster's lack of endothermy and the apparent ineffectiveness of huddling has made necessary the mother's nearly continual presence in the nest as a source of heat during the first week postpartum (Leonard, 1974). Thus, as we move toward a more complete understanding of the thermal context of infancy among different species, it is important that we consider the behavioral and physiological capabilities of individual infants within a huddle as well as the thermal contributions of the mother (Leon, 1986).

HOMEOSTASIS, PATHOLOGY, AND THE REALITY OF SEPARATE SYSTEMS

We saw above that when early investigators attempted to document the "development of homeothermy" (e.g., Brody, 1943), there was an implicit assumption that adult standards of thermoregulatory success could be safely applied to infants. This assumption, however, is mistaken. A better appreciation of thermoregulatory capacities is obtained when thermal challenges are scaled appropriately to the physical properties of the animal, when heat loss and heat gain mechanisms are considered separately, and when thermoregulatory success is evaluated using physiological and behavioral measures. Thus, while BAT may once have been considered a generic source of heat that contributes to the maintenance of body temperature, it is becoming apparent that in some species BAT may play a transitional thermoregulatory role during development, providing heat to the vital thoracic organs during cold exposure. In other words, BAT may be most appropriately conceptualized as an ontogenetic adaptation (Hall & Oppenheim, 1987) that is suited to the needs of developing animals; its early peak activity and involution prior to weaning (unless pups are reared in a cold environment; Nedergaard *et al.*, 1986) attests to a specific role during early development.

The thermoregulatory system is composed of multiple heat loss and heat gain mechanisms that are controlled at many levels of the neuraxis (Blumberg *et al.*, 1995; Satinoff, 1978). Peripheral feedback is evident; for example, as described earlier, young guinea pigs possess neural elements in the cervical spinal cord that are responsible for the initiation of shivering and that are suppressed by heat produced by overlying BAT (Brück & Wünnenberg, 1970). This "meshed" organization of effector mechanisms, in which feedback from one heat gain mechanism suppresses the initiation of a second heat gain mechanism, may be a general feature of homeostatic systems. The challenge for the developmental study of thermal homeostasis is to understand how these multiple mechanisms develop individually and how they become integrated such that, under normal conditions, they exhibit coordinated activation. Such a level of understanding, however, will only emerge from focused developmental studies of the responses of specific effector systems to well-defined environmental challenges. Measurement of a single outcome variable (e.g., rectal temperature) is no longer sufficient.

In this chapter and elsewhere, it is commonplace to read of a "thermoregulatory system" or a "cardiovascular system." Although these systems are discussed separately for ease of communication, one has to wonder whether we can advance our understanding of homeostasis while retaining this convenient fiction. In this regard, Satinoff's (1978) observation that most thermoregulatory mechanisms originally evolved as components of other regulatory processes is particularly significant. For example, temperature regulation and cardiovascular regulation rely on the use of circulatory adjustments to control heat loss through the skin and blood flow to metabolically active organs, respectively. Thus, to what "system" does vasomotor activity belong?

In addressing this question, consider Bartholomew's (1964) observation that "biologists, because of our limitations, divide ourselves into categories of specialization and then pretend that these categories exist in the biological world. As everyone knows, organisms are functionally indivisible and cannot be split into the conventional compartments of morphology, physiology, behaviour, and genetics" (p. 8). It should give us pause that over three decades later, Blessing (1997) was moved to make a similar point, encouraging us to critically reconsider the concept that the nervous system and its functions can be divided into independent parts: autonomic versus somatic, central versus peripheral, physiological versus behavioral. Therefore, using similar logic, we must wonder whether the tendency to focus our attention on one homeostatic system at a time will continue to be an effective approach to the study of homeostasis. That approximately 50% of the thermosensitive preoptic hypothalamic neurons examined by Boulant and Silva (1987) were also sensitive to other stimuli (e.g., glucose, osmotic pressure) provides a rational justification for investigations that are explicitly interactionist. Fortunately, such investigations are becoming more common (e.g., Pleschka & Gerstberger, 1994).

The benefits of an interactionist perspective are especially apparent from studies implicating developmental thermoregulatory deficits in the origins of such pathological conditions as obesity and hypertension. For example, researchers have developed a variety of mouse and rat strains whose offspring become obese (e.g., Krief & Bazin, 1991; Lowell *et al.*, 1993; Pelleymounter *et al.*, 1995). Investigations of these strains have helped lead to the identification of a fat cell-produced protein called leptin that provides feedback information to the hypothalamus and, by doing so, promotes either fat storage (when leptin levels are low) or fat wastage (when leptin levels are high). Wastage of fat is accomplished through heat production by brown adipose tissue. These and other findings are providing important new insights into the functional interplay between metabolism and feeding.

The various obese strains differ as to which part of the leptin–hypothalamic–BAT system are disrupted (Seeley & Schwartz, 1997). For example, *ob/ob* mice do not produce leptin, and *db/db* mice do not express the leptin receptor. Regardless of the actual mechanism that triggers obesity, however, it is clear that the precursors to obesity are present in the newborn and infant. For example, although onset of obesity in the Zucker *fa/fa* rat (a strain that, like the *db/db* mouse, expresses a defective leptin receptor; Chua *et al.*, 1996) becomes visually apparent when young are 4–5 weeks of age, thermal and metabolic disturbances can be detected as early as the first week postpartum (Krief & Bazin, 1991). Specifically, BAT thermogenesis is attenuated in *fa/fa* rats as early as 2 days of age (Planche, Joliff, & Bazin, 1988), and reduced body temperatures are found even among huddling pups (Schmidt, Kaul, & Carlisle, 1984). Significantly, daily pharmacological activation of BAT thermogenesis from 8 to 16 days of age is sufficient to prevent the onset of obesity in *fa/fa* rats (Charon, Dupuy, Marie, & Bazin, 1995). Therefore, it appears that thermoregulatory responses in the cold during infancy affect the future development of such related responses as energy storage and food intake.

Although BAT thermogenesis factors into the infant's overall energy balance, it does not follow that BAT thermogenesis exists to modulate energy balance. Nonetheless, as discussed above, in the context of leptin's modulatory effects on energy balance in adults, BAT thermogenesis is interpreted as a mechanism for "wasting" or "burning off" excess fat. For the infant, however, whose main concern is rapid growth and development, there is no such thing as "excess energy." Therefore, we wondered whether leptin plays a role in modulating BAT thermogenesis in non-obese rat strains under conditions of thermal challenge when energy use and conservation are in conflict. To do this, we revisited the phenomenon of starvation-induced suppression of thermogenesis during cold exposure, first investigated more than two decades ago by Bignall and his colleagues (e.g., Bignall *et al.*, 1975). When 4- to 5-day-old rats were isolated in an incubator away from maternal care for 18 hr, during which time leptin levels likely declined (Dessolin *et al.*, 1997), administration of leptin was sufficient to disinhibit BAT thermogenesis during cold exposure (Blumberg, Deaver, & Kirby, 1999). Therefore, the modulation of BAT thermogenesis by leptin in adults may emerge from a thermoregulatory role during infancy.

Like the Zucker rat for the study of obesity, the spontaneously hypertensive rat (SHR) has become the most widely used rat model of hypertension. This strain is characterized by high adult blood pressures due in large part to neurogenically induced vasoconstriction that, by 3 months of age, is irreversible (Yamori, 1984). Because, as noted above, vasoconstriction of peripheral resistance vessels is a mechanism that is important for maintaining blood pressure, we wondered whether infant SHRs exhibit deficient thermoregulatory responses during cold exposure in relation to pups of the normotensive control strain and, if so, whether they also exhibited deficiencies in their ability to maintain cardiac rate. As expected, infant SHRs were indeed less able to maintain interscapular temperature and oxygen consumption during cold exposure and, most importantly, also exhibited significant decreases in cardiac rate in comparison with their control strain (Kirby, Sokoloff, Perdomo, & Blumberg, 1999). In addition, the SHRs were less sensitive than the normotensive control strain to pharmacologic activation of BAT using a selective β_3-adrenoceptor agonist, suggesting that the SHR's thermoregulatory deficiencies were due in part to a diminished capacity to stimulate BAT. These findings raise the possibility that the SHR has been selected in part for developmental deficits in thermoregulation and that extreme demands placed on vasomotion in the service of heat retention in the infant reduce the range of vasomotion available to the adult for regulation of blood pressure (see Fregly, 1994, for a discussion of the induction of hypertension in adult rats after chronic cold exposure).

It is intriguing and instructive that our understanding of the most important animal models of obesity and hypertension has benefited substantially by considering the developmental mechanisms that contribute to these pathologies. It is particularly interesting that, for both of these animal models, thermoregulatory deficits have been detected during the first week postpartum. Such early deficits may initiate a cascade of events, for example, competing demands on multifunction effector mechanisms, that eventually lead to the adult pathology of interest. Clearly, more research is necessary to explore the possibility that the adult regulatory range of individual effector mechanisms is calibrated during development within a context that reflects the immediate functional needs of the young animal.

CONCLUSIONS AND FUTURE DIRECTIONS

After a half-century of investigations of the thermoregulatory competence of altricial infants, it had become commonplace to refer to them as poikilotherms (Blumberg & Sokoloff, 1998; Nedergaard *et al.*, 1986). The use of this term reflected a common belief that individual newborn rats lack the size and/or insulation necessary to retain the heat produced by shivering or nonshivering mechanisms (Satinoff, 1996). However, as we have seen, recent work demonstrates that isolated infant rats during moderate cold challenges exhibit responses that are consistent with thermoregulatory success, not failure.

With this new appreciation of the contributions of endothermy to the adaptation of infant rats to cold challenge, we are now in a position to reassess the contributions of endothermy to the group regulatory processes of the huddle. Although this work is only beginning, it appears that BAT thermogenesis plays a central role in making the rat huddle work as a thermoregulatory unit. In contrast, huddling infant golden hamsters are more behaviorally active than huddling rats, a characteristic that may be due to their lack of BAT thermogenesis. Additional comparative studies will be needed to determine those features of early development that are most directly influenced by the presence or absence of endothermy.

Although this chapter has focused largely on the contributions of endothermy to infant thermoregulation, it must be stressed that infant animals exhibit simple orienting responses toward warmth that are essential for the maintenance of the huddle. Indeed, the ability to thermoregulate behaviorally is fundamental and ubiquitous, being found in diverse animals including lice (Fraenkel & Gunn, 1961), worms (Mori & Ohshima, 1995), and reptiles and mammals (Hammel, Caldwell, & Abrams, 1967; Satinoff, 1996). In animals that cannot produce heat endogenously, behavioral thermoregulation is the only means by which the internal thermal environment can be regulated. In a recent exciting development, researchers working with a nematode worm showed that selective manipulation of a thermosensory neuron or of the interneurons that mediate the worm's behavioral attraction to warmth and cold resulted in predictable disruption of thermotaxic behavior (Mori & Ohshima, 1995). Most interesting was their finding that two identifiable interneurons appear to mediate the worm's thermoregulatory behavior, and that thermotaxis results from a balance between these ''cryophilic'' and ''thermophilic'' interneurons. This work is significant for its focus on an aspect of behavior that is so essential for survival but is, as is readily acknowledged, so poorly understood (Thomas, 1995).

Our fundamental lack of knowledge concerning the rules and mechanisms that guide behavioral thermoregulation in newborn mammals is especially apparent. Although excellent descriptive studies of behavioral thermoregulation exist (e.g., Hull & Hull, 1982; Kleitman & Satinoff, 1982; Leonard, 1982), relatively few studies have examined the behavioral rules and neural mechanisms involved in thermal orientation behaviors. Such studies are needed, and they will benefit from considering the possibility that BAT thermogenesis, perhaps by influencing the rate of heat loss (Leonard, 1982), contributes to species differences in the rapidity and accuracy of behavioral responding to cold exposure. Thus, broad experimental approaches that incorporate the methods and perspectives of multiple disciplines will bring us closer to a comprehensive understanding of the myriad mechanisms and responses by which infant mammals adjust successfully to thermal challenges.

Acknowledgment

Preparation of this chapter was supported by National Institute of Mental Health Grant MH50701.

REFERENCES

Alberts, J. R. (1978). Huddling by rat pups: Group behavioral mechanisms of temperature regulation and energy conservation. *Journal of Comparative and Physiological Psychology, 92,* 231–245.
Alberts, J. R. (1987). Early learning as ontogenetic adaptation. In N. A. Krasnegor, E. M. Blass, M. A. Hofer,

& W. P. Smotherman (Eds.), *Perinatal development: A psychobiological perspective* (pp. 11–37). New York: Academic Press.

Alberts, J. R., & Cramer, C. P. (1988). Ecology and experience: Sources of means and meaning of developmental change. In E. M. Blass (Ed.), *Handbook of behavioral neurobiology*, Vol. 8, pp. 1–39. New York: Plenum Press.

Allin, J. T., & Banks, E. M. (1971). Effects of temperature on ultrasound production by infant albino rats. *Developmental Psychobiology, 4,* 149–156.

Allin, J. T., & Banks, E. M. (1972). Functional aspects of ultrasound production by infant albino rats (*Rattus norvegicus*). *Animal Behaviour, 20,* 175–185.

Amir, S. (1990). Stimulation of the paraventricular nucleus with glutamate activates interscapular brown adipose tissue thermogenesis in rats. *Brain Research, 508,* 152–155.

Baker, M. A. (1979). A brain-cooling system in mammals. *Scientific American, 240,* 130–139.

Bartholomew, G. A. (1964). The roles of physiology and behaviour in the maintenance of homeostasis in the desert environment. *Symposia of the Society for Experimental Biology, 18,* 7–29.

Bignall, K. E., Heggeness, F. W., & Palmer, J. E. (1975). Effect of neonatal decerebration on thermogenesis during starvation and cold exposure in the rat. *Experimental Neurology, 49,* 174–188.

Blessing, W. W. (1997). Inadequate framework for understanding bodily homeostasis. *Trends in Neuroscience, 20,* 235–239.

Block, B. A. (1986). Structure of the brain and eye heater tissue in marlins, sailfish, and spearfishes. *Journal of Morphology, 190,* 169–189.

Blumberg, M. S. (1997). Ontogeny of cardiac rate regulation and brown fat thermogenesis in golden hamsters (*Mesocricetus auratus*). *Journal of Comparative Physiology B, 167,* 552–557.

Blumberg, M. S., & Alberts, J. R. (1990). Ultrasonic vocalizations by rat pups in the cold: An acoustic by-product of laryngeal braking? *Behavioral Neuroscience, 104,* 808–817.

Blumberg, M. S., & Lucas, D. E. (1996). A developmental and component analysis of active sleep. *Developmental Psychobiology, 29,* 1–22.

Blumberg, M. S., & Sokoloff, G. (1997). Dynamics of brown fat thermogenesis in week-old rats: Evidence of relative stability during moderate cold exposure. *Physiological Zoology, 70,* 324–330.

Blumberg, M. S., & Sokoloff, G. (1998). Thermoregulatory competence and behavioral expression in the young of altricial species—Revisited. *Developmental Psychobiology, 33,* 107–123.

Blumberg, M. S., & Sokoloff, G. (2001). Do infant rats cry? *Psychological Review, 108,* 83–95.

Blumberg, M. S., & Stolba, M. A. (1996). Thermogenesis, myoclonic twitching, and ultrasonic vocalization in neonatal rats during moderate and extreme cold exposure. *Behavioral Neuroscience, 110,* 305–314.

Blumberg, M. S., Efimova, I. V., & Alberts, J. R. (1992a). Ultrasonic vocalizations by rat pups: The primary importance of ambient temperature and the thermal significance of contact comfort. *Developmental Psychobiology, 25,* 229–250.

Blumberg, M. S., Efimova, I. V., & Alberts, J. R. (1992b). Thermogenesis during ultrasonic vocalization by rat pups isolated in a warm environment: A thermographic analysis. *Developmental Psychobiology, 25,* 497–510.

Blumberg, M. S., Knoot, T. G., & Kirby, R. F. (Submitted). Neural and hormonal control of arterial pressure during cold exposure in unanesthetized week-old rats.

Blumberg, M. S., Schalk, S. L., & Sokoloff, G. (1995). Pontine and basal forebrain transections disinhibit brown fat thermogenesis in neonatal rats. *Brain Research, 699,* 214–220.

Blumberg, M. S., Sokoloff, G., & Kirby, R. F. (1997). Brown fat thermogenesis and cardiac rate regulation during cold challenge in infant rats. *American Journal of Physiology, 272,* R1308–R1313.

Blumberg, M. S., Deaver, K., & Kirby, R. F. (1999). Leptin disinhibits BAT thermogenesis in infants after maternal separation. *American Journal of Physiology, 276,* R606–R610.

Blumberg, M. S., Sokoloff, G., & Kent, K. J. (1999). Cardiovascular concomitants of ultrasound production during cold exposure in infant rats. *Behavioral Neuroscience, 113,* 1274–1282.

Blumberg, M. S., Sokoloff, G., & Kent, K. J. (2000). A developmental analysis of clonidine's effects on cardiac rate and ultrasound production in infant rats. *Developmental Psychobiology, 36,* 186–193.

Blumberg, M. S., Sokoloff, G., Kirby, R. F., & Kent, K. J. (2000). Distress vocalizations in infant rats: What's all the fuss about? *Psychological Science, 11,* 78–81.

Booth, D. A. (1980). Conditioned reactions in motivation. In F. M. Toates & T. R. Halliday (Eds.), *Analysis of motivational processes* (pp. 3–21). New York: Academic Press.

Boulant, J. A., & Silva, N. L. (1987). Interactions of reproductive steroids, osmotic pressure, and glucose on thermosensitive neurons in preoptic tissue slices. *Canadian Journal of Physiology and Pharmacology, 65,* 1267–1273.

Brody, E. B. (1943). Development of homeothermy in suckling rats. *American Journal of Physiology, 139,* 230–232.

Brück, K., & Hinckel, P. (1996). Ontogenetic and adaptive adjustments in the thermoregulatory system. In M. J. Fregly & C. M. Blatteis (Eds.), *Handbook of physiology* (pp. 597–611). Oxford: Oxford University Press.

Brück, K., & Wünnenberg, B. (1970). "Meshed" control of two effector systems: nonshivering and shivering thermogenesis. In J. D. Hardy, A. P. Gagge, & J. A. J. Stolwijk (Eds.), *Physiological and behavioral temperature regulation* (pp. 562–580). Springfield, IL: Thomas.

Cannon, B., Jacobsson, A., Rehnmark, S., & Nedergaard, J. (1996). Signal transduction in brown adipose tissue recruitment: Noradrenaline and beyond. *International Journal of Obesity, 20* (Supplement 3), S36–S42.

Cannon, W. B. (1932). *The wisdom of the body.* New York: W. W. Norton.

Caputa, M. (1984). Some differences in mammalian versus avian temperature regulation: Putative thermal adjustments to flight in birds. In J. R. S. Hales (Ed.), *Thermal physiology* (pp. 413–417). New York: Raven Press.

Charon, C., Dupuy, F., Marie, V., & Bazin, R. (1995). Effect of the β-adrenoceptor agonist BRL-35135 on development of obesity in suckling Zucker (*fa/fa*) rats. *American Journal of Physiology, 268*, E1039–E1045.

Chua, S. C., Jr., Chung, W. K., Wu-Peng, S., Zhang, Y., Liu, S.-M., Tartaglia, L., & Leibel, R. L. (1996). Phenotypes of mouse diabetes and rat fatty due to mutations in the *ob* (leptin) receptor. *Science, 271*, 994–996.

Conklin, P., & Heggeness, F. W. (1971). Maturation of temperature homeostasis in the rat. *American Journal of Physiology, 220*, 333–336.

Crawford, E. C., Jr. (1972). Brain and body temperatures in a panting lizard. *Science, 177*, 431–433.

Daly, M. (1976). Behavioral development in three hamster species. *Developmental Psychobiology, 9*, 315–323.

Dawkins, M. J. R., & Scopes, J. W. (1965). Nonshivering thermogenesis and brown adipose tissue in the human new-born infant. *Nature, 206*, 201–202.

Dessolin, S., Schalling, M., Champigny, O., Lönnqvist, F., Ailhaud, G., Dani, C., & Ricquier, D. (1997). Leptin gene is expressed in rat brown adipose tissue at birth. *FASEB Journal, 11*, 382–387.

Dotta, A., & Mortola, J. P. (1992). Effects of hyperoxia on the metabolic response to cold of the newborn rat. *Journal of Developmental Physiology, 17*, 247–250.

Fairfield, J. (1948). Effects of cold on infant rats: Body temperatures, oxygen consumption, electrocardiograms. *American Journal of Physiology, 155*, 355–365.

Fraenkel, G. S., & Gunn, D. L. (1961). *The orientation of animals.* New York: Dover.

Freeman, P. H., & Wellman, P. J. (1987). Brown adipose tissue thermogenesis induced by low level electrical stimulation of hypothalamus in rats. *Brain Research Bulletin, 18*, 7–11.

Fregly, M. J. (1994). Induction of hypertension in rats exposed chronically to cold. In K. Pleschka & R. Gerstberger (Eds.), *Integrative and cellular aspects of autonomic functions: Temperature and osmoregulation* (pp. 201–208). Paris: John Libby Eurotext.

Gerrish, C. J., & Alberts, J. R. (1996). Environmental temperature modulates onset of independent feeding: Warmer is sooner. *Developmental Psychobiology, 29*, 483–495.

Girardier, L., & Seydoux, J. (1986). Neural control of brown adipose tissue. In P. Trayhurn & D. G. Nicholls (Eds.), *Brown adipose tissue* (pp. 122–151). London: Edward Arnold.

Goslinga, H. (1984). *Blood viscosity and shock: The role of hemodiluation, hemoconcentration and defibrination.* New York: Springer-Verlag.

Gramsbergen, A., Schwartze, P., & Prechtl, H. F. R. (1970). The postnatal development of behavioral states in the rat. *Developmental Psychobiology, 3*, 267–280.

Guyton, A. C., & Hall, J. E. (1996). *Medical physiology.* Philadelphia: Saunders.

Hahn, P. (1956). The development of thermoregulation. III. The significance of fur in the development of thermoregulation in rats. *Physiologia Bohemoslovenica, 5*, 428–431.

Hall, W. G., & Oppenheim, R. W. (1987). Developmental psychobiology: Prenatal, perinatal, and early postnatal aspects of behavioral development. *Annual Review of Psychology, 38*, 91–128.

Hammel, H. T., Jackson, C. D., Stolwijk, J. A. J., Hardy, J. D., & Strømme, S. B. (1963). Temperature regulation by hypothalamic proportional control with an adjustable set point. *Journal of Applied Physiology, 18*, 1146–1154.

Hammel, H. T., Caldwell, F. T., Jr., & Abrams, R. M. (1967). Regulation of body temperature in the blue-tongued lizard. *Science, 156*, 1260–1262.

Hård, E., Engel, J., & Lindh, A. (1988). Effect of clonidine on ultrasonic vocalization in preweaning rats. *Journal of Neural Transmission, 73*, 217–237.

Heim, T., & Hull, D. (1966). The blood flow and oxygen consumption of brown adipose tissue in the new-born rabbit. *Journal of Physiology, 186*, 42–55.

Heinrich, B. (1987). Thermoregulation by winter-flying endothermic moths. *Journal of Experimental Biology, 127,* 313–332.

Heinrich, B. (1993). *The hot-blooded insects: Strategies and mechanisms of thermoregulation.* Cambridge, MA: Harvard University Press.

Hissa, R. (1968). Postnatal development of thermoregulation in the Norwegian lemming and the golden hamster. *Annales Zoologici Fennici, 5,* 345–383.

Hofer, M. A., Brunelli, S. A., & Shair, H. (1994). Potentiation of isolation-induced vocalization by brief exposure of rat pups to maternal cues. *Developmental Psychobiology, 26,* 81–95.

Hogan, J. A. (1980). Homeostasis and behavior. In F. M. Toates & T. R. Halliday (Eds.), *Analysis of motivational processes* (pp. 3–21). New York: Academic Press.

Houk, J. C. (1988). Control strategies in physiological systems. *FASEB Journal, 2,* 97–107.

Hsieh, A. C. L., Carlson, L. D., & Gray, G. (1957). Role of the sympathetic nervous system in the control of chemical regulation of heat production. *American Journal of Physiology, 190,* 247–251.

Hull, J., & Hull, D. (1982). Behavioral thermoregulation in newborn rabbits. *Journal of Comparative and Physiological Psychology, 96,* 143–147.

Hull, D., & Segall, M. M. (1965). The contribution of brown adipose tissue to heat production in the new-born rabbit. *Journal of Physiology, 181,* 449–457.

Jans, J., & Leon, M. (1983). Determinants of mother–young contact in Norway rats. *Physiology and Behavior, 30,* 919–935.

Johanson, I. B. (1979). Thermotaxis in neonatal rat pups. *Physiology and Behavior, 23,* 871–874.

Johnston, D. W. (1971). The absence of brown adipose tissue in birds. *Comparative Biochemistry and Physiology, 40A,* 1108–1108.

Jouvet-Mounier, D., Astic, L., & Lacote, D. (1970). Ontogenesis of the states of sleep in rat, cat, and guinea pig during the first postnatal month. *Developmental Psychobiology, 2,* 216–239.

Kehoe, P., & Harris, J. C. (1989). Ontogeny of noradrenergic effects on ultrasonic vocalizations in rat pups. *Behavioral Neuroscience, 103,* 1099–1107.

Kinder, E. F. (1927). A study of nest-building activity of the albino rat. *Journal of Experimental Zoology, 147,* 117–161.

Kirby, R. F., & Blumberg, M. S. (1998). Maintenance of arterial pressure in infant rats during moderate and extreme thermal challenge. *Developmental Psychobiology, 32,* 169–181.

Kirby, R. F., Sokoloff, G., Perdomo, E., & Blumberg, M. S. (1999). Thermoregulatory and cardiovascular responses of infant SHR and WKY rats to cold exposure. *Hypertension, 33,* 1465–1469.

Kleitman, N., & Satinoff, E. (1982). Thermoregulatory behavior in rat pups from birth to weaning. *Physiology and Behavior, 29,* 537–541.

Krief, S., & Bazin, R. (1991). Genetic obesity: Is the defect in the sympathetic nervous system? A review through developmental studies in the preobese Zucker rat. *Proceedings of the Society for Experimental Biology and Medicine, 198,* 528–538.

Leon, M. (1986). Development of thermoregulation. In E. M. Blass (Ed.), *Handbook of behavioral neurobiology* (Vol. 8, pp. 297–322). New York: Plenum Press.

Leonard, C. M. (1974). Thermotaxis in golden hamster pups. *Journal of Comparative and Physiological Psychology, 86,* 458–469.

Leonard, C. M. (1982). Shifting strategies for behavioral thermoregulation in developing golden hamsters. *Journal of Comparative and Physiological Psychology, 96,* 234–243.

Lovegrove, B. G., Heldmaier, G., & Ruf, T. (1991). Perspectives of endothermy revisited: The endothermic temperature range. *Journal of Thermal Biology, 16,* 185–197.

Lowell, B. B., SSusulic, V., Hamann, A., Lawitts, J. A., Himms-Hagen, J., Boyer, B. B., Kozak, L. P., & Flier, J. S. (1993). Development of obesity in transgenic mice after genetic ablation of brown adipose tissue. *Nature, 366,* 740–742.

Lyman, C. P., & Blinks, D. C. (1959). The effect of temperature on the isolated hearts of closely related hibernators and non-hibernators. *Journal of Cellular and Comparative Physiology, 54,* 53–63.

Lyman, C. P., & Chatfield, P. O. (1950). Mechanism of arousal in the hibernating hamster. *Journal of Experimental Zoology, 114,* 491–516.

Miczek, K. A., Weerts, E. M., Vivian, J. A., & Barros, H. M. (1995). Aggression, anxiety and vocalizations in animals: GABA$_A$ and 5-HT anxiolytics. *Psychopharmacology, 121,* 38–56.

Mori, I., & Ohshima, Y. (1995). Neural regulation of thermotaxis in *Caenorhabditis elegans. Nature, 376,* 344–348.

Mortola, J. P. (1983). Comparative aspects of the dynamics of breathing in newborn mammals. *Journal of Applied Physiology, 54,* 1229–1235.

Mortola, J. P. (1987). Dynamics of breathing in newborn mammals. *Physiological Reviews, 67,* 187–243.

Mrosovsky, N. (1990). *Rheostasis: The physiology of change.* Oxford: Oxford University Press.

Nedergaard, J., Connolly, E., & Cannon, B. (1986). Brown adipose tissue in the mammalian neonate. In P. Trayhurn & D. G. Nicholls (Eds.), *Brown adipose tissue* (pp. 152–213). London: Edward Arnold.

Newkirk, K. D., Silverman, D. A., & Wynne-Edwards, K. E. (1995). Ontogeny of thermoregulation in the Djungarian hamster (*Phodopus campbelli*). *Physiology and Behavior, 57,* 117–124.

Ogilvie, D. M., & Stinson, R. H. (1966). The effect of age on temperature selection by laboratory mice (*Mus musculus*). *Canadian Journal of Zoology, 44,* 511–517.

Okon, E. E. (1971). The temperature relations of vocalization in infant Golden hamsters and Wistar rats. *Journal of Zoology* (London), *164,* 227–237.

Pelleymounter, M. A., Cullen, M. J., Baker, M. B., Hecht, R., Winters, D., Boone, T., & Collins, F. (1995). Effects of the *obese* gene product on body weight regulation in *ob/ob* mice. *Science, 269,* 540–543.

Planche, E., Joliff, M., & Bazin, R. (1988). Energy expenditure and adipose tissue development in 2- to 8-day-old Zucker rats. *International Journal of Obesity, 12,* 352–360.

Pleschka, K., & Gerstberger, R. (1994). *Integrative and cellular aspects of autonomic functions: Temperature and osmoregulation.* Paris: John Libbey Eurotext.

Pohl, H. (1965). Temperature regulation and cold acclimation in the golden hamster. *Journal of Applied Physiology, 20,* 405–410.

Richter, C. P. (1943). Total self regulatory functions in animals and human beings. *Harvey Lectures, 38,* 63–103.

Rothwell, N. J., Stock, M. J., & Thexton, A. J. (1983). Decerebration activates thermogenesis in the rat. *Journal of Physiology, 342,* 15–22.

Satinoff, E. (1978). Neural organization and evolution of thermal regulation in mammals. *Science, 201,* 16–22.

Satinoff, E. (1991). Developmental aspects of behavioral and reflexive thermoregulation. In H. N. Shair, G. A. Barr, & M. A. Hofer (Eds.), *Developmental psychobiology: New methods and changing concepts* (pp. 169–188). Oxford: Oxford University Press.

Satinoff, E. (1996). Behavioral thermoregulation in the cold. In M. J. Fregly & C. M. Blatteis (Eds.), *Handbook of Physiology* (pp. 481–505). Oxford: Oxford University Press.

Schmidt, I., Kaul, R., & Carlisle, H. J. (1984). Body temperature of huddling newborn Zucker rats. *Pflüger's Archive, 401,* 418–420.

Schmidt-Nielsen, K. (1981). Countercurrent systems in animals. *Scientific American, 244,* 118–128.

Schmidt-Nielsen, K. (1991). *Animal physiology: Adaptation and environment.* Cambridge: Cambridge University Press.

Schneider-Picard, G., & Girarder, L. (1982). Postnatal development of sympathetic innervation of rat brown adipose tissue reevaluated with a method allowing for monitoring flavoprotein redox state. *Journal of Physiology* (Paris), *78,* 151–157.

Schoenfeld, T. A., & Leonard, C. M. (1985). Behavioral development in the Syrian golden hamster. In H. I. Siegel (Ed.), *The hamster: Reproduction and behavior* (pp. 289–321). New York: Plenum Press.

Seeley, R. J., & Schwartz, M. W. (1997). The regulation of energy balance: Peripheral hormonal signals and hypothalamic neuropeptides. *Current Directions in Psychological Science, 6,* 39–44.

Shaddy, R. E., Tyndall, M. R., Teitel, D. F., Li, C., & Rudolph, A. M. (1988). Regulation of cardiac output with controlled heart rate in newborn lambs. *Pediatric Research, 24,* 577–582.

Shibata, M., Benzi, R. H., Seydoux, J., & Girardier, L. (1987). Hyperthermia induced by prepontine knife-cut: Evidence for a tonic inhibition of non-shivering thermogenesis in anesthetized rat. *Brain Research, 436,* 273–282.

Smith, R. E. (1961). Thermogenic activity of the hibernating gland in the cold-acclimated rat. *Physiologist, 4,* 113.

Smith, R. E. (1964). Thermoregulatory and adaptive behavior of brown adipose tissue. *Science, 146,* 1686–1689.

Smith, R. E., & Horwitz, B. A. (1969). Brown fat and thermogenesis. *Physiological Reviews, 49,* 330–425.

Smith, R. E., & Roberts, J. C. (1964). Thermogenesis of brown adipose tissue in cold-acclimated rats. *American Journal of Physiology, 206,* 143–148.

Sokoloff, G., & Blumberg, M. S. (1997). Thermogenic, respiratory, and ultrasonic responses of week-old rats across the transition from moderate to extreme cold exposure. *Developmental Psychobiology, 30,* 181–194.

Sokoloff, G., & Blumberg, M. S. (1998). Active sleep in cold-exposed infant Norway rats and Syrian golden hamsters: The role of brown adipose tissue thermogenesis. *Behavioral Neuroscience, 112,* 695–706.

Sokoloff, G., Blumberg, M. S., & Adams, M. M. (2000). A comparative analysis of huddling in infant Norway rats and Syrian golden hamsters: Does endothermy modulate behavior? *Behavioral Neuroscience, 114,* 585–593.

Sokoloff, G., Kirby, R. F., & Blumberg, M. S. (1998). Further evidence that BAT thermogenesis modulates cardiac rate in infant rats. *American Journal of Physiology, 274,* R1712–R1717.

Spiers, D. E., & Adair, E. R. (1986). Ontogeny of homeothermy in the immature rat: Metabolic and thermal responses. *Journal of Applied Physiology, 60,* 1190–1197.

Sundin, U., Herron, D., & Cannon, B. (1981). Brown fat thermoregulation in developing hamsters (*Mesocricetus auratus*): A GDP-binding study. *Biology of the Neonate, 39,* 141–149.

Taylor, P. M. (1960). Oxygen consumption in new-born rats. *Journal of Physiology, 154,* 153–168.

Tazawa, H., & Nakagawa, S. (1985). Response of egg temperature, heart rate and blood pressure in the chick embryo to hypothermal stress. *Journal of Comparative Physiology B, 155,* 195–200.

Teitel, D. F., Sidi, D., Chin, T., Brett, C., Heymann, M. A., & Rudolph, A. M. (1985). Developmental changes in myocardial contractile reserve in the lamb. *Pediatric Research, 19,* 948–955.

Thomas, J. H. (1995). Some like it hot. *Current Biology, 5,* 1222–1224.

West, M. J., King, A. P., & Arberg, A. A. (1988). The inheritance of niches: The role of ecological legacies in ontogeny. In E. M. Blass (Ed.), *Handbook of behavioral neurobiology* (Vol. 9, pp. 41–62). New York: Plenum Press.

Wiener, N. (1961). *Cybernetics or control and communication in the animal and the machine.* Cambridge, MA: MIT Press.

Williams, G. C. (1966). *Adaptation and natural selection.* Princeton, NJ: Princeton University Press.

Winslow, J. T., & Insel, T. R. (1991). The infant rat separation paradigm: A novel test for novel anxiolytics. *Trends in Pharmacological Sciences, 12,* 402–404.

Yamori, Y. (1984). Development of the spontaneously hypertensive rat (SHR) and of various spontaneous rat models, and their implications. In W. de Jong (Ed.), *Handbook of hypertension, Volume 4: Experimental and genetic models of hypertension* (pp. 224–239). New York: Elsevier.

Youmans, W. B., Murphy, Q. R., Turner, J. K., Davis, L. D., Briggs, D. I., & Hoye, A. S. (1963). Activity of abdominal muscles elicited from the circulatory system. *American Journal of Physical Medicine, 42,* 1–70.

Youmans, W. B., Tjioe, D. T., & Tong, E. Y. (1974). Control of involuntary activity of abdominal muscles. *American Journal of Physical Medicine, 53,* 57–74.

Zhao, J., Unelius, L., Bengtsson, T., Cannon, B., & Nedergaard, J. (1994). Coexisting b-adrenoceptor subtypes: Significance for thermogenic process in brown fat cells. *American Journal of Physiology, 267,* C969–C979.

6

Development of Behavior Systems

JERRY A. HOGAN

The purpose of this chapter is to present a general framework for studying the development of behavior. The thesis to be defended here is that the building blocks of behavior are various kinds of perceptual, central, and motor components, all of which can exist independently. The study of development is primarily the study of changes in these components themselves and in the connections among them.

I begin the chapter by explaining my conception of a behavior system. The basic concepts that I use are generally derived from classical ethological theory as set forth, for example, by Tinbergen (1951). There are, however, a number of differences in the way I define and use these concepts, and these differences are discussed where appropriate. The bulk of the chapter is devoted to the presentation and discussion of examples showing how behavior systems develop. Many of these examples are based on my own work on chickens, but I also show how the behavior systems of chickens can be considered to be typical of behavior systems in other species. One such system is the language system in humans, and one section of the chapter is devoted to showing how the general framework presented here can be applied to the development of human language. Finally, I discuss a number of general issues, including the distinction between causal and functional classification of behavior systems, the relevance of functional considerations to causal analyses, and whether any general principles of development emerge from the data.

THE CONCEPTION OF A BEHAVIOR SYSTEM

No two occurrences of behavior are ever identical, and it is therefore necessary to sort behavior into categories in order to make scientific generalizations. These categories can be defined in different ways (e.g., structurally, causally, or functionally; Hinde, 1970, Chapter 2; Hogan, 1994a) and at different levels of complexity

JERRY A. HOGAN   Department of Psychology, University of Toronto, Toronto, Canada M5S 3G3.

Developmental Psychobiology, Volume 13 of *Handbook of Behavioral Neurobiology*,
edited by Elliott Blass, Kluwer Academic / Plenum Publishers, New York, 2001.

(e.g., individual muscle movements, limb movements, or acts; Gallistel, 1980). The concept of a behavior system is defined here structurally, and the level to be analyzed corresponds to the complexity indicated by the terms *feeding behavior, aggressive behavior, play behavior,* and so on. These terms can be considered names for behavior systems as a whole, but our analysis begins with a consideration of the parts of which these systems are constructed.

Three kinds of parts can be distinguished: motor parts, perceptual parts, and central parts. All of these parts are viewed as corresponding to structures within the central nervous system. For this reason, the word *mechanism*[*] is used in the rest of this chapter in references to these parts. Each motor mechanism, perceptual mechanism, or central mechanism is conceived of as consisting of some arrangement of neurons (not necessarily localized) that is able to act independently of other such mechanisms. These mechanisms are here called *behavior mechanisms* for two reasons. First, the actual neural connections, their location, and their neurophysiology are not of direct interest in the study of behavior. Second, the activation of a behavior mechanism results in an event of behavioral interest: a particular perception, a specific motor pattern, or an identifiable internal state.

Behavior mechanisms can be connected with one another, and the organization of these connections determines the nature of the behavior system. In order to make the discussion more concrete, I shall use the feeding system of a chicken as my example.

MOTOR MECHANISMS

We say a chicken is feeding when it walks about looking at the ground, when it scratches at the substrate, and when it pecks and swallows small objects. Walking, scratching, pecking, and swallowing are all easily recognizable motor patterns and can be viewed as reflecting the motor mechanisms of the feeding system. Three points here are worthy of mention.

First, although the behavior patterns of walking and so on are easily recognizable, there is considerable variation between different instances of the "same" pattern. In a practical sense, this variation does not usually interfere with the identification of a pattern, and that is sufficient for our present purpose. The second point is essential. What we observe is only a reflection or manifestation of the motor mechanisms of the system. The motor mechanisms themselves are groups of neurons located inside the central nervous system of the animal; activation of a motor mechanism is responsible for coordinating the muscle movements that we actually see. Finally, the concept of a motor mechanism is clearly related to the ethological concept *Erbkoordination* (Lorenz, 1937) or *fixed action pattern* (Hinde, 1970; Tinbergen, 1951) but is meant to be much broader in scope and to encompass all types of coordinated movements.

PERCEPTUAL MECHANISMS

Corresponding to the motor mechanisms on the efferent side of a behavior system are perceptual mechanisms on the afferent side. Perceptual mechanisms solve

---

[*]The word *mechanism* usually connotes analysis at a molecular level. Nonetheless, as I discuss in detail elsewhere (Hogan, 1994a, pp. 9–10), the dictionary definition of mechanism merely implies cause, and is agnostic with respect to the level of analysis. I use the word mechanism to emphasize the fact that the perceptual, motor, and central units that are discussed in this chapter are causal concepts at the behavioral level.

the problem of stimulus recognition and are often associated with particular motor mechanisms. In the feeding system of a chicken, there must be perceptual mechanisms for recognizing the objects at which the bird pecks, for what it swallows, and for the type of environment in which the bird scratches. There must also be perceptual mechanisms that are sensitive to changes in the chick's internal state consequent to its behavior. Particular perceptual mechanisms may be restricted to a single sensory modality, but frequently integrate information from several modalities.

Perceptual mechanisms are inherently more difficult to study than motor mechanisms because the output of a perceptual mechanism can be "seen" only after it has activated some motor mechanism. Thus, there are always more steps where variation can occur. The general method used to study perceptual mechanisms is to present stimuli that vary along different dimensions and to ascertain which combination of characteristics is most effective in bringing about certain responses.

The concept *perceptual mechanism* is clearly related to concepts such as *releasing mechanism* (Baerends & Kruijt, 1973; Lorenz, 1937; Tinbergen, 1951); *Sollwert*, or *comparator mechanism* (Hinde, 1970; von Holst, 1954); *cell assembly* (Hebb, 1949); and *analyzer* (Sutherland, 1964). However, as with the term *motor mechanism, perceptual mechanism* is meant to encompass all types of stimulus recognition mechanisms, including such "cognitive" mechanisms as *ideas* and *memories* (Hogan, 1994a).

CENTRAL MECHANISMS

The final part of a behavior system to be considered is the central mechanism. This part is responsible for integrating the input from various perceptual mechanisms and coordinating the output to the various motor mechanisms associated with it. In many cases, it is also responsible for the timing and activation of the whole behavior system. It is the central mechanism that usually corresponds to the name we give to a behavior system: a hunger mechanism, an aggression mechanism, a sexual mechanism, and so on. The concept *central mechanism* is clearly related to the neurophysiologic concepts *central excitatory mechanism* (Beach, 1942), *central motive state* (Stellar, 1960), or *neural center* (Doty, 1976), but it will be used here in a still more general sense. Central mechanisms do not differ in any basic way from motor or perceptual mechanisms; they are distinguished separately because of their function of coordinating motor, perceptual, and motivational mechanisms.

BEHAVIOR SYSTEMS

We can now return to the concept *behavior system* and define it as an organization of perceptual, central, and motor mechanisms that act as a unit in some situations. A pictorial representation of this definition is shown in Figure 1.

The first part of the definition is structural and is basically similar to Tinbergen's (1951, p. 112) definition of an instinct; it is also similar to the *functional* organization of von Holst and von St. Paul (1960). Hierarchical organization is also implied in this part of the definition, and it is thus related to conceptions of Tinbergen (1951), Baerends (1976), and Gallistel (1980); see also Hogan (1981). Further, as we shall see, there are various levels of perceptual and motor mechanisms, and the connections among them can become very complex. A diagram such as Figure 1, if expanded to encompass all the facts that are known, would soon become unmanageable. In the extreme, it would become congruent with a wiring diagram of the brain. The main function of such a diagram, and of the concept of a behavior system, is to direct our thinking into particular pathways.

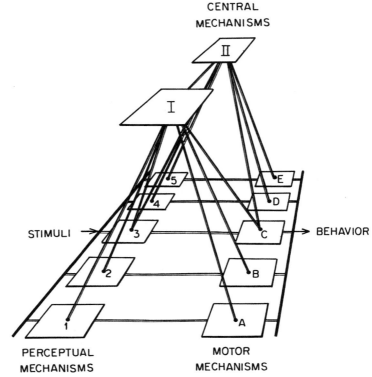

Figure 1. Conception of behavior systems. Stimuli from the external world are analyzed by perceptual mechanisms. Output from the perceptual mechanisms can be integrated by central mechanisms and/or channeled directly to motor mechanisms. The output of the motor mechanisms results in behavior. In this diagram, central mechanism I, perceptual mechanisms 1, 2, and 3, and motor mechanisms A, B, and C form one behavior system; central mechanism II, perceptual mechanisms 3, 4, and 5, and motor mechanisms C, D, and E form a second behavior system. Systems 1–A, 2–B, and so on can also be considered less complex behavior systems. (From Hogan, 1988.)

The second part of the definition of a behavior system is causal: at present, the only method for determining behavioral structure is through causal (or motivational) analysis. In discussing the development of behavior systems, we shall be interested in both structural and causal (motivational) aspects.

THE DEVELOPMENT OF BEHAVIOR SYSTEMS

In a very real sense, the development of behavior begins at conception and continues until death. Nonetheless, much can be understood about the development of behavior systems by considering only the period between birth (hatching) and maturity, and that is what I shall do here. The thesis of this chapter is that perceptual, central, and motor mechanisms are the building blocks out of which complex behavior is formed, and that a developmental analysis requires looking for the factors causing the development of the building blocks themselves as well as for the way connections among these building blocks become established.

In some cases, the building blocks and/or their connections appear for the first time "prefunctionally" (Schiller, 1949); that is, functional experience is not necessary for their development. A building block (e.g., the pecking motor mechanism) is functional when its associated response (i.e., pecking) occurs in its adaptive context (i.e., grasping small objects). If the pecking response occurs in its normal form before the chick has ever grasped an object, the development of the pecking motor mechanism can be said to occur prefunctionally: experience grasping an object is not necessary for the development of a normal pecking response.

It should be noted that saying that a behavior mechanism develops prefunctionally implies only that particular kinds of experience play no role in its development; there is no implication about the role of other kinds of experience. For example, the development of the pecking motor mechanism in the chick may well be influenced by events associated with beak movements that occur in the egg before hatching or with head and beak movements that occur during hatching. The pecking motor mechanism would nonetheless still be regarded as appearing prefunctionally. This concept is discussed in greater detail later.

Even in cases in which behavior develops prefunctionally, developmental questions arise. I begin with an example of such a system. I then consider several examples of how individual behavior mechanisms develop and, finally, some examples of the development of more complex systems.

THE "GUSTOFACIAL REFLEX": A PREFUNCTIONALLY DEVELOPED SYSTEM

Steiner (1979) showed that newborn human infants have at least three gustofacial reflexes. A sweet stimulus to the tongue elicits a "smile" reaction, a sour stimulus elicits a "pucker" reaction, and a bitter substance elicits a "disgust" reaction. The identification of these reactions by even inexperienced observers is highly reliable. In terms of the concepts discussed above, we can posit that the newborn infant has three perceptual mechanisms for particular tastes (a sweet, a sour, and a bitter mechanism) and three motor mechanisms (a smile, a pucker, and a disgust mechanism). These mechanisms and the specific connections between them are formed prefunctionally, that is, before the consequences of ingesting sweet, sour, or bitter substances have been experienced and before any social (or other) reactions to these facial expressions can have been perceived. Nonetheless, there are many questions of developmental interest that can be asked about these results.

With respect to the motor mechanisms, there is a large literature on the form and development of human facial expressions. Ekman and Friesen (see Ekman, 1982) devised a facial action coding system which analyzes all human adult facial expressions as combinations of about 50 basic action units, and Oster (1978) reported that almost all of these discrete action units can be identified in the facial movements of newborn infants. In this system, the smile, pucker, and disgust patterns discussed by Steiner consist of particular combinations of the basic action units. One can ask how these motor patterns are organized, how they change as the infant grows older, and what experience is necessary for the changes to occur. Thelen (1985) used this framework of hierarchical organization of coordinative structures for understanding the development of motor mechanisms in general, and I return to some of her ideas in a later section.

The perceptual mechanisms that recognize sweet, sour, and bitter are probably the basic perceptual units, and developmental interest would focus on connections between them and other behavior mechanisms rather than on the development of the perceptual mechanisms themselves. Some of these connections develop before birth, and may depend on specific experiences of the fetus. These would include possible effects of tasting and swallowing amniotic fluid or feedback from movements of facial or other muscles. We are not concerned in this chapter with such prenatal experiences, but it is important to realize that there is a complex developmental history before the emergence of even a prefunctionally developed system.

Other connections develop after birth. For example, many adults will smile at the taste of coffee (a bitter substance). In such a case, presumably neither the perceptual nor the motor mechanism has changed over time. What has changed is the connection between them. Further, the change is not simply one in which the bitter mechanism becomes attached to the smile mechanism, because other bitter substances still elicit a disgust expression. Identification of the changes that occur and the experience that is necessary requires experimental analysis (Rozin, 1984), but this type of formulation of the problem makes that analysis easier to tackle.

A related question has to do with connections between the motor mechanisms and higher level coordinative structures. People smile not only in response to sweet tastes, but also in response to a wide range of stimuli associated with the hunger, sexual, parental, and other systems. How does the smile mechanism become attached to these various systems? This question also requires experimental analysis (e.g., Blass, Ganchrow, & Steiner, 1984), and several examples of this type are considered below.

DEVELOPMENT OF PERCEPTUAL MECHANISMS

Two of the most studied examples of behavior development, song learning and imprinting in birds, are both cases that involve a perceptual mechanism that develops independently of connections with central and motor mechanisms. Several aspects of these studies seem worthwhile to mention here. The development of food recognition mechanisms serves as a final example.

SONG RECOGNITION MECHANISMS.   Some time ago, Thorpe (1958, 1961) showed that the male chaffinch, *Fringilla coelebs*, had to learn to sing its species-specific song, and that this learning occurred in two stages. First, the young bird had to hear the normal song (or, within limits, a similar song); later, it learned to adjust its vocal output to match the song it had heard when it was young. Similar results have also been found for the white-crowned sparrow, *Zonotrichia leucophrys* (Konishi, 1965; Marler, 1970a), though not necessarily for other species of song birds (Hultsch, 1993; Logan, 1983; Marler, 1976). The first, or memorization, stage of learning involves the development of a perceptual mechanism, and that is discussed here; the second, or selection, stage involves the development of a motor mechanism, and that is discussed later. There have been many reviews of the bird song literature (e.g., Nelson, 1997; see also chapters by DeVoogd and by West & King in this volume), and only highly selected aspects are mentioned in this chapter.

Konishi (1965) and Marler (1976, 1984) proposed that the results of studies of song learning imply the existence of an auditory template, which was conceived of as a sensory mechanism that embodies species-specific information. The normal development of the template requires auditory experience of the proper sort at the proper time. In our terms, the template becomes a song-recognition (perceptual) mechanism that is partially formed at hatching. One question that has been asked is whether there is one or many templates. Originally it was thought that the young bird memorized a single song and that later variation in produced song came about because of mismatches during the selection stage. More recently (Marler & Peters, 1982; Nelson, 1997), it has become clear that the bird memorizes a variety of species-specific songs when young, but only one (or one subset) of these is selected later for production. How this choice is made is not known, but it now appears that the template used in the development of song production is stored in a different part of the brain from the other song memories (Bolhuis, Zijlstra, den Boer-Visser, & van der Zee, 2000; Jarvis & Nottebohm, 1997).

A second question concerns constraints on the kinds of experience that can affect development. Thorpe (1961) found that chaffinches would learn to sing normal or rearranged chaffinch songs heard when young, but exposure to songs of other species resulted in songs no different from those sung by birds raised in auditory isolation. The range of stimuli that affect development turns out to depend crucially on such factors as the species, the age at which the bird is exposed, the previous experience of the bird, and the conditions under which the bird is exposed (Nelson, 1997). There are no easy causal generalizations.

A third question concerns the processes that are involved in development. In the memorization stage, it is often assumed that mere exposure to an adequate stimulus is sufficient for perceptual learning to occur. In a restricted sense, this is probably true, but what makes a stimulus adequate often depends critically on the conditions under which the bird is exposed: For example, in many cases, memorization is more likely to occur when exposure occurs during social interaction with another bird (Baptista & Gaunt, 1997; Baptista and Petrinovich, 1984; Clayton, 1994; Nelson, 1997; Petrinovich, 1985), though the mechanism through which social interaction has these effects remains an open question (Houx & ten Cate, 1998, 1999).

PARENT AND PARTNER RECOGNITION MECHANISMS. Many species of birds do not recognize conspecifics on the basis of their song. These species have analogous perceptual mechanisms that analyze visual or other sensory input. The development of such perceptual mechanisms has usually been studied in the context of imprinting. This concept, as originally elaborated by Lorenz (1935/1970), was primarily concerned with the process by which early experience affects development. Lorenz proposed that "through imprinting, the bird acquires a schema of the conspecific animal" (p. 133). He also noted that the young of some species such as the curlew, *Numenius arquata*, require no visual experience in order to recognize members of their own species, whereas the young of other species such as the greylag goose, *Anser anser*, direct all their species-typical social behaviors to the first moving object they see. Imprinting was relevant only to the acquired aspects of the schema. In our terms, we would say that most, and perhaps all, species have a preassigned perceptual mechanism that serves a species recognition function. In a species such as the curlew, this perceptual mechanism develops prefunctionally. In a species such as the greylag goose, this perceptual mechanism requires various kinds of experience for its development. Bolhuis (1996) and van Kampen (1996) also analyzed imprinting with an emphasis on the development of perceptual mechanisms. The question of whether there is a single perceptual mechanism for species recognition or independent mechanisms for parent recognition, partner recognition, etc., will be considered in the general discussion.

It turns out that all the problems relating to the development of song recognition mechanisms mentioned above are also applicable to the development of the perceptual mechanisms studied in imprinting. For example, work of Horn and his colleagues has shown that there are two independent perceptual mechanisms involved in parent recognition in young chicks. One, called a learning mechanism, is concerned with specific details of the imprinting stimulus, and the other, called a predisposition, responds to generalized characteristics of fowl such as the head and eyes (for reviews see Bolhuis, 1991; Bolhuis & Honey, 1998; Horn, 1985). The outputs of these two mechanisms must summate, perhaps in a higher level perceptual mechanism, to determine whether a particular object is recognized as the parent.

A related example comes from the work of ten Cate (1986), who demonstrated a case of double imprinting. Young zebra finches, *Taeniopygia guttata*, that are exposed early in life to both zebra and Bengalese finches, *Lonchura striata*, may later court both species. A stable preference is formed for both these species over other, similar species to which the zebra finches were not exposed when young. Further experiments (ten Cate, 1987) investigated what kind of internal representation (perceptual mechanism) is necessary to account for this phenomenon. Ten Cate found that doubly imprinted males courted a zebra finch/Bengalese finch hybrid female more than they courted pure-bred females of either species. He concluded that a single, combined representation is sufficient to account for a male's courtship preferences.

The problem of constraints on what experience can be effective and at what stage of development of a perceptual mechanism has been studied extensively in the imprinting literature, and the general conclusions are the same as for memorization of songs (Bolhuis, 1991, 1996; ten Cate, 1994). The case of the developing predisposition is especially interesting because it illustrates an important aspect of prefunctionality. Young, dark-reared domestic chicks have a predisposition to approach objects resembling adult conspecifics, but only if they receive certain kinds of nonspecific experience such as an opportunity to run in a wheel (Johnson & Horn, 1988); and, this nonspecific experience must occur during a restricted time after hatching to be effective (Johnson, Davies, & Horn, 1989). Thus, the predisposition develops prefunctionally because the chick approaches specific visual stimuli even though it has not had any visual experience; nonetheless, this effect is not seen if the chick does not have other kinds of experience. A similar example is provided by the development of the auditory recognition mechanism of the species' maternal call in Peking ducklings (Gottlieb, 1980).

The problem of the processes through which experience has its effects has also received much attention. As with song memorization, simple exposure to an adequate stimulus can be sufficient for imprinting to occur, but social interaction with the imprinting stimulus can enhance its effects (ten Cate, 1984). More recently, van Kampen and Bolhuis (1991) showed that simultaneous exposure to an auditory stimulus and a visual stimulus results in enhanced learning about each stimulus separately. Further experiments by Bolhuis and Honey (1994) supported the conclusion that conjoint exposure to a visual and an auditory stimulus leads to the formation of an integrated memory of them.

FOOD RECOGNITION MECHANISMS. The work of Steiner (1979) suggests that newborn infants have well-developed perceptual mechanisms for recognizing sweet, sour, and bitter. Most substances that humans (and other animals) treat as food, however, are recognized on the basis of more complex properties and require specific experience for recognition to develop (for review see Hogan, 1973, 1977). I will here discuss two examples of how food recognition mechanisms develop in chicks and cats. I should emphasize that I am considering "recognition mechanisms" in a strictly (behaviorally) causal sense. That is, stimuli that activate a food recognition mechanism, for example, are those stimuli that the animal treats as food. We infer that an animal is treating a stimulus as food from the occurrence of behavior that belongs to the hunger system. Such stimuli may or may not be nutritious and could even be poisonous.

Newly hatched chicks peck at a wide variety of objects, although, even at the first opportunity, certain colors and shapes are preferred (Fantz, 1957; Hess, 1956). These preferences need not be a reflection of an undeveloped food recognition mechanism, however, for at least two major reasons. First, pecking is a component of aggressive, sexual, and grooming behavior as well as of feeding behavior, and the stimuli that release and direct pecking in these various contexts are quite different. Second, chickens continue to peck a wide variety of objects throughout their lives, even after the objects toward which they direct their feeding, grooming, aggressive, and sexual behavior have become quite specific. Thus, one could view these early preferences as being due to a perceptual mechanism directly connected to the pecking mechanism in the same way that the various taste mechanisms are connected to specific motor mechanisms in infants. This "independent" pecking might be regarded as serving an exploratory function, and it also has many of the characteristics of play.

The putative food recognition mechanism in newly hatched chicks must be largely unspecified because of the very wide range of stimuli that are characteristic of items that chicks will come to accept as food. Certain taste and tactile stimuli are more acceptable than others (for review see Hogan, 1973), but these stimuli can be effective only after the chick has the stimulus in its mouth. In some cases, taste and tactile feedback seem to be sufficient to cause an item to become recognized as food. For example, as early as 1–2 days of age, a chick that has eaten one mealworm will treat all subsequent mealworms as food. Presumably, the taste of the mealworm is sufficient for subsequent visual recognition to occur because a second mealworm will be accepted immediately after the first, and thus long before any effects of digestion could be expected to play a role (Hogan, 1966). Taste is also sufficient for a chick to develop visual recognition of a stimulus to be rejected: a 1-day-old chick will learn to reject a distasteful cinnabar caterpillar in just one trial (Morgan, 1896; see also Hale & Green, 1979). The fact that mealworms can come to be recognized as food (i.e., are avidly ingested) and other insects can come to be rejected as food before nutritive factors gain control of pecking on day 3 (see below) is evidence that the food recognition mechanism is independent of the central mechanism of the developing hunger system.

The food recognition mechanism also develops under the influence of the long-term (1–2 hr) effects of ingestion. Experiments by Hogan-Warburg and Hogan (1981) provide evidence that chicks gradually learn to recognize food particles as a result of the reinforcing effects of food ingestion. In these experiments, visual stimuli from the food gained significant control over the chicks' behavior after one substantial food meal, though oral stimuli gained control of ingestion more slowly.

The development of food recognition in young kittens is similar in many ways to that in chicks (Baerends-van Roon & Baerends, 1979). Kittens begin ingesting their first solid food at about 4 weeks of age. Some items are immediately recognized as food, whereas others require various kinds of experience before being accepted (or rejected) as food. Fish odor appears to be attractive to all cats, even those with no experience of fish. Fish is ingested as early as a kitten is able to eat solid food, but the main problem for the kitten is learning how to catch a fish. This topic is discussed in the next section. Mouse odor, on the other hand, does not appear to have an inherent attractiveness for cats. Mice become recognized as food only after a kitten has eaten a mouse. This can happen if a mother cat presents a dead (and opened) mouse. It can also happen if a kitten attacks and bites a live mouse by itself. It is not yet possible to say whether the taste of the mouse is sufficient experience for its subsequent recognition as food (as in the chicks) or whether nutritional effects of digestion are necessary. The Baerends' did observe that a shrew may be caught and ingested by a naive kitten, but it is vomited within 15–20 min. Thereafter, kittens may catch and "play" with shrews, but they never ingest them. This finding suggests that the effects of digestion may be the critical experience for food recognition to develop. Such observations also indicate considerable independence of catching and eating behavior, a topic discussed later.

In a functional sense, the nutritional effects of ingestion should be the ultimate factor determining which objects are recognized as food. But sometimes other factors override the effects of nutrition and lead to the development of a food recognition mechanism that is maladaptive. Two observations made on chicks' food preferences are relevant here (Hogan, 1971). First, many chicks that were fed meal-worms on the first few days after hatching died at about 6 or 7 days. These chicks could generally be characterized as mealworm fanatics because of their excited, positive behavior toward mealworms. These mealworm fanatics never learned to eat the regular chicken food that literally surrounded them, and they apparently died of starvation. Second, many chicks that were raised on a mixture of chicken food and aquarium gravel also died at about 6 days of age, also apparently of starvation. In this case, the gravel seemed to be an exceptionally good releasing stimulus for pecking and swallowing. Both these examples suggest that factors other than the nutritional effects of ingestion can play an important role in the development of food recognition.

DISCUSSION.   The development of perceptual mechanisms illustrates most of the problems encountered in the development of behavior systems in general. First, the postulation of a template or schema implies a kind of modularity in the brain in that a certain part of the brain is preassigned a specific function. Second, there are constraints on the kinds of experience that can affect development and on the age or stage of development at which this experience can be effective. Third, there is the problem of developmental processes: Are the effects of experience direct or indirect? Is mere exposure sufficient, or is some sort of reinforcement necessary? Finally, there is much variability between species in the role played by experience and the types of constraints encountered. These problems are all interrelated, and I return to them in the general discussion.

One generalization about perceptual mechanisms per se is that the evidence supports their existence in at least three functional levels of organization: feature recognition, object recognition, and function recognition. Feature recognition mechanisms discriminate among various sizes, shapes, colors, smells, tastes, and so on. This is presumably the level at which the gustofacial reflex is organized in human infants. The reason for distinguishing between object recognition and function recognition is that objects with similar properties, such as food crumbs and sand, mealworms and cinnabar caterpillars, or mice and shrews, are easily recognized (after appropriate experience) as being food or nonfood, whereas other objects with greatly disparate properties, such as grain, insects, fish, and the leaves of various plants, are easily included in the food category. Similarly, a mockingbird, *Mimus polyglottos*, mimics very accurately the songs of many different species (Baylis, 1982). Therefore, it must have a number of perceptual mechanisms for recognizing each different song. Further, the various songs that the mockingbird has learned are combined into an overall song that has species-specific characteristics (Logan, 1983), so there must also be an additional perceptual mechanism at a higher level of organization. Ten Cate's (1994) imprinting results with zebra finches tell the same story.

DEVELOPMENT OF MOTOR MECHANISMS

Many motor mechanisms develop prefunctionally. For instance, young chicks show normal locomotion and pecking movements almost immediately after hatching. Within the next few days, ground scratching and various grooming movements appear. Kruijt (1964) showed that the proper functioning of these and other movements in the posthatching situation is not a necessary causal factor for their development. Of course, prehatching conditions obviously influence the development of these movements, though the processes responsible for behavioral organization remain largely unknown (Oppenheim, 1974). Studies of the responses of young rat pups to electrical stimulation of the medial forebrain bundle provide an additional example (for review see Moran, 1986). Three-day-old rat pups show a number of organized behavior patterns such as licking, pawing, gaping, and lordosis in response to such stimulation. These patterns are not seen in their normal functional context until later in development. Thus, these motor mechanisms must also be organized prior to their functioning.

It should be emphasized here that, although motor patterns are visible to an observer, motor mechanisms are not. An example should make this point clear. Kuo (1967) noted that chicks that developed with their yolk sac in an abnormal position were often crippled when they hatched. He interpreted these results to mean that the development of normal walking movements required functional experience in the egg: the legs had to push actively against the yolk sac for normal development to occur. Such experience is indeed necessary for the development of normal joints (Drachman & Sokoloff, 1966), and without properly functioning joints, a chick cannot move normally. Nonetheless, the movements of a crippled chick cannot provide evidence for whether or not the motor mechanism for walking has developed normally. Such evidence certainly does not contradict the conclusion of Hamburger (1973) that the neural patterning underlying the walking movements of a chick develops without functional experience (Lehrman, 1970). (For a related example concerning human locomotion, see Thelen & Fisher, 1982.)

Perhaps the best studied case of how a motor mechanism develops on the basis of functional experience is the development of bird song, and this provides my first example. I then consider the development of some displays in birds; experience is effective in a surprising way in this example. Finally, I discuss aspects of the development of behavior sequences in dustbathing of chickens, grooming in rats, and prey catching in cats. These examples all give insight into the development of more complex behavior systems.

SONG LEARNING. As seen above, in many species the young bird forms an auditory image of the song it will learn to sing. Learning to sing the song does not happen until later, when the internal state (e.g., the level of testosterone) is appropriate. At this point, it appears that the bird learns to adjust its motor output to match the image it has previously formed. This adjustment must involve the bird's hearing itself because deafened birds never learn to produce any song that approaches normal song (Konishi, 1965).

Experiments by Stevenson (1967) showed that hearing its species-specific song could serve as a reinforcer for an operant perching response in male chaffinches. On the basis of these results, Hinde (1970) suggested that song learning might involve matching the sounds produced by the young bird with the stored image: sounds that matched the image would be reinforced, whereas other sounds would be extinguished. In this way, a normal song could develop in much the same way as an experimenter originally trains a rat to press a lever (Skinner, 1953).

In most species, three stages in the production of song can be distinguished: subsong, plastic song, and crystallized song (Thorpe, 1961). During the subsong phase, the bird essentially babbles, and slowly adjusts its production to match phrases and songs it heard during the memorization stage; it may also invent new combinations of phrases during this phase. These changes presumably come about in the manner suggested by Hinde. In the plastic song phase, the bird may be singing a number of songs that resemble songs it previously heard. Which of these songs becomes chosen as the crystallized song depends, in many species, on the songs it hears from other birds at this time. In some species, the bird selects a similar song (which probably accounts for the occurrence of local dialects) and in other species, the bird selects a dissimilar song. In either case, a selection is made from songs already developed (Marler & Peters, 1982; Nelson, 1997). The selection process presumably also involves some kind of reinforcement, often provided by the behavior of conspecifics. A particularly interesting example is the song of the brown-headed cowbird, *Molothrus ater*. Males of this species increase their performance of those songs that are associated with a wing stroke display given by the females (West & King, 1988).

DISPLAYS. A display is a behavior pattern that is adapted to serve as a signal to a conspecific (Tinbergen, 1952). The mechanism controlling the display is thus the motor counterpart of species-recognition perceptual mechanisms discussed above. Displays are often complex, yet they typically develop prefunctionally. For example, waltzing is a courtship display in chickens that essentially involves the male's circling a female in a characteristic posture. Kruijt (1964) showed that the form of this display can be derived from components of behavior that belong to the aggression and escape systems, and that these systems are activated when waltzing first appears. Nonetheless, waltzing appears even in animals that are reared in social isolation, so social experience cannot be a necessary causal factor in its development.

One example of a display in which social experience has been implicated as a causal factor in its development is the "oblique posture with long call" of the black-headed gull, *Larus ridibundus*. Groothuis (1992) raised gulls to the age of 1 year either in social isolation, in small groups of 2–4 individuals, or in large groups of 12. Black-headed gulls are colonial breeders, and large groups are the normal social environment for the developing young. All the birds raised in large groups, 50% of the birds raised in small groups, and 35% of the birds raised in social isolation developed the normal display in the first year. Of particular interest is that about 40% of the birds raised in small groups developed an aberrant display in which the head was held in an abnormal posture. Further experience in large groups for more than 1 year had no effect on the form of this aberrant display. This result contrasts with the finding that all of the isolated and other birds that originally showed only fragmentary forms of the display subsequently developed a normal display when placed together in a large group. A separate experiment (Groothuis & Meeuwissen, 1992) showed that isolated birds that were injected with testosterone at 10 weeks of age all developed a normal display within a few days of injection.

One process underlying the development of this display may be the same as that suggested by Hinde for the development of bird song. There may be some sort of template sensitive to proprioceptive feedback from the display that selects out the correct forms from all the transitional forms that normally occur. Such a cognitive structure that recognizes proprioceptive feedback has been proposed to explain the results of experiments on imitation by human infants (Field, Woodson, Greenberg, & Cohen, 1982). Nonetheless, the results from the testosterone experiments, in which essentially no transitional forms were seen, do not support such a process in the gulls. Further, the fact that many isolated birds developed a normal display means that social experience cannot be a necessary causal factor for normal development. However, the aberrant displays that developed in some birds raised in small groups suggest that social interactions can be of importance in special circumstances. Groothuis (1992, 1994) discusses several hypotheses to explain these results, one of which is that abnormal experience encountered in the small groups could have distorted normal development. I return to this idea in the discussion.

DUSTBATHING AND GROOMING.   Dustbathing in the adult fowl consists of a sequence of coordinated movements of the wings, feet, head, and body that serve to spread dust through the feathers. It occurs regularly, and bouts of dustbathing last about ½ hr (Vestergaard, 1982). When dust is available, dustbathing functions to remove excess lipids from the feathers and to maintain good feather condition (van Liere & Bokma, 1987). The sequence of behaviors in a dustbathing bout begins with the bird pecking and raking the substrate with its bill and scratching with its feet. These movements continue as the bird squats down and comes into a sitting position. From time to time, the bird tosses the dusty substrate into its feathers with vertical movements of its wings and also rubs its head in the substrate. It then rolls on its side and rubs the dust thoroughly through its feathers. These sequences of movement may be repeated several times. Finally, the bird stands up, shakes its body vigorously, and then switches to other behavior. A diagram of the dustbathing behavior system is shown in Figure 2.

Dustbathing does not appear fully formed in the young animal. Rather, individual elements of the system appear independently, and only gradually do these elements become fixed in the normal adult form. Pecking is seen on the day of hatching, but the other motor components appear gradually over the first 7 or 8 days posthatch (Kruijt, 1964). Vestergaard, Hogan, and Kruijt (1990) asked whether the rearing environment influenced the organization of the motor components. They observed small groups of chicks that were raised either in a normal environment containing sand and grass sod or in a poor environment in which the floor was covered with wire mesh. A comparison of the dustbathing motor patterns of 2-month-old birds raised in the two environments showed surprisingly few differences. The form and frequency of the individual behavior patterns as well as the temporal organization of the elements during extended bouts of dustbathing developed almost identically in both groups. There were some differences in the microstructure of the bouts that could be related to the presence or absence of specific

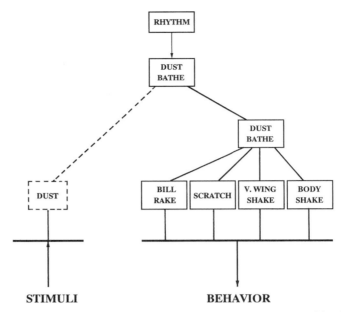

Figure 2. The dustbathing system of a young chick. Boxes represent putative cognitive (neural) mechanisms: a perceptual mechanism responsible for recognizing dust; a central dustbathing mechanism responsible for integrating input from the perceptual mechanism and other internal influences; several motor mechanisms responsible for specific actions as well as a higher level motor mechanism responsible for the patterning of individual actions during dustbathing bouts. Solid lines indicate mechanisms and connections among them that develop prefunctionally; dashed lines indicate mechanisms and connections that develop as the result of specific functional experience. (Adapted from Vestergaard *et al.*, 1990 with permission. Copyright 1990 by E. J. Brill.)

feedback, but the motor mechanisms and their coordination developed essentially normally in chicks raised in a dustless environment (see also van Liere, 1992). Clearly, the experience of sand in the feathers removing lipids or improving feather quality is not necessary for the integration of the motor components of dustbathing into a normal coordinated sequence.

More recently, Larsen, Hogan, and Vestergaard (2000) studied in detail the development of dustbathing behavior sequences in chicks from hatching to 3 weeks of age. They found that the individual behavior elements, as soon as they appeared, were incorporated into the normal adult sequence structure; this occurred even though the form of the elements themselves is not yet fixed. These results support the conclusion that separate mechanisms are responsible for the form of the individual behavior elements and for the organization of these elements into recognizable sequences as shown in Figure 2. A similar conclusion was reached by Berridge (1994) on the basis of results on the development of grooming sequences in young mice (Fentress, 1972; Fentress & Stilwell, 1973) and young rats and guinea pigs (Colonnese, Stallman, & Berridge, 1996). In fact, Berridge and Fentress (1986) and Berridge, Fentress, and Parr (1987) called certain sequences of grooming movements "syntactic chains" to emphasize the rules controlling natural action sequences (see also the chapter by Fentress in this volume). I return to these ideas in the section on human language.

PREY CATCHING.   My final example of the development of sequences of motor patterns is the prey-catching behavior of cats (Baerends-van Roon & Baerends, 1979). Locomotion, pouncing, angling (with one paw), and biting are the basic motor patterns out of which effective prey-catching develops, and all these behaviors can be seen, prefunctionally, by the time the kitten is about 4 weeks old. The way these behaviors become integrated depends primarily on the type of prey being caught. If a mouse is the prey, locomotion and biting are sufficient to catch and kill, whereas with larger prey, pouncing is necessary in addition. If a fish is the prey, angling and biting are the necessary motor patterns. The evidence suggests that the "correct" behavior sequences are selected on the basis of the effects of the behavior. In other words, an operant shaping process can account for all the results, with the proviso that the basic elements—locomotion, pouncing, angling, and biting—are not themselves shaped. This conclusion is supported by the fact that the course of development can vary considerably among individuals even though the final result is quite stereotyped.

An important difference between prey-catching sequences and many of the other examples discussed above is that prey-catching sequences in cats do not "crystallize." That is, functional experience continues to be effective in shaping new sequences. For example, kittens that have developed proficient fish-catching behavior can subsequently learn to catch mice, although there is some interference from the previous learning in that such kittens take longer to learn to kill the mouse with a bite than do naive kittens. Learning to catch a fish after the kitten has already developed mouse-catching behavior turns out to be considerably more difficult. The primary problem here is that older kittens have a stronger tendency to avoid getting wet than younger kittens. If the fear of water can be overcome, the fish-catching sequence can be easily acquired. This last example indicates an important problem in the study of development: It seems that certain cases of learning may be irreversible when, actually, indirect factors (such as fear of water in this case) obscure the fact that functional experience can still have direct effects on development.

DISCUSSION.   The development of motor mechanisms illustrates the same problems previously discussed with respect to perceptual mechanisms: modularity, constraints, processes, and species differences. The emphases are somewhat different in that most motor mechanisms (with the notable exception of bird song) and even many motor sequences (such as dustbathing in chicks and grooming in rats) develop prefunctionally, whereas almost all perceptual mechanisms are directly influenced by functional experience. Further, when functional experience is relevant, mere exposure is probably sufficient for perceptual mechanisms, whereas some sort of reinforcement is usually necessary for motor mechanisms and motor sequences. These topics are considered again in the general discussion.

Another similarity between perceptual and motor mechanisms is the existence of different functional levels of organization. In the case of motor mechanisms, there is the level of the individual motor pattern and the level of motor pattern integration. The results reviewed above for dustbathing, grooming, and prey catching all provided evidence that these levels are independent. In many ways, the level of motor pattern integration is especially interesting because it provides the basis for the temporal patterning, or syntax, of functionally related behaviors. As Lashley (1951) pointed out, all skilled acts, including human language, seem to involve the same problems of serial ordering. In a later section, I will attempt to show how the framework developed here can be useful in understanding the development of human language.

In a recent discussion of motor development, Thelen (1995) proposes a dynamic theory in which "repeated cycles of perception and action can give rise to emergent new forms of behavior without preexisting mental or genetic structures" (p. 93). She opposes her theory to one in which the brain matures and movements appear when the appropriate level of maturity is reached. Although I would maintain that there are always preexisting structures, in some ways her approach is similar to the one taken in this chapter. For example, we have seen that feedback from the performance of a song or display can affect the form of future performances. What is very different is that her basic units of action lie at a lower functional level of organization than is considered here, and she explicitly considers factors that I call prefunctional and do not analyze further. With respect to the latter, her analysis is similar to Kuo's analysis of walking in chicks discussed earlier.

DEVELOPMENT OF CONNECTIONS BETWEEN CENTRAL AND MOTOR MECHANISMS

We can say that a central and a motor mechanism are connected when the occurrence of a behavior varies directly with the presence of factors known to affect the central mechanism. Consider the behavior of pecking and the central hunger mechanism in a chicken. If the amount of pecking varies directly with the amount of food deprivation, then we have evidence that hunger and pecking are connected. On the other hand, if variations in food deprivation have no effect on the amount of pecking, we have evidence that hunger and pecking are not connected, that is, are independent mechanisms. The developmental problem is how we get from the state of independence to the state of connectedness. There are many examples of situations in which a motor mechanism becomes connected to a particular central mechanism, including examples from the operant conditioning literature. I shall discuss some of these later, but I begin with some examples of the development of normal feeding behavior.

HUNGER. A surprising fact about the feeding behavior of many neonatal animals is that their early feeding movements are relatively independent of motivational factors associated with food deprivation. Hinde (1970, pp. 551ff.) reviewed a variety of evidence from studies on kittens, puppies, lambs, and human infants that show that the amount of suckling by a young animal is very little influenced by the amount of food it obtains. More recently, a series of studies on the development of feeding in chicks and in neonatal rats has been published, and these are reviewed briefly here.

A chick begins pecking within a few hours of hatching, but its nutritional state does not influence pecking until about 3 days of age (Hogan, 1971). When chicks were 1 or 2 days old, 5 hr of food deprivation did not influence the subsequent rate of pecking at food, whereas by the time the chicks were 4 or 5 days old, 5 hr of food deprivation led to a large increase in pecking at food. A very similar change in the control of feeding has been found in rat pups (Blass, Hall, & Teicher, 1979; Cramer & Blass, 1983; Hall & Williams, 1983). Before the age of about 2 weeks, the occurrence of behaviors such as nipple search and nipple attachment, as well as the amount of suckling itself, was not influenced by food (i.e., maternal) deprivation of as long as 22 hr. After 2 weeks, however, deprived pups attached to the nipple more quickly and suckled longer than nondeprived pups. Similarly, when tested in a spatial discrimination task in a Y maze, nutritive suckling provided a greater incentive than nonnutritive suckling only after the pups were older than 2 weeks (Kenny, Stoloff, Bruno, & Blass, 1979).

The developmental question with respect to these results is: How do the motivational factors associated with food deprivation come to control feeding behavior? For chicks, early experiments (for review see Hogan, 1977) led to the hypothesis that it is the experience of pecking followed by swallowing that causes the connection between the central hunger mechanism and the pecking mechanism to be formed. In other words, a chick must learn that pecking is the action that leads to ingestion; once this association has been formed, nutritional factors can directly affect pecking. Subsequent experiments have shown that the association of pecking with ingestion is indeed the necessary and sufficient condition for pecking to become integrated into the hunger system (Hogan, 1984). Experiments on the development of pecking in ring doves, *Streptopelia roseogrisea*, also indicate that experience is necessary for hunger to gain control of pecking; though in this case, the necessary experience apparently involves interaction with the parents as well as with food (Balsam & Silver, 1994; Balsam, Graf, & Silver, 1992; Graf, Balsam & Silver, 1985).

Similar experiments with rat pups have not been done, though the problem with mammals in general is more complex because the suckling response drops out altogether at weaning and is replaced by different behaviors (Hall & Williams, 1983). Hall and his colleagues have shown that, under special conditions, rat pups ingest food away from the mother very soon after birth, but these experiments have not asked the same questions being asked here (see also Johanson & Terry, 1988). In the study by Kenny *et al.* (1979), the infant rats received their nourishment through intragastric feeding between days 12 (before their ingestion was influenced by hunger) and day 17. When tested at day 17, motivational control of their ingestion was the same as in normally reared pups, which implies that experience eating solid food is not necessary for motivational control to develop. However, there are some results from guinea pigs that are also relevant (Reisbick, 1973). Guinea pigs normally begin ingesting solid food within 1 day of birth, and Reisbick found that experience of ingesting and swallowing was necessary before the guinea pigs showed evidence of discriminating between nutritious and nonnutritious objects. These results are very similar to the results from the chicks and have been discussed in more detail elsewhere (Hogan, 1977).

OPERANT CONDITIONING.   A second source of evidence for the development of connections between central and motor mechanisms is the operant conditioning literature. The process of reinforcement, in general, can be regarded as influencing the development of connections between central and motor mechanisms. For example, the response of lever pressing is an easily recognizable motor pattern in a rat. Reinforcing lever pressing with food leads to a connection of the motor mechanism for lever pressing with the hunger system, and reinforcing with water leads to a connection with the thirst system.

Schiller (1949) reported the results of studies of problem solving by chimps. He noted that many of the behavior patterns used by his chimps to procure food that was placed out of reach were apparently the same manipulative patterns that had first appeared spontaneously and prefunctionally. The patterns included "weaving," "poking and sounding," and "joining sticks." Schiller suggested that these patterns could be considered operant responses that were used to solve the problem, and that they were reinforced when the chimp was successful. In the terminology used here, we could say that the originally independent motor mechanisms responsible for the various observed behavior patterns became connected to the hunger system as a result of operant reinforcement. The test for "connection" here, as elsewhere, is to see if the occurrence of a behavior varies directly with the presence of factors known to affect the central mechanism: Do hungry chimps engage in these behaviors more than sated chimps? Schiller's results suggest that they do.

One question that arises from these results is whether any motor mechanism can be attached to the hunger system using food reinforcement. Rice (1978) tried to affect the occurrence of shrill calls and twitters in young chicks by using food reinforcement, but he was unsuccessful. Shettleworth (1975), using golden hamsters, *Mesocricetus auratus*, also asked whether various behavior patterns could be influenced by food reinforcement. In one set of experiments, she observed animals in their home cages and in an unfamiliar environment both when deprived and when not deprived of food. In another set of experiments, she reinforced animals with food when they performed various behavior patterns, including scrabbling, digging, rearing, face washing, scratching, and scent marking. She found that food reinforcement was effective in increasing the occurrence of scratching, digging, and rearing, but that it had very little effect on the occurrence of face washing, scratching, and scent marking. The first three patterns all increased in frequency in hungry hamsters, and the latter three decreased in frequency. Thus, behavior patterns that belonged to the hamster's hunger system—when the criterion used is a positive correlation with food deprivation—could be influenced by food reinforcement, whereas behavior patterns that belonged to other systems could not. These results all indicate a considerable degree of inflexibility with respect to which motor patterns can become connected to which central mechanisms.

DEVELOPMENT OF CONNECTIONS BETWEEN PERCEPTUAL AND CENTRAL MECHANISMS

We can say that a perceptual mechanism and a central mechanism are connected when a stimulus that activates the perceptual mechanism can lead to the occurrence of the set of behaviors known to belong to the central mechanism. For instance, an egg recognition mechanism is connected to the incubation system in many birds because the presentation of an egg (or other appropriate stimulus) can lead to approach, retrieval, and settling on the nest. The evidence necessary to show that a perceptual mechanism is in fact connected to a central mechanism is to show that the presentation of an adequate stimulus has the same effect on the central mechanism (priming effects; Hogan & Roper, 1978, p. 231) as a direct manipulation of the relevant internal factors, for example, by deprivation or the injection of hormones.

DUSTBATHING.   Functional experience plays an essential role in the development of the perceptual mechanism for recognizing dust and of the connection between it and the central mechanism (dashed lines in Figure 2). Evidence for the role of experience in the development of the dust recognition mechanism itself is reviewed by Sanotra, Vestergaard, Agger, and Lawson (1995). Factors responsible for the connection between the dust recognition mechanism and the central mechanism are reviewed here.

Young chicks can be seen engaging in dustbathing movements on almost any surface that is available, ranging from hard ground and stones to sand and dust. In fact, Kruijt (1964) found that making the external situation as favorable as possible for dustbathing was insufficient for releasing the behavior. This result implies that early dustbathing may be controlled exclusively by internal factors (see below). It also implies that the connection between the dust recognition perceptual mechanism and the central mechanism is not formed until well after the motor and central mechanisms are functional.

Vestergaard and Hogan (1992) found that early dustbathing is most likely to occur in whatever substrate is pecked at most. They point out that pecking is a movement that functions as exploratory, feeding, dustbathing, and later aggressive behavior. They suggest that perceptual mechanisms specific to each system develop gradually out of exploratory pecking on the basis of functional experience. It remains to be determined whether removal of lipids, the sensory feedback from the substrate in the feathers, or facilitation of the dustbathing behavior itself is the crucial factor.

Other evidence shows that early experience can lead to stable preferences for particular stimuli (Petherick, Seawright, Waddington, Duncan, & Murphy, 1995; Vestergaard & Baranyiova, 1996). As an extreme example, Vestergaard and Hogan (1992) raised birds on wire mesh but gave them regular experience on a substrate covered with coal dust, white sand, or a skin of junglefowl feathers. In choice tests given at 1 month of age, some of the birds that had had experience with junglefowl feathers were found to have developed a stable preference for dustbathing on the feathers. This example is important because it shows how a system can develop abnormally. It also suggests that the pecking associated with dustbathing may be a cause for "feather pecking," a common pathological condition in which some hens pull out the feathers of their cage mates, which is seen in many commercial groups of fowl (Johnsen, Vestergaard, & Nørgaard-Nielsen, 1998; Vestergaard, Kruijt, & Hogan, 1993).

HUNGER.   The results with the chicks discussed above show that a mealworm recognition mechanism can become connected to the motor mechanism for pecking at least 1 day before nutrition (i.e., the central "hunger" mechanism) gains control of pecking. The evidence indicates that the ingestion of mealworms remains semiindependent of hunger, probably throughout life: satiated chicks avidly ingest many mealworms, and the ingestion of a substantial number of mealworms, at least in the first week after hatching, has no effect on the amount of other food subsequently ingested (Hogan, 1971). This semiindependence of mealworm ingestion and hunger is probably the same phenomenon as the semiindependence of suckling and hunger in rats and prey catching and hunger in cats and most other predators.

Evidence that a food recognition mechanism becomes connected to the central hunger mechanism comes from the fact that food particles develop incentive value between 3 and 5 days posthatching (Hogan, 1971); development of incentive value probably reflects the same process involved in the development of food recognition discussed above (Hogan-Warburg & Hogan, 1981). More direct evidence of perceptual mechanisms' becoming connected to central mechanisms is provided by several examples from the learning literature.

CLASSICAL CONDITIONING.   There are now numerous examples of complex, species-typical behaviors that become released by previously neutral stimuli that develop their effectiveness by means of a classical conditioning procedure. For instance, Adler and Hogan (1963) paired the presentation of a weak electric shock with a mirror to a male Siamese fighting fish, *Betta splendens*, and showed that full aggressive display could be conditioned to the shock. In a similar way, Farris (1967) conditioned the courtship behavior of Japanese quail, *Coturnix japonica*, to a red light. Moore (1973) showed that a small lighted key followed consistently by food elicited a food peck in a pigeon; when followed consistently by water, it elicited a drinking peck. Blass *et al.* (1984) were able to condition the ingestive behaviors of head orientation and sucking in human infants (which are unconditioned responses to the oral delivery of a sucrose solution) to gentle forehead stroking. These and many other cases exemplify the development of connections between a perceptual mechanism and a set of behaviors as a result of a classical conditioning procedure. They do not, however, distinguish between a connection between a perceptual mechanism and a central mechanism or directly between a perceptual mechanism and a complex motor mechanism.

There are also cases, however, where a connection between a perceptual mechanism and a central mechanism is directly implicated. Wasserman (1973) looked at the behavior of young chicks tested in a cool environment. The chicks were trained by being exposed to a lighted key for several seconds and then to presentation of heat from a heat lamp. After several pairings of the light and the heat, the chicks began to approach the key when it lighted up and showed pecking and "snuggling" movements to it. These behaviors were never shown to the heat lamp itself (which was suspended above the chicks, out of reach). Pecking and snuggling movements are behaviors shown by young chicks when soliciting brooding from a mother hen (Hogan, 1974). Wassermann's results imply that the recognition mechanism for the lighted key becomes connected to a thermoregulatory system in the young chick (Sherry, 1981), and that the presentation of this stimulus to a cold chick elicits brooding solicitation movements.

More recently, similar examples have multiplied. The systems include hunger, aggression, sex, and fear, in species ranging from insects through fish and birds to mammals including humans (for reviews see Domjan & Holloway, 1998; Fanselow & De Oca, 1998; Timberlake, 1994). What many of these examples show is that previously neutral stimuli can, as a result of classical conditioning procedures, develop control of entire behavior systems. The studies by Fanselow and his colleagues, using the species-specific defense reactions of rats as their behavior, have even shown that the conditioned stimulus has its effects through the same neural structures as the unconditioned stimulus.

DEVELOPMENT OF CONNECTIONS AMONG PERCEPTUAL, CENTRAL, AND MOTOR MECHANISMS

The previous sections have presented evidence about the effects of various kinds of experience on the development of connections between pairs of building blocks. The principles of development that emerge from those results are sufficient to allow us to understand much of the development of more complex systems. As we shall see, however, some new principles seem also to be involved in these more complex cases. A review of some examples of the development of dustbathing, hunger, aggressive, sexual, and play systems will illustrate how these principles operate, and the final section will consider how they could be applied to the study of human language acquisition.

DUSTBATHING. The dustbathing system of an adult chicken consists of a perceptual mechanism that recognizes dust, and a central mechanism that integrates internal motivational factors with signals from the perceptual mechanism and a circadian clock (Hogan, 1997), and controls the motor mechanisms that constitute dustbathing (Figure 2). As seen above, the perceptual mechanism itself and its connection with the central mechanism require specific functional experience for their development, whereas the motor components of dustbathing as well as the temporal organization of those components develop prefunctionally. Some evidence is also available for the development of the central mechanism.

In adult fowl, the occurrence of dustbathing varies directly with the length of time a bird has been deprived of the opportunity to dustbathe; it also occurs primarily in the middle of the day (Vestergaard, 1982). In young chicks, as soon as dustbathing is seen, at 1 week of age, it is controlled by the effects of dust deprivation. Hogan, Honrado, and Vestergaard (1991) found that deprivation effects could be demonstrated as early as 8 days of age (the age that complete dustbathing bouts first appear) and that they did not change over at least a 4-week period. No specific experience was necessary for the motivational factors associated with dust deprivation to gain control of dustbathing (see also Vestergaard, Damm, Abbot, & Bildsøe, 1999). Similarly, Hogan and van Boxel (1993) found that a daily rhythm, with most dustbathing occurring in the middle of the day, was seen in chicks as young as 14 days of age. These results suggest that the central mechanism for dustbathing and the connection between the central mechanism and the motor mechanisms also develop prefunctionally.

HUNGER.   The hunger system of an adult chicken consists of various perceptual mechanisms that serve a food recognition function, motor mechanisms that function to locate and ingest food, and a central mechanism that integrates signals from the physiological mechanisms concerned with nutrition and modulates signals from the perceptual mechanisms and to the motor mechanisms (Figure 3). We have

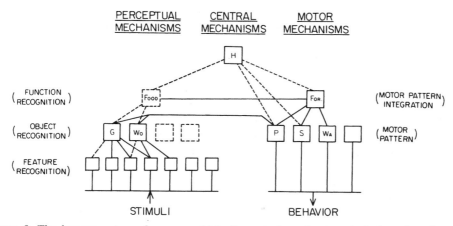

Figure 3. The hunger system of a young chick. Perceptual mechanisms include various feature-recognition mechanisms (such as of color, shape, size, and movement), object-recognition mechanisms (such as of grainlike objects [G], wormlike objects [Wo], and possibly others), and a function-recognition mechanism (Food). Motor mechanisms include those underlying specific behavior patterns (such as pecking [P], ground scratching [S], walking [Wa], and possibly others) and an integrative motor mechanism that could be called foraging (For). There is also a central hunger mechanism (H). Solid lines indicate mechanisms and connections that develop prefunctionally; dashed lines indicate mechanisms and connections that develop as the result of specific functional experience. (From Hogan, 1988.)

seen above how the perceptual mechanisms develop and what experience is necessary for the central mechanism to develop its modulating function. With respect to motor mechanisms, the previous discussion has focused entirely on pecking. There are, however, several other motor mechanisms that are normally associated with the hunger system, such as those controlling ground scratching and locomotion. As with dustbathing, both the individual motor mechanisms of the system and the integration of these mechanisms into effective foraging behavior appear prefunctionally. Unlike with dustbathing, however, the integration of the motor mechanisms disintegrates in the absence of effective functional experience (Hogan, 1971). Hogan (1988) reviewed the evidence that suggests that new connections are formed between the central hunger mechanism and individual motor mechanisms on the basis of the specific experience of the individual chick (dashed lines between H and P and between H and S in Figure 3), and that these new connections effectively block the expression of the original prefunctional connections.

The general picture that emerges from all the data is summarized in Figure 3. A young chick has a number of feature recognition perceptual mechanisms, an undeveloped food recognition mechanism, an independent central hunger mechanism, an integrated complex of motor mechanisms, and some connections between the perceptual and motor mechanisms; these mechanisms are available prefunctionally. The food recognition mechanism develops (perhaps simultaneously with a number of object recognition mechanisms) under the influence of experience with certain tastes or positive nutritious aftereffects of ingestion. The food recognition mechanism probably has connections to the motor mechanisms prefunctionally. A connection between the central hunger mechanism and the complex of motor mechanisms develops as a result of the experience of pecking followed by swallowing, and between the central hunger mechanism and the food recognition mechanism as a result of experience of the nutritive aftereffects of ingesting particular particles (the incentive value of food crumbs). More specific connections develop between the central hunger mechanism and particular motor mechanisms on the basis of nutritive feedback as well. These specific connections are in evidence especially when the chick is hungry, but the original prefunctional connections among perceptual mechanisms and motor mechanisms remain operative and can be seen especially when the chick is not hungry.

It should be noted that the picture for the development of prey catching in kittens is not essentially different from the picture just presented for chicks. Although the individual behavior patterns used in prey catching (pouncing, angling, biting) are originally independent in the sense that the precise ordering of components is not determined, nonetheless these behavior patterns do not occur at random. The specific patterns of these components that develop with respect to particular stimuli can be considered subsystems of the sort that chicks develop with respect to mealworms or to grainlike objects. These subsystems in kittens—a mouse-catching system or a fish-catching system—also have a relationship to the central hunger mechanism that is very similar to the relationship between hunger and pecking or ground scratching in chicks.

A final point is that the development of a hunger system can be greatly influenced by factors that are basically irrelevant to feeding or nutrition. The factor that has been mentioned here is fear, with respect to both the development of recognition of mealworms as food in chicks and the development of fish catching in kittens. A chick that is too afraid of a mealworm will never pick one up (Hogan, 1965), and a kitten that is too afraid of water will never learn to catch a fish. Such indirect motivational factors play an even more important role in the development of social behavior, as will be seen below.

AGGRESSION. The aggression system of an adult chicken consists of perceptual mechanisms that serve an "opponent" recognition function, various motor mechanisms that are used in fighting (including those that control threat display, leaping, wing flapping, kicking, and pecking), and a central mechanism that is sensitive to internal motivational factors (such as testosterone) and that coordinates the activation of the motor mechanisms. Kruijt (1964) showed that fighting develops out of hopping, which is a locomotory pattern that is not initially released by or directed toward other chicks. While hopping, chicks sometimes bump into each other by accident, and in the course of several days, hopping gradually becomes directed toward other chicks. Frontal threatening starts to occur, and by the age of 3 weeks, pecking and kicking are added to aggressive interactions. Normal well-coordinated fights are not seen until 2–3 months.

The various behavior patterns comprising adult fights can be seen to occur, independently, in the 1- to 2-week-old chick, well before their integration into fighting behavior. This means that functional social experience could be a necessary factor guiding development. This is, however, not the case. In other experiments, Kruijt (1964) raised chicks in social isolation for the first week of life and then placed them together in pairs. Many of these chicks showed aggressive behavior toward each other within seconds. Further, the fights that developed were characteristic of the fights of 1-month-old, socially raised chicks. Such results suggest that the organization of the motor components of the aggression system as well as the connections between the central and motor mechanisms develop prefunctionally, and that the occurrence of aggressive behavior requires only the proper motivational state. Similar results and conclusions apply to the development of aggressive displays in gulls (Groothuis, 1994). In this way, the aggression system is more like the dustbathing system than it is like the hunger system.

Whether functional social experience ever affects the organization of the motor mechanisms of the aggression system in chickens as it does in gulls remains an open question. Males raised by Kruijt in social isolation for more than 1 year still showed reasonably normal aggressive patterns, and the abnormalities that were seen could be accounted for in terms of interference from other systems such as fear. Nonetheless, social experience could be necessary for fighting to develop a high degree of effectiveness. One method of testing this idea is to see whether chickens can be "trained" to fight by appropriate tutors. Kuo (1967) reported that such methods are effective in training various breeds of dogs to fight.

The opponent recognition perceptual mechanism must be partially formed prefunctionally because a chick as young as 2.5 days old will respond with frontal threat and aggressive pecks to the stimulus of a 6-cm green wooden triangle moved directly in front of it (Evans, 1968). Likewise, socially isolated chicks showed fully coordinated aggressive behavior when confronted with another chick at the age of 1 week. But isolated chicks of the same age can also direct aggressive behavior to a light bulb hanging in the cage, and older isolated males often come to direct their aggressive behavior to their own tails (Kruijt, 1964). Presumably the complete development of the perceptual mechanism depends on the proper experience at the proper time (just like the templates for song learning in many species), but the experiments necessary to explore this idea have not yet been done.

The development of normal aggressive behavior in kittens does require specific functional experience. Baerends-van Roon and Baerends (1979) described early attack behavior, which included most of the same behavior patterns previously discussed with respect to prey catching, including pouncing and biting. These patterns are apparently the same when originally directed to either a prey or another kitten, but they become modified in different way as a result of feedback from the opponent. In particular, the force of the pounce, the extension of the claws, and the strength of the bite all become reduced after a nestmate responds in kind. The occurrence of "play" behavior, especially in the period from 4 to 8 weeks, seems to provide the kitten with essential experience for the development of normal social behavior. Two kittens that were raised in social isolation (after weaning at 7 weeks) showed either unrestrained attack or total avoidance when confronted with a normally reared cat at the age of several months. These two cats also showed abnormal maternal behavior when they later had their own litters. The Baerends' suggested that normal development requires a proper balance of attack and escape motivation.

SEX.   The sex system of a normal adult rooster consists of perceptual mechanisms that serve a "partner" recognition function; motor mechanisms for locomotion, copulation (which includes mounting, sitting, treading, pecking, and tail lowering), and various displays, such as waltzing, wing flapping, tidbitting, and cornering; and a central mechanism that is sensitive to internal motivational factors such as testosterone and that coordinates the activation of the motor mechanisms. In small groups of junglefowl, Kruijt (1964) saw mounting and copulatory trampling (treading) on a model in a sitting position as early as 3–4 days, but such behavior was not common until weeks later. Full copulation with living partners did not occur before the males were 4 months old.

Many of the components of the copulatory sequence, including mounting, sitting, and pecking, are seen independently in young chicks, and there is ample opportunity for social experience to influence the occurrence and integration of these components. As with aggression, however, several lines of evidence suggest that the motor mechanisms are already organized soon after hatching, and that their expression merely requires a sufficiently high level of motivation. For example, Andrew (1966) was able to elicit well-integrated mounting, treading, and pelvic lowering in socially isolated domestic chicks as young as 2 days old by using the stimulus of a human hand moved in a particular manner. Andrew also found that injection of testosterone greatly increased the number of chicks that responded sexually in his tests during the first 2 weeks (see also Groothuis, 1994, for similar evidence on the expression of sexual displays in gulls, and Williams, 1991, for evidence on the expression of sexual behavior in rat pups). Further, junglefowl males that had been raised in social isolation for 6–9 months copulated successfully with females within so few encounters that it was clear that the motor mechanisms had been integrated before testing (Kruijt, 1962).

The occurrence of the courtship displays presents a somewhat different picture. For example, waltzing is first seen at 2–3 months of age, when it always appears in the context of fighting. As already mentioned, the form of the display seems to develop independently of social experience. The factors controlling the occurrence of waltzing, however, seem to be largely determined by social experience. Waltzing to a female often has the effect that the female crouches, and a crouching female is the signal for mounting and copulation. Experiments reported by Kruijt (1964) showed that the frequency of waltzing increased when mating was contingent on its occurrence and decreased when mating was not allowed. This finding suggests that, in normal development, the switch that is seen from the occurrence of waltzing in a fighting context to a sexual context may require the experience of the display followed by copulation. This interpretation is also supported by the behavior of the males that were socially isolated for 6–9 months. These animals did not show waltzing (or the other displays) before mating with the female, but they often showed displays before attacking her. Thus, copulation seems to be the reinforcer that causes the motor mechanism for waltzing to become attached to the central coordinating mechanism for sex.

Tidbitting is a display that consists in ground pecking directed to edible or inedible objects and/or ground scratching, accompanied with high, rhythmically repeated calls. It develops out of the pecking and calling that accompany "food running" (Kruijt, 1964), which can be seen in young chicks as early as 2 days. Tidbitting is especially interesting because it serves a courtship function in males but a parental function in females. In all three contexts, it serves to attract conspecifics from a distance: food-running chicks attract other chicks and the mother hen, tidbitting males attract females, and food-calling (tidbitting) mother hens attract their chicks. As with waltzing, the form of the tidbitting display does not depend on social experience because it is seen in both chicks and adults that have been raised in social isolation. The causal factors controlling food running are complex and include escape, hunger, and possibly aggression (Hogan, 1966). Andrew (1966) reported that testosterone injections did *not* increase the occurrence of "juvenile tidbitting," whereas they did increase copulatory behavior. Nonetheless, in adult males sexual factors play a primary role in the occurrence of tidbitting (Kruijt, 1964; van Kampen, 1997; van Kampen & Hogan, 2000), and in adult females, parental factors play a primary role (Sherry, 1977). Somewhat surprisingly, sexual factors are not implicated in the response of females to a tidbitting male (van Kampen, 1994). All these results imply that the motor mechanism for tidbitting develops new connections with central mechanisms in the course of development. Unfortunately, there have been no experiments to determine what kind of experience is necessary for the switch in causal factors to occur (but see Moffatt & Hogan, 1992).

The development of the perceptual mechanisms of the sex system and their connections to the central sex mechanism seems to be much more susceptible to the effects of experience than the development of the motor mechanisms. For example, junglefowl chicks become sexually dimorphic at about 1 month of age. By about 2 months, young males begin to show incomplete sexual behavior toward other animals, but such behavior is directed equally toward males and females. Only gradually, as a result of specifically sexual experience, does sexual behavior become directed exclusively to females (Kruijt, 1964). As seen above, the development of the partner recognition mechanism has been studied intensively for many years in the context of sexual imprinting, and there is extensive evidence documenting the influence of both prefunctional and functional factors (Bischof, 1994; Bolhuis, 1991).

Much of the work of Harlow and his students and of Hinde and his students on the development of social behavior in rhesus monkeys, *Macaca mulatta*, is also relevant to this discussion (see, for example, Harlow & Harlow, 1965; Hinde, 1977; Sackett, 1970). The parallels between the development of chicken behavior and monkey behavior are remarkable, and many of the points made in the previous discussion could have been illustrated just as easily by reference to the monkey results. There are also important parallels between the work discussed above and the development of human social behavior (see, for example, Rutter, 1991), but a discussion of these is beyond the scope of this chapter.

PLAY. The topic of play has been discussed extensively in the context of development. Here, I briefly present some ethological ideas about the causation of play, and I show how they complement the behavior system framework. A more general treatment of play that includes a discussion of problems caused by the confusion of cause and function is given by Martin and Caro (1985) and in the chapter by Burghardt in this volume.

The first important idea was expressed by Lorenz (1956): "It seems characteristic of 'play' that instinctive movements are thus performed independently of the higher patterns into which they are integrated when functioning 'in serious'" (p. 635). In other words, the motor mechanisms are activated independently of an activation of the central mechanisms. Insofar as Kruijt's (1964) analysis of junglefowl development is correct, this is precisely the case in newly hatched chicks. As the animal grows older and causal factors for the central mechanisms grow stronger, the independence of motor and central mechanisms decreases, and one might expect the frequency of play to decline, which generally happens. Nonetheless, the analysis of the hunger system suggests that, even when particular motor patterns such as pecking and/or ground scratching become integrated into the system, these same movements can occur independently, especially when the causal factors that activate the central mechanism (i.e., the level of hunger) are weak.

Similar results have also been seen in other species. Lorenz (1956) described the behavior of a young raven that showed a wide array of "playful" movements toward a strange object when not hungry, but that immediately tried to eat such an object if it was hungry. Likewise, Schiller's (1949) chimpanzees showed a playful manipulation of objects, especially when not hungry. The motor patterns of the raven and the chimps under these circumstances could be recognized as being similar to motor patterns belonging to various adult behavior systems.

Once various behavior systems have developed, it may be that play ceases. This, of course, is not true in many species. Morris (1956) suggested that play occurs when central mechanisms are switched off: "The mechanisms of mutual inhibition and sequential ordering mechanisms are not switched on and as a result there is no control over the types and sequences of motor patterns in the usual sense" (p. 643). Switching off central mechanisms would effectively return the animal to a very early stage of development, in which the appearance of play would again be expected. A more elaborate version of this idea was suggested by Baerends-van Roon and Baerends (1979) and was based on their observations of kittens. They proposed that, in cats at least, a central play mechanism exists that, when activated, inhibits other central mechanisms, and "play" could thus appear. Thus, species-typical patterns of play can be understood as being due to a differential inhibition of central mechanisms. Further, when play occurs, its causation remains the same as Lorenz originally suggested: the independent activation of motor mechanisms.

HUMAN LANGUAGE. In this section, I hope to show how it is possible to consider human language to be a behavior system that is similar in many respects to the behavior systems we have already considered. Human language, of course, is vastly more complex than dustbathing or feeding in chickens, but, as a biological system, both the organization and development of language should share many of the principles governing these simpler systems. There is an enormous literature on language and its development, and only very restricted aspects can be considered here.

To begin, it is necessary to identify the building blocks of the language system: What are the perceptual, motor, and central mechanisms comprising the system? As we shall see, which building blocks are chosen depends on one's definition of language. For my present purposes, I will start with the perceptual and motor mechanisms that recognize and produce the sounds in a language. I will also restrict my discussion to specific speech sounds (i.e., phonemes) as opposed to other vocal aspects of language such as prosody (Locke, 1993, 1994; Locke & Snow, 1997).

It has been known for some time that human infants as young as 1 month are able to perceive phonetic distinctions categorically in a similar way to normal adults (Eimas, Siqueland, Jusczyk, & Vigorito, 1971; Eimas, Miller, & Jusczyk, 1987). More recent evidence demonstrates that these perceptual categories can be altered by linguistic experience. For example, in a cross-cultural study of 6-month-old American and Swedish infants, Kuhl, Williams, Lacerda, Stevens, and Lindblom (1992) found the two groups exhibited a language-specific pattern of phonetic perception to native- and foreign-language vowel sounds. Of particular interest is that these effects of experience are seen by 6 months of age, that is, before the infant itself begins producing speech sounds. Further, by 1 year of age, infants no longer respond to speech contrasts that are not used in their native language, even those that they did discriminate at earlier ages (Werker & Tees, 1992). Thus, the perceptual mechanisms responsible for speech perception in infants are both highly structured at birth and highly malleable in that they are shaped, instructively and selectively, by exposure to the linguistic environment (Kuhl, 1994).

Normal infants begin to babble between 6 and 10 months (for review see Locke, 1993). The initial sounds produced by the infant are species specific (i.e., are similar in infants raised in different linguistic environments), and include phonemes not found in its native language. As the child grows older, the distribution of sounds comes more nearly to approximate the distribution in its linguistic environment, and the nonnative sounds drop out. The mechanism by which these changes occur has not really been adequately analyzed, but it presumably involves a process of matching vocal output to the previously developed perceptual mechanisms (templates) by auditory feedback (Marler, 1976).

These results for the development of the perceptual and motor mechanisms that recognize and produce speech sounds involve the same problems of modularity, constraints, and processes that we have seen before, especially with respect to the changes that occur in the development of bird song. These parallels have been noted for many years (e.g., Lenneberg, 1967; Marler, 1970b), and continue to provide mutual insights into the development of both systems at both a behavioral and neural level (Doupe & Kuhl, 1999; Hauser, 1996; Snowdon & Hausberger, 1997).

Speech sounds, however, are only one aspect of normal spoken language. Sounds become combined into words (morphemes: units of meaning), and one can ask whether the phonemes or the morphemes are the basic units of the language system. Words can always be broken down into their constituent sounds, but there is now considerable evidence that infants learn utterances (words or short phrases) as a whole during the first 2 years with respect to both perception and production (Jusczyk, 1997; Locke & Snow, 1997). It is only later that children are able to break utterances down into smaller sound packets. For these and other reasons, Locke

(1994) argues specifically that phonemes are not the basic building blocks of human language. We might then conclude that words are the basic units of the language system, but first we must consider what the words represent.

Birds do not sing randomly. They sing when the appropriate internal and external factors are active. In most cases, this is when the sexual and/or aggression behavior systems are activated.* Humans speak in comparable circumstances, but the range of circumstances in which humans speak is very much broader. In fact, cognitive psychologists (Shelton & Caramazza, 1999) have proposed that humans possess a semantic system that receives input from spoken and written words (phonological and orthographic input lexicons) and responds with output of spoken and written words (phonological and orthographic output lexicons). Our speech (or writing) thus expresses the state of our semantic system. Of course, our semantic system is much more complex than the sexual and aggression systems of songbirds (and much of the field of cognitive psychology is devoted to understanding the organization of the semantic system and the mechanisms of lexical access to it), but it seems certain that the principles of organization and development are similar. In the present context, some of Shelton and Caramazza's (1999) conclusions are particularly interesting. They reviewed studies of language processing following brain damage and found results that "broadly support a componential organization of lexical knowledge—the semantic component is independent of phonological and orthographic form knowledge, and the latter are independent of each other" (p. 5). In my terms, their language system can be considered to have a central semantic mechanism with perceptual mechanisms for recognizing words and motor mechanisms for producing words.

One important component of language has not been discussed: Words can be combined into sentences, and it is at this level of analysis that the concepts of grammar and syntax are generally used. It is also this level of organization to which the concept "language instinct" (Pinker, 1994) has been applied. The operation of grammatical structures is not normally apparent until some time after the age of 2 years, when words become recombined into novel utterances that follow particular rules (Locke & Snow, 1997). One set of rules, called generative grammar, was proposed by Chomsky (1965). These rules have been reasonably successful in describing the types of sentences produced by native speakers of English, but there has been great controversy about how these rules develop in the child, particularly whether specific kinds of linguistic experience are necessary (see, e.g., Tomasello, 1995). The details of this controversy need not concern us here, except to say that most of the issues are the same as we have already met in describing the development of grooming (Berridge, 1994) and dustbathing (Larsen *et al.*, 2000) sequences, some of which are considered further in the general discussion.

As a final point, there is a long history of authors' proposing uniqueness for the human species on the basis of aspects of the syntax found in our language. In a recent discussion of this issue, Kako (1999) continues this debate. He points out that in its "most generic form, syntax is defined as a set of rules for assembling units of any type into larger units" (p. 1) He then discards this definition because he finds it too broad, and proposes instead a set of four structural properties that define the core of syntax. Once again, the details of his proposal need not concern us here. It is

---

*Note that this interpretation assumes that the perceptual and motor mechanisms for song have become attached to the sexual and aggression systems. We have not considered these systems in a songbird, and have not asked how these connections develop.

my opinion, however, that the generic definition has many advantages if one is interested in looking for the similarities, rather than the differences, among systems (Lashley, 1951). It must be true, by definition, that human language is unique because all species are unique as are the specific behavior systems possessed by each species.

These results suggest that the human language system comprises three basic sets of components at two major levels of organization, and that these components develop largely independently. The sensory–motor components correspond to the perceptual and motor mechanisms depicted in Figure 1 (with additional connections between them and the central mechanisms), whereas the semantic (meaning) and syntax components correspond to two separate central mechanisms.

This general conception is also supported by the results of studies of deaf children. For example, deaf children born to deaf parents who communicate using sign language do not babble vocally; rather, such children babble with their hands (Pettito & Marentette, 1991). Manual babbling occurs at about the same age that vocal babbling occurs in children with normal hearing who have been raised in a vocal environment. Further, the development of sign language proceeds in much the same way as the development of vocal language with respect to both structure and use. Goldin-Meadow (1997) found that the same general rules apply and that they appear at the same age. These results all suggest that the language system can use auditory–vocal units or visual–manual units equally well. Studies of the neural organization of language (Hickok, Bellugi, & Klima, 1998) are also consistent with this interpretation.

One can ask, finally, whether this conception of human language as a behavior system actually furthers our understanding of language and its development. I think it does in at least two important ways. First, by breaking the system up into its components, the study of the pieces becomes more tractable. There has already been considerable success in comparing the development of bird song and human speech (Doupe & Kuhl, 1999), and the development of grooming sequences may provide a useful model for some aspects of the development of syntax. Further, insofar as these components are the "natural" pieces of the system, it becomes easier to understand how the system could have evolved (Hauser, 1996; Pinker, 1994). A second important reason is that development of all three sets of components requires both functional and nonfunctional experience and involves the same problems of modularity, constraints, and processes that have appeared before: Solutions to these problems in one system should easily generalize to other systems.

DEVELOPMENT OF INTERACTIONS AMONG BEHAVIOR SYSTEMS

A basic tenet of ethological theory is that various behaviors of an animal—and often the most interesting ones—are the expression of the activation of not just a single behavior system, but of the interaction of two or more systems that are activated simultaneously. This conflict hypothesis was proposed by Tinbergen (1952) and has been discussed by Kruijt (1964), Baerends (1975), and Groothuis (1994). Two major studies have directly addressed the development of interactions among systems—those of Kruijt (1964) in chickens and Groothuis (1994) in black-headed gulls. I will restrict my discussion here to the behavior of chickens.

Kruijt's (1964) results show that the major behavior systems of escape, aggression, and sex develop in chickens in that order. Further, activation of a system already developed inhibits the expression of systems that are just beginning to develop.

Thus, a young chick that shows frontal threatening and jumping to another chick may immediately stop this early aggressive behavior if it bumps into the other too hard. As the chicks grow older and the causal factors for aggression become stronger, however, such escape stimuli no longer stop aggressive behavior. Rather, attack and escape begin to occur in rapid alternation, and various irrelevant movements start to appear during fighting. Likewise, early sexual behavior is immediately interrupted if either the attack or the escape system is activated, but later, behavior containing components of attack, escape, and sex can be seen simultaneously. As seen above, there is evidence that the basic organization of these major systems is formed prefunctionally and that their expression merely requires a sufficiently high level of causal factors. The gradual appearance of more complex interactions can be interpreted as reflecting changes in the strength of causal factors (i.e., motivational changes) rather than changes in the connections among central mechanisms (i.e., developmental changes).

The fact that another member of the species is the adequate stimulus for activating the escape, aggression, and sex systems means that all these systems must normally be activated when a conspecific is present. Kruijt (1964) points out that the precise state of activation of these systems at any moment depends on the previous history of the male and on the appearance, distance, and behavior of the other bird. He suggests that the appearance of smooth, typical adult courtship behavior depends on an increasing activation and mutual inhibition of the attack and escape systems, and the relationship between attack and escape is stabilized by the activation of the sexual system. He posits a stabilizing factor in order to explain why the adult animals do not constantly switch quickly from performing one type of behavior to performing another.

The stabilizing influence of sex on the agonistic systems of escape and aggression is not merely a consequence of increasing hormone levels as the animals grow older. Experience also plays a major role. Kruijt (1964) found that junglefowl males reared from hatching in social isolation for more than 9 months showed serious and apparently irreversible abnormalities in their courtship and sexual behavior. To a large extent, these abnormalities could be characterized as switching too quickly among escape, aggressive, and sexual behavior. In other words, the stabilizing influence of sex was present only after experience of the sort that would occur during normal early development. Kruijt also found that as little as 2½ months of normal social experience immediately after hatching was sufficient to obviate the effects of subsequent social isolation for periods of at least 16 months. These results are difficult to interpret because during the first 2½ months of life chicks show essentially no sexual behavior. Thus, they could not be learning anything specific about sexual behavior. Instead, it would seem that the experience a chick gains during normal encounters early in life provides the information necessary for it to stabilize its agonistic systems, and that normal sexual behavior can only occur if the agonistic systems are already stabilized.

Results of the Baerends', mentioned above, also support this interpretation. Their kittens that were raised in isolation from peers showed either unrestrained attack or complete avoidance when confronted with a normal kitten, and this pattern was also seen later in a sexual situation. Rhesus monkeys that were raised in isolation from peers also showed inadequate sexual behavior when adult (Harlow & Harlow, 1962). However, in both the cats and the monkeys, a particularly "good" partner was able to compensate for the behavioral deficiency in the isolation-reared animals (Harlow & Suomi, 1971; Novak & Harlow, 1975). The description of these encounters suggests that the sexual behavior system itself had not developed abnormally, but that abnormal fear or aggression interfered with the performance of sexual behavior. The conclusion that can be drawn from these studies is that well-integrated interactions among behavior systems are necessary for the normal, well-coordinated behavior we see in adult animals, and that functional experience is necessary for such integration to occur.

How a stabilizing influence develops has not been studied. We have seen that some of the experiences of a normally raised young chick, such as bumping into other chicks or being pecked at as a result of pecking another chick, are not necessary for normal aggressive behavior to develop. Such early social experiences might, however, be crucial for developing a normal attack–escape relationship. I shall return to this problem in the discussion.

General Discussion

Structure, Cause, Function, and Development

In this chapter, I have defined a behavior system in terms of its structure. Other investigators define a behavior system in terms of its functional characteristics (e.g., Timberlake, 1994). There may often be a close correspondence between systems defined in structural and functional terms, but this is by no means always the case; it is very easy for confusion to arise. For example, a structural definition of sexual behavior would include a description of the perceptual mechanisms that analyze stimuli and activate a central sexual coordinating mechanism plus a description of the motor patterns that occur when the central mechanism is activated. A functional definition of sexual behavior would emphasize reproduction, that is, those behaviors that lead to successful propagation of the species. It should be clear that many animals, including humans, engage in sexual behavior by the structural definition when the behavior definitely will have no reproductive function. Further, courtship behaviors in many species are necessary for successful reproduction, even though the courtship behaviors themselves can be considered to belong to nonsexual behavior systems such as fear and aggression (Baerends, 1975; Tinbergen, 1952). Another example would be the language system which could be defined in terms of its communication function (Hauser, 1996). It should be emphasized that one type of definition is not inherently better or worse than the other type: which type is most useful depends on the questions being asked (Hogan, 1994a).

Development implies changes in the structure of behavior, both changes in the organization of the behavior mechanisms themselves as well as changes in the connections among behavior mechanisms. To this point, I have only considered the causes of changes in behavioral structure. In this section, I will briefly discuss some examples of functional questions. Since I believe that there is no necessary relation between cause and function (Hogan, 1994a), it might seem that there is nothing to be gained toward understanding causal mechanisms by asking functional questions. In theory, this should be true. In practice, however, the problems that an animal must solve in order to survive provide the selection pressures that are responsible for evolution by natural selection, and it turns out that the evolutionary solutions to these problems sometimes use causal mechanisms that are related to the function that the behavior serves. I shall consider first some examples of functional questions that do not increase our understanding of development, and then some examples that do.

IS DEVELOPMENT SELECTED?   It is almost a truism that natural selection should operate at all stages of development, and not only on the adult outcome, because any developmental process that reduces the probability of reaching adulthood will be very strongly selected against, all other things being equal. Nonetheless, a genotype with advantageous consequences at a particular stage of development can be selected for only if its consequences in the adult do not reduce the fitness of the individual possessing it. What this means is that, at any particular stage of development, behavior may be far from optimal: it need only be good enough to bring the animal to adulthood.

This line of reasoning also leads to other conclusions. For example, it seems intuitively obvious that the best mechanism for regulating a particular outcome would be one that is directly sensitive to the outcome. Thus, the best mechanism for regulating nutrition, say, would be one that could directly sense the state of nutrition. This is another way of saying that an optimal mechanism should be based on a simple, direct relationship between cause and function. But as we have seen, development is an extremely complex process and one in which optimal solutions may be the exception rather than the rule. It follows that development is opportunistic in the sense that any available means will be used to produce an acceptable end. Two examples should make this point clearer.

We have seen above that pecking in newly hatched chicks is not controlled by factors related to nutrition. When it became clear that experience was necessary for nutritional control to develop, it seemed reasonable to look at the effects of various kinds of nutritional experience on the occurrence of pecking. That approach turned out not to be the key to solving the puzzle because the necessary experience was not nutritional, but an association between the act of pecking and the effects of swallowing any solid object. These results were surprising (and took a long time to discover) because of our preconceptions about the relationships between the causes and functions of behavior. We intuitively feel that, when behavior changes in an adaptive direction, the cause of the change should be related to factors associated with the adaptation. Thus, when pecking changes in such a way that relatively more nutritive items are ingested, we infer that something about nutrition was responsible for the change. But in this case, our inference was wrong. Pecking behavior to food and sand during the test changes for reasons that are completely unrelated to nutrition.

A second example is provided by the analysis of Hall and Williams (1983) of the relationship between suckling and other ingestive behavior in rats. Suckling and eating are both behaviors that function to provide nutritive substances to rats, suckling normally for the first 3 weeks after birth and eating thereafter. In their search for the causal mechanisms underlying ingestion, Hall and his colleagues originally assumed that these mechanisms would be similar in both newborn and older animals. In fact, after many years of work, their results showed that the causal mechanisms controlling suckling are largely independent of the mechanisms controlling eating. Their analysis suggests that both systems coexist simultaneously, and that only one system is expressed at a time. Hall and Williams (1983, p. 250) concluded: "Such findings for suckling illustrate the general difficulty in determining the relationship between adaptive behavior of infancy and functionally similar representations in adulthood." Subsequently, Hall and Browde (1986) made similar studies of infant mice and discovered that the causal factors underlying eating are considerably different from those in rats. Thus, the study of the development of feeding behavior in chicks, rats, and mice shows that mechanisms for change have evolved that lead to an adaptive result, but that these mechanisms often bear little resemblance to our prior ideas of what they should be.

ADAPTATIONS FOR DEVELOPMENT. The problems that an animal has to solve for survival put selection pressures on the causal mechanisms for the behavior that can evolve. It is for this reason that functional thinking can help us to understand causal mechanisms that we have discovered, and in some cases, it may direct our attention to seeking causal mechanisms that we would not otherwise have thought of. Oppenheim (1981) provided many examples and an excellent discussion of this issue. Here, I shall make a few functional comments about the case of nutrition, which we have already considered from a causal perspective.

In almost all species of animal, the method of acquiring nutrition changes—willy-nilly, at least once, and often two or more times—in the course of the animal's lifetime. In mammals, for example, nutrition is provided to the fetus via the placenta and, after birth, first by suckling and later by eating. Suckling, as a motor mechanism, exists before birth and after weaning, but it is not expressed then. Thus, at some stage in development, suckling must be "switched in" to provide nutrition and later must be "switched out." Similarly, in birds, the yolk sac provides nutrition in the egg and for some time after hatching; then the young bird may receive food from its parents by gaping; and finally, it feeds itself using some sort of pecking movement. Here, too, something must regulate when gaping is used and when pecking is used. This, then, is the problem the animal must solve. How does it do it?

The causal answer to this question is probably different for every species. We have seen at least a partial causal answer for chicks in the results we have obtained from pecking. But these results raise several obvious functional questions, two of which can be considered here. First, why should pecking not be controlled by nutritional factors at hatching? Second, why should experience be necessary for pecking to become integrated into the hunger system?

One can imagine that, if pecking were originally controlled primarily by the chick's nutritional state, pecking might not occur at all until the yolk reserves were exhausted. Such a chick would not have as much experience with its world as a chick that had engaged in exploratory pecking during the first few days. Given that the control of pecking must shift sometime between hatching and the time when pecking is necessary for providing nutrients, there is no particular reason that experience should not provide the timing of the shift. On the other hand, there is one important reason that experience should provide the timing: Birds can hatch early or late with respect to their overall stage of development (and mammals can be born prematurely or past term). Endogenous timing of the switch in causal factors to or from pecking or suckling would be disastrous if, for example, a 1-week premature baby could not suckle in its first week, or if a baby could not be weaned early if its mother's milk supply were interrupted. In general, it seems certain that experiential factors provide a more reliable timing cue than endogenous factors could provide in most cases where a switch between methods of acquiring nutrition occurs.

In this context, it is useful to return to the concept of a play system and to consider what function it may serve. We have seen that the essence of the concept is that motor mechanisms have a chance to be "free" of influence from central mechanisms. Such freedom may give the motor mechanisms an opportunity to become incorporated into other central mechanisms. One can imagine that such a flexible system would be useful during development, especially in cases where something may have gone wrong, and in which the so-called normal connections would not function optimally. Similar functions for play have been suggested before, but one problem with such explanations is that adult behavior develops equally well in individuals that vary greatly in the amount of play they exhibit (Martin & Caro, 1985). Here we can see a function for the developmental situation described by Groothuis (1994). Endogenous factors are sufficient to determine the development of particular behavior systems (or particular motor mechanisms), but during development the possibility exists for experience to bring about a somewhat different outcome. Under normal conditions of development, either endogenous factors or play could provide alternative pathways to reach the same result. Only under special conditions (such as those provided to the gulls by Groothuis) would the different pathways lead to different results.

THE CONCEPT OF PREFUNCTIONAL.   It should be clear by now that it is quite possible to discuss the causal development of behavior without using the word *innate*. Nonetheless, it must also be clear that I have used the word *prefunctional*, defined as developing without the influence of functional experience, in many places where others would have said *innate*. In some ways, this is how Lorenz (1961, 1965) suggested the term *innate* be used, though he was not always consistent in his use (Lehrman, 1970). Nonetheless, there are still some problems with using a functional definition, and I shall briefly mention two of them here. First, it may be useful to indicate why I think the concept *prefunctional* is necessary at all.

Lehrman (1970) pointed out that one important reason for the controversy between him and Lorenz was that the two were interested in different problems: Lehrman was interested in studying the effects of all types of experience on all types of behavior at all stages of development, whereas Lorenz was interested only in studying the effects of functional experience on behavior mechanisms at the stage of development at which they begin to function as modes of adaptation to the environment. In other words, Lehrman used a causal criterion to determine what was interesting to study, whereas Lorenz used a functional criterion. These two criteria are equally legitimate (Hogan, 1994a), but the functional criterion used by Lorenz corresponds to the way most people think about development. In fact, it is logically consistent to talk about behavior development that is prefunctional (or innate) versus behavior development that is learned when the criterion is the absence or presence of functional experience. (I prefer the word *prefunctional* to the word *innate* because the latter has too many additional meanings.) I think it is important to show how behavior that can be classified as prefunctional still presents interesting developmental problems that can be investigated in a causal framework. That is one of the things I have tried to do in this chapter.

It is also important to see some of the difficulties inherent in using a functional definition. Perhaps the most important of these is that the function of a behavior is not always obvious. For example, if the function of pecking is viewed as being to provide nutrition, pecking becomes integrated into the hunger system prefunctionally; if the function of pecking is viewed as being to bring about ingestion, then pecking becomes integrated into the hunger system through functional experience. In either case, the causal process is the same. Similar problems arise when there are alternative routes to reaching the same end, as in the development of the oblique posture in the black-headed gull.

A related problem is that the function of behavior can change over the course of time. Sometimes, this change is due to changes in the environment and sometimes to changes in the behavior mechanisms themselves. This means that, at best, the concept *prefunctional* is only relative: It can usefully be used to describe situations with respect only to the particular function that the investigator has in mind.

SOME PRINCIPLES OF DEVELOPMENT

The process of development is extremely complex, to a large extent because so many interdependent events occur simultaneously (Hogan, 1978; Kuo, 1967). Unfortunately, it is not possible to comprehend all the important variables at the same time, so that various sorts of distinctions and simplifications must be made in order to further our understanding (Bateson, 1999). The basic simplification that has been made in this chapter is that of describing behavior in terms of motor, central, and perceptual mechanisms and the connections among them. These mechanisms are conceived of as structural units of behavior of a particular magnitude and complexity. Development is viewed as changes in these underlying behavioral mechanisms and their connections. This conception makes it possible to see a clear analogy between behavior development and the development of specialized cells and tissues in the embryo (e.g., Slack, 1991; Waddington, 1966). I begin this section with a brief overview of the development of the nervous system and then discuss the problems of modularity, constraints, and processes that have been identified above. Finally, I consider some special aspects of the development of social behavior.

DEVELOPMENT OF THE NERVOUS SYSTEM.   Brown, Hopkins, and Keynes (1991) divided brain development at the cellular level into four major stages: (1) genesis of nerve cells (proliferation, specification, and migration), (2) establishing connections (axon and dendritic growth, and synapse formation), (3) modifying connections (nerve cell death and reorganization of initial inputs), and (4) adult plasticity (learning and nerve growth after injury). Stages 3 and 4 are the most relevant to our question.

During fetal development, many more nerve cells are formed than will be found in the adult brain. These nerve cells all send out axons and establish connections with target cells (other neurons and muscle cells), but a large proportion of them die before the synapses become functional. The mechanisms underlying this process involve electrical activity in the nerve cells and their targets, but they are still not fully understood (Oppenheim, 1991). It is thought that neuronal death may serve to eliminate errors in the initial pattern of connections. The axons of the cells that remain are often found to have more extensive branches and to contact more postsynaptic cells than they will in the adult. The mechanisms that bring about axonal remodeling—that is, the elimination and reorganization of these terminal branches—also involve activity in the neurons. In brief, it has been shown that specific spatial and temporal patterns of electrical activity in both the nerve cells and their target cells are necessary for functional connections to form between them: "cells that fire together wire together" (Shatz, 1992, p. 64).

The process of axonal remodeling occurs both pre- and postnatally, and it is essentially irreversible. Once the axons have established functional connections with other neurons or muscles, those connections appear to be a permanent part of neural organization. The mechanisms that are responsible for adult plasticity involve facilitation or inhibition of synaptic transmission and the growth of dentritic spines which presumably correlate with the formation of new synapses (Bolhuis, 1994; Brown *et al.*, 1991; see also DeVoogd, 1994, for a discussion of neurogenesis in adult birds). Whether these changes are reversible remains a matter of conjecture.

The work of Hubel and Wiesel established that visual stimulation plays a vital role in the development of the mammalian visual system (for reviews of the early work, see Blakemore, 1973; Wiesel, 1982). They showed, for example, that normal development of the connections between cells of the lateral geniculate nucleus and the visual cortex in the cat requires binocular visual stimulation soon after the kitten's eyes open. Allowing a kitten to see with only one eye at a time during the critical period results in most cortical cells being responsive to stimulation from one eye only, whereas binocular stimulation results in most cortical cells being responsive to stimulation from both eyes. These results were interpreted in terms of the eyes competing for control of cells in the cortex and are an example of axonal remodeling. These results were important because they showed that the organization of a sensory system was actually driven by stimulation from the environment. They also provided a model for how the perceptual mechanisms underlying bird song learning and filial and sexual imprinting might develop (Bischof, 1994; Bolhuis, 1994; DeVoogd, 1994).

The neural activity responsible for axonal remodeling in the visual cortex is triggered by stimuli originating in the environment after the kitten is born and has opened its eyes. More recently, other investigators have asked whether neural activity is also necessary for neural connections to form *in utero*, and, if so, how this activity is instigated. Shatz (1992) and her collaborators, for example, have looked at axonal remodeling in the lateral geniculate nucleus of the cat, which occurs before birth. They found that the same kind of action-potential activity is necessary for developing normal connections from the retina to the lateral geniculate as is later necessary for normal connections to form in the cortex. Rather than being instigated by stimulation from the external world, however, the neural activity was caused by patterns of spontaneous neural firing. How these waves of activity are generated remains to be discovered.

These two cases of axonal remodeling illustrate the difference between development based on functional experience (organization of the visual cortex) and development that occurs prefunctionally (organization of the lateral geniculate nucleus). What is important in the present context, however, is that the mechanisms for synaptic change are the same before and after birth, and it is irrelevant for the connection being formed whether the neural activity arises from exogenous or endogenous sources. In fact, the same connection can be formed in either way. Some behavioral examples will be used to illustrate this point in the section on processes.

MODULARITY.   An important assumption made in this chapter is that particular parts of the central nervous system subserve particular functions, and that, by the time behaviorally interesting events are occurring, these parts, or modules, are preassigned. This means that, at the particular stage of development under consideration, the range of possibilities for further development of a particular behavior mechanism are so restricted that only special (i.e., already determined) kinds of experience can have a developmental effect on that mechanism. In practice, this means that, by the time of birth (or hatching), the central nervous system is already highly differentiated, with the general organization of pathways and connections already determined. By this stage of development, reversing the functions of major parts of the brain is generally impossible in the sense just discussed. Under these circumstances, it seems justified to speak of the song-recognition perceptual mechanism or the ground-scratching motor mechanism or the aggression central mechanism as prefunctionally developed units of behavioral structure subject to further (but quite restricted) differentiation on the basis of subsequent experience.

It should be realized, however, that, if we follow the development of any behavior mechanism backward in time, we can always find a stage in which the nerve cells making up the behavior mechanism could have subserved a different behavior mechanism under somewhat different conditions. If we go back still further, we will find a stage when the cells could have become something other than nerve cells, and so on. At the time of birth—an arbitrary time I have chosen for convenience—a particular set of nerve cells may have differentiated to the point where, if they survive, they will be the cells that mediate mate recognition, and in this sense, they are preassigned that function. But they are preassigned only from the point of view of future development. There has been much recent discussion about the meaning of modularity with respect to higher cognitive functions (e.g., a language module). This is not the forum to comment on this issue except to say that the principles involved are the same as I have just discussed (Karmiloff-Smith, 1992, 1998).

CONSTRAINTS.   Constraints on development actually arise as an interaction between the structures (modules) available at any given time and the processes that can lead to changes in those structures. Two such constraints that are ubiquitous in discussions of development will be considered here: irreversibility and critical periods.

Insofar as behavior mechanisms can be regarded as preassigned, they illustrate the problem of the irreversibility of development. When Waddington (1966) discussed the question of whether the differentiation of cells is reversible, his answer was, "it depends." It depends on what cell, in what animal, at what stage of development, and so on. This is already an important point because similar reasoning shows that it is nonsense to ask a question such as: Is imprinting irreversible? One can only begin to answer such a question after specifying the species, the particular imprinting procedures, the stage of development, and so on.

More important, Waddington specified some of the processes that are responsible for the irreversibility of cell differentiation. For example, some or all of the genetic material may have been "used up" or may have otherwise disappeared in the course of the development of the cell, or the genetic material may still be present, but, for various reasons, cannot be accessed. The most frequent reason for irreversibility, however, seems to be that "development involves such a complicated network of processes that it would be an extremely long and tricky process to unravel them. One could, in theory, take an automobile, dismantle it, and build the pieces up again with a little modification into two motorcycles, but it wouldn't be easy; and it is something like this that we are asking a differentiated cell to do when we try to persuade it to lose its present differentiation and develop into something else" (Waddington, 1966, pp. 54–55). Processes with similar characteristics seem certain to underlie cases of behavior irreversibility.

The best-documented cases of total irreversibility involve motor mechanisms for bird song, as exemplified by the "crystallization" of song in the chaffinch (Thorpe, 1961) and the white-crowned sparrow (Marler, 1970a). The perceptual mechanisms, or templates, on which these songs are based are probably also fixed irreversibly once they have developed, although here the evidence is somewhat controversial (Baptista & Gaunt, 1997; Nelson, 1997). Many of the courtship and agonistic displays seen especially in birds, such as waltzing in chickens (Kruijt, 1964) or the oblique posture in the black-headed gull (Groothuis, 1994), are probably also fixed irreversibly once they have developed. These cases probably all involve axonal remodeling and are analogous to the case of cell differentiation, in which the genetic material either disappears or becomes inaccessible during the course of development.

Here it is useful to emphasize the distinction between perceptual and motor mechanisms themselves, and the various connections that may exist between them: Even though a perceptual or motor mechanism has crystallized, there are still possibilities for alternative pathways among them. The concept of imprinting, for example, implies a change in a perceptual mechanism as a result of experience. In some species, such a change may be irreversible, but subsequent experience may lead to additional pathways being formed between other perceptual mechanisms and the sexual behavior system, and these new connections may mask the original imprinting. A rather difficult experimental analysis would be necessary to investigate this possibility. We have seen, however, a case such as this on the motor side of the hunger system in chickens: An original connection between pecking and ground scratching was masked, but not destroyed, by later experience.

The most common reason that behavior changes are apparently irreversible is probably the same reason that most cell differentiation is irreversible: So many events would have to be undone (or compensated for) that change becomes almost impossible. A very simple case where changes could still be made was training a kitten to catch fish after it had already learned to catch mice (Baerends-van Roon & Baerends, 1979). Here there were two problems. One was an indirect, motivational problem: A fear of water inhibited any attempt to catch the fish. Once the fear of water could be overcome, the kitten faced a direct, developmental problem: rearranging motor mechanisms in a different sequence. In this case, rearrangement was possible, although with some interference from the original learning.

A more complex case is the sexual behavior of male junglefowl raised in social isolation. Here, subtle aspects of the integration of the aggression and escape systems seem to be permanently missing. Because this integration plays a determining role in permitting sexual behavior to occur, these effects of social isolation are effectively irreversible, even though the copulatory motor patterns remain intact. The fact that some consequences of normal social experience during the first few weeks of life are sufficient for the development of relatively normal adult behavior implies that axonal remodeling-type processes are involved. In effect, it could be that various perceptual neurons are competing for connections with the attack and escape systems, and that a stable attack–escape balance depends on the pattern of connections that finally develops. This is a speculative suggestion, but it does fit in well with what is known about the development of neural connections at earlier stages. Such a suggestion also implies that no new principles of development are required to understand the development of behavior system interaction. It remains to be seen whether some sort of "therapy" could be devised to cope with this problem—as was possible in the cats and monkeys raised in social isolation—but that is an empirical matter.

The fact that development is not reversible (except as discussed above) means that constraints of various sorts are inherent in developing systems. The most commonly discussed constraint is a "critical" or "sensitive" period that corresponds to the embryological concept of competence (Waddington, 1966). In essence, these concepts refer to the fact that the developing system is especially susceptible to particular external influences at particular stages of development. This topic has often been a matter of controversy, especially with respect to the factors responsible for the beginning and the end of the period (see Bateson & Hinde, 1987, for an excellent discussion of sensitive periods). Nonetheless, the previous discussion should make clear that probably all aspects of development are associated with critical periods. At each stage of development, the animal is different from what it was; it is only to be expected that the effects of the "same" experience will be different in the different stages (Schneirla, 1956; Schneirla, Rosenblatt, & Tobach, 1963). The factors that are responsible for the beginning and ending of these periods are probably different in every case.

PROCESSES.   What are the processes of behavior development? There is not yet any answer to this question, but I think several points are worth making. To begin, it seems very unlikely that the biochemical processes responsible for altering the structure of behavioral mechanisms and their connections are different before and after a particular behavior begins to function. If this is indeed the case, a number of results I have discussed become more easily understandable.

A first example is provided by the results of Groothuis (1992, 1994). He found that the oblique posture in the black-headed gull developed normally when a gull was reared either in social isolation or in large social groups, but that it sometimes developed abnormally when a gull was raised with only two or three peers. One can suppose that under circumstances of social isolation, endogenously produced patterns of neural firing provide the information necessary to develop the normal connections in the motor mechanism responsible for the form of the display, prefunctionally. When peers are present, functional social experience provides the information. Performance of precursors of the display often leads to reactions by the other gulls. These reactions, in turn, provide additional neural stimulation which could interfere with endogenously produced patterns that thus lead to different (abnormal) connections being formed in the motor mechanism. If these connections require repeated stimulation to form, the probability that the average experience will be "correct" is greater in a large group than in a small group, where the effects of the behavior of one abnormal individual companion would be relatively greater (see the results from groups of songbirds raised in isolation from adult song, Marler, 1976). This line of reasoning suggests that functional and prefunctional "experience" provide alternative routes for the control of behavior system development, a suggestion that can also account for some of the results for the development of the aggression system in chickens reviewed above and for the results of play in several species (Martin & Caro, 1985).

As a second example, one of the interesting aspects of the perceptual phase of song learning in birds is the very large differences among species with respect to what kind of experience is needed for an adequate template to develop. At one extreme, a male cowbird raised in social isolation will develop a normal species' song (King & West, 1977), whereas a chaffinch or white-crowned sparrow raised similarly will develop a song that at best contains only a few species-specific elements (Marler, 1976; Thorpe, 1961). On the other hand, the time at which hearing the species' song is effective for learning is much more restricted in the white-crowned sparrow than it is in the chaffinch. Likewise, if socially isolated males are played variants of the typical species' song, or indeed songs of other species or even pure tones, some species are able to learn only the song of their own species, whereas other species are able to learn a much wider range of sound patterns. Similar species differences are also characteristic of the range of stimuli to which young birds will imprint and the time at which these stimuli are effective (Lorenz, 1935). In all cases, however, a perceptual mechanism develops that serves a species-recognition function.

One way to understand how so many apparently different ways can lead to a similar functional outcome is to suppose that once certain kinds of structural change have occurred in the development of a perceptual mechanism, further change is no longer possible (crystallization, irreversibility). It then follows that the timing of triggering events becomes crucial in determining which events will affect development. In a particular species of songbird, for example, one can imagine that, if genetically triggered events occur in the perceptual mechanism for song recognition before the young bird can hear, then the perceptual mechanism is fixed, prefunctionally, in that species, and posthatching experience can no longer have an effect. If the triggering events are delayed, however, the posthatching experience of the bird can provide the trigger. In this way, the same type of perceptual mechanism can be used for either "innate" or "learned" song recognition.

The timing of events that trigger irreversible changes in developing behavior mechanisms can also explain some apparent differences between perceptual and motor mechanisms. It is noteworthy that, with the exception of bird song and human language, the motor mechanisms of the behavior systems discussed above all develop prefunctionally, whereas all the perceptual mechanisms require at least some functional experience in order to achieve the normal adult form. This fact might suggest that there are some fundamental differences in the causal factors responsible for the development of perceptual and motor mechanisms. Such a conclusion is unlikely to be true because in both cases the organization of neural or neuromotor connections depends on particular spatiotemporal patterns of neural activity that can be generated either endogenously or exogenously. Prior to birth, most of the causal factors would be endogenous, although external stimulation may play a role in some cases (e.g., the auditory system in ducks; Gottlieb, 1978). After birth, both internal and external factors could be important. The fact that most of the motor mechanisms we have considered develop prefunctionally very likely reflects the fact that motor mechanisms generally become organized earlier in development than perceptual mechanisms (Hogan, 1994b).

It is tempting to speculate that development of behavior mechanisms that involve the elimination and reorganization of terminal axon branches (axonal remodeling) is essentially irreversible. The critical period then becomes the time at which the axonal remodeling occurs; it would depend on all the factors that can affect the timing of the remodeling. The production of new synapses continues to occur throughout life and could modulate the structure of behavior mechanisms after the critical period has passed.

These ideas have some similarities to proposals by Bateson (1987) and Greenough, Black, and Wallace (1987), though there are some important differences as well. The latter authors distinguish between experience-expectant and experience-dependent information storage based upon the functional requirements of particular brain systems: "Experience-expectant information storage refers to incorporation of environmental information that is ubiquitous in the environment and common to all species members, such as the basic elements of pattern perception.... Experience-dependent information storage refers to incorporation of environmental information that is idiosyncratic, or unique to the individual, such as learning about one's specific physical environment or vocabulary" (p. 539). They also suggest that experience-expectant processes depend on selection or pruning of overproduced synaptic connections (i.e., axonal remodeling, as discussed above), whereas experience-dependent processes depend on formation of new synaptic connections.

With respect to the type of environmental information stored, the development of all the perceptual and motor mechanisms we have described would be classified as experience-expectant, and therefore, according to Greenough *et al.*, dependent on synapse pruning. As seen above, however, the mechanisms responsible for axonal remodeling are the same regardless of whether the information comes from the environment or from endogenous processes. Thus, the word "experience" would have to be used broadly so as to include all information originating outside the specific brain structure itself (Lehrman, 1970; Schneirla, 1965). It does not seem that this use of the word was intended. A further problem with this classification arises with respect to the environmental information stored during imprinting, which is common to all members of the species in the natural environment. As Bolhuis (1994) discusses, the development of perceptual mechanisms during imprinting must also involve experience-dependent processes.

Greenough *et al.* also suggest that their categories offer a new view of phenomena that have previously been labeled critical or sensitive periods. Instead of viewing these phenomena as due "to the brief opening of a window, with experience influencing development only while the window is open," their approach "allows consideration of the evolutionary origins of a process, its adaptive value for the individual, the required timing and character of experience, and the organism's potentially active role in obtaining appropriate experience for itself" (p. 539). This view proposes a functional explanation for a causal phenomenon, which leads to all the problems discussed above (see also Hogan, 1994a). The proposal I have made above includes all experience-expectant processes, but is considerably broader and becomes congruent with the putative neural mechanism underlying it.

Finally, if it is true that the processes responsible for altering the structure of behavioral mechanisms and their connections do not differ before and after a particular behavior begins to function, it follows that the processes responsible for learning are no different from the processes responsible for development in general. In other words, the same structural change can be triggered by different events, for example, by genes or by the experience of "reinforcement." The important point is that the change itself cannot be classified as genetic or learned because it could have been triggered either way. I have discussed elsewhere (Hogan, 1994c) how a consideration of the structures that are changing can provide a good basis for classifying different types of learning.

THE ROLE OF EARLY SOCIAL EXPERIENCE. In a discussion comparing the development of social and non-social behavior systems (Hogan, 1994b), I concluded that both kinds of systems develop according to the same rules, and that there appear to be no systematic differences between them. The question then arises why topics such as imprinting and bonding have assumed such an important role in the developmental literature. My answer is that this interest is related to Lorenz' original conception of imprinting. He defined imprinting as "the acquisition of the object of instinctive behavior patterns oriented towards conspecifics" (Lorenz 1935/1970, p. 124). In terms of the concepts used in this chapter, we would say that imprinting refers to the development of a perceptual mechanism (or schema) that is responsible for species recognition and that is connected to all (or many of) the social behavior systems in the animal. The reason imprinting is so important is that Lorenz' definition implies that a single perceptual mechanism serves a number of different behavior systems and that this perceptual mechanism develops irreversibly very early in life.

Current evidence from imprinting studies is usually interpreted to mean that the object-recognition mechanisms for filial and sexual behavior develop separately (Bolhuis, 1991, 1996). Lorenz showed in his studies of jackdaws that the objects of the various functional systems he discussed (infant, sexual, social) might be different and might develop at different periods in the animal's life. Thus, the implication of Lorenz' definition may generally not be true. Nonetheless, the idea that early experience has far-reaching, general effects on later social behavior has remained influential and is supported by a wide variety of evidence (e.g., Bowlby, 1991; Hofer, 1996; Rutter, 1991). The question is, how do these effects come about if the perceptual mechanisms of the various social behavior systems develop independently?

One suggestion is likely to be widely applicable. Hofer (1987, 1996) and his colleagues studied the processes of early social attachment in young rats and their responses to separation from their mothers. Their results show that separation has extensive effects on the young rats' behavior, similar to (though not as dramatic as) the effects of maternal separation on the behavior of young rhesus monkeys (Harlow & Harlow, 1962; Hinde, 1977). Hofer analyzed these effects into two components. The first involves the formation of an attachment system, which has similarities to the one proposed by Bowlby (1991) for human infants and to the filial system implicated in imprinting studies in birds. This system develops as the young rat learns the characteristics of its mother; when the infant is separated from her, it shows distress reactions, and shows relief when it is later returned to her. If one substitutes an alternative "caregiver" for the mother, such as an inanimate object or another rat pup, Hofer's results show that the attachment system still seems to function normally.

The novel aspect of Hofer's analysis is the second component: the behavioral and physiological effects that occur during long-term separation from the mother are shown to depend on specific aspects of the mother–infant interaction. Hofer isolated a number of regulators including body warmth, tactile and olfactory stimulation, stimulation peculiar to the suckling situation, etc. Many of the specific effects of these factors have been described by Fleming and Blass (1994). A real mother provides all the necessary regulators, but alternative caregivers do not. Under such circumstances, various behavioral and physiological abnormalities will develop.

Kraemer (1992) interpreted the development of primate social attachment in similar terms to those of Hofer. He pointed out that a young rhesus monkey may become attached to an abusive mother or to a peer, and that such young monkeys can be seen in many ways to have a normally developed attachment system. But such monkeys also develop abnormally in many other ways. Kraemer provided evidence that absence of an adequate caregiver leads to aberrant development of brain biogenic amine systems which are implicated in the control of sensorimotor integration and emotion: "If the attachment process fails, or if the caregiver is incompetent as a member of the species, the developing infant will also fail to regulate its social behavior and may be dysfunctional in the social environment" (p. 493). It seems likely that similar processes determine the attack–escape relationship with respect to the development of sexual behavior in chickens as discussed above: A young chick can become imprinted on an inanimate object and develop a normal filial system, but the inanimate object does not provide the conditions for normal agonistic behavior to develop.

Lorenz' and Hofer's theories are the same in that both postulate that a representation of the imprinting object or caregiver (perceptual mechanism) is formed early in ontogeny. In Lorenz' theory, that representation controls a number of social behavior systems, and long-term effects are seen because each system matures at its own time in the life of the animal. In Hofer's theory, the representation controls only the attachment system; long-term effects are seen because the object to which the animal is attached provides the necessary conditions for various biochemical and neural changes that are indispensable for normal development of other systems.

CONCLUSIONS

The development of behavior systems is a very complex process, involving intricate interactions of external and internal causal factors with the genes and their products at every stage. Yet the principles involved in this process seem relatively simple. Specific patterns of neural activity are responsible for the formation of the basic behavioral mechanisms and many of the connections between them, probably through the mechanism of synapse pruning or axonal remodeling. Later stimulation causes the formation of new synapses which probably underlie the modification of behavioral mechanisms and the formation of new connections between them; new synapses are probably also important in the development of new representations (cognitive structures). These neural processes are, in fact, sufficient for understanding a wide range of developmental phenomena including critical periods and irreversibility. Yet, understanding the neural mechanisms that determine development tells us nothing about how a particular system will develop in a particular animal. The development of any specific system and of its interactions with other systems will need to be studied in each case.

One of the remarkable things about development is how normal most individuals become in spite of large variations in the experiences to which they are exposed. Waddington (1966) coined the term *canalization* to express this fact with respect to the morphology of the animal, and we have seen a similar picture with respect to behavior. The basic structure of the perceptual, central, and motor mechanisms, as well as the basic interconnections among these units, develops, by and large, prefunctionally. The experience of the individual is, of course, important, often in very unexpected ways, but typically, the basic structure of behavior is extraordinarily stable. Nonetheless, development, especially of social behavior, sometimes goes seriously wrong. Such disturbed development can often be traced to peculiarities in the social experience of the young animal, especially to periods of social deprivation. In general, the development of non-functional behavior is due to a combination of structural and motivational causes.

Structural causes for abnormal behavior include the development of aberrant behavior mechanisms and the development of anomalous connections among behavior mechanisms. For example, a chick that is force-fed and is not allowed to peck in its first 2 weeks after hatching is later unable to peck at food when hungry, presumably because the motor mechanism for pecking remains independent of the central mechanism for hunger (Hogan, 1977); or, the partner recognition mechanism may develop with the image of the wrong species or of a member of the same sex, and interspecific courtship or homosexual behavior would be seen. However, structural aberrations probably account for only a small proportion of developmental problems. Most disturbed development probably results from motivational causes such as an abnormally high activation of particular behavior systems or atypical interaction among behavior systems. For example, excessively fearful animals have general difficulties in learning new tasks, like the older kittens learning to catch fish, and in expressing normal social behavior; and the inadequate integration of fear and aggression is probably the main reason for problems in the expression of sexual behavior, as seen in isolated roosters, cats, and monkeys. In all these cases, the basic behavioral structure is present, but the more subtle interactions among behavior systems are missing. It is, of course, sometimes difficult to distinguish between structural and motivational aberrations. Nonetheless, the causal analysis of the development of behavior systems, as discussed in this chapter, provides a framework within which to attack these problems.

REFERENCES

Adler, N. T., & Hogan, J. A. (1963). Classical conditioning and punishment of an instinctive response in *Betta splendens*. *Animal Behaviour, 11*, 351–354.

Andrew, R. J. (1966). Precocious adult behaviour in the young chick. *Animal Behaviour, 14*, 485–500.

Baerends, G. P. (1975). An evaluation of the conflict hypothesis as an explanatory principle for the evolution of displays. In G. P. Baerends, C. Beer, & A. Manning (Eds.), *Function and evolution in behaviour* (pp. 187–227). London: Oxford University Press.

Baerends, G. P. (1976). The functional organization of behaviour. *Animal Behaviour, 24*, 726–738.

Baerends, G. P., & Kruijt, J. P. (1973). Stimulus selection. In R. A. Hinde & J. G. Stevenson-Hinde (Eds.), *Constraints on learning* (pp. 23–49). London: Academic Press.

Baerends-van Roon, J. M., & Baerends, G. P. (1979). The morphogenesis of the behaviour of the domestic cat, with a special emphasis on the development of prey-catching. *Verhandelingen der Koninklijke Nederlandse Akademie van Wetenschappen, Afd. Natuurkunde, Tweede Reeks* [*Proceedings of the Royal Netherlands Academy of Sciences, Section Physics, Second Series*], Part 72. Published monograph, 116 pp.

Balsam, P. D., & Silver, R. (1994). Behavioral change as a result of experience: toward principles of learning and development. In J. A. Hogan & J. J. Bolhuis (Eds.), *Causal mechanisms of behavioral development* (pp. 327–357). Cambridge: Cambridge University Press.

Balsam, P. D., Graf, J. S., & Silver, R. (1992). Operant and Pavlovian contributions to the ontogeny of pecking in ring doves. *Developmental Psychobiology, 25*, 389–410.

Baptista, L. F., & Gaunt, S. L. L. Social interaction and vocal development in birds. In C. T. Snowdon & M. Hausberger (Eds.), *Social influences on vocal development* (pp. 23–40). Cambridge: Cambridge University Press.

Baptista, L. F., & Petrinovich, L. (1984). Social interaction, sensitive phases and the song template hypothesis in the white-crowned sparrow. *Animal Behaviour, 32*, 172–181.

Bateson, P. P. G. (1987). Imprinting as a process of competitive exclusion. In J. P. Rauschecker and P. Marler (Eds.), *Imprinting and cortical plasticity* (pp. 151–168). New York: Wiley.

Bateson, P. P. G. (1999). Foreword. In J. J. Bolhuis & J. A. Hogan (Eds.), *The development of animal behaviour: a reader* (pp. ix–xi). Oxford: Blackwell.

Bateson, P. P. G., & Hinde, R. A. (1987). Developmental changes in sensitivity to experience. In M. H. Bornstein (Ed.), *Sensitive periods in development* (pp. 19–34). Hillsdale, NJ: Erlbaum. [Reprinted in Bolhuis & Hogan (1999)]

Baylis, J. R. (1982). Avian vocal mimicry: Its function and evolution. In D. E. Kroodsma, E. H. Miller, & H. Ouellet (Eds.), *Acoustic communication in birds* (Vol. 2, pp. 51–83). New York: Academic Press.

Beach, F. A. (1942). Analysis of factors involved in the arousal, maintenance, and manifestation of sexual excitement in male animals. *Psychosomatic Medicine, 4*, 173–198.

Berridge, K. C. (1994). The development of action patterns. In J. A. Hogan & J. J. Bolhuis (Eds.), *Causal mechanisms of behavioral development* (pp. 147–180). Cambridge: Cambridge University Press.

Berridge, K. C. & Fentress, J. C. (1986). Contextual control of trigeminal sensorimotor function. *Journal of Neuroscience, 6*, 325–330.

Berridge, K. C., Fentress, J. C., & Parr, H. (1987). Natural syntax rules control action sequence of rats. *Behavioural Brain Research, 23*, 59–68.

Bischof, H.-J. (1994). Sexual imprinting as a two-stage process. In J. A. Hogan & J. J. Bolhuis (Eds.) *Causal mechanisms of behavioral development* (pp. 82–97). Cambridge: Cambridge University Press.

Blakemore, C. (1973). Environmental constraints on development in the visual system. In R. A. Hinde & J. Stevenson-Hinde (Eds.), *Constraints on learning* (pp. 51–73). Cambridge: Cambridge University Press.

Blass, E. M., Hall, W. G., & Teicher, M. H. (1979). The ontogeny of suckling and ingestive behaviors. *Progress in Psychobiology and Physiological Psychology*, 8, 243–299.

Blass, E. M., Ganchrow, J. R., & Steiner, J. E. (1984). Classical conditioning in newborn humans 2–48 hours of age. *Infant Behavior and Development*, 7, 125–134.

Bolhuis, J. J. (1991). Mechanisms of avian imprinting: A review. *Biological Reviews*, 66, 303–345.

Bolhuis, J. J. (1994). Neurobiological analyses of behavioural mechanisms in development. In J. A. Hogan & J. J. Bolhuis (Eds.) *Causal mechanisms of behavioral development* (pp. 16–46). Cambridge: Cambridge University Press.

Bolhuis, J. J. (1996). Development of perceptual mechanisms in birds: Predispositions and imprinting. In C. F. Moss & S. J. Shettleworth (Eds.), *Neuroethological studies of cognitive and perceptual processes* (pp. 158–184). Boulder, CO: Westview Press. [Reprinted in Bolhuis & Hogan (1999)]

Bolhuis, J. J., & Hogan, J. A. (Eds.). (1999). *The development of animal behaviour: A reader.* Oxford: Blackwell.

Bolhuis, J. J., & Honey, R. C. (1998). Imprinting, learning and development: From behavior to brain and back. *Trends in Neurosciences*, 21, 306–311.

Bolhuis, J. J., Zijlstra, G. G. O., den Boer-Visser, A. M., & van der Zee, E. A. (2000). Localized neuronal activation in the zebra finch brain is related to the strength of song learning. *Proceedings of the National Academy of Sciences of the USA*, 97, 2282–2285.

Bowlby, J. (1991). Ethological light on psychoanalytical problems. In P. Bateson (Ed.), *The development and integration of behaviour* (pp. 301–313). Cambridge: Cambridge University Press.

Brown, M. C., Hopkins, W. G., & Keynes, R. J. (1991). *Essentials of neural development.* Cambridge: Cambridge University Press.

Chomsky, N. (1965). *Aspects of the theory of syntax.* Cambridge, MA: MIT Press.

Clayton, N. S. (1994). The influence of social interactions on the development of song and sexual preferences in birds. In J. A. Hogan & J. J. Bolhuis (Eds.), *Causal mechanisms of behavioral development* (pp. 98–115). Cambridge: Cambridge University Press.

Colonnese, M. T., Stallman, E. L., & Berridge, K. C. (1996). Ontogeny of action syntax in altricial and precocial rodents: Grooming sequences of rat and guinea pig pups. *Behaviour*, 133, 1165–1195.

Cramer, C. P., & Blass, E. M. (1983). Mechanisms of control of milk intake in suckling rats. *American Journal of Physiology*, 245, R154–R159.

DeVoogd, T. J. (1994). The neural basis for the acquisition and production of bird song. In J. A. Hogan & J. J. Bolhuis (Eds.), *Causal mechanisms of behavioral development* (pp. 49–81). Cambridge: Cambridge University Press.

Domjan, M., & Holloway, K. S. (1998). Sexual learning. In G. Greenberg and M. M. Haraway (Eds.), *Comparative psychology: A handbook* (pp. 602–613). New York: Garland.

Doty, R. W. (1976). The concept of neural centers. In J. C. Fentress (Ed.), *Simpler networks and behavior* (pp. 251–265). Sunderland, MA: Sinauer.

Doupe, A. J., & Kuhl, P. K. (1999). Birdsong and human speech: Common themes and mechanisms. *Annual Review of Neuroscience*, 22, 567–631.

Drachman, D. B., & Sokoloff, L. (1966). The role of movement in embryonic joint development. *Developmental Biology*, 14, 401–420.

Eimas, P. D., Siqueland, P., Jusczyk, P., & Vigorito, J. (1971). Speech perception in infants. *Science*, 171, 303–306.

Eimas, P. D., Miller, J. L., & Jusczyk, P. W. (1987). On infant speech perception and the acquisition of language. In S. Harnad (Ed.), *Categorical perception.* (pp. 161–195). Cambridge: Cambridge University Press.

Ekman, P. (1982). *Emotion in the human face*, 2nd ed. Cambridge: Cambridge University Press.

Evans, R. M. (1968). Early aggressive responses in domestic chicks. *Animal Behaviour*, 16, 24–28.

Fanselow, M. S., & De Oca, B. M. (1988). Defensive behaviors. In G. Greenberg & M. M. Haraway (Eds.), *Comparative psychology: A handbook* (pp. 653–665). New York: Garland.

Fantz, R. L. (1957). Form preferences in newly hatched chicks. *Journal of Comparative and Physiological Psychology*, 50, 422–430.

Farris, H. E. (1967). Classical conditioning of courting behavior in the Japanese quail (*Coturnix c. japonica*). *Journal of the Experimental Analysis of Behavior*, 10, 213–217.

Fentress, J. C. (1972). Development and patterning of movement sequences in inbred mice. In J. Kiger (Ed.), *The biology of behavior* (pp. 83–132). Corvallis: Oregon State University Press.

Fentress, J. C., & Stilwell, F. P. (1973). Grammar of a movement sequence in inbred mice. *Nature*, 224, 52–53.

Field, T. M., Woodson, R., Greenberg, R., & Cohen, D. (1982). Discrimination and imitation of facial expressions by neonates. *Science, 218,* 179–181.

Fleming, A. S., & Blass, E. M. (1994). Psychobiology of the early mother-young relationship. In J. A. Hogan & J. J. Bolhuis (Eds.), *Causal mechanisms of behavioral development* (pp. 212–241). Cambridge: Cambridge University Press.

Gallistel, C. R. (1980). *The organization of action.* Hillsdale, NJ: Erlbaum.

Goldin-Meadow, S. (1997). The resilience of language in humans. In C. T. Snowdon & M. Hausberger (Eds.), *Social influences on vocal development* (pp. 293–311). Cambridge: Cambridge University Press.

Gottlieb, G. (1978). Development of species identification in ducklings: IV. Change in species-specific perception caused by auditory deprivation. *Journal of Comparative and Physiological Psychology, 92,* 375–387.

Gottlieb, G. (1980). Development of species identification in ducklings—VI: Specific embryonic experience required to maintain species-typical perception in Peking ducklings. *Journal of Comparative and Physiological Psychology, 94,* 579–587. [Reprinted in Bolhuis & Hogan (1999)]

Gottlieb, G. (1997). *Synthesizing nature–nurture: prenatal roots of instinctive behavior.* Mahwah, NJ: Erlbaum.

Graf, J. S., Balsam, P. D., & Silver, R. (1985). Associative factors and the development of pecking in the ring dove. *Developmental Psychobiology, 18,* 447–460.

Greenough, W. T., Black, J. E., & Wallace, C. S. (1987). Experience and brain development. *Child Development, 58,* 539–559.

Groothuis, T. G. G. (1992). The influence of social experience on the development and fixation of the form of displays in the black-headed gull. *Animal Behaviour, 43,* 1–14.

Groothuis, T. G. G. (1994). The ontogeny of social displays: Interplay between motor development, development of motivational systems and social experience. In J. A. Hogan & J. J. Bolhuis (Eds.), *Causal mechanisms of behavioral development* (pp. 183–211). Cambridge: Cambridge University Press.

Groothuis, T. G. G., & Meeuwissen, G. (1992). The influence of testosterone on the development and fixation of the form of displays in two age classes of young black-headed gulls. *Animal Behaviour, 43,* 189–208.

Hale, C., & Green, L. (1979). Effect of initial pecking consequences on subsequent pecking in young chicks. *Journal of Comparative and Physiological Psychology, 93,* 730–735.

Hall, W. G., & Browde, J. A. (1986). The ontogeny of independent ingestion in mice: Or, why won't infant mice feed? *Developmental Psychobiology, 19,* 211–222.

Hall, W. G., & Williams, C. L. (1983). Suckling isn't feeding, or is it? A search for developmental continuities. *Advances in the Study of Behavior, 13,* 219–254.

Hamburger, V. (1973). Anatomical and physiological basis of embryonic motility in birds and mammals. In G. Gottlieb (Ed.), *Behavioral embryology* (pp. 52–76). New York: Academic Press.

Harlow, H. F., & Harlow, M. K. (1962). Social deprivation in monkeys. *Scientific American, 207,* 136–146. [Reprinted in Bolhuis & Hogan (1999)]

Harlow, H. F., & Harlow, M. K. (1965). The affectional systems. In A. M. Schrier, H. F. Harlow, & F. Stollnitz (Eds.), *Behavior of nonhuman primates* (Vol. 2, pp. 287–334). New York: Academic Press.

Harlow, H. F., & Suomi, S. J. (1971). Social recovery by isolation-reared monkeys. *Proceedings of the National Academy of Sciences of the USA, 68,* 1534–1538.

Hauser, M. D. (1996). *The evolution of communication.* Cambridge, MA: MIT Press.

Hebb, D. O. (1949). *The organization of behavior.* New York: Wiley.

Hess, E. H. (1956). Natural preferences of chicks and ducklings for objects of different colors. *Psychological Reports, 2,* 477–483.

Hickok, G., Bellugi, U., & Klima, E. S. (1998). The neural organization of language: Evidence from sign language aphasia. *Trends in Cognitive Sciences, 2,* 129–136.

Hinde, R. A. (1970). *Animal behaviour* (2nd ed.). New York: McGraw-Hill.

Hinde, R. A. (1977). Mother–infant separation and the nature of inter-individual relationships: Experiments with rhesus monkeys. *Proceedings of the Royal Society of London B, 196,* 29–50. [Reprinted in Bolhuis & Hogan (1999)]

Hofer, M. A. (1987). Early social relationships: A psychobiologist's view. *Child Development, 58,* 633–647.

Hofer, M. A. (1996). On the nature and consequences of early loss. *Psychosomatic Medicine, 58,* 570–581.

Hogan, J. A. (1965). An experimental study of conflict and fear: An analysis of behavior of young chicks to a mealworm. Part I: The behavior of chicks which do not eat the mealworm. *Behaviour, 25,* 45–97.

Hogan, J. A. (1966). An experimental study of conflict and fear: An analysis of behavior of young chicks to a mealworm. Part II: The behavior of chicks which eat the mealworm. *Behaviour, 27,* 273–289.

Hogan, J. A. (1971). The development of a hunger system in young chicks. *Behaviour, 39,* 128–201.

Hogan, J. A. (1973). How young chicks learn to recognize food. In R. A. Hinde & J. G. Stevenson-Hinde (Eds.), *Constraints on learning* (pp. 119–139). London: Academic Press.

Hogan, J. A. (1974). Responses in Pavlovian conditioning studies. *Science, 186,* 156–157.

Hogan, J. A. (1977). The ontogeny of food preferences in chicks and other animals. In L. M. Barker, M. Best, and M. Domjan (Eds.), *Learning mechanisms in food selection.* Waco, TX: Baylor University Press.

Hogan, J. A. (1978). An eccentric view of development: A review of *The Dynamics of Behavior Development* by Zing-Yang Kuo. *Contemporary Psychology, 23,* 690–691.

Hogan, J. A. (1981). Hierarchy and behavior: A review of *The Organization of Action* by C. R. Gallistel. *Behavioral and Brain Sciences, 4,* 625.

Hogan, J. A. (1984). Pecking and feeding in chicks. *Learning and Motivation, 15,* 360–376.

Hogan, J. A. (1988). Cause and function in the development of behavior systems. In E. M. Blass (Ed.), *Handbook of behavioral neurobiology,* Volume 9, *Developmental psychobiology and behavioral ecology* (pp. 63–105). New York: Plenum Press.

Hogan, J. A. (1994a). The concept of cause in the study of behavior. In J. A. Hogan & J. J. Bolhuis (Eds.), *Causal mechanisms of behavioral development* (pp. 3–15). Cambridge: Cambridge University Press.

Hogan, J. A. (1994b). Development of behavior systems. In J. A. Hogan & J. J. Bolhuis, (Eds.) *Causal mechanisms of behavioral development* (pp. 242–264). Cambridge: Cambridge University Press.

Hogan, J. A. (1994c). Structure and development of behavior systems. *Psychonomic Bulletin and Review, 1,* 439–450.

Hogan, J. A. (1997). Energy models of motivation: A reconsideration. *Applied Animal Behaviour Science, 53,* 89–105.

Hogan, J. A., & Roper, T. J. (1978). A comparison of the properties of different reinforcers. *Advances in the Study of Behavior, 8,* 155–255.

Hogan, J. A., & van Boxel, F. (1993). Causal factors controlling dustbathing in Burmese red junglefowl: Some results and a model. *Animal Behaviour, 46,* 627–635.

Hogan, J. A., Honrado, G. I., & Vestergaard, K. S. (1991). Development of a behavior system: Dustbathing in the Burmese red junglefowl (*Gallus gallus spadiceus*): II. Internal factors. *Journal of Comparative Psychology, 195,* 269–273.

Hogan-Warburg, A. J., & Hogan, J. A. (1981). Feeding strategies in the development of food recognition in young chicks. *Animal Behaviour, 29,* 143–154.

Horn, G. (1985). *Memory, imprinting, and the brain.* Oxford: Oxford University Press.

Houx, B. B., & ten Cate, C. (1998). Do contingencies with tutor behaviour influence song learning in zebra finches? *Behaviour, 135,* 599–614.

Houx, B. B., & ten Cate, C. (1999). Song learning from playback in zebra finches: Is there an effect of operant contingency? *Animal Behaviour, 57,* 837–845.

Hultsch, H. (1993). Tracing the memory mechanisms in the song acquisition of nightingales. *Netherlands Journal of Zoology, 43,* 155–171.

Jarvis E. D., & Nottebohm, F. (1997). Motor-driven gene expression. *Proceedings of the National Academy of Sciences, 94,* 4097–4102.

Johanson, I. B, & Terry, L. M. (1988). Learning in infancy: A mechanism of behavioral change in development. In E. M. Blass (Ed.), *Handbook of behavioral neurobiology,* Volume 9, *Developmental psychobiology and behavioral ecology* (pp. 245–281). New York: Plenum Press.

Johnsen, P. F., Vestergaard, K. S., & Nørgaard-Nielsen, G. (1998). Influence of early rearing conditions on the development of feather pecking and cannibalism in domestic fowl. *Applied Animal Behaviour Science, 60,* 25–41.

Johnson, M. H., & Horn, G. (1988). Development of filial preferences in dark-reared chicks. *Animal Behavior, 36,* 675–683.

Johnson, M. H., Bolhuis, J. J., & Horn, G. (1985). Interaction between acquired preferences and developing predispositions during imprinting. *Animal Behaviour, 33,* 1000–1006.

Johnson, M. H., Davies, D. C., & Horn, G. (1989). A critical period for the development of a predisposition in the chick. *Animal Behaviour, 37,* 1044–1046.

Jusczyk, P. W. (1997). Finding and remembering words: Some beginnings by English-learning infants. *Current Directions in Psychological Science, 6,* 170–174.

Kako, E. (1999). Elements of syntax in the systems of three language-trained animals. *Animal Learning and Behavior, 27,* 1–14.

Karmiloff-Smith, A. (1992). *Beyond modularity: A developmental perspective on cognitive science.* Cambridge, MA: MIT Press.

Karmiloff-Smith, A. (1998). Development itself is the key to understanding developmental disorders. *Trends in Cognitive Sciences, 2,* 389–398.

Kenny, J. T., Stoloff, M. L., Bruno, J. P., & Blass, E. M. (1979). The ontogeny of preferences for nutritive over nonnutritive suckling in the albino rat. *Journal of Comparative and Physiological Psychology, 93,* 752–759.

King, A. P., & West, M. J. (1977). Species identification in the North American cowbird: Appropriate responses to abnormal song. *Science, 195,* 1002–1004.

Konishi, M. (1965). The role of auditory feedback in the control of vocalizations in the white-crowned sparrow. *Zeitschrift für Tierpsychologie, 22,* 770–783.

Kraemer, G. W. (1992). A psychobiological theory of attachment. *Behavioral and Brain Sciences, 15,* 493–511.

Kruijt, J. P. (1962). Imprinting in relation to drive interactions in Burmese red junglefowl. *Symposia of the Zoological Society, London, 8,* 219–226.

Kruijt, J. P. (1964). Ontogeny of social behaviour in Burmese red junglefowl (*Gallus gallus spadiceus*). *Behaviour* (Supplement 12) pp. 1–201.

Kuhl, P. K. (1994). Learning and representation in speech and language. *Current Opinion in Neurobiology, 4,* 812–822.

Kuhl, P. K., Williams, K. A., Lacerda, R., Stevens, K. N., & Lindblom, B. (1992). Linguistic experience alters phonetic perception in infants by 6 months of age. *Science,, 255,* 606–608.

Kuo, Z. Y. (1967). *The dynamics of behavioral development.* New York: Random House. [Excerpt reprinted in Bolhuis & Hogan (1999)]

Larsen, B. H., Hogan, J. A., & Vestergaard, K. S. (2000). Development of dustbathing behavior sequences in the domestic fowl: The significance of functional experience. *Developmental Psychobiology, 36,* 5–12.

Lashley, K. S. (1951). The problem of serial order in behavior. In L. A. Jeffress (Ed.), *Cerebral mechanisms in behavior* (pp. 112–136). New York: Wiley.

Lehrman, D. S. (1970). Semantic and conceptual issues in the nature–nurture problem. In L. R. Aronson, E. Tobach, D. S. Lehrman, & J. S. Rosenblatt (Eds.), *Development and evolution of behavior* (pp. 17–52). San Francisco: Freeman. [Reprinted in Bolhuis & Hogan (1999)]

Lenneberg, E. H. (1967). *Biological foundations of language.* New York: Wiley.

Locke, J. L. (1993). *The child's path to spoken language.* Cambridge, MA: Harvard University Press.

Locke, J. L. (1994). The biological building blocks of spoken language. In J. A. Hogan & J. J. Bolhuis (Eds.), *Causal mechanisms of behavioral development* (pp. 300–324). Cambridge: Cambridge University Press.

Locke, J. L., & Snow, C. (1997). Social influences on vocal learning in human and nonhuman primates. In C. T. Snowdon & M. Hausberger (Eds.), *Social influences on vocal development* (pp. 274–292). Cambridge: Cambridge University Press.

Logan, C. A. (1983). Biological diversity in avian vocal learning. In M. D. Zeiler & P. Harzem (Eds.), *Advances in analysis of behavior,* Volume 3, *Biological factors in learning.* (p. 143–176). Chichester, UK: Wiley.

Lorenz, K. (1935). Der Kumpan in der Umwelt des Vogels. *Journal für Ornithologie, 83,* 137–213, 289–413 [Translation (1970), Companions as factors in the bird's environment. *In Studies in animal and human behaviour* (Vol. 1, pp. 101–258) London: Methuen]. [Excerpt reprinted in Bolhuis & Hogan (1999)]

Lorenz, K. (1937). Über die Bildung des Instinktbegriffes. *Naturwissenschaften, 25,* 289–300, 307–318, 324–331.

Lorenz, K. (1956). Plays and vacuum activities. In *L'instinct dans le comportement des animaux et de l'homme* (pp. 633–637). Paris: Masson.

Lorenz, K. (1961). Phylogenetische Anpassung und adaptive Modifikation des Verhaltens. *Zeitschrift für Tierpsychologie, 18,* 139–187.

Lorenz, K. (1965). *Evolution and modification of behavior.* Chicago: University of Chicago Press. [Excerpt reprinted in Bolhuis & Hogan (1999)]

Marler, P. (1970a). A comparative approach to vocal learning: Song development in white-crowned sparrows. *Journal of Comparative and Physiological Psychology* (Monograph Supplement), *71,* 1–25.

Marler, P. (1970b). Birdsong and speech development: Could there be parallels? *American Scientist, 58,* 669–673.

Marler, P. (1976). Sensory templates in species-specific behavior. In J. C. Fentress (Ed.), *Simpler networks and behavior* (pp. 314–329). Sunderland, MA: [Reprinted in Bolhuis & Hogan (1999)]

Marler, P. (1984). Song learning: Innate species differences in the learning process. In P. Marler & H. S. Terrace (Eds.), *The biology of learning* (pp. 289–309). Berlin: Springer.

Marler, P., & Peters, S. (1982). Subsong and plastic song: Their role in the vocal learning process. In D. E. Kroodsma & E. H. Miller (Eds.), *Acoustic communication in birds* (Vol. 2, pp. 25–50). New York: Academic Press.

Martin, P., & Caro, T. M. (1985). On the functions of play and its role in behavioral development. *Advances in the Study of Behavior, 15,* 59–103.

Moffatt, C. A., & Hogan, J. A. (1992). Ontogeny of chick responses to maternal food calls in the Burmese red junglefowl (*Gallus gallus spadiceus*). *Journal of Comparative Psychology, 106,* 92–96.

Moore, B. R. (1973). The role of direct Pavlovian reactions in simple instrumental learning in the pigeon.

In R. A. Hinde & J. G. Stevenson-Hinde (Eds.), *Constraints on learning.* (pp. 159–188). London: Academic Press.

Moran, T. H. (1986). Environmental and neural determinants of behavior in development. In E. M. Blass (Ed.), *Handbook of behavioral neurobiology,* Volume 8, *Developmental psychobiology and developmental neurobiology* (p. 99–128). New York: Plenum Press.

Morgan, C. L. (1896). *Habit and instinct.* London: Arnold.

Morris, D. (1956). [Discussion following Lorenz] In *L'instinct dans le comportement des animaux et de l'homme* (pp. 642–643). Paris: Masson.

Nelson, D. A. (1997). Social interaction and sensitive phases for song learning: A critical review. In C. T. Snowdon & M. Hausberger (Eds.), *Social influences on vocal development* (p. 7–22). Cambridge: Cambridge University Press.

Novak, M. & Harlow, H. F. (1975). Social recovery of monkeys isolated for the first year of life: I. Rehabilitation and therapy. *Developmental Psychology, 11,* 453–465.

Oppenheim, R. W. (1974). The ontogeny of behavior in the chick embryo. *Advances in the Study of Behavior, 5,* 133–172.

Oppenheim, R. W. (1981). Ontogenetic adaptations and retrogressive processes in the development of the nervous system and behaviour: A neuroembryological perspective. In K. J. Connolly & H. F. R. Prechtl (Eds.), *Maturation and development: Biological and psychological perspectives* (pp. 73–109). Philadelphia: Lippincott. [Reprinted in Bolhuis & Hogan (1999)]

Oppenheim, R. W. (1991). Cell death during development of the nervous system. *Annual Review of Neuroscience, 14,* 453–501.

Oster, H. (1978). Facial expression and affect development. In M. Lewis & L. A. Rosenblum (Eds.), *The development of affect* (pp. 43–76). New York: Plenum Press.

Petherick, J. C., Seawright, E., Waddington, D., Duncan, I. J. H., & Murphy, L. B. (1995). The role of perception in the causation of dustbathing behaviour in domestic fowl. *Animal Behaviour, 49,* 1521–1530.

Petitto, L. A., & Marentette, P. F. (1991). Babbling in the manual mode: Evidence for the ontogeny of language. *Science, 251,* 1493–1496.

Petrinovich, L. (1985). Factors influencing song development in the white-crowned sparrow (*Zonotrichia leucophrys*). *Journal of Comparative Psychology, 99,* 15–29.

Pinker, S. (1994). *The language instinct.* New York: Morrow.

Reisbick, S. H. (1973). Development of food preferences in newborn guinea pigs. *Journal of Comparative and Physiological Psychology, 85,* 427–442.

Rice, J. C. (1978). Effects of learning constraints and behavioural organization on the association of vocalizations and hunger in Burmese red junglefowl chicks. *Behaviour, 67,* 259–298.

Rozin, P. (1984). The acquisition of food habits and preferences. In J. D. Matarazzo, S. M. Weiss, J. A. Herd, N. E. Miller & S. M. Weiss (Eds.), *Behavioral health: A handbook of health enhancement and disease prevention* (pp. 590–607). New York: Wiley.

Rutter, M. (1991). A fresh look at "maternal deprivation." In P. Bateson (Ed.), *The development and integration of behaviour* (pp. 331–374). Cambridge: Cambridge University Press.

Sackett, G. P. (1970). Unlearned responses, differential rearing experiences, and the development of social attachments by rhesus monkeys. In L. A. Rosenblum (Ed.), *Primate behavior* (pp. 111–141). New York: Academic Press.

Sanotra, G. S., Vestergaard, K. S., Agger, J. F., & Lawson, L. G. (1995). The relative preferences for feathers, straw, wood-shavings and sand for dustbathing, pecking and scratching in domestic chicks. *Applied Animal Behaviour Science, 43,* 263–277.

Schiller, P. H. (1949). Manipulative patterns in the chimpanzee. In C. H. Schiller (Ed.), *Instinctive behavior* (pp. 264–287). New York: International Universities Press.

Schneirla, T. C. (1956). Interrelationships of the "innate" and the "acquired" in instinctive behavior. In *L'instinct dans le comportement des animaux et de l'homme* (pp. 387–452). Paris: Masson.

Schneirla, T. C. (1965). Aspects of stimulation and organization in approach/withdrawal processes underlying vertebrate behavioral development. *Advances in the Study of Behavior, 1,* 1–74.

Schneirla, T. C., Rosenblatt, J. S., & Tobach, E. (1963). Maternal behavior in the cat. In H. Rheingold (Ed.), *Maternal behavior in mammals* (pp. 122–168). New York: Wiley.

Shatz, C. J. (1992). The developing brain. *Scientific American, 267,* 60–67. [Reprinted in Bolhuis & Hogan (1999)]

Shelton, J. R., & Caramazza, A. (1999). Deficits in lexical and semantic processing. Implications for models of normal language. *Psychonomic Bulletin and Review, 6,* 5–27.

Sherry, D. F. (1977). Parental food-calling and the role of the young in the Burmese red junglefowl (*Gallus g. spadiceus*). *Animal Behaviour, 25,* 594–601.

Sherry, D. F. (1981). Parental care and development of thermoregulation in red junglefowl. *Behaviour, 76*, 250–279.

Shettleworth, S. J. (1975). Reinforcement and the organization of behavior in golden hamsters. *Journal of Experimental Psychology: Animal Behavior Processes, 1*, 56–87.

Skinner, B. F. (1953). *Science and human behavior.* New York: Macmillan.

Slack, J. M. W. (1991). *From egg to embryo.* Cambridge: Cambridge University Press.

Snowdon, C. T., & Hausberger, M. (Eds.) (1997). *Social influences on vocal development.* Cambridge: Cambridge University Press.

Steiner, J. E. (1979). Human facial expressions in response to taste and smell stimulation. *Advances in Child Development and Behavior, 13*, 257–295.

Stellar, E. (1960). Drive and motivation. In J. Field, H. W. Magoun, & V. E. Hall (Eds.), *Handbook of physiology* (Section 1, Vol. 3, pp. 1501–1528). Washington, DC: American Physiological Association.

Stevenson, J. G. (1967). Reinforcing effects of chaffinch song. *Animal Behaviour, 15*, 427–432.

Sutherland, N. S. (1964). The learning of discriminations by animals. *Endeavour, 23*, 148–152.

ten Cate, C. (1984). The influence of social relations on the development of species recognition in zebra finch males. *Behaviour, 91*, 263–285.

ten Cate, C. (1986). Sexual preferences in zebra finch males exposed to two species. I. A case of double imprinting. *Journal of Comparative Psychology, 100*, 248–252.

ten Cate, C. (1987). Sexual preferences in zebra finch males exposed to two species. II. The internal representation resulting from double imprinting. *Animal Behaviour, 35*, 321–330.

ten Cate, C. (1994). Perceptual mechanisms in imprinting and song learning. In J. A. Hogan & J. J. Bolhuis (Eds.), Causal mechanisms of behavioral development (pp. 116–146). Cambridge: Cambridge University Press.

Thelen, E. (1985). Expression as action: A motor perspective on the transition from spontaneous to instrumental behaviors. In G. Zivin (Ed.), *The development of expressive behavior* (pp. 221–267). Orlando, FL: Academic Press.

Thelen, E. (1995). Motor development: A new synthesis. *American Psychologist, 50*, 79–95.

Thelen, E., & Fisher, C. M. (1982). Newborn stepping: An explanation for a "disappearing reflex." *Developmental Psychology, 18*, 760–775.

Thorpe, W. H.(1958). The learning of song patterns by birds, with especial reference to the song of the chaffinch (*Fringilla coelebs*). *Ibis, 100*, 535–570.

Thorpe, W. H. (1961). *Bird song.* Cambridge: Cambridge University Press.

Timberlake, W. (1994). Behavior systems, associationism, and Pavlovian conditioning. *Psychonomic Bulletin and Review, 1*, 405–420.

Tinbergen, N. (1951). *The study of instinct.* Oxford: Oxford University Press.

Tinbergen, N. (1952). Derived activities: Their causation, biological significance, origin and emancipation during evolution. *Quarterly Review of Biology, 27*, 1–32.

Tomasello, M. (1995). Language is not an instinct. *Cognitive Development, 10*, 131–156.

van Kampen, H. S. (1994). Courtship food-calling in Burmese red jungle fowl. I. The causation of female approach. *Behavior, 131*, 261–275.

van Kampen, H. S. (1996). A framework for the study of filial imprinting and the development of attachment. *Psychonomic Bulletin and Review, 3*, 3–20.

van Kampen, H. S. (1997). Courtship food-calling in Burmese red junglefowl: II. Sexual conditioning and the role of the female. *Behaviour, 134*, 775–787.

van Kampen, H. S., & Bolhuis, J. J. (1991). Auditory learning and filial imprinting in the chick. *Behaviour, 117*, 303–339.

van Kampen, H. S. & Hogan J. A. (2000). Courtship food-calling in Burmese red jungle fowl: III. Factors influencing the male's behavior. *Behavior, 137*, 1191–1209.

van Liere, D. W. (1992). The significance of fowls' bathing in dust. *Animal Welfare, 1*, 187–202.

van Liere, D. W., & Bokma, S. (1987). Short term feather maintenance as a function of dustbathing in laying hens. *Applied Animal Behaviour Science, 18*, 197–204.

Vestergaard, K. S. (1982). Dust-bathing in the domestic fowl: Diurnal rhythm and dust deprivation. *Applied Animal Ethology, 8*, 487–495.

Vestergaard, K. S., & Baranyiova, E. (1996). Pecking and scratching in the development of dust perception in young chicks. *Acta Veterinaria Brno, 65*, 133–142.

Vestergaard, K. S., & Hogan, J. A. (1992). The development of a behavior system: Dustbathing in the Burmese red junglefowl. III. Effects of experience on stimulus preference. *Behaviour, 121*, 215–230.

Vestergaard, K. S., Hogan, J. A., & Kruijt, J. P. (1990). The development of a behavior system: Dustbathing in the Burmese red junglefowl: I. The influence of the rearing environment on the organization of dustbathing. *Behaviour, 112*, 99–116.

Vestergaard, K. S., Kruijt, J. P., & Hogan, J. A. (1993). Feather pecking and chronic fear in groups of red junglefowl: Its relations to dustbathing, rearing environment and social status. *Animal Behaviour, 45*, 1127–1140.

Vestergaard, K. S., Damm, B. I., Abbot, U. K., & Bildsøe, M. (1999). Regulation of dustbathing in feathered and featherless domestic chicks: The Lorenzian model revisited. *Animal Behaviour, 58*, 1017–1025.

von Holst, E. (1954). Relations between the central nervous system and the peripheral organs. *British Journal of Animal Behaviour, 2*, 89–94.

von Holst, E., & St. Paul, U. von. (1960). Vom Wirkungsgefüge der Triebe. *Naturwissenschaften, 47*, 409–422. [Translation (1963), On the functional organisation of drives. *Animal Behaviour, 11*, 1–20]

Waddington, C. H. (1966). *Principles of development and differentiation.* New York: Macmillan. [Excerpt reprinted in Bolhuis & Hogan (1999)]

Wasserman, E. A. (1973). Pavlovian conditioning with heat reinforcement produces stimulus-directed pecking in chicks. *Science, 181*, 875–877.

Werker, J. F., & Tees, R. C. (1992). The organization and reorganization of human speech perception. *Annual Review of Neuroscience, 15*, 377–402.

West, M. J., & King, A. P. (1988). Female visual display affects the development of males song in the cowbird. *Nature, 334*, 244–246.

Wiesel, T. N. (1982). Postnatal development of the visual cortex and the influence of the environment. *Nature, 299*, 583–591.

Williams, C. L. (1991). Development of a sexually dimorphic behavior: Hormonal and neural controls. In H. N. Shair, G. A. Barr, & M. A. Hofer (Eds.), *Developmental psychobiology: New methods and changing concepts* (pp. 206–222) New York: Oxford University Press.

7

The Development and Function of Nepotism

Why Kinship Matters in Social Relationships

WARREN G. HOLMES

INTRODUCTION: CONCEPTUAL ISSUES AND TERMINOLOGY

Each year since 1904, the Carnegie Hero Fund Commission has presented a medal and financial award to civilians who voluntarily risked their own lives to save another person's. The Hero Fund came about because, according to its founder, Andrew Carnegie, "I do not expect to stimulate or create heroism by this fund, knowing well that heroic action is impulsive; but I do believe that, if the hero is injured in his bold attempt to serve or save his fellows, he and those dependent upon him should not suffer pecuniarily" (Carnegie, 2000). One type of hero who is ineligible for an award is the hero who saves a member of his or her own immediate family. Why would the Commission exclude from the list of deserving heroes those who save a close genetic relative? I suggest that the reason is quite straightforward: we humans take for granted that our kin will come to our aid in times of need; self-sacrifice for a relative like a child, a sibling, or a niece is part of our "nature" and thus does not merit pecuniary reward (Burnstein, Crandall, & Kitayama, 1994). Still, why should this be so? One of my aims in this chapter is to explain the ubiquitous and deeply ingrained nature of nepotism that permeates human and nonhuman social relationships. My analysis may help provide a deeper biological understanding of why saving a friend or stranger might merit financial reward, whereas saving a close relative is "natural" and therefore would not.

My general purpose in this chapter is to examine nepotism, the preferential treatment of genetic relatives, from both developmental and functional perspec-

WARREN G. HOLMES   Psychology Department, University of Michigan, Ann Arbor, Michigan 48109-1109.

Developmental Psychobiology, Volume 13 of *Handbook of Behavioral Neurobiology*, edited by Elliott Blass, Kluwer Academic / Plenum Publishers, New York, 2001.

tives,* a task that is difficult, in part, because there are few frameworks that incorporate both developmental and functional thinking about social behavior (West, King, & Arberg, 1988). One risk I face in this endeavor is that developmental and functional accounts are fundamentally different kinds of explanations for the same trait. That is, the terms "proximate" and "ultimate" represent different ways the same trait can be analyzed and explained (Tinbergen, 1963). The distinction between proximate explanations, which emphasize forces that operate during the lifetime of an individual, and ultimate explanations, which emphasize evolutionary forces that operate on individuals across generations, has proven useful in behavioral research (Alcock & Sherman, 1994; but see Dewsbury, 1994). By explicitly identifying two levels of analysis and by recognizing that explanations at one level are the immediate manifestations of and complement rather than compete with explanations at the other level (e.g., Maynard Smith *et al.*, 1985), one can improve communication among investigators who work at different levels (Lehrman, 1970) and produce more complete explanations for behavior than would otherwise be possible. By examining both the ontogeny and function of nepotism, I hope to offer a more inclusive and satisfying explanation for preferential treatment of kin than would be possible if my analysis proceeded at a single level.

PROXIMATE AND ULTIMATE ISSUES: THE POPE'S NEPHEWS AND THE MEANING OF NEPOTISM

Clear definitions of terms help establish a common framework for analyzing and explaining behavior. Such definitions can also sharpen the distinction among different hypotheses offered to explain the same phenomenon. Although good definitions share several common features such as clarity, precision, and specificity, terms can often be defined in more than one way and different investigators may offer different definitions for the same term (e.g., see the multiple meanings of "adaptation" in Rose & Lauder, 1996, or of "constraints" in Antonovics & van Tienderen, 1991). Thus, I begin my developmental and functional analysis of nepotism by defining and discussing some terms that will be critical to my ideas.

At first blush, the dictionary meaning of nepotism is straightforward: "favoritism that is shown to or patronage that is granted to relatives" (*American Heritage Dictionary*, 3rd ed.). However, the etymological background of nepotism reveals a deeper complexity that coheres with a functional analysis. Sixtus IV, a 15th century pope, brought legal status to the tradition of providing special favors to "Il Nipotismo," the Pope's "nephews," who were actually sons sired by Sixtus. In the eyes of many, the Pope's actions produced "one of the greatest mischiefs that oppresseth the Church [because] to favour his kindreds' interest, he had forgot himself, and the Church, thinking of nothing but of how to advance them [Sixtus' nephews] to their satisfaction" (Leti, 1669, p. 41). Sixtus lavished such wealth on his "nephews" that people were led to say, "Rome had as many Popes as Sixtus had Nephews" (Leti, 1669, p. 42). "Nephew" became a euphemism for the son of a putatively celibate clergyman and a relative to whom special favors were granted.

I use nepotism to refer to favoritism shown to relatives, in accord with the above definition from the American Heritage Dictionary, but to contextualize my mean-

---

*The term "functional" is used in different ways by investigators seeking answers to different kinds of questions. By functional, I mean the consequences of a behavior that, in present environments, increase an individual's likelihood of surviving and reproducing successfully.

ing, I must flesh out "favoritism" and "relatives," with the help of proximate and ultimate thinking. (Also see Alexander's, 1979, pp. 43–58, and Sherman's, 1980a, discussions of the meaning of nepotism.) Favoritism implies that one individual, a donor, bestows a material or psychological benefit upon another individual, a recipient. Examples of kin favoritism, which I will use interchangeably with nepotism, include sharing valuable resources with relatives (Koenig & Mumme, 1987; Parker, Waite, & Decker, 1995), providing social or psychological support to relatives during conflict (Chapais, 1992; Packer, Gilbert, Pusey, & O'Brien, 1991), and protecting relatives from harm (Hailman, McGowan, & Woolfenden, 1994; Hoogland, 1983), among many other possibilities. However, for these examples to represent kin favoritism at the ultimate level, the "material or psychological benefit" must increase either the survival or reproductive success of these related *recipients*, which, in turn, will enhance the genetic survival of the donor (Alexander, 1979). Female vervet monkeys, *Cercopithecus aethiops*, for instance, sometimes care for the young of other females. These allomothers care selectively for their younger (maternal) siblings more often than they care for unrelated infants and allomaternal care does have beneficial reproductive consequences for some of the individuals involved (Figure 1). Mothers with allomaternal daughters benefit by having shorter interbirth intervals than mothers who do not receive help, which probably enhances mothers' lifetime production of young (Fairbanks, 1990). Allomaternal daughters benefit from providing care because, perhaps as a result of "maternal practice," they are more successful rearing their own first-born infant than are less experienced allomothers. Interestingly, we do not know whether infant recipients actually benefit from allomaternal care. (Lee, 1987, reports that calf survival in African elephants, *Loxodonta africana*, is enhanced by allomaternal care from female family members). Silk (1999) examines various functional explanations for allomaternal care in nonhuman primates, as does Manson (1999), who refers to "infant handling."

As I have just suggested, behavior that "bestows a material or psychological

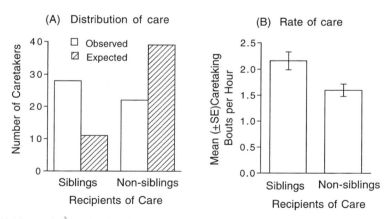

Figure 1. Evidence that captive female vervet monkeys direct their allomaternal care preferentially to their younger siblings rather than nonsiblings. (A) The number of females that provide allomaternal care (e.g., grooming, holding, carrying) to their younger siblings or nonsiblings compared with the distribution of care expected based on the number of younger siblings and nonsiblings available to each allomaternal female (chance) and (B) the mean rate of caretaking bouts that allomaternal females direct to their younger siblings or nonsiblings. (Adapted from Fairbanks, 1990, with permission. Copyright 1990 Academic Press.)

benefit upon another individual" at the proximate level may not always represent nepotism at the ultimate level. For instance, in various species of birds and mammals, sexually mature adults forego reproduction and help relatives, usually their parents, rear the helpers' younger siblings, which suggests that helpers engage in nepotism (Emlen, 1991; Solomon & French, 1997; helping is not always directed to relatives though; Cockburn, 1998). Avian helpers do enhance the production of young in several species, but in some instances, parents with helpers do not rear more young than those without helpers (Cockburn, 1998; Koenig & Mumme, 1987). Helping behavior such as feeding nestlings that are helpers' younger siblings could be considered kin favoritism at the proximate level but not at the ultimate level unless helping caused more young to be reared successfully, thereby improving, indirectly, the helper's fitness (details below). Similarly, the antipredator warning calls given in many species of mammals are hypothesized to alert other conspecifics, especially kin, of impending danger (Sherman, 1977), but I know of no data which demonstrate that alarm calls actually increase the survival of individuals that hear the calls, although I certainly believe that they do.

In summary, definitions and frameworks for the study of nepotism depend on the currencies that compose the material or psychological benefit bestowed by donors upon recipients and, ultimately, upon themselves. In this light, proximate and ultimate definitions of nepotism are complementary and not competing alternatives. If one is interested in the evolutionary origin or subsequent modification of kin favoritism over generations, then one must assess for donors and recipients the fitness correlates or consequences of putative acts of favoritism. However, given the difficulties of conceptualizing and measuring fitness (R. Dawkins, 1982), there are pragmatic reasons to design empirical studies of kin favoritism based on its proximate definition. Most of the examples I present below will portray nepotism from a proximate perspective.

WHO IS A "RELATIVE" AND WHAT IS "FITNESS"?

Like "favoritism," the meaning of "relatives" in definitions of nepotism also differs depending on one's conceptual framework. In a functional framework, relatives are individuals who share a *recent* common ancestor, which means that relatives are individuals who share genes identical by immediate descent. For example, cousins are relatives who share copies of genes that resided in their recent common ancestors, aunts or uncles. The phrase "identical by immediate descent" is critical because whereas all humans share in common over 95% of *all* of their genes, relatives share genes *over* and *above* the background of 95% because they have a *recent* ancestor(s) in common (R. Dawkins, 1989; Grafen, 1984). To quantify genetic relatedness, evolutionary scientists use the coefficient of relatedness, r, which ranges from 0.0 to 1.0 and represents the proportion of genes which are identical by descent between two individuals or the probability that two individuals share a copy of the same allele at a given locus due to recent common ancestry. Bennett (1987) reviews various ways in which relatedness has been measured.

In a functional framework, there are two categories of relatives, *descendant kin* such as parents and offspring, and *nondescendant kin* such as siblings, cousins, and aunts. (It is sometimes more helpful to refer to "lineal kin" rather than descendant kin; for example, parents are lineal kin to their offspring rather than descendant kin.) Recognition of these two categories of kin was crucial to Hamilton (1964) when he developed the concept of inclusive fitness. In population genetics models, Hamilton showed that an individual could have a *direct effect* on its genetic representation in future generations by producing offspring and an *indirect effect* on its genetic representation by assisting nondescendant relatives in their reproductive efforts. Thus, inclusive fitness is calculated as the number of offspring produced by an individual due to its own efforts *plus* how the individual affects the production of offspring by nondescendant relatives where all offspring are weighted by the appropriate coefficient of relatedness (Figure 2). [It is easy to misunderstand the concept of inclusive fitness for several reasons, as discussed by R. Dawkins (1979), M. S. Dawkins (1995) and Grafen (1984).]

Three terms are often used in association with Hamilton's (1964) rule and inclusive fitness, two of which originated in Brown (1980, 1987) and one in Maynard Smith (1964). Brown proposed terms to identify the two routes by which an individual could affect the transmission of its DNA between generations. "Direct fitness" refers to the production of descendant kin (e.g., offspring) due to an individual's own efforts and "indirect fitness" refers to the effects of the individual on the production of nondescendant kin (e.g., siblings, nieces). (Also see West-Eberhard's, 1975, p. 6, discussion of "classical fitness" and the "kinship component" of inclusive fitness.) Maynard Smith (1964) introduced the term "kin selection" to explain the evolutionary process that enhanced the survival of close relatives, including both descendant and nondescendant kin.

Unfortunately, some authors have incorrectly taken kin selection to refer only to indirect-fitness effects, despite the clarity of Brown's (1980, 1987) framework and Maynard Smith's (1964) explicit statement that the evolution of the placenta and of parental care were due to kin selection, as was the evolution of sibling favoritism. For example, Shields (1980) questioned Dunford's (1977) and Sherman's (1977) arguments that alarm calls in round-tailed ground squirrels, *Spermophilus tereticaudus*, and Belding's ground squirrels, *S. beldingi*, respectively, had evolved due to kin selection. What Shields meant to ask was whether ground squirrels' alarm calls evolved due to

$$\text{Inclusive Fitness (IF)} = \text{Direct Fitness (DF)} + \text{Indirect Fitness (IDF)}$$
$$\text{(Direct Selection)} \qquad \text{(Indirect Selection)}$$
$$\underbrace{\qquad\qquad\qquad\qquad}_{\text{Kin Selection}}$$

$$\text{IF} = \qquad (r_{p\text{-}o} \times N_p) \qquad + \qquad \Sigma\,(\,r_{d\text{-}r} \times N_{ndk})$$
$$\text{Direct} \qquad + \qquad \text{Indirect}$$
$$\underbrace{\qquad\qquad\qquad}_{\text{Kin Selection}}$$

$r_{p\text{-}o}$ = parent-offspring coefficient of relatedness (see text)

N_p = number of surviving young due to the actions of the parent

$r_{d\text{-}r}$ = donor-recipient coefficient of relatedness

N_{ndk} = number of surviving young due to the actions of nondescendant kin

Figure 2. A schematic diagram of Hamilton's (1964) concept of inclusive fitness divided into its component parts, direct fitness and indirect fitness (Brown, 1980, 1987). Direct fitness refers to an individual's production of its own offspring, whereas indirect fitness refers to how an individual affects the production of offspring by its nondescendant kin (e.g., siblings). The meaning of "kin selection," which is not synonymous with indirect fitness, and "coefficient of relationship" is discussed in the text.

direct-fitness effects associated with parental care, indirect-fitness effects associated with aid to nondescendant kin, or a combination of the two. It is possible to quantify how direct- and indirect-fitness effects are responsible for maintaining a trait in a population (Vehrencamp, 1979), and it might be valuable to know the degree to which the evolutionary maintenance of alarm calling is due to direct- and indirect-fitness effects (e.g., Blumstein, Steinmetz, Armitage, & Daniel 1997). For example, both direct- and indirect-fitness effects maintain alarm calls in *S. beldingi* (Sherman, 1977, 1985) whereas direct-fitness effects alone maintain calling in 13-lined ground squirrels, *S. tridecemlineatus* (Schwagmeyer, 1980). However, "a unit of inclusive fitness gained through aid to [non-descendant] relatives is exactly equivalent to a unit of classical [direct] fitness in so far as prediction of the results of reproductive competition among individuals is concerned" (West-Eberhard, 1975, p 14). This perspective on inclusive fitness emphasizes the additive nature of direct- and indirect-fitness effects on intergenerational gene transmission and reduces the likelihood that these two kinds of effects will be pitted against each other as if both were not part of the same whole. For an example of inclusive-fitness logic and the value of differentiating direct fitness from indirect fitness, see the exchange between Hauber and Sherman (1998) and Blumstein and Armitage (1998) on the function and targets of alarm calls of yellow-bellied marmots, *Marmota flaviventris*.

To summarize, using Hamilton's (1964) inclusive fitness theory, Maynard Smith's (1964) term kin selection, and Brown's (1987) framework of direct and indirect selection, I suggest that parental care is a form of nepotism, assisting nondescendant relatives is nepotism, and kin selection (inclusive fitness) includes both direct- and indirect-fitness effects, the relative impact of which will vary from one situation to another (Figure 2). In this chapter on the ontogeny and function of nepotism, I stress the development of social relationships between *nondescendant kin* rather than descendant kin because an extensive literature already exists on the development of parent–offspring relationships (e.g., Gubernick & Klopfer, 1981; Krasnegor & Bridges, 1990), whereas in the vast majority of organisms little has been published on the ontogeny of social relationships between nondescendant kin.

Conditions Necessary for the Evolution of Nepotism

For nepotism to have evolved by natural selection, there must have been heritable variation in the tendency to treat relatives favorably and nepotistic behavior must have produced greater inclusive fitness gains for the actor than nonnepotistic forms of behavior. For nepotism to be maintained by selection in a population, (1) relatives must be available for social interactions (the opportunity condition), (2) relatives must be distinguishable from nonrelatives (the kin-recognition condition), and (3) individuals must be able to enhance the survival and/or reproductive success of their relatives, and therefore their own inclusive fitness, by nepotistic acts (the fitness-enhancement condition). I will examine each of these requirements in turn.

The opportunity condition may seem so patent that it does not merit mention, but listing it explicitly highlights two features of nepotism. First, nepotism is more common in group-living than in asocial species because in many cases, several group members may be close kin (see Smuts, Cheney, Seyfarth, Wrangham, & Struhsaker, 1987, on primates; see Rubenstein & Wrangham, 1986, on birds and mammals; see Crozier & Pamilo, 1996, on social insects). Second, the opportunity condition helps explain at the proximate level why nepotistic behavior is often displayed largely or exclusively by one sex (i.e., it is sex-limited). In most species of mammals and birds, individuals typically emigrate from their birthplace before they reach sexual maturity, their natal dispersal. In mammals, natal dispersal is male-biased and in birds it is female-biased (Smale, Nunes, & Holekamp, 1997). If natal dispersal results in individuals of only one sex living among their kin as adults, then, unless kin dispersed together, nepotism, if it occurred, would be limited to the nondispersing sex. For example, among the ground-dwelling squirrels (family Sciuridae), males leave their birthplace, and alarm calls, a form of nepotism (Sherman, 1980a), are given almost exclusively by philopatric adult females (reviewed in Blumstein & Armitage, 1997; Hoogland, 1995).

The second requirement for nepotism is that kin must be recognizable (Hamilton, 1964). Kin recognition is inferred from the differential treatment of relatives based on correlates of genetic relatedness (Holmes & Sherman, 1983), and considerable evidence shows that recognition abilities exist in many taxa (Fletcher & Michener, 1987; Sherman, Reeve, & Pfennig, 1997). The word "correlates" is an important part of the definition of kin recognition for two reasons. First, it underscores the point that kin recognition does not involve an assessment of genes identical by descent per se, but rather a responsiveness to factors that, under species-typical circumstances, are reliably associated with genes identical by descent. The correlates are part of an organism's "ontogenetic niche," the suite of social and ecological factors that an individual inherits, along with its genes, that fosters adaptive developmental outcomes (West *et al.*, 1988). The correlates vary among taxa and include, for example, sharing a common nest with siblings during early development (Beecher, 1991; Holmes, 1984) or phenotypic similarity (Heth, Todrank, & Johnston, 1998; Holmes, 1986; Sun & Muller-Schwarze, 1997). Second, the word serves as a reminder that recognition could be mediated by relatives' phenotypic features such as their odors or vocalizations (direct recognition) or by nonphenotypic, spatial cues like the location of nests or burrows (indirect recognition; Waldman, Frumhoff, & Sherman, 1988). As long as the cues that mediate differential treatment are reliability correlated with kinship, selection will favor individuals that use phenotypic or nonphenotypic cues to modify their interactions with conspecifics (Sherman *et al.*, 1997).

Kin recognition and kin favoritism are not interchangeable terms. Because kin recognition refers to a process that produces *differential* treatment and kin favoritism to *preferential* treatment, evidence of kin favoritism provides evidence for kin recognition. However, differential treatment of kin could be expressed in ways that do not benefit kin. For example, kin could elicit more olfactory investigation than non-kin (Holmes, 1984) or individuals could look longer at their relatives than at nonrelatives (Wu, Holmes, Medina, & Sackett, 1980), so that evidence for kin recognition would not provide evidence for kin favoritism (Gamboa, Reeve, & Holmes, 1991). Furthermore, even if kin are recognized, this does not insure that they should be treated favorably (Sherman *et al.*, 1997). Besides donor–recipient relatedness, the direct-fitness benefits to a recipient and the direct-fitness costs to a donor must be considered to determine whether selection will favor aid to particular kin (Hamilton, 1964).

Finally, for nepotism to be maintained in a population, individuals must be able to enhance the survival or reproductive success of their relatives. Male lions, *Panthera leo*, for example, form cooperative coalitions to defend their group's territorial boundaries and chase intruding males that would displace resident males and kill cubs sired by the residents. Smaller coalitions (two or three males) sometimes include unrelated males, but larger coalitions typically comprise close relatives (e.g., brothers). Males in larger coalitions, besides maintaining residency longer than males in smaller coalitions, father more surviving offspring per capita than smaller coalitions (Packer *et al.*, 1991). One of the most extreme examples of vertebrate nepotism is seen in the aptly-named naked mole-rat, *Heterocephalus glaber*, a hairless, fossorial rodent found in eastern Africa (Sherman, Jarvis, & Alexander, 1991). Naked mole-rats live in subterranean colonies composed of 75–80 animals which are closely related to each other due to inbreeding (mean within-colony $r = 0.81$). In the typical mole-rat colony, one female and one to three males are active breeders, and other nonbreeding adults care cooperatively for developing young by delivering food, maintaining the colony's tunnel system, and protecting colony members from predators (Lacey & Sherman, 1997). Whether breeders produce young that survive until weaning depends critically on the direct and indirect care that young receive from their nonbreeding adult relatives.

INCLUSIVE FITNESS AND THE DEVELOPMENT OF NEPOTISM

Most evolutionary scientists identify Hamilton's (1964) concept of inclusive fitness (Figure 2) as the most influential extension of natural selection theory since Darwin's (1859) seminal work. Hamilton's concept generated numerous theoretical models and a large body of empirical evidence on nepotism and for the effects of genetic relatedness on mating behavior (e.g., see references in Alexander & Tinkle, 1981; Bateson, 1983; Crozier & Pamilo, 1996). However, for investigators interested in behavioral development, inclusive fitness theory has not had an equivalent impact. With the exception of mother–offspring relationships, studies of the *development* of nepotism are surprisingly rare given the amount of research that has focused explicitly on how kinship influences social behavior.

For example, the study of cooperative breeding in birds, in which adult helpers care for young they did not produce, was invigorated by applying Hamilton's (1964) rule for the evolution of altruism (Brown, 1987; Emlen, 1991; also see Solomon & French, 1997, for mammalian examples). However, almost no work has been done on the development of social relationships between helpers and immature recipients (Komdeur & Hatchwell, 1999). The evolutionary paradox posed by alloparental behavior (how could natural selection favor such apparent altruism?) has generated a lively debate about whether helping behavior is an adaptation or an unselected by-product of parental care that occurs when animals do not disperse from their natal environment (see the review in Mumme & Koenig, 1991). One way to help resolve the debate would be to study the ontogeny of alloparental behavior, including the social relationships that arise between helpers and immature recipients, to determine whether alloparental care emerges from a different developmental process than the process that generates parental care (e.g., Komdeur & Hatchwell, 1999). This is because adaptations are traits of individuals that were designed to solve specific problems related to survival or reproduction in a given environment and that conform to *a priori* design specifications (Williams, 1966).

Another reason to study the developmental basis of nepotism is to account for situations in which nepotism is expected but does not occur. The genetic benefits of providing parental care to young usually require that the provider be related to the recipient (Clutton-Brock, 1991) and, indeed, there is ample evidence for discriminative parental care (Beecher, 1991; Holmes, 1990). In several species of socially monogamous birds, however, extrapair mating occurs so that a female's eggs are fertilized by more than one male and cuckolded males care for nestlings some of which they did not sire (Kempenaers & Sheldon, 1996). A reduction in the likelihood of paternity should not always favor a reduction in paternal care, but why would males ever care for their competitors' offspring (Westneat & Sherman, 1993)? One proximate answer is that in birds, parents learn the phenotypic attributes of their offspring such as their individually distinctive calls by interacting directly with nestlings (Beecher, 1981) and the successful development of this mechanism is thwarted because related and unrelated nestlings occupy the same nest. A male could compare nestlings' phenotypes with his own to discriminate between his own young and those sired by another male (self-referent phenotype matching; Holmes, 1986; Lacy & Sherman, 1983), but selection does not seem to have favored a developmental process that could facilitate self-matching in mixed-paternity broods (Kempenaers & Sheldon, 1996). If we understood the ontogeny of nestlings' phenotypes and of parental care by cuckolded males, perhaps we could explain why selection has not produced a developmental process for paternal self-matching.

BEHAVIORAL DEVELOPMENT AND EVOLUTIONARY CHANGE

A final reason to conduct ontogenetic studies of nepotism is that they may enhance our understanding of the evolution of nepotism. Developmental processes, the outcomes of which are subject to natural selection, can constrain or promote behavioral evolution, including acts of nepotism, so that ontogenetic studies of nepotism can inform evolutionary studies of nepotism. A developmental constraint is a restriction on phenotypic outcomes caused by the nature or dynamics of a developmental process and such constraints may evolve for various reasons and take many forms (Maynard Smith *et al.*, 1985). For example, some constraints exist due to the laws of physics and chemistry, whereas others exist because complex adaptations depend on multiple alleles, so that an allelic change might reduce the cost of developing one trait but significantly increase the cost of developing another. Thus, once a developmental process has evolved, it may limit the range or direction of future evolutionary modification (e.g., Loeschcke, 1987; Nijhout & Emlen, 1998).

On the other hand, developmental processes can promote evolutionary change due to phenotypic plasticity, which is the capacity of one genotype, interacting with the environment, to yield more than one form of anatomy, physiology, or behavior (West-Eberhard, 1989; see Gordon, 1992, for a discussion of the multiple meanings of "phenotypic plasticity"). Plasticity itself is subject to natural selection (Scheiner, 1993), and developmental plasticity can contribute to both the origin and elaboration of evolutionarily novel traits (e.g., Nijhout & Emlen, 1998; reviewed in Maynard Smith *et al.*, 1985). This is because developmental plasticity results in a much greater array of phenotypes available for selection to act upon than there are genotypes. For example, brown-headed cowbirds, *Molothrus ater*, are obligate brood parasites, which has led some investigators to suggest that the ontogeny of species or mate recognition must be "innate" or "developmentally closed." Freeberg (1996) has shown, however, that males and females display differences in courtship and mating preferences depending on the population-specific social experiences they have as juveniles. These experiences have later reproductive consequences because they generate courtship patterns in males and mate preferences in females that mediate mating behavior, which, in turn, determines gene frequencies. Thus, as a result of behavioral plasticity during juvenile development, a phenotype may arise and natural selection could act on the neophenotype which would produce changes in gene frequencies (see the review in Gottlieb, 1992).

In this chapter, I will examine some empirical studies that focus explicitly on the development of social relationships between *nondescendant* kin. Because nepotism often occurs in group-living organisms (references above) and because social relationships routinely develop between group members (e.g., Hinde, 1983), an understanding of nepotistic behavior is illuminated by explaining how individuals develop social relationships with their kin. Thus, I will focus on the ontogeny of amicable or cohesive relationships between nondescendant kin that I believe provide the underpinnings for nepotistic behavior.

Background and Methods

Nepotism and the Behavioral Ecology of Belding's Ground Squirrels

The sciurid rodents (family Sciuridae) or ground-dwelling squirrels include marmots (genus *Marmota*; Barash, 1989), prairie dogs (genus *Cynonmys*; Hoogland, 1995), and about 20 species of ground squirrels (genus *Spermophilus*; Murie & Michener, 1984). Among the ground squirrels, the behavioral ecology of Belding's ground squirrels, *S. beldingi*, has been investigated intensively, and field researchers have chronicled various kinds of nepotism in two contexts. First, when terrestrial predators like coyotes, *Canis latrans*, appear in a colony, females living among certain classes of kin (mothers, daughters, or littermate sisters) are more likely than females without such kin to utter vocalizations that alert others to danger but increase caller mortality (Sherman, 1977, 1985). Second, following hibernation each spring, females establish their natal burrow in which they will rear their litter. During this period, females are less likely to chase and fight neighboring females to which they are closely related than neighbors to which they are unrelated (Sherman, 1980b, 1981a, b). Moreover, following parturition in natal burrows, related neighbors (mothers, daughters, or littermate sisters) are more likely than unrelated neighbors to cooperate and chase off infanticidal animals that enter burrows and kill unweaned young (Sherman, 1981a).

Documentation of nepotism in free-living *S. beldingi* has led me to pursue the developmental basis of kin favoritism in this species. I believe that the emergence of nepotism in *S. beldingi* illustrates aspects of an ontogenetic process that operates in several species of group-living organisms. Accordingly, I will describe some of the work my colleagues and I have done on *S. beldingi* and highlight studies on other organisms in which nepotism seems to unfold according to a similar ontogenetic process.

S. beldingi are group-living, 250- to 300-g, diurnal rodents that inhabit subalpine meadows and other open areas in the western United States (Jenkins & Eshelman, 1983). They are obligate hibernators that are typically active aboveground from May through September, depending on local conditions (Morton & Gallup, 1975). Adult (≥ 1 year old) females produce a single litter of five to seven young annually and females typically live four to six years. Adult (≥ 2 years old) males compete intensively with each other for mating opportunities during the 3-week-long mating season, which begins shortly after females emerge from hibernation, and males typically live three to four years (Sherman & Morton, 1984). Because each female mates with multiple partners during her single annual estrus (Hanken & Sherman, 1981), litters routinely include full and maternal half-siblings so I will use "litter-mates" rather than "siblings" to describe young born in the same litter. I have concentrated my developmental studies on littermates rather than other classes of kin because they are the most common nondescendant relative in *S. beldingi* (Sherman, 1981b) and because female littermates treat each other nepotistically (references above).

After mating, each female digs a natal burrow in which she bears one litter annually and which she defends against conspecifics until her young first appear above ground at 25–28 days of age (their natal emergence; Sherman, 1981a). Juveniles (young-of-the-year that have undergone natal emergence) use their natal burrow as an activity center for about 3 weeks, during which time they interact with their littermates and other juveniles and adults (Holekamp, 1983). About 3 weeks after natal emergence, males begin to leave their birthplace, almost never to return home (their natal dispersal), whereas females establish burrows within 20–40 m of their birthplace (Holekamp, 1986). Due to natal dispersal, adult males rarely encounter matrilineal kin and thus have no opportunity to treat them nepotistically. In contrast, adult females live within a nexus of maternally related female kin and interact regularly with various female relatives (mothers, daughters, and littermate sisters) and behave nepotistically, as outlined above. However, immigration and mortality patterns are such that a female's immediate neighbors include both kin and non-kin so that nepotism is not mediated by any simple proximity or length-of-association rule of thumb (Sherman, 1980b).

EARLY DEVELOPMENTAL ENVIRONMENTS AND THE ONTOGENY OF NEPOTISM

A functional analysis of nepotism (e.g., Alexander, 1979; Hamilton, 1964) is not predicated on the belief that the development of nepotism is "hard-wired" so that kin favoritism occurs inevitably when relatives are first encountered. When animals are reared in their species-typical environment, nepotism emerges as a result of a developmental process in which social favoritism seems to be shaped and narrowed by experiences that occur predictably at particular times in specifics contexts (see West *et al.*'s, 1988, discussion of development in an ontogenetic niche). In this light, many organisms have the *developmental capacity* to become universal altruists, treating kin and non-kin beneficently. As I hope to show below, what species-typical developmental processes do is to build constraints on motivational and behavioral systems that are potentially wide open so that individuals come to treat each other beneficently to the degree that they share genetic interests in common. If my developmental perspective on nepotism is correct, then, like the development of ingestive and some other major categories of behavior (Blass, 1995), the ontogeny of nepotism begins with motivational and behavioral systems that are quite malleable, with the potential to assume multiple forms, and proceeds over time to become systems that are narrowly focused and likely to generate adaptive outcomes.

In several taxa, the seeds of adult nepotism are planted during early ontogeny due to interactions that often occur only between kin (Bekoff, 1977; Michener, 1983; Periera & Fairbanks, 1993). This suggests that one could gain insights into the developmental origins of nepotism by examining the early environments in which ontogeny occurs; the earliest such environment for polycotous mammals (those that bear multiple young in a litter) is the uterus (Robinson & Smotherman, 1991). When a polycotous female mates with multiple males during the same receptive period, fetuses subsequently develop in a uterine environment that includes some individuals which are twice as closely related to each other (full sibling $r = \frac{1}{2}$) as they are to others (maternal half-sibling $r = \frac{1}{4}$). Under these conditions, which are common in ground squirrels (Schwagmeyer, 1990), nepotism could occur if blastocysts or fetuses competed more intensively with half-siblings than full siblings for uterine space or maternally-derived nutrients. I do not know whether such differential competition occurs in *S. beldingi* or in any other mammal, but intrauterine competition can be intense in polytocous species even between siblings. In pronghorn antelope, *Antilocapra americana*, for example, tube-shaped blastocysts become knotted together, which reduces by 30% the number of surviving embryos. Some embryos even produce a necrotic tip, an extension of the chorioallantois, that lances and kills the adjacent embryo (O'Gara, 1969).

The stage is set for intrauterine competition in *S. beldingi* because over 75% of litters are multiply sired (Hanken & Sherman, 1981) and more fetuses start to develop than are born (Morton & Gallup, 1975). Prenatal adjustment of litter size can be viewed through the lens of parent–offspring competition (Trivers, 1974), but the sibling competition models of Mock and Parker (1997) demonstrate that intersibling competition must also be weighed to understand litter-size adjustment. Mock and Parker's ideas are especially cogent for litters that include unequally related young in which benefits would accrue to fetuses that targeted half-siblings rather than full siblings. How fetuses might discriminate between full and maternal half-siblings and direct competition toward the latter is an unexplored question, but recent data on the impact of prenatal experience on mammalian development forces us to appreciate the richness of in utero effects on behavioral ontogeny (Lecanuet, Fifer, Krasnegor, & Smotherman, 1995).

Following hatching or birth, early development in many species occurs in holes, nests, burrows, or other sequestered places that are accessible only to certain conspecifics like parents. Such environments provide an ideal milieu for social relationships to develop between rearingmates, the young that share a common early rearing environment, which in many species are closely related (e.g., full or maternal half-siblings). First, these environments provide a relatively safe haven for developing young. Second, they include young of more or less similar size and experience whose thermal and nutritional needs are provided and they exclude unrelated conspecifics. Finally, they offer many opportunities for young to interact and monitor the consequences of their interactions as they begin to develop sensorimotor skills and social competence. As Blass (1999, p. 37) has written, "For many mammals, the rewarding, charged, yet gentle environment of the nest, therefore, provides sensory information concerning mother and litter mates that influence the infants' social and ingestive behavior in the present and in the immediate and long-term future" (also see Alberts & Cramer, 1988). On the other hand, these environments may be the seat of intense competition (e.g., Fraser & Thompson, 1991), leading to siblicide, when rearingmates experience acute resource shortages (Mock & Parker, 1997). However, when young do begin to forge cohesive relationships with their rearingmates, these early kin preferences may contribute to the subsequent appearance of adult nepotism in several species (Bekoff, 1981), including many ground-dwelling squirrels, as argued originally by Michener (1983).

S. beldingi pups fit this mold because from birth until natal emergence, they develop in an underground burrow, a preemergent environment that is inhabited only by their mother and littermates. Accordingly, I have been investigating the development of littermate preferences by manipulating the preemergent environment of captive ground squirrels and subsequently observing juveniles' early social interactions that occur in their postemergent (above-ground) environment. First, my assistants and I allowed field-mated, individually housed females to rear their offspring almost to weaning (25–27 days of age) in a "nursery building" at our research site, the Sierra Nevada Aquatic Research Laboratory (SNARL) near Mammoth Lakes, California. Then we transferred animals to outdoor enclosures at SNARL for behavioral observations. Into each 10-m^2 enclosure, we (usually) placed four unrelated (to each other) mothers and their offspring, two males and two females per litter. Each litter had its own underground nestbox, an artificial burrow (see drawings and details in Holmes, 1994). Enclosure-housed juveniles "emerged" (their natal emergence) from their artificial burrows at the same age as their free-living counterparts (ca. 27 days of age), and we recorded their behavioral interactions daily for 1–3 weeks, beginning on the day of natal emergence. Our general purpose was to determine how kinship (littermate and non-littermate pairs) and sex (male–male, male–female, and female–female pairs) influenced the development of juveniles' social relationships, including those with their preferred social partners. The pattern of early development in an isolated, kin-only environment, followed by further development in a new environment inhabited by kin and non-kin of various ages, is a common pattern in many taxa and one that contributes to the development of social preferences in several group-living species (Holmes, 1997).

The Meaning and Assessment of Social Preferences

A social preference is declared when individuals display nonrandom amicable behavior or affiliative responses to one social stimulus rather than another or when individuals direct aggressive or other forms of harmful behavior to one social stimulus rather than another (Holmes, 1988; Kummer, 1978). In many species, individuals prefer their kin as social partners, in which case social preferences are kin preferences that can be manifested in various ways. For example, in several cercopithecine monkeys, females provide behavioral support to their relatives by forming social alliances with kin that are critical to acquiring and maintaining social rank in a troop (Chapais, 1992; Pereira, 1995). In colonies of the social wasp, *Polistes fuscatus*, a resident female (one on her own nest) will often encounter other females on her nest some of which may be unrelated intruders (Gamboa, Foster, Scope, & Bitterman, 1991). Resident females are quite tolerant when they encounter their female nestmates, typically their sisters, but residents behave aggressively toward unrelated non-nestmates to prevent nest usurpation by the intruders (Figure 3). The kin preferences expressed by tadpoles of Arizona tiger salamanders, *Ambystoma tigrinum nebulosum*, can be quite dramatic and precise (Pfennig, Sherman, & Collins, 1994): in laboratory tests in which two types of prey are available simultaneously, cannibalistic larvae consume (1) more non-kin than siblings, (2) more non-kin than cousins, and (3) more cousins than siblings (Figure 3).

Dyadic play is one of the most frequent kinds of social interactions between free-living juvenile ground squirrels during their initial weeks above ground (Holekamp, 1983; Michener, 1981; Waterman, 1988), and I have used social play as a bioassay to assess juveniles' social preferences. Both the definition(s) and function(s) of mammalian play have been debated for years (see reviews in Bekoff & Byers, 1981; Fagen, 1981; Martin & Caro, 1985) and disputes continue (Bekoff & Byers, 1998; Pellis, Field, Smith, & Pellis, 1997). However, ground squirrel researchers have had little trouble

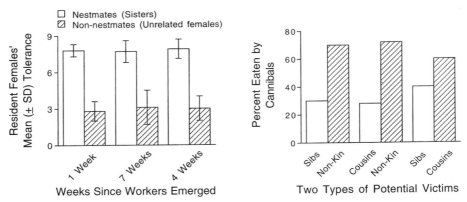

Figure 3. Evidence that female paper wasps and tiger salamander larvae both display social preferences for kin over non-kin. (A) In field tests of kin favoritism, resident female paper wasps that encounter other females on the resident's nest behave aggressively (a low tolerance score) toward unrelated nonnestmates and nonaggressively (a high tolerance score) toward related nestmates. Throughout the summer, resident females behave aggressively to non-kin, as different broods of workers emerge on a resident's nest. (Adapted from Gamboa, Foster, Scope, & Bitterman, 1991, with permission. Copyright 1991 Springer-Verlag.) (B) In laboratory tests of kin favoritism, cannibalistic tiger salamander larvae were placed in tanks with equal numbers of two types of potential prey (e.g., siblings and non-kin) that differed in how they were related to the cannibals. Cannibals preferentially consumed the less-related individuals in each of the three test conditions. (Adapted from Pfennig, Sherman, & Collins, 1994, with permission. Copyright 1994 Oxford University Press.)

defining and recognizing social play (Holmes, 1995), and they have routinely categorized it as amicable or cohesive when it occurs between juveniles during their initial weeks above ground (e.g., Michener, 1981; Waterman, 1988). (Pellis and Iwaniuk, 1999, discuss different kinds of social play in rodents, including "play fighting" and "sexual play," both of which occur in *Spermophilus*.) Dyadic play in *S. beldingi* comprises motor patterns that may include rushing at, pouncing on, or chasing; mounting; and wrestling (see drawings of *S. columbianus* play by Steiner, 1971). During daily observations of enclosure-housed juveniles, my assistants and I recorded the frequency of dyadic play bouts, in which two juveniles engaged in one or more of the motor patterns just listed. I then used play-bout frequencies per pair to assess juveniles' social preferences under the assumption that preferred social partners play together more often than nonpreferred partners, although I do not assume that there is a linear relationship between play-bout frequencies and the strength of social preferences (see Holmes, 1994, for details).

I do not view ground squirrel play as nepotistic behavior at either the proximate or ultimate level, although future work might show that play is nepotistic (for reviews of the potential short-term and long-term benefits of social play, see Bekoff & Byers, 1981; Martin & Caro, 1985; Pellis *et al.*, 1997). I do hypothesize, however, that by playing together, juveniles avail themselves of important opportunities to probe the quality and strength of a potential social relationship (Hinde, 1979, 1983). That is, play may allow newly-emergent juveniles to "test" social partners who offer different potential payoffs (Kummer, 1978; Zahavi, 1977). Play in this "testing" framework could provide a behavioral context in which close social relationships are established, monitored, and modified (Lee, 1983; Poirier & Smith, 1974). I am thus hypothesizing that play can both contribute to the formation of an amicable or cohesive relationship and reflect the existence of such a relationship once it is established. In this light, if juveniles' cohesive relationships were maintained until adulthood, they could support differential social interactions between adults and facilitate nepotism. With this logic in mind, I have proceeded to examine the frequency of social play between captive juveniles and the role of kinship in the development of social preferences, using methods described above.

QUANTIFYING SOCIAL PREFERENCES IN "GROUPS" OF GROUND SQUIRRELS

To quantify *S. beldingi* social play, I will report the mean ± 1 SEM frequency of play bouts for various types of pairs. For example, I summed all play bouts that occurred between littermates during the 10–14 days a group was observed and divided by the total number of littermate pairs in the group to calculate the mean frequency of littermate play per pair. A "group" is the set of animals that simultaneously occupied the *same* enclosure, and my assistants and I observed several different groups, as described below. Groups typically comprised four litters, with one mother and (usually) two male and two female young in each litter. Mothers in a group were unrelated to each other and usually interacted agonistically although infrequently. Comparisons of play-bout frequencies *between* groups are problematic because groups were observed under different weather conditions that affected overall rates of social interactions (Holmes & Mateo, 1998).

For *within-group* comparisons, the statistical tests I report between the frequencies of littermate and non-littermate play control for other factors that influence play such as the sex of the interactants, their relative body weights, and the appearance of the same individual in more than one pair. Because littermate and non-littermate pairs in the same group were observed simultaneously (and thus for the same amount of time), frequency data (play bouts per pair) and rate data (play bouts per pair per time unit), which I do not report, reveal the same set of play-partner preferences. For an explanation of statistical techniques (e.g., log-linear analyses), sample sizes, and significance levels, see Holmes (1994).

KIN PREFERENCES IN GROUND SQUIRRELS AND OTHER ORGANISMS

THE SOCIAL PREFERENCES OF JUVENILE BELDING'S GROUND SQUIRRELS

Our observations of several groups of ground squirrels revealed clear and easily recognizable littermate preferences. When summed over juveniles' first 10–14 days above ground, play bouts between littermates are two to four times more frequent than play bouts between non-littermates on a per-pair basis (Figure 4). For the three groups in Figure 4, we observed one or more play bouts in every possible littermate pair, whereas we did not observe a single play bout in 28– 41% of all possible non-littermate pairs. For play bouts that lasted at least 5 sec, the duration of littermate bouts (22.5 ± 4.3 sec) was significantly longer than the duration of non-littermate bouts (14.3 ± 3.0 sec; W. Holmes, unpublished data from three groups). Thus, both the frequency and duration of littermate play were greater than non-littermate play.

The data in Figure 4 are based on dyads as the unit of analysis because I am interested in the development of social relationships, the nature of which is a product of (at least) two individuals (Hinde, 1979). However, I also examined play-bout frequencies using individuals as the unit of analysis and again found clear

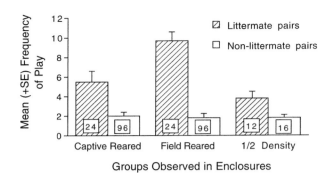

Figure 4. Evidence that juvenile Belding's ground squirrels prefer their littermates over non-littermates as play partners. (Littermates are young produced by the same mother and may be full or maternal half-siblings, which result when a female mates with multiple males.) In three groups, each observed for 10–14 days in an outdoor enclosure beginning on the day of juveniles' natal emergence, play-bout frequencies between littermates were significantly higher than between non-littermates (see Holmes, 1994, for actual statistical tests based on log-linear models). Captive-reared juveniles were born and raised to weaning in the laboratory by field-mated mothers before they were transferred to outdoor enclosures. Field-reared juveniles were live trapped in the field shortly after they first came above ground at about the time of weaning and then transferred to an enclosure with their mothers. Finally, the ½-density group was treated like the captive-reared group, but only two rather than four litters lived in an enclosure. Regardless of the preemergent rearing environment or density of juveniles in enclosures, littermates were preferred over non-littermates as social partners. The number of pairs observed are shown inside bars. (Adapted from Holmes, 1994, with permission. Copyright 1994 Academic Press.)

evidence of a littermate preference. In two intensively observed groups, I identified for each juvenile the agemate with which it played the most often and found that a littermate was the most frequent play partner for 91% of all juveniles, averaged across the two groups. This shows that the high frequency of littermate play is not simply due to a few pairs of especially active littermates.

In summary, play data from several groups of ground squirrels observed in outdoor enclosures and analyzed from different perspectives all demonstrate that juvenile *S. beldingi* prefer their littermates as social partners during juveniles' initial weeks above ground. Having described juveniles' kin preferences, I now turn to the developmental process that produces them.

THE ROLE OF EARLY EXPERIENCE AND THE DEVELOPMENT OF LITTERMATE PREFERENCES

The interactions that occur between parents and offspring and between siblings or littermates when females bear multiply sired litters are one of the richest sources of early environmental stimulation that mediate ontogeny in many vertebrates. Such interactions can have long-lasting effects via their influence on anatomy, physiology, and behavior (e.g., see the review by Laviola & Teranova, 1998). For example, young that share a common nest or burrow during early development interact repeatedly with each other and are likely to become familiar as individuals or as classes of individuals (e.g., rearingmates versus non-rearingmates). This "familiarity" is hypothesized to provide a foundation for later kin favoritism in many species (Alexander, 1990; Bekoff, 1981). In laboratory tests designed to assess food sharing in spiny mice, *Acomys cahirinus*, for example, pairs composed of siblings reared together display more cooperative feeding than pairs of unrelated young reared apart (Porter, Moore, & White, 1981). Similarly, in a huddling-preference study with cross-fostered young, unrelated spiny mice pups reared together huddled preferentially with each other, whereas siblings reared apart did not huddle preferentially with each other when they were reunited (Porter, Tepper, & White, 1981). Thus, spiny mouse kin preferences are mediated by the familiarity* that develops between rearingmates, which would be siblings in species-typical environments, rather than by some other correlate of genes identical by descent. When familiarity mediates preferential treatment of kin, it may be that familiar individuals (e.g., related rearingmates) are attracted to each other, but it may also be that unfamiliar individuals (e.g., unrelated non-rearingmates) are unattractive and thus aversive. In the eusocial wasp, *P. fuscatus*, for example (Figure 3), the ontogeny of kin preferences for nestmates entails the development of behavioral intolerance of unfamiliar non-nestmates rather than the behavioral tolerance of familiar nestmates (Gamboa, Reeve, Ferguson, & Wacker, 1986).

Interactions between rearingmates in a common burrow also direct the development of social preferences in juvenile *S. beldingi*, as was revealed when pairs of young were switched reciprocally between mothers (M) shortly after birth (e.g., two of M A's offspring, Aa and Ab, were fostered to M B and simultaneously two of M B's offspring, Ba and Bb, were fostered to M A). I cross-fostered 5-day-old pups between litters and later recorded play bouts between enclosure-housed juveniles (Holmes, 1990, explains why *S. beldingi* mothers readily accept alien young introduced into their nestbox). Familiar juveniles, those that shared a nestbox as pups after being cross-fostered, played together significantly more often during the first weeks above ground than unfamiliar juveniles, those that were reared as pups in different nestboxes (Figure 5). (See Holmes & Sherman, 1982, for related results from pups switched within a few hours of birth.)

The cross-fostering results (Figure 5) demonstrate that sharing a natal burrow prior to emerging above ground influences the development of social preferences in *S. beldingi* juveniles. In nature, an ontogenetic link between familiarity and social preferences for kin will be adaptive only to the degree that familiarity is accurately correlated with kinship. If, for example, multiple mating by females produces broods or litters of full and maternal-half siblings, all of which are reared together, then an ontogenetic process based *only* on a rearingmate versus non-rearingmate distinction (i.e., familiar versus unfamiliar) will produce a less than ideal outcome because individuals would treat full and maternal-half siblings as if they were equally related (but see Keller's, 1997, review on why this "less-than-ideal outcome" might be adaptive).

As we have seen (Figure 5), familiarity established between rearingmates does influence social preferences in *S. beldingi*, but something besides familiarity must be guiding social development because non-littermates reared together played significantly less often than littermates reared together (5.9 ± 0.9 versus 8.9 ± 1.0 bouts per pair, respectively). In addition, littermates reared apart played significantly more often than non-littermates reared apart (3.3 ± 0.4 versus 2.4 ± 0.3 bouts per pair, respectively; W. G. Holmes, unpublished data; also see the 18-day-old results in

---

*"Familiarity" in the kin recognition and nepotism literature is usually used as a descriptive term to indicate that individuals have associated directly with each other before their social discrimination abilities were examined. Later in the chapter, I discuss "familiarity" as an explanatory rather than a descriptive construct (see p. 308).

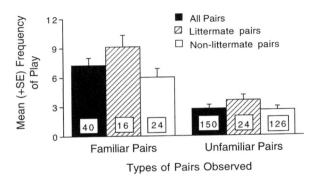

Figure 5. Evidence from one group of young ground squirrels cross-fostered between litters at 5 days of age that being reared together affects the development of social preferences. Pairs of individuals that were reared together after cross-fostering (familiar pairs) displayed more frequent bouts of play than pairs of individuals that were reared apart (unfamiliar pairs) whether members of a pair were related (littermates) or unrelated, which demonstrates an effect of "familiarity" on the development of juveniles' social preferences. There was also a "relatedness" effect on play-bout frequencies, which is explained in the text. The number of pairs observed is shown inside bars.

Figure 9). It may be that a juvenile's social preferences are influenced not only by familiarity but also by the phenotypic similarity between a juvenile and its potential social partners. This phenotypic-resemblance hypothesis derives from earlier research (Holmes, 1986) in which I cross-fostered newborns so that they were reared with different numbers of related (to them) and unrelated young. Results from subsequent paired-encounter tests of cross-fostered, *unfamiliar* females suggested that how one female treated another depended on the phenotypic similarity (probably olfactory) between the two individuals.

Although familiarity established during early ontogeny can have potent effects on social relationships, it is important to appreciate that familiarity *describes* the outcome of a developmental process, but it does not *explain* the outcome. For example, if an associative learning process is responsible for the "familiarity effect" in young ground squirrels (Figure 5), "familiarity" does not specify the instrumental contingencies that might shape pups' preferences for their littermates. How might "familiarity" serve as an explanatory construct rather than a descriptive label? Body temperature is one of the most potent modulators of mammalian infant behavior (Blumberg & Sokoloff, 1998), and I hypothesize that ground squirrel pups' efforts to meet their thermal needs is the first step in the developmental process that generates the "familiarity" which subsequently mediates juveniles' littermate preferences. My (untested) hypothesis is based on developmental studies by Alberts and his associates of huddling behavior in captive Norway rats (*Rattus norvegicus*).

The huddle is the early postnatal environment for pups in many species of polytocous mammals. It is a cluster of nestmates that are in continuous direct contact with each other and often with their mother. In a compelling series of studies on the development of affiliative bonds in infant rats, Alberts and his associates reported that pups actively participate in forming, maintaining, and regulating their huddle (Alberts, 1978). In doing so, pups experience thermotactile and olfactory stimulation that starts to produce filial preferences for the mother, and perhaps for nestmates, by allowing contingencies to be established between warmth and various odors encountered in the huddle (the nutritive rewards of suckling are not necessary to produce filial preferences, but may be sufficient; Brunjes & Alberts, 1979).

Filial preferences mediated by olfactory cues could originate in pups by "mere exposure" to the cues, a general, nonassociative effect that produces familiarity (Sluckin, 1964). However, for filial preferences to develop in rats, pups must be exposed to odors *in association with* thermotactile stimulation. Exposure to odors while isolated from other nest stimuli, including mother and nestmates, does not produce filial preferences. This verifies that filial preferences begin to develop via contingency-based learning and that simple familiarization does not explain their ontogenetic origins (Alberts & May, 1984). Like rat pups, *S. beldingi* neonates develop in a littermate huddle attended by their mother, and their thermoregulatory abilities develop in this context (Maxwell & Morton, 1975). Thus, it may be that littermate preferences in *S. beldingi* begin to develop due to the thermotactile and olfactory experiences that pups have when huddled together in their natal burrow. For rat pups, the associative learning process for filial preferences starts during the first 2 weeks after birth; for ground squirrel pups, I do not know when it begins. For both rats (Panksepp, Siviy, & Normansell, 1984) and ground squirrels (see above), subsequent interactions, especially social play in the periweaning period, contribute to the further development of juveniles' social preferences.

Maternal Effects and the Development of Kin Preferences

Due to either direct- or indirect-fitness effects, individual A might have a reproductive interest in the development of a social relationship between individuals B and C. For instance, if A were related to either B or C, A might attempt to facilitate or hinder social interactions between B and C (de Waal, 1996). In many vertebrates, "individual A" is likely to be a mother, given that mothers provide postnatal care for their young and in some species mother–offspring associations continue even after offspring can survive without direct maternal assistance (Clutton-Brock, 1991; Gubernick & Klopfer, 1981; Pereira & Fairbanks, 1993). When mothers are present during early ontogeny, they could influence the development of nepotism by facilitating or preventing their offsprings' interactions, as a function of the relatedness of potential social partners. For instance, Small and Smith (1981) recorded how often captive rhesus macaque, *Macaca mulatta*, mothers resisted attempts by other troop members to touch, grab or carry a mother's infant (collectively, "infant grabs"). Mothers are about twice as likely to resist attempted grabs initiated by an infant's paternal half-sibling than by age- and sex-matched juveniles that are unrelated to the infant, even though both classes of monkeys attempt infant grabs at similar rates. It is not known whether these maternal interventions affect infant social development, but other studies demonstrate a powerful maternal effect.

For example, In rhesus macaque groups on Cayo Santiago, mothers stay near their infants and allow or actively thwart attempts by other monkeys to interact with their infant (Berman, 1982). The result of maternal intervention is that the affiliative network of an infant comes to match its mother's, including the predominance of maternally related females in the network. de Waal (1996) proposes that young females copy the social preferences of their mothers through a process of cultural learning that depends on infants' access to various social partners, which is regulated by mothers (Berman, Rasmussen, & Suomi, 1997). Given the evidence that some monkeys can recognize not only their own relatives but also those of other group members (e.g., Cheney & Seyfarth, 1999), it would be an understatement to say that the development of affiliative relationships could be subject to a host of influences from the many players that are part of an infant's social environment. It may be easiest for investigators to detect direct interventions by adults into infants' developing social relationships, but indirect effects, even if they are hard to assess, may also be quite important. In cooperatively breeding white-fronted bee-eaters, *Merops bullockoides*, for example, fathers may actively disrupt attempts by their sons to disperse from their natal group and establish breeding relationships. The indirect consequence of father–son disruptions is that sons end up remaining in their natal group and developing caregiving relationships with their younger siblings (Emlen & Wrege, 1992).

Belding's ground squirrel mothers also figure prominently in the development of their offsprings' social relationships, as was revealed by a "mothers-absent" experiment. Holmes and Mateo (1998) allowed mothers to rear their offspring almost to weaning and then transferred 4 litters, each in its own nestbox, to outdoor enclosures, leaving mothers behind so that 16 juveniles occupied an enclosure. Juveniles living in the permanent absence of mothers did not display play-partner preferences for littermates (Figure 6): littermate and non-littermate play occurred equally often on a per-pair basis, although other aspects of juveniles' growth and behavioral development were not compromised by the absence of mothers. Because kin favoritism depends on kin recognition (see above section, Conditions Necessary for the Evolution of Nepotism), we wondered whether removing mothers had disrupted the development of littermate recognition, which could explain why littermates were not preferred play partners. The failed-recognition hypothesis predicts no differential treatment in any kind of social interaction, but this prediction was not supported because, in the absence of mothers, juveniles investigated (nose–body contact) their littermates more often than non-littermates, which verifies that they could discriminate between related and unrelated agemates (Figure 6; Holmes & Mateo, 1998).

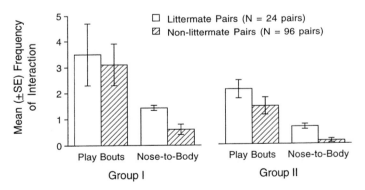

Figure 6. Evidence from two replicate groups of juvenile ground squirrels that littermates are not preferred social partners when mothers are permanently absent from the postweaning environment even though juveniles can still discriminate between littermates and non-littermates. That is, play-bout frequencies between littermates do not differ significantly from those between non-littermate pairs in either replicate group, whereas frequencies of nose-to-body contact do differ between littermates and non-littermates. These results suggest that mothers must be present after the natal emergence of their litter for their offspring to develop littermate preferences and that when mothers are absent, the lack of a littermate preference is not due to an inability to recognize littermates. (Adapted from Holmes & Mateo, 1998, with permission. Copyright 1998 Academic Press.)

Earlier in this chapter, I suggested that evidence for kin recognition was not necessarily evidence for kin favoritism because differential treatment (recognition) was not always reflected by preferential treatment (favoritism). The behavior of juvenile ground squirrels living without mothers provides empirical support for the distinction between recognition and favoritism. The fact that juveniles recognized their littermates but did not prefer them as social partners (Figure 6) points to different ontogenetic processes for these two outcomes. (Holmes and Mateo, 1998, explain why it is unlikely that littermate preferences had developed in juveniles that failed to express them because mothers were absent.)

How did *S. beldingi* mothers influence social development? In a study of maternal effects on the ontogeny of alarm-call responses, Mateo and Holmes (1997) distinguish between direct maternal effects, in which mothers actively orient their behavior toward their young, and indirect maternal effects, in which mothers' normal behavior inadvertently affects their young, that is, a mother's behavior is not contingent on the presence or activities of her young. As described earlier in this section, rhesus macaque mothers actively control who has physical access to their infants, which is an example of a direct maternal influence. Similarly, spotted hyena, *Crocuta crocuta*, mothers sometimes break up intensely aggressive interactions between their offspring and other agemates (Jenks, Weldele, Frank, & Glickman, 1995). In contrast, Holmes and Mateo (1998) found no evidence that ground squirrel mothers controlled social access to their offspring. For example, we never saw a mother chase away an unrelated juvenile that attempted to approach one of her offspring, we never saw a mother intervene in an ongoing juvenile–juvenile social interaction, causing its termination, nor did we ever see a mother try to sequester her young or otherwise control their movements so as to restrict social access. We were thus unable to discern any direct effect by mothers on the development of littermate preferences.

What our observations suggest is that *S. beldingi* mothers have an indirect effect on their juvenile offsprings' social development by how mothers' presence affects nocturnal exposure to non-littermates. In groups that include mothers in enclosures, about 85% of the average juvenile's nocturnal burrowmates are its littermates and, during the day, littermates are preferred play partners. However, in groups that lack mothers, only about 20% of nightime burrowmates are littermates, the percentage expected by chance, and littermates are not preferred play partners during the day, that is, littermate and non-littermate play occur equally often (Figure 7). Although there is a clear association between the presence/absence of mothers and the relatedness of nocturnal burrowmates, the proximate causes of this relationship remain elusive (Holmes & Mateo, 1998). For example, mothers when present do not actively determine where juveniles sleep by, for example, herding their offspring into a burrow or acting as gatekeepers, inspecting each juvenile as it retires for the night. Neither do mothers act as a "social magnet," drawing their offspring into a burrow when mothers retire for the night. Whatever constellation of factors determines nightime burrow use (e.g., prior familiarity with a burrow, the presence of mother and/or littermates), juveniles that inhabit a common burrow at night cannot help but associate directly with each other. Ground squirrels' social preferences depend on with whom they share a burrow prior to natal emergence (Figure 5) and, as I show below (see section, Temporal Aspects in the Development of Kin Preferences), their preferences are also affected by the identities of their post-emergent social companions. The presence of a recently-emergent juvenile's mother, even if she takes no direct action, ensures that the juvenile's primary social

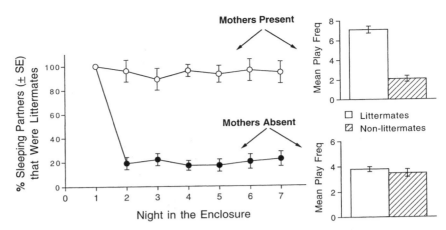

Figure 7. Evidence from ground squirrels that when mothers are permanently absent from their off-springs' postemergent environment, juveniles are much more likely to share burrows at night with non-littermates and fail to display play-partner preferences for their littermates. The sleeping partner data show for the average juvenile the percentage of its nighttime burrowmates that are littermates during the first 7 nights in an enclosure, depending on whether mothers were present in or permanently absent from enclosures. (Adapted from Holmes, 1994, with permission. Copyright 1994 Academic Press.)

companions, including nocturnal burrowmates, will be its littermates rather than non-littermates, which contributes to the consolidation of littermate preferences.

I believe that indirect maternal effects on the development of kin preferences are quite common in group-living species with extended maternal care and that these indirect effects will often be difficult to document due to their subtle nature. For example, vervet mothers occasionally intervene directly in their infant's play with other agemates. However, if a mother is simply near her offspring, her infant plays less often and terminates play bouts more often than if the mother is not nearby (Govindarajulu, Hunte, Vermeer, & Horrocks, 1993). Thus, the mere physical presence of mothers can influence social development (Pereira & Altmann, 1985) even if we cannot yet pinpoint how indirect maternal effects are mediated.

TEMPORAL ASPECTS IN THE DEVELOPMENT OF KIN PREFERENCES

WHEN ARE KIN PREFERENCES FIRST EXPRESSED? The defining attribute of a developing system is its always changing nature due, in part, to a series of recursive interactions between a developing organism and its environment. If developing systems continually change, then proximate explanations for ontogeny must address temporal issues. One of the most basic questions about developmental "timing" is when during ontogeny does an individual first display a particular behavior. For example, bank swallow, *Riparia riparia*, parents appear unable to discriminate between their own and alien young in the parents' nest until nestlings are 16–17 days of age (Beecher, Beecher, & Hahn, 1981; see Holmes, 1990, pp. 450–453, for a discussion of the difference between when parent–offspring recognition *is* first manifested from when it *could be* first manifested). In contrast to bank swallow parents, human mothers can discriminate shortly after birth between their infant's odor and that of another age- and sex-matched infant (Porter, Cernoch, & McLaughlin, 1983).

To determine when *free-living S. beldingi* first display littermate preferences, one would have to wait not only until natal emergence (recall that litters develop in isolated burrows), but an additional 3–5 days because this is when juveniles first encounter non-littermates (W. G. Holmes & J. M. Mateo, unpublished data). By observing *captive* juveniles, however, I was able to discover whether littermate preferences were already present at natal emergence because burrows in the 10-m² enclosures are close enough together that juveniles could encounter related and unrelated agemates on their first day above ground (day 1). Unfortunately, juveniles do not play often enough on day 1 to permit inferential statistical tests, but the pattern of play-bout frequencies during juveniles' initial days above ground demonstrates that they are biased to play preferentially with their littermates from day 1 (Figure 8; also see Holmes, 1997). Thus, it appears that by the time of their natal emergence juveniles have acquired a littermate preference as a result of repeated interactions between young that shared a natal burrow. I will amend this suggestion below.

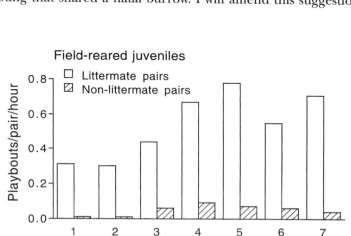

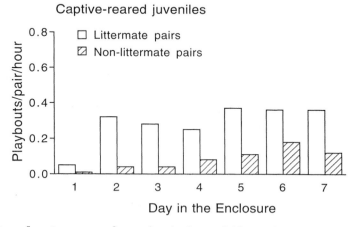

Figure 8. Evidence from two groups of ground squirrels, one field-reared and the other captive-reared, which suggests that juveniles prefer their littermates as play partners from the first day (day 1) juveniles emerge from their natal burrow. The mean values show rates of littermate and non-littermate play on each of the first 7 days that juveniles were observed in outdoor enclosures. Overall play rates differ between the two groups because young were reared under different environmental conditions and observed under different weather conditions.

WHEN DOES CRUCIAL EXPERIENCE OCCUR? Another fundamental question about the temporal nature of development is when during ontogeny do particular experiences occur that are necessary to produce the phenotype of interest (e.g., Immelmann & Suomi, 1981). In some species of birds, for example, filial preferences are critically dependent on experiences that occur shortly before the preferences are expressed by young birds (e.g., see Bolhuis's, 1991, review on imprinting); in contrast, sexual preferences depend on experiences that occur several months before the preferences are expressed, although sexual preferences may also be affected by adults' experiences (e.g., Kruijt & Meeuwissen, 1993). Because newly-emergent juvenile ground squirrels prefer their littermates as play partners on day 1 (Figure 8), it seems that preemergent rather than postemergent experiences may be largely responsible for these preferences. This raises the question of when the crucial experiences occur and how long they last. Although littermates share a burrow for up to 4 weeks days before natal emergence, in some instances relatively short-term social experience is more likely to channel behavioral development if it occurs at a particular point in time, the so-called "sensitive period," rather than at some other time (Bateson, 1979). To investigate sensitive periods, researchers often switch young between rearing environments, for instance, from one nest to another, when social sensitivity is hypothesized to be high (e.g., Beecher *et al.*, 1981; Holmes & Sherman, 1982; Kruijt & Meeuwissen, 1993).

In the development of *S. beldingi* littermate preferences, field data suggest that a sensitive period exists shortly before natal emergence (see Figure 6 in Holmes & Sherman, 1982). To examine this possibility with captive ground squirrels, I cross-fostered young 9 days before natal emergence (18 days of age) or 2 days before natal emergence (25 days of age). I reasoned that if experience just before natal emergence were crucial to the establishment of littermate preferences, then switching 25-day-old young (late cross-fostering) should compromise the development of littermate preferences, regardless of rearing history, whereas switching 18-day-old young (early cross-fostering) should not threaten social development. The cross-fostering results support my reasoning (Holmes, 1997). Early-fostered young display play-partner preferences that differ with pairs' relatedness (littermates or non-littermates) and pairs' postfostering rearing experience (reared together or reared apart). In contrast, late-fostered young play indiscriminately with respect to both relatedness and post-fostering rearing experience (Figure 9). It is possible that late fostering simply disrupts in a nonspecific fashion all aspects of development, but this seems unlikely because late-fostered young grow at the same rate as early-fostered young and display species-typical sex differences in play-bout frequencies (see below). Additional work is needed to characterize more fully how the temporal nature of rearing experience influences juvenile social development, but the cross-fostering results (Figure 9) suggest that preemergent young experience heightened social sensitivity just before their initial aboveground appearance (Holmes, 1997). The timing of this social sensitivity makes good functional sense. The sensory and motor systems of periemergent young are relatively well developed (Maxwell & Morton, 1975), which facilitates opportunities for littermates to interact and monitor the consequences of their interactions.

When considering how temporal variation in social sensitivity might affect the development of social preferences, including those for kin, one must appreciate that sensitivity may depend on both temporal *and* contextual features of the environment. A classic example of temporal variation in social sensitivity is the sensitive period for filial imprinting in precocial birds (Bateson, 1979). This developmental process is sometimes described as preprogrammed and endogenously driven, but, more often than not, it is one that is responsive to an organism's external environment (Bateson, 1981), especially the social makeup of the environment. For instance, in an imprinting study with mallards, *Anas platyrhynchos*, 24-hr old ducklings were briefly exposed to a model hen (or some other object) and 1 day later given a choice between the familiar model and an unfamiliar redhead duck, *Aythya americana*, model. If, after ducklings were exposed to the imprinting model, they were housed with their siblings, then, during subsequent tests, ducklings preferred the familiar model. However, if ducklings were housed alone after exposure to the imprinting model, then they did not prefer the familiar model during subsequent tests (Lickliter & Gottlieb, 1985). For ducklings to display species-typical imprinting, they had to interact with their siblings after they were exposed to the imprinting model, which demonstrates that temporal sensitivity and social context interact to generate a filial preference. In fact, posthatching social experience with siblings may actually interfere with the maintenance of an imprinted maternal preference unless ducklings have ongoing exposure to their mother (Lickliter & Gottlieb, 1986).

Let me return now to whether young *S. beldingi* emerge from their natal burrows with fully-developed littermate preferences. That littermate play is more frequent than non-littermate play beginning on day 1 above ground (Figure 8) suggests that the developmental process which generates littermate preferences has been completed by natal emergence. However, this interpretation is challenged by results from the "no-moms" experiment (see Figure 6). If littermate preferences were fully formed concomitant with natal emergence, then why did juveniles fail to display them when mothers were permanently absent? By manipulating juveniles' social experience on their initial days above ground, Holmes and Mateo (1998) found that the littermate bias with which juveniles emerge must be solidified by aboveground interactions with littermates to ensure an enduring preference. We permanently removed all mothers and restricted juveniles' movements by placing 1-m² wire-mesh covers over burrow entrances so that juveniles could interact only with their litter-

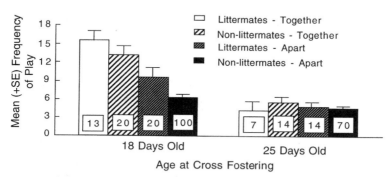

Figure 9. Evidence that social experience a few days before young first come above ground is crucial to the development of littermate preferences in juvenile ground squirrels. The mean values show the frequency of play bouts between juveniles during their first 20 days aboveground. Young were cross-fostered between litters at 18 days of age (about 9 days before natal-emergence age) or 25 days of age (about 2 days before natal-emergence age). Only juveniles in the former group display play-partner preferences. "Together" and "apart" refer to whether young occupied the same or different nestboxes, respectively, after cross-fostering. The number of pairs observed, shown inside bars, differs between groups because different numbers of juveniles were originally placed in enclosures. (Reprinted from Holmes, 1997, with permission. Copyright 1997 Academic Press.)

mates on days 1–3. Then we removed the wire-mesh covers and gave juveniles access to both littermates and non-littermates for 1 week. During this week, littermates played together significantly more often than non-littermates in the continued absence of mothers (Holmes & Mateo, 1998). The littermate-only experience on days 1–3 highlights the crucial nature of early aboveground interactions in the ontogeny of littermate preferences.

The potency of early aboveground interactions on *S. beldingi* social development is also reflected in juveniles' daily rates of play. Figure 8 shows that the ratio of littermate to non-littermate play, averaged over juveniles' first 7 days above ground, is more than three times higher for field-reared juveniles (11.5 littermate:non-littermate bouts, on average) than captive-reared juveniles (3.5 littermate:non-littermate bouts, on average). Field-reared young (the top panel in Figure 8) were not trapped and placed in an enclosure until juveniles had been above ground for 2 or 3 days, during which time they played frequently, but only with their littermates. In the field, play between newly-emergent juveniles is restricted largely to littermates due to juveniles' tendency to remain near their natal burrow, making littermates the only agemates available for social interactions (W. G. Holmes & J. M. Mateo, unpublished data). In contrast, captive-reared young on their initial days above ground were able to play with littermates and non-littermates both of which were spatially accessible due to burrow proximity in enclosures. Thus, the critical nature of early aboveground social experience (days 1–3) on the ontogeny of kin preferences in *S. beldingi* is reflected in the clear bias that field-reared juveniles had for littermates over non-littermate play partners. These juveniles, in contrast to captive-reared young, inhabited a "pure-kin" environment during their first few days aboveground before non-littermates were incorporated into their social world. It is worth noting that captive-reared and field-reared young differed in other ways (e.g., during rearing, only captive-reared young were handled to monitor growth rates, and captive-reared young weighed more than field-reared young at weaning) that may have influenced social development, although it is not clear how these differences might have affected littermate:non-littermate play ratios.

SEX DIFFERENCES IN THE DEVELOPMENT OF NEPOTISM

SEX DIFFERENCES IN NEPOTISM AND NATAL DISPERSAL.   In several species, members of one sex are much more likely to behave nepotistically than members of the other sex. In cercopithicene primates (macaques, baboons, and vervets), for example, females but not males provide social support to matrilineally related female kin during competitive interactions (Chapais, 1992). In some species of cooperatively breeding birds, the adult helpers that feed and protect their younger siblings (or half-siblings) are much more likely to be males than females (Brown, 1987; Cockburn, 1998). The most distinctly sex-limited examples of nepotism occur in eusocial insects (members of the order Hymenoptera, which include many species of ants, bees, and wasps) in which haplodiploid genetic sex determination (females are diploid and males haploid) results in unusually high degrees of relatedness between females (e.g., sister–sister $r = \frac{3}{4}$). In several of these species, females behave nepotistically in various ways, including foregoing direct reproduction and investing preferentially in their sisters, whereas males rarely behave nepotistically despite living among their close kin (reviewed in Crozier & Pamilo, 1996), although the link between genetic relatedness and nepotism in eusocial insects is not always clear (e.g., Breed, Welch, & Cruz, 1994).

A common proximate reason for sex-limited nepotism is the greater likelihood that members of one sex will disperse from their birthplace, the result of which is that only members of the philopatric sex have opportunities to interact with and assist their kin. Under the natal-dispersal hypothesis, nepotistic behavior should be more common in female mammals and male birds than in opposite-sexed individuals because female mammals and male birds are usually philopatric relative to opposite-sexed conspecifics that disperse (Smale *et al.*, 1997). There are, of course, examples, like lions, in which natal dispersal is sex-limited (Pusey & Packer, 1987) and yet adults of both sexes behave nepotistically to nondescendant kin (Packer *et al.*, 1991; but see Grinnell, Packer, & Pusey, 1995, on kinship in males). Thus, there is not a universal association between sex-limited dispersal and sex-limited nepotism (Cockburn, 1998).

There are at least 10 proximate hypotheses to explain why natal dispersal occurs in mammals (Holekamp, 1986; Smale *et al.*, 1997), but not all try to explain sex differences in dispersal tendencies. Moreover, those hypotheses that do grapple with why members of only one sex are sedentary do not necessarily try to explain sex-limited nepotism. Members of the philopatric sex may have opportunities to treat nondescendant kin nepotistically, but whether they do will depend, at the ultimate level, on the fitness consequences of nepotism and, at the proximate level, on the social relationships they have developed with each other. Just because members of the philopatric sex live among their nondescendant relatives it does not necessarily mean that they will treat them preferentially (e.g., Schwagmeyer, 1980). On the other hand, if some members of the sex that typically emigrates did not do so, then they might behave nepotistically if, at the proximate level, kin favoritism were mediated by a simple "availability-of-kin" rule (i.e., if kin are encountered and recognized, treat them nepotistically).

S. BELDINGI SEX DIFFERENCE IN LITTERMATE PREFERENCES. As I noted earlier, the nepotistic behavior of free-living Belding's ground squirrels, which includes antipredator alarm calls, territorial defense of natal burrows, and defense of unweaned young, is restricted to adult females, as it is in most sciurid rodents (Barash, 1989; Hoogland, 1995; Murie & Michener, 1984). If I am correct that social play both reflects and further contributes to cohesive relationships between related juveniles to influence adult nepotism, then one might expect that play-bout frequencies should vary with both the relatedness and the sex of the interactants. "If we assume ... that the social interactions of juveniles serve some developmental function for adult social behavior, then to the degree that there are sex differences in the adult social behavior ... so too should there be sex differences in the form of social play that contribute to the development of adult social behavior" (Meaney, Stewart, & Beatty, 1985, p. 34; but see Pellis *et al.*, 1997 for another view of sex differences in play). There are sex differences in *S. beldingi* play (Nunes *et al.*, 1999): male–male pairs play significantly more often than male–female and female–female pairs (Holmes, 1994). However, littermates play more often than non-littermates in all three sex-of-pair combinations. In fact, the largest within-sex disparity in play frequency occurs between male littermate and male non-littermate pairs (Figure 10). This also holds for the proportion of all social interactions that are scored as play (W. G. Holmes, unpublished data).

A developmental account of *S. beldingi* nepotism requires an explanation for its sex-limited nature, which is not obvious in the play-bout frequencies of juveniles since members of both the nepotistic (females) and nonnepotistic (males) sex

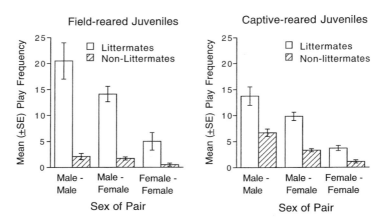

Figure 10. Evidence from two groups of enclosure-housed juvenile ground squirrels that both males and females prefer their littermates as play partners even though as adults only females behave nepotistically. The mean values show play-bout frequencies from a group of field-reared and a group of captive-reared young as a function of the sex of the interactants.

prefer littermates as play partners (Figure 10). However, I have quantified social play at the molar level, combining specific play motor patterns into play bouts, rather than quantifying the more molecular motor patterns themselves. When social play is dissected in greater detail, it may become necessary to recharacterize sex differences, including the underlying mechanisms responsible for them, as shown for sex differences in play fighting in laboratory rats (Pellis *et al.*, 1997).* Regardless of the mechanisms that produce sex differences in play, however, if the quality or quantity of play by *S. beldingi* juveniles does have important consequences for adult social behavior, then these consequences may differ for males and females (Nunes *et al.*, 1999). For females, for example, play may enhance the development of social relationships with littermates; for males, play may contribute to motor training and the acquisition of social skills that are important in adulthood when males compete intensively for mating opportunities (Meaney *et al.*, 1985). But to account for male play, this reasoning must be extended to explain why males play preferentially with their littermates. Littermates are more similar in age/experience and body weight than non-littermates, and this phenotypic similarity may make male littermates the most appropriate social partners if males are developing their competitive skills by engaging in frequent social play (e.g., Thompson, 1996).

I cannot yet forge a strong link between sex differences in juvenile play (Figure 10) and sex differences in adult social behavior, which means that I cannot yet offer a compelling *developmental* explanation for why adult female *S. beldingi* behave nepotistically and males do not. On the other hand, perhaps we already have a more complete developmental explanation for sex-limited nepotism than we realize: for juvenile male ground squirrels, natal dispersal precludes the kind of ongoing interactions with littermates that may be needed to sustain the social preferences which male juveniles manifest during their initial weeks above ground. There are prece-

*There are sex differences in *S. beldingi* social play. For example, males initiate play more often than females, males play more often than females (usually in male–male dyads), and males display some play motor patterns (e.g., mounting) more often than females. However, none of these differences provides a proximate insight into why females and not males behave nepotistically as adults.

dents in the developmental literature for the notion that appropriate stimulation in a particular context is necessary to sustain behavior over time in young animals. Weaning in laboratory rats, for example, usually begins at 15–17 days of age and typically ends by about 35 days of age (Thiels, Cramer, & Alberts, 1988). The maintenance of suckling during the weaning period depends, in part, on the act of suckling itself; suckling is sustained until 50–70 days of age if juveniles are housed with a succession of preweaning litters and their dams (Pfister, Crammer, & Blass, 1986). When juvenile male *S. beldingi* remain housed with littermates well past the usual age of natal dispersal, they continue to express littermate preferences (W. G. Holmes & J. M. Mateo, unpublished data). As I suggested above, a sex difference in ground squirrel nepotism may develop because of a sex difference in natal dispersal (males leave, females stay) that, under species-typically circumstances, denies males the kind of ongoing stimulation necessary to sustain the littermate preferences they express as young juveniles.

Concluding Comments

Competition, Not Cooperation, between Kin

My focus in this chapter has been on the development of cohesive social relationships between closely related nondescendant kin, and I have tried to link developmental explanations for early kin preferences with functional explanations for adult nepotism. Although our own close kin are frequent recipients of favoritism, anyone who has struggled with a recalcitrant child, a behind-the-times parent, or a know-it-all sibling is well aware that relatives can, at times, be trying. At the ultimate level of analysis, Hamilton's (1964) rule acknowledges that kin, no matter how closely related, should not *always* be treated beneficently. Mock and Parker (1997), for instance, document and explain examples from many taxa on the "dark side" of sibling relationships that often result in siblicide. What Hamiltonian logic suggests is that individuals are expected to have evolved to include not only genetic relatedness in their nonconscious social calculus, but also the direct-fitness benefits to recipients and direct-fitness costs to themselves before engaging in nepotistic acts. (At the proximate level, of course, such assessments are mediated by the immediate correlates of relatedness and the short-term currencies of reproductive benefits and costs.) For example, to explain why newborn piglets have specialized incisors that they use as weapons in neonatal competition with littermates (Fraser & Thompson, 1991) or why female prairie dogs often kill their own younger siblings, nieces, and nephews (Hoogland, 1985), one must consider reproductive benefits and costs and not just degrees of relatedness. As Mock and Parker (1998, p. 1) explain, whenever close kin experience acute competition for the same limited resource "natural selection can be expected to have forged some rather nasty behavioural resolutions." The ontogeny of sibling competition warrants just as much scrutiny as does the ontogeny of sibling favoritism.

Development and the Conditional Nature of Social Relationships

In my developmental analysis of nepotism, I have focused on the ontogeny of amicable or cohesive relationships between relatives rather than on the ontogeny of nepotistic relationships per se. I have done this because, in a Darwinian framework, favoritism to kin should be a conditional rather than a universal feature of an individual's behavior, as noted in the prior paragraph. This perspective forces us to realize that relatives will sometimes, but not always, be targets for favoritism. Accordingly, it would be misleading to label a relationship as "nepotistic" if two relatives cooperated some of the time and competed selfishly at other times. Perhaps the relationships that often develop between kin should be labeled "strategically nepotistic" to stress their conditional nature.

Another perspective which I have advocated in this chapter is that kin favoritism emerges from a developmental history rather than springing up, fullblown, overnight. In many cases, individuals must interact with each other over time to assess and shape a social relationship that subsequently may include nepotistic behavior. In the ground squirrels that I have been studying, for example, nepotism occurs between adults and not juveniles, but, in accord with Michener's (1983) early ideas, I hypothesize that adult nepotism is grounded in juvenile experience and the social relationships that develop between related young. Under these circumstances, which I believe characterize many vertebrates, one must work to understand a *developmental process* in juveniles to explain a *developmental outcome*, nepotistic behavior, in adults. In this light, I must emphasize that in this chapter I have documented the social preferences of *juvenile* ground squirrels during their first few weeks after natal emergence. Although we know that the littermate preferences of captive juveniles remain intact at the end of the summer (W. G. Holmes & J. M. Mateo, unpublished data), we have yet to determine whether these juvenile preferences are linked directly to adult nepotism.

At the beginning of this chapter, I alluded to the Hero Fund created by Andrew Carnegie to encourage acts of heroism in the face of life-threatening emergencies (Johnson, 1996). For evolutionary scientists, inclusive-fitness logic offers a functional or ultimate-level explanation for why a hero who risks his or her life to save a close relative is excluded from the Hero Fund: self-sacrifice for a relative can enhance a hero's inclusive fitness and thus generates its own genetic reward. For developmental scientists and for investigators who want to link developmental and functional explanations, much work remains to be done to explain the ontogeny of nepotism, including why we routinely expect assistance from our close relatives, and why, in turn, we stand ready to assist them.

Acknowledgments

First and foremost, I wish to thank the group of dedicated, hard-working undergraduate assistants who contributed enormously to my research on Belding's ground squirrels, including J. Ashley, A. Bell, M. Blocker, L. Cablk, M. Clancy, J. Davis, D. King, J. Osborn, M. Palopoli, S. Speizer, L. Starr, and M. Wolf. I conducted my *S. beldingi* research at the Sierra Aquatic Research Laboratory and want to extend my deep thanks to D. Dawson, S. Roripaw, and other staff members at SNARL for all the help and expertise they provided. Drafts of this chapter were improved considerably by suggestions from J. Mateo, J. Mitani, and members of the latter's behavioral ecology seminar. I thank E. Blass for his incisive comments on various developmental issues that I consider in this chapter and for identifying examples which show that many developmental processes seem to be designed to constrain and direct phenotypes toward adaptive ends. My research was supported by funds from the National Institute of Mental Health (MH 43861) and two units at the University of Michigan: The Office of the Vice President for Research and the Horace H. Rackham School of Graduate Studies.

Alberts, J. R. (1978). Huddling by rat pups: Group behavioral mechanisms of temperature regulation and energy conservation. *Journal of Comparative and Physiological Psychology, 92*, 231–245.

Alberts, J. R., & Cramer, C. P. (1988). Ecology and experience: Sources of means and meaning of developmental change. In E. M. Blass (Ed.), *Handbook of behavioral neurobiology*, Volume 9, *Developmental psychobiology and behavioral ecology* (pp. 1–39). New York: Plenum Press.

Alberts, J. R., & May, B. (1984). Nonnutritive, thermotactile induction of filial huddling in rat pups. *Developmental Psychobiology, 17*, 161–181.

Alcock, J., & Sherman, P. (1994). The utility of the proximate–ultimate dichotomy in ethology. *Ethology, 96*, 58–62.

Alexander, R. D. (1979). *Darwinism and human affairs*. Seattle, WA: University of Washington Press.

Alexander, R. D. (1990). Epigenetic rules and Darwinian algorithms. *Ethology and Sociobiology, 11*, 241–303.

Alexander, R. D., & Tinkle, D. W. (Eds.). (1981). *Natural selection and social behavior*. New York: Chiron Press.

Antonovics, J., & van Tienderen, P. H. (1991). Ontoecogenophyloconstraints? The chaos of constraint terminology. *Trends in Ecology and Evolution, 6*, 166–168.

Barash, D. P. (1989). *Marmots: Social behavior and ecology*. Stanford, CA: Stanford University Press.

Bateson, P. (1979). How do sensitive periods arise and what are they for? *Animal Behaviour, 27*, 470–486.

Bateson, P. (1981). Control of sensitivity to the environment during development. In K. Immelmann, G. W. Barlow, L. Petrinovich, & M. Main (Eds.), *Behavioral development* (pp. 432–453). Cambridge: Cambridge University Press.

Bateson, P. (Ed.). (1983). *Mate choice*. Cambridge: Cambridge University Press.

Beecher, M. D. (1981). Development of parent–offspring recognition in birds. In R. K. Aslin, J. R. Alberts, & M. R. Petersen (Eds.), *Development of perception* (Vol. 1, pp. 45–65). Orlando, FL: Academic Press.

Beecher, M. D. (1991). Successes and failures of parent–offspring recognition in animals. In P. G. Hepper (Ed.), *Kin recognition* (pp. 94–124). Cambridge: Cambridge University Press.

Beecher, M. D., Beecher, I. M., & Hahn, S. (1981). Parent–offspring recognition in bank swallows (*Riparia riparia*): II. Development and acoustic basis. *Animal Behaviour, 29*, 95–101.

Bekoff, M. (1977). Mammalian dispersal and the ontogeny of individual behavioral phenotypes. *American Naturalist, 111*, 715–732.

Bekoff, M. (1981). Mammalian sibling interactions: Genes, facilitative environments, and the coefficient of familiarity. In D. J. Gubernick & P. H. Klopfer (Eds.), *Parental care in mammals* (pp. 307–346). New York: Plenum Press.

Bekoff, M., & Byers, J. A. (1981). A critical reanalysis of the ontogeny and phylogeny of mammalian social and locomotor play: An ethological hornet's nest. In K. Immelmann, G. W. Barlow, L. Petrinovich, & M. Main (Eds.), *Behavioral development* (pp. 296–337). Cambridge: Cambridge University Press.

Bekoff, M., & Byers, J. A. (1998). *Animal play: Evolutionary, comparative, and ecological perspectives*. Cambridge: Cambridge University Press.

Bennett, B. (1987). Measures of relatedness. *Ethology, 74*, 219–236.

Berman, C. M. (1982). The ontogeny of social relationship with group companions among free-ranging infant rhesus monkeys I. Social networks and differentiation. *Animal Behaviour, 30*, 149–162.

Berman, C. M., Rasmussen, K. L. R., & Suomi, S. J. (1997). Group size, infant development and social networks in free-ranging rhesus monkeys. *Animal Behaviour, 53*, 405–421.

Blass, E. M. (1995). Ontogeny of ingestive behavior. In A. Morrison & S. Fluharty (Eds.), *Progress in psychobiology and physiological psychology* (Vol. 16, pp. 1–51). New York: Academic Press.

Blass, E. M. (1999). The ontogeny of human face recognition: Orogustatory, visual and social influences. In: Rochat, P. (Ed.), *Early social cognition* (pp. 35–65). Hillsdale, NJ: Erlbaum.

Blumberg, M. S., & Sokoloff, G. (1998). Thermoregulatory competence and behavioral expression in the young of altricial species—revisited. *Developmental Psychobiology, 33*, 107–123.

Blumstein, D. T., & Armitage, K. B. (1997). Alarm calling in yellow-bellied marmots: I. The meaning of situationally variable alarm calls. *Animal Behaviour, 53*, 143–171.

Blumstein, D. T., & Armitage, K. B. (1998). Why do yellow-bellied marmots call? Animal Behaviour, 56, 1053–1055.

Blumstein, D. T., Steinmetz, J., Armitage, K. B., & Daniel, J. C. (1997). Alarm calling in yellow-bellied marmots: II. The importance of direct fitness. *Animal Behaviour, 53*, 173–184.

Bolhuis, J. J. (1991). Mechanisms of avian imprinting: A review. *Biological Reviews, 66*, 303–345.

Breed, M. D., Welch, C. K., & Cruz, R. C. (1994). Kin discrimination within honey bee (*Apis mellifera*) colonies: An analysis of the evidence. *Behavioral Processes, 33*, 25–40.

Brown, J. L. (1980). Fitness in complex avian social systems. In H. Markl (Ed.), *Evolution of social behavior: Hypotheses and empirical tests* (pp. 115–128). Weinheim, Germany: Verlag Chemie.

Brown, J. L. (1987). *Helping and communal breeding in birds*. Princeton, NJ: Princeton University Press.

Brunjes, P. C., & Alberts, J. R. (1979). Olfactory stimulation induces filial preferences for huddling in rat pups. *Journal of Comparative and Physiological Psychology, 93*, 548–555.

Burnstein, E., Crandall, C., & Kitayama, S. (1994). Some neo-Darwinian decision rules for altruism: Weighing cues for inclusive fitness as a function of the biological importance of the decision. *Journal of Personality and Social Psychology, 67*, 773–789.

Carnegie. (2000). Carnegie Hero Fund Commission website, http://www.carnegiehero.org/history. shtml, revision 19 December 2000.

Chapais, B. (1992). The role of alliances in social inheritance of rank among female primates. In A. Harcourt & F. de Waal (Eds.), *Coalitions and alliances in humans and other animals* (pp. 29–59). New York: Oxford University Press.

Cheney, D. L., & Seyfarth, R. M. (1999). Recognition of other individuals' social relationships by female baboons. *Animal Behaviour, 58*, 67–75.

Clutton-Brock, T. (1991). *The evolution of parental care*. Princeton, NJ: Princeton University Press.

Cockburn, A. (1998). Evolution of helping behavior in cooperatively breeding birds. *Annual Review of Ecology and Systematics, 29*, 141–177.

Crozier, R. H., & Pamilo, P. (1996). *Evolution of social insect colonies*. Oxford: Oxford University Press.

Darwin, C. (1859). *On the origin of species*. London: Murray.

Dawkins, M. S. (1995). *Unravelling animal behaviour* (2nd ed.). Essex, England: Longman Scientific & Technical.

Dawkins, R. (1979). Twelve misunderstandings of kin selection. *Zeitschrift für Tierpsychologie, 51*, 184–200.

Dawkins, R. (1982). *The extended phenotype*. Oxford: Freeman.

Dawkins, R. (1989). *The selfish gene*, 2nd ed. New York: Oxford University Press.

Dewsbury, D. A. (1994). On the utility of the proximate–ultimate distinction in the study of animal behavior. *Ethology, 96*, 63–68.

Dunford, C. (1977). Social system of round-tailed ground squirrels. *Animal Behaviour, 25*, 885–906.

Emlen, S. T. (1991). Evolution of cooperative breeding in birds and mammals. In J. R. Krebs & N. B. Davies (Eds.), *Behavioural ecology*, 3rd ed. (pp. 301–337). London: Blackwell Scientific.

Emlen, S. T., & Wrege, P. H. (1992). Parent–offspring conflict and the recruitment of helpers among bee-eaters. *Nature, 356*, 331–333.

Fagen, R. (1981). *Animal play behavior*. New York: Oxford University Press.

Fairbanks, L. A. (1990). Reciprocal benefits of allomothering for female vervet monkeys. *Animal Behaviour, 40*, 553–562.

Fletcher, D. J. C., & Michener, C. D. (1987). *Kin recognition in animals*. Chichester, England: Wiley.

Fraser, D., & Thompson, B. K. (1991). Armed sibling rivalry among suckling piglets. *Behavioral Ecology and Sociobiology, 29*, 9–15.

Freeberg, F. M. (1996). Assortative mating in captive cowbirds is predicted by social experience. *Animal Behaviour, 52*, 1129–1142.

Gamboa, G. J., Reeve, H. K., Ferguson, I. D., & Wacker, T. L. (1986). Nestmate recognition in social wasps: The origin and acquisition of recognition odours. *Animal Behaviour, 34*, 685–695.

Gamboa, G. J., Foster, R. L., Scope, J. A., & Bitterman, A. M. (1991). Effects of stage of colony cycle, context, and intercolony distance on conspecific tolerance by paper wasps (*Polistes fuscatus*). *Behavioral Ecology and Sociobiology, 29*, 87–94.

Gamboa, G. J., Reeve, H. K., & Holmes, W. G. (1991). Conceptual issues and methodology in kin-recognition research: A critical discussion. *Ethology, 88*, 109–127.

Gordon, D. M. (1992). Phenotypic plasticity. In E. Fox Keller & E. A. Lloyd (Eds.), *Keywords in evolutionary biology* (pp. 255–262). Cambridge, MA: Harvard University Press.

Gottlieb, G. (1992). *Individual development and evolution*. New York: Oxford University Press.

Govindarajulu, P., Hunte, W., Vermeer, L. A., & Horrocks, J. A. (1993). The ontogeny of social play in a feral troop of vervet monkeys (*Cercopithecus aethiops sabaeus*): The function of early play. *International Journal of Primatology, 14*, 701–719.

Grafen, A. (1984). Natural selection, kin selection and group selection. In J. R. Krebs & N. B. Davies (Eds.), *Behavioural ecology*, 2nd ed. (pp. 62–84). Sunderland, MA: Sinauer.

Grinnell, J., Packer, C., & Pusey, A. E. (1995). Cooperation in male lions: Kinship, reciprocity or mutualism? *Animal Behaviour, 49*, 95–105.

Gubernick, D. J., & Klopfer, P. H. (Eds.). (1981). *Parental care in mammals*. New York: Plenum Press.

Hailman, J. P., McGowan, K. J., & Woolfenden, G. E. (1994). Role of helpers in the sentinel behavior of the florida scrub jay (*Aphelocoma c. coerulescens*). *Ethology, 97*, 119–140.

Hamilton, W. D. (1964). The genetical evolution of social behaviour. *Journal of Theoretical Biology, 7,* 1–52.

Hanken, J., & Sherman, P. W. (1981). Multiple paternity in Belding's ground squirrel litters. *Science, 212,* 351–353.

Hauber, M. E., & Sherman, P. W. (1998). Nepotism and marmot alarm calling. *Animal Behaviour, 56,* 1049–1052.

Heth, G., Todrank, J., & Johnston, R. E. (1998). Kin recognition in golden hamsters: Evidence for phenotype matching. *Animal Behaviour, 56,* 409–417.

Hinde, R. A. (1979). What do we mean by a relationship? In R. A. Hinde (Ed.), *Towards understanding relationships* (pp. 14–39). London: Academic Press.

Hinde, R. A. (Ed.). (1983). *Primate social relationships.* Oxford: Blackwell Scientific.

Holekamp, K. E. (1983). *Proximal mechanisms of natal dispersal in Belding's ground squirrel (Spermophilus beldingi beldingi).* Ph.D. thesis, University of California, Berkeley.

Holekamp, K. E. (1986). Proximal causes of natal dispersal in Belding's ground squirrels (*Spermophilus beldingi*). *Ecological Monographs, 56,* 365–391.

Holmes, W. G. (1984). Sibling recognition in thirteen-lined ground squirrels: Effects of genetic relatedness, rearing association, and olfaction. *Behavioral Ecology and Sociobiology, 14,* 225–233.

Holmes, W. G. (1986). Kin recognition by phenotype matching in female Belding's ground squirrels. *Animal Behaviour, 34,* 38–47.

Holmes, W. G. (1988). Kinship and the development of social preferences. In E. M. Blass (Ed.), *Handbook of behavioral neurobiology,* Volume 9, *Developmental psychobiology and behavioral ecology* (pp. 389–413). New York: Plenum Press.

Holmes, W. G. (1990). Parent–offspring recognition in mammals: A proximate and ultimate perspective. In N. A. Krasnegor & R. S. Bridges (Eds.), *Mammalian parenting* (pp. 441–460). New York: Oxford University Press.

Holmes, W. G. (1994). The development of littermate preferences in juvenile Belding's ground squirrels. *Animal Behaviour, 48,* 1071–1084.

Holmes, W. G. (1995). The ontogeny of littermate preferences in juvenile golden-mantled ground squirrels: Effects of rearing and relatedness. *Animal Behaviour, 50,* 309–322.

Holmes, W. G. (1997). Temporal aspects in the development of Belding's ground squirrels' litter-mate preferences. *Animal Behaviour, 53,* 1323–1336.

Holmes, W. G., & Mateo, J. M. (1998). How mothers influence the development of littermate preferences in Belding's ground squirrels. *Animal Behaviour, 55,* 1555–1570.

Holmes, W. G., & Sherman, P. W. (1982). The ontogeny of kin recognition in two species of ground squirrels. *American Zoologist, 22,* 491–517.

Holmes, W. G., & Sherman, P. W. (1983). Kin recognition in animals. *American Scientist, 71,* 46–55.

Hoogland, J. L. (1983). Nepotism and alarm calling in the black-tailed prairie dog (*Cynomys ludovicianus*). *Animal Behaviour, 31,* 472–479.

Hoogland, J. L. (1985). Infanticide in prairie dogs: Lactating females kill offspring of close kin. *Science, 230,* 1037–1040.

Hoogland, J. L. (1995). *The black-tailed prairie dog.* Chicago: University of Chicago Press.

Immelmann, K., & Suomi, S. J. (1981). Sensitive phases in development. In K. Immelmann, G. W. Barlow, L. Petrinovich, & M. Main (Eds.), *Behavioral development: The Bielefeld Interdisciplinary Project* (pp. 395–430). Cambridge: Cambridge University Press.

Jenkins, S. H., & Eshelman, B. D. (1984). *Spermophilus beldingi. Mammalian Species, 221,* 1–8.

Jenks, S. M., Weldele, M. L., Frank, L. G., & Glickman, S. E. (1995). Acquisition of matrilineal rank in captive spotted hyaenas: Emergence of a natural social system in peer-reared animals and their offspring. *Animal Behaviour, 50,* 893–904.

Johnson, R. C. (1996). Attributes of Carnegie medalists performing acts of heroism and of the recipients of these acts. *Ethology and Sociobiology, 17,* 355–362.

Keller, L. (1997). Indiscriminate altruism: Unduly nice parents and siblings. *Trends in Ecology and Evolution, 12,* 99–103.

Kempenaers, B., & Sheldon, B. C. (1996). Why do male birds not discriminate between their own and extra-pair offspring? *Animal Behaviour, 51,* 1165–1173.

Koenig, W. D., & Mumme, R. L. (1987). *Population ecology of the cooperatively breeding acorn woodpecker.* Princeton, NJ: Princeton University Press.

Komdeur, J., & Hatchwell, B. J. (1999). Kin recognition: Function and mechanism in avian societies. *Trends in Ecology and Evolutionary Biology, 14,* 237–241.

Krasnegor, N. A., & Bridges, R. S. (Eds.). (1990). *Mammalian parenting.* New York: Oxford University Press.

Kruijt, J. P., & Meeuwissen, G. B. (1993). Consolidation and modification of sexual preferences in adult male zebra finches. *Netherlands Journal of Zoology, 43,* 68–79.

Kummer, H. (1978). On the value of social relationships to non-human primates: A heuristic scheme. *Social Science Information, 17*, 687–705.

Lacey, E. A., & Sherman, P. W. (1997). Cooperative breeding in naked mole-rats: Implications for vertebrate and invertebrate sociality. In N. G. Solomon & J. A. French (Eds.), *Cooperative breeding in mammals* (pp. 267–301). Cambridge: Cambridge University Press.

Lacy, R. C., & Sherman, P. W. (1983). Kin recognition by phenotype matching. *American Naturalist, 121*, 489–512.

Laviola, G., & Terranova, M. L. (1998). The developmental psychobiology of behavioural plasticity in mice: The role of social experiences in the family unit. *Neuroscience and Biobehavioral Reviews, 23*, 197–213.

Lecanuet, J.-P., Fifer, W. P., Krasnegor, W. P. F., & Smotherman, W. P. (Eds.). (1995). *Fetal development: A psychobiological perspective.* Hillsdale, NJ: Erlbaum.

Lee, P. C. (1983). Play as a means for developing relationships. In R. A. Hinde (Ed.), *Primate social relationships* (pp. 82–89). Oxford: Blackwell Scientific.

Lee, P. C. (1987). Allomothering among African elephants. *Animal Behaviour, 35*, 278–291.

Lehrman, D. S. (1970). Semantic and conceptual issues in the nature–nurture problem. In L. R. Aronson, E. Tobach, D. S. Lehrman, & J. S. Rosenblatt (Eds.), *Development and evolution of behavior* (pp. 17–52). San Francisco: Freeman.

Leti, G. (1669). *Il nipotismo di Roma, or, The history of the popes nephews: From the time of Sixtus IV, anno 1471, to the death of the late Pope Alexander VII, anno 1667.* London: John Starkey.

Lickliter, R., & Gottlieb, G. (1985). Social interactions with siblings is necessary for the visual imprinting of species-specific maternal preference in ducklings. *Journal of Comparative Psychology, 99*, 371–378.

Lickliter, R., & Gottlieb, G. (1986). Training ducklings in broods interferes with maternal imprinting. *Developmental Psychobiology, 19*, 555–566.

Loeschcke, V. (Ed.). (1987). *Genetic constraints on adaptive evolution.* Berlin: Springer-Verlag.

Manson, J. H. (1999). Infant handling in wild *Cebus capucinus*: Testing bonds between females? *Animal Behaviour, 57*, 911–921.

Martin, P., & Caro, T. M. (1985). On the functions of play and its role in behavioral development. In J. S. Rosenblatt, C. Beer, M.-C. Busnel, & P. J. B. Slater (Eds.), *Advances in the study of behavior* (Vol. 15, pp. 59–103). New York: Academic Press.

Mateo, J. M., & Holmes, W. G. (1997). Development of alarm-call responses in Belding's ground squirrels: The role of dams. *Animal Behaviour, 54*, 509–524.

Maxwell, C. S., & Morton, M. L. (1975). Comparative thermoregulatory capabilities of neonatal ground squirrels. *Journal of Mammalogy, 56*, 821–828.

Maynard Smith, J. (1964). Group selection and kin selection. *Nature, 201*, 1145–1147.

Maynard Smith, J., Burian, R., Kauffman, S., Alberch, P., Campbell, J., Goodwin, B., Lande, R., Raup, D., & Wolpert, L. (1985). Developmental constraints and evolution. *Quarterly Review of Biology, 60*, 265–287.

Meaney, M. J., Stewart, J., & Beatty, W. W. (1985). Sex differences in social play: The socialization of sex roles. In J. S. Rosenblatt, C. Beer, M.-C. Busnel, & P. J. B. Slater (Eds.), *Advances in the study of behavior* (Vol. 15, pp. 1–58). New York: Academic Press.

Michener, G. R. (1981). Ontogeny of spatial relationships and social behaviour in juvenile Richardson's ground squirrels. *Canadian Journal of Zoology, 59*, 1666–1676.

Michener, G. R. (1983). Kin identification, matriarchies, and the evolution of sociality in ground-dwelling sciurids. In J. F. Eisenberg & D. G. Kleiman (Eds.), *Advances in the study of mammalian behavior* (pp. 528–572). Shippensburg, PA: American Society of Mammalogists.

Mock, D. W., & Parker, G. A. (1997). *The evolution of sibling rivalry.* Oxford: Oxford University Press.

Mock, D. W., & Parker, G. A. (1998). Siblicide, family conflict and the evolutionary limits of selfishness. *Animal Behaviour, 56*, 1–10.

Morton, M. L., & Gallup, J. S. (1975). Reproductive cycle of the Belding ground squirrel (*Spermophilus beldingi beldingi*): Seasonal and age differences. *Great Basin Naturalist, 35*, 427–433.

Mumme, R. L., & Koenig, W. D. (1991). Explanations for avian helping behavior. *Trends in Ecology and Evolution, 6*, 343–344.

Murie, J. O., & Michener, G. R. (1984). *The biology of ground-dwelling squirrels.* Lincoln: University of Nebraska Press.

Nijhout, H. F., & Emlen, D. J. (1998). Competition among body parts in the development and evolution of insect morphology. *Proceedings of the National Academy of Sciences of the USA, 95*, 3685–3689.

Nunes, S., Muecke, E. M., Anthony, J. A., & Batterbee, A. S. (1999). Endocrine and energetic mediation of play behavior in free-living Belding's ground squirrels. *Hormones and Behavior, 36*, 153–165.

O'Gara, B. (1969). Unique aspects of reproduction in the female pronghorn (*Antilocapra americana*). *American Journal of Anatomy, 125*, 217–232.

Packer, C., Gilbert, D. A., Pusey, A. E., & O'Brien, S. J. (1991). A molecular genetic analysis of kinship and cooperation in African lions. *Nature, 351,* 562–565.

Parker, P. G., Waite, T. A., & Decker, M. D. (1995). Kinship and association in communally roosting black vultures. *Animal Behaviour, 49,* 395–401.

Panksepp, J., Siviy, S., & Normansell, L. A. (1984). The psychobiology of play: theoretical and methodological perspectives. *Neuroscience and Biobehavioral Reviews, 8,* 465–492.

Pellis, S. M., Field, E. F., Smith, L. K., & Pellis, V. C. (1997). Multiple differences in the play fighting of male and female rats. Implications for the causes and functions of play. *Neuroscience and Biobehavioral Reviews, 21,* 105–120.

Pellis, S. M., & Iwaniuk, A. N. (1999). The roles of phylogeny and sociality in the evolution of social play in muroid rodents. *Animal Behaviour, 58,* 361–373.

Pereira, M. E. (1995). Development and social dominance among group-living primates. *American Journal of Primatology, 37,* 143–175.

Pereira, M. E., & Altmann, J. (1985). Development of social behavior in free-living nonhuman primates. In E. S. Watt (Ed.), *Nonhuman primate models for human growth and development* (pp. 217–309). New York: Liss.

Pereira, M. F., & Fairbanks, L. A. (Eds.). (1993). *Juvenile primates.* New York: Oxford University Press.

Pfenning, D. W., Sherman, P. W., & Collins, J. P. (1994). Kin recognition and cannibalism in polyphenic salamanders. *Behavioral Ecology, 5,* 225–232.

Pfister, J. F., Cramer, C. P., & Blass, E. M. (1986). Suckling in rats extended by continuous living with dams and their preweanling litters. *Animal Behaviour, 34,* 415–420.

Poirier, F. E., & Smith, E. O. (1974). Socializing functions of primate play. *American Zoologist, 14,* 275–287.

Porter, R. H., Moore, J. D., & White, D. M. (1981). Food sharing by sibling vs nonsibling spiny mice (*Acomys cahirinus*). *Behavioral Ecology and Sociobiology, 8,* 207–212.

Porter, R. H., Tepper, V. J., & White, D. M. (1981). Experiential influences on the development of huddling preferences and "sibling" recognition in spiny mice. *Developmental Psychobiology, 14,* 375–382.

Porter, R. H., Cernoch, J. M., & McLaughlin, F. J. (1983). Maternal recognition of neonates through olfactory cues. *Physiology and Behavior, 30,* 151–154.

Pusey, A. E., & Packer, C. (1987). The evolution of sex-biased dispersal in lions. *Behaviour, 101,* 275–310.

Robinson, S. R., & Smotherman, W. P. (1991). Fetal learning: Implications for the development of kin recognition. In P. G. Hepper (Ed.), *Kin recognition* (pp. 308–334). Cambridge: Cambridge University Press.

Rose, M. R., & Lauder, G. V. (Eds.). (1996). *Adaptation.* San Diego, CA: Academic Press.

Rubenstein, D. I., & Wrangham, R. W. (Eds.). (1986). *Ecological aspects of social evolution.* Princeton, NJ: Princeton University Press.

Scheiner, S. M. (1993). Genetics and evolution of phenotypic plasticity. *Annual Review of Ecology and Systematics, 24,* 35–68.

Schwagmeyer, P. L. (1980). Alarm calling behavior of the thirteen-lined ground squirrel, *Spermophilus tridecemlineatus. Behavioral Ecology and Sociobiology, 7,* 195–200.

Schwagmeyer, P. L. (1990). Ground squirrel reproductive behavior and mating competition: A comparative perspective. In D. A. Dewsbury (Ed.), *Contemporary issues in comparative psychology* (pp. 175–196). Sunderland, MA: Sinauer.

Sherman, P. W. (1977). Nepotism and the evolution of alarm calls. *Science, 197,* 1246–1253.

Sherman, P. W. (1980a). The meaning of nepotism. *American Naturalist, 116,* 604–606.

Sherman, P. W. (1980b). The limits of ground squirrel nepotism. In G. W. Barlow & J. Silverberg (Eds.), *Sociobiology: Beyond nature/nurture?* (pp. 505–544). Boulder, CO: Westview Press.

Sherman, P. W. (1981a). Reproductive competition and infanticide in Belding's ground squirrels and other animals. In R. D. Alexander & D. W. Tinkle (Eds.), *Natural selection and social behavior* (pp. 311–331). New York: Chiron Press.

Sherman, P. W. (1981b). Kinship, demography, and Belding's ground squirrel nepotism. *Behavioral Ecology and Sociobiology, 8,* 251–259.

Sherman, P. W. (1985). Alarm calls of Belding's ground squirrels to aerial predators: Nepotism or self-preservation? *Behavioral Ecology and Sociobiology, 17,* 313–323.

Sherman, P. W., & Morton, M. L. (1984). Demography of Belding's ground squirrels. *Ecology, 65,* 1617–1628.

Sherman, P. W., Jarvis, J. U. M., & Alexander, R. D. (Eds.). (1991). *The biology of the naked mole-rat.* Princeton, NJ: Princeton University Press.

Sherman, P. W., Reeve, H. K., & Pfennig, D. W. (1997). Recognition systems. In J. R. Krebs & N. B. Davies (Eds.), *Behavioral ecology,* 4th ed., (pp. 69–96). Oxford: Blackwell Scientific.

Shields, W. M. (1980). Ground squirrel alarm calls: Nepotism or parental care? *American Naturalist, 116,* 599–603.

Silk, J. B. (1999). Why are infants so attractive to others? The form and function of infant handling in bonnet macaques. *Animal Behaviour, 57,* 1021–1032.

Sluckin, W. (1964). *Imprinting and early learning.* Chicago: Aldine.

Smale, L., Nunes, S., & Holekamp, K. E. (1997). Sexually dimorphic dispersal in mammals: Patterns, causes, and consequences. In P. J. B. Slater, J. S. Rosenblatt, C. T. Snowdon, & M. Milinski (Eds.), *Advances in the study of behavior* (Vol. 26, pp. 181–250). New York: Academic Press.

Small, M. F., & Smith, D. G. (1981). Interactions with infants by full siblings, paternal half-siblings, and nonrelatives in a group of rhesus macaques (*Macaca mulatta*). *American Journal of Primatology, 1,* 91–94.

Smuts, B. B., Cheney, D. L., Seyfarth, R. M., Wrangham, R. W., & Struhsaker, T. T. (Eds.). (1987). *Primate societies.* Chicago: University of Chicago Press.

Solomon, N. G., & French, J. A. (Eds.). (1997). *Cooperative breeding in mammals.* Cambridge: Cambridge University Press.

Steiner, A. L. (1971). Play activity of Columbian ground squirrels. *Zeitschrift für Tierpsychologie, 28,* 247–261.

Sun, L., & Muller-Schwarze, D. (1997). Sibling recognition in the beaver: A field test for phenotype matching. *Animal Behaviour, 54,* 493–502.

Thompson, K. V. (1996). Play-partner preferences and the function of social play in infant sable antelope, *Hippotragus niger. Animal Behaviour, 52,* 1143–1155.

Tinbergen, N. (1963). On aims and methods of ethology. *Zeitschrift für Tierpsychologie, 20,* 410–433.

Trivers, R. L. (1974). Parent–offspring conflict. *American Zoologist, 14,* 249–264.

Vehrencamp, S. L. (1979). The role of individual, kin, and group selection in the evolution of sociality. In P. Marler & J. G. Vandenbergh (Eds.), *Handbook of behavioral neurobiology,* Volume 3, *Social behavior and communication* (pp. 351–394). New York: Plenum Press.

de Waal, F. B. M. (1996). Macaque social culture: Development and perpetuation of affiliative networks. *Journal of Comparative Psychology, 110,* 147–154.

Waldman, B., Frumhoff, P. C., & Sherman, P. W. (1988). Problems of kin recognition. *Trends in Ecology and Evolution, 3,* 8–13.

Waterman, J. M. (1988). Social play in free-ranging Columbian ground squirrels, *Spermophilus columbianus. Ethology, 77,* 225–236.

West, M. J., King, A. P., & Arberg, A. A. (1988). The inheritances of niches: The role of ecological legacies in ontogeny. In E. M. Blass (Ed.), *Handbook of behavioral neurobiology,* Volume 9, *Developmental psychobiology and behavioral ecology* (pp. 41–62). New York: Plenum Press.

West-Eberhard, M. J. (1975). The evolution of social behavior by kin selection. *Quarterly Review of Biology, 50,* 1–33.

West-Eberhard, M. J. (1989). Phenotypic plasticity and the origins of diversity. *Annual Review of Ecology and Systematics, 20,* 249–278.

Westneat, D. F., & Sherman, P. W. (1993). Parentage and the evolution of parental behavior. *Behavioral Ecology, 4,* 66–77.

Williams, G. C. (1966). *Adaptation and natural selection.* Princeton, NJ: Princeton University Press.

Wu, H. M. H., Holmes, W. G., Medina, S. R., & Sackett, G. P. (1980). Kin preference in infant *Macaca nemestrina. Nature, 285,* 225–227.

Zahavi, A. (1977). The testing of a bond. *Animal Behaviour, 25,* 246–247.

8

Play

Attributes and Neural Substrates

GORDON M. BURGHARDT

Play being the all important business of childish life, and all play consisting more or less in acts, it is play then, above all, that we must seek for the beginnings of voluntary activity. (Compayré, 1902; p. 151)

INTRODUCTION

Play is an important aspect of neurobehavioral development. While many agree (e.g., Byers, 1998; Fagen, 1981), the lack of attention to play has prompted one eminent neuroscience researcher to write, "It is sad that play research has not been of greater interest for neuroscientists ... the modern search for the mythological 'fountain of youth' should focus as much on the neurobiological nature of mental youthfulness and play as on ways to extend longevity" (Panksepp, 1998, p. 281). Underscoring the dearth of work in this area, a major compendium on cognitive neuroscience (Gazzaniga, 1995) lacks a single index reference to play, curiosity, or even exploration! To add another small voice to the call to make play a fruitful neurobiological pursuit, the present review develops and extends themes of play that I have broached previously in this format (Burghardt, 1988), but more specifically addresses neurobiological correlates. The evolutionary neural model of play suggested below in broad strokes is presented here in a condensed, verbal manner. It is an uneasy union of a developmental proximate scheme (the mechanisms and ontogenic processes of play in individuals) with an evolutionary scheme (how play originated in ancestral species and diversified, gaining new functions facilitating behavioral, emotional, and cognitive complexity). Rather than focus on either

GORDON M. BURGHARDT   Departments of Psychology and Ecology & Evolutionary Biology, University of Tennessee, Knoxville, Tennessee 37996-0900.

Developmental Psychobiology, Volume 13 of *Handbook of Behavioral Neurobiology*, edited by Elliott Blass, Kluwer Academic / Plenum Publishers, New York, 2001.

proximate or ultimate mechanisms alone, I will attempt to integrate them. It seems to me that compartmentalization of play research is one reason for the slow progress in understanding play in mechanistic, developmental, or evolutionary terms. Although there has been much relevant work on animal play in the last decade (Bekoff & Byers, 1998; see also Burghardt 1998a; Pellis & Iwaniuk, 2000), there remains a substantial need for broader conceptual integration in neuroscience to guide future behavioral research. This chapter is a step in that direction.

To accomplish my goals, I will first present criteria for recognizing play that lessens the ambiguities in earlier definitions. Definitional clarity and rigor are essential for progress in understanding play as a phenomenon as well as in charting its taxonomic distribution and revealing its physiological underpinnings. Second, in the context of surplus resource theory (SRT), I will review some neurobiological findings that relate to playfulness, including the status of *behavioral* play, including play with objects and conspecifics, as a precursor to *mental* play involving imagination and ideas.

Although the broader theoretical interpretations as well as the empirical research on which they are based will be preliminary, partial, and speculative, the importance of studying play is that it is a likely precursor and concomitant of mammalian cognitive, emotional, and behavioral complexity. This complexity arguably dwarfs that which is found in any other group of animals. Nonetheless, the failure of various attempts to establish the adaptiveness of play, let alone even to define play, has resulted in a conundrum that may be best resolved within a comparative, evolutionary perspective (Burghardt, 1984). Thus, my original impetus for studying play was to try to understand why many mammals, but virtually no nonavian (ectothermic) reptiles, were considered playful (e.g., Fagen, 1981). I contrasted living reptiles with mammals in an attempt to compare behavioral development in these important radiations of vertebrates (Burghardt, 1984, 1988). The major premise was that to understand playfulness, we had carefully to observe animals that apparently did not play (at least in obvious ways) as well as those animals that play a great deal. I argued that physiological (i.e., metabolic) factors and the presence or absence of parental care were major determinants of behavioral development including play. These traits, in turn, affected the evolutionary trajectories in behavior, cognition, and emotion of mammals and reptiles. However, what kinds of roles can play have in these phenomena?

PLAY: EVOLUTIONARY DETRITUS OR EVOLUTIONARY PUMP?

For over a century, claims have been made concerning the positive, perhaps crucial, role of play in the mental life and behavior of human and even nonhuman animals (Burghardt, 1999). These claims, derived from theory, have remained controversial and largely unsupported empirically (Martin & Caro, 1985; Power, 2000; Smith, 1988, 1996). Such support that has been advanced has often been correlational or experimentally flawed. Claims have been made that play facilitates learning (including reading and mathematics), imagination, socialization, behavioral flexibility, mental agility, and creativity (e.g., Fromberg & Bergen, 1998; Hartley, Frank, & Goldenson, 1952). Evolutionary arguments based on selected comparative data often accompany these assertions (Fagen, 1984). In spite of the lack of experimental support (Power, 2000), such claims have certainly been viewed as plausible by many who have studied play. Even those most responsible for disconfirming experiments in children find it difficult to completely dismiss these putative benefits (e.g., P. K. Smith, 1996). I accept the premise that play has an important role in the behavioral, social, emotional, cognitive, physiological, and development realms in the lives of many animals including humans. However, this role is likely to be multifaceted, variable, and involve often complex, indirect, and subtle processes in which immediate benefits, not long-term delayed benefits, are most important (Martin & Caro, 1985).

One way in which this multifaceted process is expressed is that the origins of play, those *primary processes* or setting conditions through which "play" first evolved in ancient vertebrates and their modern descendants, need to be empirically and theoretically distinguished from derived *secondary processes*. These latter may currently provide, or in an earlier evolutionary time period did provide, physiological, behavioral, cognitive, or emotional advantages to those animals engaging in play. Consider a developmental analogy with eating. Here the primary processes are regulatory and available at birth: neonatal animals, including mammals, do not need to learn to identify basic foods or to chew and swallow them. Physiological mechanisms such as blood sugar level control initiation and termination of feeding bouts. Specific appetites, feeding in response to social cues, and learning to obtain or handle difficult food items would be secondary processes. Similarly, primary processes for play are the proximate conditions, such as lowered thresholds for eliciting behavior due to stimulus deprivation, that underlie "playlike" performances. Such primary processes provide an essential scaffolding upon which the highly diverse and complex structures of mammalian play can be built through the selective actions of secondary processes providing important cognitive, emotional, and motivational functions (Burghardt, 1998a; Spinaka, Newberry, & Behoff, in press). For example, the exaggerated movements, role reversals, and testing of the individuals or objects being played with may allow learning of the reactions of other entities as well as aiding control of the player's own body. The distinction is similar to that underlying the process of behavioral ritualization, whereby autonomic, thermoregulatory, defensive, or feeding responses become incorporated into social displays (Burghardt, 1973; Grier & Burk, 1992). This distinction allows us to see that the surplus energy theory popularized by Spencer (1872) and the recapitulation theory of play popularized by G. S. Hall (1904) both dealt with primary processes in play, but did not explicitly address the derived (secondary) functions of play that Groos (1898) and his modern followers contend are the major reasons for play (Burghardt, 1998a, b). In brief, the surplus energy theory held that play occurred when well-fed "higher" animals had an excess of energy and needed to "blow it off" through vigorous activity. The recapitulation view held that juvenile play, which resembled adult "serious" behavior, was a necessary biological developmental process animals, including humans, went through that recapitulated behavior patterns once necessary for survival in earlier times. The practice theory held that play was important solely as a means to perfect adult behavior in species with long periods of immaturity and extensive parental care. The tendency of most early, and even some current, writers on play to confuse proximate mechanisms and adaptive value led to much needless controversy and, especially, a neglect of the former (Burghardt, 1984). Play can thus be viewed as both a product and cause of evolutionary change; that is, playful activities may be a source of enhanced behavioral and mental functioning as well as a byproduct or remnant of prior evolutionary events. It is probably a mistake to think that play originally evolved in order to provide such advantages, and this mistake may have hindered a more accurate and scientifically supported analysis of play.

If the distinction between primary and secondary processes is appropriate, we need to ground the phenomena of play in a phylogenetic context. What are the processes leading to playfulness throughout vertebrate evolution? The answer to this question is best sought before the search for putative consequences of play. Elsewhere, I present an extended treatment of animal play, and the historical, definitional, theoretical, and comparative aspects are explored in greater depth and with supporting data and citations (Burghardt, in press). By focusing on the origins and evolution of play, I think it is possible to outline a testable model of how playful activities arose and how they shaped the behavior patterns of animals, including facilitating complex mental processing. I will present a way of looking at play as a potential source of the rapid evolution of behavioral and mental complexity in endothermic vertebrates and as a resource for ontogenetic change in behavior that can be both conservative and novel.

DEFINING PLAY

In order to evaluate the role of play in creating or influencing the animal mind, we must first identify the phenomena to be considered playful. This is not a trivial exercise, especially when seeking the origins of behavioral phenotypes. Sometimes even the most accepted and clear examples of play can be problematic. The traditional categories of play are locomotor, object (including predatory), and social. These frequently overlap and are not distinct (Burghardt, 1998b; Fagen, 1981). Thus, in order to compare play across species we must have some way of isolating play behavior using a definition that is not specific to any one kind of play (such as social play). Another important problem is that both the proximate causes and adaptive functions of play may differ for different kinds of play, differ throughout ontogeny, and differ across species. These issues, along with legitimate difficulties in defining, describing, and analyzing playful behavior, have contributed to the lack of serious and careful scientific attention to the role of play in development and evolution. Thus, we first need to agree on how to recognize play.

Play is primarily an activity, a dynamic process of interaction. In this respect, however, play is not unique. After reading many treatments and definitions of play, one will probably be tempted to throw up one's hands and assert that "play just cannot be usefully defined." Fagen (1981) has gathered together many of these unsuccessful definitions. According to Wilson (1975, p. 164), "No behavioral concept has proved more ill-defined, elusive, controversial, and even unfashionable." R. W. Mitchell (1990, p. 197) states that "Play is the hobgoblin of animal behavior, mischievously tempting us to succeed in what, judging from the number of failed attempts, seems a futile task: defining play." Gandelman (1992, p. 215) agrees: "It seems unlikely that an adequate behavioral definition of play can ever be formulated." No prior approach, in other words, has been successful in listing the minimal criteria necessary to set play apart from other phenomena in a precise manner (Burghardt, 1999). Since, for my purposes, I needed to identify possible candidates for play in a wide diversity of species, I extracted from the literature on animal and human play, and my own 35 years of behavioral study, a set of criteria and arranged them in a formal and definitive manner. My goal was to provide criteria for play that could accommodate those approaching behavioral phenomena from different methodological and conceptual orientations (physiological, cognitive, ethological, etc.). It is important to note that none of these is alone sufficient. A more complete discussion of these criteria and their rationale is presented elsewhere (Burghardt, in press).

The *first criterion* for recognizing play is that the performance of the behavior is not completely functional in the form or context in which it is expressed; that is, it includes elements, or is directed toward stimuli, that do not contribute to current survival. The critical term is "not completely functional," instead of "purposeless," nonadaptive, or having a "delayed benefit."

The *second criterion* for recognizing play is that the behavior is spontaneous, voluntary, intentional, pleasurable, rewarding, or autotelic ("done for its own sake"). Here only one term of these often overlapping concepts needs to apply. Note that this criterion also accommodates any subjective concomitants of play (having fun, enjoyable), but does not make this essential for recognizing play. It should be noted, however, that laughter is a major indicator of joyful affect and has ancient evolutionary roots in brain evolution (Panksepp, 2000). Even rats may laugh when tickled (Panksepp & Burgdorf, 2000).

The *third criterion* for recognizing play is that it differs from the "serious" performance of ethotypic behavior in at least one respect: it is incomplete (generally through inhibited or dropped final elements), exaggerated, awkward, precocious, or involves behavior patterns with modified form, sequencing, or targeting. Notice that this is a "structural" or descriptive definition. It also implicitly acknowledges, but does not require, that play may be found only in juveniles.

The *fourth criterion* for recognizing play is that the behavior is performed repeatedly in a similar, but not rigidly stereotyped, form, during at least a portion of the animal's ontogeny. This criterion might seem a bit counterintuitive since the apparent freedom, flexibility, and versatility of play have been so often noted. However, repetition of patterns of movement is found in all play and games in human and nonhuman animals. It is also useful in distinguishing curiosity responses to novel stimuli from the play actions that may follow initial exploratory behavior.

The *fifth criterion* for recognizing play is that it is initiated when the animals are adequately fed, healthy, and free from acute or chronic stress (e.g., predator threat, harsh microclimate, social instability) or intense competing systems (e.g., feeding, mating, resource competition, nestbuilding): In other words, the animals are in a "relaxed field." This is an essential ingredient of play, as play is one of the first types of behavior to drop out when animals are hungry, threatened, or under inclement environmental conditions.

While these criteria do not provide a crisp, one-line definition, they do seem to cover every accepted example of play and to exclude, with consistent reasoning, much that is problematic and controversial. Play is not limited to social behavior, or juvenile behavior, or behavior that involves special signals. It excludes stereotyped, abnormal behavior as well as exploration. Although much behavior called play is vigorous and energetic from a human perspective, this also is not required. In applying individual criteria, the apparent fuzziness of the boundaries in some cases reflects the fact that some aspects of play and nonplay lie on a continuum. Nonetheless, my claim is that, if all the above criteria are met, the behavior deserves to be called play for the purpose of subsequent analysis. This approach does not prejudge whether, as I believe, play is a distinct but heterogeneous category, or whether all play is best viewed as a manifestation of a homologous process. The five criteria also make no assumptions about the function of play. The criteria also avoid the controversy of whether or not there is a separate motivation underlying play (S. L. Hall, 1998). Time will tell if new observations can be handled by these five criteria, although I have not encountered any that cannot.

For convenience sake, I will refer to the five criteria as:

1. Not completely functional
2. Endogenous component
3. Structural/temporal difference
4. Repeated
5. Relaxed field

Keeping in mind the nuances underlying each word, a one-sentence definition could then read as follows: *Play is repeated, incompletely functional behavior differing from more adaptive versions structurally, contextually, or ontogenetically, and initiated voluntarily when the animal is in a relaxed or unstressed setting.*

Applying the Play Criteria

As an example that satisfies the five criteria, the description of play fighting in rat pups by Panksepp, Siviy, and Normansell (1984, p. 466) is exemplary. When two pups "are placed together in a nonthreatening environment, they rapidly begin to exhibit vigorous fighting: animals chase and pounce on each other, sometimes unilaterally, sometimes mutually with rapid role reversals. They repeatedly poke and nip each other, often at the nape of the neck but also on the ventral surface when one animal is pinned." I chose this example not only because it concisely satisfies all five criteria , but also because play fighting in rats is the type of play most well studied experimentally and physiologically and discussed extensively below.

A problem, however, is that many descriptions of putative play are difficult to interpret because information that would enable us to apply the five criteria are absent. Thus, a side benefit of these criteria is to point out what kinds of information we need to know. Documentation and independent observation by several qualified investigators is also important in firmly accepting claims about play for some species.

I have found these five criteria to be particularly valuable in evaluating claims and evidence for play in problematic groups such as ants, fishes, monotremes, birds, and nonavian reptiles. The criteria also help guard against the anthropomorphism that has been endemic to the study of animal play since its inception. However, the application of *critical anthropomorphism* (Burghardt, 1985) can be used to retain the motivational and hedonic aspects of play without losing an objective and public assessment of play. These latter aspects of play will become clarified as physiological and neural measures of internal states become more refined.

As noted above, there is no implication that all behaviors satisfying the play criteria have the same causal (neurobiological) mechanisms or that they are homologous. A species' normal behavior, ecology, ontogenetic development, metabolism, neural organization, and phylogeny influence the manner in which play is expressed. In fact, it is because play is almost certainly a heterogeneous category that arises repeatedly in vertebrate evolution (Burghardt, 1998a) that we need clear criteria to help us sort out the processes and variables influencing playfulness and point the way to experiments and critical phyletic comparisons.

THE PHYLOGENETIC ORIGINS OF PLAY

323

PLAY:
ATTRIBUTES AND
NEURAL
SUBSTRATES

Play has been viewed as primarily a property of mammals. Griffiths (1978) even listed play as one of the defining attributes of mammals. He argued that since monotremes play [actually, the platypus (Fagen, 1981), but even this is controversial], monotremes should be considered true mammals and not "quasi-mammals." Birds have been acknowledged as engaging in play (reviews in Fagen, 1981; Ficken, 1977; Ortega & Bekoff, 1987). However, a clear discontinuity was generally recognized in that play was considered common in endothermic vertebrates (especially mammals) and absent in the traditional "cold-blooded" vertebrates: fishes, amphibians, and nonavian reptiles (reptiles from here on). As someone who has spent decades studying reptile behavioral ontogeny and being impressed with the degree of complexity and learning in their behavior, I had to agree that I had seen nothing that appeared to be playful (Burghardt, 1984).

However, my colleagues and I have recently presented data establishing that behavior that would readily be called play if found in mammals can be identified in turtles and perhaps other reptiles (Burghardt, Ward, & Roscoe, 1996; Kramer & Burghardt, 1998; Krause, Burghardt, & Lentini, 1999; review in Burghardt, 1998a). These include batting around balls, swimming through hoops, engaging in precocious courtship, and other typical play responses. Elsewhere, I review the suggestive, but frequently poorly documented evidence for play in other nontraditional animals (Burghardt, in press). One can make a case that play with objects also occurs in some squamate reptiles and crocodilians, fish, and perhaps even in some invertebrates (Burghardt, 1999; Mather & Anderson, 1999). Consider the comparison of play with flying. Wings in birds, bats, and flies have differing genetic, physiological, and phylogenetic substrates, although they have a similar functional role in flying. Flying, of course, is much easier to recognize than play. Wings may be for flying, but we also readily recognize wings in flightless birds, such as rheas and ostriches, and consider functional and nonfunctional wings as homologous. Play, as a label, shares some interesting similarities with flying as a label. For example, object play in cats may share much more in common with predation than object play shares with social play wrestling (S. L. Hall, 1998; Pellis *et al.*, 1988).

METABOLISM, DEVELOPMENT, AND STIMULATION AS SETTING FACTORS

Why is play, even if very modest and incipient versions are found in ectothermic vertebrates, primarily a phenomenon of mammals? I have developed a list of life history and physiological factors that appeared to be different between "typical" reptiles and mammals. These have been incorporated in an evolutionary/developmental perspective called surplus resource theory (SRT). Developed in two chapters (Burghardt, 1984, 1988), many of the core ideas have been extended and elaborated by Coppinger and Smith (1989) and Barber (1991). SRT incorporates physiology (e.g., activity metabolism, thermoregulation), life history (e.g., parental care, altriciality, food niche, ontogenetic shifts), and psychological factors (stimulus deprivation, exploration).

The first suite of traits involves energetic differences between reptiles and mammals. Reptiles typically are metabolically constrained from performing vigorous, energetically expensive behaviors, especially those of any substantial duration (Bennett, 1982). This constraint is due to their low resting and maximal metabolic rates, limited aerobic capacity, and long recovery times after anaerobic expenditures. Reptiles have about 10% the metabolic rate of a comparably sized mammal (although most reptiles are much smaller than mammals). Both metabolic rates and typical body sizes are further reduced in amphibians. Small body size leads to greater heat loss and time and energy efforts to behaviorally thermoregulate in ectotherms or obtain calories in endotherms. Taken together, these factors constrain the performance, and selection for, vigorous playlike behavior with no or limited current function in ectotherms. Very small mammals also appear to play less than larger species in a lineage (Burghardt, 1988). Countering this trend is the fact that play occurs more often, at least in mammals, in juveniles that have not yet reached adult size. In these cases, the animals, such as rodent pups, have grown fur and become better able to thermoregulate, and secondary derived processes may underlie playful behavior. Phylogenetic factors and other traits such as type of parental care and quality of diet complicate the role of metabolism in play. Among primates, folivorous species play less than frugivorous or omnivorous ones (Fagen, 1981).

The second suite of characteristics involve developmental factors, almost all of which revolve around the limited parental care and attendant precocity of juvenile nonavian reptiles. With the exception of all crocodilians, most neonatal reptiles are not cared for by their parents (Shine, 1988). Consequently, they must devote their activities, immediately upon birth or hatching, toward obtaining food, finding shelter, avoiding predation, and growing rapidly. This limits safe time or opportunity for practicing or perfecting behaviors (motor abilities, perceptual–motor coordination, social skills) to be used in an uncertain future. Most young reptiles are very small, and selection will have shaped abilities that enhanced juvenile survival, such as remaining quiet and inconspicuous. Such behaviors are, of course, incompatible with vigorous play. There is also no source of high-fat, high-protein nourishment at little cost, as in juvenile birds and mammals. Although nonavian reptiles are highly precocial and *superficially* resemble, and behave like, miniature adults, they nonetheless may learn many things as they go about their serious activities. Reptiles also grow as rapidly as food resources allow; unlike mammals and birds, there is no necessary age-defined period of youth. Young reptiles, unlike young mammals, do not get fat; they just grow as fast as food resources permit. This characteristic of reptiles is termed indeterminate growth. Reptiles convert a higher percentage of assimilated energy to biomass and thus can grow on a smaller ration than a mammal. Given plentiful food, however, endotherms can add weight at a higher rate than can an ectotherm amniote; this might be a major benefit of endotherm parental care. Barber (1991) advanced a "brown fat" hypothesis in which vigorous play is a means for well-provisioned young mammals to avoid obesity, a view very akin to the Spencer's surplus energy formulation. This could be a primary process as well, but no experimental evidence supports the hypothesis. In any event, obesity is not a problem for wild reptiles, although many captive reptiles of the larger species are often fed excessively rich or abundant food, leading to overweight, lethargic adult animals. How a "play to avoid being fat" mechanism evolved when, for example, reducing caloric intake would seem to be simpler is not clear, though activity-induced thermogenesis may be involved. The last life history characteristic I will mention is that reptiles typically have large litter and clutch sizes with both smaller neonates and higher juvenile mortality than endotherms. Thus, any delayed benefits of play would be less worth a current risk.

Third, animals in stimulus-deprived (boring) environments would be most likely to engage in behavior to relieve sensory and response deprivation and to increase arousal. Such boredom might be expected in the well-provisioned and protected environments (e.g., nests, burrows) provided by endothermic parents. This factor may be the critical one in the consistent finding that well-cared-for captive animals play much more than their wild counterparts. If juveniles are buffered from the demands of survival and the species possesses a complex repertoire of evolved and active behavior patterns, then they have a behavioral resource to draw on when deprived of stimulation; threshold lowering and reorganization of behavior sequences into less precise adult-like forms could result. Since the simplistic views of a unitary nonspecific arousal system are no longer accepted, being replaced with a conception of functionally different neural systems mediated by monoaminergic and cholinergic neurotransmitters "contributing to different forms of behavioral activation" (Robbins & Everitt, 1995, p. 703), the plausibility of such processes has increased.

Information on these three kinds of factors (degree of precocity, metabolism, and environmental stimulation) helps explain why some groups of animals play and others do not. As a guide to identifying behavior in nonavian reptiles that could be candidates for traditional-appearing play, I subsequently (Burghardt, 1988) used the above and related physiological, psychological, and life history contrasts to predict that mammalian or avian-like play in reptiles should be rare and occur only in specific contexts in which those factors facilitating play in mammals and birds are also present. They also allow us to predict in what mammalian groups we would expect to find the most complex play and the most time spent in play. A series of predictions was supported that suggest that the three factors listed above are involved in facilitating play. For example, less energy is needed for locomotion in water than on land and such energetic considerations would suggest that one would expect that aquatic mammals would be particularly playful—and they are. The first play we confirmed in reptiles was in aquatic turtles (Burghardt *et al.*, 1996; Kramer & Burghardt, 1998), supporting predictions made a decade earlier (Burghardt, 1988).

Certain kinds of complex ecological and social interactions and operations upon the environment seem to characterize the more playful species, and ecology is a fourth major component in the occurrence of play. For example, object play is found most often in active predators, scavengers, extractive foragers, and generalist feeders that rely on manipulation of limbs and mouth. Reptiles do not have the rich repertoire of possible movements of limbs and face seen in many mammals. The lack of parental care may have prevented the evolution of the social bonding and affiliation in reptiles to the extent seen in birds and mammals and thus social play is less likely than object and locomotor play. Interestingly, it is in those reptiles with more variable diets and active foraging techniques, such as soft-shelled turtles (Burghardt, 1998a), in which more exploration, curiosity, and, sometimes, play, are found. The same probably holds in fishes as well (Burghardt, in press).

A major premise of this approach is that the initial advantages of incipient playlike behavior did not involve any particular functions such as perfecting later behavior, increasing endurance, or facilitating behavioral flexibility. Play can evolve independently whenever physiological (including neural), life history, metabolic, ecological, and psychological conditions, in conjunction with a species' behavioral repertoire, reach a threshold level. In mammals, however, play may have been both a cause and consequence of the evolving disparity between the periods of juvenile dependence and adult responsibilities. More specifically, according to the model advanced here, the advent of parental care led to the deterioration of some neonatal response systems through less precise functional motor patterns, the lowering of stimulus thresholds, and the broadening of the effective stimuli. In addition, the increased aerobic metabolic capacities resulting from endothermy modified or even reorganized developmental processes so that incipient play and other experiential avenues were not only available to mammals, but may well have had to be exploited by them for continued survival by replacing lost, suppressed, or maturationally delayed response systems. For example, if natural selection is continually honing predatory skills so that less successful juvenile predators starve or are otherwise less fit compared to more skilled predators, then once selection is removed the mechanisms for capturing prey should become less precise. A prediction would be that domesticated cats removed for generations from preying on live animals would be less competent hunters as juveniles than would nondomesticated small cats. Among garter snakes (*Thamnophis*), we have shown that species specializing on aquatic prey capture fish more efficiently than species that are prey generalists or earthworm specialists. Experience, however, can make up for some of the deficits in terrestrial prey specialists (Burghardt & Krause, 1999; Halloy & Burghardt, 1990).

It is in this parental care transition that the need to distinguish between primary and secondary processes in play becomes critical. A secondary process of play derived from the primary processes outlined above would be supported if research showed that vigorous rough and tumble play of young rats and dogs enhances adult performance, promotes socialization, or increases behavioral flexibility. However, it is precisely because the more primary processes of play have been ignored that predictions made from secondary processes have fared poorly. An example of an important primary process derived from SRT would be the role of metabolic rate or parental care in production of "surplus" behavioral "mutants" that could in turn be selected ontogenetically and phylogenetically. An apparent secondary process derived from a detailed consideration of primary processes is based on the recent claim by Byers and Walker (1995) of a correlation between the onset of vigorous motor play with the age at which permanent long-term changes occur in the muscular and cerebellar systems of several species of domesticated animals. If play is essential or even useful in establishing these permanent physiological systems, then a secondary process has been established. If the play behavior is a mere accompaniment to this developmental process with no causal role, then it is but another primary process.

The study of the role of play in development should initially focus on the primary processes leading to behavior satisfying the five criteria for play. I postulate that increased endurance, functional endothermy, parental care, and lack of sufficient external stimulation facilitate incipient playlike behavior. As the trends favoring this incipient play expanded, play acquired functions, including those underlying greater behavioral, social, and cognitive complexity through the evolution of secondary processes. Some writers have argued that play is a "random process generator" or a means of creating "adaptive variability," and similar views abound (e.g., Sutton-Smith, 1999). A more nuanced version is Fagen's (1974) comparison of the variability seen in play to that found in genetic systems. That is, in play we see similarities to chromosomal inheritance in the sense that in play sequences we can find *recombination, fragmentation, translocation,* and *duplication.* Such variation can be the raw materials for natural selection to operate upon insofar as the behavioral variation also has some inherited component. Recent molecular genetics methods allowing overexpressed alleles and the elimination of loci (knockouts) allow further genetic metaphors. However, play is far from random and seems to operate within tight boundaries. Pellis (1993) has documented how play fighting can differ among closely related species with little or no overlap.

Thus, the challenge is not just to state that play creates novel behavioral phenotypes, but to uncover the actual behavioral processes underlying such behavioral variation. I view these processes as of two main types. The initial secondary processes were those that maintained the instinctive skills of animals under threat of deterioration due to the lack of natural selection acting to maintain the precision of the predatory, defensive, and social skills necessary for survival in the absence of parental or other protection. Support for this is seen in the far less skillful initial foraging and predatory behavior of many juvenile mammals as compared to turtles, lizards, snakes, and crocodilians. Later, the neural and physiological changes resulting from this experienced-based learning and plasticity opened up new possibilities for cognitive and emotional complexity in many mammals and some birds over nonavian reptiles. For example, could play among littermates help establish or consolidate kin recognition cues that may later facilitate recruitment to aid in defense against predators in social mobbing species (see chapter by Holmes in this volume)? How might these more advanced secondary processes have operated? Were they in fact tertiary processes? Could they have jump-started a cognitive revolution? In the remainder of this chapter, I will outline a scenario based on recent advances in neuroscience.

PLAY: AN OVERLOOKED CRITICAL COMPONENT OF DEVELOPMENTAL PSYCHOBIOLOGY

The early emphasis by biologists and psychologists on the instinctive nature of much animal and human behavior was largely lost in psychology with the arrival of behaviorism. Virtually all early theories of play, even when they conflicted with each other, included a role for instinctive or innate components. That this became unacceptable even among play theorists is shown in the review by E. D. Mitchell and Mason (1934), who were at pains to accept the then ascendant "modern" anti-instinct perspective (e.g., Bernard, 1924), stating "Out of habits and attitudes arise motives and desires. These are the drives that lead to play" (E. D. Mitchell & Mason, 1934, p.70). Nevertheless, ethologists (e.g., Tinbergen, 1951) and some important developmental psychobiologists such as Carmichael (1954), Sperry (1951), and McGraw (1943) continued to argue that there was considerable prewiring and endogenous maturation of neural systems that could not be explained by naive environmentalism, just as current evolutionary psychologists are claiming that many sex and gender differences cannot be explained by naive sex-role social constructivism (e.g., Buss, 1999). Today we know that phenotypic expression of behavior patterns is a complex epigenetic outcome of interactions and feedback occurring at many levels from gene allele to protein synthesis to behavioral practice to social experience. If so, play may have a more subtle, yet profound, role in behavioral ontogeny and phylogeny that we are only beginning to appreciate.

One of the outcomes of the focus of the early ethologists on species-typical behavior patterns and their development was that the individual motor components ("fixed action patterns") were often functional at birth or after a period of postnatal maturation with minimal experiential input (Tinbergen, 1951). This was somewhat less true of the sequential organization of such patterns as demonstrated by Eibl-Eibesfeldt's studies of nest building in rats and nut opening in squirrels (for review see Burghardt, 1973). The perceptual control of behavior was much more malleable, however. It is no accident that studies of operant and classical conditioning, habituation, imprinting, and even plasticity in snake behavior focus on the stimulus and contextual cues controlling behavior rather than on the topographic forms of behavior. Nevertheless, most of the research on the functions of play focused on the behavioral organization rather than the context of play (Martin & Caro, 1985). Recently, with the advent of cognitive ethology, more emphasis on the role of play in communication, intentionality, flexibility, and ability to anticipate behavioral responses of others is appearing in research (e.g., Allen & Bekoff, 1998; Bekoff & Allen, 1998; Biben, 1998; Pellis, & Pellis, 1998b). However, an early type of secondary process in play may have been the performance of basic species-typical motor activities by young animals. In altricial species, as compared to precocial species, a more drawnout process would be expected, as there would be a greater window of time for incompletely functional performances to be influenced by playlike experiences. However, even in human beings, the motor components of species-typical behavior (facial expressions, reaching, grasping, biting) may be only modestly influenced by experiences, although the contextual deployment and perceptual–motor control may be influenced to a considerable degree.

A good test case in which the first type of secondary process (maintaining "instinctive behavior") in play may be relevant is learning to walk in human children. McGraw (1943) emphasized the importance of maturation: infants do not learn to walk, they walk when ready. A similar view has been expressed on the onset of play fighting in rats (Ikemoto & Panskepp, 1992). Thelen (1995) confirmed that human infants do indeed have many precocial locomotor skills, such as coordinated stepping movements at 1 month of age if held on a treadmill. Nevertheless, she argues, these congenitally highly complex behavior patterns still need much experience in order to develop fully and advocates that walking and other behaviors should be viewed in a "multicausal" framework. Although compelling evidence for this is hard to come by for human locomotion, I am willing to concede that practice is as important for perfecting this skill in children as it seems to be for predatory behavior in some young snakes (Burghardt & Krause, 1999; Halloy & Burghardt, 1990).

Thelen (1995, p. 85) argues that "new views of motor development emphasize strongly the roles of exploration and selection in finding solutions to new task demands. This means that infants must assemble adaptive patterns from modifying their current movement dynamics." Although Thelen's ideological position is to overthrow accepted genetic and neural inputs in behavioral development, a softer version in which the adaptive deployment and altering of much motor behavior may involve considerable ontogenetic input ("learning") has value. Aiding this process are motivational factors (as when kids keep trying to stand alone, crawl up stairs, cross a room to get to a parent or toy), accompanied by the joy of success, which facilitates the learning of a novel means of accomplishing a behavioral outcome. Central to Thelen's view is that action and perception are linked and "that each component in the developing system is both cause and product" and that "cognition is emergent from the same dynamic processes as those governing early cycles of perception and action" (Thelen, 1995, p. 94). Even "higher order mental activities, including categorization, concept formation, and language, must arise in a self-organized manner from the recurrent real time activities of the child just as reaching develops from cycles of matching hand to target" (Thelen, 1995, p. 94). Thus, here is a conceptual model for how sophisticated motor activities, often involved in play, are derived from the same kind of processes as are involved in the development of advanced mental abilities.

Applying the five criteria of play including the incomplete functionality, repetition, endogenous motivation, and structural/temporal factors all speak to "learning to walk" and other behavioral tasks in altricial mammals as more than superficially playlike. The task now is how to forge this link between the two kinds of secondary processes, and Thelen (1995) provides no clear means of going beyond a vague multicausal ("nonlinear") framework and the importance of the largely innate locomotor system in her emphasis on behavioral refinement and deployment. We need analytical research that separates and evaluates the various putative influences. For example, Meer, Weel, and Lee (1995) showed that spontaneous arm movements in newborn (10-24-day-old) infants are more frequent when the infants could see their arms. Furthermore, visual information is used to counteract external forces (weights, strings) on their limbs in maintaining a reference position. The authors speculate that such early developmental processes may be critical in the development of more purposeful reaching and grasping skills 3 months or so later. Playlike behavior may be important in many of these developmental transitions in human and nonhuman species. Comparable experimental work with animals in which controlled and long-term manipulation of sensory–motor integration during social, locomotor, and object play is carried out could be an entrée to the analytical work that has been so lacking in play research.

An exception is Piagetian and neo-Piagetian developmental research in children and nonhuman primates (Parker & McKinney, 1999), where much, though not all, research finds that four periods of cognitive development (sensorimotor, pre-operations, concrete operations, and formal operations) build on one another sequentially, as do several stages within each period. Play, as characterized above, is an integral and important component in cognitive development (Piaget, 1962). For example, the sensorimotor intelligence series involves a sequence of stages including circular and secondary circular reactions. The latter include repeated actions on the environment such as repeatedly striking a mobile or shaking a rattle. These are found in many species. In this approach, pretend play does not occur until the symbolic preoperations subperiod.

Play behavior, as is true of many other behavior patterns, has several sequential components that may have different causal bases. Even a circular reaction can be partitioned. For example, if an object such as a ball is introduced into a kitten's environment and play with it ensues, we observe the following: orientation to the object (initiation), running over to the object (approach), hitting it with a paw, perhaps several times in a row (engagement), leaving the ball, and switching to another activity (termination). Such sequences may occur repeatedly in bouts. Recently, Willingham (1998, 1999) put forth a general model of motor skill learning that postulates a specific set of four processes involved in motor control of just the first part of the above sequence. Consider the process of reaching for and moving an object such as when a cat reaches out at a dead mouse and pulls it to her mouth. This entails a "decision" to perform the act (strategic process), a translation of the spatial locations of cat, paw, and object into an "egocentric space" (perceptual–motor integration), a sequential ordering such that the paw is moved to the object (the "goal location") before being retrieved (sequencing), and finally the translation of the above into appropriate muscle firing (dynamic process). In each of these four stages, learning of different kinds may be involved in enhanced behavioral performance. Play could be a means of ensuring that behavior patterns be both practiced and deployed in effective manners either to maintain skills removed from the operation of natural selection due to parental care or to enhance performance beyond that which could possibly be anticipated by innate wiring. Even the most congenitally perfected behaviors (predation) may still have to be employed in widely varying environments to widely disparate objects. Perhaps object play in kittens is so exaggerated or otherwise so different from real predation that such "unrealistic" practice interferes with predatory behavior. If so, then object play has no secondary function regarding predation and exists for other reasons (e.g., stimulus deprivation). Evidence suggests, however, that predatory play in cats is linked to motivational systems involving predation (S. L. Hall, 1998; Pellis *et al.*, 1988).

RELATIONSHIPS AMONG PLAYFULNESS, BRAIN SIZE, AND INTELLIGENCE

As a major goal of this chapter is to discuss the neural substrates of play, a brief discussion of the rich history of speculation on brain size, play, and intelligence is needed. The role of play in cognitive performance is often advanced by pointing out that play is found most commonly in those birds and mammals that have large brains and are considered "intelligent" (Fagen, 1981). Play and curiosity could thus be marks of intelligence as well as necessary for the development of a sophisticated mind. The view that play is most common in large-brained animals is rarely evaluated rigorously; what we find when we do so has not been very encouraging.

First, we need to determine if species with larger brains are more intelligent, not an easy task even if we had a clear understanding of intelligence in nonhuman animals. Such a putative relationship has been at the heart of much comparative psychology; the problem was the lack of a general measure of intelligence (speed of learning/complexity of problems solved) that could be used across species with widely differing morphological, ecological, social, and phylogenetic attributes. Thus a more limited emphasis on behavioral specializations of an ethological and ecological nature became popular (Burghardt, 1973). Nevertheless, the general pattern of increased relative brain to body size ratios among those mammals and birds considered most intelligent or adaptable in behavior is hard to ignore. Efforts to develop a comparative measure of intelligence that can cut across species are still underway; a recent one claimed to be promising for primates from prosimians to *Homo* is the "transfer index" (Beran, Gobson, & Rumbaugh, 1999). The correlation with brain size is high enough (0.83) that the authors used it to predict the intelligence of extinct hominid species. Brain size, body size, and metabolic rate are all highly correlated, so let us accept that animals with larger brains are, in some important senses, smarter or more cognitively adept than others. Does this relate to play?

Certainly, crude correlations appear impressive. Primates and aquatic mammals have large brains and are considered highly playful, and chimps seem to play much more than prosimians (Fagen, 1981). Among birds, the "big-brained" species in the parrot and corvid families play extensively (Fagen, 1981; Heinrich & Smolker, 1998; Ortega & Bekoff, 1987). Sloths, insectivores, and many less playful mammals and birds have smaller brains. The best evidence for fish object play is in mormyrid fish (Meder, 1958; Meyer-Holzapfel, 1960; G. M. Burghardt, unpublished observations) that have brains larger, relatively, than most mammals (Butler & Hodos, 1996) and may even exceed the brain/body weight ratio of adult human beings (Bullock, 1977)! Furthermore, species with more altricial young often play more, or more complexly, than even close relatives that are more precocial (Burghardt, 1988).

The problem is that variation in many other features of life history, physiology, ecology, and phylogeny render suspect general impressions that overall brain size is the most important factor. The application of the newer methods of comparative analysis that control for phylogeny and look at variation in brain size and playfulness within a narrow taxonomic lineage is rare. This is due to the lack of quantitative and descriptive data on play in many species closely related to a particularly well-studied species as well as the difficulties in comparing the details of play records gathered using different methods and terminology. In some studies, social play is not labeled as such but is pooled under "other social" or not even recognized. In other cases, social play is more commonly described as "play" than object or solitary locomotor/rotational play (e.g., Panksepp, 1998).

Some recent studies do provide some initial data in three disparate groups of mammals: marsupials, primates, and rodents. Byers (1999) looked at a composite measure of play in marsupials from most families and did find a positive relationship with overall brain size. However, he did not use a comparative method that controlled for lack of independence across phylogeny and he compared average brain size and play levels using family means. Some families were represented by only one species, and Iwaniuk, Nelson, and Pellis (2001) analyzed data that are more complete and found that the correlation with overall brain size does not hold. However, a positive relationship of play with neocortex size may exist in marsupials.

In a study with 26 species of muroid rodents using the method of independent contrasts, Iwaniuk *et al.* (2001) also found no significant relationship between brain size and a relative measure of complexity of social and locomotor play. Sufficient data were not available for evaluating various brain regions, however. They also performed a study on 20 species of primates for which they had both play data (complexity of adult play) and measures of overall brain size as well as measures of the relative size of the cerebellum, neocortex, and nonvisual cortex. No significant relationships of any brain measure and play were found. More extensive work might still uncover a relationship. In a study of seven primate species at the Edinburgh Zoo, Lewis (2000) looked at social, solitary, and object play and applied the method of independent contrasts, finding that only social play was positively and significantly related to neocortex ratio. This preliminary study needs critical replication with more species and in controlled settings.

Studies of play in birds using comparative methods are not yet available. However, the occurrence of feeding innovations in birds is related to increasing relative forebrain size (Lefebre, Whittle, Lascaris, & Finkelstein, 1997). Interestingly, although altricial birds have larger brains as adults than do more precocial species, developmental mode was not a significant factor, although phylogeny was. It would be worthwhile to compare feeding innovations and object play in this sample.

Even if play is more frequent and diverse in animals with larger telencephalons (or parts thereof), we still have to decide whether this is a feature supporting enhanced mental abilities, maintaining instinctive responses that are now far more plastic than ancestral states, or is a currently epiphenomenal or atavistic phenomenon. SRT suggests that the later two views cannot be rejected out of hand (the surplus energy and recapitulation play theories; Burghardt, 1998a). On the other hand, play could evolve secondary processes that allow behavior and psychological abilities to rapidly alter. A hallmark of mammalian behavioral evolution is the rapid diversification of behavior, telencephalon, and genome in a relatively short (geologically speaking) time span as compared to many ectothermic vertebrates and invertebrates, whose core behavior patterns and abilities may have changed little over millennia in spite of often rapid microevolutionary adaptation . For example, physical and behavioral adaptations involving milk production and its delivery and ingestion are restricted to modern mammals. Facial expressions and vocal repertoire are also much more complex in all mammals as compared to the extant reptilian sister groups (squamates, tuatara, turtles).

There are some problems, however, with the putative relationship between play and cognition. Dogs raised in social isolation appear to be more social and curious than normally reared dogs, especially to novel situations; they are less neophobic. Actually, it has been shown that such dogs are rather stupid; even the most trivial items amuse them persistently (Melzack & Thompson, 1956). There is also considerable evidence that habituation, rapid boredom with stimuli, is a good measure of intelligence and can be used with preverbal infants with considerable success (Colombo, 1993). If becoming bored is a sign that one has mastered what there is to know or do with an object, then choosing stimuli that are more complex might be the way to maintain play. This is exactly what Piagetian mastery play is directed toward. Thompson (1998) advocated a similar process that she terms self-assessment; animals play to perform and "practice" a skill until it becomes mastered or too easy (habituated, boring). However, curiosity killed the cat. Rodents, fish, and tadpoles that stray too far from secure retreats for food have been shown to be more at risk of predation (e.g., Sih, 1992). Thus, prudent behavior has some advantages over the exuberant "joy of life" excitement many view as the essence of play. Clearly, some ecological contexts might have facilitated play in some contexts and opposed it in others. It is for this very reason that a simple unitary functional explanation of play is unlikely to be discovered.

Another telling argument against a simple brain size and intelligence relationship with play can be found by using domestication as a model system for testing surplus resource theory. Many of the processes that I postulate as having occurred in juvenile mammals with the onset of parental care and the consequent buffering from the demands of life are also found in domesticated species. Price (1984) documented these processes completely independently of the theory outlined here. The match is remarkable. The retention of juvenile morphology and behavior into adulthood is seen particularly clearly in dogs. Litter sizes increase and many behavioral skills shown in wild populations (wolves) show deterioration in domesticated forms (dogs) that have considerably smaller brains (Coppinger & Coppinger, 1998). The pattern of increasing playfulness as brain size decreases may be quite common. Kruska (1987, 1988) showed that in many mammals, domestic populations have brain sizes 5% less than wild populations after only a few generations. This is shown in both a rodent (bank vole) and carnivore (polecat). Ferrets were domesticated from wild stock (polecats) 2500 years ago and now have brain sizes 30% smaller than wild animals. Generally, the detrimental effects of domestication on brain size are greater in groups with larger brains (e.g. carnivores) than those with smaller brains (e.g. rodents). It is the former that have relatively larger telencephalons and it is the telencephalon, particularly the neocortex, that shows the largest reduction (Kruska, 1988). It would seem that a careful study of play behavior in wild animals undergoing domestication might be a most useful method to see how buffering animals from the harsher aspects of existence may change the amount, type, and frequency of play as well as cognitive capacities. It is important to be critically anthropomorphic (Burghardt, 1985), as it is all too easy to conclude that the more affectionate or compliant dog is smarter than the more elusive, unpredictable, high-strung wolf or devious coyote.

Perhaps the smarter animals were actively selected against in domestication. This appears unlikely, as in the most careful study of the process by Belyaev and colleagues (Trut, 1999). Merely selecting silver foxes, *Vulpes vulpes*, for tameness alone led, after 10–35 generations, to animals that showed marked changes in coat color and ear and tail morphology, more rapid sensory development (e.g., eyes open earlier), and delayed onset of fear and the plasma corticosteroid surge. The foxes thus have a longer period for bonding with humans and other animals and showed much more interest and "friendliness" to people (which seems to include playful interactions) than the initial stock population. Although I could not find any data on brain size, cranial height and width were both reduced in the selected foxes (Trut, 1999).

Play is often considered juvenile behavior and thus its retention in adults, including humans, is due to neoteny or retained childlike characteristics. While this may be true for some species, such as dogs, neo-Piagetian analyses suggest that certain kinds of play, especially in primates, reflect continued brain and cognitive development rather than the retention of juvenile characteristics (Parker & McKinney, 1999). Nevertheless, although overall brain size or neocortex ratios may not be directly correlated with the amount, complexity, or intensity of play, the neural underpinnings of play need to be addressed. However, this endeavor must not be approached with the view that play is solely a process meant to enhance the cognitively oriented secondary process of play. Superficially similar play behavior may have different evolutionary histories and neural substrates.

Is Play Behavior Localized in the Brain?

Does play have its own neural underpinnings, a play circuit (or 'module') as suggested by Panksepp (1998) and advocated for most behavioral systems by evolutionary psychologists (Buss, 1999)? Or is play a product of a special set of circumstances that could arise in almost any neurobehavioral system and is this distributed rather than localized? Indirect evidence supporting the modularity view comes from studies indicating a specific motivation for play of certain types (Rasa, 1984). Contrary evidence comes from those who see play as derived from incipient behavior (intention movements; Lorenz, 1981), the emergence of "prefunctional" behaviors in ontogeny (Hogan, 1988; Kruijt, 1964), or conflicts between behavior systems (Pellis *et al.*, 1988). We are far from an answer, but in the model I am developing, the former would have evolved from the latter and thus we might expect both specific and more diffuse neural underpinnings for play of different types, different phases of the sequence, and in different species. That being said, research reviewed below, although not definitive, is suggestive even if it often does not involve specific measures of play as dependent variables.

The Triune Brain

McLean's (1985, 1990) writings on the triune brain and the comparative perspective it provides have been influential, though controversial (e.g., Butler & Hodos, 1996). In brief, the forebrain of mammals is divided into three parts. The first of these is the "reptilian brain," variously referred to as the R-complex, basal ganglia, extrapyramidal motor system, or striatopallidal complex, in which essential "instinctive" behaviors such as feeding, fighting, dominance displays, and mating are controlled along with fear, anger, and exploration (Panksepp, 1998). All mammals inherited these brain structures from their reptilian ancestors. The second part is the "old mammalian" brain, variously referred to as the limbic system, visceral brain, or emotional brain, composed of brain areas in which emotions involving parental and filial affectionate behavior and social bonding/distress are controlled and elaborated. The third part is the "neomammalian" brain, variously referred to as the neocortex, somatic cognitive nervous system, or thalamic–neocortical axis, which, while influenced by the older forebrain systems, is the place where higher cognitive functions such as perception, reasoning, and logical thought are deployed. Panksepp (1998) refers to these three brain regions as being involved in *innate behavioral knowledge, affective knowledge,* and *declarative knowledge,* respectively. The congruence of these three brain regions with the classical 19th century division of psychological phenomena into motivational, emotion, and cognitive areas is probably not coincidental. Although the comparative sweep is certainly too simplistic (Butler & Hodos, 1996) and incongruent with modern studies of reptile behavior and neuroanatomy, MacLean does provide a broad integrative canvas and testable speculations for modeling the neural substrates of behavior. Damasio (1999) provides an updated neural model of the interconnections among emotions, feelings, cognition, and consciousness in the organization of brain and behavior. He classifies emotions as primary (e.g., fear, anger) and secondary (e.g., grief at the death of a friend, joy that your team won the game) (Damasio, 1995). Primary emotions are innate and control is centered in the limbic system, particularly the amygdala, whereas secondary emotions have learned components, controlled by input from the prefrontal cortex, that interact with the innate, less conscious, processes. Building on this distinction, Sutton-Smith (in press) argued that play is a separate tertiary emotion involving virtual contexts in which the primary and secondary emotions can be experienced in a less risky or nonserious setting. This may be true of much play at the derived level but neural evidence is unavailable to test this interesting hypothesis. The evidence presented below shows that the "reptilian brain" is, however, critically involved in many of the interrelated processes involving emotion, cognition, and the brain (Cabanac, 1999).

MacLean (1985) derives play from family life, and in that sense, his view is compatible with the emphasis on parental care in providing an environmental context in which social play, as characterized above, could evolve. For MacLean, however, play evolved along with the limbic system and is a distinctly mammalian attribute. However, the frequent occurrence of social play in many birds and in a few reptiles (and perhaps fish) appears to make this dependence suspect. However, the limbic system and associated structures do have homologues and precursors in reptilian and even anamniote vertebrates (Butler & Hodos, 1996), and nonavian reptiles have more parental care than was previously suspected (Shine, 1988). Thus, precursors or convergent evolution of play is to be expected according to SRT. The problem has been that if play is seen as primarily a higher cognitive function most prevalent in "higher" animals, especially mammals with lots of neocortex, then the restriction of play to such animals is not only obvious but negates any reason to look elsewhere. If, however, play is derived from instinctive motor patterns and has endogenous motivational/emotional components, then the "reptile" brain is where we need to look for neural input into the primary processes of play, especially nonsocial play, as well as the beginnings of its secondary evolution.

This view, that nonmammalian brains provide essential information for understanding the evolution and function of advanced mammalian brains, is a key message in a provocative review of mammalian brain evolution by Deacon (1990). He argues that claims about progressive evolutionary novelties in brain structures are the result of misunderstandings of homologies, development, functional constraints, as well as the consequences of increasing brain size. "With the differentiation of new neural circuits from ancestral circuits and the elaboration of corresponding new functional adaptations we can expect to trace functional homologies in the form of underlying functional similarities and constraints. Even in extreme cases in which neural structure is co-opted for new adaptations that are radically different than the ancestral function, the underlying homologies will likely exert a major organizing influence on the form and range of variability of the new function" (Deacon, 1990, p. 638). In practice, the changes involved in the nervous system to alter species-specific behavior need not be very great and can be based on alterations in receptor type, receptor distribution, and neurotransmitter inputs. A recent review concludes that "rather subtle changes in the nervous system can cause large and important changes in the behavior of an organism.... even single neurons can show dramatic alterations with very small changes in parameters such as the density of ion channels.... modulatory inputs to neural networks seem to be a very plastic trait in CNS evolution.... The nervous system is organized in such a way that it can accept these changes and incorporate them to generate novel species-specific behavior" (Katz & Harris-Warrick, 1999, p. 633). Thus, the essentially conservative nature of the vertebrate brain and the cascading behavioral changes derived from small alterations in the nervous system support the derivation of play, language, and "higher" mental functions from evolutionarily old systems that may be retained largely intact (cf. Wilczynski, 1984). Small changes in the developmental and selective contexts in which behavior is expressed can lead to rapid changes in behavioral phenotypes and cognitive abilities that mark mammalian evolution. Thus the derivation of secondary and even tertiary processes of play from primary processes is compatible with comparative neuroanatomy as the isomorphism between the triune brain and taxonomic position breaks down.

STUDIES ON BRAIN REGIONS AND PLAY

Brain ablation, drugs, hormones, social isolation, and other experimental methods have been used to understand the neurobiology of play, and extensive reviews are available (Panksepp, 1998; Panksepp *et al.*, 1984; Pellis & Pellis, 1998a; Vanderschuren, Niesink, & Van Ree, 1997)). The vast majority of this work has focused on play fighting in rodents. With the limitations of such taxonomic and play-type bias acknowledged, as well as interpretive problems in localizing brain function, the findings allow us to begin to answer some of the questions about the neural substrates of play. Most of the evidence implicates the forebrain, particularly the telencephalon. In the diencephalon, some hypothalamic lesions that disrupt sexual and aggressive behavior (e.g., medial preoptic) did not affect play at all, whereas other hypothalamic areas (e.g., ventromedial) decreased play initiation and maintenance, perhaps due to increased aggression (Panksepp *et al.*, 1984; Vanderschuren *et al.*, 1997). Thalamic lesions interfering with somatosensory transmission disrupt play bouts (e.g., reducing pinning), but not play initiation. Immediately it is seen that even in the play fighting of but one species, the neural control may be complex and stage (initiation, engagement)-specific.

In the following sections, various neural (brain and neurochemical) substrates are discussed and their relevance to specific features of play identified. These often speculative links lead to predictions that can be tested with lesion, stimulation, and brain imaging studies. However, the larger message might be that many parts of the nervous system are involved, and these vary across play types and among aspects of a given type of play. For example, the motor systems underlying play may be more fixed than the perceptual and contextual aspects of play. While the motor patterns used in play may become refined, the integration of sensory and motor systems may be more heavily affected due to the unpredictable consequences of interacting with objects, social partners, and even one's own body (e.g., bouncing on hard or soft ground). These aspects of play are shared with other behavior systems, of course, but play may exaggerate, push, or refine other developmental systems.

BASAL GANGLIA. The basal ganglia or striatopallidal complex (Butler & Hodos, 1996) is composed of dorsal and ventral striatopallidal complexes and include structures such as the caudate nucleus, putamen, globus pallidus, nucleus accumbens, olfactory tubercle, and substantia innominata. This complex, especially the nucleus accumbens, has output to the substantia nigra (at least in mammals) and ventral tegmental area (VTA). The basic organization of the basal ganglia was already established in ancestral tetrapods and did not evolve later with amniote vertebrates as believed until recently (Marín, Smeets, & González, 1998). Siviy (1998) reviewed the neurobiology of play, studied primarily through use of "pinning" behavior in play-fighting laboratory rats (see also Panksepp, 1998; Panksepp *et al.*, 1984; Vanderschuren *et al.*, 1997). Basal ganglia lesions interfere with motor patterning, including play fighting in rats (Siviy, 1998). The nucleus accumbens and caudate putamen seem particularly involved in social play. Furthermore, the basal ganglia are prominent in reptiles (Butler & Hodos, 1996) and are involved in both instinctive behavior patterns (Greenberg, Fort, & Switzer, 1988) and learning (Punzo, 1985). However, the "fast-paced social interaction, in which the intentions of the play partners must be constantly evaluated … might require more flexibility than the basal ganglia are capable of, hence an increased dependence upon limbic structures in mammalian play" (Siviy, 1998, p. 222). Supporting this is that videos of object play that we have recorded in fish, aquatic turtles, and Komodo monitors (e.g., G. M. Burghardt, Chiszar, Murphy, Romano, & Walsh, in preparation) look much more mammal-like when viewed at four to eight times normal speed.

Exploratory or appetitive "seeking" behavior (Panksepp, 1998) is also influenced by the basal ganglia (medial pallium). While exploration is not itself play according to our five criteria, it does overlap with play, especially early in ontogeny, and could well be an essential precursor of play. Exploratory behavior, especially of biologically significant stimuli and novel environments, is common in fish and reptiles (Burghardt, 1984). In humans, brain imaging studies have shown that the ventral striatum responds to novel information, such as subtle visual sequence changes, and this process occurs without awareness (Berns, Cohen, & Mintun, 1997). A recent study on maze learning in rats (Jog, Kubota, Connolly, Hillegaart, & Graybiel, 1999) traced the changes in neuronal firing in the sensorimotor (caudal) striatum that occurred during learning and habit formation and found particularly marked changes at the beginning and end of the maze run, with some neurons firing at both start and goal attainment. The authors speculate that dysfunctions of basal ganglia systems may occur where organisms may have difficulty either initiating or stopping actions, or in switching from one behavioral sequence to another. Such phenomena, they claim, may underlie repetitive stereotypical behavior in degenerative neural diseases such as Parkinson's disease.

Based on findings such as the above, I would predict that in social play interference with the nucleus accumbens and caudate putamen would disrupt the patterning of the behavior and that interference with the caudal striatum would interfere with the initiation or cessation of play bouts. The triggering response to novel objects in the environment might be controlled by the ventral striatum.

LIMBIC SYSTEM.   The limbic system is composed of the hippocampus, dentate gyrus, subicular and entorhinal cortices, cingulate gyrus, amygdala, and septum, but is also involved with the striatum and diencephalon. The limbic system is associated with emotional responses more positive than the rage and fear associated with the striatopallidal complex and may modulate the latter. For example, ablation of the septum or amygdala increases levels of irritable and aggressive behavior (Panksepp, 1998).

However, the concept of a structurally or functionally integrated limbic system is increasingly being questioned (Durant, 1985; LeDoux, 1991) as are component structures such as the amygdala (Swanson and Petrovich, 1998). In fact, the amygdala may play important roles in both integrating somatic and visceral processes (Blessing, 1997) and associated learning such as fear conditioning (LeDoux, 1991).

Although positive emotions are hard to decipher in the rather facially inexpressive ectothermic reptiles (Bowers & Burghardt, 1992), recent reports suggest that they do respond emotionally to stimuli, in contrast to amphibians and fish (for review see Cabanac, 1999). Social play seems to depend on aspects of the limbic system in that affiliation is required for the physical contact and "bonding" seen among play partners. This affiliation depends on the amygdala and cingulate gyrus in both rodents and primates (Pellis & Pellis, 1998a). Furthermore, in rats it has been shown that the amygdala is a target tissue for androgen-induced increased play fighting in male rats. Lesioning the amygdala in juvenile rats did not affect social play in females, but reduced the male level to the female level, implicating a role for perinatal androgens (Meaney, Dodge, & Beatty, 1981; Vanderschuren et al, 1997). Damasio (1999) presents data, primarily from humans, that bilateral impairment of the amygdala reduces both fear in threatening situations and the ability to even learn to fear dangerous situations. Does this imply that female rats are normally less stressed or competitive in play than males? Are any reward components derived from simulating "fear" in play fighting impaired and thus play frequency in males declines?

Play is marked by sequences of behavior derived from ethotypic responses. Graybiel (1995) concludes that the basal ganglia and limbic system act as a goal attainment system with the former responsible for the establishment and execution of motor patterns and the latter with the recognition of goals and evaluation of behavioral outcomes. The basal ganglia and limbic system are tightly interconnected in mammals and thus tie the motivation and emotion regions of the brain with both effector systems and more cognitive neocortical (e.g., prefrontal cortex) processing (Groenewegen, Wright, & Beijer, 1996; Graybiel, 1997).

In contrast with the basal ganglia, then, which are implicated in the motor performance of both object and social play, mammalian studies suggest that the amygdala is essential for the assessment of play partners in social play and the ability to make appropriate decisions, including role reversals and self-handicapping, and to attain species-typical outcomes such as pinning. Except for emydid turtles, however, there is little evidence of social play in reptiles (Burghardt, 1998a)

Learning that takes place through locomotor and object play might be consolidated in the hippocampus as there is increasing evidence that the hippocampus is involved in spatial learning, not just affective responses (Sherry, Jacobs, & Gaulin, 1992). In fact, mice raised in enriched environments with many objects with which to interact had 15% more neurons in the dentate gyrus than controls and showed better spatial learning in a water maze (Kempermann, Kuhn, & Gage, 1997). A related genetic study is discussed below. In rats, enriched environments lead to similar improvements on the same task as well as greater levels of nerve growth factor (involved in dendrite formation) in the hippocampus (Pham, Söderström, Winblad, & Mohammed, 1999). Thus, a secondary process for object and locomotor play may be increased neuron growth that may then be deployed for enhanced learning in other settings that may not involve play at all. It also has been established that neurogenesis occurs in adult brains of birds and mammals (Gross, 2000) and also in adult ectothermic reptiles (Font *et al.*, 1995, 1997).

Similar convergences in the neural substrates of play systems almost certainly exist. For example, evidence reviewed below suggests that the prefrontal cortex might be particularly important in the motor performance of voluntary activity generated in the basal ganglia and limbic system as well as be involved in spatial working memory (Halgren & Marinkovic, 1995; Beiser & Houk, 1998).

NEOCORTEX.   The most complex neural expressions of behavioral and cognitive plasticity are found in the neocortex, but even here motor and perceptual processes are heavily represented. In mammals, the dorsal pallium becomes expanded into the six-layered neocortex or isocortex (Butler & Hodos, 1996) and this becomes the most voluminous part of the brain in many species. It has been clear for almost two decades that cortical representations in adult animals are not fixed entities, but are dynamic and are continually modified by experience (Buonomano & Merzenich, 1998).

Kolb and Whishaw (1998, p. 47) provide a comprehensive review of brain plasticity and state that in "studies of the effects of housing on rats in enriched environments, we consistently see changes in young animals in overall brain weight on the order of 7%–10% after 60 days.... It would be difficult to estimate the total number of increased synapses, but it is probably on the order of 20% in the cortex, which is an extraordinary change!" Other studies show comparable results of enrichment in rodents. Even in other contexts, brain plasticity is manifested. "One of the most intriguing questions in behavioral neuroscience concerns the manner in which the brain, and especially the neocortex, can modify its function throughout one's lifetime. Taken all together, the evidence discussed above makes a strong case for a relationship between brain plasticity and behavioral change. Indeed, it is now clear that experience alters the synaptic organization in species as diverse as fruit flies and humans" (Kolb & Whishaw, 1998, p. 58). Can play be responsible for much of this change? Enrichment experiences for mammals as diverse as rodents, bears, and primates involve presenting stimuli in which play responses are elicited. It must be pointed out, however, that what passes for enrichment in captive animal studies is more often a pale compensation for the deprived state in which almost all such animals are kept (Burghardt, 1996). A careful study of the brains of the same undomesticated species in both its natural habitat and typical standard and "enriched" captive conditions has, to my knowledge, not yet been performed.

Interestingly, play fighting in rats is not dependent on having a cortex, as studies of decorticate rodents have repeatedly shown complex play fighting in these animals (Gandelman, 1992; Panksepp, 1998; Panksepp, Normansell, Cox, & Siviy, 1994; Pellis, Pellis, & Whishaw, 1992). Although decorticate rats initiate and defend against playful attacks as well as do normal rats, the pattern of defense is altered so that the pinning frequency is lower (Pellis *et al.*, 1992). Sexual, feeding, and aggressive behaviors are also shown by decorticate rats. However, such rats do show abnormal sexual behavior as adults and these deficiencies are similar to those found in rats raised in isolation as juveniles, and thus deprived of opportunities for social play (and other types of interactions as well). These findings suggest that social play as juveniles may refine adult behavior, although they do not shape it. A social play bout has both offensive and defensive moves and feints that are frequently very quick (Pellis & Pellis, 1998a). As noted in the preceding section, experiences involving object play can significantly increase neuron density in areas other than the neocortex. The developmental changes in play, such as the frequent loss of vigorous play fighting as animals mature, may be related to prefrontal lobe maturation. Evidence suggests that in rats, the prefrontal lobes modulate the output of mesolimbic and nigrostriated systems (Whishaw *et al.*, 1992). If the two systems underlie different components of social play, then the lower level of prefrontal control may allow more rapid fluctuations in chasing, nape contact and pinning (S. Siviy, pers. comm., 2000).

Based on energetic grounds, I earlier predicted that very small animals in a lineage would play less than larger species, using the difference between and mice and rats as an example (Burghardt, 1988, prediction 2). Pellis and Iwaniuk (2000) suggest that the difference might also be due to the lesser amount of postnatal growth found in mice as compared to rats. This observation supports prediction 12, that more altricial animals should play more, having a longer period of juvenile dependence. The argument that prenatal motility is functionally continuous with play also is compatible with the greater amount of play postnatally in more altricial species (Bekoff, Byers, & Bekoff, 1980). Parker and McKinney (1999) claim the reverse, however, that the more precocial mammals grow more slowly, have a longer period of parental care, and have larger brains. Thus, they should play more. This contradiction can be tested with more comparative data, but may not be a general principle, as the claim is supported only by data from selected primates. Primate neonates are born more precocial (e.g., eyes open, fur developed, grasp reflexes strong and effective) than many other mammals, but the arboreal habitat and nomadic foraging of primates, whereby females travel with young rather than depositing them in nests, render any such correlation suspect. Humans are secondarily altricial in this view, a consequence of the larger brain and resulting head size that makes earlier delivery a mechanical necessity.

Another factor is brain density. Generally, larger species of mammals have larger brains, but both relative brain size and the absolute density of neurons in the cortex decreases (Abeles, 1991). Thus in mice, cortical neuron density has been calculated at 142,500 neurons/mm^3, whereas in rats, it is 105,000/mm^3. In larger mammals, the change is even more dramatic, with cats at 30,800/mm^3, "monkey" at 21,500/mm^3, and humans at 10,500/mm^3. The lower density means larger cells and particularly that more synaptic connections are able to be formed. Although such estimates are fraught with methodological difficulties, Abeles (1991) calculated that in the mouse, each cortical neuron receives 8000 synapses, in the monkey, 20,000 synapses, and in a human, 40,000 synapses. It is clear that with increased neocortex volume in larger brains, both the increased connections possible for each neuron and the total space in which synapse formation can occur raise many more opportunities for experiential effects to create interconnections that are more complex. The sheer number of such connections dwarfs the number of estimated genes in an animal's DNA, especially in the larger brained species. Recently, electron microscopy has been employed to document that long-term potentiation, an important neurophysiological process in learning, actually induces new functional synapses between axons and dendritic spines (Toni, Buchs, Nikonenko, Bron, & Muller, 1999).

The areas of the neocortex that might be most involved in play are the motor cortex, premotor cortex, and the prefrontal cortex, as discussed below.

CEREBELLUM. Although most research has involved the role of the forebrain (especially the telencephalon) in play, the cerebellum, forming the roof of the hindbrain, may also be an important region. The cerebellum is found in all vertebrates. Although the cerebellum is dwarfed by the cerebrum of mammals and of birds, it is, with few exceptions, enlarged in them relative to anamniotes. Recent research has expanded its traditional role in balance and orientation to being a major vertebrate brain site for the control of motor learning, conditioning, and accurate precise movements, including the tracking of both external moving objects and the animal's own movements (Paulin, 1993; Bell, Cordo, & Harnad, 1996). Evidence based on functional magnetic resonance imaging (fMRI) shows that the cerebellum is involved in "predicting the specific sensory consequences of movements" (Blakemore, Wolpert, & Frith, 1999, p. 448). What was found is that activity in the somatosensory cortex was related to activity in the cerebellum when self-produced movements led to tactile stimulation, but not when the tactile stimulation was externally produced. In addition, the cerebellar cortex is well suited for the task of incorporating patterned behavioral organization into motor and learning systems in the brain (Houk, Buckingham, & Barto, 1996). These ideas are expanded and refined in Braitenberg, Heck, and Sultan (1997) and the commentaries therein. As Courchesne (1997, p. 249) concludes, "In sum, the cerebellum plays an anticipatory role in the learning and smooth coordination of diverse neurobiological features." It is thus not surprising that the cerebellum might be involved in play. Interestingly, it is in the weakly electric mormyrid fish (e.g., elephant nose fish), that process many and complex electrical signals, that object play involving fine motor control has been described. Here the cerebellum is huge, even overlying the entire forebrain (Butler & Hodos, 1996). All these observations support the prediction that the cerebellum is involved in the rapid and refined movements involved in active play of all types.

The role of the cerebellum in play ontogeny has been the subject of recent theory. In a review of the functions of play, especially the exercise or "getting in shape" theories, Byers and Walker (1995) point out that none of 16 different physiological exercise effects posited for play, such as increased oxygen-carrying capacity in the blood, increased blood volume, and increased endurance, have permanent effects. Once a long distance runner shifts to the couch, one can't keep the fitness accrued by prior exercise in the bank, let alone amass interest! By looking at the ages when play peaks in three domestic species (mice, rats, and cats), Byers and Walker found that these play peaks coincided closely with the ages that permanent experience-dependent synaptic development occurred in the cerebellum. Although this is a correlation based on a small sample size, the idea that play is associated with a sensitive period for neural integration is an important one (Byers, 1998). A recent study evaluating limited social play deprivation in rats found a decrement in amount of social interaction in adulthood when rats were not able to play during a period comparable to the period that Byers and Walker (1995) argue is the critical period for cerebellar synapse formation (Berg *et al.*, 1999). However, synaptic plasticity in the cerebellum has been found in adult rats given daily exercise in a complex 3 dimensional maze: this was not found if rats had comparable amounts of activity in a running wheel (Kleim *et al.*, 1998). Certainly the cerebellum is involved in play activities and may also be involved in the refining of movements performed in play. Play, which in its typical guise involves vigorous and quick movements in limbs, torso, and head in response to rapidly changing visual stimulation, may involve the cerebellum. The only work I am aware of on play in cerebellum impaired animals are the unpublished studies by Panksepp's laboratory on rats with total cerebellectomies. The rats grew up fine but were "spastically uncoordinated". Although they tried to play when put together, they were "all over the place". Thus the motivational system was unimpaired, although the execution system was drastically affected (Panksepp, pers. comm., 2000). The best example of social playlike behavior in a nonavian reptile is the precocial courtship of young emydid turtles, which have a complex courtship involving vigorous movement (Kramer & Burghardt, 1998). It is possible that precocial courtship aids a turtle in the three-dimensional maneuvering needed for successful adult courtship (Burghardt, 1998). Postnatal neurogenesis in the cerebellum has been found in an emydid species (*Trachemys scripta*) that shows precocial courtship (Enrique Font, pers. comm., 2000).

Nonetheless, although the relationship of play with degree of cerebellar development is intriguing, there are some questions. One is that different kinds of play (social, object, locomotor) wax and wane at different times. Second, changes in many other parts of the brain may also be taking place at the same time rapid cerebellar changes are occurring, and it is most likely that play, being such a diverse phenomenon, may be involved in brain development over many parts of the nervous system. The evidence reviewed above on extensive cortical development due to enrichment suggests that rather permanent changes in synapse connections might be occurring in other brain areas. Furthermore, positive feedback loops for information processing and working memory involved in sustaining activity in prefrontal cortex neurons and the dentate nucleus in the cerebellum, as well as trans-striatal loops through the basal ganglia to the prefrontal cortex, have also been described (Beiser & Houk, 1998).

In addition, one of the features of play is its organization in bouts of intermittent activity. This noncontinuous activity is not likely to provide the cardiovascular benefits posited for intense sustained activity. Recently work throughout the phylogenetic tree has established that start and stop activity is very common and, in fact, appears to be a more energy-efficient strategy than continuous locomotion. In mammals, including human beings, the muscle enzyme pyruvate dehydrogenase has been identified as involved in the mechanism underlying the relative efficiency of intermittent locomotion (Pennisi, 2000)

In conclusion, the specific neural circuits involved in any type of play are still little known; specifically, there is no definitive evidence for neural circuits limited to play (Panksepp, 1998). There is now need for brain imaging studies to localize patterns of brain activity accompanying play and related phenomena, although doing so without immobilization is not yet possible. Assessing glucose metabolism or multiple EEG recording may be more promising initially, although behaviorally controlled gene expression is the most exciting prospect for the future (Jarvis *et al.*, 2000). An initial attempt in a c-fos gene activation study found elevated activity in some brain areas after a play bout (N. S. Gordon, S. Kollch-Walker, H Akil, and J. Panksepp, submitted). In the hindbrain large significant increases were found in the deep and dorsolateral tectum. In the midbeam only the ventromedial hypothalamus showed significant activation. In the forebrain, dorsal and ventral striatum and somatosensory cortex were significantly elevated. Interestingly, there was no change in the cingulate or amygdala.

Criterion 5 (relaxed field) for the identification of play is based on the recognition that true play does not begin when animals are stressed. In a quantitative study of social play in squirrel monkeys, *Saimiri sciureus*, cortisol levels were inversely related to the amount of play, especially when separated animals were reintroduced after an interval (Biben & Champoux, 1999). Is there a mechanism by which this effect is mediated? Maier and Watkins (1998, p. 97) argue that "stressors activate circuitry that evolved to mediate defense against infection and produce sickness behavior." Developing a model to represent how cytokines communicate with the brain and immune system bidirectionally, they present evidence on the mechanisms underlying the changes in behavior, cognition, and emotion that occur in sickness. Stress-related chemicals such as interleukin-1, released from the peripheral nervous system, cross into the brain, where there are receptors for them. Such chemical input to the brain may then lead to further production of the same chemical in the brain, especially in the hypothalamus and hippocampus, the latter a major area for experience-induced memory systems. Although Maier and Watkins do not explicitly mention play, the lack of play at developmentally appropriate species-characteristic periods could be an early indicator of stress. Although blockade of interleukin brain receptors reduces symptoms of sickness to infection, the role of such blockade on play has not, to my knowledge, been studied. Nonetheless, evidence suggests that sickness induces a motivational reorganization of behavioral priorities such that essential ones such as nest building and retrieving pups still occur, but less critical behaviors are replaced with those more amenable to recuperation (Aubert, 1999). These findings lend support to Chiszar's (1985) view that juvenile playfulness may communicate information to parents about the competence of their offspring, especially if "competence" is expanded to include current conditions and state of the animal.

The possibilities are even more far-reaching. Recent work shows that maternal responses by rat mothers to their offspring differ in the amounts of grooming, licking, and arched-back nursing. Offspring from mothers with high rates of these maternal behaviors are less fearful when placed in a novel environment, are less easily stressed, and show high rates of maternal behavior, all of which are transferred to their offspring (Francis, Diorio, Liu, & Meaney, 1999). Cross-fostering studies showed that the transmission was nongenetic. Does maternal behavior affect social play behavior as well? This has not been studied in this paradigm, although social (and thus play) deprivation at the age of 4 and 5 weeks in rats causes defects in adulthood (Berg *et al.*, 1999; Hol, Berg, van Ree, & Spruijt, 1999).

A number of studies have looked at neurochemicals and social play (for reviews see Panksepp, 1998; Siviy, 1998; Vanderschuren *et al.*, 1997). Methods include use of both antagonists and agonists. Although the results are not totally consistent, given the complex nature of just social play even in rodents, this is not surprising. In fact, many of the neurological and neurochemical studies in this area may need to be reconsidered as more sophisticated ways of looking at the topography and sequential organization of rodent play fighting are incorporated into future work (Pellis & Pellis, 1998). Some patterns do appear, however. Parts of the basal ganglia and limbic system have considerable serotonergic and endorphin receptors. Noradrenergic system blockade increases play and agonists decrease it. Serotonergic blockade either increases or has no effect on play. Agonists are inconsistent and show reversals at different doses. Cholinergic system blockade reduces play (Siviy, 1998), but others report mixed effects (Panskepp, Siviy, & Normansell, 1984; Vanderschuren *et al.*, 1997).

343

PLAY:
ATTRIBUTES AND
NEURAL
SUBSTRATES

Low levels of morphine increase social play. This is particularly seen in an increase in a rat's pinning frequency or dominance in play fighting; opiate antagonists reduce both levels of social play and relative dominance during such play (Panskepp, 1998). Interestingly, amphetamines and methylphenidate (Ritalin), which increase levels of catecholamines, greatly reduce social play and play initiation or soliciting while increasing attention and exploration. This provides a means of validating the separation of play and exploration (Panskepp, 1998). The reduction in play may, however, have little to do with the effect on catecholamines (Vanderschuren *et al.*, 1997).

Dopamine is the neurotransmitter most often implicated in play and is heavily concentrated in the basal ganglia and associated structures implicated in play. In fact, the substantia nigra and ventral tegmental area (involved in motor systems) have perhaps the highest concentrations of dopamine in the brain (Butler & Hodos, 1996). Dopamine is involved in reward, pleasure, arousal, and motor patterning systems of motivated behavior. Exploiting the sequential organization of play sequences (i.e., initiation, engagement, and termination) may be useful in clarifying the role of dopamine in play. This literature cannot be explored in depth here, but it is important to note that the studies of Siviy and others suggest that dopamine antagonists reduce play, but agonists do not always facilitate play. To resolve this inconsistency, Siviy (1998) proposes that dopamine might be involved in the initiation of play through anticipated reward systems and summarizes suggestive experimental evidence. He speculates that

> Stimuli which predict a playful experience would result in increased activity in the dopaminergic mesolimbic pathway. This would result in an increased release of dopamine in mesolimbic terminal areas, such as the prefrontal cortex and nucleus accumbens, resulting in energization of the animal and behavior patterns that would increase the probability of a playful interaction. Because of the diffuse nature of noradrenergic, serotonergic and opioid pathways, these systems are likely to exert a more modulatory influence on how the play bout will unfold. Increased noradrenergic activity may enhance the ability of a rat to focus its attention on the task at hand (i.e., playing), while increased opioid activity may enhance the pleasure associated with playing. For all this to happen, serotonin levels must also be low. (Siviy, 1998, p. 232)

Play with objects usually involves novelty, and responses to novel objects may be processed differently than responses to novel environments (Besheer, Jensen, & Bevins, 1999; Orofino, Ruarte, & Alvarez, 1999). Dopamine subtypes are involved in different aspects of novel object learning and place conditioning (Besheer *et al.*, 1999). Novelty seeking involves an appetitive phase and is related to stimulus seeking and even risky behavior in people. Depue and Collins (1999) analyzed individual differences in people in extraversion, which involves positive affect, optimism, sensation seeking, sociability, assertiveness, high activity, and being energetic (and many similar terms derived from various converging personality typologies). They summarize these terms with the label *positive incentive motivation*, relate the studies on humans to nonhuman mammal studies, and claim that approach, exploration, and engagement is the final general behavioral path. Activity of the medial orbital cortex and parallel cortical networks that integrate complex cognitive processes guide "incentive motivated behavior through the environment" (p. 498). The central modulatory system here is again the VTA dopamine projection system; individual differences in the VTA dopamine system are shown to modulate reward, incentive, volitional, and coordinated locomotor behavior. Although play is never mentioned, they conclude that "human incentive motivation is associated with both positive, emotional feelings such as elation and euphoria and motivational feelings of desire, wanting, craving, potency, and self-efficacy" (p. 504). Note that these dopamine-oriented terms feature some of the key attributes of play developed in the five criteria, notably criteria 2 and 5.

Depue and Collins also suggest a model that could be used to test these ideas in regards to play. Different inbred strains of mice differ genetically in the number of dopamine neurons produced prenatally. Consistent with the theory, C57BL6 mice have more dopamine neurons than DBA2 mice and show more novelty-induced exploratory behavior than DBA2 mice. Rats would be the prime animals to test since different strains also vary in their dopamine concentrations and playfulness (Siviy, Baliko, & Bowers, 1997). Recently, sensation seeking and risk taking behavior in adolescent human beings has been linked to differential maturation of dopamine terminal regions (Spear 2000a, b). As such behavior is part of 'risky play' (drug use, extreme sports, sex, gambling), we may be closing in on a predictive understanding of otherwise difficult behavior to fathom. Testosterone levels may also be involved and thus explain some gender differences.

Dopamine is also involved in the sensitization of behavior and other experience-dependent processes in development. Synaptic plasticity in behavioral sensitization "occurs in part through the interaction of DA and glutamate in the VTA region" Depue and Collins (1999, p. 508). Behaviorally, high and low responders to novelty differ in the former having more dopamine receptors. One feature of this model is that extraversion as a social response is associated with solitary responses to novel stimuli. The later is akin to object play. Thus, this model helps bring together the often heterogeneous types of play and helps explain their association.

A recent experiment using transgenic mice overexpressing the N-methyl-D-aspartate (NMDA) receptor 2B in the neocortex and hippocampus showed that such mice as well as controls explored and interacted with novel objects (Tang et al., 1999). When a new novel object was introduced, both groups retained the discrimination of novel from habituated stimuli for 1 hr, but only the transgenic animals retained the discrimination for 1 or 3 days. This suggests that small genetic changes in neuroreceptors can greatly affect the utilization of the exact same experience in later behavior. Such an ability could, if adaptive, be readily enhanced by natural selection. The transgenic animals also showed faster water maze learning and both more rapid fear conditioning and subsequent extinction. Physiological evidence from brain slices showed greater synaptic potentiation in the transgenic mice that might underlie the differences in learning. The NMDA receptor "does serve as a graded molecular switch for gating the age-dependent threshold for synaptic plasticity and memory formation" (Tang et al., 1999, p. 633).

A provocative review of the neural basis of learning has argued that dopaminergic neurons with fluctuating outputs "signal changes or errors in the predictions of future salient and rewarding events" (Schultz, Dayan, & Montague, 1997, p. 1593). This view is based on evidence that cues attached to learned predictable cues already associated with responses or stimuli are not themselves learned, as they add no new information. Thus, if a rat learns a conditioned response to a visual cue, a subsequent auditory cue added to the highly predictable visual cue will not become itself a learned cue. Only if there are deviations or probabilities will the learning to the new cue take place. The authors argue that dopamine neurons in the ventral tegmental area and substantia nigra, brain areas long associated with processing of rewarding stimuli, are thus pivotally involved in target structures such as frontal cortex, striatum, and nucleus accumbens. The fluctuating output of these dopamine neurons is correlated with error signals and "the delivery of this signal to target structures may influence the processing of predictions and the choice of reward-maximizing actions" (Schultz *et al.*, 1997, p. 1597). They specifically note that these mechanisms should operate on "ethologically important" time scales (p. 1597). Note also, given criterion 4 for play, that "dopamine neurons emit an excellent appetitive error (teaching) signal" and that dopamine neurons are also involved in responses to novel stimuli and decrease (habituate?) with repeated stimulation. There is also evidence supporting the role of mesolimbic dopamine in aiding formation of learned associations between contextual stimuli and aversive or rewarding events, which can underlie compulsive behavior and addictions (Spanagel & Weiss, 1999). Thus, the case for dopamine as an important chemical in all kinds of play is further enhanced, as well as a possible linkage with compulsive and addictive behavior. All mammals show a decline in play fighting after puberty. A mesolimbic dopamine system has been found to underlie the age-related modulation of social play in rats (Siviy, 1998). From a comparative perspective, new data are establishing the ubiquity of catecholamine systems, especially dopaminergic ones, in the brains of all vertebrates and developmental studies are beginning (Smeats & González, 2000).

This review establishes that the basal ganglia, VTA, prefrontal cortex, and dopamine systems are critically involved in reward, anticipation, memory, and goal orientation seen in the often fast paced, contextually sensitive, and anticipatory responses (Pellis & Pellis, 1998a, b) of locomotor, object, and social play. However, these systems are also involved in many nonplay activities, suggesting that the pathways involved may overlap, converge, or reinforce a variety of behavior systems. Thus the hypothesis that play originated in the initiation and execution of instinctive behavior sequences, in which motor performance was itself rewarding, and through which repeated performance could enhance performance through practice in changing contexts and experience-based modifications of sequences, is plausible. The difficulty in isolating specific long-term benefits through play may be due to the fact that play may provide only one of several ways to enhance behavioral and cognitive performance, a facilitator not essential in every case. Play may be the preferred method if animals are in good conditions characterizing the "relaxed field" criterion, but relying just on this easily inhibited activity would not be adaptive: play as a more incremental system to enhance behavior might be more readily evolved. Nevertheless, although the exact neural systems involved in behavior and learning are only imperfectly understood, play, as characterized here, may be a key concept in integrating the often narrowly focused neuroscience research with the development and evolution of behavior and the landmarks of higher mental performance.

INTERNALIZING MOTOR PATTERNS: FROM ACTION TO IMAGINATION

In people, including children, play fighting and rough and tumble play are little studied. With the exception of organized sports, this is actually true of physical activity play in general (Pellegrini & Smith, 1998). Outside of the movements (e.g., circular reactions) of infants, the literature on human play generally focuses on sociodramatic play, pretend play, imaginary play, verbal play, creativity, and fantasy

(Burghardt, 1999), which may all be summarized as mental play. The putative importance of these kinds of play for cognitive and social–emotional development is emphasized by postulating internal processes that are indirectly measured through behavioral output such as artistic production, building block structures, role playing, ability at board games, etc. These types of play are favored and encouraged because they are thought to enhance cognitive ability and social/emotional skills (Johnson, Christie, & Yawkey, 1999). The neglect of physical "free" play as valuable for everyone is shown by the reduction in recess time in elementary schools in the United States.

If, however, the evolutionary origins and ontogenetic precursors of "mental play" are in active motor play, then the neglect of active motor play in preschool through elementary grades might actually impair some areas of cognitive and social performance of children and adults. Pellegrini and Smith (1998) argue for a continuity of physical play from the rhythmic stereotypes of infants (aids in control of specific movements) to exercise play (which may have some incidental cognitive benefits) to rough and tumble play (learning of social dominance roles). These are immediate rather than deferred benefits. Unfortunately, as these authors know, the research documenting such deferred benefits of play is controversial and difficult to carry out both in human and nonhuman animals (Martin & Caro, 1985).

It is difficult to show that the skills or enhanced physical condition derived through motor play could *only* be accomplished by physical activity in play (Burghardt, 1984). Play, as a product of surplus resources, may be, in terms of physical condition and prowess, only one of several means to this end. As shown above, the endogenous motivational and emotional concomitants of play (criterion 2) might serve as mechanisms to stimulate activity in animals, including humans. Perhaps much motoric play is more important for cognitive and emotional development than it is for motor development; if so, we need to examine the relationship between physical and mental play by searching for neurological connections between the two processes. In the following sections, some suggestive evidence is presented that this link can, and is, being made.

Play involves activity with the self, objects, other species of animals, and especially conspecifics. The vigorous activities of play in which feints, exaggerated movements, and other activities critical to predicting and anticipating actions of objects or other organisms are prominent features of play (see criterion 3, structural difference). Visual-motor integration may be heavily involved, although, of course, other sensory systems may be at play as well. Can these aspects of play be linked in any way to imagination?

BASAL GANGLIA AND NEOCORTEX

A body of work is accumulating on the relations among motoric acts, mental imagery, and the brain (Jeannerod, 1994; Kosslyn & Sussman, 1995). Circuits in the basal ganglia are involved in both motor learning and cognitive behavior (Graybiel, 1997; Jog *et al.*, 1999), suggesting that the links between motor performance and cognitive processes are either very ancient or slight modifications from ancestral vertebrate systems (Katz & Harris-Warrick, 1999). Striatal projection neurons depend on afferent and loop circuits in both neocortex, especially premotor, parietal, and prefrontal cortex, and the thalamus (Jog *et al.*, 1999).

Particularly intriguing is a series of papers from a research group in Parma, Italy, that suggest a mechanism whereby a more direct cognitive link may be made.

Initially this research group showed that neurons in area F5 of the premotor cortex (near the arcuate sulcus) fired when monkeys (*Macaca nemestrina*) performed a goal-directed action, such as reaching for and grasping a piece of food. However, they also showed that some of these same neurons fired when the monkeys just *observed* the experimenter doing the same thing (Gallese, Fadiga, Fogassi, & Rizzolatti, 1996). The authors termed these neurons "mirror neurons" and found several types. Some of the neurons were specific to single actions by the monkey or the observed experimenter such as grasping, holding, manipulating, and placing. Other mirror neurons were specific to two or three actions combined such as grasping and placing, placing and holding, or grasping, placing, and holding. It is interesting to note that this area of the premotor cortex has been considered homologous with Broca's area in the human brain. The authors argue that such mirror neuronal systems in humans could be involved in the recognition of both actions and sounds. Further experiments showed that the mirror neurons did not fire when the observed experimenter reached for food that was not present or if he was in darkness and could not be seen.

To test their findings in humans, in which electrophysiologic recording from single neurons deep in the brain was not feasible, the motor cortex was stimulated by transcranial magnetic stimulation (Fadiga, Fogassi, Pavesi, & Rizzolatti, 1995). Electromyographs or motor evoked potentials (MEPs) were recorded in four hand muscles under several conditions. The authors found that MEPs only increased when the subjects saw an experimenter grasping objects or observed the experimenter tracing geometrical figures with his arm (e.g., Greek letters). Increases were not found when the subjects just looked at the three-dimensional objects or had to signal verbally when a light dimmed. Comparable responses were recorded when the subjects themselves executed the behavior they observed the experimenter performing. Similar results were obtained when positron emission tomography was used to localize the brain regions involved (Rizzolatti *et al.*, 1996). Mirror neurons might underlie imitation, and some evidence suggests that play in animals facilitates imitation of novel behavior (Miklosi, 1999).

These and other studies suggest that mental imagery, physical movements, and perception can be linked in areas of the brain even down to involvement of the same neuron (Jeannerod, 1994). Thus "imagery is a bridge between perception and motor control" (Kosslyn & Sussman, 1995, p. 1040). This may be a common phenomenon. The same striatal neurons in rats may fire at both the start and end of a maze run (Jog *et al.*, 1999). With accumulating evidence on the great dendritic (synaptic) changes that can rapidly occur in the brain, practice and repetition of similar but not identical behavior during play may subserve a role in the shifting mental states involved in anticipating, predicting, and controlling one's own behavior in relation to external stimuli.

MENTAL ACTIVITY

The above studies model how we could move from play as a motoric, active response to stimuli to active play involving pretense and make believe. What about behavior that seems largely divorced from physical actions such as imagination, fantasy, creativity, and complex social assessments? These phenomena may be more closely linked than our theories suggest. Pretense, fantasy, and creativity also involved exaggerated, novel, even distorted and seemingly inappropriate responses. Many people need to draw, sound out, or act out their ideas to "test them." Talking to oneself is commonplace, and doing it soundlessly may be the essence of "thinking," which is often the rehearsal of different actions and possible outcomes. Is there any evidence for merely thinking affecting motor behavior?

Motor imagery is the process of imagining behavioral actions. When people are asked to imagine movements, the neurophysiological responses in various brain regions resemble those made when such movements are actually executed (Kosslyn & Sussman, 1995; Yágüez, Canavan, Lange, & Hömberg, 1999). This also happened with mental rotation tasks in humans and in monkeys when preparing to move an arm in a specific way (Kosslyn & Sussman, 1995). When people were asked to imagine writing a letter and then actually wrote the letter, the same areas in the cerebellum, prefrontal cortex, and supplementary motor cortex were activated (Kosslyn & Sussman, 1995).

Is it possible that there is a functional link between the mental imagery and physical performance in that the former enhances the latter? Intriguing support is provided by several experiments that demonstrate the role of mental imagery in enhanced motor abilities (Lovell & Collins, 1997; Yaguez et al., 1998; Page, 2000).

If merely imagining activity is functional and thus evolutionarily adaptive, the next step is mentally to rehearse different actions. Linking these mental rehearsals with possible outcomes is a hallmark of creativity, innovation, and social adeptness. At a less exalted level, this rehearsal necessitates anticipating the future. Those organisms producing more of these mental options should then have, through natural selection, an advantage in producing more of the innovations that "take." It is far too early to accept such a process on the scientific evidence to date.

However, if the ideas have some merit, we would expect to be able to trace the operation of such mechanisms in a comparative fashion. Although this has not yet been done, initial studies in human adults and children seem promising (Frith & Frith, 1999). Functional magnetic resonance imaging and positron emission tomography both indicated that the medial prefrontal cortex and superior temporal sulcus are involved in interpreting the mental states of characters in stories or cartoons. The authors, in trying to explain the origin of such mental imagination, suggest that they arise from the following four "preexisting abilities that are relevant to mentalizing," all of which have a demonstrated neural basis (Frith & Frith, 1999, p. 1693). These include (1) distinguishing animate from inanimate entities, (2) following the gaze of another individual, (3) representing goal-directed actions, and (4) distinguishing actions of self from those of others.

There is considerable similarity between these four abilities and the four processes involved in basic motor control developed by Willingham (1999) and described above. Willingham presents a neuroanatomic locus for each of the four processes he described including (from "higher" to "lower"), in turn, the dorsal frontal cortex, premotor and posterior parietal cortex, basal ganglia and supplementary motor cortex, and spinal interneurons. There might indeed be similar neural involvement in motor representations and the highest cognitive abilities attributed to imagination, role playing, fantasy, and other types of "mental" play. Moreover, dopamine may be the link among them all. Previc (1999) recently attributed the origins of human intelligence and advanced cognitive abilities to an expanded dopaminergic system. This system, he argues, underlies six of the most important cognitive skills in which we excel including motor planning, cognitive flexibility, temporal analysis/sequencing, and generativity. He also suggests that considering dopamine may offer a new way of looking at the advanced cognitive accomplishments of some nonprimate mammals and birds who have differing cortical anatomies and evolutionary histories. It might also be useful to look at other vertebrates that may have relatively advanced cognitive abilities. Brain size is not necessarily related to play (Iwaniuk *et al.*, 2001). Rather, play may have secondarily provided expanded opportunities for behavioral flexibility, plasticity, and cognitive decision making through the operation of natural selection on behavior, and thus the nervous system, instigated by the evolution of parental care, endothermy, and increased metabolic rates, as outlined in the section, Metabolism, Development, and Stimulation as Setting Factors.

The ultimate transmutation of motor activity to imaginary events may lie in dreaming. Panksepp (1998) argued for the similarity of the play and dream systems neurologically. Interestingly, young birds may actually refine and learn aspects of their songs while they sleep (Dave, Yu, & Margoliash, 1998). Apparently, during sleep, neurons in two brain areas important to song learning both respond to the songs, while only one of the areas did so in birds that were awake. Subsequently it was shown that "... 'spontaneous' activity of these neurons during sleep matches their sensorimotor activity, a form of song 'replay'" (Dave & Margoliash, 2000). Much more needs to be done in this area, but this pioneering study does suggest a playlike role for the ultimate relaxed field.

CONCLUSIONS

This chapter has ranged widely in the search for neural processes that may underlie playful behavior. The view of play and its origins outlined here suggests the episodic development throughout vertebrate evolution of playlike behavior under suitable ecological and physiological conditions. Using the five criteria for play, there is compelling evidence that play evolved independently in several lines of fishes, turtles, squamate reptiles, crocodilians, birds, and marsupials, as well as virtually all families of placental mammals (Burghardt, in press). Nonetheless, play is clearly most common in mammals and birds, groups with high metabolic rates, endothermy, parental care, and relatively large brains, although the specific relationships among these variables is not well established.

Play is derived from instinctive behavior patterns whose patterning and motivation are controlled by the basal ganglia of the telencephalon and structures in the diencephalon. The rapid rate of evolutionary changes in endothermic animals, especially mammals, in genome size, brain size, and behavioral complexity are remarkable and little understood (Gottlieb, 1992). Play may have been a major engine in this rapid cascade of evolutionary change that led to increased cognitive complexity. This may occur by moving initially playful responses to more serious endeavors and functions, so that the once playful behavior has been transformed and "fixed" so that it has been transformed to being outside the realm of play using the criteria developed above. The thesis here is that after a period of evolutionary reorganization in behavioral ontogeny accentuated by the lengthening of parental care (Burghardt, 1988), play can facilitate rapid behavioral and mental development by providing altered phenotypes for natural selection to prune and shape. It can also reflect deteriorated or developmentally stalled behavior as found in some domesticated species (Burghardt, 1984).

Just as play is not a simple process, neither are the neural substrates of play simple. Ultimately, physiological information will supplant or refine the five play criteria. Play, especially locomotor and object play and the target-directed aspects of social play involve the cerebellum. The basal ganglia are critical in the motivational, structural, and learned aspects of play, explaining why so much play resembles movements found in adult "serious" behavior. The limbic system appears largely responsible for the emotional and affiliative aspects of play. The role of dopamine is important as a facilitator of play, but it may also be a factor in extreme deviations such as dangerous risk-taking behavior, stereotypies, and even addictions. The role of the neocortex may be to provide the animal with a greater ability to utilize the information gathered during interactions with the world, provide neural resources for rapid synapse formation, and derive novel ways of dealing with environmental challenges. From here it is not too far to thought processes, planning, and especially creativity as consequences of playful activities. But much play may still be nonfunctional, evolutionary detritus, or byproducts of developmental retardation. Thus, the paradoxical nature of play endures.

Acknowledgments

The work reported here was partially supported by grants from the National Science Foundation and from the Science Alliance and Research Office of the University of Tennessee. Elliott Blass made many valuable suggestions, including the use of ingestion as a model for primary and secondary process. I also thank Julie Albright, Paul Andreadis, Matthew Bealor, Marc Behoff, Enrique Font, Neil Greenberg, Mark Krause, Matthew Lanier, Jaak Panksepp, Serge Pellis, and Steve Siviy for reading earlier drafts and many others for helpful discussions, references, and suggestions.

References

Abeles, M. (1991). *Corticonics*. Cambridge: Cambridge University Press.

Allen, C., & Bekoff, M. (1998). *Species of mind*. Cambridge, MA: MIT Press.

Aubert, A. (1999). Sickness and behavior in animals: A motivational perspective. *Neuroscience and Biobehavioral Reviews, 23*, 1029–1036.

Barber, N. (1991). Play and energy regulation in mammals. *Quarterly Review of Biology, 66*, 129–147.

Beiser, D. G., & Houk, J. C. (1998). Model of cortical-basal ganglionic processing: Encoding the serial order of sensory events. *Journal of Neurophysiology, 79*, 3168–3188.

Bekoff, M., & Allen, C. (1998). Intentional communication and social play: How and why animals negotiate and agree to play. In M. Bekoff & J. A. Byers (Eds.), *Animal play: Evolutionary, comparative, and ecological perspectives* (pp. 97–114). Cambridge: Cambridge University Press.

Bekoff, M., & Byers, J. A. (Eds.). (1998). *Animal play: Evolutionary, comparative, and ecological perspectives*. Cambridge: Cambridge University Press.

Bekoff, M., Byers, J. A., & Bekoff, A. (1980). Prenatal motility and postnatal play: Functional continuity? *Developmental Psychobiology, 13*, 225–228.

Bell, C., Cordo, P., & Harnad, S. (1996). Controversies in neuroscience IV: Motor learning and synaptic plasticity in the cerebellum: Introduction. *Behavioral and Brain Sciences, 19*, v–vi.

Bennett, A. F. (1982). The energetics of reptilian activity. In C. Gans & F. H. Pough (Eds.), *Biology of the Reptilia* (Vol. 13, pp. 155–199). London: Academic Press.

Beran, M. J., Gobson, K. R., & Rumbaugh, D. M. (1999). Predicting hominid intelligence from brain size. In M. C. Corballis & S. E. G. Lea (Eds.), *The descent of mind: Psychological perspectives on hominid evolution* (pp. 88–97). Oxford: Oxford University Press

Berg, C. L. v. d., Hol, T., Everts, H., Koolhaas, J. M., van Ree, J. M., & Spruijt, B. M. (1999). Play is indispensable for an adequate development of coping with social challenges in the rat. *Developmental Psychobiology, 34*, 129–138.

Bernard, L. L. (1924). *Instinct: A study in social psychology*. London: George Allen and Unwin.

Berns, G. S., Cohen, J. D., & Mintun, M. A. (1997). Brain regions responsive to novelty in the absence of awareness. *Science, 276*, 1272–1275.

Besheer, J., Jensen, H. C., & Bevins, R. A. (1999). Dopamine antagonism in a novel-object recognition and a novel-object place conditioning preparation with rats. *Behavioural Brain Research, 103*, 35–44.

Biben, M. (1998). Squirrel monkey play fighting: Making the case for a cognitive training function of play. In M. Bekoff & J. A. Byers (Eds.), *Animal play: Evolutionary, comparative, and ecological perspectives* (pp. 161–182). Cambridge: Cambridge University Press.

Biben, M., & Champoux, M. (1999). Play and stress: Cortisol as a negative correlate of play in *Saimiri*. In S. Reifel (Ed.), *Play and culture studies* (Vol. 2, pp. 191–208). Stamford, CT: Ablex.

Blakemore, S.-J., Wolpert, D. M., & Frith, C. D. (1999). The cerebellum contributes to somatosensory cortical activity during self-produced tactile stimulation. *NeuroImage, 10*, 448–459.

Blessing, W. W. (1997). Inadequate frameworks for understanding bodily homestasis. *Trends in Neuroscience, 20*, 235–239.

Bowers, B. B., & Burghardt, G. M. (1992). The scientist and the snake: Relationships with reptiles. In H. Davis & D. Balfour (Eds.), *The inevitable bond* (pp. 250–263). Cambridge: Cambridge University Press.

Braitenberg, V., Heck, D., & Sultan, F. (1997). The detection and generation of sequences as a key to cerebellar function: Experiments and theory. *Behavioral and Brain Sciences, 20*, 229–277.

Bullock, T. H. (1977). *Introduction to nervous systems*. San Francisco: Freeman.

Buonomano, D. V., & Merzenich, M. M. (1998). Cortical plasticity: From synapses to maps. *Annual Review of Neuroscience, 21*, 149–186.

Burghardt, G. M. (1973). Instinct and innate behavior: Toward an ethological psychology. In J. A. Nevin & G. S. Reynolds (Eds.), *The study of behavior: Learning, motivation, emotion, and instinct* (pp. 322–400). Glenview, IL: Scott, Foresman.

Burghardt, G. M. (1984). On the origins of play. In P. K. Smith (Ed.), *Play in animals and humans* (pp. 5–41). Oxford: Blackwell.

Burghardt, G. M. (1985). Animal awareness: Current perceptions and historical perspective. *American Psychologist, 40*, 905–919.

Burghardt, G. M. (1988). Precocity, play, and the ectotherm–endotherm transition: Superficial adaptation or profound reorganization? In E. M. Blass (Ed.), *Handbook of behavioral neurobiology*, Volume 9 *Developmental psychobiology and behavioral ecology* (pp. 107–148). New York: Plenum.

Burghardt, G. M. (1998a). The evolutionary origins of play revisited: Lessons from turtles. In M. Bekoff & J. A. Byers (Eds.), *Animal play: Evolutionary, comparative, and ecological perspectives* (pp. 1–26). Cambridge: Cambridge University Press.

Burghardt, G. M. (1998b). Play. In G. Greenberg & M. Haraway (Eds.), *Comparative psychology: A handbook* (pp. 757–767). New York: Garland Press.

Burghardt, G. M. (1999). Conceptions of play and the evolution of animal minds. *Evolution and Cognition, 5*, 115–123.

Burghardt, G. M. (in press). *The genesis of animal play: Testing the limits*. Cambridge, MA: MIT Press.

Burghardt, G. M., & Krause, M. A. (1999). Plasticity of foraging behavior in garter snakes (*Thamnophis sirtalis*) reared on different diets. *Journal of Comparative Psychology, 113*, 277–285.

Burghardt, G. M., Ward, B., & Rosccoe, R. (1996). Problem of reptile play: Environmental enrichment and play behavior in a captive Nile soft-shelled turtle (*Trionyx triunguis*). *Zoo Biology, 15*, 223–238.

Buss, D. M. (1999). *Evolutionary psychology*. New York: Allyn and Bacon.

Butler, A. B., & Hodos, W. (1996). *Comparative vertebrate neuroanatomy: Evolution and adaptation*. New York: Wiley-Liss.

Byers, J. A. (1998). Biological effects of locomotor play: Getting into shape, or something more specific? In M. Bekoff & J. A. Byers (Eds.), *Animal play: Evolutionary, comparative, and ecological perspectives* (pp. 205–220). Cambridge: Cambridge University Press.

Byers, J. A. (1999). The distribution of play behaviour among Australian marsupials. *Journal of Zoology, London, 247*, 349–356.

Byers, J. A., & Walker, C. (1995). Refining the motor training hypothesis for the evolution of play. *American Naturalist, 146*, 25–40.

Cabanac, M. (1999). Emotion and phylogeny. *Journal of Consciousness Studies, 6*, 176–190.

Carmichael, L. (1954). The onset and early development of behavior. In L. Carmichael (Ed.), *Manual of child psychology*, 2nd ed. (pp. 60–185). New York: Wiley.

Chiszar, D. (1985). Ontogeny of communicative behaviors. In E. S. Gollin (Ed.), *The comparative development of adaptive skills: Evolutionary implications* . Hillsdale, NJ: Erlbaum.

Colombo, J. (1993). *Infant cognition: Predicting later intellectual functioning*. London: Sage.

Compayré, G. (1902). *Development of the child in later infancy*. New York: D. Appleton.

Coppinger, R., & Coppinger, L. (1998). Differences in the behavior of dog breeds. In J. Serpell (Ed.), *Genetics and the behavior of domestic animals* (pp. 167–202). New York: Academic Press.

Coppinger, R. P., & Smith, C. K. (1989). A model for understanding the evolution of mammalian behavior. In H. Genoways (Ed.), *Current mammalogy* (Vol. 2, pp. 335–374). New York: Plenum Press.

Courchesne, E. (1997). Prediction and preparation: Anticipatory role of the cerebellum in diverse neurobehavioral functions. *Behavioral and Brain Sciences, 20,* 248–249.

Damasio, A. R. (1995). *Descartes error: Emotion, reason, and the human brain.* New York: Avon.

Damasio, A. R. (1999). *The feeling of what happens.* New York: Harcourt Brace.

Dave, A. S., Yu, A. C., & Margoliash, D. (1998). Behavioral state modulation of auditory activity in a vocal motor system. *Science, 282,* 2250–2254.

Deacon, T. W. (1990). Rethinking mammalian brain evolution. *American Zoologist, 30,* 629–705.

Depue, R. A., & Collins, P. F. (1999). Neurobiology of the structure of personality: Dopamine, facilitation of incentive motivation, and extraversion. *Behavioral and Brain Sciences, 22,* 491–568.

Durant, J. R. (1985). The science of sentiment: The problem of the cerebral localization of emotion. *Perspectives in Ethology, 6,* 1–31.

Fadiga, L., Fogassi, L., Pavesi, G., & Rizzolatti, G. (1995). Motor facilitation during action observation: A magnetic stimulation study. *Journal of Neurophysiology, 73,* 2608–2611.

Fagen, R. M. (1974). Selective and evolutionary aspects of animal play. *American Naturalist, 108,* 850–858.

Fagen, R. (1981). *Animal play behavior.* New York: Oxford University Press.

Fagen, R. (1984). Play and behavioural flexibility. In P. K. Smith (Ed.), *Play in animals and humans* (pp. 159–173). Oxford: Blackwell.

Ficken, M. S. (1977). Avian play. *Auk, 94,* 573–582.

Font, E., García-Verdugo, J. M., Desfilis, E. & Pérez-Cañellas, M. (1995). Neuron-glia interrelations during 3-acetylpyridine-induced degeneration and regeneration in the adult lizard brain. In A. Vernadakis & B. Roots (Eds.), *Neuron-glia interrelations during plasticity: II. Plasticity and regeneration* (pp. 275–302). Totowa, NJ: Humana Press.

Font, E., Desfilis, E., Pérez-Cañellas, M., Alcántara, S., & García-Verdugo, J. M. (1997). 3-Acetylpyridine-induced degeneration and regeneration in the adult lizard brain: A qualitative and quantitative analysis. *Brain Research, 754,* 245–259.

Francis, D., Diorio, J., Liu, D., & Meaney, M. J. (1999). Nongenomic transmission across generations of maternal behavior and stress responses in the rat. *Science, 286,* 1155–1158.

Frith, C. D., & Frith, U. (1999). Interacting minds—A biological basis. *Science, 286,* 1692–1695.

Fromberg, D. P., & Bergen, D. (Eds.). (1998). *Play from birth to twelve and beyond.* New York: Garland Press.

Gallese, V., Fadiga, L., Fogassi, L., & Rizzolatti, G. (1996). Action recognition in the premotor cortex. *Brain, 119,* 593–609.

Gandelman, R. (1992). *The psychobiology of behavioral development.* New York: Oxford University Press.

Gazzaniga, M. S. (Ed.). (1995). *The cognitive neurosciences.* Cambridge, MA: MIT Press.

Gottlieb, G. (1992). *Individual development and evolution: The genesis of novel behavior.* New York: Oxford University Press.

Graybiel, A. M. (1995). Building action repertoires: Memory and learning functions of the basal ganglia. *Current Opinion in Neurobiology, 5,* 733–711.

Graybiel, A. M. (1997). The basal ganglia and cognitive pattern generators. *Schizophrenia Bulletin, 23,* 459–469.

Greenberg, N., Font, E., & Switzer, R. (1988). The reptilian striatum revisited. In W. K. Schwerdtfeger and W. J. A. J. Smeets (Eds.), *The forebrain of reptiles: Current concepts of structure and function* (pp. 162–177). Basal: Karger-Verlag.

Grier, J. M., & Burk, T. (1992). *Biology of animal behavior,* 2nd ed. St. Louis, MO: Mosby Year Book.

Griffiths, M. (1978). *The biology of the monotremes.* New York: Academic Press.

Groenewegen, H. J., Wright, C. I., & Beijer, A. V. J. (1996). The nucleus accumbens: Gateway for limbic structures to reach the motor system? *Progress in Brain Research, 107,* 485–511.

Groos, K. (1898). *The play of animals* New York: D. Appleton.

Gross, C. G. (2000). Neurogenesis in the adult brain: Death of a dogma. *Nature Reviews Neuroscience, I,* 67–73.

Halgren, E., & Marinkovic, K. (1995). Neurophysiological networks integrating human emotions. In M. S. Gazzaniga (Ed.), *The cognitive neurosciences* (pp. 1137–1151). Cambridge, MA: MIT Press.

Hall, G. S. (1904). *Adolescence: Its psychology and its relations to physiology, anthropology, sociology, sex, crime, religion and education.* New York: D. Appleton.

Hall, S. L. (1998). Object play in adult animals. In M. Bekoff & J. A. Byers (Eds.), *Animal play: Evolutionary, comparative, and ecological perspectives* (pp. 45–60). Cambridge: Cambridge University Press.

Halloy, M., & Burghardt, G. M. (1990). Ontogeny of fish capture and ingestion in four species of garter snakes (*Thamnophis*). *Behaviour, 112,* 299–318.

Hartley, R. E., Frank, L. K., & Goldenson, R. M. (1952). *Understanding children's play.* New York: Columbia University Press.

Heinrich, B., & Smolker, R. (1998). Play in common ravens (*Corvus corax*). In M. Bekoff & J. A. Byers

(Eds.), *Animal play: Evolutionary, comparative, and ecological perspectives* (pp. 27–44). Cambridge: Cambridge University Press.

Hogan, J. A. (1988). Cause and function in the development of behavior systems. In E. M. Blass (Ed.), *Handbook of behavioral neurobiology*, Volume 9, *Developmental psychobiology and behavioral ecology*, pp. 63–106). New York: Plenum Press.

Hol, T., Berg, C. L. v. d., van Ree, J. M., & Spruijt, B. M. (1999). Isolation during the play period in infancy decreases adult social interaction in rats. *Behavioural Brain Research, 100*, 91–97.

Houk, J. C., Buckingham, J. T., & Barto, A. G. (1996). Models of the cerebellum and motor learning. *Behavioral and Brain Sciences, 19*, 368–383.

Ikemoto, S. & Panksepp, J. (1992). The effects of early social isolation on the motivation for social play in rats. *Developmental Psychobiology, 25*, 261–274.

Iwaniuk, A. N., Nelson, J. E., & Pellis, S. M. (2001). Do big-brained animals play more? Comparative analyses of play and relative brain size in mammals. *Journal of Comparative Psychology, 115*, 29–41.

Jarvis, E. D., Ribeiro, S., da Silva, M. L., Ventura, D., Vielliard, J., & Mello, C. V. (2000). Behaviorally driven gene expression reveals song nuclei in hummingbird brain. *Nature, 406*, 628–632.

Jeannerod, M. (1994). The representing brain: Neural correlates of motor intention and imagery. *Behavioral and Brain Sciences, 17*, 187–245.

Jog, M. S., Kubota, Y., Connolly, C. I., Hillegaart, V., & Graybiel, A. M. (1999). Building neural representations of habits. *Science, 286*, 1745–1749.

Johnson, J. E., Christie, J. F., & Yawkey, T. D. (1999). *Play and early childhood development*. New York: Longman.

Katz, P. S., & Harris-Warrick, R. M. (1999). The evolution of neuronal circuits underlying species-specific behavior. *Current Opinion in Neurobiology, 9*, 628–633.

Kempermann, G., Kuhn, H. G., & Gage, F. H. (1997). More hippocampal neurons in adult mice living in an enriched environment. *Nature, 386*, 493–495.

Kleim, J. A., Swain, R. A., Armstrong, K. A., Napper, R. M. A., Jones, T. A., & Greenough, W. T. (1998). Selective synaptic plasticity within the cerebellar cortex following complex motor skill learning. *Neurobiology of Learning and Memory, 69*, 274–289.

Kolb, B., & Whishaw, I. Q. (1998). Brain plasticity and behavior. *Annual Review of Psychology, 49*, 43–64.

Kosslyn, S. M., & Sussman, A. L. (1995). Role of imagery in perception: Or, there is no such thing as immaculate perception. In M. S. Gazzaniga (Ed.), *The cognitive neurosciences* (pp. 1035–1042). Cambridge, MA: MIT Press.

Kramer, M., & Burghardt, G. M. (1998). Precocious courtship and play in emydid turtles. *Ethology, 104*, 38–56.

Krause, M. A., Burghardt, G. M., & Lentini, A. (1999). Improving the lives of captive reptiles: Object provisioning in Nile soft-shelled turtles (*Trionyx triunguis*). *Lab Animal, 2*, 38–41.

Kruijt, J. P. (1964). Ontogeny of social behaviour in Burmese red jungle fowl (*Gallus gallus spadiceus*). *Behaviour Supplement, 12*, 1–201.

Kruska, D. (1987). How fast can total brain size change in mammals? *Journal für Hirnforschung, 28*, 59–70.

Kruska, D. (1988). Mammalian domestication and its effect on brain structure and behavior. In H. J. Jerison and I. Jerison (Eds.), *Intelligence and Evolutionary Biology*. NATO ASI series, Vol. G17, pp. 211–250. Berlin: Springer-Verlag.

Le Doux, J. E. (1991). Emotion and the limbic system concept. *Concepts in Neuroscience, 2*, 169–199.

Lefebre, L., Whittle, P., Lascaris, E., & Finkelstein, A. (1997). Feeding innovations and forebrain size in birds. *Animal Behaviour, 53*, 549–560.

Lewis, K. P. (2000). A comparative study of primate play behaviour: Implications for the study of cognition. *Folia Primatologica, 71*, 417–421.

Lorenz, K. Z. (1981). *The foundations of ethology*. New York: Springer-Verlag.

Lovell, G., & Collins, D. The mediating role of sex upon the relationship between mental imagery ability and movement acquisition rate. *Journal of Human Movement Studies, 32*, 187–210.

MacLean, P. (1985). Brain evolution relating to family, play and the separation call. *Archives of General Psychiatry, 42*, 405–417.

MacLean, P. (1990). *The triune brain in evolution*. New York: Plenum Press.

Maier, S. F., & Watkins, L. R. (1998). Cytokines for psychologists: Implications of bidirectional immune-to-brain communication for understanding behavior, mood, and cognition. *Psychological Review, 105*, 83–107.

Marín, O., Smeets, W. J. A. J., & González, A. (1998). Evolution of the basal ganglia in tetrapods: A new perspective based on recent studies in amphibians. *Trends in Neuroscience, 21*, 487–494.

Martin, P., & Caro, T. M. (1985). On the function of play and its role in behavioral development. *Advances in the Study of Behavior, 15*, 59–103.

Mather, J. A., & Anderson, R. C. (1999). Exploration, play, and habituation in octopuses (*Octopus dofleini*). *Journal of Comparative Psychology, 113*, 333–338.

McGraw, M. B. (1943). *The neuromuscular maturation of the human infant.* New York: Columbia University Press.

Meaney, M. J., Dodge, A. M., & Beatty, W. W. (1981). Sex-dependent effects of amygdaloid lesions on the social play of pubertal rats. *Physiology and Behavior, 26*, 467–472.

Meder, E. (1958). Gnathonemus petersii (Günter). *Zeitschrift für Vivaristik, 4*, 161–171.

Meer, A. L. H. v. d., Weel, F. R. v. d., & Lee, D. N. (1995). The functional significance of arm movements in infants. *Science, 267*, 693–695.

Melzack, R., & Thompson, W. R. (1956). Effects of early experience on social behavior. *Canadian Journal of Psychology, 10*, 82–90.

Meyer-Holzapfel, M. (1960). Über das Spiel bei Fischen, insbesondere beim Tapirrüsselfisch (*Mormyrus kannume* Forskål). *Zoologische Garten, 25*, 189–202.

Miklosi, A. (1999). The ethological analysis of imitation. *Biological Review, 74*, 347–374.

Mitchell, E. D., & Mason, B. S. (1934). *The theory of play.* New York: A. S. Barnes.

Mitchell, R. W. (1990). A theory of play. In M. Bekoff & D. Jamieson (Eds.), *Interpretation and explanation in the study of animal behavior*, Volume 1, *Interpretation, intentionality, and communication* (pp. 197–227). Boulder, CO: Westview Press.

Orofino, A. G., Ruarte, M. B., & Alvarez, E. O. (1999). Exploratory behaviour after intra-accumbens histamine and/or histamine antagonists injection in the rat. *Behavioural Brain Research, 102*, 171–180.

Ortega, J. C., & Bekoff, M. (1987). Avian play: Comparative evolutionary and developmental trends. *Auk, 104*, 338–341.

Page, S. J. (2000). Imagery improves upper extremity motor function in chronic stroke patients: A pilot study. *Occupational Therapy Journal of Research, 20*, 200–215.

Panksepp, J. (1998). *Affective neuroscience.* New York: Oxford University Press.

Panksepp, J. (2000). the riddle of laughter. *Current Directions in Psychological Science, 9*, 183–186.

Panksepp, J., & Burgdorf, J. (2000). 50-khz chirping (laughter?) in response to conditioned and unconditioned tickle-induced reward in rats: Effects of social housing and genetic variables. *Behavioral Brain Research, 115*, 25–38.

Panksepp, J., Siviy, S., & Normansell, L. (1984). The psychobiology of play: Theoretical and methodological perspectives. *Neuroscience and Biobehavioral Reviews, 8*, 465–92.

Panksepp, J., Normansell, L., Cox, J. F., & Siviy, S. M. (1994). Effects of neonatal decortication on the social play of juvenile rats. *Physiology and Behavior, 56*, 429–443.

Parker, S. T., & McKinney, M. L. (1999). *Origins of intelligence: The evolution of cognitive development in monkeys, apes, and humans.* Baltimore, MD: Johns Hopkins University Press.

Paulin, M. G. (1993). The role of the cerebellum in motor control and perception. *Brain, Behavior and Evolution, 41*, 39–50.

Pellegrini, A. D., & Smith, P. K. (1998). Physical activity play: The nature and function of a neglected aspect of play. *Child Development, 69*, 577–598.

Pellis, S. M. (1993). Sex and the evolution of play fighting: A review and model based on the behavior of muroid rodents. *Play Theory and Research, 1*, 55–75.

Pellis, S. M., & Iwaniuk, A. N. (2000). Comparative analyses of the role of postnatal development on the expression of play fighting. *Developmental Psychobiology, 36*, 136–147.

Pellis, S. M., & Pellis, V. C. (1998a). Play fighting of rats in comparative perspective: A schema for neurobehavioral analysis. *Neuroscience and Biobehavioral Reviews, 23*, 87–101.

Pellis, S. M., & Pellis, V. C. (1998b). Structure–function interface in the analysis of play fighting. In M. Bekoff & J. A. Byers (Eds.), *Animal play: Evolutionary, comparative, and ecological perspectives* (pp. 115–140). Cambridge: Cambridge University Press.

Pellis, S. M., O'Brien, D. P., Pellis, V. C., Teitelbaum, P., Wolgin, D. L., & Kennedy, S. (1988). Escalation of feline predation along a gradient from avoidance through "play" to killing. *Behavioral Neuroscience, 102*, 760–777.

Pellis, S. M., Pellis, V. C., & Whishaw, I. Q. (1992). The role of the cortex in play fighting by rats: Developmental and evolutionary implications. *Brain, Behavior and Evolution, 39*, 270–284.

Pennisi, E. (2000). In nature, animals that stop and start win the race. *Science, 288*, 83–85.

Pham, T. M., Söderström, S., Winblad, B., & Mohammed, A. H. (1999). Effects of environmental enrichment on cognitive function and hippocampal NGF in the non-handled rats. *Behavioural Brain Research, 103*, 63–70.

Piaget, J. (1962). *Play, dreams and imitation in chilhood.* New York: Norton.

Power, T. G. (2000). *Play and exploration in children and animals.* Mahwah, NJ: Erlbaum.

Previc, F. H. (1999). Dopamine and the origins of human intelligence. *Brain and Cognition, 41*, 299–350.

Price, E. O. (1984). Behavioral aspects of domestication. *Quarterly Review of Biology, 59,* 1–32.

Punzo, F. (1985). Neurochemical correlates of learning and role of the basal forebrain in the brown anole, *Anolis sagcei* (Lacertilia:Iguanidae). *Copeia, 1985,* 409–414.

Rasa, O. A. E. (1984). A motivational analysis of object play in juvenile dwarf mongooses (*Helogale undulata rufula*). *Animal Behaviour, 32,* 579–589.

Rizzolatti, G., Fadiga, L., Matelli, M., Bettinardi, V., Paulesu, E., Perani, D., & Fazio, F. (1996). Localization of grasp representations in humans by PET: 1. Observation versus execution. *Experimental Brain Research, 111,* 246–252.

Robbins, T. W., & Everitt, B. J. (1995). Arousal systems and attention. In M. S. Gazzaniga (Ed.), *The cognitive neurosciences* (pp. 703–720). Cambridge, MA: MIT Press.

Schultz, W., Dayan, P., & Montague, P. R. (1997). A neural substrate of prediction and reward. *Science, 275,* 1593–1599.

Sherry, D. F., Jacobs, L. F., & Gaulin, S. J. C. (1992). Spatial memory and adaptive specialization of the hippocampus. *Trends in Neuroscience, 15,* 298–303.

Shine, R. (1988). Parental care in reptiles. In C. Gans & R. B. Huey (Eds.), *Biology of the Reptilia,* Volume 16, *Ecology B* (pp. 275–329). New York: Liss.

Sih, A. (1992). Prey uncertainty and the balancing of antipredator and feeding needs. *American Naturalist, 139,* 1052–1069.

Siviy, S. M. (1998). Neurobiological substrates of play behavior: Glimpses into the structure and function of mammalian playfulness. In M. Bekoff & J. A. Byers (Eds.), *Animal play: Evolutionary, comparative, and ecological perspectives* (pp. 221–242). Cambridge: Cambridge University Press.

Siviy, S. M., Baliko, C. N., & Bowers, S. (1997). Rough-and-tumble play behavior in Fischer-344 and buffalo rats: Effects of social isolation. *Physiology and Behavior, 61,* 597–602.

Smeets, W. J. H. J., and González, H. (2000). Catecholamine systems in the brain of vertebrates: New perspectives through a comparative approach. *Brain Research Reviews, 33,* 308–379.

Smith, P. K. (1988). Children's play and its role in early development: A reevaluation of the 'play ethos'. In A. D. Pellegrini (Ed.), *Psychological bases for early education* (pp. 207–226). Chichester, England: Wiley.

Smith, P. K. (1996). Play, ethology, and education: A personal account. In A. D. Pellegrini (Ed.), *The future of play theory* (pp. 3–21). Albany, NY: SUNY Press.

Spanagel, R., & Weiss, F. (1999). The dopamine hypothesis of reward: Past and current status. *Trends in Neuroscience, 22,* 521–527.

Spear, L. P. (2000a). Neurobehavioral changes in adolescence. *Current Directions in Psychological Science, 9,* 111–114.

Spear, L. P. (2000b). The adolescent brain and age-related behavioral manifestations. *Neuroscience and Biobehavioral Reviews, 24,* 417–463.

Spencer, H. (1872). *Principles of psychology,* Vol. 2, 2nd ed. New York: D. Appleton.

Sperry, R. W. (1951). Mechanisms of neural maturation. In S. S. Stevens (Ed.), *Handbook of experimental psychology* (pp. 236–303). New York: Wiley.

Spinka, M., Newberry, R. C., & Behoff, M. (in press). Mammalian play: Can training for misfortune be fun? *Quarterly Review of Biology.*

Sutton-Smith, B. (1999). Evolving a consilience of play definitions: Playfully. In S. Reifel (Ed.), *Play and culture studies* (Vol. 2, pp. 239–256). Stamford, CT: Ablex.

Sutton-Smith, B. (in press). Play as a tertiary emotion. In J. L. Roopnarine (Ed.), *Play and culture studies* (Vol. 4). Stamford, CT: Ablex.

Swanson, L. W., & Petrovich, G. D. (1998). What is the amygdala? *Trends in Neuroscience, 21,* 323–331.

Tang, Y.-P., Shimizu, E., Dube, G. R., Rampon, C., Kerchner, G. A., Zhuo, M., Liu, G., & Tslen, J. Z. (1999). Genetic enhancement of learning and memory in mice. *Nature, 401,* 63–69.

Thelen, E. (1995). Motor development: A new synthesis. *American psychologist, 50,* 79–95.

Thompson, K. V. (1998). Self assessment in juvenile play. In M. Bekoff & J. A. Byers (Eds.), *Animal play: Evolutionary, comparative, and ecological perspectives* (pp. 183–204). Cambridge: Cambridge University Press.

Tinbergen, N. (1951). *The study of instinct.* Oxford: Clarendon Press.

Toni, B., Buchs, P.-A., Nikonenko, I., Bron, C. R., & Muller, D. (1999). LTP promotes formation of multiple spine synapses between a single axon terminal and a dendrite. *Nature, 402,* 421–425.

Trut, L. N. (1999). Early canid domestication: The farm-fox experiment. *American Scientist, 87,* 160–169.

Vanderschuren, L. J. M. J., Niesink, R. J. M., & Van Ree, J. M. (1997). The neurobiology of social play behavior in rats. *Neuroscience and Biobehavioral Reviews, 21,* 309–326.

Whishaw, I. Q., Fiorino, D., Mittleman, G., & Castaneda, E. (1992). Do forebrain structures compete for behavior expression? Evidence from amphetamine-induced behavior, microdialysis, and caudate-accumbans lesions in medial frontal cortex damaged rats. *Brain Research, 576,* 1–11.

Wilczynski, W. (1984). Central neural systems subserving a homoplasous periphery. *American Zoologist, 24,* 755–763.

Willingham, D. B. (1998). A neuropsychological theory of motor skill learning. *Psychological Review, 105,* 558–584.

Willingham, D. B. (1999). The neural basis of motor-skill learning. *Current Directions in Psychological Science, 6,* 178–182.

Wilson, E. O. (1975). *Sociobiology: The new synthesis.* Cambridge, MA: Belknap Press.

Yágüez, L., Nagel, D., Hoffman, H., Canavan, A. G. M., Wist, E., & Hömberg, V. (1998). A mental route to motor learning in improving trajectorial kinematics through imagery training. *Behavioral Brain Research, 90,* 95–106.

Yágüez, L., Canavan, A. G. M., Lange, H. W., & Hömberg, V. (1999). Motor learning by imagery is differentially affected in Parkinson's and Huntington's diseases. *Behavioural Brain Research, 102,* 115–127.

9

Emerging Psychobiology of the Avian Song System

TIMOTHY J. DEVOOGD AND CHRISTINE LAUAY

This chapter builds on the excellent review of avian song development by Bottjer and Arnold (1986) that appeared in an earlier volume of this series (Blass, 1986). We first summarize the knowledge and perspectives on the neurobiology of song development and production of that time. We then present an overview of the major advances that have occurred since the earlier review, and finish with a description of the research avenues that we feel are likely to yield significant advances in our understanding of the neurobiology of bird song.

STATE OF KNOWLEDGE AT THE TIME OF THE BOTTJER AND ARNOLD (1986) REVIEW

SONG ACQUISITION

Already by the mid 1980s, there had been substantial research on the nature of birdsong and how it was acquired. Much of this early work was carried out or inspired by Peter Marler. Research, usually involving rearing birds in the laboratory with varying degrees of sensory and social deprivation, demonstrated that many of the components of the song of oscine birds (usually called the songbirds) were learned by hearing renditions of adult song (reviewed by Marler, 1970, 1976, 1984). In the limited number of species that had been studied closely, this learning was only possible during a restricted sensitive period shortly after fledging. In some species, for example, swamp sparrows (Marler & Peters, 1982), the period of auditory learning from adult models could be months in advance of song practice and production.

TIMOTHY J. DEVOOGD AND CHRISTINE LAUAY   Department of Psychology, Cornell University, Ithaca, New York 14853.

Developmental Psychobiology, Volume 13 of *Handbook of Behavioral Neurobiology*, edited by Elliott Blass, Kluwer Academic / Plenum Publishers, New York, 2001.

Even in these species, some aspects of the species-typical song were innate and did not require learning from adult models. Thus, song and swamp sparrows that had been isolated from conspecific song would still sing as adults, and their song resembled normal song in segmentation, duration, and frequency range (Marler & Sherman, 1983). This "isolate" song would often contain sound units (called notes or syllables) that, although simple or distorted, could still be recognized as similar to syllables in normal songs.

This innate bias toward the conspecific song was also evident in choice experiments. Juvenile male swamp and song sparrows presented with both conspecific and heterospecific song during the sensitive period would almost always choose to copy the conspecific models, and juvenile males presented only with heterospecific song often would either acquire the sound elements most like ones found in conspecific song or would structure the elements acquired into a pattern typical of conspecific song (reviewed by Marler & Peters, 1989). Konishi (1965) and Marler (1970, 1976) suggested that such results could be explained by the existence of an innate template for characteristics of conspecific song which guides selection of a song model and which is modified by learning particular songs or supplemented by an additional template formed by the learning. These behavioral observations and the hypotheses which summarized them generated a large number of experimental studies on how general the phenomena were as well as on the neural bases for the learning, as will be described below.

SONG EXPRESSION

Complex songs in the oscine subfamily are produced entirely or predominantly by males. Song is produced in two broad contexts: as an agonistic signal to other males and to attract a female or maintain a pair bond with a female. Species differ in the degree to which song is used in one context or the other (data reviewed by Kroodsma & Byers, 1991; Morton, 1996; Smith, 1991). Behavioral research prior to the earlier volume in this series had shown that sounds making up the song of an individual male are idiosyncratic (within species and dialect constraints) and are usually produced in a highly stereotyped manner (for example, Waser & Marler, 1977). As even the most casual observer knows, birdsong is vigorous and ubiquitous in spring, at least for birds that reproduce in temperate zones. The observations that song is produced seasonally and is associated with reproduction suggested that it was influenced by gonadal steroids. This conclusion was strengthened by observations that adult male songbirds that have been castrated decrease song production or cease singing altogether, but resume when given exogenous testosterone (reviewed by Harding, 1986).

In summary, song production is a natural learned behavior. In most songbirds that have been studied, song is stable once acquired (for example, Petrinovich, 1988), it is readily (even exuberantly) expressed, and it is steroid dependent. Thus, as argued in the earlier review (Bottjer & Arnold, 1986), song provides exceptional possibilities for understanding neuroendocrine integration, organization of complex motor behaviors, and the neural bases for motor learning.

SONG SYSTEM NEUROBIOLOGY

Nottebohm and colleagues (Nottebohm, Stokes, & Leonard, 1976; Nottebohm, Kelley & Paton, 1982) used tract-tracing and lesions to identify the structures that have become known as the song system (Figure 1). Briefly, the high vocal center (HVC) in neostriatum projects to robustus archistriatalis (RA) in archistriatum, which in turn projects to the tracheosyringeal portion of cranial nerve nucleus twelve (nXIIts), the brainstem nucleus that ennervates the avian vocal organ, the syrinx. Bilateral lesions of any of these in males permanently ends further song production (Nottebohm *et al.*, 1976). HVC also projects rostrally to Area X, which projects via the medial part of the dorsolateral thalamus (DLM) to the lateral part of the magnocellular nucleus of the anterior neostriatum (lMAN), which in turn projects to RA and back to Area X (Bottjer, Halsema, Brown, & Miesner, 1989; Nixdorf-Bergweiler, Lips, & Heinemann, 1995; Nottebohm *et al.*, 1976; Okuhata & Saito, 1987). Adult birds with lesions of Area X or lMAN can produce acoustically normal song (Bottjer, Miesner, & Arnold, 1984; Sohrabji, Nordeen, & Nordeen, 1990). However, if either Area X or lMAN is lesioned in juvenile birds before song crystallization, the structure of the song that is eventually produced is severely distorted (Bottjer *et al.*, 1984; Scharff & Nottebohm, 1991; Sohrabji *et al.*, 1990). Based on these observations, the song system until recently was viewed as consisting of two subsystems: a caudal production system that, with the help of auditory feedback, maintains the connectivity required for producing a song in a stereotyped manner, and a rostral system that is essential for song acquisition but of little known use thereafter (Bottjer & Arnold, 1986). Both systems originate from HVC, which receives auditory information from the less specialized telencephalic auditory area, Field L. Recent anatomic and electrophysiological data to be discussed below indicate a continuing role of the rostral system in singing. In this original formulation, only song production was related to anatomy—little attention was paid to neural

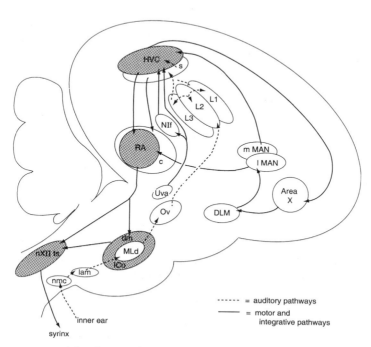

Figure 1. Song system nuclei and connections as understood at the time of the Bottjer & Arnold (1986) review. Areas known to be essential for song production are shaded.

correlates either of song processing in females or of song perception and classification more generally. These qualities are essential attributes of song use in a naturalistic context and recent findings are beginning to describe the brain regions responsible for them, as will be described below.

Dramatic anatomic differences between males and females exist throughout this network, especially for species in which females are unable to sing. HVC, RA, nXIIts, and Area X are much smaller in females than in males. In zebra finches, for example, there is a five-fold difference between the sexes in the volume of HVC and RA (Nottebohm & Arnold, 1976). In species in which females sing, the anatomic differences can be much smaller (Brenowitz, Arnold, & Levin, 1985; DeVoogd, Brenowitz, & Arnold, 1988). In large part, the sex differences are due to differences in the number of neurons in the nuclei (Gurney, 1981; Konishi & Akutagawa, 1985). However, at least in RA, the neurons that persist in females have shorter dendrites and have fewer synapses than neurons in males (DeVoogd & Nottebohm, 1981a). In species such as canaries or red-winged blackbirds that sing and reproduce seasonally, adult song system morphology is not fixed. Nuclei from both the rostral and caudal divisions of the song system are larger in the spring than in the fall (Kirn, Clower, Kroodsma, & DeVoogd, 1989; Nottebohm, 1981). Increased survival of new neurons contributes to the spring increase in the volume of HVC (reviewed by Nottebohm, 1987). The increases in the volumes of RA and nXIIts are associated with dendritic growth and the addition of up to 40% more synapses (Clower, Nixdorf, & DeVoogd, 1989; DeVoogd, Nixdorf, & Nottebohm, 1985; reviewed by DeVoogd, 1991, 1994).

Song System Neuroendocrinology

Neurons in most of the nuclei of the song system concentrate gonadal steroids (reviewed by DeVoogd, 1991, 1994). In nXIIts, it is likely that every single neuron has androgen receptors. Many neurons in HVC, RA, and lMAN also have androgen receptors. In addition, many cells in HVC and lMAN have estrogen receptors. The functions of these receptors were (and are) unclear, but they could plausibly be the means by which systemic information about reproductive state, as indicated by blood steroid levels, ensures that singing coincides with reproduction and is less likely to occur at other times. The levels of gonadal steroids can be as much as two orders of magnitude higher in the spring than in the fall (Wingfield & Farner, 1993), and may induce many of the changes in morphology of song system nuclei that occur with the change in seasons. In fact, many of the seasonal changes in song system anatomy can be mimicked by administering gonadal steroids (DeVoogd & Nottebohm, 1981b; Nottebohm, 1980).

Song System Physiology

In order to understand how the song system produces a stereotyped, learned song, we must understand the physiology of the neurons that comprise this system. The first experiments on song system physiology were published just prior to the Bottjer and Arnold (1986) review. Many neurons in HVC are responsive to sound bursts in general (in anesthetized birds). In one of the more exciting early discoveries concerning the song system, Margoliash (1983) found that neurons in HVC of males are selectively responsive to conspecific song, and many neurons are particularly tuned to the bird's own song. Less was known of the physiology of song production. McCasland and Konishi (1981) had showed that neurons in HVC are active during song production. Little was known of how these characteristics develop or change across seasons.

SONG SYSTEM DEVELOPMENT

Songbirds in general are very altricial at hatching, but develop quickly thereafter. Much of the telencephalon is undifferentiated at hatching; song system nuclei become apparent in zebra finches between 10 and 15 days after hatching and establish their connections over the ensuing 2–3 weeks (Bottjer, Glaessner, & Arnold, 1985; Konishi & Akutagawa, 1985; Kirn & DeVoogd, 1989). In males, the volumes of the telencephalic nuclei asymptote at adult levels in the second month after hatching (Kirn & DeVoogd, 1989), except for lMAN. This structure paradoxically shrinks in volume in the second month after hatching, eventually attaining a volume about half of its initial maximum (Bottjer *et al.*, 1985).

Sex dimorphisms in song system anatomy appear during weeks 3–6 after hatching, in large part resulting from cell death and neuronal regression in females (Kirn & DeVoogd, 1989). During this interval, large numbers of steroid receptors appear in telencephalic song system nuclei (Arnold, Nottebohm, & Pfaff, 1976); they, too, become dimorphic as greater numbers are formed or retained in males (Arnold & Saltiel, 1979; E. J. Nordeen, Nordeen, & Arnold, 1987; K. W. Nordeen, Nordeen, & Arnold, 1986). Giving estradiol to newly hatched females results in substantial masculinization of each of these dimorphic features. If coupled with later administration of testosterone, treated females acquire and produce a male-like stereotyped song (Gurney, 1981; Gurney & Konishi, 1980). These findings led to the belief that the process of normal sexual differentiation of the song system is like brain differentiation in mammals: estradiol in juvenile males acts to stimulate and create brain circuits and localized hormonal receptors such that testosterone at puberty can induce a male-typical behavior (song) that is not possible for females. As will be described below, this model is incorrect and is being substantially revised to account for recent detailed observations on steroid synthesis, action, and metabolism in the brains of developing finches.

PROGRESS IN THE LAST 15 YEARS

The last 15 years have seen several highly significant shifts in research on the avian song system. Viewed broadly, most of these involve an increasing emphasis on understanding song and the song system in a more natural context. Thus, observations on song system development are being related to natural acquisition of song. Considerable research is examining the process of song system differentiation in normal developing males and females. Neuronal physiology is being studied in awake, behaving animals. Also, immediate-early gene activation is now being used to map brain areas involved in production and reception of song in more naturalistic social contexts (reviewed by Ball & Gentner, 1998). These will be described in more detail below.

SINGING BEHAVIORS

One of the more interesting developments in the study of song acquisition has been the increasing recognition of the importance of context in learning (reviewed

TIMOTHY J.
DEVOOGD AND
CHRISTINE LAUAY

by Eens, 1997). This can be seen at many levels. Young starlings learn twice as many types of phrases when exposed to a live tutor as when exposed to tapes of song (Chaiken, Bohner, & Marler, 1993). Young nightingales learn songs more completely and more accurately if from a live tutor than simply from a tape (reviewed by Todt & Hulsch, 1996). This occurs even if the live "tutor" is their human caretaker wearing a tape recorder! If young zebra finches cannot see their tutor as well as hear him, they will not learn his song (reviewed by Slater, 1989). Starlings will mimic human words, but do so only if hand-reared by humans, and without contact with other starlings (West, Stroud, & King, 1983). Zebra finches preferentially copy songs of the adult males with which they have the most interactions (Clayton, 1987; Williams, 1990). Even in species that will learn song from a tape recorder, such as the white-crowned sparrow, sounds that are heard late in the sensitive period are more likely to be learned if they come from a live tutor than from a tape recorder (reviewed by Baptista, Bell, & Trail, 1993). It is not clear which aspects of social context confer these patterns of selectivity. One contributor is likely to be level of arousal or attention (Hultsch, Schleuss, & Todt, 1999). Another could be the capacity of the juvenile to interact with the tutor. For example, Adret (1993) showed that juvenile male zebra finches will learn a song from a tape if their actions turn it on.

Social context can also influence song content in older birds. Many bird species learn and practice more sounds as juveniles than will finally be used in their songs (reviewed by Nelson, Marler, & Morton, 1996). In some of these species, selecting the components of the final song from the larger set that has been practiced is an active process that takes place in the first reproductive season. Thus, sounds that are similar to those produced by older resident males are retained and others are inhibited to create the adult song. In white-crowned sparrows, the male then retains this song without major change throughout adult life. Recent evidence suggests that the sounds that an individual has practiced but not incorporated into the mature song are inhibited rather than being totally lost (Benton, Nelson, Marler, & DeVoogd, 1998). Marler (1990) suggested that this process of shaping the content of song in response to interactions with adult conspecifics is a distinct phase of song learning, termed action-based learning (reviewed by Nelson, 1997). In fact, Nelson (1998) suggests that late learning from tutors in white-crowned sparrows, as described above, should be understood not as an extended sensitive period, but as an instance of action-based learning in which the opportunity to interact with the tutor leads to expression of sounds by the juvenile that were acquired earlier. The social cues that lead to modifications of song need not even be vocal. In brown-headed cowbirds, young males will acquire song characteristics specific to a dialect region of the cowbirds' range simply from exposure to adult females from that region. Apparently, gestures by the females cue the young males about sounds that are especially stimulatory, which they then retain in song while dropping sounds that are less effective in eliciting interest from the females (reviewed by West & King, 1996).

Context also influences song expression in adulthood. Thus, song production, especially in species that breed in temperate zones, is substantially constrained to times leading up to and including annual phases of reproduction (Morton, 1996). However, one of the exciting advances of the last decade has been the repeated demonstration that song production and content are also affected by many more precise or subtle contextual factors. Several examples indicate the range of these interactions.

First, social context affects the likelihood of singing. Thus, male juncos sing vigorously when unmated and less often when mated (Ketterson, Nolan, Wolf, &

Ziegenfus, 1992). This change can occur very quickly—in wood warbler species, males rapidly increase song production if their mate is removed and return to the lower rate as soon as she is returned or they re-pair (reviewed by Spector, 1992; Staicer, 1996). Such results suggest that song can be primarily an advertisement used to attract a mate or to solicit extrapair copulations rather than a behavior that helps maintain a pair bond.

Second, an individual may produce different song variants in different social contexts. For example, juncos produce short-range song vigorously when courting, but very little when their mate is incubating (Titus, 1998). Male zebra finches sing both undirected songs and songs directed at a female. These two sorts of song appear to differ in meaning—the former seem to be more of an advertising signal to females generally, whereas the latter are precopulatory signals directed at a mate (Zann, 1996). As will be described below, the two forms of song also differ in their neural effects.

Third, a number of bird species have several distinct classes of song within their repertoires that appear to be used with different categories of individuals (reviewed by Kroodsma, 1996). In the North American Parulinae warblers, for example, one song class tends to be used for agonistic interactions with other males and varies regionally and another class is similar across the geographic range of the species and tends to be used for interactions with females (reviewed by Spector, 1992). Both song classes are learned, but learning of one permits regional and/or individual variation whereas learning the other supports geographic homgeneity. How this dichotomous learning or the specificity of use is acquired is largely unexplored.

Finally, song perception is integrated with other forms of knowledge. For example, males of species from several avian families show enhanced responses to the song of a neighbor when it is played from an inappropriate location over the response to the song when it comes from the neighbor's territory (reviewed by Stoddard, 1996; Naguib & Todt, 1998). Such data suggest that these species recognize individuals on the basis of song content and source. Also, female zebra finches in reproductive condition respond to song with a rise in estrogen level, but only if the song comes from a male in a particular location with respect to the nest (Tchernichovski, Schwabl, & Nottebohm, 1998). All of the experiments mentioned here indicate the integration of song production or perception with a variety of other actions on the basis of diverse forms of information, from endocrine to visual to auditory to what might even be called cognitive maps. This has significant implications for the study of the neurobiology of song, as will be described below.

SONG SYSTEM NEUROANATOMY

Clearly the behavioral data described above indicate that brain areas responsible for song production must interact with brain areas involved in many other aspects of perception, cognition, and action. Yet in early studies on the song system, anatomic data available appeared to show that song nuclei communicated primarily with each other and did not have major connections with other brain regions (except for Field L, the source of auditory input) (Nottebohm *et al.*, 1976, 1982). Consequently, many early studies treated the song system as a discrete entity almost independent from processing in other brain regions. Several studies in recent years have found novel neural pathways whereby the song system may interact with these other systems (Figure 2). Song system nucleus RA is connected to areas in the brain stem that are involved in respiration (Vicario, 1993; Wild, 1997a), which in turn

TIMOTHY J.
DEVOOGD AND
CHRISTINE LAUAY

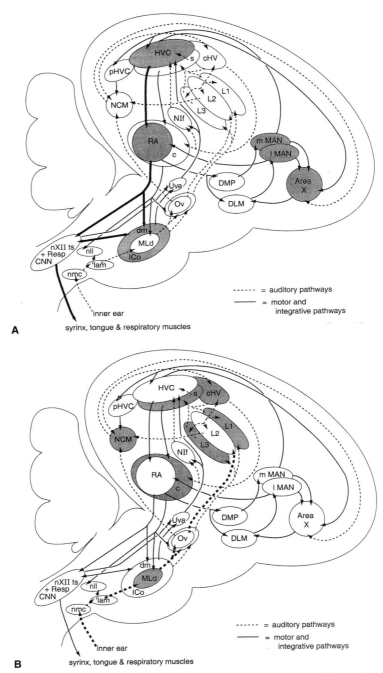

Figure 2. Song system nuclei and pathways as currently understood. Dashed lines indicate projections that appear to be primarily auditory; solid lines indicate projections that are primarily motor or integrative. (A) Bold lines indicate the primary song production pathways. Shaded areas show immediate early gene activation in birds that are *producing* song. (B) Bold lines indicate the primary auditory reception pathways. Shaded areas show immediate early gene activation in birds that are *hearing* song.

project both directly and via the dorsomedial portion (DM) of the intercollicular nucleus to nucleus uvaeformis (UVA) of the diencephalon (Reinke & Wild, 1998; Striedter & Vu, 1998) and then on to HVC (Nottebohm *et al.*, 1982). This loop might be used to control vocal intensity and to coordinate the interplay between singing and inspiration on a scale of fractions of 1 sec (Suthers, 1997; Wild, 1997b). There is a projection from the hypothalamus to the dorsomedial nucleus of the posterior thalamus (DMP) that in turn projects to medial MAN (Foster, Mehta, & Bottjer, 1997), which could be used to transmit neuronal signals about motivational and reproductive state to the song system (Riters & Ball, 1999). There are extensive, complex interconnections between the original song system nuclei, high auditory regions, and zones adjacent to the song and auditory areas, such as caudal hyperstriatum ventrale (cHV), HVC shelf, and paraHVC, whose functions are not understood (Foster & Bottjer, 1998). In addition, novel feedback loops have been discovered between song system nuclei (MAN → Area X; RA → DMP → mMAN; DM → Uva) (Nixdorf-Bergweiler, Lips, & Heinemann, 1995; Striedter & Vu, 1998; Vates, Vicario, & Nottebohm, 1997) and from song system areas to auditory nuclei (RAcup → ovoidalis, MLd) (Mello, Vates, Okuhata, & Nottebohm, 1998). Several of these projections are bilateral (DMP → mMAN, inspiratory centers → Uva), in contrast to prior observations that song system nuclei projections were exclusively ipsilateral. In addition, rostral hyperstriatum accessorium, a somatosensory area, projects to Field L and RA, as well as zones that are afferent to HVC and lMAN (Wild & Williams, 1999). These findings underscore on a neural level what has long been clear on a behavioral level: song does not occur in isolation but, is integrated with respiration, context, ongoing behaviors, and endocrine state.

Less is known of the topography within nuclei and major projections. The brainstem nucleus nXIIts has subdivisions that project to different muscles of the syrinx (Vicario & Nottebohm, 1988). The lateral central and medial portions of lMAN, respectively, tend to project to ventromedial, ventrolateral, and dorsal portions of RA (Johnson, Sablan, & Bottjer, 1995; Iyengar, Viswanathan, & Bottjer, 1999). Similarly, there is a rough topography in the descending projections of RA: a dorsal region of RA projects to DM respiratory nuclei, a medial region to rostral nXIIts, and a ventral region to caudal nXIIts (Vicario, 1994). However, there is no evident pattern in the projection from HVC to RA (Foster & Bottjer, 1998; Vicario, 1994). Nowhere in the song system is there the highly precise mapping that is found in mammalian sensory and motor cortices. One of the major anatomic puzzles remaining to be understood in this system is how these connections are organized so as to produce highly patterned, stereotyped vocalizations and the extent to which this organization is determined by vocal experience. It is likely that these very detailed structure–function relations will become clearer as more precise tract-tracing techniques coupled with physiology are applied to this system, as has recently been done by Reinke and Wild (1998).

Song System Neurophysiology

Much of the research in neurophysiology of the song system in recent years has yielded incremental advances rather than revolutionary ones. Thus, a prominent feature of neuronal physiology in several song system nuclei is responsiveness to song, and in many cases, enhanced responsiveness to the bird's own song (reviewed by Margoliash, 1997). This selectivity has been closely studied by manipulating the songs that are presented to the birds. In HVC, neurons can be found that are highly sensitive to the timing or relative amplitude characteristics of song (Theunissen & Doupe, 1998). Neurons can be found in HVC as well that are sensitive to pairs of syllables from the song or to the harmonics of a particular sound from song (Margoliash & Fortune, 1992). Neurons selective for song are also found in the rostral nuclei Area X and lMAN, the thalamic nucleus DLM (Doupe & Konishi, 1991; Maekawa & Uno, 1996), as well as the HVC afferent structure, nucleus interface of the neostriatum (Janata & Margoliash, 1999). Neurons in L1 and L3 of the high auditory area, Field L, are responsive to harmonic complexes like those found in song (Lim & Kim, 1997), although these neurons are not specifically selective to the bird's own song (Janata & Margoliash, 1999). A few neurons in L1 and L3 respond more to normal song than to song with the syllable order reversed (Lewicki & Arthur, 1996). However, in spite of these advances, much remains to be done in describing how auditory signals are transformed from relatively little specificity as they enter Field L to the highly precise song-specific responses in the song system. Furthermore, virtually nothing is known of whether these auditory signals are used to monitor and maintain the acoustic structure of the bird's own song or to evaluate the songs of others. Substantial advances in our understanding of processing in these auditory areas would be likely if conventional analytic techniques such as lesions, electrophysiology, or immediate-early gene expression were applied to birds behaving in complex or naturalistic contexts.

Neurons in all of the major song system nuclei of the forebrain are activated during song production (Hessler & Doupe, 1999; Yu & Margoliash, 1996). Activity in HVC neurons precedes the onset of vocalization. The pattern of activity shown by individual neurons often differs between syllables and is consistent enough across repetitions of a syllable that it is possible to predict which syllable is about to be produced (Yu & Margoliash, 1996). However, activity of HVC neurons that is associated with a particular note differs when that note appears in different syllables, suggesting that HVC activity is directing the production and sequencing of syllables, rather than notes. RA neurons also show increased activity that predicts singing. In contrast to the HVC results, neurons showed firing patterns that were predictive for particular notes, regardless of where the note appeared (Yu & Margoliash, 1996), suggesting that RA directs the motor coordinations associated with each song note. These conclusions were supported by studies indicating that stimulating HVC or RA during song has distinct disruptive effects. Stimulating HVC causes the bird to start its song again at the beginning, whereas stimulating RA causes distortion of the sounds currently being produced, although the overall timing and production of song do not change (Vu, Mazurck, & Kuo, 1994; Vu, Schmidt, & Mazurek, 1998).

Several recent discoveries are likely to change our perspective on the function of song system nuclei. The first is the discovery that the physiology of neurons in RA is substantially different in an animal that is awake and behaving as opposed to an animal that is asleep or anesthetized (Dave, Yu, & Margoliash, 1998; Schmidt & Konishi, 1998). In a finch that is awake, RA neurons do not change their activity in response to auditory stimuli, including the bird's own song. When a bird is asleep, however, RA neurons change their pattern of activity in response to sound, and in particular show pronounced activity in response to the bird's own song (Dave *et al.*, 1998). This neuronal activity in RA resembles the responses previously observed in birds that were anesthetized (Margoliash, 1983). Thus, while audition probably does not immediately influence the actions of RA and nXIIts in producing song, it may participate in reinforcing memories for song production during sleep in a way similar to that proposed for the mammalian hippocampus (Skaggs & McNaughton, 1996).

The second is that the activity in lMAN and Area X associated with singing differs depending on social context (Hessler & Doupe, 1999). In both nuclei, the level of neural activity is lower when the bird is producing directed song than when it is producing undirected song; the neuronal activity pattern of directed song shows less variation from one song rendition to the next. This result, together with the data on immediate-early gene activation (below), suggests that the rostral nuclei of the song system participate in song production by adults, not just in song acquisition. Furthermore, these areas are informed of the social context for singing, as shown by the differences in firing in different contexts. Though this has dramatic effects on physiology, the songs produced in these two contexts are almost identical, suggesting a dissociation within the brain between the act and its meaning.

SONG SYSTEM NEUROENDOCRINOLOGY

The general framework in which the neuroendocrinology of the adult song system is understood has not changed substantially in the last decade. Steroid receptors are present in several song system nuclei. In canaries and zebra finches, males have a greater proportion of cells with the receptors than do females. The receptor levels are similar between the two sexes in several wren species in which females sing (reviewed by Brenowitz, 1997). The receptors are believed to respond to the rises in gonadal steroid level associated with reproduction so as to synchronize song with other reproductive behaviors (reviewed by Schlinger, 1997). This has been studied experimentally by manipulating steroid levels or their effects in adults. Thus, males with low testosterone either because of castration or because of gonadal regression during autumn typically sing less than reproductive males. Frequently, females are treated with steroids and are assessed almost as models for male neuro-endocrine function. This has been done with female zebra finches, which never normally sing, and female canaries, which sing occasionally but produce a less stable song with fewer syllable types than males (Weichel, Schwager, Heid, Guttinger, & Pesch, 1986). Testosterone and its metabolites have no apparent activational effect on the song system in female zebra finches. However, if testosterone or estradiol is given during a critical period early in development (discussed below), subsequent testosterone treatment as adults will induce song (Grisham & Arnold, 1995; Schlinger & Arnold, 1991; Simpson & Vicario, 1991). In contrast, female canaries produce a song with male-typical structure and temporal pattern when treated with testosterone as adults (Nottebohm, 1980; Vallet, Kreutzer, & Gahr, 1996). In fact, their songs contain sound types that are normally only produced by males and which elicit high levels of sexual displays in untreated normal females (Vallet *et al.*, 1996). Thus, provided that a base neural substrate is present in adulthood, the song system remains sensitive to hormonal manipulation in females such that testosterone not only can increase the incidence of singing behavior, but can also lead to the expression of novel, highly potent song components, as it does in males.

FUNCTION OF SONG SYSTEM NUCLEI

Perhaps the most exciting advances in recent years in our understanding of song and the song system have come from studies on immediate-early genes (IEGs) and their products. These are genes that are turned on by high levels of neural activity including those often associated with plastic changes in neuronal functioning. The IEGs, in turn, activate other genes concerned with producing proteins

TIMOTHY J.
DEVOOGD AND
CHRISTINE LAUAY

needed, for example, for structural changes (Dubnau & Tully, 1998). As might be expected, singing is associated with IEG expression in HVC and RA (Figure 2A) (Jarvis & Nottebohm, 1997; Kimpo & Doupe, 1997). In fact, in HVC, IEG expression occurs specifically in the neurons that project to RA (Kimpo & Doupe, 1997). Somewhat surprisingly, singing is also associated with an increase in IEG mRNA in Area X, lMAN, and mMAN as well as in HVC (Jarvis & Nottebohm, 1997; Jin & Clayton, 1997). Expression of the mRNA in HVC and Area X is almost linearly related to the amount of singing the bird did in the ½ hr prior to sacrifice in both juveniles and adults (Jin & Clayton, 1997). These neuronal responses have been observed both in the lab (Jarvis & Nottebohm, 1997; Jarvis *et al.*, 1998; Kimpo & Doupe, 1997) and in the wild (Jarvis, Schwabl, Ribeiro, & Mello, 1997). In zebra finches, there are dramatic differences in the IEG expression associated with directed and undirected song. There is as much as five times more expression of the IEG mRNA in lMAN and 10 times more in Area X when the bird was alone and producing undirected song than when it was singing directed song to a female (Jarvis, Scharff, Grossman, Ramos, & Nottebohm, 1998). These results, like the physiological results above, indicate that social context has a major influence on song production. They also indicate that the rostral nuclei and Area X in particular, though not essential for song production per se, are activated by singing in ways that suggest that they may play a role in relating production to context.

In dramatic contrast, merely hearing a song is not associated with activation of the conventional nuclei of the song system, but rather with activation of subdivisions L1 and L3 of the auditory area Field L, zones adjacent to HVC and RA, as well as two newly described areas of the telencephalon named NCM and cHV (Figure 2B) (Jarvis & Nottebohm, 1997; Mello & Clayton, 1994; Mello, Vicario, & Clayton, 1992). Indeed, there are distinct patterns of activation in NCM of canaries to hearing different song syllables (Ribeiro, Cecchi, Magnasco, & Mello, 1998), patterns that seem topographic in that they show an orderly distribution of responses related to between- and within-class variation in the acoustic structure of the syllables. Research using tract-tracing in zebra finches indicates that the zones near HVC and RA as well NCM and cHV are reciprocally connected with Field L (Vates, Broom, Mello, & Nottebohm, 1996; Vates *et al.*, 1997) and with auditory nuclei elsewhere in the brain (Mello *et al.*, 1998). Such results suggest that brain areas involved in song perception and in song production are physically distinct. As discussed above, Area X and lMAN, once thought to be primarily perceptual, are unambiguously activated by producing but not by hearing song. Furthermore, the dramatic sex differences of song system nuclei seem constrained to the structures that are activated by producing song. In adult canaries, the pattern of IEG activation in response to *hearing* song is similar in males and females (Mello & Ribeiro, 1998). Neuronal firing in NCM increases when the bird hears conspecific song, although the response is general and not specific to a particular song as it is in HVC (Stripling, Volman, & Clayton, 1997). Both the physiological response (Chew, Vicario, & Nottebohm, 1996; Stripling *et al.*, 1997) and the IEG response (Mello, Nottebohm, & Clayton, 1995) habituate with repeated presentation of a particular song. Intriguingly, this response (and its habituation) does not differ between adult male and female zebra finches (Chew *et al.*, 1996). Using repeated song presentation followed by IEG measures, it was possible to demonstrate simultaneous habituation to a variety of conspecific songs and retention of the decreased response for at least 24 hr. These results suggest that brain region NCM may form memories of songs and function in individual recognition on the basis of song in both sexes. They make it possible to begin studying the neural bases for song perception and use, and how these are affected by experience and continuing interactions with conspecifics.

SONG SYSTEM DEVELOPMENT

In most adult songbirds, the song system is sexually dimorphic in structure and function. In males, it acquires and maintains learned motor elaborations of innate song structures, and its neurons come to respond very specifically to the bird's own song. How is this differentiation and precision achieved? Numerous studies have documented correlations across development, although the relevance of many such correlations for song acquisition or expression are unclear. For example, expression of nitric oxide synthase (an enzyme used in making nitric oxide, which has been implicated in plastic changes) is high in song system nuclei in juveniles and subsequently decreases (Wallhausser-Franke, Collins, & DeVoogd, 1995). Catecholamine levels decrease across development for several song system nuclei (Harding, Barclay, & Waterman, 1998). Binding of melatonin (Gahr & Kosar, 1996) and androgens (reviewed by Schlinger, 1997) increases across development, a change that is much greater for males than for females. In each experiment of this kind, the investigators suggest ways in which the biological changes that they observed might cause or be the substrate for a particular characteristic of avian song. However, two further sorts of experiments are necessary for such comments to be more than hopeful speculation. First, experiments are needed in which the neurobiology is altered. If the behavior depends on this attribute of neurobiology, it, too, should be altered. Conversely, experiments are needed in which the behavioral feature is suppressed or altered. If this feature is influencing the neurobiology, development of the neurobiological feature should also be altered. Ideally, both sorts of manipulation should be quantitative and should be assessed at multiple developmental time points. In recent years, many students of the song system have used one or the other of these approaches. However, interpretation of the experiments is complex, as it is often difficult to change a particular behavioral or neurobiological attribute and it is even more difficult to change one attribute without affecting others. The summary below will indicate the approaches for doing so that have been tried and the results they have obtained.

How Does the Brain (and Song System) Become Male or Female? By the time that songbirds are adults, the motor nuclei of the song system are large in males and small in females in most species that have been studied. There are correspondingly large sex differences in neuroendocrinology, neurochemistry, and neurophysiology. What are the mechanisms by which these sex differences develop? In the classic theory of sexual differentiation in mammals, female brain organization and body form are seen as the developmental "default." In genetic males, testicular tissue in the embryo secretes testosterone, which acts to masculinize and defeminize the brain, peripheral tissue, and subsequent behavior. Thus, the mammalian pattern of sexual differentiation was believed to be steroid dependent, and the presence of testicular secretions was believed to induce the male form. Inherent in this theory is the dual potentiality of the brain of either genetic sex to become either male or female, which would be completely determined by the presence during development of the appropriate gonadal hormones.

Initial research findings in the song system were seen as consistent with or supporting the mammalian model. Estradiol, a metabolite of testosterone, has been shown to produce sexual differentiation in RA and MAN by preventing cell loss (Konishi & Akutagawa, 1988; K. W. Nordeen, Nordeen, & Arnold, 1987). Arnold and Saltiel (1979) compared autoradiograms of male and female zebra finches injected with tritiated testosterone. They found sex differences in the number of hormone-accumulating cells in HVC and MAN, such that females had significantly fewer cells than did males. It seemed plausible that sex differences in the sensitivity of neural populations to hormones may underlie the anatomic dimorphisms. Further evidence for an early organizational effect of hormones coupled with a later activational effect comes from studies of hormonal manipulation in female zebra finches. If female nestlings are treated with estradiol on the first day of life, telencephalic nuclei (MAN, HVC, RA) are permanently masculinized (Gurney, 1981; Gurney & Konishi, 1980). If females were also treated with dihydrotestosterone (DHT), then masculinization was enhanced (Gurney, 1981). Thus, estradiol and DHT were thought to be complementary in their effects, where estradiol increases neuron size and DHT increases neuron number. These findings were compatible with the theory that, like mammals, the testes of the young male zebra finch secrete testosterone which is metabolized into estradiol and DHT. The latter two each have separate actions that contribute to the masculine pattern of development. Later androgen treatment during adulthood would then complete the masculinization of the song system (Gurney, 1981).

More recent studies argue against this hypothesis and the generality of the mammalian model. For example, Jacobs, Grisham, and Arnold (1995) found that estrogen treatment accompanied by DHT treatment did not have any additional masculinizing effect over that seen with estrogen alone. Schlinger and Arnold (1991) treated male and female nestlings with either DHT or flutamide, an antiandrogen. They found that DHT masculinized the female song system, but had no effect in males. Surprisingly, antiandrogen treatment failed to demasculinize any measures in the male song system. In fact, the volume and neuron number was hypermasculinized in the telencephalic nucleus RA as a result of flutamide treatment (Schlinger & Arnold, 1991). In addition, Wade, Schlinger, Hodges, and Arnold (1994) induced testes formation in female zebra finches by using a powerful aromatase inhibitor during embryogenesis. The presence of a testis was not sufficient to masculinize the female's song system. Thus, hormonal manipulations can only partially masculinize females, and removal of these hormones from genetic males does not result in the female phenotype. The song system does not seem to have the equipotentiality seen in mammals. Factors influencing differentiation other than gonadal hormones must be investigated to resolve this conundrum.

Applying the mammalian theory of sexual differentiation to songbirds also conflicted with an earlier body of research on the sexual differentiation of secondary (nongonadal) sexual characteristics in non-songbird species. Gonadectomized chickens or ducks developed masculine physical characteristics regardless of their genetic sex (Wolff, 1959). In ducks, removal of the ovary resulted in a masculine pattern of development of the phallic organ and syrinx (the avian vocal organ). Furthermore, estrogen treatment halted the masculine development of these organs both *in vitro* and *in vivo* (Taber, 1964). It was later found that estrogen receptors are expressed in the syrinx, and estrogen acts directly on the syrinx to prevent masculine development (Takahashi & Noumura, 1987). These results suggest that the masculine development of nongonadal characteristics occurs in the absence of ovarian secretions, and that feminine development requires estrogen secretions by the embryonic ovary. It seemed, then, that the differentiation of secondary sex characteristics in both vertebrate taxa was controlled by gonadal hormones, but in birds, ovarian estrogens inhibit masculine development, whereas in mammals testicular secretions promote masculine development (Arnold, Wade, Grisham, & Jacobs, 1996).

How can the masculinizing action of estradiol on the song system in zebra finches be reconciled with its demasculinizing action on other physical characteristics? Perhaps both patterns coexist within a given species. The sexual differentiation of copulatory behaviors may be demasculinized by estrogen, while sexual differentiation of the song system may be masculinized by estrogen. Adkins-Regan and Ascenzi (1987) found that estradiol treatment in male zebra finch chicks resulted in less masculine copulatory behavior compared to control males. The coexistence hypothesis, however, must explain how one hormone can be responsible for masculinizing some brain regions and demasculinizing others in the same animal (Arnold *et al.*, 1996). Perhaps there are alternating critical periods in development during which gonadal steroids initiate either masculinization or demasculinization. Such a scenario would require that males have high levels of estrogen during the masculinizing period and females have high levels of estrogen during the demasculinizing period. This is unlikely, as there is no evidence for an alternating pattern of estrogen levels between the sexes for the first 2 months after hatching (Adkins-Regan, Abdelnabi, Mobarak, & Ottinger, 1990; Schlinger & Arnold, 1992) and the critical periods for both the masculinizing effect of estrogen on the female song system and the demasculinizing effects of estrogen on sexual behavior are in the first week posthatch (Adkins-Regan, Mansukhani, Seiwert, & Thompson, 1994). Furthermore, the role of estrogens in sexual differentiation has been questioned because the dose of estradiol required for masculinization of the female song system is far above the normal physiological range (Adkins-Regan *et al.*, 1990) and may not be a physiological effect. The only possible way in which the hormone-organizing theory can still contribute to differentiation is if differences in steroid levels in eggs early in incubation set in motion some aspects of sexual differentiation. This seems unlikely, as fadrozole (an antiestrogen) treatment on embronic day 5 induces the formation of testicular tissue in females, but has no effect on the song system (Wade & Arnold, 1996).

As a result of the paradoxes in studies on song system differentiation and the ways in which song system observations are incompatible with prior theories, researchers have begun to look elsewhere for the mechanisms of sexual differentiation. Arnold (1996) reviewed evidence that gonadal differentiation in mammals is initiated by the action of gene products, and proposed that the sexual differentiation of neural tissue could also be initiated through nonhormonal gene products rather than indirectly via gonadal hormones. In contrast to the universal actions of circulating steroids, such a process would require that the nonsecreted protein(s) responsible for differentiation must be synthesized within the tissue that is under the developmental control of the protein. There is evidence that the transcript for the SRY gene believed responsible for testicular differentiation in mammals is expressed in the brain and other nongonadal tissue in some mammals (Harry, Koopman, Brennan, Graves, & Renfree, 1995). Furthermore, molecular response components that tie the protein to the initial steps of differentiation must also be present. Considerable research is needed to evaluate this hypothesis. For example, the sex-specific genes would have to be identified and shown to be present and functioning in the appropriate tissue at the correct times in development. The expression of these genes would also have to be shown to be independent of hormones, and sex reversal must occur if those genes are knocked out (Arnold *et al.*, 1996).

TIMOTHY J.
DEVOOGD AND
CHRISTINE LAUAY

Regardless of the mechanism of initial differentiation, it is clear that multiple processes participate in making the song system so distinct between males and females. Androgen receptors are expressed early in development in many of the motor nuclei of the song system (reviewed by Arnold, 1996, 1997). More are found in HVC of males than of females from posthatching day 9, the earliest age at which they are observed. This sex difference appears even brain slices *in vitro*, without any exogenous steroid (Gahr & Metzdorf, 1999). Thus from very early development, it is possible that circulating steroids may have different effects in males and females because of differences in the number and location of receptors. There are also high levels of estrogen receptors in HVC of males and females early in development (K. W. Nordeen *et al.*, 1987), but the number decreases over time in females (Gahr & Konishi, 1988). The time course of the decrease closely resembles the progressively diminished masculinizing effect of estrogen on the female song system in this species (Konishi & Akutagawa, 1988).

Development of the anatomy of song system nuclei has been carefully described in zebra finches. Initial growth of song system nuclei is rapid. All major nuclei can be distinguished by 2 weeks of age and are similar in the two sexes (except that Area X is not discrete in females) (Bottjer *et al.*, 1985; Kirn & DeVoogd, 1989). Major projections between nuclei are present by 3 weeks (Foster & Bottjer, 1998; Iyengar, Viswanathan, & Bottjer, 1999). In females, initial growth of HVC and RA is similar to that seen in males. However, in females, these song production nuclei then shrink back to a fraction of their earlier size as a result of high rates of cell death (Kirn & DeVoogd, 1989). The mechanism whereby masculinization reduces developmental cell death is unclear. Augmented addition of newly generated neural cells in males may contribute to the sexual differentiation of HVC and Area X (E. J. Nordeen & Nordeen, 1989). Alternatively, aspects of cell loss in females may be active—a nonestrogenic ovarian factor may inhibit the proliferation and/or survival of cells in HVC (Hidalgo, Barami, Iverson, & Goldman, 1995).

There is evidence that sex differences in one brain region can cascade to affect others. In zebra finches, adult males have a greater number of lMAN neurons that project to RA than do females (E. J. Nordeen, Grace, Burek, & Nordeen, 1992). Many of these projection neurons accumulate androgens. Both sexes form this projection early in development. However, many of these neurons then die in females unless the females are treated with estradiol. Perhaps hormone treatment of females stimulates masculinization through the retention of androgen-concentrating lMAN → RA projection neurons. Neurons in RA would be spared due to the stimulation or activation provided by these projection neurons, which could help account for the larger size of both of these structures in the male bird (E. J. Nordeen *et al.*, 1992).

Another example of such a cascade effect is the estrogen-related addition of neurons to HVC. Giving estrogen to developing female zebra finches increases the number of neurons in the nucleus, including androgen-accumulating cells that project to RA (E. J. Nordeen & Nordeen, 1989). Thus, estradiol serves to establish the motor pathway for song production, masculinizes the number of androgen-accumulating cells within HVC, and is also likely to enhance neuron survival in RA (E. J. Nordeen & Nordeen, 1988, 1989).

Glial cells could be an intermediate agent in the process of differentiation. Sex differences in neuronal survival are preceded by differential gliogenesis (E. J. Nordeen & Nordeen, 1996). An increase in the number of glial cells results in greater retention of neurons (E. J. Nordeen, Voelkel, & Nordeen, 1998). The greater number of glia cells in males could regulate local concentrations of active steroid metabolites, leading to higher local concentrations of estradiol in males (Nordeen & Nordeen, 1996), which in turn could affect connectivity and survival as indicated above.

Finally, neurotrophic factors may also influence differentiation. Lesioning lMAN in young male zebra finches normally leads to substantial cell death in RA. This is prevented by infusing RA with one of several neurotrophic growth factors (Johnson, Hohmann, DiStefano, & Bottjer, 1997). Infusion of the growth factor brain-derived neurotrophic factor (BDNF) in HVC in adult female canaries likewise prevents cell death (Rasika, Alvarez-Buylla, & Nottebohm, 1999). BDNF and its mRNA are expressed in HVC in young zebra finches (Akutagawa & Konishi, 1998; Dittrich, Feng, Metzdorf, & Gahr, 1999). Levels of the mRNA rise across development in males but not females, and the level is modulated by increasing or decreasing levels of estradiol (Dittrich *et al.*, 1999). Together, these data are consistent with a model in which stimulation of steroid receptors in HVC leads to synthesis of a trophic factor that is then transported to RA and released, thereby directly or indirectly preserving neurons.

To sum up, development of a dimorphic song system now appears far more complicated, and far more interesting, than it did 15 years ago. Yes, steroid receptors are present and steroids affect them, but it is now clear that sex differences in steroid level cannot explain the initiation of differentiation. Research on this question in the song system may lead to the discovery of novel mechanisms of differentiation that involve dimorphic transcription in the central nervous system of initiating factors. Yes, differential cell death is an obvious means by which the male and female song systems differentiate. However, it is hard to determine whether this is a cause of the other sex differences in this system or a consequence. Dissociating these various developmental influences only will not give much insight into how the song system forms, but may tell us about the development of integrated neural systems more broadly.

How Do Juvenile Birds Learn Song?   Young birds typically have a sensitive period during which they are open to learning a song. This early learning shapes the function of the song system and the song that the birds will later produce. This can be seen in the exquisite precision with which neurons in the song system respond to the bird's own song. Which aspects of innate structure guide the birds in selecting a tutor (or form the "innate" template for song)? What mechanisms are used for the learning (and where)? Do they occur in females as well as in males? Which developmental changes in the nervous system encode and preserve the learning, and which are merely coincidental? Why does the sensitive period end? Here, too, while the questions are clear, the explanations remain incomplete and tentative.

Much of the research on the development of the song system has focused on males and has looked at females primarily in order to define the critical times and events that lead to organization of a male system (reviewed by Arnold, 1997). Thus, little is known of the developmental events that are critical for a female to be able to recognize and classify song. In males, the gross volume of HVC, RA, and Area X increase throughout development such that they attain adult volumes during the sensitive period for auditory song learning (Nixdorf-Bergweiler, 1996). In contrast, lMAN attains a maximal volume by about day 20 in males and then decreases in volume and neuron number by more than 40% over the next 3 weeks (Bottjer & Sengelaub, 1989; Nixdorf-Bergweiler, 1996). This initial growth and later regression in lMAN occurs in both sexes such that throughout development the nucleus appears similar in males and females (Nixdorf-Bergweiler, 1996). Nothing is known of the developmental anatomy or connectivity of NCM, cHV, the "shelf" below HVC, or the "cup" adjacent to RA, the regions that show an increase in IEG expression in adult males and females after hearing song (Ball & Gentner, 1998).

The neurobiology of sensitive period learning is not understood in males, much less in females. Clearly, young birds are highly selective in choosing which environmental sounds they will accept as suitable models for song. Some discrimination relies on social cues (Petrinovich & Baptista, 1987). However, much of the basis for acceptance depends on sound structure, which has been studied in only a few species (Marler, 1991). The selectivity suggests the existence of filtering or discriminatory circuits that gate what can be learned by the song system and are formed without reference to environmental information. Several studies suggest that juvenile birds can recognize and respond to conspecific sounds at ages far earlier than the sensitive period for song learning (Clemmons, 1995; Leonard, Horn, Brown, & Fernandez, 1997). By 3 or 4 weeks after hatching, both male and female white-crowned sparrows increase their calling selectively in response to conspecific song, and many neurons in HVC and the HVC "shelf" below it are activated by phrase types that comprise song (Whaling, Solis, Doupe, Soha, & Marler, 1997). Such data suggest that aspects of auditory selectivity are formed in the song system as it develops, prior to the start of the sensitive period for song learning. If gating circuitry that relays only song-like sounds into the song system is formed early in development (perhaps in the projections between Field L and NCM and cHV, see Figure 2), it would help explain both the selectivity of song learning and the speed and accuracy with which acceptable models are acquired and transformed into vocal output. As indicated above, such mechanisms are plausible for females as well as males. There are hints that females, too, have a sensitive period for song learning (Nelson, Marler, Soha, & Fullerton, 1997). Female zebra finches have clear song preferences as adults (Houtman, 1992; S. A. Collins, Hubbard, & Houtman, 1994) and can be trained to discriminate songs of conspecific males (Cynx & Nottebohm, 1992). However, no systematic research has studied the neurobiology of song learning in females.

The early existence of circuitry that can recognize essential features of conspecific song could also explain the behavioral effects of song deprivation in species that have sensitive periods for song acquisition. Male birds that do not hear an acceptable song model during development eventually form a simple song which includes many species-typical characteristics (Marler, 1970; Marler & Sherman, 1983; Price, 1979). In zebra finches, this deprived song often incorporates maternal call notes and converges with the songs of other young males being reared in the same environment (C. E. Collins, Airey, Guzman, Tremper, & DeVoogd, under revision; Price, 1979; Volman & Khanna, 1995). Thus, learning of simple conspecific sounds will occur if complex ones are not present.

Many song birds acquire the sound elements that they will use in song during a posthatching sensitive period which normally ends well before reproduction (Marler, 1976, 1984, 1991). It is unclear what causes the end of the sensitive period, although several factors are likely to be involved (and may in fact be interrelated). Young male zebra finches that have been isolated from song are still able to learn complex song features in the third month and even the fourth month after hatching, well after the sensitive period would have ended in normally reared birds (Eales, 1985; Morrison & Nottebohm, 1993). This extension of the sensitive period is less likely in group-housed birds, perhaps because they have been able to learn simple sounds from each other (Jones, Ten-Cate, & Slater, 1996). These results suggest that either the social experience or acquisition of a satisfactory song may terminate behavioral plasticity. Raising testosterone levels before the normal rise of puberty can result in an abnormally simple song (Bottjer & Hewer, 1992; Korsia & Bottjer, 1991), suggesting that puberty may also play a role in ending the sensitive period.

Little is known of the neural mechanism(s) responsible for sensitive-period song learning. We have found that rearing male zebra finches without experience of adult models for song results in retaining large numbers of dendritic spines in lMAN from day 30 to day 55, in contrast to a decrease found in control males (Wallhausser-Franke, Nixdorf-Bergweiler, & DeVoogd, 1995; Nixdorf-Bergweiler, Wallhauser-Franke, & DeVoogd, 1995). Similar decreases in spine density have been found with early auditory learning in chicks (Bock & Braun, 1998, 1999; Wallhausser & Scheich, 1987). Together, these results suggest a selective model for early learning in which networks are built up in advance of learning, which then involves retaining synapses that are used and eliminating those that are not. This model is attractive for explaining attributes of early sensitive-period learning such as its speed, selectivity, fidelity, and endurance. Further study is needed, however, because only one experiment has linked synaptic selection to the song system.

In a key discovery on song acquisition, Bottjer *et al.* (1984) showed that lesioning lMAN in male zebra finches that had not yet acquired their songs caused a profound impairment in subsequent acquisition. It is likely that this effect results not just from the loss of the lMAN input to RA, but from premature maturation of the RA synapses from HVC and concomitant loss of plasticity in this pathway (Kittelberger & Mooney, 1999). Thus, recent experiments on the substrate for song learning have focused on the actions of the projection from lMAN to RA, with the hypothesis that, at least in development, this pathway is used as a comparator or reinforcer, and sustains or strengthens RA circuits that participate in correct motor coordinations. This perspective was supported by the discovery that most of the lMAN $\rightarrow$ RA synapses use N-methyl-D-aspartate (NMDA) receptors, whereas the HVC $\rightarrow$ RA synapses use Alpha-amino-3-hydroxyl-5-methyl-4 isoxazole proprionic acid (AMPA) as well as NMDA glutamate receptors (Mooney & Konishi, 1991; Stark & Perkel, 1999). NMDA receptors are only activated when stimulated by their neurotransmitter at the same time as depolarization in the postsynaptic ending. When activated, they can induce a variety of lasting changes in the structure and physiology of the synaptic ending, as might be needed to encode learning. The lMAN synapses form early and therefore are especially influential in RA over the interval when song learning is beginning (Mooney, 1992). NMDA receptors also occur at high levels within lMAN itself early in development (Boettiger & Doupe, 1998); in contrast to their persistence in RA, the number of NMDA receptors in lMAN decreases as the bird matures (Aamodt, Kozlowski, Nordeen, & Nordeen, 1992; Basham, Sohrabji, Singh, Nordeen, & Nordeen, 1999).

The role of NMDA receptors in song learning has been studied experimentally in several elegant studies in the Nordeen labs. Injecting an NMDA antagonist systemically (Aamodt, Nordeen, & Nordeen, 1996) blocks acquisition of song in young males tutored every other day if it is done just before exposure to a conspecific tutor, but not if done on nontutoring days (reviewed by K. W. Nordeen & Nordeen, 1997). Much of this effect occurs with injections of the antagonist directly into lMAN before tutoring (Basham, Nordeen, & Nordeen, 1996). In many systems, protein kinase C (PKC) is a cytoplasmic intermediate for plastic changes in synaptic function. Sakaguchi and Yamaguchi (1997) found that this enzyme has a peak in expression in RA during the sensitive period which does not occur in birds that are deafened or raised without access to a tutor. Further work is needed to determine if PKC rises when a tutor is provided or if interfering with PKC results in impaired song production.

The exquisite sensitivity with which neurons in HVC and lMAN respond to the bird's own song develops rapidly during the sensitive period for song learning (Doupe, 1997). By 60 days, neurons in lMAN respond preferentially to the tutor song, the bird's own developing song, or both over reversed song or the songs of other conspecifics (Solis & Doupe, 1997). The two sorts of song seem to influence these neurons separately: if the syrinx has been denervated such that the bird is only able to produce a distorted copy of the tutor's song, both the copy and the original stimulate lMAN neurons (Solis & Doupe, 1999). Thus, it appears that neurons in the song system form connections such that they are sensitive to conspecific song. The bird's experience of particular songs and of designing an idiosyncratic song during its sensitive period for learning then shapes the sensitivity of the neurons and results in the lasting precision of the mature system.

PLASTICITY IN THE SONG SYSTEM OF ADULT BIRDS

Many songbirds are reproductive seasonally. In temperate zones, the reproductive phase is in the spring and is signaled by increasing day length. Typically, males in these species sing vigorously during the time leading up to and including reproduction and sing little or not at all at other times. Much of the recent research on song system plasticity in adults has described features of the song system that differ between birds in these two behavioral states. For example, under decreased day length, dendrites in RA become shorter and lose 40% of their spines (Hill & DeVoogd, 1991), the rate of neuronal death in HVC increases by fourfold (Kirn & Schwabl, 1997), the proportion of HVC neurons that contains androgen receptors drops by half (Soma, Hartman, Wingfield, & Brenowitz, 1999), the density of astrocytes decreases substantially in HVC (Kafitz, Guttinger, & Muller, 1999), and the number of synapses and the number of vesicles per synapse in nXIIts decrease (Clower *et al.*, 1989). Taken together, the seasonal changes are pervasive, making it almost impossible to determine which cause the variation in behavior, which are caused by it, and which are independent of it. Indeed, given that many species retain all or most of their song from one season to the next, it is useful to examine what does *not* change. There is little if any seasonal change in the volume of lMAN, which also has much smaller sex differences than other song system nuclei (although few species have been studied) (Brenowitz, Nalls, Wingfield, & Kroodsma, 1991; Hamilton, King, Sengelaub, West, 1997). In preliminary observations, Benton, Cardin, and DeVoogd (1998) found that dendrites of certain HVC cells that project to Area X do not differ in morphology between female canaries on short days, long days, or long days plus testosterone. Two studies have looked closely at effects of lesions of the rostral nuclei in adults. Lesioning Area X impairs zebra finches in discriminations between their own song and that of a conspecific, but not in discriminating songs of two canaries (Scharff, Nottebohm, & Cynx, 1998). Lesioning lMAN in white-crowned sparrows can cause the birds to introduce new syllables into song (Benton, Nelson, 1998). Taken together, these studies suggest that the rostral circuit of the song system is less plastic in adulthood than the caudal, and therefore more suited to retaining access to the crystallized song and to shaping song structure, both at the start of the reproductive season and throughout it.

Many of the effects of varying day length can be caused by experimentally varying levels of gonadal steroid (reviewed by Ball, 1993; DeVoogd, 1994). Several recent studies indicate, however, that there are independent effects on song system nuclei of changing photoperiod and steroid level (Bernard & Ball, 1997; Bernard, Wilson, & Ball, 1997; Gulledge & Deviche, 1998; Tramontin, Wingfield, & Brenowitz, 1999). Intriguingly, social context plays a role here as well, moving male white-crowned sparrows from short days to long days results in greater HVC growth if the males are placed with reproductive females than if not (Tramontin *et al.*, 1999). Again, it is difficult to separate cause from effect, as the males housed with females also sang more often than males housed with other males.

The neurobiology underlying another form of adult behavioral plasticity, recognition of individual conspecifics, has received little attention. Thus, territorial male songbirds often become familiar with the songs of neighbors and even come to know which song should come from which site (reviewed by Stoddard, 1996). Females in some species are able to compare the songs of many males in selecting a mate (Buchanan & Catchpole, 1997; Hasselquist, Bensch, & Von-Schantz, 1996). Females may learn the songs of their mates (Hausberger, Richard-Yris, Henry, Lepage, & Schmidt, 1995; Miller, 1979). Both male and female starlings can recognize novel songs of an individual after experience with the song of that individual (Gentner & Hulse, 1998). Clearly, some kinds of song learning and comparison can continue throughout life. However, very little is known of which brain areas are used for these abilities. Female zebra finches treated with estradiol normally show courtship displays in response to conspecific, but not heterospecific song. If, in addition, cHV is lesioned in the females, they show the displays to both (MacDougall-Shackleton, Hulse, & Ball, 1998a). Similar effects occur in female canaries following lesions of HVC (Brenowitz, 1991; Del Negro, Gahr, Leboucher, & Kreutzer, 1998). Such data suggest that the females either cannot inhibit inappropriate responses or can no longer discriminate between the songs of the two species. However, nothing is known of which brain structures are needed in either sex to form or retain memories for particular conspecific songs.

Brain–Behavior Relations in the Song System

Assessing brain–behavior relations at multiple levels of comparison is one of the wonderful possibilities of song system neurobiology. Thus, it is possible identify individual differences in behavior and relate them to song system neurobiology. One can also evaluate sex differences in song-related behaviors across species and then assess whether there are parallel differences in song system morphology. Finally, it is possible to measure differences in the content or use of song across songbird species and relate this to species differences in song system structure and physiology. In each case, the approach may provide a novel insight into function. Finding a close relation between HVC anatomy and amount of song learning, for example, should indicate that a function of HVC is to carry out the learning. Stated like this, it is clear that the approach has the same limitations as any other correlational approach. However, finding strong associations (as well as finding none where associations were expected) has led to new insights about function of the song system as a whole.

Individual differences in singing have been studied in relatively few species. Nottebohm, Kasparian, and Pandazis (1981) found that in reproductive adult male canaries, the volumes of HVC and RA were significantly correlated to the number of syllables in the song repertoire. These relations were not found in a smaller sample of red-winged blackbirds (Kirn *et al.*, 1989), a species with a small number of song types. Marsh wrens are a species that, like canaries, selects from a repertoire of sounds in producing a song. As in the canary, HVC volume in marsh wrens is significantly related to repertoire size (Canady, Kroodsma, & Nottebohm, 1984). Recently, we found a similar relation in sedge warblers between HVC volume and the number of sounds in their song repertoires as recorded on their territories (Airey, Buchanan, Catchpole, Szekely, & DeVoogd, 2000). Although no significant relation has been found in zebra finches between the volume of HVC and the total number of sound elements in song (MacDougall-Shackleton, Hulse, & Ball, 1998b, but see Airey & DeVoogd, 2000), there is a significant relation between HVC volume and the number of *learned* elements (those that clearly match sounds produced by tutors) (Ward, Nordeen, & Nordeen, 1998).

As each of these studies were correlational, it was not possible to determine whether differential song learning led to differential song system anatomy or whether differences in anatomy (due to genetic or developmental factors) led to differences in the amount of learning that was possible. By manipulating opportunity for learning, Brenowitz, Lent, and Kroodsma (1995) found that amount of song learning does not appear to affect volumes of song system nuclei. Marsh wrens tutored with either 45 song types (similar to normal mean repertoire size of 54) or 5 song types grew up to produce songs with repertoires similar to those with which they were tutored. The volumes of HVC and RA did not differ between these groups (Brenowitz *et al.*, 1995). However, within the group tutored with the large repertoire, the volumes of both RA and HVC tended to vary with the size of the repertoire that was acquired. These results support the second possibility, that individual differences in song system anatomy set limits to learning. Thus, under natural conditions, in which many tutors and songs are present, healthy birds should learn up to their anatomic capacity and therefore would show significant correlations between repertoire and HVC volume, as reported for unmanipulated marsh wrens, sedge warblers, canaries, and zebra finches—and the experimental group of marsh wrens above that had a large array of songs to learn. It is possible that species that learn a single song or that do not use a repertoire of sounds use individual variation in HVC differently, but too few such species have been studied even for reasonable speculation. In the repertoire species, little is known of how the volume of HVC sets limits to learning. The marsh wrens that learned the small repertoires have fewer spine synapses in HVC than do those that learned the large repertoires (Airey, Kroodsma, & DeVoogd, 2000), although overall volume of HVC did not differ (Brenowitz *et al.*, 1995). This suggests that restricting the amount of song learning below this built-in capacity results in augmented loss of synapses.

Research that was described earlier indicates that females pay attention to song and, in many species, select males as mates on the basis of their song. The results mentioned above showing that individual differences in song system anatomy are reflected in differences in song suggest that when females choose a male with an elaborate song, they are also choosing one with a relatively large HVC. Evolutionary theory suggests that such choices occur for good reasons, either to enhance immediate fitness (by choosing a mate with good resources or helpful behaviors) or to enhance ultimate fitness (by choosing a mate with genes that will benefit the progeny). Recent research indicates that female choice on the basis of song may enhance both sorts of fitness (Airey, Castillo-Juarez, Casella, Pollak, & DeVoogd, 2000). Airey *et al.* measured the volumes of all major song production nuclei and of the telencephalon in more than 100 adult male zebra finches from 38 families.

Variation in the sizes of each of the song nuclei is correlated with overall telencephalon size. This suggests that when a female chooses a male with a large HVC, on average, she is also choosing one with a large brain and all of the potential augmented capacities that this may bring. (A caveat: while many evolutionary and neurobiological theorists would suggest that individuals with a larger brain have advantages in mating or survival over less endowed conspecifics, there is very little unambiguous support for this from life history.) This research also found that variation in the volumes of HVC, RA, and nXIIts is heritable, with heritability values as large as or larger than those found for such traits as brain or body size. This suggests that when a female chooses a male with a large HVC, she is also ensuring that her sons will have a large HVC. Finally, variation in the sizes of different song nuclei is correlated, but only for nuclei that are monosynaptically connected. Thus, sexual selection may occur for a trait (like a large song repertoire) for which one of the nuclei is most responsible. However, because of the associations between nuclei, this will have effects on the anatomy (and therefore function) of other nuclei as well.

Only one study has related individual differences in song-related behaviors in females to neuroanatomy. Hamilton *et al.* (1997) found that lMAN is larger in female cowbirds that most reliably distinguish high-potency conspecific songs from low-potency ones than in less discriminating females. Individual differences between females in the anatomy of song system nuclei are not confined to cowbirds or to lMAN. They have been found in each of the species in which female anatomy has been measured. As described earlier, we now know that song reception is processed in brain regions such as NCM and cHV that have not yet been quantified in males, much less in females. Clearly, further work is needed to determine how individual differences in anatomy in females are related to individual differences in song perception and in how perception is used by the females.

In species in which males do all or virtually all of the singing, most song system nuclei are much larger in males than in females (reviewed by Ball, Casto, & Bernard, 1994). In species in which females produce some song, the differences are much smaller (Brenowitz & Arnold, 1986). However, the sex differences in volume do not relate simply to differences in song learning or production because even in species in which males and females do not differ substantially in singing, sex differences in the volumes of most of the song production nuclei still exist (DeVoogd, Houtman, & Falls, 1995; Gahr, Sonnenshein, & Wickler, 1998).

Finally, there are immense differences across the more than 4000 songbird species in the composition, content, and use of song. Comparative study offers the opportunity to relate the multiple dimensions along which song differs between species to song system anatomy and thereby infer which anatomic features are most associated with a particular behavioral characteristic as well as how these traits may have evolved (reviewed by DeVoogd & Szekely, 1998). However, relatively little of this has been done. A careful scan of the studies summarized above indicates that most work has been done on zebra finches, with almost all of the rest on a smattering of temperate-zone finch and sparrow species.

Explicitly comparative study requires a large enough number of species to be statistically meaningful and an analytic approach in which degree of relatedness between species is factored in. Using one such approach, the method of independent contrasts, in a sample of males from 41 species in eight families, DeVoogd, Krebs, Healy, and Purvis (1993) found that the volume of HVC with respect to telencephalon volume was significantly correlated with the number of songs typically sung by males of the species (Figure 3). The volume of Area X was not related

TIMOTHY J.
DEVOOGD AND
CHRISTINE LAUAY

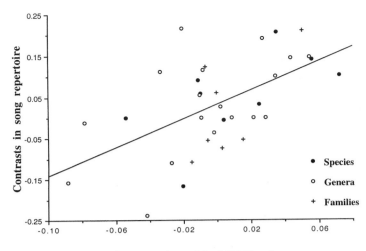

Contrasts in residual HVC volume

Figure 3. Increases in the volume of HVC tend to be associated with increases in the song repertoire typically attained by males of a species. The method of independent contrasts was used in order to control statistically for varying degrees of relatedness in this sample of 41 oscine species. In this sample, the contrasts among species and genera contribute more to the association than the contrasts among families, suggesting that close brain-behavior associations are most evident in instances of recent evolutionary divergence. (From DeVoogd *et al.*, 1993. Copyright 1993 The Royal Society.)

to any of the song measures. This association suggests that the anatomy of HVC in particular reflects capacity for learning, and that this assignment has existed for the millions of generations since these species had common ancestors. Ideally, such a result would indicate that one could select for study a species in which behaviors related to the scientific question being examined are especially prominent and be confident that the results obtained would apply to songbirds more generally. Furthermore, these results would suggest that it might be possible to determine functions of other brain nuclei by comparing song acquisition and expression in diverse species in which the nuclei are especially large to species in which they are especially small. However, examining such a diverse phylogeny is also likely to prevent seeing meaningful brain–behavior relations. Song composition varies so much across songbirds that it is not possible to determine measures that apply to all (how would one quantify the amount of learning in the "song" of a crow, for example?) It is also possible that the neural substrate for song production or perception could have become qualitatively different in lineages in which the use of song has diverged substantially. These theoretical limitations can be reduced by studying instead a more restricted phylogeny in which the use and form of song differ quantitatively and applying the findings to birds closely related to that lineage. In a closely related phylogeny of eight reed warbler species, Szekely, Catchpole, DeVoogd, Marchl, and DeVoogd (1996) found a similar association between the number of song syllables in the repertoire of the typical male and the relative volume of HVC, but not the volumes of RA, Area X, or lMAN. Once again, the neural association with motor learning is focal and not spread across the song system. Furthermore, applying parsimonious models of evolutionary change to these anatomic measures indicates that HVC volume has increased across evolutionary time in some lineages of this phylogeny and decreased in others. Under these models, other nuclei have shifted as well, but in different patterns. Such findings suggest that changes in different song system nuclei affect distinct attributes of song. If so, then selection based on a particular song attribute would act on heritable components of variation in the nucleus most related to that attribute.

AND WHERE DO WE GO FROM HERE?

As we enter the new century, it is common in many endeavors to make predictions. The paragraphs below are our evaluation of domains that appear to offer exciting opportunities for deeper understanding of song behavior and its neurobiology. A common theme is research directed at greater integration of song and the song system with other aspects of avian biology. There are major gaps in our understanding of several levels of integration in the research reviewed above. Little has been done relating evolution and inheritance of song attributes to neurobiology. Much needs to be done as well in relating song and the song system to other behavioral systems such as those involved in other forms of aggressive or reproductive behaviors. Zebra finches, the songbird in which most of the songbird research has been carried out, are a colonial species with a dominance hierarchy, in which pairs are monogamous and cooperate in raising young (Zann, 1996). As argued above, song plays a role in establishing and maintaining dominance, forming pairs, and instructing young. Virtually no research has tried to determine how song system brain areas contribute to these behaviors. Even the simplest experiments looking for changes in dominance or pairing status as a result of lesioning each of the song system nuclei have not been done. Finally, there is very little research on the interplay between the actions of one individual (and the brain systems responsible for these actions) and the actions and brain systems of another individual. The rationale for selecting these directions will be discussed below.

We need to understand the neurobiology of innateness. There are many constraints on song perception and production that are independent of the peripheral receptors and effectors and either do not require experience or interact with experience in restricted ways. Young males can be highly selective about the sounds they learn and from whom they are willing to learn them. Songs contain species-typical features even in the absence of access to a song model. Females can be very consistent in assessing which songs are potent or attractive, whether or not they know the singer. How are such qualities encoded in the brain and passed on within a species? Research described above suggests that there are ways in which the physiology of song system neurons is selective for song-like features almost from the time that the cells can be monitored. However, no research has described the anatomy that gives rise to this selectivity within a species or how it differs between species that develop very different forms of song.

Do bird parents pass on to their progeny other neural features related to song and to fitness (for example, capacity for learning)? Research described above suggests that aspects of song system morphology are heritable. It is not clear from this research how evolution of song system anatomy is constrained. In other words, if an elaborate song is attractive to females, why doesn't every male produce one? Perhaps song system anatomy is so closely tied to the anatomy of other brain systems that enhanced song system structure cannot be expressed without also changing other brain regions. Perhaps, too, elaboration of song is constrained by the particular environment in which it is produced as well as by the need to maintain markers to

TIMOTHY J.
DEVOOGD AND
CHRISTINE LAUAY

individual identity. The developmental mechanisms that might produce these restraints are not understood.

Far too little is known of the pattern and meaning of interspecies variation in song system anatomy and physiology. The song system cannot be understood as a modal zebra finch pattern to which minor variation is added for other species. This lack of a more general perspective cannot be fixed by studying any single species that seems to be different is some aspect of its use of song (discussed in detail by DeVoogd & Szekely, 1998). Information gained from comparative study can make use of the immense natural experiment of the speciation of oscine birds to provide insights into structure–function relations across many different dimensions of song organization, content, and use. In addition, as argued above, it can show how present neurobiology has been selected and shaped across evolutionary time.

Song is one component in a variety of aggressive and agonistic behaviors. It can be directed or nondirected, and its intensity can be modulated. It can easily lead to aggression or mate guarding, depending on the context. Little is known of how the song system is connected to or integrated with neural regions responsible for these other behaviors. This is certainly not a research issue confined to avian neurobiology. How systems of behavior are managed and integrated is an unresolved issue in many domains of behavioral neuroscience (Newman, 1999). However, because song and the neuroanatomy that controls it are so accessible, the larger questions of brain system organization and integration may also be tractable in songbirds, and have not yet been systematically studied.

Natural song is produced in complex social contexts. It is a medium of communication and the reaction of the audience affects subsequent vocalizations and other activities. Only recently has research begun to address the neurobiology of song *reception* (Jarvis & Nottebohm, 1997; Mello & Ribeiro, 1998; Ribeiro *et al.*, 1998; Stripling *et al.*, 1997). Much remains to be done on how signals from other birds are processed in the brain and how they influence the song system and ultimately behavior. For example, substantial behavioral data indicate that a bird adapts his song to the song of neighbors. He learns and remembers the songs of his neighbors, and his response to song differs in predictable ways depending on whether the song is familiar or not and from an appropriate location or not. It is not known how these streams of information are put together in the brain. Furthermore, in many species, females choose males on the basis of their songs. In others, female vocal preferences drive male song acquisition and performance. In fact, female choice based on song is believed to have driven speciation in several lineages. Very little has been done on the neural basis of normal song perception in females, even though this is key to this process and must be at least as complicated and interesting as the male system. Perhaps female acuity has been one of the factors leading to action-based learning in which young and therefore presumably less attractive males shape their song to resemble more closely that of the older adults. Clearly, female reception and evaluation of song is rapid and accurate, but not even the gross anatomic brain regions responsible are definitely known.

Finally, research on the avian song system has been impressive in that many of the phenomena that were first documented in this system have been beacons for novel understanding of neurobiology in other taxa, especially mammals. The observations that major brain structures differ between males and females, that large physiological shifts occur in adult neuroanatomy and neural function, that rapid anatomic synaptic alteration occurs in response to shifts in circulating steroids, and that adult neurogenesis is not only possible but widespread exemplify the comparative and intrinsic value of research on the avian song system. Current discoveries in the song system that point to nongenomic steroid receptors and differential brain development as a result of direct readout of chromosomal sex may one day be seen as additional instances of insight into general principles in neurobiology. It is almost certain that research in the next decade will use this remarkable system to uncover still more general principles of brain evolution, development, and integration.

Acknowledgments

This work was supported by NIH grants HD21033 and MH56093.
The authors thank Dr. Sarah Newman for composing and providing Figures 1 and 2.

REFERENCES

Aamodt, S. M., Kozlowski, M. R., Nordeen, E. J., & Nordeen, K. W. (1992). Distribution and developmental change in [3H] MK-801 binding within zebra finch song nuclei. *Journal of Neurobiology, 23,* 997–1005.

Aamodt, S. M., Nordeen, E. J. & Nordeen, K. W. (1996). Blockade of NMDA receptors during song model exposure impairs song development in juvenile zebra finches. *Neurobiology of Learning and Memory, 65,* 91–98.

Adkins-Regan, E., & Ascenzi, M. (1987). Social and sexual behaviour of male and female zebra finches treated with oestradiol during the nestling period. *Animal Behaviour, 35,* 1100–1112.

Adkins-Regan, E., Abdelnabi, M., Mobarak, M., & Ottinger, M. A. (1990). Sex steroid levels in developing and adult male and female zebra finches (*Poephila guttata*). *General and Comparative Endocrinology, 78,* 93–109.

Adkins-Regan, E., Mansukhani, V., Seiwert, C., & Thompson, R. (1994). Sexual differentiation of brain and behavior in the zebra finch: Critical periods for effects of early estrogen treatment. *Journal of Neurobiology, 25,* 865–877.

Adret, P. (1993). Operant conditioning, song learning and imprinting to taped song in the zebra finch. *Animal Behaviour, 46,* 149–159.

Airey, D. C., & DeVoogd, T. J. (2000). Variation in song complexity and HVC volume are significantly related in zebra finches. *Neuroreport, 10,* 2339–2344.

Airey, D. C., Buchanan, K. L., Catchpole, C. K. Szekely, T., & DeVoogd, T. J. (2000). Song complexity, sexual selection and a song control nucleus (HVC) in the brains of European sedge warblers. *Journal of Neurobiology, 43,* 244–253.

Airey, D. C., Castillo-Juarez, H., Casella, G., Pollak, E. J., & DeVoogd, T. J. (2000). Variation in the volume of zebra finch song control nuclei is heritable: Developmental and evolutionary implications. *Proceedings of the Royal Society, Series B, 267,* 2099–2104.

Airey, D. C., Kroodsma, D. E., & DeVoogd, T. J. (2000). Learning a larger song repertoire increases spine density in a songbird telencephalic control nucleus. *Neurobiology of Learning and Memory, 73,* 274–281.

Akutagawa, E., & Konishi, M. (1998). Transient expression and transport of brain-derived neurotrophic factor in the male zebra finch's song system during vocal development. *Proceedings of the National Academy of Sciences of the USA, 95,* 11429–11434.

Arnold, A. P. (1996). Genetically triggered sexual differentiation of brain and behavior. *Hormones and Behavior, 30,* 495–515.

Arnold, A. P. (1997). Sexual differentiation of the zebra finch song system: Positive evidence, negative evidence, null hypotheses, and a paradigm shift. *Journal of Neurobiology, 33,* 572–584.

Arnold, A. P. & Saltiel, A. (1979). Sexual difference in pattern of hormone accumulation in the brain of a songbird. *Science, 205,* 702–704.

Arnold, A. P., Nottebohm, F., & Pfaff, D. W. (1976). Hormone accumulating cells in vocal control and other brain regions of the zebra finch (*Poephila guttata*). *Journal of Comparative Neurology, 165,* 487–512.

Arnold, A. P., Wade, J., Grisham, W., & Jacobs, E. C. (1996). Sexual differentiation of the brain in songbirds. *Developmental Neuroscience, 18,* 124–136.

Ball, G. F. (1993). The neural integration of environmental information by seasonally breeding birds. *American Zoologist, 33*, 185–199.

Ball, G. F., & Gentner, T. Q. (1998). They're playing our song: Gene expression and birdsong perception. *Neuron, 21*, 271–274.

Ball, G. F., Casto, J. M., & Bernard, D. J. (1994). Sex differences in the volume of avian song control nuclei: Comparative studies and the issue of brain nucleus delineation. *Psychoneuroendocrinology, 19*, 485–504.

Baptista, L. F., Bell, A., & Trail, P. W. (1993). Song learning and production in the white-crowned sparrow: Parallels with sexual imprinting. *Netherlands Journal of Zoology, 43*, 17–33.

Basham, M. E., Nordeen, E. J., & Nordeen, K. W. (1996). Blockade of NMDA receptors in the anterior forebrain impairs memory acquisition in the zebra finch. *Neurobiology of Learning and Memory, 66*, 295–304.

Basham, M. E., Sohrabji, F., Singh, T. D., Nordeen, E. J., & Nordeen, K. W. (1999). Developmental regulation of NMDA receptor 2B subunit mRNA and ifenprodil binding in the zebra finch anterior forebrain. *Journal of Neurobiology, 39*, 155–167.

Benton, S., Cardin, J. A., & DeVoogd, T. J. (1998). Lucifer yellow filling of Area X-projecting neurons in the high vocal center of female canaries. *Brain Research, 799*, 138–147.

Benton, S., Nelson, D. A., Marler, P., & DeVoogd, T. J. (1998). Anterior forebrain pathway mediates seasonal song expression in adult male white-crowned sparrows (*Zonotrichia leucophrys*). *Behavioural Brain Research, 96*, 135–150.

Bernard, D. J., & Ball, G. F. (1997). Photoperiodic condition modulates the effects of testosterone on song control nuclei volumes in male European starlings. *General and Comparative Endocrinology, 105*, 276–283.

Bernard, D. J., Wilson, F. E., & Ball, G. F. (1997). Testis-dependent and -independent effects of photoperiod on volumes of song control nuclei in American tree sparrows (*Spizella arborea*). *Brain Research, 760*, 263–169.

Blass, E. M. (Ed.). (1986). *Handbook of behavioral neurobiology*, Volume 8, *Developmental psychobiology and developmental neurobiology*. New York: Plenum Press.

Bock, J., & Braun, K. (1998). Differential emotional experience leads to pruning of dendritic spines in the forebrain of domestic chick. *Neural Plasticity, 6*, s17–27.

Bock, J., & Braun, K. (1999). Blockade of N-methyl-D-aspartate receptor activation suppresses learning-induced synaptic elimination. *Proceedings of the National Academy of Sciences of the USA, 96*, 2485–2490.

Boettiger, C. A., & Doupe, A. J. (1998). Intrinsic and thalamic excitatory inputs onto songbird LAMN neurons differ in their pharmacological and temporal properties. *Journal of Neurophysiology, 79*, 2615–2628.

Bottjer, S. W., & Arnold, A. P. (1986). The ontogeny of vocal learning in songbirds. In E. M. Blass (Ed.), *Handbook of behavioral neurobiology*, Volume 8, *Developmental psychobiology and developmental neurobiology* (pp. 129–161). New York: Plenum Press.

Bottjer, S. W., & Hewer, S. J. (1992). Castration and antisteroid treatment impair vocal learning in male zebra finches. *Journal of Neurobiology, 23*, 337–353.

Bottjer, S. W., & Sengelaub, D. R. (1989). Cell death during development of a forebrain nucleus involved with vocal learning in zebra finches. *Journal of Neurobiology, 20*, 609–618.

Bottjer, S. W., Miesner, E., & Arnold, A. P. (1984). Forebrain lesions disrupt development but not maintenance of song in passerine birds. *Science, 224*, 901–903.

Bottjer, S. W., Glaessner, S. L., & Arnold A. P. (1985). Ontogeny of brain nuclei controlling song learning and behavior in zebra finches. *Journal of Neuroscience, 5*, 1556–1562.

Bottjer, S. W., Halsema, K. A., Brown, S. A., & Miesner, E. A. (1989). Axonal connections of a forebrain nucleus involved with vocal learning in zebra finches. *Journal of Comparative Neurology, 279*, 312–326.

Brenowitz, E. A. (1991). Altered perception of species-specific song by female birds after lesions of a forebrain nucleus. *Science, 251*, 303–305.

Brenowitz, E. A. (1997). Comparative approaches to the avian song system. *Journal of Neurobiology, 33*, 517–531.

Brenowitz, E. A., & Arnold, A. P. (1986). Interspecific comparisons of the size of neural song control regions and song complexity in duetting birds: Evolutionary implications. *Journal of Neuroscience, 6*, 2875–2879.

Brenowitz, E. A., Arnold, A. P., & Levin, R. A. (1985). Neural correlates of female song in tropical duetting birds. *Brain Research, 343*, 104–112.

Brenowitz, E. A., Nalls, B., Wingfield, J. C., & Kroodsma, D. E. (1991). Seasonal changes in avian song nuclei without seasonal changes in song repertoire. *Journal of Neuroscience, 11*, 1367–1374.

Brenowitz, E. A., Lent, K., & Kroodsma, D. E. (1995). Brain space for learned song in birds develops independently of song learning. *Journal of Neuroscience, 15*, 6281–6286.

Buchanan, K. L., & Catchpole, C. K. (1997). Female choice in the sedge warbler *Acrocephalus schoeno-baenus*: Multiple cues from song and territory quality. *Proceedings of the Royal Society of London, Series B, 264*, 521–526.

Canady, R. A., Kroodsma, D. E., & Nottebohm, F. (1984). Population differences in complexity of a learned skill are correlated with the brain space involved. *Proceedings of the National Academy of Sciences of the USA, 81*, 6232–6234.

Chaiken, M., Bohner, J., & Marler, P. (1993). Song acquisition in European starlings (*Sturnus vulgaris*): A comparison of the songs of live-tutored, tape-tutored, untutored and wild-caught males. *Animal Behaviour, 46*, 1079–1090.

Chew, W. J., Vicario, D. S., & Nottebohm, F. (1996). A large capacity memory system that recognizes the calls and songs of individual birds. *Proceedings of the National Academy of Sciences of the USA, 93*, 1950–1955.

Clayton, N. S. (1987). Song tutor choice in zebra finches. *Animal Behaviour, 35*, 714–722.

Clemmons, J. R. (1995). Development of a selective response to an adult vocalization in nestling black-capped chickadees. *Behaviour, 132*, 1–20.

Clower, R. P., Nixdorf, B. E., & DeVoogd, T. J. (1989). Synaptic plasticity in the hypoglossal nucleus of female canaries: Structural correlates of season, hemisphere and testosterone treatment. *Behavioral Neural Biology, 52*, 63–77.

Collins, C. E., Airey, D. C., Guzman, J. R., Tremper, K. A,. & DeVoogd, T. J. Group-raised, song-deprived zebra finches exhibit abnormal song structure, but normal social behavioral development. *Journal of Comparative Psychology*, under revision.

Collins, S. A., Hubbard, C., & Houtman, A. M. (1994). Female mate choice in the zebra finch: The effect of male beak colour and male song. *Behavioral Ecology and Sociobiology, 35*, 21–25.

Cynx, J., & Nottebohm, F. (1992) Role of gender, season and familiarity in discrimination of conspecific song by zebra finches. *Proceedings of the National Academy of Sciences of the USA, 89*, 1368–1371.

Dave, A. S., Yu, A. C., & Margoliash, D. (1998). Behavioral state modulation of auditory activity in a vocal motor system. *Science, 282*, 2250–2254.

Del Negro, C., Gahr, M., Leboucher, G., & Kreutzer, M. (1998). The selectivity of sexual responses to song displays: Effects of partial chemical lesion of the HVC in female canaries. *Behavioural Brain Research, 96*, 151–159

DeVoogd, T. J. (1991). Endocrine modulation of the development and adult function of the avian song system. *Psychoneuroendocrinology, 16*, 41–66.

DeVoogd, T. J. (1994). Interactions between endocrinology and learning in the avian song system. *Annals of the New York Academy of Sciences, 743*, 19–43.

DeVoogd, T. J., & Nottebohm, F. (1981a). Sex differences in dendritic morphology of a song control nucleus in the canary: A quantitative Golgi study. *Journal of Comparative Neurology, 196*, 309–316.

DeVoogd, T. J., & Nottebohm, F. (1981b). Gonadal hormones induce dendritic growth in the adult brain. *Science, 214*, 202–204.

DeVoogd, T. J. & Szekely, T. (1998). Causes of avian song: Using neurobiology to integrate proximal and ultimate levels of analysis. In I. Pepperberg, A. Kamil, & R. Balda (Eds.), *Animal cognition in nature* (pp. 337–380), New York: Academic Press.

DeVoogd, T. J., Nixdorf, B., & Nottebohm, F. (1985). Formation of new synapses related to acquisition of a new behavior. *Brain Research, 329*, 304–308.

DeVoogd, T. J., Brenowitz, E. A., & Arnold, A. P. (1988). Small sex differences in song control dendrites are associated with minimal differences in song capacity. *Journal of Neurobiology, 19*, 199–209.

DeVoogd, T. J., Krebs, J. R., Healy, S. D., & Purvis, A. (1993). Evolutionary correlation between repertoire size and a brain nucleus amongst passerine birds. *Proceedings of the Royal Society of London, Series B, 254*, 75–82.

DeVoogd, T. J., Houtman, A., & Falls, B. (1995). White-throated sparrow morphs that differ in song production rate also differ in the anatomy of some song-related brain areas. *Journal of Neurobiology, 28*, 202–213.

Dittrich, F., Feng, Y., Metzdorf, R., & Gahr, M. (1999). Estrogen-inducible, sex-specific expression of brain-derived neurotrophic factor mRNA in a forebrain song control nucleus of the juvenile zebra finch. *Proceedings of the National Academy of Sciences of the USA, 96*, 8241–8246.

Doupe, A. J. (1997). Song- and order-selective neurons in the songbird anterior forebrain and their emergence during vocal development. *Journal of Neuroscience, 17*, 1147–1167.

Doupe, A. J., & Konishi, M. (1991). Song-selective auditory circuits in the vocal control system of the zebra finch. *Proceedings of the National Academy of Sciences of the USA, 96*, 11339–11343.

Dubnau, J., & Tully, T. (1998) Gene discovery in *Drosophila*: New insights for learning and memory. *Annual Review of Neuroscience, 21*, 407–444.

Eales, L. A. (1985). Song learning in zebra finches: Some effects of song model availability on what is learnt and when. *Animal Behaviour, 33,* 1293–1300.

Eens, M. (1997). Understanding the complex song of the European starling: An integrated ethological approach. *Advances in the Study of Behavior, 26,* 355–434.

Foster, E. F., & Bottjer, S. W. (1998). Axonal connections of the high vocal conter and surrounding cortical regions in juvenile and adult male zebra finches. *Journal of Comparative Neurology, 397,* 118–138.

Foster, E. F., Mehta, R. P., & Bottjer, S. W. (1997). Axonal connections of the medial magnocellular nucleus of the anterior neostriatum in zebra finches. *Journal of Comparative Neurology, 382,* 364–381.

Gahr, M., & Konishi, M. (1988). Developmental changes in estrogen-sensitive neurons in the forebrain of the zebra finch. *Proceedings of the National Academy of Sciences of the USA, 85,* 7380–7383.

Gahr, M., & Kosar, E. (1996). Identification, distribution and developmental changes of a melatonin binding site in the song control system of the zebra finch. *Journal of Comparative Neurology, 367,* 308–318.

Gahr, M., & Metzdorf, R. (1999). The sexually dimorphic expression of androgen receptors in the song nucleus hyperstriatalis ventrale pars caudale of the zebra finch develops independently of gonadal steroids. *Journal of Neuroscience, 19,* 2628–2636.

Gahr, M., Sonnenshein, E., & Wickler, W. (1998). Sex difference in the size of the neural song control regions in a dueting songbird with similar song repertoire size of males and females. *Journal of Neuroscience, 18,* 1124–1131.

Gentner, T. Q., & Hulse, S. H. (1998). Perceptual mechanisms for individual vocal recognition in European starlings, *Sturnus vulgaris. Animal Behaviour, 56,* 579–594.

Grisham, W., & Arnold, A. P. (1995). A direct comparison of the masculinizing effects of testosterone, androstenedione, estrogen, and progesterone on the development of the zebra finch song system. *Journal of Neurobiology, 26,* 163–170.

Gulledge, C. C., & Deviche, P. (1998). Photoperiod and testosterone independently affect vocal control region volumes in adolescent male songbirds. *Journal of Neurobiology, 36,* 550–558.

Gurney, M. (1981). Hormonal control of cell form and number in the zebra finch song system. *Journal of Neuroscience, 1,* 658–673.

Gurney, M. E., & Konishi, M. (1980). Hormone induced sexual differentiation of brain and behavior in zebra finches. *Science, 208,* 1380–1383.

Hamilton, K. S., King, A. P., Sengelaub, D. R., & West, M. J. (1997). A brain of her own: A neural correlate of song assessment in a female songbird. *Neurobiology of Learning and Memory, 68,* 325–332.

Harding, C. F. (1986). The role of androgen metabolism in the activation of male behavior. *Annals of the New York Academy of Sciences, 474,* 371–378.

Harding, C. F., Barclay, S. R., & Waterman, S. A. (1998). Changes in catecholamine levels and turnover rates in hypothalamic, vocal control, and auditory nuclei in male zebra finches during development. *Journal of Neurobiology, 34,* 329–346.

Harry, J., Koopman, P., Brennan, F., Graves, J., & Renfree, M. (1995). Widespread expression of the testis-determining gene SRY in a marsupial. *Nature Genetics, 11,* 347–349.

Hasselquist, D., Bensch, S., & Von-Schantz, T. (1996). Correlation between male song repertoire, extra-pair paternity and offspring survival in the great reed warbler. *Nature, 381,* 229–232.

Hausberger, M., Richard-Yris, M.-A., Henry, L., Lepage, L., & Schmidt, I. (1995). Song sharing reflects the social organization in a captive group of European starlings (*Sturnus vulgaris*). *Journal of Comparative Psychology, 109,* 222–241.

Hessler, N. A., & Doupe, A. J. (1999). Social context modulates singing-related neural activity in the songbird forebrain. *Nature Neuroscience, 2,* 209–211.

Hidalgo, A., Barami, K., Iverson, K., & Goldman, S. A. (1995). Estrogens and non-estrogenic ovarian influences combine to promote the recruitment, and decrease the turnover of new neurons in the adult female canary brain. *Journal of Neurobiology, 27,* 470–487.

Hill, K. M., & DeVoogd, T. J. (1991). Decreased daylength results in dendritic shrinkage and loss of dendritic spines in a brain nucleus involved in seasonal production of song. *Behavioral and Neural Biology, 56,* 240–250.

Houtman, A. (1992). Female zebra finches choose extra-pair copulations with genetically attractive males. *Proceedings of the Royal Society of London, Series B, 249,* 3–6.

Hultsch, H., Schleuss, F., & Todt, D. (1999). Auditory–visual stimulus pairing enhances perceptual learning in a songbird. *Animal Behaviour, 58,* 143–149.

Iyengar, S., Viswanathan, S. S., & Bottjer, S. W. (1999) Development of topography within song control circuitry of zebra finches during the sensitive period for song learning. *Journal of Neuroscience, 19,* 6037–6057.

Jacobs, E. C., Grisham, W., & Arnold, A. P. (1995). Lack of a synergistic effect between estradiol and dihydrotestosterone in the masculinization of the zebra finch song system. *Journal of Neurobiology, 27,* 513–519.

Janata, P., & Margoliash, D. (1999). Gradual emergence of song selectivity in sensorimotor structures of the male zebra finch song system. *Journal of Neuroscience, 19,* 5108–5118.

Jarvis, E. D., & Nottebohm, F. (1997). Motor-driven gene expression. *Proceedings of the National Academy of Sciences of the USA, 94,* 4097–4102.

Jarvis, E. D., Schwabl, H., Ribeiro, S., & Mello, C. V. (1997). Brain gene regulation by territorial singing behavior in freely ranging songbirds. *Neuroreport, 3,* 2073–2077.

Jarvis, E. D., Scharff, C., Grossman, M. R., Ramos, J. A., & Nottebohm, F. (1998). For whom the bird sings: Context-dependent gene expression. *Neuron, 21,* 775–788.

Jin, H., & Clayton, D. F. (1997). Localized changes in immediate-early gene regulation during sensory and motor learning in zebra finches. *Neuron, 19,* 1049–1059.

Johnson, F., Sablan, M. M., & Bottjer, S. W. (1995). Topographic organization of a forebrain pathway involved with vocal learning in zebra finches. *Journal of Comparative Neurology, 358,* 260–278.

Johnson, F., Hohmann, S. E., DiStefano, P. S., & Bottjer, S. W. (1997). Neurotrophins suppress apoptosis induced by deafferentation of an avian motor-cortical region. *Journal of Neuroscience, 17,* 2101–2111.

Jones, A. E., Ten-Cate, C., & Slater, P. J. B. (1996). Early experience and plasticity of song in adult male zebra finches (*Taeniopygia guttata*). *Journal of Comparative Psychology, 110,* 354–369.

Kafitz, K. W., Guttinger, H. R., & Muller, C. M. (1999). Seasonal changes in astrocytes parallel neuronal plasticity in the song control area HVc of the canary. *Glia, 27,* 88–100.

Ketterson, E. D., Nolan, V., Wolf, L., & Ziegenfus, C. (1992) Testerone and avian life histories: Effects of experimentally elevated testosterone on behavior and correlates of fitness in the dark-eyed junco (*Junco hyemalis*). *American Naturalist, 140,* 980–999.

Kimpo, R. R., & Doupe, A. J. (1997). FOS is induced by singing in distinct neuronal populations in a motor network. *Neuron, 18,* 315–325.

Kirn, J. R., & DeVoogd, T. J. (1989). The genesis and death of vocal control neurons during sexual differentiation in the zebra finch. *Journal of Neuroscience, 9,* 3176–87.

Kirn, J. R., & Schwabl, H. (1997). Photoperiod regulation of neuron death in the adult canary. *Journal of Neurobiology, 33,* 223–231.

Kirn, J. R., Clower, R. P., Kroodsma, D. E., & DeVoogd, T. J. (1989). Song-related brain regions in the red-winged blackbird are affected by sex and season but not repertoire size. *Journal of Neurobiology, 20,* 139–163.

Kittelberger, J. M., & Mooney, R. (1999). Lesions of an avian forebrain nucleus that disrupt song development alter synaptic connectivity and transmission in the vocal premotor pathway. *Journal of Neuroscience, 19,* 9385–9398.

Konishi, M. (1965). The role of auditory feedback in the control of vocalization in the white-crowned sparrow. *Zeitschrift für Tierpsychologie, 22,* 584–599.

Konishi, M., & Akutagawa, E. (1985). Neuronal growth, atrophy and death in a sexually dimorphic song nucleus in the zebra finch brain. *Nature, 315,* 145–147.

Konishi, M., & Akutagawa, E. (1988). A critical period for estrogen action on neurons of the song control system in the zebra finch. *Proceedings of the National Academy of Sciences of the USA, 85,* 7006–7007.

Korsia, S., & Bottjer, S. W. (1991). Chronic testosterone treatment impairs vocal learning in male zebra finches during a restricted period of development. *Journal of Neuroscience, 11,* 2362–2371.

Kroodsma, D. E. (1996). Ecology of passerine song development. In D. E. Kroodsma & E. H. Miller (Eds.), *Ecology and evolution of acoustic communication in birds* (pp. 3–19), Ithaca, NY: Cornell University Press.

Kroodsma, D. E., & Byers, B. E. (1991). The functions(s) of bird song. *American Zoologist, 31,* 318–328.

Leonard, M. L., Horn, A. G., Brown, C. R., & Fernandez, N. J. (1997). Parent–offspring recognition in tree swallows, *Tachycineta bicolor. Animal Behaviour, 54,* 1107–1116.

Lewicki, M. S., & Arthur, B. J. (1996). Hierarchical organization of auditory temporal context sensitivity. *Journal of Neuroscience, 16,* 6987–6998.

Lim, D., & Kim, C. (1997). Emerging auditory response interactions to harmonic complexes in field L of the zebra finch. *Auris Nasus Larynx, 24,* 227–232.

MacDougall-Shackleton, S. A., Hulse, S. H., & Ball, G. F. (1998a). Neural bases of song preferences in female zebra finches (*Taeniopygia guttata*). *NeuroReport, 9,* 3047–3052.

MacDougall-Shackleton, S. A., Hulse, S. H., & Ball, G. F. (1998b). Neural correlates of singing behavior in male zebra finches (*Taeniopygia guttata*). *Journal of Neurobiology, 36,* 421–430.

Maekawa, M., & Uno, H. (1996). Difference in selectivity to song note properties between the vocal nuclei of the zebra finch. *Neuroscience Letters, 218,* 123–126.

Margoliash, D. (1983). Acoustic parameters underlying the responses of song-specific neurons in the white-crowned sparrow. *Journal of Neuroscience, 3,* 1039–1057.

Margoliash, D. (1997). Functional organization of forebrain pathways for song production and perception. *Journal of Neurobiology, 33,* 671–693.

Margoliash, D., & Fortune, E. S. (1992). Temporal and harmonic combination-sensitive neurons in the zebra finch's HVc. *Journal of Neuroscience, 12,* 4309–4326.

Marler, P. (1970). A comparative approach to vocal learning: Song development in white-crowned sparrows. *Journal of Comparative and Physiological Psychology, 71/2,* 1–25.

Marler, P. (1976). Sensory templates in species-specific behavior. In J. Fentress (Ed.), *Simpler networks and behavior* (pp. 314–329). Sunderland, MA: Sinauer.

Marler, P. (1984). Song learning: Innate species differences in the learning process. In P. Marler & H. S. Terrace (Eds.), *The biology of learning* (pp. 289–309). Berlin: Springer-Verlag.

Marler, P. (1990). Song learning: The interface between behaviour and neuroethology. *Philosophical Transactions of the Royal Society of London, Series B, 329,* 109–114.

Marler, P. (1991). Differences in behavioural development in closely related species: Birdsong. In P. Bateson (Ed.), *The development and integration of behaviour: Essays in honour of Robert Hinde* (pp. 41–70). Cambridge: Cambridge University Press.

Marler, P., & Peters, S. (1982). Long-term storage of learned birdsongs prior to production. *Animal Behaviour, 30,* 479–482.

Marler, P., & Peters, S. (1989). Species differences in auditory responsiveness in early vocal learning. In R. J. Dooling & S. H. Hulse (Eds.), *The comparative psychology of audition: Perceiving complex sounds* (pp. 243–273). Hillsdale, NJ: Erlbaum.

Marler, P., & Sherman, V. (1983). Song structure without auditory feedback: Emendations of the auditory template hypothesis. *Journal of Neuroscience, 3,* 517–531.

McCasland, J. S., & Konishi, M. (1981). Interaction between auditory and motor activities in an avian song control nucleus. *Proceedings of the National Academy of Sciences of the USA, 78,* 7815–7819.

Mello, C. V., & Clayton, D. F. (1994). Song-induced ZENK gene expression in auditory pathways of songbird brain and its relation to the song control system. *Journal of Neuroscience, 14,* 6652–6666.

Mello, C. V., & Ribeiro, S. (1998). ZENK protein regulation by song in the brain of songbirds. *Journal of Comparative Neurology, 393,* 426–438.

Mello, C. V., Nottebohm, F., & Clayton, D. F. (1995). Repeated exposure to song leads to a rapid and persistent decline in an immediate early gene's response to that song in the zebra finch telencephalon. *Journal of Neuroscience, 15,* 6919–6925.

Mello, C. V., Vicario, D. S., & Clayton, D. F. (1992). Song presentation induces gene expression in the songbird forebrain. *Proceedings of the National Academy of Sciences of the USA, 89,* 6818–6822.

Mello, C. V., Vates, G. E., Okuhata, S., & Nottebohm, F. (1998). Descending auditory pathways in the adult male zebra finch (*Taeniopygia guttata*). *Journal of Comparative Neurology, 395,* 137–160.

Miller, D. B. (1979). The acoustic basis of mate recognition by female zebra finches (*Taeniopygia guttata*). *Animal Behaviour, 27,* 376–380.

Mooney, R. (1992). Synaptic basis for developmental plasticity in a birdsong nucleus. *Journal of Neuroscience, 12,* 2464–2477.

Mooney, R., & Konishi, M. (1991). Two distinct inputs to an avian song nucleus activate different glutamate receptor subtypes on individual neurons. *Proceedings of the National Academy of Sciences of the USA, 88,* 4075–4079.

Morrison, R. G., & Nottebohm, F. (1993). Role of a telencephalic nucleus in the delayed song learning of socially isolated zebra finches. *Journal of Neurobiology, 24,* 1045–1064.

Morton, E. S. (1996). A comparison of vocal behavior among tropical and temperate passerine birds. In D. E. Kroodsma & E. H. Miller (Eds.), *Ecology and evolution of acoustic communication in birds* (pp. 258–268). Ithaca, NY: Cornell University Press.

Naguib, M., & Todt, D. (1998). Recognition of neighbors' song in a species with large and complex song repertoires: The thrush nightingale. *Journal of Avian Biology, 29,* 155–160.

Nelson, D. A. (1997). Social interaction and sensitive phases for song learning: a critical review. In C. T. Snowdon & M. Hausberger (Eds.), *Social interaction and development* (pp. 7–22). Cambridge: Cambridge University Press.

Nelson, D. A. (1998). External validity and experimental design: The sensitive phase for song learning. *Animal Behaviour, 56,* 487–491.

Nelson, D. A., Marler, P., & Morton, M. L. (1996). Overproduction in song development: An evolutionary correlate with migration. *Animal Behaviour, 51,* 1127–1140.

Nelson, D. A., Marler, P., Soha, J. A., & Fullerton, A. L. (1997). The timing of song memorization differs in males and females: A new assay for avian vocal learning. *Animal Behaviour, 54,* 587–597.

Newman, S. W. (1999). The medial extended amygdala in male reproductive behavior: A node in the mammalian social behavior network. *Annals of the New York Academy of Sciences, 877,* 242–257.

Nixdorf-Bergwieler, B. E. (1996). Divergent and parallel development in volume sizes of telencephalic song nuclei in male and female zebra finches. *Journal of Comparative Neurology, 375*, 445–456.

Nixdorf-Bergwieler, B. E., Lips, M. B., & Heinemann, U. (1995). Electrophysiological and morphological evidence for a new projection of LMAN neurones towards area X. *Neuroreport, 6*, 1729–1732.

Nixdorf-Bergwieler, B. E., Wallhausser-Franke, E., & DeVoogd, T. J. (1995). Regression in neuronal structure during song learning in songbirds. *Journal of Neurobiology, 27*, 204–215.

Nordeen, E. J., & Nordeen, K. W. (1988). Sex and regional differences in the incorporation of neurons born during song learning in zebra finches. *Journal of Neuroscience, 8*, 2869–2874.

Nordeen, E. J., & Nordeen, K. W. (1989). Estrogen stimulates the incorporation of new neurons into avian song nuclei during adolescence. *Developmental Brain Research, 49*, 27–32.

Nordeen, E. J., & Nordeen, K. W. (1996). Sex difference among nonneuronal cells precedes sexually dimorphic neuron growth and survival in an avian song control nucleus. *Journal of Neurobiology, 30*, 531–542.

Nordeen, E. J., Nordeen, K. W., & Arnold, A. P. (1987). Sexual differentiation of androgen accumulation in the zebra finch brain through selective cell loss and addition. *Journal of Comparative Neurology, 259*, 393–399.

Nordeen, E. J., Grace, A., Burek, M. J., & Nordeen, K. W. (1992). Sex-dependent loss of projection neurons involved in avian song learning. *Journal of Neurobiology, 23*, 671–679.

Nordeen, E. J., Voelkel, L., & Nordeen, K. W. (1998). Fibroblast growth factor-2 stimulates cell proliferation and decreases sexually dimorphic cell death in an avian song control nucleus. *Journal of Neurobiology, 37*, 573–581.

Nordeen, K. W., & Nordeen, E. J. (1997). Anatomical and synaptic substrates for avian song learning. *Journal of Neurobiology, 33*, 532–548.

Nordeen, K. W., Nordeen, E. J., & Arnold, A. P. (1986). Estrogen establishes sex differences in androgen accumulation in zebra finch brain. *Journal of Neuroscience, 6*, 734–738.

Nordeen, K. W., Nordeen, E. J., & Arnold, A. P. (1987). Estrogen accumulation in zebra finch song control nuclei: Implications for sexual differentiation and adult activation of song behavior. *Journal of Neurobiology, 18*, 569–582.

Nottebohm, F. (1980). Testosterone triggers growth of brain vocal control nuclei in adult female canaries. *Brain Research, 189*, 429–437.

Nottebohm, F. (1981). A brain for all seasons: Cyclical anatomical changes in song-control nuclei of the canary brain. *Science, 214*, 1368–1370.

Nottebohm, F. (1987). Plasticity in adult avian central nervous system: Possible relation between hormones, learning and brain repair. In Plum, F. (Ed.), *Handbook of physiology, Section 1, The nervous system. Part V, Higher functions of the brain, Part 1* (pp. 85–108). Bethesda, MD: American Physiology Society.

Nottebohm, F., & Arnold, A. P. (1976). Sexual dimorphism in vocal control areas of the song bird brain. *Science, 194*, 211–213.

Nottebohm, F., Stokes, T. M., & Leonard, C. M. (1976). Central control of song in the canary. *Journal of Comparative Neurology, 165*, 457–468.

Nottebohm, F., Kasparian, S., & Pandazis, C. (1981). Brain space for a learned task. *Brain Research, 213*, 99–109.

Nottebohm, F., Kelley, D. B., & Paton, J. A. (1982). Connections of vocal control nuclei in the canary telencephalon. *Journal of Comparative Neurology, 207*, 344–357.

Okuhata, S., & Saito, N. (1987). Synaptic connections of thalamo-cerebral vocal nuclei of the canary. *Brain Research Bulletin, 18*, 35–44.

Petrinovich, L. (1988). Individual stability, local variability and the cultural transmission of song in white-crowned sparrows (*Zonotrichia leucophrys nuttalli*). *Behavior, 107*, 208–240.

Petrinovich, L., & Baptista, L. F. (1987). Song development in the white crowned sparrow: Modification of learned song. *Animal Behaviour, 35*, 961–974.

Price, P. (1979). Developmental determinants of structure in zebra finch song. *Journal of Comparative and Physiological Psychology, 93*, 260–277.

Rasika, S., Alvarez-Buylla, A., & Nottebohm, F. (1999). BDNF mediates the effects of testosterone on the survival of new neurons in an adult brain. *Neuron, 22*, 53–62.

Reinke, H., & Wild, J. M. (1998). Identification and connections of inspiratory premotor neurons in songbirds and budgerigar. *Journal of Comparative Neurology, 391*, 147–163.

Ribeiro, S., Cecchi, G. A., Magnasco, M. O., & Mello, C. V. (1998). Toward a song code: Evidence for a syllabic representation in the canary brain. *Neuron, 21*, 359–371.

Riters, L. V., & Ball, G. F. (1999) Lesions to the medial preoptic area affect singing in the male European starling (*Sturnus vulgaris*). *Hormones and Behavior, 36*, 276–286.

Sakaguchi, H., & Yamaguchi, A. (1997). Early song deprivation affects the expression of protein kinase C in the song control nuclei of the zebra finch during a sensitive period of song learning. *NeuroReport, 8,* 2645–2650.

Scharff, C., & Nottebohm, F. (1991). A comparative study of the behavioral deficits following lesions of various parts of the zebra finch song system—Implications for vocal learning. *Journal of Neuroscience, 11,* 2896–2913.

Scharff, C., Nottebohm, F., & Cynx, J. (1998). Conspecific and heterospecific song discrimination in male zebra finches with lesions in the anterior forebrain pathway. *Journal of Neurobiology, 36,* 81–90.

Schlinger, B. A. (1997). Sex steroids and their actions on the birdsong system. *Journal of Neurobiology, 33,* 619–631.

Schlinger, B. A., & Arnold, A. P. (1991). Androgen effects on the development of the zebra finch song system. *Brain Research, 561,* 99–105.

Schlinger, B. A., & Arnold, A. P. (1992). Plasma sex steroids and tissue aromatization in hatchling zebra fiches: Implications for the sexual differentiation of singing behavior. *Endocrinology, 130,* 289–299.

Schlinger, B. A., & Arnold, A. P. (1991). Androgen effects on the development of the zebra finch song system. *Brain Research, 561,* 99–105.

Schmidt, M. F., & Konishi, M. (1998). Gating of auditory responses in the vocal control system of awake songbirds. *Nature Neuroscience, 1,* 513–518.

Simpson, H. B., & Vicario, D. S. (1991). Early estrogen treatment alone causes female zebra finches to produce learned, male-like vocalizations. *Journal of Neurobiology, 22,* 755–776.

Skaggs, W. E., & McNaughton, B. L. (1996). Replay of neuronal firing sequences in rat hippocampus during sleep following spatial experience. *Science, 271,* 1870–1873.

Slater, P. J. B. (1989). Bird song learning: Causes and consequences. *Ethology, Ecology and Evolution, 1,* 19–46.

Smith, W. J. (1991). Singing is based on two markedly different kinds of signaling. *Journal of Theoretical Biology, 152,* 241–253.

Sohrabji, F., Nordeen, E. J., & Nordeen, K. W. (1990). Selective impairment of song learning following lesions of a song control nucleus in juvenile zebra finches. *Neural Behavioral Biology, 53,* 51–63.

Solis, M. M., & Doupe, A. J. (1997). Anterior forebrain neurons develop selectivity by an intermediate stage of birdsong learning. *Journal of Neuroscience, 17,* 6447–6462.

Solis, M. M., & Doupe, A. J. (1999). Contributions of tutor and bird's own song experience to neural selectivity in the songbird anterior forebrain. *Journal of Neuroscience, 19,* 4559–4584.

Soma, K. K., Hartman, V. N., Wingfield, J. C., & Brenowitz, E. A. (1999). Seasonal changes in androgen receptor immunoreactivity in the song nucleus HVc of a wild bird. *Journal of Comparative Neurology, 409,* 224–236.

Spector, D. A. (1992). Wood-warbler song systems. A review of paruline singing behaviors. In D. M. Power (Ed.), *Current ornithology* (Vol. 9, pp. 199–238). New York: Plenum Press.

Staicer, C. A. (1996) Honest advertisement of pairing status: Evidence from a tropical resident wood-warbler. *Animal Behaviour, 51,* 375–390.

Stark, L. L., & Perkel, D. J. (1999). Two-stage, input-specific synaptic maturation in a nucleus essential for vocal production in the zebra finch. *Journal of Neuroscience, 20,* 9107–9116

Stoddard, P. K. (1996). Vocal recognition of neighbors by territorial passerines. In D. E. Kroodsma & E. H. Miller (Eds.), *Ecology and evolution of acoustic communication in birds* (pp. 356–374). Ithaca NY: Cornell University Press.

Striedter, G. F., & Vu, E. T. (1998). Bilateral feedback projections to the forebrain in the premotor network for singing in zebra finches. *Journal of Neurobiology, 34,* 27–40.

Stripling, R., Volman, S. F., & Clayton, D. F. (1997). Response modulation in the zebra finch neostriatum: Relationship to nuclear gene regulation. *Journal of Neuroscience, 17,* 3883–3893.

Suthers, R. A. (1997). Peripheral control and lateralization of birdsong. *Journal of Neurobiology, 33,* 632–652.

Székely, T., Catchpole, C. K., DeVoogd, A., Marchl, Z., & DeVoogd, T. J. (1996). Evolutionary changes in a song control area of the brain (HVC) are associated with evolutionary changes in song repertoire among European warblers (*Sylviidae*). *Proceedings of the Royal Society of London, Series B, 263,* 607–610.

Taber, E. (1964). Intersexuality in birds. In C. N. Armstrong, & J. J. Marshall (Eds.), *Intersexuality in vertebrates including man* (p. 285). New York: Academic Press.

Takahashi, M. M., & Noumura, T. (1987). Sexually dimorphic and laterally asymmetric development of the embryonic duck syrinx: Effect of estrogen on *in-vitro* cell proliferation and chondrogenesis. *Developmental Biology, 121,* 417–422.

Tchernichovski, O., Schwabl, H., & Nottebohm, F. (1998). Context determines the sex appeal of male zebra finch song. *Animal Behaviour, 55,* 1003–1010.

Theunissen, F. E., & Doupe, A. J. (1998). Temporal and spectral sensitivity of complex auditory neurons in the nucleus HVc of male zebra finches. *Journal of Neuroscience, 18,* 3786–3802.

Titus, R. C. (1998) Short-range and long-range songs: Use of two acoustically distinct song classes by dark-eyed juncos. *Auk, 115,* 386–393.

Todt, D., & Hultsch, H. (1996). Acquisition and performance of song repertoires: Ways of coping with diversity and versatility. In D. E. Kroodsma & E. H. Miller (Eds.), *Ecology and evolution of acoustic communication in birds* (pp. 79–96). Ithaca, NY: Cornell University Press.

Tramontin, A. D., Wingfield, J. C., & Brenowitz, E. A. (1999). Contributions of social cues and photoperiod to seasonal plasticity in the adult avian song control system. *Journal of Neuroscience, 19,* 476–483.

Vallet, E., Kreutzer, M., & Gahr, M. (1996). Testosterone induces sexual release quality in the song of female canaries. *Ethology, 102,* 617–628.

Vates, G. E., Broome, B. M., Mello, C. V., & Nottebohm, F. (1996). Auditory pathways of caudal telencephalon and their relation to the song system of adult male zebra finches (*Taenopygia guttata*). *Journal of Comparative Neurology, 366,* 613–642.

Vates, G. E., Vicario, D. S., & Nottebohm, F. (1997). Reafferent thalamo-"cortical" loops in the song system of oscine songbirds. *Journal of Comparative Neurology, 380,* 275–290.

Vicario, D. S. (1993). A new brainstem pathway for vocal control in the zebra finch song system. *Neuroreport, 4,* 983–986.

Vicario, D. S. (1994). Motor mechanisms relevant to auditory–vocal interactions in songbirds. *Brain Behavior and Evolution, 44,* 265–278.

Vicario, D. S., & Nottebohm, F. (1988). Organization of the zebra finch song control system: I. Representation of syringeal muscles in the hypoglossal nucleus. *Journal of Comparative Neurology, 271,* 346–354.

Volman, S. F., & Khanna, H. (1995). Convergence of untutored song in group-reared zebra finches (*Taeniopygia guttata*). *Journal of Comparative Psychology, 109,* 211–221.

Vu, E. T., Mazurek, M. E., & Kuo, Y.-C. (1994). Identification of a forebrain motor programming network for the learned song of zebra finches. *Journal of Neuroscience, 14,* 6924–6934.

Vu, E. T., Schmidt, M. F., & Mazurek, M. E. (1998). Interhemispheric coordination of premotor neural activity during singing in adult zebra finches. *Journal of Neuroscience, 18,* 9088–9098.

Wade, J., & Arnold, A. P. (1996). Functional testicular tissue does not masculinize development of the zebra finch song system. *Proceedings of the National Academy of Sciences of the USA, 93,* 5264–5268.

Wade, J., Schlinger, B. A., Hodges, L., & Arnold, A. P. (1994). Fadrozole: A potent and specific inhibitor of aromatase in the zebra finch. *General and Comparative Endocrinology, 94,* 53–61.

Wallhausser, E., & Scheich, H. (1987). Auditory imprinting leads to differential 2 deoxyglucose uptake and dendritic spine loss in the chick rostral forebrain. *Developmental Brain Research, 31,* 29–44.

Wallhausser-Franke, E., Collins, C. E., & DeVoogd, T. J. (1995). Developmental changes in the distribution of NADPH-diaphorase containing neurons in telencephalic nuclei of the zebra finch song system. *Journal of Comparative Neurology, 356,* 345–354.

Wallhausser-Franke, E., Nixdorf-Bergweiler, B. E., & DeVoogd, T. J. (1995). Song isolation prevents the early loss of spine synapses in song system nucleus l-MAN of zebra finches. *Neurobiology of Learning and Memory, 64,* 25–35.

Ward, B. C., Nordeen, E. J., & Nordeen, K. W. (1998). Individual variation in neuron number predicts differences in the propensity for avian vocal imitation. *Proceedings of the National Academy of Sciences of the USA, 95,* 1277–1282.

Waser, M. S., & Marler, P. (1977). Song learning in canaries. *Journal of Comparative and Physiological Psychology, 91,* 1–7.

Weichel, K., Schwager, G., Heid, P., Guttinger, H. R., & Pesch, A. (1986). Sex differences in plasma steroid concentrations and singing behaviour during ontogeny in canaries (*Serinus canaria*). *Ethology, 73,* 281–294.

West, M., & King, A. (1996). Eco-gen-actics: A systems approach to the ontogeny of avian communication. In D. E. Kroodsma & E. H. Miller (Eds.), *Ecology and evolution of acoustic communication in birds* (pp. 20–38). Ithaca, NY: Cornell University Press.

West, M. J., Stroud, A. N., & King, A. P. (1983). Mimicry of the human voice by European starlings: The role of social interaction. *Wilson Bulletin, 95,* 635–640.

Whaling, C. S., Solis, M. M., Doupe, A. J., Soha, J. A., & Marler, P. (1997). Acoustic and neural bases for innate recognition of song. *Proceedings of the National Academy of Sciences of the USA, 94,* 12694–12698.

Wild, J. M. (1997a). Neural pathways for the control of birdsong production. *Journal of Neurobiology, 33,* 653–670.

Wild, J. M. (1997b). Functional anatomy of neural pathways contributing to the control of song production in birds. *European Journal of Morphology, 35,* 303–325.

Wild, J. M,. & Williams, M. N. (1999). Rostral wulst of passerine birds: II Intratelencephalic projections to nuclei associated with the auditory and song systems. *Journal of Comparative Neurology, 413*, 520–534.

Williams, H. (1990). Models for song learning in the zebra finch: Fathers or others? *Animal Behaviour, 39*, 745–757.

Wingfield, J. C., & Farner, D. S. (1993). Endocrinology of reproduction in wild species. In D. S. Farner, J. R. King, & K. C. Parkes (Eds.), *Avian biology* (Vol. 9, pp. 163–327). San Diego, CA: Academic Press.

Wolff, E. (1959). Endocrine function of the gonad in developing vertebrates. In A. Gorbman (Ed.), *Comparative endocrinology* (pp. 568–569). New York: Wiley.

Yu, A. C., & Margoliash, D. (1996). Temporal hierarchical control of singing in birds. *Science, 273*, 1871–1875.

Zann, R. A. (1996). *The zebra finch: A synthesis of field and laboratory studies.* Oxford: Oxford University Press.

10

The Development of Action Sequences

John C. Fentress and Simon Gadbois

Introduction

> The Brain functions for action. (Edelman, 1987, p. 63)

Many years ago, Lashley (1951) challenged behavioral neuroscientists to examine movement properties and the serial order of behavior more specifically. This was in many respects a logical follow-up of Hebb's (1949) concerns with the *organization* of behavior, according to which, individual behavioral and brain properties must be isolated, but also examined within broader contexts of expression (Fentress, 1999). For example, Hebb devised concepts of *cell assembly* and *phase sequence* to help behavioral neuroscientists evaluate the fact that *all* behavior is organized in time. Movement is a directly observable manifestation of this dynamic ordering in brain and behavior (Berridge & Whishaw, 1992; Fentress, 1990, 1992; Golani, 1992; Kelso, 1997; Thelen & Smith, 1994). As such, quantitative analyses of movement can provide fundamental insights into brain–behavior organization, including the developmental profiles that occur dynamically across levels and time frames of operation. As stated by Churchland and Sejnowski (1992, p. 178), "Our brains are dynamical, not incidentally or in passing, but essentially, inevitably, and to their very core." As will become clear in this chapter, we agree with this position, and believe that action dynamics can provide fundamental insights into processes at both ontogenetic and integrative time frames of organization.

The dynamics of movement and brain organization, however, provide funda-

John C. Fentress   Department of Psychology and Neuroscience, Dalhousie University, Halifax, Nova Scotia, Canada B3H 3J5, and Departments of Psychology and Biology, University of Oregon, Eugene, Oregon 97403.   Simon Gadbois • Department of Psychology, Dalhousie University, Halifax, Nova Scotia, Canada B3H 3J5.

Developmental Psychobiology, Volume 13 of *Handbook of Behavioral Neurobiology*, edited by Elliott Blass, Kluwer Academic / Plenum Publishers, New York, 2001.

mental conceptual as well as analytical problems. Two basic problems are as follows: (1) How does one find natural divisions in observed streams of events that will also clarify rules of relation among individual movement and circuit properties? (2) How does one deal both with the flexibilities and constraints of integrated action? In addition to the obvious importance of these questions for basic research, they are critical to the evaluation of movement disorders. In humans, movement disorders can attack brain and behavior at many different levels of organization (e.g., see Poizner et al., 1995, on limb apraxia; Rapoport, 1991, on obsessive compulsive disorders). Marsden and Fahn (1994), Rossenbaum (1991), and Weiner and Lang (1989) provide comprehensive surveys of the rich variety of human movement disorders, their origins, and their neural substrates.

As stressed by Bernstein (1967), Golani (1992), and others, movement emerges from postural change. In Thach's (1996, p. 415) terms, "Moving the skeleton is an engineer's nightmare." This is in part because the nervous system has to deal with changing peripheral forces both in moment-to-movement action sequencing and across development (Thelen & Smith, 1994). To simplify these complex issues, it is clear that all movement has three essential properties, also highlighted in our previous review (Fentress & McLeod, 1986). First, movement is relational. To understand movement organization, it is necessary to examine relations not only between organism and environment, but also among different actions, limb segment properties, and so forth. Second, movement, by definition, is dynamic. Movement is thus *change* in (multiple) relations. Movement thus involves many potential changes in postural relations within the organism and between the organism and its environment. Static abstractions of movement dynamics (such as "acts") thus have limited utility for the understanding of process and underlying mechanism. Third, movement is multilayered. This means that individual movement properties are arranged together into higher-order functional units. Hierarchical models of brain have a long history, starting at least from the classic work of Sherrington (1906) (for example, see Arbib, 1995; Cordo & Harnad, 1994; Fentress, 1990; Georgopoulos, 1991; Rosenbaum, 1991) and remain conceptually important (with modifications).

Development provides an increasingly appreciated and powerful analytical tool with which to examine movement (for example, Bekoff, 1992; Berridge, 1994; Bradley & Smith, 1988; Fentress, 1972, 1992; Fentress & Bolivar, 1996; Foster, Sveistrup, & Woollacott, 1996; McLeod & Fentress, 1997; Pellis, Castaneda, McKenna, Tran-Nguyen, & Whishaw, 1993; Robinson & Smotherman, 1992; Sachs, 1988; Smotherman & Robinson, 1996; Thelen & Smith, 1994). Development makes it possible to trace the differentiation of individual movement properties and their organization into higher order functional units (Fentress, 1992; Fentress & McLeod, 1986; Thelen & Smith, 1994). Many human movement disorders have both a genetic and epigenetic base (Durr & Brice, 1996; Foster et al., 1996; Gerlai, 1999; Klintsova, Matthews, Goodlett, Napper, & Greenough, 1997; Marsden & Fahn, 1994; H. Poizner et al., 1995; Weiner & Lang, 1989), and thus the understanding of both normal movement assembly and its disassembly in disease is essential. There are, however, relatively few research programs that combine analyses of movement ontogeny with movement integration. The recent advent of powerful genetic probes in laboratory animals has opened an opportunity (Crawley & Paylor, 1997; Durr & Brice, 1996; Goldowitz & Eisenman, 1992; Jacobson & Anagnostopoulos, 1996; Lederhendler, 1997). Yet, the precision of analyses at cellular and biochemical levels has not been matched with equal precision of analysis at the behavioral level. As stated by Nelson (1997) in a recent review of transgenic/knockout mutants, "Presumably, subtle behavioral changes of knockout mice await discovery" (p. 190). He argues, further, that, "behavioral researchers must provide ethograms, or complete behavioral assessments, of various *Mus* strains to accomplish the important goal of mapping the genetic engines driving the mechanisms of behavior" (p. 192).

Accordingly, one goal of this chapter is to provide quantitative analyses of movement ontogeny and control in mutant mice with neurological disorders. These analyses will clarify both the subtleties and broader profiles of movement organization. Basic observations of ongoing movement will be documented for both control and mutant animals. This provides a direct comparison, in time, of points where the movements of control and mutant animals diverge. Experimental *perturbations* applied to specific movement phases during ontogeny will clarify the relative autonomy versus contextual sensitivity of defined movement operations. This will also be achieved through analyses of a very different behavior, food caching in a very different mammalian genus, the wild canids (e.g., foxes, coyotes, wolves). We shall, in conclusion, offer some broadly sketched models of interactive and self-organizational processes that may apply equally to moment-by-moment performance and the developmental substrates of performance. Our hope is that these concrete examples and broadly sketched model systems will help guide future research into both action sequence organization and the ontogenetic processes that make this performance possible.

MOVEMENT ANALYSIS

INTRODUCTION

Historically, there have been two basic approaches to movement analysis. The first divides movement into conceptually discrete "acts" and asks how these acts are sequenced (in a linear fashion) through time. The second divides movement into its various dynamic profiles and operations. This permits evaluation of the (nested) processes through which movement is formed. The first method has the advantage of relative simplicity, at the expense of understanding movement dynamics. The second has the advantage of understanding dynamics, but often at the expense of cataloguing functional units in movement (for reviews see Fentress, 1999; Fentress & McLeod, 1986). The work by us and our colleagues has attempted to synthesize these two approaches (e.g., Berridge, 1994; Berridge, Fentress & Parr, 1987; Bolivar, Danilchuk, & Fentress, 1996; Bolivar, Manley, & Fentress, 1996; Coscia & Fentress, 1993; Gadbois & Fentress, 1997a, b; Golani & Fentress, 1985; McLeod, 1996; McLeod & Fentress, 1997). *The challenge is to link systems and mechanisms perspectives into a common framework.*

CONTEXT

Movement streams occur within the context of dynamic balances among intrinsic and extrinsic states. Contextual evaluations can clarify movement control. For example, in early studies (Fentress, 1968a, b; 1972) parameters of rodent grooming behavior could be understood by examining the animal's environment, time since presentation of independent variables, and the temporal association of grooming with other actions, such as locomotion and immobility (reviewed in Fentress 1990, 1992). At a more refined level, individually defined actions, or movement operations (as in face grooming strokes), can be understood in terms of their sequential associations (Berridge, 1994; Berridge, *et al.*, 1987; Coscia & Fentress, 1993; Cromwell & Berridge, 1996; Fentress & Bolivar, 1996). These studies allow one to examine motivational parameters in movement as well as patterns of temporal organization in movement. In doing so, it is possible to determine the extent to which individual movements or movement properties are modular (intrinsically ordered versus responsive to their more broadly defined surround). A critical lesson is that individual movements as well as movement sequences can change their rules of control with context (Prochazka, 1989). While movement modules provide a useful analytical framework, the lesson of context is that these modules are not entirely autonomous in their operations. They are influenced to varying degrees by extrinsic factors. Modules can be recruited to the service of a number of different behaviors. Biting (pecking) or holding to the ground appear in both feeding or killing sequences. It is the balance of interactive and self-organizing operations that is critical in the selection and appearance of particular encapsulated modules. Interestingly, the striatal mediation of grooming sequences studied by Berridge and his colleagues (e.g., Cromwell & Berridge, 1996) fits well with both the idea of action chunking, including issues related to our discussion of super-segmental events in prosody (Graybiel, 1998), and also somatosensory processing that is dynamically ordered and only partially compartmentalized (Brown, Feldman, Divac, Hand, & Lidsky, 1994). We know little of how these rules change in development.

HIERARCHICAL STRUCTURE

Our early work on rodents demonstrated a facial grooming "grammar" defined by rule-governed changes in sequential order of movement components that could in turn be defined by higher order movement "units." These units are sequentially ordered (Fentress, 1972; Fentress & Stilwell, 1973). This basic perspective has been refined by subsequent analyses of movement phases and their temporal associations (Aldridge, Berridge, Herman, & Zimmer, 1993; Berridge *et al.*, 1987; Golani & Fentress, 1985). The hierarchical perspective has also allowed us to dissociate movements of varying degrees of complexity, such as rodent grooming versus swimming actions (Bolivar, Danilchuk, & Fentress, 1996; Bolivar, Manley, & Fentress, 1996; Fentress, 1992; Fentress & Bolivar, 1996). At each of these levels of movement, behavioral operations and their neural control follow partially distinct rules. This clarifies hierarchical operations within the nervous system. An important, but often neglected point in neurobiological studies of movement is that higher order movement configurations can be *simpler* (economically described) and more invariant in form than are their lower-order constituents. For example, mice can maintain relatively *invariant* contact pathways between forepaws and face through a variety of individual (and compensatory) movements of trunk, head, and forelimbs (Golani & Fentress, 1985). These *"relational invariants"* (Fentress, 1990) were suggested by many early workers (e.g., Hebb's "motor equivalence"). However, the implications of higher order constraints in action are less frequently appreciated. Recent models of motor "schemas" are beginning to address this (e.g. reviews in Arbib, 1995; Cordo & Harnad, 1994; Graybiel, 1998; Hayes, Davidson, Keele, & Rafal, 1997; Ivry, 1996; Keele, Davidson, & Hayes, 1996; Kelso, 1997; Prochazka, Clarac, Loeb, Rothwell, & Wolpaw, 2000; Rosenbaum, 1991; Stein, Grillner, Selverston, & Stuart, 1997; Thelen & Smith, 1994).

CENTRAL PROGRAMMING

Early models of movement were based upon reflex chains (Sherrington, 1906). Our work and that of many others has indicated the importance of central programming (see reviews in the previous paragraph). However, these *central programming* models must be viewed critically (e.g., see the recent discussion in Prochazka *et al.*, 2000, as well as reviews in Stein *et al.*, 1997). In our laboratory, we have shown that the relative balance between central and peripheral movement control shifts with expressive contexts and movement dynamics (Berridge & Fentress, 1987a, b; Fentress, 1990, 1992). Strongly activated movement patterns become more stereotyped and ballistic (reviewed in Fentress, 1988, 1990). This indicates that central programming is a *relative* concept. Central and peripheral states interact through dynamic operations. These operations separate and combine. The search for localized central pattern generators must therefore also acknowledge extrinsic factors. Even at the level of invertebrate movement, early "central pattern generator" models were flawed through excessive inference of rigid autonomy. It is now known that these generators, in all species thus far investigated, contain "multifunctional" units (Getting, 1988). Circuits and their elements are modulated by extrinsic variables (Marder & Weimann, 1992). Dynamic and distributed network models of movement control are necessary. (For recent reviews that include action patterns for a number of vertebrate and invertebrate systems as well as specific neural systems, see Stein *et al.*, 1997.)

Movement Ontogeny

Movement is progressively assembled, disassembled, and reassembled during ontogeny (for examples, see Bekoff, 1992; Bradley & Smith, 1988; Fentress, 1992; Fentress & McLeod, 1986; Foster *et al.*, 1996; Golani, 1992; Golani & Fentress, 1985; McCrea, Stehouwer, & van Hartesveldt, 1997; McLeod, 1996; Robinson & Smotherman, 1992; Sachs, 1988; Thelen & Smith, 1994). Our early work showed that many rodent actions emerge in a relatively normal form with an anterior–posterior sequence. This is true even when basic sources of sensory input ("experience") are removed from birth (Fentress, 1972, 1992). For example, early amputation of the forelimbs in infant mice does not prevent the establishment of basic grooming movements such as shoulder rotations and even appropriately timed (but now abortive) licking movements of the tongue. Others and we have also shown that grooming movements are both species- and strain-specific (Berridge, 1994; Fentress & Bolivar, 1996). Our recent work with *weaver* (Bolivar, Danilchuk, & Fentress, 1996; Bolivar, Manley, & Fentress, 1996) and *jimpy* (Bolivar, 1996; see also Fentress & Bolivar, 1996) neurological mutant mice has clarified the role of single genes in movement ontogeny and control and this will be reviewed below. Ultimately, movement is an *epigenetic* phenomenon. Thus the differentiation and expression of movement (*plus movement control circuits*) also depends upon factors extrinsic to the genes. We have demonstrated that postural support (such as an infant rodent "grooming chair") can lead to the expression of movements that are normally not seen until later postnatal ages. Many early movements have also been traced prenatally (Robinson & Smotherman, 1992; Smotherman & Robinson, 1996). The developmental progression of movement follows predictable timetables that can be defined separately for limb segment kinematics, limb trajectories, and higher order movement configurations (Fentress, 1992; Golani, 1992; Golani & Fentress, 1985). One should expect that both genetic manipulations and experience could modify these developmental timetables. Current research on this topic is inadequate.

Accordingly, this chapter updates and expands a review on motor development written for a previous edition of the present *Handbook* (Fentress & McLeod, 1986). In that chapter, we took a broad ethological perspective, which we shall also do here. In addition, we shall offer the analogy to prosody in human speech as a means for stressing both the segmental and supersegmental nature of natural movement patterns, and add information relevant to the neural mechanisms associated with action sequencing and their development. We now put our main focus on the sequencing of actions that are in important ways both separate and joined together, the development of these separable and combined actions, and certain aspects of their underlying central controls. This is a more specific focus on issues of pattern formation than were addressed in the previous chapter. Details about issues of spontaneous versus elicited actions in early development were addressed more fully in the previous chapter. In both chapters, we argue that developmental analyses can help segregate the mature functional sequences by identifying components as they appear in a sequence developmentally and can clarify the consequence of their appearance in terms of changing in timing, phase, order, etc. (see the chapter by Hogan in this volume), and that there is the need for clear taxonomic dissections of action that allow for the construction of models that can guide future research at the level of both systems properties in behavior and the dissection of mechanism.

INTERIM SUMMATION: THE SALIENCE OF TAXONOMY

Action sequences can be dissected in many ways. Once these action sequences are dissected into their components, it then becomes possible to ask how these components are arranged together in time and across levels of organization. Ontogenetic analyses can be of great value here, for one can trace the progressive assembly, disassembly, and reassembly of action profiles defined by specific criteria. These evaluations in turn can help clarify the organization of underlying brain structures, changing peripheral and environmental demands, and alterations in function. At this point, new technologies in genetic and epigenetic analyses may help bridge the richness of ethological studies of natural action sequences with a deeper understanding of mechanisms of origin and control. We argue that in each case, components must be examined in their broader operational, or systems, contexts as a guard against the incompleteness of unidirectional reductionism. A first question is how we define components in such a way that they can be merged within more global dynamic processes that make integrated action, including its ontogeny, possible.

The next section of this chapter offers speech prosody as useful scaffolding for action sequence analysis more generally. We shall then give two primary illustrations: an ethological evaluation of food caching in wild canids, and more restricted laboratory analyses of grooming sequences in inbred mice, including strains with defined neurological disorders. We next offer a framework of interactive/self-organizing ("*ISO*") systems that we believe can help bridge studies of action sequence performance with issues of underlying ontogeny and brain function. In this way we believe that mechanistic and systems perspectives can be brought into closer register than exists in much of the literature today.

PROSODY IN ACTION SEQUENCING

We have argued that action sequences involve relational, dynamic, and multi-leveled events. In many streams of action (e.g., bird song and human speech) one can isolate components that are distinctive and also repeated. These stream "segments" or "acts" clearly suggest a degree of modularity in organization. At the same time, there are cases where the components are less clearly defined, and there are rules of combination that cross over individual segmental boundaries. To aid the reader (and ourselves!) we have made a series of cartoon representations that clarify some of the taxonomic issues we have offered in previous sections of this chapter. The cartoons are not intended to be exhaustive. Rather, they highlight issues that are germane to all forms of action sequence analysis. The reason for analogy to prosody in human speech is that linguists have been equally concerned with the combination of segmental and supersegmental properties (e.g., Jusczyk & Aslin, 1995; Ramus & Mehler, 1999). The term prosody refers to supersegmental aspects of human speech, such as rhythm, emphasis, tempo, and the like. Thus, when we speak with another person, we not only articulate individual phonemes, words, phrases, and so forth, but also connect, or overlay, these speech components into a melody that has rules that are defined in their own terms. At still higher levels of organization, we refer to the meanings or functions of prosodic variations. This is analogous to most traditional definitions of action: food acquisition, mating, fighting, facial recognition, and the like (Blass, 1999). While we recognize that analogies between complex human activities such as speech and animal actions have their limitations, it is our view that there are a number of underlying issues, or at least concerns, that are held in common. The developmental approach allows the tracking of these components and their integration as new elements are folded into the pattern as they become available. The following cartoons of action relationships allow us to conceptualize some of the relationships and their changes over time.

A Brief Expansion on Linguistic Definitions

We offer this brief expansion for the reader uninitiated in linguistic analysis. While it is relevant to the materials that follow, some readers may prefer to skip to the next section.

Syntagmatic Analysis.   This considers linear sequences of events and studies of their combinations (concatenations, combinations of events/actions), i.e., the patterns of transitions between events. This approach is concerned with the configuration of the sequences, their intrinsic organization, and the set of rules or syntax that governs the sequential configuration. This is analogous to the most common approaches to sequence analysis in ethology.

Paradigmatic (Associative) Analysis.   This is used when a contrastive/ comparative approach is taken, i.e., when the whole set of alternative events is considered. A paradigmatic approach would focus on the selection ["choice" of specific movements over others in the potential substitutions (one movement for another)]. The approach focuses on the set of movements (actions) available to and used by the species, age group, population, etc.

A recent review of these conceptualizations, developed may years ago by Ferdinand de Saussure, 1857–1913, can be found in Crystal (1997). In a developmental perspective, it may be useful to specify if the search for changes in time focuses on syntagmatic principles (e.g., sequential analysis) or paradigmatic principles (the actual repertoire of the individual or age group or the ethogram of the species).

PROSODIC FEATURES. More relevant to this chapter are the "prosodic" features of behavior, i.e. the features of behavior parallel to the actual syntax, semantics, etc. This approach contributes to our understanding of behavior in two ways.

1. The serial, sequential patterns of change (or stability) in a sequence of actions is not sufficient to capture the essence of the flow of behavior. Sequential patterns, if restricted to the combination of events (actions, movement), may not be found in most sequences under investigation until spatiotemporal factors are taken into consideration. The duration of events, gaps between them, and overlaps each deserve consideration. This refers to the metrics of behavior in time and space. Prosody and music offer an analogy to this problem: patterns in rhythm and "melody," for instance, could be found without any "syntagmatic" or syntactic patterns.

2. In linguistics, and more specifically in phonology, this approach is illustrated by the distinction between segmental (phonemic) and supersegmental (prosodic) theories of speech (segmental and supersegmental phonology). The first approach tends to look at the discrete units of speech (phones or phonemes). In the sequential analysis (Bakeman & Gottman, 1986; Gottman & Roy, 1990) of behavior, this is the main approach: units (movements, actions) are defined and their configuration is analyzed (e.g., AABBBBABBABBBBABBABABABBBBBAB). The second deals with features extending over more than one segment (e.g., pitch). This perspective seeks patterns that proceed and progress beyond and above the behavioral units (specific actions or movements). Terms such as "behavioral fluctuations" or "modulations" are representative.

The purpose of this brief section has been to suggest to the reader that action sequences potentially contain a remarkable richness. The extent to which this richness can be found in all species, or all stages of ontogeny, is a topic that is remarkably little explored. In the cartoons that follow, we shall try to provide a visual representation of some potential issues that will then be expanded in our review of canid caching and rodent grooming.

SCHEMATIC REPRESENTATIONS OF (POTENTIAL) ACTION PROPERTIES

Figure 1 represents a very simple case of only two actions, each presumed to be discrete. The filled and open circles represent these two actions (or action properties). In an action sequence, there are only two possible alternatives: alternations between the two actions or repeats (perseverations) of a single action. In our subsequent example of canid caching, we shall concentrate upon just two actions, tamping food into the ground and scooping dirt over the food. As we shall see, even here,

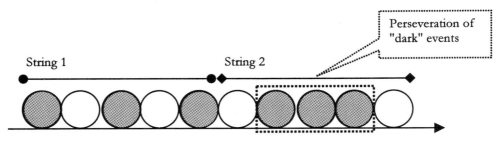

Figure 1. Two basic concepts in behavioral sequences: perseveration and alternation of actions (Berridge *et al.*, 1987).

the overall sequence of events may have surprising properties, such as the ability to predict the duration of the entire sequence by looking at the arrangement of elements at the beginning of the sequence (Gadbois, Fentress, & Harrington, n.d.). In the case of rodent grooming, we shall examine a sequence that is richer in its component structure, but still find patterns of alternation between acts and perseverations of a given act to be germane to a full analysis (Berridge *et al.*, 1987). In each case, one can ask the following question: (1) Do the elements within the sequence become more clarified (fully differentiated from one another) in ontogeny? (2) Do the rules of sequential connectivity change with time? (3) Are these rules sensitive to extrinsic perturbations or are they more "ballistic" in their control? (4) Do different parts of the overall sequence vary in their internal properties, such as sensitivity to disruption? (5) Do sequences tend to start and end with one or the other elements? (6) Can one discern "higher order" patterning in the sequences (i.e., are they ergodic or more complexly organized into nested stochastic patterns)? Note that many other properties of action are omitted from this simple figure, such as timing, emphasis, overlap, rhythms, and the like.

Figure 2 schematizes two of these additional properties, overlaps and gaps. The overlaps suggest that specific properties of a particular action may cooccur to differing degrees (Golani, 1992). The movements may be imperfectly discrete, as in the phenomenon of "co-articulation" in human speech. In coarticulated speech, the precise pronunciation of a given phoneme can vary (within limits) as a function of which specific sounds have preceded it and which will follow it. Sounds can blend, but the human auditory system has the ability to decode that blending. In both canid caching and rodent grooming, we have found evidence for the occasional blending of individual action properties, which is a method for increased efficiency in production (sort of like doing two things at once). Think how strange early computer simulations of human speech were; the sounds were too discrete to gain an overall sense of speech flow. Interestingly, we have found that in young animals, and for certain neurologically damaged animals, action elements lose their ability to be coarticulated, giving the overall sequence a rather "rough" and "uncoordinated appearance" (reviews in Fentress, 1990, 1992). The inclusion of gaps in this figure adds one dimension of time. Action properties may be separated by various intervals. A simple analysis of the sequencing of acts or action properties loses this property. Again, we have found that in young animals, both canids and mice (as well as some neurologically damaged mice), the overall flow of action sequencing appears to be broken by these gaps. To cite one example, Fentress (1972) found that swimming movements in young mice often involved longer limb cycle times, not so much

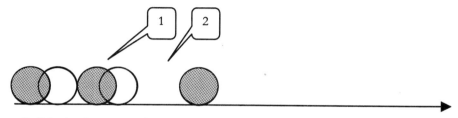

Figure 2. Behavioral events may be discrete, but they are often overlapping with each other (1) or are temporally spaced (2).

because the individual movement velocities were slower than in adult animals, but because there were momentary pauses between cycles. Perceptually we could not pick this up without the aid of high-speed films. In a sense, the gaps are the opposite of coarticulation, or blending, between movement properties, where individually defined events share certain timing properties. To make the point of Figure 2 intuitively obvious, a mere listing of the notes in a Beethoven or Mozart composition would fail to capture the dynamics of performance that bring these compositions alive. Sequencing and timing work together both for listening pleasure and for the generation of action sequences. Changes in both sequencing and timing properties during ontogeny are thus potentially critical. (It is hard to put these two together, for mathematical reasons that we shall not go into here.)

Figure 3 is a cartoon representation of overshadowing among action properties, and also encapsulation. These are temporal and spatial "invasions" that go beyond the previous figure. They suggest action properties that are competing for time and space. Overshadowing (part 1 of Figure 3) describes the spatiotemporal "takeover" of one element or property by another; here the white circle "takes over" the gray one. This can be due in part to the fact that nervous systems have a limited capacity for multiple actions, and thus one action can interrupt the execution of another, or at least aspects of another. Properties of fighting might take over, or supersede, properties of feeding, for example. At more micro levels of organization, a given property of motor performance might block the expression of an alternative and simultaneously activated property of motor performance. Indeed, Sherrington's (1906) classic studies of reflex integration were based largely on this notion. Thus, if limb withdrawal and scratching movements were simultaneously activated, limb withdrawal usually blocked scratching. This can be understood mechanistically from the perspective of lower thresholds and asymmetric reciprocal inhibition. We can, of course, understand it functionally, for withdrawal implies the potential to escape harm, while scratching is often an activity that can be postponed without harm. The second aspect of Figure 3 is conceptually related, but distinct. Here the open circle represents a relatively short duration and/or intense action property that is temporally shared with the longer duration closed-circle event. Thus, as demonstrated beautifully by the work of Golani and his colleagues, movement properties can be invaded or shared with one another to varying degrees (Golani, 1992). In both our canid and rodent research, we found many cases where protracted changes of state can temporally block ongoing actions that are then resumed. If, for example, wolves or other canids are caching food and are presented with a major disturbance (such as the presence of noisy student observers or the presence of heavy equipment) the caching becomes suspended until the disturbance has ceased. At this point, it is very

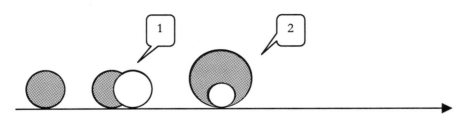

Figure 3. Behavioral events may both overshadow one another (1) or encapsulate one another in time (2).

common to see the animals basically pick up where they left off, and go back to the caching sequence. Interestingly, our impression from incompletely analyzed data records suggests that the animals might often perform the previously suspended caching actions more vigorously, in analogy to Sherrington's studies of reflex "rebound." This would be an interesting area to pursue further, in a developmental context. Actions might, for example, block each other at one level of organization but also share facilitatory circuits at another level of organization. There may also be dynamic changes in the balance of facilitation versus blocking both for performance details and during ontogeny. We shall return to this theme below in the context of interactive/self organizing (ISO) systems. We also have incompletely analyzed observations that when rodents are engaged in a variety of actions, from feeding to grooming, relatively protracted disturbances can stop them (i.e., they take avoidance responses), but they then return to their previous activities when the disturbances are terminated. Here our impression is that the animals will often run through the first part of the sequence rather rapidly and then, reaching where they had been, they return to a more typical rate of performance. A bird may momentarily terminate its feeding behavior when given the opportunity to mate, only to return to feeding after mating, often with an apparently increased "vigor," and so forth. Again, Sherrington (1906) anticipated such observations with his descriptions of "rebound" at the spinal level. Give a foot pinch during scratching behavior and the scratching is suspended until the (for our purposes, protracted) pinch terminates. Scratching then rebounds (returns to) its formal scratching state, sometimes even with apparently greater intensity, a topic that we shall return to later in this chapter. [The supplementary question, rarely studied ontogenetically, to our knowledge, is whether something such as a *mild* pinch could facilitate responses to a scratch stimulus. There are reasons to suspect this may be so (Fentress, 1990, 1992, 1999). We hope to make the relevance of this question clear by the end of the chapter.]

Figure 4 uses different vertical and horizontal dimensions to represent action properties. The vertical dimension symbolizes actions that can and often do vary widely in their intensity (amplitude, strength) of expression. Thus, one animal might push or shove another animal gently and slowly or with considerable force. There are obviously many potential dimensions to the concept of intensity in action, and to our knowledge, these have not yet been dissected carefully within a developmental context. The cartoon is intended to point toward future research that is needed. The cartoon also contains ellipses that vary along their horizontal dimension, representing duration. Temporal properties of individual actions, or the substrates of action, can vary in both their individual extent and rules of overlap. Further, for action sequences as a whole, temporal domains, as in the timing of a

Figure 4. Varying the size and shape of sets can translate the concepts of intensity, amplitude, or magnitude (vertical; 1) and duration (horizontal; 2).

musical composition, can overlay the rules involved with the articulation of expression for individual properties within the overall action sequence. As pointed out by Thelen and Smith (1994), among others, developmental changes in the temporal properties of movement have been largely neglected. Since movement is fundamentally a dynamic (time-defined) domain (Ivry, 1996; Kelso, 1997), this is very unfortunate. As Ivry (and others) have suggested, "The fourth dimension of our world, time, has tended to be the forgotten stepchild in many theories of perception and motor control" (Ivry, 1996, p. 851). We do not find ourselves in agreement with any suggestion that dynamic (timing) properties of movement preclude the analysis of mechanism (e.g., Kelso, 1997; Thelen & Smith, 1994). Indeed, we take the opposite stance, as articulated by workers such as Ivry, that different timing mechanisms and other specific operations can be isolated within the nervous system and may indeed have a different ontogenetic course. For example, Ivry argued that cerebellar mechanisms may be primarily concerned with short-term timing events, whereas structures within the basal ganglia may be primarily concerned with longer term (as in supersegmental "prosodic") timing events. Since cerebellar and basal ganglia circuit properties develop with somewhat separable ontogenetic timetables, there is a wonderful but opportunity to use timing to isolate these different mechanisms. We shall return to this topic below. Whatever the details of our conclusions, timing is critical. That is the major point of Figure 4.

Thus far we have acted as if action or system boundaries can always be defined easily and without ambiguity. This is not always the case, and Figure 5 represents just a small sample of the reasons. First, it is not always clear what dimensions we should use to define an action. By some criteria, one action can appear to be distinct from another, but by other criteria, the separation is less clear. In mouse grooming movements, for example, actions that are quite distinct in their form can share common timing properties. Should we use *timing* or *form*, or some other criteria, to make our separations? In other cases, actions that are similar in their overall form vary distinctively in time, amplitude, direction of action, apparent function, and so on. We can now ask the question in a more complex manner. Are we talking about the relative amplitude of actions, their direction toward the environment, their kinematic details, or some inconsistent *combination* of each of these dimensions, right up to the level of human perceptual, categorical, and interpretative biases (Edelman, 1987; Golani, 1992)? Part 1 of Figure 5 is simply meant to imply that there may be dimensional ambiguity. If licking and certain face washing movements appear to share common timing mechanisms, to what extent are we justified in separating them? If certain tamping and scooping movements in canid caching share kinematic properties, to what extent are we justified in distinguishing them? Here

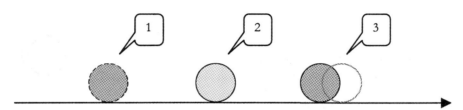

Figure 5. Ambiguity reflected in unclear boundaries or identity (undefined beginnings and ends of events or uncertainty in event identity): (1) dimensional ambiguity (dashed boundaries); (2) categorical ambiguity ("white or gray"?); (3) transitional ambiguity (underdefined boundaries between two events).

we wholeheartedly agree with Golani that one has to try to assemble a multidimensional picture of action. We also agree fully with Edelman (1987) that this is a matter of the taxonomic schemes we employ: is a red square more distinct from a green square or a red circle? It is foolish to operate as if one scheme is necessarily superior to another. The key question is how *animal* nervous systems are making their categorizations in action and perception. It is simply too easy to apply superficial judgements on the situation. That can degrade our analyses of performance mechanism, and also ontogeny. It can lead to a perspective of unitary and static modules that are simply unprincipled (Thelen & Smith, 1994). Part 1 of Figure 5 is simply designed to alert us to this problem, which can quickly become complex and also a major source of confusion among workers with different analytical perspectives. This leads directly into part 2 of Figure 5, where we move to higher order interpretations of a behavioral act or sequence of acts. At this point, we have moved from simple description to interpretation, and it is important to keep this distinction as clear as possible. How do we decide, for example, if the bites given by one young wolf to another are "playful" or "aggressive"? These are not trivial questions, but unless we understand their underlying ambiguities, we can become very superficial in our analyses. Or, to take an example offered earlier, suppose a young mouse reaches rhythmically toward its face *as if* grooming, but fails to make sophisticated contact with its forepaws and its face. Is it grooming? Or, finally, think of the case of a young male rat showing copulatory patterns with another young rat; is this mating? The list could be extended indefinitely. The third part of Figure 5 is meant to depict ambiguous rules of transition, such as often found in early ontogeny. When do children on a playground stop playing and start fighting? When do young wolves move from play actions to attack? These are not trivial questions, and ones that we cannot address fully here. But they need to be recognized.

Figure 6 is designed to remind us that within an overall action sequence there may be saltatory patterns where "recurrent patterns" are buried within a stream of actions that are less clearly ordered in their sequencing. The reason for including this figure is that overall sequential (stochastic, Markovian) analyses fail to highlight the possible existence of special phases within the sequence (e.g., Berridge *et al.*, 1987). In other words, action systems can tighten and loosen, or even change in a qualitative manner, their rules of sequential dependence as the overall sequence progresses. We have found this to be true both for canid food-caching patterns and rodent grooming. One needs to look at specific sequence phases as they occur within the overall sequence. Tightening and loosening rules of connectivity among action components within an overall sequence can reflect prosodic rules for the sequence as a whole. We shall subsequently look at this from two perspectives: (1) changes in the tightness of patterning of individually defined "events" and (2) changes in the actual assembly of these event packages during ontogeny and performance. To make this clear, the former refer to sequential predictability of action properties, such as in relatively relaxed versus stereotyped phases of rodent grooming, and the second refer to changes in the assembly of individual action properties, such as when

Figure 6. Very deterministic patterns (marked by horizontal line and alternation between shaded and white circles) can be interspersed with action sequences that are more or less random.

we use individual letters to make different words, individual words to make different phrases, etc. Each of these perspectives offers the important suggestion that both segmental and supersegmental issues are at play; thus our analogy to prosody in human speech.

Note that thus far we have cartooned actions as if there are indeed singular (even if complex!) events that can be placed upon a linear string. This is obviously a gross simplification. Action sequences are made up of numerous parallel as well as sequential events. To stand and walk, one must have to keep balance, move the legs in conjunction with the torso, know where one is going, know the reasons for going there, and so forth. Action sequences involve an orchestrated combination of both simultaneous and sequential events. We do not have time or space in this chapter to cover all of these. Figure 7, however, will be used to highlight some of the parallel processes discussed more fully below.

RELATIONAL INVARIANTS IN ACTION.   There is one point that we should make clear before giving more concrete examples of some of the issues that we have but sketched above. *Action sequences involve coordinated combinations of action elements (and or processes).* It is often an error to isolate individual properties of action as if they operate in full independence of one another. For ontogenetic analyses, it is equally important to examine how individual action properties become differentiated (refined) and how they become integrated (combined). The reason for this is that effective behavioral performance often involves what we here call "relational invariances" in movement. This is not a new idea. One can trace it back to Hebb and to Lashley, who spoke of motor equivalences in action. But it is critical, and is especially germane to developmental analyses and levels of brain function. We can each pick up a glass of water through many different details of motor coordination, and then once having picked up the water, drink it through even more details of motor coordination. In brief, we put together integrated *packages* of behavior. The individual details may change, as would the notes of a transposed melody or the words in a sentence that was trying to make a point. It is our view that both the component structure and the rules by which components are combined are at the heart of developmental analysis. Any reader of this book can achieve the same goal through a variety of means. That is the idea of motor equivalence. There are "higher order" invariances in behavior that come about through variable means. How the nervous system accomplishes this remains a major mystery. How ontogeny both refines individual movement properties and establishes rules for their combination so that higher order functions can be achieved remains an equal mystery. We do not pretend to have the answers to such mysteries, but we offer two primary examples

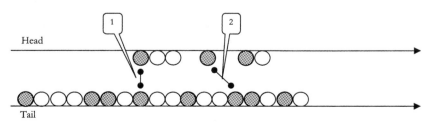

Figure 7. Parallelism with synchrony (1) or asynchrony (2); e.g., the parallel activity of head and tail in a feeding animal.

that illustrate them: canid caching behavior and rodent grooming. We then offer thoughts as to how neural structures might be differentially involved at different levels of organization and how the combination of genetic and epigenetic analysis might help us understand these issues more fully.

INTERIM SUMMARY: PROSODY AND ACTION SEQUENCING

In the present section, we attempted to present, in simplified cartoon form, just some of the many issues that confront investigators of action sequence analysis. We first tried to show how even simple sequences, involving two discrete actions, can offer many opportunities for research (Figure 1). We then made brief note of how temporal factors can lead to both overlapping and temporally disconnected properties of action (Figure 2). From there we looked briefly at the still more complex temporal issues of overshadowing and encapsulation in movement (Figure 3). We then tried to cartoon in a more explicit fashion how both intensity and temporal factors can affect our analyses (Figure 4). We next tried to delve somewhat more deeply into action category and action complex analyses (Figure 5). Two items we raised, albeit briefly, were specific phase sequence properties (Figure 6) and the fact that action sequencing involves a number of parallel as well as sequential operations within the nervous system (Figure 7). In each case, we oversimplified the story. For example, many actions do indeed occur in parallel as well as sequentially, and these rules almost certain change during ontogeny. For those concerned either with ontogenetic analyses at the behavioral level or the relation of brain mechanisms to different parameters of movement, these are important considerations. We fully appreciate that they still represent oversimplifications. Our analogy to prosody in human speech, where supersegmental and segmental analyses join, is perhaps the best overall summary of where we are coming from and where we want to head. If one takes these figures and the ideas that underlie them, then the richness for future research opportunities should be apparent. We know that the figures are abstract. We thus provide two illustrations, canid caching and rodent grooming, that may make them more explicit. We then offer a framework of "interactive/self-organizing" (ISO) processes that we believe apply both to the moment-to-moment performance of behavior and to its ontogeny. Clearly, one of our major challenges is to link these time frames of performance and ontogeny together within a more comprehensive framework than we have at present. A second and equally challenging opportunity is to seek ways to link current neurobiological investigations with analyses of un-restricted action sequences such as studied by ethologists for several decades. We shall thus conclude our chapter with ideas on how this might be done.

CANID CACHING BEHAVIOR

Food caching is a common behavior found in dogs, related canids (such as wolves, coyotes, and foxes), and many but not all carnivores. It involves a rich but relatively stereotyped sequence of actions that is species-specific in detail, but with many movements and their sequencing that can be recognized across species (Macdonald, 1976; Vander Wall, 1990). As shown by Phillips, Danilchuk, Ryon, and Fentress (1990) and Phillips, Ryon, Danilchuk, and Fentress (1991) for wolves and coyotes, respectively, the animals pick up a food item, explore for a site to bury it, dig a hole with varying combinations of single and double (alternating) paw movements, drop the food into the hole, "tamp" the food into the hole via rapid and often forceful head movements, and "scoop" dirt over the cache with the nose via horizontal head movements starting from the chest and extending outward. In young animals, these sequences are often incomplete, and may include burying sticks and other objects during play.

Lorenz (1982) provided a lovely description of the richness of food caching behavior and its phylogenetically constrained origins: "If ... the observer of captive animals sees the way a young wolf or dog carries a bone to behind the dining room drapes, lays it down there, scrapes violently for a while next to the bone, pushes the bone with his nose to the place where all the scraping was done and then, again with his nose now squeaking along the surface of the parquetry flooring, shoves the nonexistent earth back into the hole that has not been dug and goes away satisfied, the observer knows quite a lot about the phylogenetic program of the behavior pattern" (pp. 48–49). This suggests interesting analogies to other systems, such as rat drinking systems that are relatively ballistic, i.e., run as of a sequence independent of actual outcome (Blass & Hall, 1976).

We recently became interested in broadening our comparative analysis of caching behavior, including its super-segmental ("prosodic") structure and development. Canids are scatter hoarders as opposed to larder hoarders (Morris, 1962; Tinbergen, 1972; Vander wall, 1990), i.e., they spread their caches or storage sites over fairly large areas. Larder hoarders tend to store all their food items in a single central location, often in a shelter, the den or burrow, with or without satellite locations (e.g., chambers). An example of a larder hoarder is the American pine marten (*Martes americana*). In the first month of data collection for this study (Gadbois *et al.*, n.d.), two individuals of this species were tested for food caching. The martens each cached food (larder style), but never engaged in the active and complex caching patterns exhibited by wolves, coyotes, and red foxes.

Our most recent data concern red foxes, considered more "primitive" in terms of their social and perhaps other forms of behavior than are members of the genus *Canis* (dog, wolf, coyote, and four species of jackals) (Wayne, 1993). The groups are genetically distinct. To facilitate comparison with our two-element "prosody" analyses (above) we shall here concentrate on the final phase of caching, "tamping" (T) and "scooping" (S). Tamps and scoops can alternate or show perseveration (repeats of the same act) (Figure 1). Thus a fox sequence might be of the form TTTTSSSTSTTSSSTTTSSTTTSSTS. There can also be overlaps and gaps in the sequence (Figure 2). For example, during transitions between tamps and scoops, the animals may show an intermediate pattern, such as a scoop that terminates with a downward push of the snout. Similarly, one action may overshadow another, such as when an initial scooping action is replaced by tamping movements, or the reverse. Scoops tend to be of longer duration than tamps, and one occasionally sees what is a protracted scoop that encapsulates a tamp (Figure 3). While our data are incompletely analyzed, it is clear that tamping and scooping movements can vary in their amplitude and duration (Figure 4). As illustration, the overall tamping and scooping sequence can vary in its duration (and number of elements). The duration of individual tamps at the beginning of a short caching sequence tends to be longer than is the duration of initial scoops. For long caching sequences, the relationship tends to be reversed (Table 1). While tamps and scoops tend to be distinctive in their morphology, there were occasions where a clear taxonomic distinction could not be made (Figure 5). This was particularly true for young animals. Within the overall *tamp/scoop sequence*, there were often periods where clear saltatory or recurrent patterns could be found, and other periods were the sequencing of the two actions was more random (determined by in-house software program) (see Figure 6). Finally, tamps and scoops are clearly not restricted to neck movements, but involved parallel and often complexly timed adjustments in other body postures (Figure 7).

Thus this apparently simple sequence between two actions revealed a surprising richness. We became particularly interested in how the overall sequence duration (and number of elements) varied as a function of species (wolf, coyote, red fox) and changed with age. For example, caching sequences are much more rapid and stereotyped in foxes than in coyotes or wolves. That is, foxes have a "machine-like" appearance to them that is distinctive. Wolves are by many criteria the most flexible in the movement details of their caching behavior. While our developmental data are incomplete, it appears that young animals will often exhibited truncated sequences. These are often interrupted by playful running and jumping. Food (or sticks) may not be buried at all.

There were clear supersegmental properties of the overall caching sequence that we then sought to explore more fully. Even within individuals, some sequences were orchestrated with relatively rapid movements and others with more deliberate, slow movements. One of the most interesting features that we sought to understand was variation in sequence duration. To do this, we combined repetitions of a single action, tamp or scoop, into what we define as a segment. Thus, the sequence TTTSST contains three segments, two made up of tamps (three elements and one element) and one made up of scoops (two elements). We then calculated segment length, total number of segments, and total number of elements.

For each species, tamps almost always initiate the entire sequence. Remarkably, if one takes the ratio of tamps to scoops in the initial segment of each (which in the above example would be 3:2), this ratio can be used to predict quite accurately the total sequence duration. Thus, the animals let us know early in the tamp/scoop sequence how long that sequence will last, including the number and combination of elements overall. Tamps initiated sequences over 95% of the time, and scoops tended to terminate the sequences, however long the sequences were. If we now define sequence length in terms of number of perseverative and alternating segments (where TTT or SSS would be a single segment), we can plot T1-to-S1 ratios and predict with surprising accuracy total sequence length (Figure 8). Overall, the foxes appeared to be the most "mechanical" (predictable) in their behavior, and we still have not completed a definitive analysis of how these patterns might change ontogenetically for any of the species. There are more subtle rhythmical and amplitude ("prosodic") modulations within the sequence that are still being analyzed.

Early reports can be found in Gadbois and Fentress (1997a, b). Gadbois and Fentress (1997a) failed to find a conclusive difference in young fox kits (from 3 to 6 months old) in either the sequential structure of perseverations or the temporal structure of the entire tamp–scoop phase. However, we do have preliminary data

TABLE 1. DURATION OF TAMPS AND PROTRACTED SCOOPS IN A SEQUENCE OF FOOD CACHING ACTION IN FOXES

	1	2	3	4	5	6	7	8	9	10	11	12	13	14
T1 (sec)	0	4.4	4.5	4.3	4.5	5.8	5.5	5.3	5.2	5.3	6.0	6.0	7.4	9.0
S1 (sec)	0	8.1	7.2	4.2	5.0	3.5	3.7	2.3	3.7	2.1	1.0	1.4	2.0	1.0

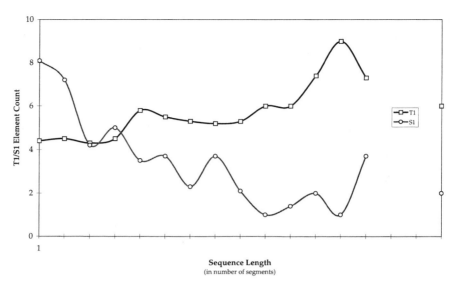

Figure 8. The change in ratio of initial number of tamps and scoops in the first sequence of each as a predictor of the total length of the caching sequence. See text for further details.

that the young animals were often less effective in actually burying the food, and were likely to misdirect their movements away from the cached food target. Often, food items would be ignored altogether or pushed out of the caching hole without correction. It is as if the different segments of the overall sequence were running off more independently than in adult animals.

The misdirection of movements was reported much earlier by Fentress (1967) for pouncing behavior in a young wolf. A toy "prey" (rubber bone) was pulled by a string either in front of the animal or to one side of the animal. Often when the object was pulled to the side of the wolf cub, he would make prey-catching pounces, but in the direction of his initial body orientation! It is as if the directional tuning of the behavior developed later than the basic prey jump. Thus, there is often a fragmentation of action sequences that occur in a more fully integrated style later in ontogeny. A challenge for future research is to see how developmental processes of movement differentiation and movement integration operate together. We shall discuss this further with respect to rodent grooming in the next section.

INTERIM SUMMARY AND EXTENSION

We have demonstrated that caching is a relatively stereotyped but also richly nested sequence of movements. In adult animals, these movements are often surprisingly persistent, even when environmental circumstances for their expression, such as snow versus hard ground, vary (Phillips *et al.*, 1990, 1991; cf. quote by Lorenz cited earlier). In young canids. the actions are less well integrated and have a less "ballistic" appearance. During rapid caching sequences, we have also found the animals to be less responsive to extrinsic disturbances than they are under other circumstances. The central phases of the caching sequence, which we can predict by the methods outlined above, appear to be least easily disrupted. Thus the sequence as a whole has rules that are not apparent from simple element-by-element analyses.

To return to our earlier analogies with human prosody, it is clear that both segmental and supersegmental properties of action sequencing deserve evaluation. Simple statistical (e.g., stochastic) analyses of overall tamp/scoop sequences would have failed to reveal rules by which these segments are joined through more global (supersegmental) properties that "glue" the sequence as a whole together. We suspect that issues of sequencing and timing, along with examination of subtle variations in form and emphasis across action sequences, will offer new opportunities to untangle clusters of ontogenetic variables that may operate with different time courses and involve different central nervous system mechanisms. The potential connection to central nervous system mechanisms, and forms of future analysis, can be seen more clearly in the next section on rodent grooming behavior.

As we shall show with rodent grooming, rapid and stereotyped movements in the middle of a grooming sequence are often the most "ballistic" in their organization. There are numerous comparisons, such as the work by Blass and Hall (1976) which showed that the lick pattern in drinking could not be broken until the rats had moved into a state of cellular overhydration. At that point, it was very easily broken. A hypothesis this suggests, to which we shall return at the end of the chapter, is that control boundaries within behavioral sequences may change their focus, such as sensitivity to extrinsic factors, with variations in performance and age. In canid caching, we have suggested that rapid and stereotyped performance is less easily disrupted than more "leisurely" performance, and that young animals are often more easily distracted in their "playful" caching than are adults. This suggests that in addition to describing ongoing sequences of natural behavior, the application of specific perturbations during specific sequence phases at specific ages may provide a powerful analytical tool. We shall return to this theme.

RODENT GROOMING BEHAVIOR

GENERAL PICTURE OF RODENT GROOMING

Rodent grooming has provided a rich source of developmental data relevant to the considerations outlined above. As we began to show many years ago (Fentress, 1972; Fentress & Stilwell, 1973), it is possible to analyze grooming sequences into basic stroke types and to determine "grammatical" rules for the temporal organization of these stroke types. Further, it is possible to determine hierarchical rules of sequential organization, such as ABC ABC ACD ADE BCD, where each stroke type is represented by a letter. Groups of letters can be thought of as words, and rearrangements of the letters form phrases. This in turn allows us to ask whether there is a clear developmental sequence in, e.g., the differentiation of letters, their combination into words, and phrases—each grouping being viewed as a more complex level of organization. Again analogy to human speech proves useful.

Visualize a young mouse grooming. Infant mice cannot support themselves stably in a sitting posture until they are about 10 days of age. How can it groom? Normally, newborn and young mice do not groom extensively (Golani & Fentress, 1985). One might think this is simply a matter of immature brain development, yet rudimentary grooming has been observed prenatally (Smotherman & Robinson, 1996). So our studies of action sequence development must certainly extend beyond thoughts of a brain in a dish. Other factors need to be considered [see Bateson & Martin (1999), Gottlieb, Wahlsten, & Lickliter (1998), and Thelen & Smith (1994) for recent thoughtful discussions of related issues with respect to the development of action patterns in both animals and humans].

First, our young mouse needs postural stability. It must sit up before it can make grooming movements in anything like an adult pattern. This demands a large measure of muscular and vestibular control. Infant mice do not groom in an adult pattern. They do not have the strength or balance to do so. Yet, if one places them in a high-chair, with sufficient body support, they can and do groom, in a progressively coherent and rich manner. So, to use the terminology of Hogan (1988; and Chapter 6 in the present volume), the basic neural circuits are available prefunctionally, but are masked by other constraints. In this instance, they require special (supportive or promoting) circumstances for their expression. It turns out that this is a very common phenomenon in both animal and human development; at some level, circuit properties may be established long before they are used in behavioral expression. This supports the hypothesis of Blass and others (chapters in this volume) that there can be "core" operations in development that are at least relatively independent of experience, such as through practice. In some sense, they can thus be viewed as "maturational." However, it is important to recall that embryological "experiences" can occur at all levels of organization, ranging from organism–environment interactions to interactions among cells and even at the biochemical level. Differentiation itself would be impossible without these distinctive sources of "experience" (Atchley & Hall, 1991; Wolpert, 1998). The ancient nature–nurture issue gets pushed back, but it does not disappear, and probably never will.

Even in cellular terms, "One of the biggest stumbling blocks in understanding the development and evolution of morphology has been the extrapolation from simple changes during ontogeny to seemingly complex alterations in morphology" (Atchley & Hall, 1991, p. 116). It is the establishment of which units to look at and which processes that may glue these units together, whether morphological or functional, that continues to be one of our major challenges. This is why we stress in this chapter both the importance of looking at individual aspects of action pattern formation and the way these become clustered as development proceeds. Studies of behavioral ontogeny and embryology more generally remain in need of establishing closer ties.

Still, it is a striking fact that many of the basic circuits that underlie the control of behavior emerge more or less independently of experiences that we might expect to be essential, such as practice. As an additional concrete illustration, Hall and Bryan (1980) demonstrated precocial self-feeding in rat pups when the animals were put into a warm testing chamber. As a human example, Thelen and Ulrich (1991) showed that infants who were posturally supported and placed on a treadmill performed alternating stepping movements normally seen only considerably later in ontogeny (approximately 1 year). Thus, brains are not isolated ontogenetic "bathtubs" in any sense, but both express and change their properties in dynamic ways as a function of experiences throughout ontogeny and environmental supports that are provided at any given stage of ontogeny. Action sequence analyses can help dissect both where and when these changes and sensitivities to environmental supports occur.

Golani and Fentress (1985) attempted to tease out some of the parameters of early supported grooming by describing three movement parameters: the kinematics for individual limb segments, the trajectories of the forepaws as they reached toward the face, and the contacts established between the forepaws and the face. We used the basic technique of supporting infant mice in a "chair" surrounded by mirrors so that we could also score their behavior in three dimensions. In broad terms, prior to 100 hr postnatally, we observed a rich variety of forelimb movements that certainly appeared like grooming "attempts" but which led to variable forelimb trajectories and ineffective contact movements across the face. In the second 100 hr, the young mice restricted movement of their heads (thus reducing the challenge of having the forepaws find a moving target), tightened several kinematic parameters, and also reduced the variability, as well as limited the extent, of their forepaw-to-face contacts. That is, by 4 days they were now clearly "grooming," but only the anterior parts of their face. Later, both the movements and the head became reelaborated in a coordinated manner, leading to a rich variety of strokes previously described by Fentress and Stilwell (1973).

So the question in some sense remains: "Where" did these improvements come from? It is well known that many rodent brain circuits are established postnatally, and its possible that some of these diverse timetables contributed to the different grooming parameters. One possibility was that relatively early-developing brainstem and cerebellar circuits contributed to the tighter coordination of grooming at 4 days, and that later-developing striatal (basal ganglia) circuits contributed to the elaboration of distinct stroke types and their sequentially patterning (see Fentress and Bolivar, 1996, for more extended review).

When an infant mouse is placed in a grooming posture and activated by a light pinch of the tail, elaborate movements are elicited that, while not fully functional grooming, appear very grooming-like, and in this way parallels the rat findings. Place the same infant mouse in another posture, activate it again, and it performs predictably different and distinct actions. Too strong of a tail pinch precludes grooming even if the animal is in a correct grooming posture. The mouse is more likely to orient directly to its tail (Fentress, 1972, 1990). Thus, activational and patterning mechanisms are potentially separable. There is the further interesting possibility that one can model how two "distinct" systems, such as tail pinch and grooming, might interact, a point to which we shall return below. What our observations did make clear is that early grooming and other action patterns reflect posture, support, activation, and certainly a number of other factors.

Both the sequencing and timing of grooming movements appear to reflect maturational events that are in some sense independent of experience. Thus, Berridge (1994) used allometric analyses of movement duration and body size for guinea pigs and rats to argue that there is a consistency between neural timing circuits and eventual body size that can change across ontogeny. It is important to be clear on this point: there are stabilities in movement timing that persist even when body size changes dramatically. Young guinea pigs and adult rats are approximately the same size, yet the timing of grooming movements at developmental times where body size is similar are species-specific. The link between maturational and epigenetic factors in grooming deserves much further investigation. For example, what if particular movements were blocked, such as via collars, during development? Or, what if grooming in early life was activated such as through the application of ammonia fumes? Or could practice in other motor skills such as swimming generalize in the sense that they have consequences in the production of grooming? We do not know the answer to any of these questions. We do know, however, that a variety of factors such as postural support can have major implications for the production of grooming and a variety of other behavioral patterns (Thelen & Smith, 1994).

JOHN C.
FENTRESS AND
SIMON GADBOIS

At this point, we recognized there was an opportunity to utilize neurological mutant mice to assist us in our developmental dissections. There are a variety of mutant mice that have distinctive patterns and time courses of cellular degeneration. Our laboratory decided to look at supported swimming movements (Bolivar, Manley, & Fentress, 1996) as well as grooming (Bolivar, Danilchuk, & Fentress, 1996) movements in *weaver* mice. These mice have well-documented early disorders of the cerebellum (with granule cell death; Bayer *et al.*, 1996) and a more protracted degeneration in nigrostriatal dopamine production (Roffler-Tarlov, Martin, Grabiel, & Kauer, 1996; Tepper & Trent, 1993). There is growing, but incomplete evidence that the cerebellum's primary contribution to movement is in the production of finely coordinated patterns of muscular activity (Thach, 1996), whereas the striatum contributes importantly to such things as the initiation of movement, and the coordination of higher order sequences (Graybiel & Kimura, 1995; Jackson & Houghton, 1995). [See also Berridge & Whishaw, 1992, and Cromwell & Berridge (1996) for experimental support (via lesion studies) of this distinction with specific reference to grooming.]

Our choice of employing a support swim test was based upon two premises. First, we could study individual limb movement properties and coordination between the limbs at an early age, without the confounds of gravity (body support). This would permit us to compare swimming coordination with locomotor ataxia, long known to be pronounced by the second postnatal week (Fentress & Bolivar, 1986). Second, it is well known that there are systematic changes in rodent swimming "style" that have an ontogenetic timetable (Bekoff & Trainer, 1979; Fentress, 1972). Shortly after birth the animals use the forelimbs only, then all four limbs, and then the hind limbs only. We could thus ask whether this timetable was altered.

The general procedure was to place each animal in a warm swimming bath, suspended by a rubber band, starting on postnatal day 3 and continuing through the third postnatal week. Our primary goal was to seek any early differences between the control and mutant animals, especially where their behavioral performance might diverge during ontogeny. As in our other rodent studies, single-frame analyses of cine and video records allowed for a precise description of movement parameters. The *weaver* mice did display a relative slowness in limb movements during the swim as well as a moderate delay in switching between less mature and more mature swimming styles. Interestingly, their coordination when supported in the swimming task remained essentially normal, by which time they had clear signs of cerebellar-type ataxia during locomotion. It was not possible, however, to disintangle possible cerebellar and striatal contributions to this task. What was clear is that mice with early and specific neurological disorders could be distinguished as early as the third postnatal day. We also learned that providing animals with environmental supports can mask as well as reveal central nervous system effects (see the above comments on early supported grooming).

Following an earlier study by Coscia and Fentress (1993) on grooming ontogeny in *weaver* versus control mice, in which clear problems in the balance in the mutant animals along with truncated grooming bouts were revealed, we decided to examine their grooming ontogeny in more detail. By looking at grooming in both pre-swim and post-swim conditions, we could in principle disintangle issues of basic motor coordination from higher order "bout" structure involving sequences of strokes, from issues of activation (via the use of pre- and post-swim measures of grooming).

In simpler terms, we could separate issues of activation and pattern in grooming ontogeny. This is of interest in that it is well known that striatal deficits, as in Parkinson's disease, can affect both the activation and organization of movement. To our knowledge this had never been studied ontogenetically in an animal model with known central nervous system deficits. The pattern of results was both complex and clear. First, the mean time spent grooming in the *weaver* mice was less through the analyzed period from 11 to 19 days postnatal. Second, from day 15 the *post-swim weaver* mice spent more time grooming than did the *pre-swim* controls. Third, in *post-swim* but not *pre-swim* mice, both the mutant and control mice showed progressive increases in total grooming. This indicates clearly that some of the overall grooming deficits in *weaver* mice could be alleviated through activation, but that levels of activation were also important for controls.

We then decided to examine the number of grooming bouts. At postnatal day 17 there was a striking divergence between mutant and control animals. In control animals, swimming had little effect upon grooming bout numbers, but greatly increased bout lengths. In the mutant animals, swimming had a major effect on bout number at the time striatal damage becomes marked, but the bouts remained truncated. In conclusion, therefore, we could say that (1) activation can improve the initiation of grooming in the *weaver* mice, but (2) their ability to complete the complex sequence of strokes found in controls never approached the control level. Further, each of these effects exhibited an ontogenetic time course that closely corresponds to nigrostriatal effects found at cellular and biochemical levels of analysis. Analysis of grooming stroke types, which are normally sequential arranged, also indicated that the mutant mice somehow got "lost" in the full grooming sequence, and that deficit could be moderately reduced in an age dependent manner by the experience of grooming. This provided an important control for otherwise undisclosed issues of balance loss. The truncation in grooming movements was at the most only partially correlated with balance loss, a conclusion proposed earlier by Coscia and Fentress (1993).

The first lesson from this rather complex pattern of results is that in ontogenetic analyses of action sequences it is important to take multiple measures of action (as the different measures reveal different patterns). The second lesson is that it is important to test the animals under more than one context. The exciting possibility is that judicious use of neurological mutant mice may indeed provide insights into behavioral distinctions that can in turn be related to known central nervous system disorders. Prosody reminds us that both sequential and temporal information can help clarify behavior (Bressers *et al.*, 1995)

MORE ON ACTIVATION AND SELF-ORGANIZATION

In very early studies of grooming in adult voles, Fentress (1968a, b) provided evidence that moderate levels of "fear" due to an overhead model predator could facilitate the probability, speed, and duration of grooming movements. This grooming most often occurred once the "predator" stimulus had passed, and with a predictable time course. Here we noted that moderate swimming could "activate" grooming in mutant mice. Also noted briefly above was that in young mice, a moderate tail pinch could generate grooming *as long as* the animals were in a correct posture. Less formally, we observed that rodents often groom when there are signs of ambivalence in what to do. One can place an animal in a small container, for example, and generate excessive grooming. If the animal is overly stressed, it will try to escape. These observations clearly indicate that in the performance of action sequences there can be a dynamic interplay between external activating events and grooming performance.

We have also shown that early and late phases of grooming are much more easily interrupted than are rapid and stereotyped phases (Fentress, 1988), which suggests that in addition to activation of the behavior from often multiple and surprising sources, once a behavior such as grooming gets going the behavior often becomes more self-organized, that is, more tightly focused and less prone to disruption from external sources. In an unpublished study by E. Buckle in our laboratory, it was found that mice with neonatal lesions of the striatum that in some sense mimic the effects found in *weaver* mice were much more sensitive to laboratory noises than were control animals, in that these noises would cause them to terminate their grooming. These animals also rarely entered the full stereotyped phase of grooming, thus again suggesting a dynamic interplay between strongly activated behavior and sensitivity to environmental events (Figure 9).

The dynamic structure of grooming responsive to external perturbations appears to be richer than simply being more "self-organized" (under central control) during the middle phases of grooming. Ongoing research indicates that there are certain phases within the overall sequence in which sequential rules change (e.g., going from ABCABCABC to something like ADEADEADE). During these "phase transitions," the animals momentarily increase their sensitivity to external events. Thus, there are at least two processes. The first can be sketched as increased stereotypy during approximately the middle of the overall sequence. The second involves predictable shifts in sequential patterning, during which time the animals are momentarily more sensitive to events in the external environment. The overall pattern is sketched in Figure 10. This again shows why prosody (or speech or music) analogues to animal behavior can be so useful: there are individual events and different combinations of events to which we must be alert. Some of these event

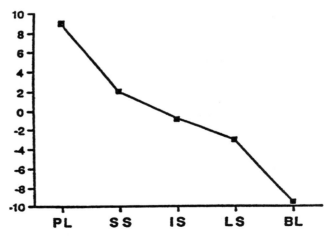

Figure 9. The degree to which control and dopamine lesioned mice completed their grooming sequences. The proportions of paw licking (PL) to small facial strokes (SS) to intermediate (IS) to long (LS) to full body licking (BL) are altered. This strongly suggests that the striatum is involved in action sequence completion. (Ordinate = scaled change from control proportions.)

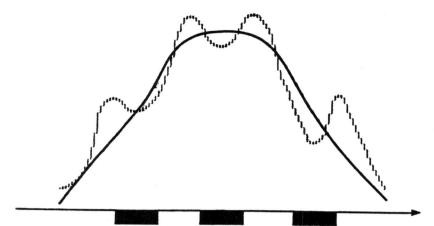

Figure 10. Changes in threshold to interruptions in grooming by extraneous events. There are two profiles. (1) During the middle phases of grooming the threshold for disruption increases; and (2) across this overall phase, there are "phase transitions" during which for brief times, the animals are increasingly sensitive to environmental perturbations.

combinations appear to serve as action "chunks" and others appear to cross over even these relatively defined "chunks" in performance.

As a developmental issue, it is interesting to ask whether the ontogeny of grooming phases, or syntax, depends upon the prior perfection of individual movements or whether there is a degree of independence between these levels of control. Indeed, Colonnese, Stallman, and Berridge (1996) provided data that in young rats, but less obviously in precocial guinea pigs, these levels of grooming organization can develop somewhat independently. In guinea pigs, action components were adult-like from postnatal day 1. The grooming syntax was also adult-like. By contrast, in rats, individual movements were often abnormal at the end of the first week of postnatal life (see Golani & Fentress, 1985, for mice), but the authors also observed clear evidence for "syntactic chains" (prosodic structure). The syntactic chains were not always perfect, and the authors had to take into account omissions, extraneous actions, and hybridization across actions. Their overall finding, however, supports the hypothesis that in action sequencing different levels of organization are somewhat independent in both their operation and ontogeny. The differentiation (refinement) of individual action properties and their integration (combination) *each* deserves evaluation as a developmental event. Since it is likely that different neural circuits contributed differentially to these various levels of organization, future genetic and epigenetic analyses offer considerable promise at both the mechanistic and systems levels of organization.

PROMISES AND CAUTIONS IN THE USE OF MUTANT ANIMALS

There are both promises and cautions in the future use of neurological mutant animals in such studies. Mutant mice are being "manufactured" at an accelerating rate and with increased specificity of neural operations affected. For example, there are now mutant mice that have cerebellar dysfunctions that affect different specific cell types (e.g., Purkinje cells versus granule cells) with quite distinctive ontogenetic timetables. Similarly, specific receptor knockouts have been created (e.g., DIA dopamine receptors in the striatum). Chimeric animals and the almost certainty of inducible and highly localized "knockout" functions will provide the future behavioral scientist with potentially valuable tools not previously available. But as Nelson (1997) has argued, the value of these animals in linking ontogenetic brain and behavioral functions will certainly depend upon sophisticated behavioral analyses. A main point of this chapter is that action sequence analyses across levels and time frames of organization provide one valuable tool. With the almost certain advent of miniaturized recording and imaging techniques in the near future, the opportunities look promising indeed. We also believe that many of the issues of action sequence analysis found in other more strictly ethological studies, such as in the canid caching behavior reviewed previously, will be reflected in issues that can be pursued at complementary levels in the laboratory.

One has to be careful at this time not to jump on any bandwagons, however. The senior author has gone to several recent "mutant mouse" meetings where a rather extreme form of reductionistic enthusiasm is apparent, almost as of a "one gene/one behavior" quality. Knockouts are also lesions, and developmental lesions that result in a variety of compensatory consequences. These are difficult to trace. Genes have developmental effects that often cross over traditional neurobiological categories (such as cerebellum versus striatum). Of course, that is the way our genes help construct us, and thus judicious use of these genetic alterations, *in a carefully evaluated ontogenetic context*, can help clarify the way nature does indeed work.

There is a last caution that we would like to emphasize, one that has only been pursued directly in a few laboratories, but with often striking results. Genes do not operate in a vacuum. Behavior is an *epigenetic* event.

We have made recent unpublished chance observations on Huntington's mice that emphasize the importance of developmental context in evaluations of mutant animals. The particular mice employed (in the Dalhousie laboratory of my colleagues H. Robinson and E. Donovan-Wright) are of interest because they have rather specific deficits of striatal function that lead to progressive Huntington's-type symptoms (thus far incompletely documented). The mice are extremely fragile, and many would often die a few weeks after birth. However, Dr. Donovan-Wright noted that when the mice were handled, their symptoms were greatly delayed and their life spans were extended dramatically. The laboratory of W. Greenough at the University of Illinois has also worked with cerebellar-deficit mice and has reported impressive results indicating that skill training but not just mere exercise also reduced the severity of behavioral symptoms (Kleim *et al.*, 1998). Pham, Soderstom, Winblad, and Mohammed (1999) recently found that environmental enrichment can be similarly effective in reducing the symptoms of mice with genetic predisposition to have deficits in hippocampal nerve growth factor. Careful analyses of behavioral ontogeny in mutant mice under various conditions of "early experience" could thus provide extremely important insights into the classic nature–nurture debates, and at a much more refined level than previously possible.

We offer a few additional suggestions: (1) First and foremost we suggest that analyses of action sequence development and various levels of organization observed under a variety of specific contexts offer powerful tools. (2) A combination of observations on unperturbed (ongoing) behavior can be combined with measures of specific perturbations (e.g., loud clicks) to test the animals' reactivity. This will help dissociate central and peripheral contributions to integrated behavior. (3) In addition to the use of environmental manipulations during different phases of development, it should in principle be possible to attempt pharmacologic and related internal manipulations to (a) try to produce symptoms in control animals, and (b) to "rescue" genetically compromised animals. (4) More specific ontogenetic questions can be asked about differential experience. For example, could one attempt either to enhance or diminish specific opportunities for specific movement types, such as individual phases of grooming? One could also ask whether there are influences that cross over behavioral categories, such as whether swimming practice affects grooming ontogeny. We do not yet understand how experiences at specific stages of ontogeny generalize. Would we, for example, expect generalization in movements that are similar in their form, or that serve common functions even though they differ in their form? The answers might vary with animals that have different genetic predispositions that affect different central nervous system structures. This is a greatly truncated list of possibilities. There is an enormous range of exciting and totally unexplored future opportunities, plenty to keep many laboratory workers happy for a long time.

ACTION SEQUENCES AS INTERACTIVE/SELF-ORGANIZING (ISO) SYSTEMS

In this chapter, we have broadened the concept of action beyond one of simple motor control. Rather, we see action as involving the whole organism, and in that sense agree with Hebb (1949, p. xv): "it is a truism by now that the part may have properties that are not evident in isolation." Thus, animal and human movements, while obviously involving the motor system, are organized through their connections with sensory, cognitive, motivational, and other fundamental processes. The challenge thus becomes one of looking at basic components of action within their broader expressive contexts. When one does so, it becomes clear that underlying processes exhibit a dynamic balance between two extremes: they shift in their relative sensitivity to extrinsically defined processes and relative autonomy (modularity). Neither excessive modularity nor excessive connectivity can work. Here we shall suggest more explicitly our view that action sequences indeed exhibit a dynamic balance between systems that are *both* interactive and self-organizing (Fentress, 1976). Indeed, recent neurobiological studies of integrative networks suggest the value of this same perspective. We argue that the same holds for developing systems at all levels. This allows the organism and its underlying subsystems to both maintain essential integrity (stability) and adapt to changing circumstances.

In neuroscience, the issue is often expressed as the "binding problem": how do separate functions communicate, and in doing so change their own rules of operation, which changes their rules of connectivity, and so forth? In developmental biology, the issue can be thought of in terms of the balance between mosaic and regulatory properties (Wolpert, 1998). What we are thus trying to do is to show how compartmentalized "mechanistic" and global "systems" perspectives can be brought together, each informing the other (Edelman, Gall, & Cowan, 1990). An ultimate goal is to bring both levels and time frames of analysis together, to seek parallels between behavior and neuroscience, as well as integrative and developmental perspectives. While we are a long way from that goal, we offer the interactive/self-organizing framework as a way in which it might be pursued. Prosody served as our initial analogy, for it reminded us that in action sequences, there are both components of action that can be isolated and rules of supersegmental connectivity.

Hierarchical models have of course attempted to capture this, but we believe that most such models are placed within a framework that is insufficiently fluid.

AN APPROACH TO INTERACTION/SELF-ORGANIZING SYSTEMS

Rather than speaking in simple dichotomies, such as interaction versus self-organization, we believe it is more useful to speak of a dynamic balance among polarities of emphasis. Thus, in the history of motor systems analysis, there has been a dichotomy between sensory-driven "reflexes" and "central motor programs"; yet at the intact organism level this dichotomy has proven increasingly difficult to maintain (Prochazka *et al.*, 2000). What often happens is that an animal will be responsive to sensory events, and upon responding, block out further response to these same or other sensory events. We shall return to this theme shortly. The classic nature–nurture dichotomy runs into similar problems. Thus, in a dynamically balanced manner developmental processes may for a while depend heavily upon their surround, only later to become relatively independent of this same surround, a point elegantly reviewed by Gottlieb *et al.* (1998).

In the above review, we have suggested that in the organization of action sequences, animals may often be highly responsive to extrinsic events to initiate behavior, but in cases where the resulting actions are rapid (intense) and stereotyped, they may become relatively oblivious to these same extrinsic events. During phase transitions, where the sequencing rules change, there are also momentary increases in sensitivity to external events (Figure 10). It is, in brief, possible that systems are set into play by interactions with their surround, but these events do not literally drive the details of behavior. Rather, they allow the systems to then self-organize relatively independently of strict guidance from the very events that set them into action. In development, there are many well-documented cases of processes of differentiation being set into motion by communication between cellular and higher groupings and then self-organizing more or less independently of these very same extrinsic events (Wolpert, 1998).

To the extent this is true, how can one visualize what is going on? A number of years ago, Fentress suggested the idea of shifting boundaries along the following lines. A system more or less at rest often has a broad sensitivity to a number of extrinsic events. These events then help the system self-organize. Once this process is underway, it is as if the system tightens its focus and indeed begins actively to inhibit the same extrinsic factors that initiated its action. Thus, systems not only become relatively opened and closed at different times, but one could conceptualize this in terms of boundaries that broaden and narrow as a function of their current activity profiles. Another way to state this is that the systems are relatively undifferentiated, nonspecific, and global under some conditions, then highly differentiated, specific, and locally controlled under others. It is obvious that the degree of localized versus specific function of particular events can vary with both time and stimulus intensity. This is because stimuli take time to build up in their intensity as well as to die down. These and related dynamics, however, are not yet well worked out (Fentress, 1999; Golani, 1992).

Self-organization could occur through a variety of mechanisms, starting with the genetic substrates of the organism. One approach is to conceptualize a system as having both latent potentialities and developmental predispositions. This is analogous to the concept of "induction" in embryology, where, for example, the juxtaposition of tissues does not in a strict sense "instruct" tissues how to differentiate, but simply "sets them on their course." In this developmental context, an especially appealing model, proposed by G. Edelman and his colleagues (e.g., Edelman, 1987; Sporns & Tononi, 1994), is that extrinsic factors "select" and "amplify" preexisting (phenotypic) potentialities. One can think about this in two ways. The first way is rather like opening a file drawer and selecting the particular folders to read that are already there. The second way is Darwinian, in the sense that certain potentialities are selected and amplified at the expense of others. Analogies to the immune system, which is fundamentally selective rather than instructive, has provided the impetus for such models.

Similar events may occur in integrated performance. Behaving organisms have momentary "sets" where they are more predisposed to certain forms of activity than to others. Inputs may select and amplify these predispositions. We hypothesize that this will occur if the inputs form a reasonably close match to specific inputs and are not too strong. In the latter case, the system would reset. It is a double threshold model in the sense that moderate inputs from a variety of sources can kick the system into action along a path that is in some sense already preset. Stronger inputs or inputs that are more "off target" can, conversely, re-set the system. Once the system, in either the developmental or integrative sense, is "kicked into action," our hypothesis is that the system becomes less sensitive to extrinsic factors for variable periods of time. In this sense it is self-organized by being relatively autonomous in its operations. One can visualize the system as now being more focused in its operations, with system boundaries that are more tightly constrained. Extraneous events are then blocked, thus protecting the trajectory of expression that the system is taking.

The value of such speculations of course depends on where there are rules for this conceptualized broadening and narrowing of a system's boundaries. The suggestion was that weakly activated systems broaden their focus, whereas strongly activated systems narrow their focus. One can think here of early ethological models that stressed appetitive behavior (open systems) and consummatory acts (more tightly constrained systems). The distinction between this model and the model of Tinbergen is that he represented breadth and narrowness as moving through a basically static hierarchy of organization. The emphasis of the Fentress model is that individual systems can change the size of their boundaries, broadening and narrowing as a function of current action states (Figure 11).

The question is how this can come about? If one visualizes systems, however defined, as being represented by a core excitation and a surround inhibition, and with a core excitatory potential having a lower threshold than the surrounding inhibitory potential, then once strongly activated, the core would shrink and the lateral inhibitory surround would spread. The core shrinks in part because of invasion of the surrounding inhibition, and this can be represented in such observations as increased stereotypy in action. The narrowed core is now strongly activated. In addition, the surrounding inhibition blocks out extrinsic influences, thus protecting the integrity of the initiated action. The implication is that elements will openly talk with one another at some times and cut off conversation at other times. This allows for, indeed demands, a dynamic rather than static form of organization.

If one then imagines both excitatory and inhibitory events to have warmup and decay functions, it is possible to take this spatial model and turn it into a temporal model. Elements within an action sequence may overlap and separate in terms of the processes that control them. One limitation of the model as initially stated is that it could lead to a strongly activated element that stays "turned on" forever! So there would have to be decay functions built into the system. Once this is done, it is indeed

JOHN C.
FENTRESS AND
SIMON GADBOIS

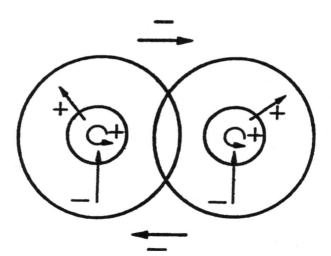

Figure 11. The intersecting circles symbolize two movement control systems. Each is depicted with an excitatory core (small circles) and surround inhibition (large circles). As discussed in the text, the relative sizes of inner excitatory core and surround inhibition vary with activation strength.

possible to generate a wide range of organized action patterns (J. C. Fentress, unpublished). There was a second limitation to the original formulation. Insufficient attention was given to the possibility that the elements that made up a system could vary from one expression to the other, or in other terms, that the same element could upon different occasions contribute to different higher order systems.

There are now numerous examples in both invertebrate and vertebrate systems that individual neural elements shuffle their individual roles and patterns of functional connectivity (e.g., Stein *et al.*, 1997). This is true both ontogenetically and as a function of the functional dynamics of individual performances of action. To cite one mammalian example, Lemon (1990) mapped the output functions of motor cortex and concluded from his detailed analyses that a given motor cortex neuron "can change its relationship with its neighbor in a behaviorally contingent manner … to move in and out of functionally different groups" (p. 347). What this clearly implies is that the circuits underlying integrated action sequencing are dynamically ordered, with relative connections and disconnections among participating neuronal elements becoming more and less potent with variations and even repetitions of the same action. In summary terms, the boundaries that separate system network properties are not static, but are dynamically organized, opening and closing to extrinsic influences in ways that remain poorly understood.

We would like to show why this is important, first in purely behavioral terms and then in neurobiological terms. One example given by Fentress (1976) was borrowed from von Holst's discussion of the magnet response upon fins in a fish. He pointed out that when the fish was in a restful state, each fin could move as if driven by more or less independent oscillators. However, as the amplitude and frequency of one fin's movements increased, this would tend to "draw in" the other fin, and the two fins would now act as if driven by a common oscillator. Fentress (1976) tried to illustrate this with the question, can we move the index fingers on each hand independently? Of course, we can move one finger and then we can move the other, so in that sense the answer is a trivial yes. Now suppose one tries to move each finger up and down so that the movement of one starts halfway down the cycle of the other finger. Very soon, one finds that the fingers refuse to behave, and they fall into an alternating pattern. Now speed up the movements, and they fall into phase, almost as if, rather than having independent instructions, or instructions that say move finger A up and finger B down, the system becomes still more tightly constrained into having one identical instruction for each. The two systems have now merged into one (see Kelso, 1997, for elaboration of these early concepts).

In brief, it is more difficult to perform two actions independently if the speed or intensity if one or more of these actions is performed in a rigorous manner. At this point, economy of control demands that one of the actions be blocked, or if they are sufficiently similar, be brought into tight register with one another. The question of similarity then becomes key. Whatever the details here, it is clear that there is a shifting in the controlling details of one or more of the actions. The control is dynamic, not static.

SUMMARY COMMENT ON POLYMORPHIC, DEGENERATE, AND DISTRIBUTED MOVEMENT CONTROL. As suggested above, but a point worth repeating is that, in both invertebrate and vertebrate movement control, important recent discoveries have led to the view that neuronal populations can shift their functional connections, that these connections can in that sense be degenerate, and that, especially in mammals, they are widespread in the nervous system. Excellent reviews can be found in Stein *et al.* (1997). This contrasts with earlier views that a single neuron served a single function. Thus, even in invertebrate systems, the classical distinction between, e.g., "command" and "central motor programming" neurons has now given way to an understanding that single cells can cross these classical boundaries, and that the individual neurons are less important than neuronal ensembles. In mammals, it has become especially clear that it is these neuronal ensembles that code specific movement properties, such as through a population vector averaging (Geogopolous, 1991). This is not to deny that different brain regions have different capacities, but there is often a remarkable shift in detail as to how these capacities are realized.

DEVELOPMENT OF ACTION SEQUENCE MODULES: THE MODEL AMPLIFIED

There is no doubt that both basic movements and more complex action sequences are in some way modular, indeed "informationally encapsulated" from early life (Blass, 1999). However, one must pay equal attention to the dynamic and distributed properties of movement, including central nervous system, muscular strength, joint biomechanics, and contingencies offered through the environment (Thelen & Smith, 1994). As Blass has pointed out, both narrowly and broadly defined components must be accounted for. In neurobiological terms, these are often characterized as local and global forms of processing (Edelman *et al.*, 1990). The challenge is to bring these local and global processing perspectives together. In this chapter, we have used the analogy to human speech prosody to suggest ways this might be accomplished.

Population codes, mentioned above, are one way this can be accomplished. From this perspective, modules are actually fluid and distributed ensembles of activity. These ensembles change their properties from movement to movement, and of course through developmental time. Precisely how this is done remains a mystery, although a number of quantitative models have been proposed (e.g.,

Sporns & Tononi, 1994). It is clear that during the production of movement, the actions of individual cells both coalesce and inhibit one another through connections that range from local microcircuits through reentrant connections that involve many parts of the central nervous system. The picture is vastly different from one that might be represented by simple static and highly localized "boxes." In motor development, basic sensory and motor circuits work together as the infant explores its world, and from such basic building blocks, there is a gradual emergence of complex action sequences. These action sequences are in turn products of attentional mechanisms, motivational states, and the like. In that sense, movement cannot be disconnected from other layers of behavioral control.

The position we suggest here is that it is best to view organizational processes in movement as involving *dynamic boundaries*. Movement systems can and do reconfigure themselves both developmentally and across moment-to-moment details of performance. Clear examples have been demonstrated with changes of cortical representation of muscle groups that occur, even in adult animals, through practice (e.g., Freil & Nudo, 1998). Muscle groups that are used together on skilled tasks become represented through larger populations of cortical neurons. Similar changes may well occur in other parts of the nervous system, but because of the distributed nature of movement representations they are more difficult to trace.

These dynamic properties of brain have been found for essentially all systems that have been subjected to empirical test. Perhaps the most straightforward pictures have come through studies of shifting boundaries in sensory receptive fields (Weinberger, 1995). Learning, attention, and motivation can cause these fields to expand and shrink. There appear to be two basic and complementary processes involved. The first is positive connectivity between elements that are contributing to the performance of a given task. The second, and equally important process, is a form of surround or lateral inhibition of elements that are involved with alternative or conflicting tasks. This inhibition grows proportionately with the intensity of task performance, thus providing a means for the activated circuits to persist in their activity so that the performance is not interrupted.

We offer, briefly, one final neurobiological illustration of the type of model we have developed here. As already noted, K. Berridge and his colleagues have provided many lines of evidence that the stereotyped phase of grooming "syntax" is related to activity in the dorsal lateral striatum, whereas cerebellar mechanisms are more involved with the fine tuning of individual movements (e.g., Aldridge *et al.*, 1993; Berridge & Whishaw, 1992). There is growing evidence that the striatum is involved in far more than motor performance, but contributes to action chunking and cognitive processes that are involved in action sequencing at several levels (e.g., Berns & Sejnowski, 1998; Graybiel, 1998). In an important series of studies, L. Brown and her colleagues (e.g., Brown *et al.*, 1994) have demonstrated how somatotopic input into striatum is organized *both* similarly to cortical topography and at other levels in terms of combinational maps. Further, there is a growing body of data that the details of these maps can change with performance, and several laboratories have offered recent models based both on core excitation and surround inhibition properties that are compatible with the more abstract ideas we have presented in this chapter (e.g. Mink, 1999; Rolls & Treves, 1998).

A number of years ago, Crick (1984) offered the metaphor of a thalamic "searchlight" that selects different patterns of activity at the cortical level, although he did not explicitly discuss the possibility that this "searchlight" could vary in its tightness of focus as well as directional tuning. The basal ganglia provide important links between the cortex and thalamus (and other brain regions), and as such may be importantly involved in selection, sequencing, and "focusing" of action sequence production. Thus, there appears to be a convergence of data and thought from many levels of investigation in both brain and behavioral sciences. With the dissection tools provided by developmental analyses, including those that combine genetic and epigenetic manipulations (such as now becoming available with neurological mutant animals), the future convergence of mechanistic and systems approaches to action sequencing is extremely promising.

PLASTICITY AND CONSTRAINT IN ACTION SEQUENCE ONTOGENY

As mentioned above, even in adult mammals, sensory and motor representations at the cortical level can show both short-term and long-term plasticity. There is almost certainly a scaffolding established early in development that constrains this plasticity, and thus a fundamental question is how this early, often prefunctional, scaffolding works together with the various forms of plasticity afforded by experience.

Our models of the possible roles of experience are certainly incomplete. Earlier views that experience somehow "stamps in" details from the environment, as if writing on a blank slate, are clearly inadequate. Equally inadequate are views that are extremely preformationist, thus neglecting the multitude of changes that experience can produce. In more embryological terms, experience does not just "instruct" as if providing nervous systems with a "mirror image" of the environment, but also serves a variety of permissive and inductive factors as the brain and its behavior mold themselves together into a unified and adaptive set of coexisting properties. Since the development of action sequences involves what Hebb (1949) referred to as "whole brain functions," the difficulties for full microanalyses are indeed deep. Thus, the study of action patterns at their own level with an eye both to the flexibilities and constraints during development remains essential to future progress.

It is not always obvious where these flexibilities and constraints are to be found. Thus, in human language, H. Neville has shown through a beautiful series of studies that brain regions that normally subserve auditory speech processing in subjects with normal hearing are not only specialized for different subfunctions, but also share fundamental circuitry with the processing of sign language in deaf subjects. Regions of the brain that process speech-specific auditory information in most individuals can thus be reconfigured to process language-relevant visual information when early developmental processes compromise auditory channels (e.g., Neville, 1996). Stein and Meredith (1993) indeed stressed, based upon neurophysiologic data, that cross-modal interactions are common both in early development and throughout life, and that functional combinations that cross individual modalities as individually studied in much neurophysiological work are essential for the understanding of adaptive performance (Golani, 1992; Thelen & Smith, 1994). In this chapter, we have tried to reach a balance between modular and more dynamic "supersegmental" concepts in brain and behavior, and to offer reasons to be alert to the fact that overly strict modular or global processing models can be misleading. A fundamental challenge is to bring these two views of local and global processing (Edeleman *et al.*, 1990; Sporns & Tononi, 1994) together. It is our conviction that developmental analyses may provide the essential analytical tool to do that. To the extent this is true, both behavioral science and neuroscience will benefit, and indeed, will have the opportunity to forge closer links than have been available in the past.

JOHN C.
FENTRESS AND
SIMON GADBOIS

A firm conclusion that can be derived from a wide range of individual research efforts conducted across many systems and at many different levels of organization is that we need a much deeper understanding of both the constraints and flexibilities in whole-animal behavior. In this chapter, we have tried to argue that the developmental organization of action sequences is a good place to begin. We have stressed the multidimensionality of these action sequences, where different facets of performance, and presumably development, operate by separable, but not totally separate rules. The exciting opportunity that lies ahead will be to clarify how developmental constraints, including their genetic foundations, work together with specific sources and consequences of experience at specific developmental phases.

It is promising that formal models of integrative function are now showing important parallels to models of brain development. Thus, von der Malsburg (1995) and others have argued that not only during performance, but also during development, circuits become specified through processes of local self-organizational and long-range lateral inhibition that preserve systems operations once they are initiated. This suggests that the connections among actions and among developmental pathways can at times be relatively independent, facilitatory, or inhibitory. Our hypothesis is that during both performance and development, system "boundaries" can be relatively open to extrinsic influences, but that these influences, at moderate levels, can "facilitate" rather than "dictate" ongoing predispositions (see also Fentress, 1999). At higher levels of input, or with inputs that are further "off target" from ongoing predispositions, a restructuring of performance or developmental trajectories may occur. In either event, once systems are strongly activated, our model is that they insulate themselves, for varying periods of time, from further extrinsic influences. This gives them a relative autonomy of action that allows them to pursue their course. If indeed there are similar operational principles that can be found in terms of performance and development, then one can expect much progress in the future in linking these diverse time frames, not to mention levels of organization, together into a more unified framework than exists at present.

We used the analogy of prosody in human speech to explore the richness of animal action sequences. This analogy reminds us that both segmental and supersegmental processes, defined in terms of action sequence elements, are involved in performance and ontogeny. This in turn reflects issues of relative autonomy (modularity) in action sequence properties and their underlying control mechanisms. If we are correct that this relative autonomy of action system properties can shift in time, in performance, and also in ontogenetic time, then there are clearly new areas of research to be explored. We have suggested just a sample of these areas. The dynamic framework of systems that are both interactive and self-organizing is offered as one method by which this might be done.

Finally, we wish to stress again that our view of action sequencing and its development cross over many traditional lines of inquiry. We noted briefly that action sequencing is intimately connected with both changing patterns of sensory, cognitive, and motivational processing. Thus, while we have concentrated upon the production of movement, a full appreciation of motor behavior must also incorporate findings from other fields (Kelso, 1997; Thelen & Smith, 1994). There have been recent successful attempts to combine cognition studies, perception, as well attention with the development of temperament and related issues that obviously play out in action sequencing (e.g., Spelke, Vishton, & von Hofsten, 1996; Posner & Rothbart,

1998). It is equally important to take both short-term and longer term changes in motivation into account (Blass, 1999). Thus, we are beginning to see not only use of improved analytical technologies for the dissection of action sequences, including their development, but also the break down of subdiscipline barriers that in many cases overly restrict the focus of investigation. Our analogy to prosody in the study of animal actions, including their development, is a continuation of this trend. The opportunities for future research are rich indeed.

Acknowledgments

Much of the research reported here was funded by the Medical Research Council of Canada and the Natural Sciences and Engineering Research Council of Canada. We are indebted to our many colleagues and friends, with special acknowledgment of E. Blass, V. Bolivar, T. Bond, F. Harrington, P. McLeod, W. Mosher, and J. Ryon. Finally, we deeply appreciate the patience and editorial skills of Elliott Blass. He took our early, and delayed (!), rough drafts and helped us complete the final draft through constructive comments over the course of preparation of this volume.

References

Aldridge, W. J., Berridge, K. C., Herman, M., & Zimmer, L. (1993). Neuronal coding of serial order: Syntax of grooming in the neostriatum. *Psychological Science, 4*, 391–395.

Arbib, M. A. (Ed.). (1995). *The handbook of brain theory and neural networks*. Cambridge, MA: MIT Press.

Atchley, W. R., & Hall, B. K. A model for development and evolution of complex morphological structures. *Biological Reviews, 66*, 101–137.

Bakeman, R., & Gottman, J. M. (1986). *Observing interaction: An introduction to sequential analysis*. Cambridge: Cambridge University Press.

Bateson, P., & Martin, P. (1999). *Design for a life: How behaviour develops*. London: Jonathan Cape.

Bayer, S. A., Wills, K. V., Wei, J., Feng, Y., Dlouy, S. R., Hodes, M. E., Verina, T., & Ghetti, B. (1996). Phenotypic effects of the weaver gene are evident in the embryonic cerebellum but not in the ventral midbrain. *Developmental Brain Research, 96*, 130–137.

Bekoff, A. (1992). Neuroethological approaches to the study of motor development in chicks: Achievements and challenges. *Journal of Neurobiology, 23*, 1486–1505.

Bekoff, A., & Trainer, W. (1979). The development of interlimb co-ordination during swimming in postnatal rats. *Journal of Experimental Biology, 83*, 1–11.

Berridge, K. C. (1994). The development of action patterns. In J. A. Hogan and J. J. Bolhuis (Eds.), *Causal mechanisms of behavioral development. Essays in honor of J. P. Kruijt* (pp. 147–180). Cambridge: Cambridge University Press.

Berridge, K. C., & Fentress, J. C. (1987a). Disruption of natural grooming chains after striatopallidal lesions. *Psychobiology, 15*, 336–342.

Berridge, K. C., & Fentress, J. C. (1987b). Deafferentation does not disrupt natural rules of action syntax. *Behavioural Brain Research, 23*, 69–76.

Berridge, K. C., & Whishaw, I. Q. (1992). Cortex, striatum and cerebellum: Control of serial order in a grooming sequence. *Experimental Brain Research, 90*, 275–290.

Berridge, K. C., Fentress, J. C., & Parr, H. (1987). Natural syntax rules control of action sequence of rats. *Behavioral Brain Research, 23*, 59–68.

Berns, G. S., & Sejnowski, T. J. (1998). A computational model of how the basal ganglia produce sequences. *Journal of Cognitive Neuroscience, 10*, 108–121.

Bernstein, N. (1967). *Coordination and regulation of movements*. New York: Pergamon Press.

Blass, E. M. (1999). The ontogeny of human infant face recognition: Orogusttory, visual, and social influences. In P. Rochat (Ed.), *Early social cognition* (pp. 35–65). Mahwah, NJ: Erlbaum.

Blass, E. M., & Hall, W. G. (1976). Drinking termination: Interactions among hydrational, orgastic, and behavioral controls in rats. *Psychological Review, 83*, 356–374.

Bolivar, V. J. (1996). *The development of movement patterns during swimming in the CNS myelin deficient jimpy mouse.* Ph.D. Thesis, Dalhousie University, Halifax, Nova Scotia, Canada.

Bolivar, V. J., Danilchuk, W., & Fentress, J. C. (1996). Separation of activation and pattern in grooming development of weaver mice. *Behavioral Brain Research, 75,* 49–58.

Bolivar, V. J., Manley, K., & Fentress, J. C. (1996). The development of swimming behavior in the neurological mutant weaver mouse. *Developmental Psychobiology, 29,* 123–137.

Bradley, N. S., & Smith, J. L. (1988). Neuromuscular patterns of stereotypic hindlimb behaviors in the first postnatal months. II. Stepping in spinal kittens. *Developmental Brain Research, 38,* 53–67.

Bressers, W. M. A., Kruk, M. R., van Erp, M. M., Willekens-Bramer, D. C., Haccou, P., & Meelis, B. (1995). Time structure of self-grooming in the rat: Self-facilitation and effects of hypothalamic stimulation and neuropeptides. *Behavioral Neuroscience, 109,* 955–964.

Brown, L. L., Feldman, S. M., Divac, I., Hand, P. J., & Lidsky, T. J. (1994). A distributed network of context-dependent functional units in the rat neostriatum. In G. Percheron (Ed.), *The basal ganglia IV* (pp. 215–227). New York: Plenum Press.

Churchland, P. S., & Sejnowski, T. J. (1992). *The computational brain.* Cambridge, MD: MIT Press.

Colonnese, M. T., Stallman, E. L., & Berridge, K. C. (1996). Ontogeny of action syntax in altricial and precocial rodents: Grooming sequences of rat and guinea pig pups. *Behaviour, 133,* 1165–1195.

Cordo, P., & Harnad, S. (Eds.) (1994). *Movement control.* Cambridge: Cambridge University Press.

Coscia, E. M., & Fentress, J. C. (1993). Neurological dysfunction expressed in the grooming behavior of developing weaver mutant mice. *Behavior Genetics, 23,* 533–541.

Crawley, J. N., & Paylor, R. (1997). A proposed test battery and constellations of specific behavioral paradigms to investigate the behavioral phenotypes of transgenic and knockout mice. *Hormones and Behavior, 31,* 197–211.

Crick, F. (1984). Function of the thalamic reticular complex: The searchlight hypothesis. *Proceedings of the National Academy of Sciences of the USA, 81,* 4586–4590.

Cromwell, H. C., & Berridge, K. C. (1996). Implementation of action sequences by a neostriatal site: A lesion mapping study of grooming syntax. *Journal of Neuroscience, 16,* 3067–3081.

Crystal, D. (Ed.). (1997). *The Cambridge encyclopedia of language,* 2nd ed.. Cambridge: Cambridge University Press.

Durr, A., & Brice, A. (1996). Genetics of movement disorders. *Current Opinion in Neurology, 9,* 290–297.

Edelman, G. M. (1987). *Neural Darwinism: The theory of group selection.* New York: Basic Books.

Edelman, G. M., Gall, W. E., & Cowan, W. M. (1990). *Signal and sense: Local and global order in perceptual maps.* New York: Wiley.

Fentress, J. C. (1967). Observations on the behavioral development of a hand-reared male timber wolf. *American Zoologist, 7,* 339–351.

Fentress, J. C. (1968a). Interrupted ongoing behaviour in voles (*Microtus agrestis* and *Clethrionomys britannicus*): I. Response as a function of preceding activity and the context of an apparently "irrelevant" motor pattern. *Animal Behaviour, 16,* 135–153.

Fentress, J. C. (1968b). Interrupted ongoing behaviour in voles (*Microtus agrestis* and *Clethrionomys britannicus*): II. Extended analysis of intervening motivational variables underlying fleeing and grooming activities. *Animal Behaviour, 16,* 154–167.

Fentress, J. C. (1972). Development and patterning of movement sequences in inbred mice. In J. Kiger (Ed.), *The biology of behavior* (pp. 83–132). Corvallis, OR: Oregon State University Press.

Fentress, J. C. (1976). Dynamic boundaries of patterned behavior: Interaction and self-organization. In P. P. G. Bateson and R. A. Hinde (Eds.), *Growing points in ethology* (pp. 135–160). Cambridge: Cambridge University Press.

Fentress, J. C. (1988). Expressive contexts, fine structure, and central mediation of rodent grooming. *Annals of the New York Academy of Sciences, 525,* 18–26.

Fentress, J. C. (1990). Organizational patterns in action: Local and global issues in action pattern formation. In G. M. Edelman, W. E. Gall, & W. M. Cowan (Eds.), *Signal and sense: Local and global order in perceptual maps* (pp. 357–382). New York: Wiley.

Fentress, J. C. (1992). Emergence of pattern in the development of mammalian movement sequences. *Journal of Neurobiology, 23,* 1529–1556.

Fentress, J. C. (1999). The organization of behavior revisited. *Canadian Journal of Experimental Psychology, 53,* 8–19.

Fentress, J. C., & Bolivar, V. J. (1996). Developmental aspects of movement sequences in mammals. In K. P. Ossenkopp, M. Kavaliers, & P. R. Sanberg (Eds.), *Measuring movement and locomotion: From invertebrates to humans* (pp. 95–114). New York: Chapman and Hall.

Fentress, J. C., & McLeod, P. J. (1986). Motor patterns in development. In E. M. Blass (Ed). *Handbook of behavioral neurobiology,* Volume 8, *Developmental psychology and developmental neurobiology* (pp. 35–97). New York: Plenum Press.

Fentress, J. C., & Stilwell, F. P. (1973). Grammar of a movement sequence in inbred mice. *Nature, 244,* 52–53.

Foster, E. C., Sveistrup, H., & Woollacott, M. H. (1996). Transitions in visual proprioception: A cross-sectional developmental study of the effect of visual flow on postural control. *Journal of Motor Behavior, 28,* 101–112.

Friel, K. M., & Nudo, R. J. (1998). Recovery of motor function after focal cortical injury in primates: Compensatory movement patterns used during rehabilitative training. *Somatosensory and Motor Research, 15,* 173–189.

Gadbois, S., & Fentress, J. C. (1997a). *Prosodic measures of a stereotyped movement sequence in canids: A new framework for developmental analysis.* Presented at the International Society for Developmental Psychobiology, 30th Annual Meeting, New Orleans, Louisiana [Abstract].

Gadbois, S., & Fentress, J. C. (1997b). *Canid caching sequences as a model for mammalian movement.* Presented at the Society for Neuroscience 27th Annual Meeting, New Orleans, Louisiana [Abstract].

Gadbois, S., Fentress, J. C., & Harrington. (In preparation). Comparative analysis of food-caching sequences in wolves (*Canis lupus*), coyotes (*Canis latrans*), and red foxes (*Vulpes vulpes*).

Georgopoulos, A. P. (1991). Higher order motor control. *Annual Review of Neurosciences, 14,* 361–377.

Gerlai, R. (1999). Targeting genes associated with mammalian behavior: Past mistakes and future solutions. In W. E. Crusio & R. T. Gerlai (Eds.), *Techniques in the behavioral and neural sciences,* Volume 13, *Handbook of molecular-genetic techniques for brain and behavior research* (pp. 381–392). Amsterdam: Elsevier.

Getting, P. A. (1988). Comparative analysis of invertebrate central pattern generators. In A. Cohen, S. Rossignol, & S. Grillner (Eds.), *Neural control of rhythmic movements in vertebrates* (pp. 101–127). New York: Wiley.

Gilbert, C. D. (1995). Dynamic properties of adult visual cortex. In M. S. Gazzaniga (Ed.). *The cognitive neurosciences* (pp. 73–90). Cambridge, MA: MIT Press.

Golani, I. (1992). A mobility gradient in the organization of vertebrate movement: The perception of movement through symbolic language. *Behavioral and Brain Sciences, 15,* 249–308.

Golani, I., & Fentress, J. C. (1985). Early ontogeny of face grooming in mice. *Developmental Psychobiology, 18,* 529–544.

Goldowitz, D., & Eisenman, L. M. (1992). Genetic mutations affecting murine cerebellar structure and function. In. P. Driscoll (Ed.), *Genetically defined animal models of neurobehavioral dysfunctions* (pp. 66–68). Boston: Birkhauser.

Gottlieb, G., Wahlsten, D., & Lickliter, R. (1998). The significance of biology for human development: A developmental psychobiological systems view. In R. H. Lerner (Ed.), *Handbook of child psychology,* Volume 1, *Theoretical models of human development* (pp. 223–273). New York: Wiley.

Gottman, J. M., & Roy, A. K. (1990). *Sequential analysis: A guide for behavioral researchers.* Cambridge: Cambridge University Press.

Graybiel, A. M. (1998). The basal ganglia and chunking of action repertoires. *Neurobiology of Learning and Memory, 70,* 119–136.

Graybiel, A. M., & Kimura, D. (1995). Adaptive neural networks in the basal ganglia. In J. C. Houk, J. L. Davis, & D. G. Beiger (Eds.), *Models of information processing in the basal ganglia* (pp. 103–116). Cambridge, MA: MIT Press.

Hall, W. G., & Bryan, T. E. (1980). The ontogeny of feeding in rats. II. Independent ingestive behavior. *Journal of Comparative and Physiological Psychology, 93,* 746–756.

Hayes, A. E., Davidson, M. C., Keele, S. W., & Rafal, R. D. (1997). *Toward a functional analysis of the basal ganglia.* Institute of Cognitive and Decision Sciences: Technical report, 97–01, Eugene, Oregon.

Hebb, D. O. (1949). *The organization of behavior.* New York: Wiley.

Hogan, D. (1988). Cause and function in the development of behavior systems. In E. M. Blass (Ed.), *Handbook of behavioral neurology,* Volume 8, *Developmental psychology and developmental neurobiology* (pp. 63–106). New York: Plenum Press.

Ivry, R. B. (1996). The representation of temporal information in perception and motor control. *Current Opinion in Neurobiology, 6,* 851–857.

Jackson, S., & Houghton, G. (1995). Sensorimotor selection and the basal ganglia. In J. C. Houck, J. L. Davis, & J. Beiser (Eds.), *Models of information processing in the basal ganglia* (pp. 337–368). Cambridge, MA: MIT Press.

Jacobson, D., & Anagnostopoulos, A. (1996). Internet resources for transgenic or targeted mutation research. *Trends in Genetics, 12,* 117–118.

Jusczyk, P. W., & Aslin, R. N. (1995). Infants' detection of sound patterns of words in fluent speech. *Cognitive Psychology, 29,* 1–23.

Keele, S. W., Davidson, M., & Hayes, A. (1996). *Sequential representation and the neural basis of motor skills.*

Institute of Cognitive and Decision Sciences: Technical Report 96–12, University of Oregon, Eugene, Oregon.

Kelso, J. A. S. (1997). *Dynamic patterns: The self-organization of brain and behavior*, Cambridge, MA: MIT Press.

Kleim, J. A., Swain, R. A., Armstrong, K. A., Napper, R. M. A., Jones, T. A., & Greenough, W. T. (1998). Selective synaptic plasticity with the cerebellar cortex following complex motor skill learning. *Neurobiology of Learning and Memory, 69,* 274–289.

Klintsova, A. Y., Matthews, J. T., Goodlett, C. R., Napper, R. M. A., & Greenough, W. T. (1997). Therapeutic motor training increases parallel fiber synapse number per Purkinje neuron in cerebellar cortex of rats given postnatal binge alcohol exposure: Preliminary report. *Alcohol Clinical and Experimental Research, 21,* 1257–1263.

Lashley, K. S. (1951). The problem of serial order in behavior. In L. A. Jeffress (Ed.), *Cerebral mechanisms in behavior* (pp. 112–136). New York: Wiley.

Lederhendler, I. I. (1997). Using new genetics tools to advance the behavioral neurosciences beyond nature versus nurture. *Hormones and Behavior, 31,* 186–187.

Lemon, R. N. (1990). Mapping the output functions of the motor cortex. In G. M. Edelman, W. E. Gall, & W. M Cowan (Eds.), *Signal and sense: Local and global order in perceptual maps* (pp. 315–355). New York: Wiley.

Lorenz, K. Z. (1982). *The foundations of ethology: The principal ideas and discoveries in animal behavior.* New York: Simon and Schuster.

Macdonald, D. W. (1976). Food caching by red foxes and some other carnivores. *Zeitschrift für Tierpsychologie, 42,* 170–185.

Marder, E., & Weimann, J. M. (1992). Modulatory control of multiple task processing in the stomatogastric nervous system. In J. Kien, C. McCrohan, & B. Winlow (Eds.), *Neurobiology of motor programme selection: New approaches to mechanisms of behavior choice* (pp. 3–19). Manchester: Manchester University Press.

Marsden, C. D., & Fahn, S. (Eds.) (1994). *Movement disorders* 3. London: Butterworth-Heinmann.

McLeod, P. J. (1996). Developmental changes in associations among timber wolf (*Canis lupus*) postures. *Behavioural Processes, 38,* 105–118.

McCrea, A. E., Stehouwer, D. J., & van Hartesveldt, C. (1997). Dopamine D1 and D2 antagonists block L-Dopa-induced air-stepping in decerebrate neonatal rats. *Developmental Brain Research, 100,* 130–132.

McLeod, P. J., & Fentress, J. C. (1997). Developmental changes in the sequential behavior of interacting timber wolf pups. *Behavioural Processes, 39,* 127–136.

Mink, J. W. (1999). Basal ganglia. In M. J. Zigmond, F. E. Bloom, S. C. Lanis, J. L. Roberts, & L. R. Squire (Eds.), *Fundamental neuroscience* (pp. 951–972). New York: Academic Press.

Morris, D. (1962). The behaviour of the green acouchi (*Myoprocta paratti*) with special reference to scatter hoarding. *Proceedings of the Zoological Society of London, 139,* 710–731.

Nelson, R. J. (1997). The use of genetic 'knockout' mice in behavioral endocrinology research. *Hormones and Behavior, 31,* 188–196.

Neville, H. J. (1996). Developmental specificity in neurocognitive development in humans. In M. Gazzaniga (Ed.), *The cognitive neurosciences* (pp. 219–231). Cambridge, MA: MIT Press.

Pellis, S. M., Castaneda, E., McKenna, M. M., Tran-Nguyen, L. T. L., & Whishaw, I. Q. (1993). The role of the striatum in organizing sequences of play fighting in neonatally dopamine-depleted rats. *Neuroscience Letters, 158,* 13–15.

Pham, T. M., Soderstrom, S., Winblad, B., & Mohammed, A. H. (1999). Effects of environmental enrichment on cognitive function and hippocampal NGF in non-handled rats. *Behavioural Brain Research, 103,* 63–70.

Phillips, D. P., Danilchuk, W., Ryon, J., & Fentress, J. C. (1990). Food-caching in timber wolves, and the question of rules of action syntax. *Behavioural Brain Research, 38,* 1–6.

Phillips, D. P., Ryon, J., Danilchuk, W., & Fentress, J. C. (1991). Food-caching in captive coyotes: Stereotypy of action sequence and spatial distribution of cache sites. *Canadian Journal of Psychology, 45,* 83–91.

Poizner, H., Clark, M. A., Merians, A. S., Macauley, B., Rothi, L., & Heilman, K. M. (1995). Joint coordination deficits in limb apraxia. *Brain, 118,* 227–242.

Posner, M. I., & Rothbart, M. K. (1998). Attention, self-regulation and consciousness. *Transactions of the Philosophical Society of London, 353,* 1915–1927.

Prochazka, A. (1989). Sensorimotor gain control: A basic strategy of motor systems? *Progress in Neurobiology, 33,* 281–307.

Prochazka, A., Clarac, F., Loeb, G. E., Rothwell, J. C., & Wolpaw, J. R. (2000). What do *reflex* and *voluntary* mean? Modern views on an ancient debate. *Experimental Brain Research, 130,* 417–432.

Ramus, F., & Mehler, J. (1999). Language identification with suprasegmental cues: A study based on speech resynthesis. *Journal of the Acoustical Society of America, 105,* 512–521.

Rapoport, J. (1991). Recent advances in obsessive-compulsive disorder. *Neuropsychopharmacology, 5*, 1–20.

Robinson, S. R., & Smotherman, W. P. (1992). Fundamental motor patterns of the mammalian fetus. *Journal of Neurobiology, 23*, 1574–1600.

Roffler-Tarlov, S., Martin, B., Grabiel, A. M., & Kauer, J. S. (1996). Cell death in the midbrain of the murine mutation weaver. *Journal of Neuroscience, 16*, 1819–1826.

Rolls, E. T., & Treves, A. (1998). *Neural networks and brain function.* Oxford: Oxford University Press.

Rosenbaum, D. A. (1991). *Human motor control.* San Diego, CA: Academic Press.

Sachs, B. D. (1988). The development of grooming and its expression in adult animals. *Annuals of the New York Academy of Sciences, 525*, 1–17.

Sherrington, C. S. (1906). *The integrative action of the nervous system.* New Haven, CT: Yale University Press.

Smotherman, W. P., & Robinson, S. R. (1996). The development of behavior before birth. *Developmental Psychology, 32*, 425–434.

Spelke, E. S., Vishton, P., & von Hofsten, C. (1996). Object perception, object-directed action, and physical knowledge in infancy. In M. S. Gazzaniga (Ed.), *The cognitive neurosciences* (pp. 165–179). Cambridge: MIT Press.

Sporns, O., & Tononi, G. (Eds.). (1994). *Selectionism and the brain.* San Diego, CA: Academic Press.

Stein, B. E., & Meredith, M. A. (1993). *The merging of the senses.* Cambridge, MA: MIT Press.

Stein, P. S. G., Grillner, S., Selverston, A. I., & Stuart, D. G. (Eds.). (1997). *Neurons, networks, and motor behavior.* Cambridge, MA: MIT Press.

Tepper, J. M., & Trent, F. (1993). *In vivo* studies of the postnatal development of rat neostriatal neurons. *Progress in Brain Research, 99*, 35–50.

Thach, W. T. (1996). On the specific role of the cerebellum in motor learning and cognition: Clues from PER activation and lesion studies in man. *Behavioral and Brain Sciences, 19*, 411–431.

Thelen, E., & Smith, L. B. (1994). *A dynamic systems approach to the development of cognition and action.* Cambridge, MA: MIT Press.

Thelen, E., & Ulrich, B. D. (1991). *Hidden skills: A dynamic systems analysis of treadmill stepping during the first year.* Chicago: University of Chicago Press.

Tinbergen, N. (1972). *The animal in its world,* Volume 1, *Field studies.* London: George Allen and Unwin.

Vander Wall, S. B. (1990). *Food hoarding in animals.* Chicago: University of Chicago Press.

von der Malsburg, C. (1995). Self-organization and the brain. In M. A. Arbib (Ed.), *The handbook of brain theory and neural networks* (pp. 840–843). Cambridge, MA: MIT Press.

Wayne, R. K. (1993). Molecular evolution of the dog family. *Trends in Genetics, 9*, 218–224.

Weinberger, N. M. (1995). Dynamic regulation of receptive fields and maps in the adult sensory cortex. *Annual Review of Neuroscience, 18*, 129–158.

Weiner W. J., & Lang, A. E. (1989). *Movement disorders: A comprehensive survey.* New York: Futura.

Wolpert, L. (1998). *Principles of development.* Oxford: Oxford University Press.

11

Selective Breeding for an Infantile Phenotype (Isolation Calling)

A Window on Developmental Processes

Susan A. Brunelli and Myron A. Hofer

Introduction

Since its origin more than 30 years ago, developmental psychobiology has had as its central focus the role of early experience in the development of behavior and physiology. Genetically heterogeneous strains have generally been used in order to assure that any effects of experimentally altered early environments would not be likely to depend upon a unique genetic contribution. There were early interests in genetic effects, however, and in some instances, specific inbred strains were used in order to study environmental effects on a particular vulnerability or predisposition. It did not take researchers long to realize that the early environment of most mammals and birds consists of a complex and changing interaction with its parents and siblings. To better understand the mysteries of the prenatal and preweaning environments, an array of novel methods and approaches have been created that continue to be effectively used by researchers in developmental psychobiology. Yet this focus has come at a cost. There have been few cross-disciplinary efforts with the field of behavior genetics. This emphasis on environmental over genetic contributions to development over the last decades could perhaps have been justified on the grounds that the early genetic contribution could not be as readily described mechanistically or varied experimentally. In the past few years this picture has

Susan A. Brunelli and Myron A. Hofer   Developmental Psychobiology, New York State Psychiatric Institute, and College of Physicians and Surgeons, Columbia University, New York, New York 10032

Developmental Psychobiology, Volume 13 of *Handbook of Behavioral Neurobiology*, edited by Elliott Blass, Kluwer Academic / Plenum Publishers, New York, 2001.

changed radically and so has the biological orientation toward genetic determinism. Development can now be studied by manipulating identified genes with known cellular effects at specific stages of development. Some forms of genetic therapy have already become a clinical reality. The direct manipulation of single genes is revolutionizing our understanding of how embryonic cellular migration and differentiation are regulated. The route from gene to behavior, however, is far more circuitous and the relevant genes have not yet been identified. Thus, we need an approach that allows us to work back from behavior to gene. We would like to be able to identify an early behavior of interest, to manipulate the expression of the major genes affecting that behavior at a specific point in development, and then analyze the important environmental (e.g., maternal) interactions through which that behavior develops.

In order to create a model system in which we can manipulate the expression of genes affecting a particular behavior during particular stages of development, we have used a novel modification of the selective breeding strategy in which the selected trait is an infantile rather than an adult behavior. This allows us in principle selectively to activate (or suppress) those genes that serve to enhance or diminish the level of a chosen behavior at a specified point in its development. We can thus explore developmental processes by manipulating age-specific gene expression, much in the way we have in the past explored development by altering the infants' environment at different ages. Early environmental events have long-term shaping effects on adult behavior, and we have reason to believe that the same will hold true for the early genetic events produced by selective breeding for an infantile trait. But this idea needs to be tested. This also will allow us to test some of our hypotheses about developmental continuities and discontinuities and the nature of what we refer to as developmental *pathways* or *trajectories* for behavior from infancy to adulthood. Will other traits be coselected and how will these be expressed later in development? Alternatively, the effects of selection for an infantile trait may be confined to infancy, resulting in an *ontogenetic adaptation* (Oppenheim, 1981), or may be confined to a simple motor trait that extends into adulthood without change in form or associated behaviors. Finally, changes in early gene expression induced by selection are likely to affect maternal behavior, both indirectly through the way altered infant behavior affects maternal responses and also more directly through the long-term effects of the differential early gene expression on the maternal behavior of adults in subsequent generations. We will present early evidence suggesting some answers to these basic questions about the mechanisms of development.

The results of artificial breeding for desirable adult traits in domesticated species played a central role in Darwin's (1859/1979) *Origin of Species*, but such rapid phenotypic changes were not thought to take place in complex habitats under conditions of natural selection until recently, when clear evidence has been gathered in vertebrate species that evolution of traits can occur within a few generations, as it does in laboratory selective breeding programs. In continuous and extensive ecological studies on finches in the Galapagos Islands over the past two decades, beak size and shape have shown systematic evolutionary changes within 1- to 2-year periods, for example in response to severe drought, necessitating a shift in feeding technique from small, soft seeds to large seeds in hard shells (Grant, 1986, pp. 184–189). In another recent study, natural populations of guppies in Trinidad were found to show rapid adaptive changes consisting of maturity at an earlier age and smaller size, more frequent litter production, and the distribution of resources to smaller offspring when environments were changed from low to high predation rates (Reznick, Shaw, Rodd, & Shaw, 1997). These differences evolved in 4–7 years and persisted when the fish were reared in a common laboratory environment. The authors calculate that these relative rates of evolution are comparable to those found with artificial selection and are 4 to 7 orders of magnitude greater than those derived from the fossil record. This increases the likelihood that the physiological, neuro-anatomic, and genetic bases for the altered traits in our study will resemble extreme variants that occur naturally and thus may resemble those individual humans diagnosed with clinical conditions that we would like to model.

In this chapter, we will explore the themes outlined above in more detail and apply this thinking to the particular selective breeding study that we have begun. First, we will describe the particular infantile behavior we have chosen, the young rat pup's calling rate when isolated from mother and siblings and familiar odors, and our reasons for viewing it as the expression of an early anxiety state. Then we will describe the general principles of selective breeding, its relation to the aims of animal model research, and the new methods now available for mapping the specific genetic loci affected. In the next section, we will explore the most novel aspect of our study, the use of selective breeding for an infantile phenotype to elucidate developmental processes, the role of early experience, and the relationship of development to evolution, in particular through the principle of heterochrony. Finally, we will describe the results of our study thus far and a perspective on future directions.

Ultrasonic Vocalizations in Infant Rats: Background

Ethology and Pharmacology of Ultrasonic Vocalization. The most consistent response of an infant that occurs *immediately* upon separation from the mother is the *separation cry* (MacLean, 1985), and consists of repeated vocalizations and diffusely agitated behavior.* This initial behavioral response to separation has been called *protest*, and is characteristic of most mammalian species, including humans (Bowlby, 1969). In rat pups, this early vocal response to separation consists of a series of high-frequency (40–70 kHz) short calls, or ultrasonic vocalizations (USV), and in experimental settings occur in isolation from mother and littermates in a novel environment. USVs also can be elicited throughout the preweaning period by stimuli such as cold (Allin & Banks, 1971; Okon, 1970a, 1971), loss of equilibrium, or rough handling (Bell, Nitschke, Gorry, & Zachman, 1971; Okon, 1970b) as well as by unfamiliar terrain and separation from familiar social companions (Noirot, 1972; Oswalt & Meier, 1975). It was for the latter reason that ultrasonic vocalizations in response to separation came to be called *distress* vocalizations by early researchers in the field (Hart & King, 1966; Nitschke, Bell, Bell, & Zachman, 1972; Smotherman, Bell, Starzec, Elias, & Zachman, 1974).

Research in many laboratories has documented that infants of a variety of species experience isolation as a stressful event, for example, as measured by central and peripheral hypothalamic–pituitary–adrenal (HPA) axis indices such as an increase in plasma cortisone or corticosterone (e.g., Gunnar, Larson, Hertsgaard, Harris, & Brodersen, 1992; Hennessy, 1997; Larson, Gunnar, & Hertsgaard, 1991; McCormick, Kehoe, & Kovacs, 1998; Stanton, Wallstrom, & Levine, 1986; Stanton, Gutierrez, & Levine, 1988; Takahashi, 1992a). Other systems involved in mediation of stress are also implicated, such as the central dopamine (Kehoe, Clash, Skipsey, & Shoemaker, 1996), and opioid systems (Blass & Brunson, 1990; Blass & Fitzgerald,

*MacLean, 1985, p. 405, Bowlby, 1969, p. 27.

1988; Blass & Kehoe, 1987; Kehoe & Boylan, 1994; Shoemaker & Kehoe, 1995), noradrenergic activity (Harvey, Moore, Lucot, & Hennessy, 1994; Kehoe & Harris, 1989), and immune function (Coe & Ericson, 1997; Coe, Rosenberg, & Levine, 1988).

Nearly two decades of pharmacologic studies have connected the USV response in rat pups *immediately* after maternal separation and isolation to the larger field of research on anxiety disorders (but also see the chapter by Blumberg in this volume for long-term changes in the regulation of other systems related to USV over time). In these pharmacologic studies, it has been USV rate, rather than other behaviors seen in isolation, that has been the most sensitive and consistent response measure, and is often the only measure affected by low doses of these compounds. Research in other species (Harvey *et al.*, 1994; Hennessy & Moorman, 1989; Hennessy & Weinberg, 1990) including primates (Hennessey, 1986; Newman, 1991) demonstrates remarkable cross-species similarities in the neural mechanisms for isolation distress calling, suggesting that they have been conserved in evolution (Newman, 1992; Scherer, 1985). Thus, investigators have viewed isolation-induced USV as "an ethologically relevant behavior" to model aspects of separation anxiety and the neurochemistry of anxiety (Brain, Kusumorini, & Benton, 1991; Hofer, 1996; Insel & Winslow, 1991b; Miczek, Tornatzky, & Vivian, 1991; Panksepp, 1998; Winslow & Insel, 1991a). MacLean (1985) and Hofer (1996) hypothesized that central modulation of USV involves systems and sites mediating fear and responsiveness to stress (Graeff, 1994).

Table 1 provides a summary of reciprocal interactions in opposing central neurochemical systems implicated in anxiety, based on pup USV responses to specific agonists, antagonists, and reuptake inhibitors. Briefly, those compounds that facilitate the actions of monoamine or excitatory systems like the N-methyl-D-aspartate (NMDA)–glycine receptor complex are implicated in increased production of USV, whereas compounds that increase efficacy of dampening systems like the benzodiazepine (BZ)–γ-aminobutyric acid (GABA) receptor complex are involved in decreasing USV rates. Receptor agonist and antagonist interactions within systems (e.g., Table 1, benzodiazepine system) show similar bidirectional control in modulating USV, and these interactions occur in lawful ways that would be predicted by animal models characterizing anxiety in adulthood (e.g., BZ–GABA compounds and serotonin compounds, D. C. Blanchard, Blanchard, Tom, & Rodgers, 1990; D. C. Blanchard, Griebel, Rodgers, & Blanchard, 1998; imipramine, BZ and serotonin compounds, R. J. Blanchard, Shepherd, Rodgers, Magee, & Blanchard, 1993; BZ, R. J. Blanchard, Griebel, Henrie, & Blanchard, 1997; File, 1992). As Table 1 shows, compounds acting as agonists at the BZ receptor invariably decrease USV rates in separated pups, whereas the BZ-receptor inverse agonists (Nutt, 1988) increase USV rates. The BZ receptor antagonist RO15-1788 reverses both agonist and antagonist actions, typical of that system (Nutt, 1988). In the serotonin (5-HT) system, of which there are multiple receptor types, 5-HT reuptake inhibitors decrease USV rates of separated pups and 5-HT reuptake enhancers increase USV rates, in accord with human clinical findings. Table 1 also shows that 5-HT$_{1A}$ agonists decrease USV and antagonists increase USV, but 5-HT$_{1A}$ effects on USV are also opposed by agonists acting at 5-HT$_{1B}$ and some 5-HT$_2$ receptor agonists, particularly those that primarily bind to 5-HT$_{1A}$ sites; 5-HT$_{2A}$ receptor agonists generally show an anxiolytic profile.

Unlike the benzodiazepine system, which has been most carefully characterized in the USV model, other studies on the serotonin and dopamine systems illustrate the fact that not all subsystems or receptor types of neurochemical systems may be involved in the regulation of USV, or that at present, their role is not fully under-

TABLE 1. EFFECTS ON 3- TO 14-DAY-OLD RAT PUP USV OF COMPOUNDS EXERTING ANXIOLYTIC/ANXIOGENIC EFFECTS IN HUMANS OR ANIMAL MODELS[a]

Receptor system	Drug	Isolated USV	Reference
Opiate			
κ-Agonist	U50,488	↑	Barr, Wang, & Carden, 1994; Carden, Barr, & Hofer, 1991; Carden, Davachi, & Hofer, 1994
κ-Antagonist	nor-Binaltorphenime	↓	S. E. Carden, unpublished
μ-Agonist	Morphine	↓	Blass & Kehoe, 1987; Barr & Wang, 1992; Carden & Hofer, 1990a; Insel, Miller, & Hill, 1988
μ-Agonist	DAMGO-enkephalin	↓	Carden, Barr, & Hofer, 1991
μ-Antagonist	NAL	↑	Barr & Wang, 1993; Blass & Fitzgerald, 1988; Blass & Kehoe, 1987
	NAL	0	Carden & Hofer, 1991
	Naloxone	0	Carden & Hofer, 1991
δ-Agonist	DPDPE-enkaphalin	↓	Carden, Barr, & Hofer, 1991
BZ-GABA_A			
BZ agonist	CDP	↓	Carden & Hofer, 1990a; Olivier, Molewijk, van der Heyden, et al., 1998
	Alprazolam	↓	Olivier, Molewijk, van der Heyden, et al., 1998; Vivian, Barros, Manitiu, & Miczek, 1997
	CGS 9896	↓	Gardner, 1985; Gardner & Budhram, 1987
	DZ	↓	Insel, Hill, & Mayor, 1986; Naito, Nakamura, & Inoue, 1998; Olivier, Molewijk, van der Heyden, et al., 1998; Vivian et al., 1997
	Oxazepam	↓	Olivier, Molewijk, van der Heyden, et al., 1998
	Premazepam	↓	Insel et al., 1986
	Suriclone	↓	Gardner & Budhram, 1987
	ZK91296	↓	Gardner & Budhram, 1987
	Zolpidem	↓	Olivier, Molewijk, van der Heyden, et al., 1998
	RO16-6028	↓	Naito, Nakamura, & Inoue, 1998
	RO23-0364	↓	Naito, Nakamura, & Inoue, 1998
BZ antagonist	Flumazenil alone	0	Insel, Hill, & Mayor, 1986
	(RO15-1788)	0	Olivier, Molewijk, van der Heyden, et al., 1998
	Flumazenil + agonist	↑	Vivian et al., 1997
BZ inverse agonist	FG-7142	↑	Insel et al., 1986
		0	Olivier, Molewijk, van der Heyden, et al., 1998
	PTZ	↑	Insel et al., 1986;
		0	Thom, 1986
GABA_Aβ site agonist	Pentobarbital	↓	Insel, Miller, Gerhard, & Hill, 1988; Vivian et al., 1997
GABA_A steroid site	Allopregnanolone (3α-hydroxy-5α-pregnan-20-one)	↓	Vivian et al., 1997; Zimmerberg, Brunelli, & Hofer, 1994

(*continued*)

TABLE 1. (*CONTINUED*)

Receptor system	Drug	Isolated USV	Reference
NMDA-glycine-glutamate complex			
Agonist	Glycine	0	Winslow, Insel, Trullas, & Skolnick, 1990
	NMDA	↑	Winslow et al., 1990
Glycine partial agonist	ACPC	↓	Winslow et al., 1990
Glycine antagonist	MDL 100,453	↓	Kehne et al., 1991
Glycine antagonist	7-CL-KYN	↓	Winslow et al., 1990
NMDA antagonist	AP7	↓	Winslow et al., 1990
NMDA antagonist	AP5	↓	Kehne et al., 1991
Chloride channel blocker	5,7-DCKA	↓	Kehne et al., 1991
	MK-801	↓	Kehne et al., 1991
	Ifenprodil	↓	Insel & Winslow, 1991b
5-HT receptors			
5-HT reuptake inhibitors	Clomiprimine	↓ 0	Winslow & Insel, 1990; Olivier, Molewijk, van der Heyden, et al., 1998
	Paroxetine	↓	Olivier, Molewijk, van der Heyden, et al., 1998
	Citalopram	↓	Olivier, Molewijk, van der Heyden, et al., 1998
	Fluvoxamine	↓	Olivier, Molewijk, van der Heyden, et al., 1998
5-HT/NA reuptake inhibitor	Imipramine	↓ then ↑ with dosage	Olivier, Molewijk, van der Heyden, et al., 1998
5-HT uptake enhancers	Tianeptine	↓	Olivier, Molewijk, van der Heyden, et al., 1998
$5-HT_1$ nonselective agonist	d,l-Propranolol	0	Winslow & Insel, 1991b
$5-HT_{1A}$ agonists	8-OH-DPAT	↓	Härd & Engel, 1988; Joyce & Carden, 1999; Olivier, Molewijk, van der Heyden, et al., 1998; Winslow & Insel, 1991b
	Flesinoxan	↓	Olivier, Molewijk, van der Heyden, et al., 1998
	Ipsapirone	↓	Olivier, Molewijk, van der Heyden, et al., 1998
	NAN-190	↓	Olivier, Molewijk, van der Heyden, et al., 1998
$5-HT_{1A}$ partial agonists	Buspirone	↓	Albinsson, Bjork, Svartengren, Klint, & Anderson, 1994; Kehne et al., 1991; Olivier, Molewijk, van der Heyden, et al., 1998; Winslow & Insel, 1991b
	BMY 7378	↓	Olivier, Molewijk, van der Heyden, et al., 1998
	MDL 73,005EF	↓	Kehne et al., 1991
$5-HT_{1A}/5-HT_{1B}$ agonist	FG5893	↓	Albinsson et al., 1994
$5-HT_{1A}$ antagonists	WAY 100	0	Olivier, Molewijk, van der Heyden, et al., 1998
	S-UH-301	↓	Olivier, Molewijk, van der Heyden, et al., 1998
	DU125530	0	Olivier, Molewijk, van der Heyden, et al., 1998

TABLE 1. (*Continued*)

439

SELECTIVE
BREEDING FOR
AN INFANTILE
PHENOTYPE

Receptor system	Drug	Isolated USV	Reference
5-HT$_{1A}$ antagonist/β-adrenergic agonist	(+/−)-pindolol	0	Albinsson *et al.*, 1994; Joyce &
		0	Carden, 1999
5-HT$_{1B/2C/Weak\ 1A}$ agonists	TFMPP	↑ then ↓ with dosage	Insel & Winslow, 1991b; Olivier, Molewijk, van der Heyden, *et al.*, 1998
	CSG12066B	↑ then ↓ with dosage	Winslow & Insel, 1991b
5-HT$_{2A}$/5-HT$_{2C}$ agonists	m-CPP	↓	Winslow & Insel, 1991b
	DOI	↓	Winslow & Insel, 1991b
5-HT$_{2A}$ antagonist	MDL 11,939	0	Kehne *et al.*, 1991
5-HT$_{2A/2C}$ antagonist	Ketanserin	↑ ↓, U-shaped from low to high doses	Olivier, Molewijk, van der Heyden, *et al.*, 1998
5-HT$_{2C/2A}$ antagonist	Ritanserin	↑	Winslow & Insel, 1991b
5-HT$_3$ agonist	Phenylbiguanide	0	Olivier, Molewijk, van der Heyden, *et al.*, 1998
5-HT$_3$ antagonists	Odansetron	0	Olivier, Molewijk, van der Heyden, *et al.*, 1998
	MDL 73147EF	0	Kehne *et al.*, 1991
NA (noradrenergic)			
NA uptake inhibitor	Amitriptyline	0	Kehne *et al.*, 1991
NAα$_2$ agonist	Clonidine	↑ prior to 20 days; 0 at 20 days	Härd, Engel, & Lindh, 1988; Kehoe & Harris, 1989
NAα$_2$ antagonist	Idazoxan	↓	Härd *et al.*, 1988
	Yohimbine	↓ prior to 20 days; 0 at 20 days	Kehoe & Harris, 1989
NAα$_1$ antagonist	Prazosin	0	Härd *et al.*, 1988
DA (dopaminergic)			
DA agonist (indirect agonist at DA, 5-HT, NE receptors)	Cocaine (acute)	↑	Barr & Wang, 1993
	Cocaine (chronic withdrawal)	↓	Barr & Wang, 1993
	D-Amphetamine	0	Kehne *et al.*, 1991
DA synthesis inhibitor	a-MT	↑	Härd & Engel, 1991
DA$_1$ antagonist	SCH 23390 (chronic)	0	Cuomo, Cagiano, Renna, De Salvia, & Racagni, 1987
DA$_2$ agonist	Sulpiride	↓	Cuomo *et al.*, 1987
DA$_2$-like antagonist	Haloperidol (acute)	0	Kehne *et al.*, 1991
	Haloperidol (chronic)	↓	Cagiano *et al.*, 1986
CCK			
CCK$_{A,B}$ agonist	CCK-8S	↓	Weller & Blass, 1988
	CCK-8SNS	0	
	CCK-8S	↓	Rex, Barth, Vagged, Daemon, & Fink, 1994
CCK$_B$ agonist	BOC-CCK-4	↑	Rex *et al.*, 1994
CCK$_A$ antagonist	Devazepide	↑	Blass & Shide, 1993; Weller & Gispan, 2000
OXT (oxytocin)			
Agonist	OXT (central)	↓	Insel & Winslow, 1990
Antagonist	OTA	↑	Insel & Winslow, 1990

(*continued*)

TABLE 1. (*CONTINUED*)

Receptor system	Drug	Isolated USV	Reference
AVP			
AVP agonist	AVP	↓ with ↑ dose	Winslow & Insel, 1993
	AVT (arginine vasotocin)	↓ with ↑ dose	Winslow & Insel, 1993
Vasopressin precursor	Hydrin 1	0	Winslow & Insel, 1993
V_1 antagonist	V_1A	0	Winslow & Insel, 1993
V_2 antagonist	V_2A	0	Winslow & Insel, 1993
CRF			
Agonist	CRF (central)	↓	Insel & Harbaugh, 1989
	CRF (central)	↑ then ↓ with higher doses	Harvey & Hennessy, 1995
	CRF (peripheral)	0	Harvey & Hennessy, 1995
Antagonist	α-Helical CRF (central)	↓	Insel & Harbaugh, 1989; Harvey & Hennessy, 1995
CRF_1 antagonist	CP-154,526	↓	Coverdale, McCloskey, Cassell, & Kehne, 1998
Cannabinoid receptor			
Agonist	CP-55,940	↓	McGregor, Dastur, McLellan, & Brown, 1996
Antagonist	SR 141716A	↓	McGregor *et al.*, 1996

[a]Abbreviations: ACPC, 1-Aminocyclopropane-carboxylic acid; AP5, D,L-aminophosphonovaleric acid; AP7, 2-amino-7-phosphonoheptanoic acid; AVP, arginine vasopressin; BOC-CCK-4, butyl-oxycarbonyl-CCK-4; BZ, benzodiazepine; CCK, cholecystokinin; CDP, chlordiazepoxide; 7-CL-KYN, 7-chlorokynurenic acid; CP 55940, (1-) *cis*-3-[2-hydroxy-4-(1,1-dimethylheptyl)phenyl]-*trans*-4-(3-hydroxypropyl)cyclehexanol; m-CPP, (1-93-chlorophenyl)piperazine; CRF, corticotropin-releasing factor; CSG12066B, 7-trifluoromethyl-4(4-methyl-piperazinyl)-pyrrolo[1,2a]quinoxaline; DA, dopaminergic; DAMGO, [D-Ala²-Nme-Phe⁴-Gly-ol]-enkephalin; 5,7-DCKA, 5,7-dichlorokynurenic acid; DOI, (±)-1-(2,5-dimethoxy-4-iodophenyl)-2-aminopropane; 8-OH-DPAT, (±)-8-hydroxy-2-(di-*n*-propylamino)tetralin; DPDPE, [D-Pen²,D-Pen²]-enkephalin; DZ, diazepam; FG-7142, *N*-methyl-beta-caboline-3-carboxamide; GABA, gamma-aminobutyric acid; 5-HT, 5-hydroxytriptamine, serotonin; hydrin-1, OXT-Gly-Lys-Arg; MDL 11,939, alpha-phenyl-1-(2-phenylethyl)-4-piperidine-methanol; MDL 73,005EF, 8-[2-(2,3-dihydro-1,4-benzodioxin-2-yl-methylamino)ethyl]-8-azaspiro[4,5]decane-7,9-dione methyl sulfonate; MDL 72,147EF, 1H-indole-3-carcoxylic acid-*trans*-octahydro-3-oxo-2,6-methano-2H-quinolizin-8-yl-ester methane sulfonate; a-MT, a-methyltyrosine; NAL, naltrexone; NMDA, *N*-methyl-D-aspartate; OTA, d(CH₂)₅ [Tyr(Me)⁹ 2,Thr⁴-NH⁹ 2,]OVT; PTZ, pentylenetetrazol; SCH 23390, (+)-R-8-chloro-2,3,4,5-tetrahydro 3-methyl-5-phenyl-1H-3-benzazepine-7-ol; SR 141716A, *N*-piperidine-1-yl-5-(4-chlorophenyl)-1-(2,4-dichloropheny)-4-(methyl-1H-pyrazole-3-carboxamide hydrochloride; TFMPP, 1-[3-fluoromethylphenyl]-piperazine; V₁A, d(CH₂)₅[Tyr-Me]AVP; V₂A, [d(CH₂)₅,D-Phe₂,Ile₄,Ala₉-NH₂]AVP.

stood. This is true for other models of anxiety in the rat, such as the plus maze, the open field, aversive conditioning paradigms, etc., in which some systems are more involved in mediating a given behavior than others (see File, 1992, for review). It is equally true in the treatment of human anxiety syndromes (Pine & Grun, 1999). This speaks to the heterogeneity of what we call "anxiety" and to the paradigms which purport to measure it in humans and animals (File, 1992). Anxiety is not a unitary construct (Archer, 1973; Belzung & Le Pape, 1994; Denenberg, 1969; File, 1992, 1995; Flaherty, Greenwood, Martin, & Leszczuk, 1998; Maier, Vandenhoff, & Crowne, 1988), and not all systems mediating anxiety in other organisms or in other contexts would be expected to mediate isolation-induced USV.

In the opiate system, 0.5–5-mg/kg doses (intraperitoneally, IP) of naltrexone had no effect on pup USV rates at any dose when pups were isolated in bare cages (Carden & Hofer, 1991), but the higher doses increased USV rates when the pup was in contact with the dam or littermates (Carden & Hofer 1990b,c). Kehoe and Blass

(1986a, b) found an effect of naloxone on calling rates with rather low doses, but in these studies, pups were tested in the presence of bedding, which reduces isolation calling rates because of familiarity. These findings can be interpreted to mean that whereas there was no intrinsic mu opiate receptor mediation of calling, when the pups' USV rates are reduced by home cage bedding or familiar conspecifics, these environmental effects may be mu receptor-mediated. Developmental effects (maturation of systems) may also be present: the alpha-2 noradrenergic agonist clonidine increases USV rates, whereas alpha-2 antagonists such as idazoxan and yohimbine decrease USV rates in infant rats in the first 2 postnatal weeks (Härd, Engel, & Lindh, 1988; Kehne et al., 1991; Kehoe & Harris, 1989). These effects are directly contrary to human clinical findings on adult anxiety and may well reflect the immaturity of presynaptic alpha-2 receptors in infancy. Examination of noradrenergic drug effects on USV later in development, however, has shown that an adult-type pattern of clonidine effects on USV may mature at 17–20 days postnatally, accounting for the discrepancies (Kehoe & Harris, 1989). Corticotropin-releasing hormone (CRH) effects on infant USV also appear to be paradoxical: CRH, known to mobilize responses to stressful stimuli in adult rats and other species (e.g., Arborelius, Owens, Plotsky, & Nemeroff, 1999), suppresses USV, whereas its antagonists, such as alpha-helical CRH, reverse that suppression (Harvey & Hennessy, 1995; Insel & Harbaugh, 1989). However, it has been shown (Brunelli, Masmela, Shair, & Hofer, 1998; Shair, Masmela, Brunelli, & Hofer, 1997; Takahashi, 1992a, b) that isolated rat pups will cease all vocalization in the presence of an adult male, a potential predator (Brown, 1986; Mennella & Moltz, 1988; Paul & Kuperschmidt, 1975), and that this is associated with increases in adrenocorticotropic hormone (ACTH) (and thus hypothalamic CRH release) over isolated controls (Takahashi, 1992a). It is likely, therefore, that central CRH plays a dose- and region-dependent role in modulating USV. Indeed, moderate CRH doses increase USV rates, whereas higher doses decrease USV rates (e.g., Hennessy et al., 1992). At high doses, CRH effects on USV may reflect a central affective state more akin to "fear" than "anxiety" (for discussions of neurobiology of defensive behaviors and related emotions, see D. C. Blanchard & Blanchard, 1988; Graeff, 1994).

In summary, benzodiazepine–GABA, a subset of opiate receptor agonists, serotonergic agonists, and reuptake inhibitors that are effective in treating human anxiety disorders (Coplan, Gorman, & Klein, 1992), and that show anxiolytic actions in many adult animal models of anxiety, reduce rat pup isolation distress vocalizations. In all, the large body of pharmacologic data provide impressive evidence that isolation-induced rat pup USV is indeed a valid model for early separation anxiety. These compounds also provide insight into the neuropharmacologic systems mediating USV, indicating the involvement of multiple neurochemical regulators on USV (Hofer, 1996).

NEUROANATOMY OF ULTRASONIC VOCALIZATIONS. Although relatively little is known about the neuroanatomy of ultrasonic vocalizations in rat pups, detailed neuroanatomic studies of both ultrasonic and audible vocalizations in adult rats, cats, dogs, and monkeys provide a general model likely to apply to infants as well. Ultrasonic vocalizations are produced by a whistle-like mechanism due to turbulence in the flow of air through the larynx (Roberts, 1975). In infant and adult rats, two divisions of the vagus nerve, the superior and inferior laryngeal nerves, innervate the laryngeal muscles; the inferior laryngeal nerve is necessary for the production of USV, as either unilateral or bilateral transection of the nerve eliminates ultrasounds, and the superior laryngeal nerve modulates production of USV via its connections with the cricothyroid muscle (Roberts, 1975; Wetzel, Kelley, & Campbell, 1980). Anatomic studies applying retrograde horseradish peroxidase have established that efferent connections to these nerves originate in the nucleus ambiguus in the rat medullary region, with the inferior laryngeal nerve arising from its dorsal formation and the superior laryngeal nerve from its ventral formation (Hinrichsen & Ryan, 1981; Wetzel *et al.*, 1980). Besides controlling vocal fold muscles, motor neurons in the nucleus ambiguous make connections with thoracic and upper lumbar ventral horn motor neurons controlling respiration (Jürgens, 1994). USV-related and expiratory-related units discharge with tonic bursts prior to and early into ultrasounds, and inspiration-related types are suppressed during ultrasound emission (Yajima & Hayashi, 1983). The nucleus retroambigualis, another medullary structure adjacent to the nucleus ambiguus, is thought to be the primary efferent pathway involved in respiratory control of vocalization (Jürgens, 1998; Zhang, Bandler, & Davis, 1995). Both of these structures receive afferent information from laryngeal and pulmonary proprioceptors via the solitary nucleus tract (Jürgens, 1998).

These motor nuclei receive massive afferent connections from the periaqueductal gray (PAG), a midbrain area implicated in both afferent and efferent control of vocalization (Jürgens, 1994). Cells in the PAG itself are active during vocalizations corresponding to activity in laryngeal and abdominal muscles, indicating direct, reciprocal projections (Larson & Kistler, 1986). Work in a variety of species including rats, bats, cats, guinea pigs, squirrel monkeys, chimpanzees, gibbons, and humans has shown that electrical or chemical (e.g., D,L-homocysteic acid, DLH) stimulation of various areas of the PAG produce naturalistic, species-typical vocalizations, generally of agonistic types; although these vocalizations are associated with PAG stimulation, they are not specifically correlated with emotional reactions (Jürgens, 1994; Zhang, Davis, Bandler, & Carrive, 1994). Large lesions of the PAG result in mutism (cat, dog, and rat), and partial lesioning affects different call types, depending upon the PAG area lesioned (Jürgens, 1994, 1998). The PAG receives both excitatory (glutamate, NMDA) and inhibitory (GABA-A, glycine, and opioid) afferent connections from limbic and forebrain structures (Beart, Summers, Stephenson, Cook, & Christie, 1990; Jürgens & Lu, 1993), and destruction of PAG eliminates stimulation-induced vocalizations from these same structures (Jürgens & Pratt, 1979). Moreover, microinjections of DLH in PAG produce firing in PAG cells which evoke coordinated excitatory and inhibitory patterns in muscle groups to produce stereotypical patterns of vocalizations in cats (e.g., howl/mew/growl) in dose-dependent patterns. Thus, recordings of respiratory rate, integrated voice, tracheal, and arterial pressure, and electromyographic (EMG) signals corresponding to cricothyroid, digastric external oblique, and diaphragm muscles all show coordinated changes in excitation patterns evoked by DLH injection in PAG (Zhang *et al.*, 1994). Thus, the PAG serves to integrate respiratory and laryngeal motor patterns involved in vocalization with sensory feedback systems (Zhang *et al.*, 1994) and this suggests that sound production circuitry is utilized by higher cortical and subcortical centers to produce respiratory and laryngeal motor patterns characteristic of different classes of vocalizations (Jürgens, 1994, 1998).

Cortical and limbic structures from which vocalizations may be elicited are those involved in affective responses: the nucleus accumbens, preoptic area, hypothalamus, midline thalamus, septum, medial amygdala, and stria terminalis (Jürgens, 1994, 1998). In these structures, vocalizations are strikingly correlated with species-typical aversive and hedonic emotional states as measured by avoidance behaviors or self-stimulation tests (Jürgens, 1998), suggesting that these structures exert primarily affective control of vocalizations. In addition, work with adult rats has implicated the dorsal hippocampus in emission of 22-kHz ultrasonic vocalizations: excitatory synaptic activity recorded in dorsal hippocampus parallels 22-kHz USV in response to foot shock, whereas pretreatment with the serotonin 1A agonist ipsa-pirone reverses these effects, suggesting that serotonin neurons originating in the dorsal raphe nucleus exert indirect control over both functions (Xu, Anwyl, De Vry, & Rowan, 1997). The anterior cingulate cortex, another forebrain structure that produces vocalizations, has been particularly implicated in isolation calls in squirrel monkeys (MacLean & Newman, 1988). Because of its connections with cortical and other forebrain structures, the cingulate cortex appears to be uniquely poised to mediate between affective and cognitive processes (Devinksy, Morell, & Voight, 1995); indeed, one of its functions is to produce volitional initiation of vocalizations (Jürgens, 1998), in production of vocalizations associated with internal states, and assessments of motivational and emotional valence to external stimuli in humans (Devinsky *et al.*, 1995). The cingulate cortex also has connections to the PAG, and PAG lesions will disrupt cingulate-stimulated calling (Jürgens & Pratt, 1979).

In summary, vocalizations, including ultrasonic vocalizations, are the product of a series of interactions of structures in the central nervous system in which afferent information about affective and internal states is relayed via PAG from forebrain and limbic structures with information going to and from brainstem, spinal, and other nuclei controlling nerves innervating muscle groups associated with vocalization. This is a simplistic view, in that more rostral brain structures like amygdala, cingulate cortex, septum, and the PAG itself have direct and multiple connections with brainstem nuclei as well as muscle groups immediately involved in vocal and respira-tory processes. Nevertheless, these interactions can be viewed as hierarchical, in that the direction of effects is from higher central control to lower structures, with the PAG acting as a central integrator of afferent and efferent signals to produce coordinated vocalizations.

GENETIC STUDIES OF ULTRASONIC VOCALIZATION IN RODENTS.   Given that fami-lies of genes code for different receptor types within a single system (Chua, 1997), the multiplicity of neurochemical regulators of USV implies that a number of genes could be involved in the production of USV in a polygenic system. Alternatively, one or two major genes could have multiple (pleiotropic) effects. The following genetic studies have examined a genetic basis for individual differences in USV rates.

Behavior–genetic studies have attempted to approximate gene influences on infant USVs in mice, using statistical modeling and cross-breeding techniques. In one of the first, Whitney, Nyby, Cable, and Dizinno (1978), studying cold-induced USVs in one strain of mice, reported that a very small number of genes affected USV rates and that these showed directional dominance, which means, by definition, that for any given pair of alleles on each gene, the dominant allele was expressed over the other (Snustad, Simmons, & Jenkins, 1997, pp. 63, 775). Thus, because dominant alleles act to increase USV rates and nondominant alleles act to decrease USV rates less powerfully, this results in higher average rates of USV in mice than if these alleles exerted additive effects (Snustad *et al.*, 1997). In subsequent studies, Hahn, Hewitt, Adams, and Tully (1987) and Hahn, Hewitt, Schanz, Weinreb, and Henry (1997) tested mice in cold (10°C), surveying genetic influences on additional parameters of USV: rate of calling, beginning and ending frequencies (in kHz) of calls, minimum versus maximum frequencies (in kHz) of calls, and on length of calls. Like Whitney *et al.* (1978), they described the effects of dominant alleles on increasing rates of USV. But there were also weak additive effects of some alleles on most parameters of USV: that is, in addition to genes showing a dominant–recessive relationship, some pairs of alleles were contributing equally and incrementally to the phenotype (Snustad *et al.*, 1997, p. 680). Because of strong dominance for high rates, Hahn *et al.* (1997) suggested that USV appears, for the most part, to be under strong directional selection for high rates under the conditions at which they tested. Roubertoux, Martin, *et al.* (1996) reported both significant dominance and additive components contributing to USV as well as evidence for interactions between genes (epistasis, meaning that two separate genes were exerting their effects on the phenotype, but necessarily in concert with one another; Snustad *et al.*, 1997, pp. 72–74, 775). Their results were consistent with the interpretation that USV was due both to multiple, independent sets of genes (polygenic inheritance; Snustand *et al.*, 1997, pp. 670–672) and to multiple forms of alleles maintained in mouse populations (polymorphisms; Snustad *et al.*, 1997, pp. 63, 627). As suggested earlier, this would be consistent with there being multiple regulators of infant USV (Hofer, 1996).

SELECTIVE BREEDING IN ANIMAL MODELS

Selective breeding is a laboratory strategy in which animals are bred for the purpose of altering the frequencies of genes underlying a particular phenotype or observable characteristic (e.g., USV isolation response) in a population by breeding for opposite extremes of that character. Over generations of breeding the average frequency of a phenotype or observable characteristic in a population is pushed in opposite directions in the two separately bred lines. As Darwin (1859/1979) noted in *The Origin of Species*, selective breeding has been practiced for thousands of years on domesticated animals to produce variety in physical and behavioral phenotypes. Indeed, it is actively practiced today in animal husbandry (e.g., Uni *et al.*, 1993). Starting in this century, selective breeding for differences in behavioral phenotypes in organisms has been used by the biological and psychological sciences to study individual differences (Hyde, 1981). By and large, geneticists have used the rapidly reproducing fruit fly, *Drosophila melanogaster*, as the organism of choice to illuminate genetic and evolutionary principles and mechanisms (e.g., Osborne *et al.*, 1997; Rutherford & Lindquist, 1998). Rodents have been preferred by those modeling genetic influences on human behavior and physiology (e.g., Roubertoux, Mortaud, *et al.*, 1996).

Small rodents also show wide ranges of variation in individual differences in behaviors that have been used to infer traits related to anxiety and stress. These traits are strongly influenced by genetic factors (e.g., Berton, Ramos, Chaouloff, & Mormède, 1997; Mathis, Neumann, Gershenfeld, Paul, & Crawley, 1995). Measures of emotionality have been selectively bred for possible genetic linkages between selected and unselected traits (DeFries, 1981). Such measures include behavioral responses in the open field in rats, defecation in Maudsley reactive/nonreactive strain rats (Blizard, 1981; Broadhurst, 1975) and mice (high/low activity, DeFries, 1981), high and low rates of avoidance conditioning (Castanon, Perez-Diaz, & Mormède, 1995), stress-induced immobility (Scott, Cierpial, Kilts, & Weiss, 1996), and defensive reactions (Naumenko *et al.*, 1989) in rats. With the advent of molecular genetic techniques, selectively bred lines have been used to detect specific genes related to behavior (e.g., Lander & Schork, 1994; Flint *et al.*, 1995). Selective breeding retains an advantage over newer selective gene manipulation techniques such as transgenes or knockouts in that it is less likely to produce pathological anomalies and allows the investigator to select for extremes of the phenotype, but still within the natural variation in the population (Lipp & Wolfer, 1999). A recent example of this is a selection study in rats which targeted the 5-HT_{1A} system via hypothermic behavioral responses to the 5-HT_{1A} agonist 8-OH-DPAT (Overstreet, Rezvani, Pucilowski, Gause, & Janowsky, 1994). Subsequent work with these selected animals has helped to delineate specific links between regional 5-HT_{1A} receptor sites in the brain, neuropharmacologic function, and animal models of depression and anxiety (File, Ouagazzal, Gonzalez, & Overstreet, 1999; Gonzalez, File, & Overstreet, 1998; Knapp, Overstreet, & Crews, 1998; Overstreet, Rezvani, Knapp, Crews, & Janowsky, 1996). Similarly, rats selectively bred for high- or low-anxiety-related behavior in the plus maze show differential responses to other anxiety tests, with the potential to discriminate among neurophysiologic and neurochemical mediators (Liebsch, Montkowski, Holsboer, & Landgraf, 1998).

CORRELATIONS IN SELECTIVE BREEDING.   The example in which selection for behavioral responses to 8-OH-DPAT has also affected behavior in anxiety tests illustrates a common phenomenon and an important concept in selective breeding studies, which is that there can be correlations between the selected trait and other, nonselected traits. Thus, if the frequency of genes for a trait (e.g., hypertension) is shown to be altered by selective breeding (by an increase in the numbers of individuals expressing that trait), and at the same time another trait (say, hyperactivity) is not selected but is also altered in a specific direction with selection, then it may be hypothesized that both traits are under the influence of the same gene or set of genes (Crabbe, Phillips, Kosobud, & Belknap, 1990). This in turn allows formulation of hypotheses about putative mechanisms common to both responses. In fact, this is one of the major goals of selective breeding studies, to examine complex relationships between phenotypes and putative relationships in gene pathways that underlie them (Deitrich, 1993).

For both humans and animals, complex phenotypes can be caused by one or two major gene effects (*pleiotropy*) in which a single gene influences multiple physiological avenues (Chua, 1997). A well-known example is the *obese* mutation of the leptin hormone in homozygous *ob/ob* mice, which leads to early-onset obesity, failure of nonshivering thermogenesis, hyperinsulinemia, and hypercorticosteronemia (Chua, 1997). On the other hand, phenotypes may reflect the action of many genes (*polygenic systems*), each of which alone has small or negligible effects (Devor, 1993). Together with so-called modifier genes which exert *genetic background effects* on gene expression (Lander & Schork, 1994) and with facilitating environmental factors, pleiotropic or polygenic gene effects generally emerge in highly complex ways to produce continuous variation in observable phenotypes (Snustad *et al.*, 1997).

In order to investigate the individual and collective contributions of these factors to their phenotypic expression, a strategy which employs multiple behavioral or physiological endpoints is used to "capture" these pathways, in much the same way that multivariate regression analyses statistically "capture" significant pathways in determining covariance among variables (Crabbe, 1999). In practice, this means that true correlations between phenotypes are determined by examining differences between lines selectively bred for high and low values of a phenotype and

SUSAN A.
BRUNELLI AND
MYRON A. HOFER

reexamining differences in replicas of these lines (Henderson, 1989). In addition, lines of animals are selectively bred for different aspects of what are thought to be correlated phenotypes. This practice of examining multiple phenotypic relationships is illustrated by a concerted program of selection for responses to alcohol which was initiated in the 1970s (McClearn, Deitrich & Erwin, 1981), and continues to the present (e.g., Carr *et al.*, 1998; Crabbe & Phillips, 1993; Crabbe, Phillips, Buck, Cunningham, & Belknap, 1999). Mice and rats have been selectively bred for multiple, differential reactions to alcohol, among them measures of preference, acute and chronic tolerance, hypothermic responses, metabolism, elimination, and ataxia. The advantage of such a program is that comparisons can be made of correlations between phenotypes from more than one experimental model examining separate systems, and differences and similarities can provide information on mechanisms underlying phenotypic effects. This multivariate approach has the ability to illuminate central and peripheral mechanisms underlying susceptibility to human alcohol addiction at many different levels (Belknap *et al.*, 1993; Crabbe, 1999; Crabbe & Belknap, 1992; Crabbe & Phillips, 1993; Deitrich, 1993; Devor, 1993; Li, Lumeng, & Doolittle, 1993; McClearn, 1993).

The correlation of phenotypes is common in selection studies and the presence of correlations suggests that one or more genes are linked (sometimes referred to as *genetically correlated*) to influence two or more phenotypes (e.g., Jones, Mills, Faure, & Williams, 1994; Sandnabba, 1996; Shen, Dorow, Huson, & Phillips, 1996). There is, however, the strong possibility that correlations between phenotypes are not linked by genetic correlations, but rather are due to chance associations among gene effects or coselection for other traits that contribute to the expression of the selected trait (Crabbe, 1999; Crusio, 1999; DeFries, 1981; Henderson, 1989). Illustrating these points, mice selected for extremes of open-field activity over 30 generations by DeFries (1981) exhibited corresponding divergence for defecation scores (large negative correlations between activity and defecation levels over generations), whereas unselected control lines remained intermediate to both selected lines. Concordance of effects in replicate lines (High$_1$/High$_2$; Low$_1$/Low$_2$; Control$_1$/Control$_2$) provided strong evidence for genetic linkage between the two phenotypes. Incidentally, over the course of time, albinism also became associated with low-activity scores in both low-activity lines, suggesting a common genetic link between coat color and activity. Nonetheless, activity and albinism were associated in one control line, but not in the other, implying that the putative correlation between coat color and activity was a chance occurrence. Such chance correlations can be caused by random *genetic drift*, the haphazard loss of allelic forms of genes that can occur in small laboratory samples (Plomin, DeFries, & McClearn, 1991; Snustad *et al.*, 1997). This highlights the necessity for multiple replicate lines in disentangling random from selection effects (Crabbe, 1999; DeFries, 1981; Henderson, 1989). Alternatively, in selecting for low activity levels in low lines, genes for albinism were coselected because they were (weakly) linked to those governing activity levels; this weak genetic effect was seen in only one replicate (DeFries, 1969; Plomin *et al.*, 1991). These two results illustrate a common phenomenon in selection studies, which is that strong genetic correlations will replicate, whereas those that are only weakly correlated may not due to lack of power to detect weak effects (Crabbe, 1999; Crusio, 1999). This can be exacerbated by the exclusion of genes simultaneously influencing two phenotypes from a given replicate [either through "*founder effects*" in which the low-frequency genes were likely lost from a small sample making up a founding population (Crabbe, 1999), or through genetic drift] and highlights the necessity for large sample sizes or multiple replications.

The classic Mendelian cross is the most widely used design to explore phenotypic correlations: in this procedure, inbred strains are crossed producing heterogeneous F_1 (heterozygous at all gene loci) and F_2 generations (gene loci independently segregating; Snustad et al., 1997). For example, two homozygous strains of rats, the spontaneously hypertensive rats (SHR) selected for adult high blood pressure values, and their control strain, the Wistar Kyoto (WKY) rat (Okamoto & Aoki, 1963), have historically shown associated differences in other neurally regulated traits (behavioral hyperactivity, salt appetite, cardiovascular reactivity). The two inbred strains were crossed to produce sequential F_1 and F_2 generations to reshuffle genes by recombining them (Snustad et al., 1997) as a test for genetic linkage of these phenotypes with blood pressure. If phenotypes again correlated in the F_2 generation, it would be evidence that these loci were segregating together, accounting for the phenotypic correlation (e.g., Castanon et al., 1995). The absence of correlations in F_2 generations between hyperactivity and blood pressure (Hendley, Atwater, Myers, & Whitehorn, 1983) and between salt appetite and blood pressure (Yongue & Myers, 1988, 1989) suggested that the phenotypes were not, in fact, genetically coupled with hypertension, although both traits may have contributed to the level of blood pressure in selected lines through known physiological pathways (Myers, 1992). In fact, Hendley and coworkers (Hendley & Ohlsson, 1991; Hendley, Wessel, & Van Houten, 1986) were able to produce two separate inbred strains derived from the SHR rat lines, those exhibiting hyperactivity without hypertension, and those exhibiting hypertension without hyperactivity, demonstrating the dissociability of these two traits. Cardiovascular reactivity to stressors continued to be associated with behavioral hyperactivity, however, suggesting that these two traits were genetically as well as physiologically linked (Knardahl & Hendley, 1990). One important caveat in studies like this is that small sample sizes, sampling error, and restricted variability may present problems of statistical power to obtain accurate correlations in F_2 generations (Crusio, 1999). In such instances, convergent evidence, as shown by Hendley and coworkers, is needed for accurate interpretation.

QUANTITATIVE TRAIT LOCI AND IDENTIFICATION OF SPECIFIC GENES.   The fundamental hypothesis of any selective breeding study is that a genetic basis will account for differentiation of "high" and "low" phenotypes produced by breeding. If a breeding program is successful in producing extremes of a phenotype, it may be hypothesized that segments of DNA that predispose animals to an extreme of a phenotype have been selectively enriched in the "high" strain, and conversely DNA segments that predispose to the opposite extreme in the "low" strain have been selectively enriched (J. A. Knowles, personal communication, 1998). Through selective breeding, the "high" and "low" strains have accumulated more "high" and "low" alleles, respectively, at a number of genetic loci scattered throughout the genome. Some of these loci differ from each other because of the random genetic drift that occurs in small breeding populations and others will differ due to the selection for the phenotype.

Techniques are available to identify the gene loci affected by selective breeding, initially at the chromosome level, but now at more molecular levels as more becomes known about rodent genomes and the selected strains specifically. Information about genes involved in constellations of phenotypes is obtained through a technique known as quantitative trait locus (QTL) mapping.

A QTL is a region of a chromosome that has been shown through genetic mapping to contain one or more of the genes that contribute to phenotypic differences (Crabbe et al., 1999). QTL mapping is based on the fact that most complex patterns of behavior are produced by one or more alleles of multiple genes, or polygenic inheritance (Lander & Schork, 1994). These allelic variations of genes are called *polymorphisms*. Through the combined use of sophisticated breeding strategies, multivariate statistical analyses, and molecular techniques, maps have been constructed that locate the positions of QTLs on the chromosomes of animal models such as mice and rats and on humans. Thus, QTL mapping provides information about the locations and numbers of genes involved underlying a phenotype (LeRoy, 1999). Ultimately, the aim of QTL mapping is to associate QTLs with established molecularly marked regions of the chromosome (microsatellite markers) with the eventual goal of identifying and cloning the gene(s) nearby accounting for the phenotype (for reviews of QTL analysis, see Crabbe, Belknap, & Buck, 1994; Crabbe *et al.*, 1999; Lander & Schork, 1994; LeRoy, 1999; Paterson, 1995; Roubertoux, Mortaud, *et al.*, 1996). While it is beyond the capacity of this chapter (or the authors) to critique the various QTL methodologies, it should be noted that as QTLs have become more common, as with all statistical inferential methods, issues and instances of false positives and false negatives have become a concern (Crusio, 1999; LeRoy, 1999).

With respect to behavior, relevant genes of mouse strains selected for extremes of behavior in the plus maze and behavior in a light/dark exploration have recently been mapped using QTL analysis (Flint *et al.*, 1995; Gershenfeld *et al.*, 1997). The first behavioral QTL analysis in the rat characterized hyperactivity in the WKHA and WKY rat strains, which show high and low activity scores, respectively, in novel environments. A hyperactivity QTL was located on chromosome 8, approximately 8 cM from the closest marker, near the dopamine transporter and Snap25, both thought to be good candidates for hyperactivity syndrome. This QTL explained 29% of the variance in the F_2 intercross of the two strains (Moisan *et al.*, 1996).

The use of identified gene segments on chromosomes also permits targeted breeding in order to observe the interactive effects of genes on different genetic and environmental backgrounds (Klöting, Berg, Kovacs, Voight, & Schmidt, 1997). As an example of the power of QTL techniques to dissect complex genetic diseases, in a series of studies, Klöting *et al.* (1997) crossed two inbred rat strains, the SHR with BB/OK, to produce first F_1 then F_2 hybrid generations. SHR show elevated blood pressures in adulthood, whereas BB/OK rats develop insulin-dependent diabetes in early adulthood. From the F2 hybrids genotyped by polymorphic markers, QTLs for blood pressure were detected on chromosomes 1, 10, 18, 20, and X. A series of backcrosses using genetically defined regions produced strains of BB/SHR rats in which the QTLs for blood pressure were represented in different combinations and expressed in blood pressure measurements. The effects of single blood pressure QTLs were modified depending upon which of these was carried by a diabetic strain, thereby demonstrating important interactions between genes, and mimicking human populations exhibiting diabetes with and without hypertension. One implication of this work is that such genetic strategies can test for interactions among QTLs against various gene background and environmental combinations (Frankel & Schork, 1996).

DEVELOPMENTAL PROCESSES

So far, the kinds of questions delineated by selective breeding studies are consistent with approaches concerned only with mechanisms of gene action on behavior without regard to development. However, as developmental psychobiologists, we view development as intrinsic to the process through which the intertwined fates of genes and their milieu are expressed in the life of an organism (Michel & Moore, 1995). So, the question of the relationship between genes and their mechanisms of action must include facilitating environments over the course of development which produce complex phenotypes.

MECHANISMS AND EFFECTS OF EXPERIENCE.   Two independent, long-term selective breeding programs have successfully bred for isolation-induced aggression in adult male mice, employing somewhat different criteria for selection. Using outbred Swiss albino mice, a selective breeding study begun by Finnish researchers (Lagerspetz 1961; cited in Sandnabba, 1996) based on multiple measures of aggression yielded high and low aggressive lines by the second selected (S_2) generation. Beginning with Institute of Cancer Research albino mice, a University of North Carolina group (Cairns, MacCombie, & Hood, 1983) reported on two replicate selection series based on measures of attack frequency and latency which diverged into high and low lines in the S_1 and S_{3-4} generations, respectively. As across-laboratory replications, these studies have reliably demonstrated the heritable nature of aggressive behavior, and using different measures, have yielded convergent data about their mediation and correlated responses. They have also asked questions regarding the development of aggression in the selected lines, elucidating gene–environment interactions.

Complementary studies by both groups have demonstrated susceptibility of male aggressive behavior to the effects of social influences during the early postweaning period. Accordingly, prior to 45 days, group rearing rather than standard postweaning social isolation abolished aggressive line differences in adulthood. Repeated testing of isolated animals during the same period lessened line differences analogous to a "dose-dependent" effect of socialization (Cairns, Hood, & Midlam, 1985; Cairns, Garièpy, & Hood, 1990). Diminished reactivity to social advances in dyadic interactions (Lewis, Garièpy, Gendreau, Nichols, & Mailman, 1994) was thought to underlie these changes in both groups. A biochemical correlate of this diminished response was a decrease in dopamine D_1 receptor density in striatum (Garièpy, Gendreau, & Lewis, 1995; Lewis *et al.*, 1994). The high-aggressive line showed greater reactivity in response to social isolation and differentially higher D_1 receptor density in striatum with isolation (Garièpy, Lewis, & Cairns, 1996). In other words, the effects of group rearing in promoting a decrease in D_1 receptor density were more profound in the high-aggressive line than for the other two lines. Hence, the rearing environment can exacerbate or reduce genetic biases among several behaviors associated with aggression, demonstrating considerable plasticity of neurobiological and functional organization during development (Garièpy *et al.*, 1996).

Even passive exposure to adult male fighting during the juvenile period can affect individual performance in adulthood (Sandnabba, 1993, 1996): prepubertal males exposed to adults from behind wire-mesh were more aggressive in adult dyadic encounters than males exposed from behind a glass container. In this instance, early experience potentiated line biases to produce higher levels of aggression in high- than in low-aggressive line males. Finally, both the Finnish and North Carolina groups demonstrated that social learning can modify characteristic levels of aggressive behavior in high- and low-aggressive strains. High- and low-aggressive lines of mice show idiosyncratic patterns of urine deposition, but social defeat by a trained fighter altered the pattern markings in high-aggressive-line mice so that they resembled those of low-aggressive-line males (Sandnabba, 1996). Similarly, the North

Carolina group found that when males from high- and low-aggressive lines were allowed to interact in a long-term social setting, such long-term interactions between lines broke down line-specific patterns of fighting. As a result, individuals formed dominance hierarchies without regard to line affiliation, and with about equal numbers from each line attaining dominance. Testosterone levels corresponded to individuals' dominance status, not line status, suggesting physiological effects of experience that were independent of genetic predispositions (Garièpy, 1994).

MATERNAL ENVIRONMENTS.   To those primarily interested in gene effects per se, the contribution of the maternal pre- or postnatal environment to the expression of a phenotype might be of interest only insofar as it must be ruled out as a non-genetic factor contributing to variability in outcome or if the phenotype turns out to be an indirect function of the dam's and not the offspring's genotype (Henderson, 1989). For those interested in development, however, the influence of the maternal environment would be of genuine interest as the center of a process of differentiation of a genotype in the development of an individual. In their review of maternal effects on offspring phenotypic expression (on which much of this discussion of maternal effects is based), Roubertoux, Nosten-Bertrand, & Carlier (1991) characterized maternal effects on offspring phenotype as those which affect variation in offspring more than the paternal effects, emanating from either (or both) maternal inheritance and maternal environment. Paternal inheritance can derive from genes on the Y chromosome which are only transmitted to male offspring. Alternatively, both maternal and paternal effects can be transmitted by genomic imprinting, in which genes are expressed differentially depending upon their maternal or paternal origin (Snustad *et al.*, 1997). Maternal inheritance effects include mitochondrial DNA (mtDNA) transferred through the maternal egg's cytoplasm.

Maternal environment effects can be classed as prenatal (cytoplasmic, uterine) or postnatal (nutritional, biobehavioral). Of the prenatal effects, maternal cytoplasmic effects can be the consequence of interactions between elements of maternal cytoplasm and offspring nuclear DNA either in the absence of or in interaction with general environmental effects on the mother. Prenatal cytoplasmic effects have been demonstrated to affect age of eye opening in inbred strains of mice (Nosten & Roubertoux, 1988).

Prenatal uterine effects have been documented across many species (e.g., Lecanuet, Fifer, Krasnegor, & Smotherman, 1995), from the effects of prenatal stress in rodents (e.g., Weinstock, 1997) to prenatal learning (Pedersen & Blass, 1981; S. R. Robinson & Smotherman, 1995), to the well-known effect of prenatal auditory stimulation on postnatal recognition of mother's voice in human infants (DeCasper & Fifer, 1980).

Maternal postnatal effects on expression of genotypes are equally common (see Myers, 1992, and below). In an early review, Broadhurst (1961) presented methods for dissecting out postnatal maternal effects, using reciprocal cross-fostering of neonates to mothers of selected or inbred lines. For instance, mouse pups with an extreme of a phenotype (say, low activity) would be cross-fostered to and reared by mothers of the opposite phenotype (high activity). If strong postnatal maternal effects were present, the activity of the low pups cross-fostered to high-activity-line mothers should more resemble the phenotype of the high strain than control pups cross-fostered within the low strain. If the phenotype of fostered high-activity pups did not differ from that of individuals of their own genetic strain, then maternal effects would not present. The last logical scenario is that if activity levels of fostered pups lay between the two selected strains, then either it would be the result of combined strong and weak pre- and postnatal effects or the infant phenotype interacting with the maternal phenotype. In that case, in evaluating postnatal effects, information about the prenatal environment would have to be obtained through methods such as grafting eggs from one strain into ovaries of females of the opposite strain, (Broadhurst, 1961; Roubertoux et al., 1992).

As a case in point, although selection for aggression in mice by the North Carolina and Finnish groups was practiced only on male fighting behavior, females in high-aggressive lines also showed higher levels of aggression. Pregnant and postparturient females from high-aggressive lines in both selection studies showed more intense maternal aggression in response to intruders than did low-aggressive or control lines (Hood & Cairns, 1988; Sandnabba, 1993). This demonstrated that the genetic influence was not specific to males, and hence not carried exclusively by the male (Y) chromosome. In a series of experiments, Roubertoux et al. (1994), Roubertoux and Carlier (1988), and Carlier, Roubertoux, and Pastoret (1991) examined this question in inbred strains of mice, using a combination of prenatal and postnatal cross-fostering methodology (noted above), and established that the inheritance of aggressive behavior in males and females was mediated at least in part through the maternal postnatal, but not prenatal environment.

In contrast, the Finnish and North Carolina groups found that in their selectively bred mouse lines, maternal postnatal cross-fostering had little effect on adult male aggressive behavior (Hood & Cairns, 1988; Lagerspetz & Wuorinen, 1965, cited in Sandnabba, 1996). Apparently, in these selected lines, female aggression is mediated through the maternal environment, whereas male aggression is not. This discrepancy in findings between fostering in inbred strains and in selectively bred lines illustrates the point made by Roubertoux, Mortaud, et al. (1996), that there may be multiple pathways through which genes and environment can achieve similar phenotypic end results in different populations.

HETEROCHRONY, DEVELOPMENT AND EVOLUTION.   One conceptualization of how development can be involved in the evolution of differences in traits either through natural selection or in selective breeding is called "heterochrony" (from the Greek, "different time"; McKinney & McNamara, 1991). Heterochrony has been defined as a change in timing (as in the onset or offset) or rate (as in speed) of developmental events (usually in the zygotic, embryonic, or infantile period) relative to the same events in ancestral forms (McKinney & McNamara, 1991). Observations of and experiments with invertebrates (insects and arthropods) demonstrate heterochrony during development, as in descriptions of the development of sea urchin species in which some species proceed through a series of early and intermediate steps in development, whereas other species skip larval stages and go directly to adult stages (Gould, 1977). One common mechanism of action producing heterochrony is the omission of stages in some cell lineages and acceleration of stages in other cell lineages at various stages of embryogenesis; that is, groups of cells responsible for differentiation into structures undergo shifts in the rates of their development (McKinney & McNamara, 1991). Some evolutionary and developmental theorists have asserted that such changes in the developmental rate of structures during ontogeny may be a critical factor in producing speciation (Gottlieb, 1987; Gould, 1977).

Recent work with mice has shown that heterochrony is also possible in mammalian genomic development. This is exemplified by a study in which a portion of a homeobox gene complex (*HoxD* enhancer) involved in specification of organs in different body parts was deleted (Zákány, Gerard, Favier, & Duboule, 1997). This induced changes in the timing of gene transcription (the transfer of genetic information from DNA to RNA; Snustad *et al.*, 1997, p. 250) controlling trunk and limb bud development, which in turn produced changes in the developmental schedule of these body parts in relation to one another (Zákány *et al.*, 1997). This study has direct implications for transcriptional heterochrony in producing variation in body phenotypes, and consequently in evolution of differences in functional changes in response to selection pressures on mammalian species.

SELECTIVE BREEDING FOR AN INFANTILE PHENOTYPE: ULTRASONIC VOCALIZATION RATES IN 10-DAY-OLD RATS

SELECTIVE BREEDING FOR USV AS AN ANIMAL MODEL FOR THE INHERITANCE OF ANXIETY

The possibility that USV rates might be an early expression of a heritable trait related to anxiety was first raised by Insel and Hill (1987) because USV rates of the Maudsley-reactive strain were significantly higher on day 5 postnatal than those of the Maudsley-nonreactive strain. These strains have been selectively bred for adult expression of "emotionality" and "nonemotionality," measured as extremes of defecation and middle square crosses in open-field tests (Blizard, 1981). In the same study, although adult subjects showed strain-characteristic levels of defecation and square crosses, no inferences could be made about continuity of behavior from infancy because longitudinal data were not available (Insel & Hill, 1987). Since rat pups from different litters vary considerably in rates of calling (Graham & Letz, 1979) and pups within litters show individual variability, it seemed likely that USV was malleable to selective breeding for differential rates of calling (Brunelli, Keating, Hamilton, & Hofer, 1996).

There are many reasons for undertaking a selective breeding program for a developmental–genetic model of anxiety. In selective breeding for USV rates in infant rats, we asked: Could ultrasonic vocalization rate responses be altered by genetic manipulation? If so, we should be able to produce lines of animals that presented heritable, familial histories of increasingly different rates of USV over generations. This segregation of lines into two distinct phenotypes would occur because gene loci will become more homozygous over generations. If USV rates were linked to anxiety, then producing selectively bred lines with the two different phenotypes would provide a means of modeling some aspects of the inheritance of familial anxiety disorders in humans (Rutter, Silberg, O'Connor, Simonoff, 1999a, b). Because they are animal models such selected lines can never reproduce human syndromes; nevertheless, the occurrence of analogous states or traits can be studied at many levels of organization, at genetic, molecular, cellular, neurochemical, and organismic levels (Dietrich, 1993; McClearn, 1993).

Second, the efficacy of animal models for research on emotional regulation has increased in the last decade based on mounting evidence that specific neurobiological mechanisms underlying basic emotional regulation in other animals (even invertebrates) are essentially (if not universally) the same as in humans, with the proviso that the more "basic" the system, the more likely the conservation across taxa (LeDoux, 1996; Roubertoux, Mortaud, *et al.*, 1996). To the extent that mechanisms underlying affective processes are common among taxa, then selective breeding for a trait thought to be related to various aspects of affect regulation should be possible. It seemed reasonable to us, therefore, in planning a selective breeding program to expect neurobiological mechanisms underlying ultrasonic vocalization responses and, indeed, the neuropharmacology of USV (Hofer, 1996; Insel, Hill, & Mayor, 1986; Miczek *et al.*, 1991) to be similar to some of the mechanisms underlying human anxiety disorders.

If the linkage with anxiety systems is true and USV is a valid early marker, then we should end with groups likely to be dissimilar on many measures of anxiety behavior that differ in neural structures as well. If USV response to isolation is a behavioral marker for an infantile form of anxiety, then genetic manipulation should produce animals that are more likely to show correlated responses in other contexts designed to elicit anxiety- or fear-like behaviors. As shown by selective breeding studies in adult rats, "emotionality," "reactivity," and other forms of behavioral differences have their counterparts in other contexts (Abel, 1991; Blizard, 1981). We could examine the convergence of these relationships in central systems implicated in pharmacology and neurochemistry of separation-induced USV in what has become a classic multivariate approach to characterizing complex, behavioral phenomena (Hinde, 1970; LeDoux, 1996). In essence, we could test whether selective breeding will produce correlated phentoypes, with the implication that these may be linked genetically.

In a selective breeding study, the underlying neurobiology might be differentiated in populations selectively bred for extremes of behavior and physiology. That is, the physiology of "high" or "low" animals should be biased in predicted directions (e.g., alterations in hypothalamo–pituitary–adrenal axis functioning; Abel, 1991; Castanon *et al.*, 1995). By producing rat strains that differ to a major degree in the USV response and correlated traits while remaining similar in other traits, we should greatly increase the signal-to-noise ratio in the underlying neuromodulator systems and brain regions that mediate these behaviors as noted in the Introduction. This should allow experimental differentiation of neural systems underlying basic processes corresponding to these extreme differences in lines selectively bred for extremes of USV rates.

If we are selecting for a "constitutional" bias, then we should be able to follow that bias throughout the lifetime of the animal. In other words, if extreme rate of USV in infants is a behavioral marker for lifelong extremes of behavioral affective regulation, it should be measurable by behavioral and physiological means. A corollary to this is that one can test the hypothesis that a behavior thought to relate to anxiety in animals is a behavioral marker for a *trait* as opposed to state induced by a specific event. This is because a selectively bred model allows longitudinal studies across contexts and throughout lifetimes of selected individuals in controlled ways. Thus, behavioral, physiological, biochemical, and ultimately genetic mechanisms, associated with USV and affect regulation could show correlated differences throughout the lifetimes of individuals.

Concomitant with this interest, we can formulate hypotheses about how selection can mirror effects of evolution on development and vice versa. As a laboratory strategy, analysis of developmental changes over many generations of selection can allow determination of the processes by which changes in a phenotype (such as aggression or USV rates) and related structures are incorporated into developmental schedules. For example, would observed changes in USV rates in development be the result of accelerated or delayed maturation, producing heterochrony? Will there be schedule changes in the development of neurobehavioral systems correlated with USV? Referred to by Cairns *et al.* (1990) as "microevolution," such changes, if carefully documented in patterns of development over time, can be verified as they evolve over the course of successive generations (for reviews of this subject, see Cairns, 1993; Cairns *et al.*, 1990).

DESIGN OF THE SELECTIVE BREEDING PROGRAM. In selective breeding for ultrasonic vocalizations we used National Institutes of Health (N:NIH) strain rats rather than the Wistar rats used for all of our previous ultrasonic vocalization studies. The N:NIH strain was developed by the National Institutes of Health specifically to provide a heterogeneous, outbred strain as a base for selective breeding. It was generated by crossing eight inbred laboratory strains from the United States and abroad (BN/SsN, MR/N, ACI/N, BUF/N, F344/N, WKY/N, WN/N, and M520/N), chosen to represent a broad range of the *Rattus norvegicus* genotype (for details of establishment of the strain, see Hansen & Spuhler, 1984). Providing a heterogeneous founding population ensures a large and normally distributed basis of alleles to respond to selection pressure, yielding the largest possible number of phenotypic responses (Hansen & Spuhler, 1984). In general, because commonly used outbred strains are offspring of a very few founding individuals which do not represent all possible allelic variants of genes, they lack sufficient variability at gene loci for selection (Crabbe, 1999; Hansen & Spuhler, 1984; Snustad *et al.*, 1997), the so-called founder effect (due to random loss of alternate gene alleles, an example of genetic drift; Snustad *et al.*, 1997, pp. 732–733). Genetic drift through loss of alleles in later generations can also occur. Inbred strains have, by definition, no genetic variability: all animals are genetically identical at all gene loci through fixation of alleles by brother–sister or father–daughter matings. In other words, we were looking for a strain of rat that provided the greatest number of polymorphisms in the population upon which selection for USV would be exerted.

We therefore obtained 25 breeding pairs of N:NIH strain rats from the NIH. From these breeding pairs, three laboratory-born progenitor generations were randomly bred in order to characterize USV rates for the strain and to examine the effects of factors which are known to influence USV (Brunelli & Hofer, 1996).

Important factors affecting USV rates are pup age and the temperature at which and the amount of time that pups spend away from the litter environment out of the thermoneutral range. Our aim was to produce animals that responded immediately with maximal vocalization rates to disruption of the social environment (e.g., odor, tactile, thermal; Hofer, Brunelli, & Shair, 1993; Hofer, Brunelli, Masmela, & Shair, 1996). We did not want to select for ultrasound production under conditions of prolonged isolation in a cool environment, which is mediated primarily by metabolic processes (Sokoloff & Blumberg, 1997; see also the chapter by Blumberg in this volume). We therefore selected pups on the basis of their USV rates experienced during the first 2 min of isolation, and because initial responses are recommended for detecting genetic differences in USV (Roubertoux, Martin, *et al.*, 1996). The age at which pups were selected would be based on factors which indicated a period in development when pups were beginning to respond principally to social stimuli signaling separation from the dam and litter (see below).

The temperature(s) at which to test pups was one of the primary environmental considerations because of potent effects on ultrasound rates (Allin & Banks, 1971; Blumberg, chapter in this volume; Blumberg, Efimova, & Alberts, 1992; Nitschke, Bell, Bell, & Zachman, 1975; Okon, 1971; Oswalt & Meier, 1975). In order to determine whether temperature variation would influence USV rate variance within and between selected lines, and indeed if selection would alter susceptibility to temperature variation in USV, our strategy was to test animals over the range of naturally-occurring ambient temperatures (~10°C range) in our laboratory (Brunelli & Hofer, 1996; Brunelli *et al.*, 1996). Body temperature measures were axillary temperature taken immediately before and after each pup's test. These measures have been used as markers of temperature regulation in our laboratory because they are highly correlated with changes in brown fat temperatures during the isolation test (Hofer & Shair, 1991, 1992).

PRELIMINARY STUDIES: PROGENITOR (PR) GENERATIONS.   Prior to selective breeding, we measured USV in a developmental study tracing emissions in 2 min of isolation at 3, 10, 15, and 18 days postnatally in the first generation (PR1) of laboratory-born N:NIH pups (Figure 1) (Brunelli *et al.*, 1996). Pups emitted their highest rates of USV at 3–4 days postnatally, went to half of that rate at 10 days, and declined a further one fourth of that rate at 15 days and to nearly zero at 18 days. Average rates of USV during isolation at room temperature were higher at 3 days than at 10 days of age, a trend found in about half of published longitudinal studies (Noirot, 1968; Noirot & Pye, 1969; Oswalt & Meier, 1975; Sales, 1979), although in other strains, USV rates were higher at 9–10 days (Allin & Banks, 1971; Graham & Letz, 1979; Nitschke *et al.*, 1975; Okon, 1971).

USV rates were found to be an environmentally stable behavioral trait in that repeated testing did not significantly affect the calling rates of either individuals or

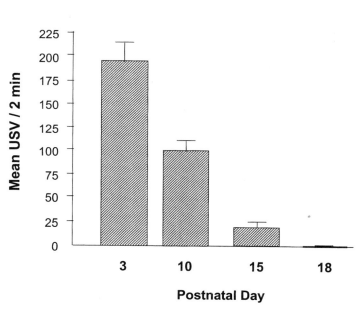

Figure 1. Average rates of USV emission for N:NIH rat pups tested on postnatal days 3, 10, 15, and 18. Litter means are the unit of analysis on each day. Based on post-hoc Tukey tests: Three-day-olds had significantly higher rates than 10-day-olds ($p < .001$), than 15-day-olds ($p < .001$) and than 18-day-olds ($p < .001$). Ten-day-olds had significantly higher rates than 15-day-olds ($p < .001$) and 18-day-olds ($p < .001$). Fifteen-day-olds and 18-day-olds did not differ. (Reprinted from Brunelli *et al.*, 1996, with permission. Copyright 1996, Wiley–Liss, Inc., a subsidiary of John Wiley & Sons, Inc.).

litters. Only at 3 days postnatal age did naturally occurring ambient temperature variations significantly affect USV responses (see Brunelli *et al.*, 1996, for details), consistent with earlier developmental studies of USV (Allin & Banks, 1971; Nitschke *et al.*, 1975; Okon, 1971; Oswalt & Meier, 1975).

The effects of litter as a factor affecting USV steadily declined over the postnatal period, suggesting that as they become older, individual pups are less susceptible to maternal and other influences that act upon the entire litter. In a regression analysis, variance in USV accounted for by litter in the PR1 generation ranged from about 75% at 3 days of age to about 15% at 18 days of age, with about 36% at 10 days of age. This was coincident with individual differences in USV responses that emerged by 10 days of age that were not simply correlations of body weight or temperature. USV rates at 10 and 15 days of age were significantly correlated, $r(51) = .51$, $p < .001$, meaning that isolation calling rates at 10 days of age were somewhat predictive of their response levels 5 days later. Rates from 3–10 days, $r(20) = .28$, and 15–18 days, $r(54) = .24$, were positively, but not significantly correlated ($ps > .10$). Findings from this study suggested that development of USV over the postnatal period in this genetically heterogeneous strain was reasonably consistent with other strains of rats tested, and pointed to the existence of individual differences traceable through the second postnatal week.

In a subsequent study, USV was measured only at 10 days of age in all three progenitor generations, PR1, PR2, PR3: 532 pups in 81 litters (Brunelli & Hofer, 1996). One purpose of the study was to determine baseline vocalization rates prior to selective breeding for high and low rates of USV at that age. This served as a foundation for selective breeding for high and low rates of USV at that age. A second purpose was to detect associations among variables measured in isolation that might suggest correlated behavioral phenotypes with USV. Evaluation by principal components factor analysis (Table 2) revealed four factors which corresponded roughly to

TABLE 2. PRINCIPAL COMPONENTS ANALYSIS OF
ASSOCIATIONS AMONG ISOLATION-INDUCED
BEHAVIORS: COMBINED PR1, PR2, PR3 GENERATIONS

Variable	Component[a]			
	1	2	3	4
Axillary T (°C)[b]	0.80	—	—	—
Rectal T (°C)[c]	0.93	—	—	—0.38
Ambient T (°C)	0.58	—	—	0.37
Weight (g)	—	0.65	—	0.90
Square cross	—	—	—	0.90
Rear	—	0.82	—	—
Defecate/urinate	—	0.39	0.31	—
Face wash	—	—	—0.71	—
USV	—	—	0.77	—
Percentage variance[d]	21.37	15.03	13.71	12.36

[a]Numbers in columns are coefficients called "loadings," which express the association of each variable to the component. Only loadings >0.30 are shown for each component.
[b]Pretest axillary temperature.
[c]Posttest rectal temperature.
[d]Percentage of variance in the entire data set accounted for by each component.

patterns indicative of (1) *thermoregulatory responses*, characterized by high positive loadings of ambient and body temperature variables without USV, (2) a mixed *maturity–fearful factor* composed of pup weight, rearing, and urinate/defecate, (3) an "*anxiety*" or "*emotionality*" factor consisting of USV, urinate/defecate (pups rarely defecated, but urination was frequent and could be counted), and face washing behavior, and (4) a mixed *locomotor–maturity* factor composed of square crosses, temperature and weight. The general features of this principal component analysis remained largely stable in the randomly bred and high lines in the 9th to 12th generations, with the low line showing some shift in variables within factors (see below).

MECHANICS OF SELECTIVE BREEDING.   Pups of the PR3 generation founded the High, Low, and Random USV lines. At PN 60–100 days, 6 offspring ($3\male$, $3\female$) from litters born to each of 18 PR2 dams (descended from each of the 25 original NIH breeders) were chosen as line breeders, based on USV rates at PN 10 days (see below). The first two ($1\male$, $1\female$) were chosen at random as Random control breeders, then littermate males and females with the highest and lowest rates of USV at PN 10, respectively, were chosen to initiate the High and Low lines. They were then bred with similarly chosen pups from other litters. Succeeding selected generations (S_1, S_2, ..., S_n) are offspring of these initial High, Random, and Low USV line breeders. The lines are maintained as closed breeding systems and mating occurs only within lines. Each closed line is made up of 18–19 families based on matrilineal descent from each of the original 25 breeding pairs. Outbreeding across families within the lines is practiced in order to minimize inbreeding, thereby minimizing genetic drift, the random fixation of alleles other than those selected for, and also to maximize the number of genes influencing USV within each line (Crabbe *et al.*, 1990; DeFries, 1981). The randomly bred line performs the added function of monitoring for the effects of genetic drift and of environmental changes across generations (Crabbe *et al.*, 1990; DeFries, 1981). In practice, this means that breeders are mated with partners from litters outside the natal families, and partners do not share parents or grandparents in common. Outbreeding also acts as a control for litter and maternal effects because it maximizes within-litter genetic heterogeneity (DeFries, 1981; Roubertoux *et al.*, 1990) allowing optimum expression of genetic-based individual differences.

The purpose of our breeding strategy, therefore, has been to maximize the overall heterozygosity of the selected lines while selecting for homozygosity at the gene loci responsible for the USV trait (DeFries, 1981; Henderson, 1989). Based on this strategy of random breeding except for the selected phenotype, when the lines are eventually mapped using molecular genetic methods, it should be possible to identify linked genes associated with USV in the selected lines (S. C. Chua, personal communication, 1999). This is because alleles should be homozygous at loci selected for in the High and Low USV lines, whereas others, not affected by selected breeding, should remain randomly segregating in both the selected lines.

RESULTS: OUTCOME OF SELECTIVE BREEDING FOR USV

Under conditions of testing noted above (for details, see Brunelli, Vinocur, Soo-Hoo, & Hofer, 1997) USV rates in Low line offspring of the first selected generation (S_1) diverged significantly from Random line controls, and the Low line has maintained significantly lower rates over all generations since (Figure 2). In the S_3

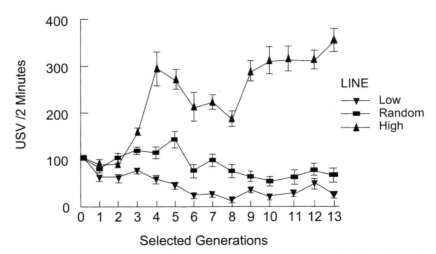

Figure 2. USV rates of 13 completed generations of selective breeding for High, Low, and Random USV lines (abscissa). The ordinate shows the average USV rate over 2 min of testing. The S_0 generation is composed of all data from the first three progenitor generations (PR1–PR3) at postnatal day 10. The S_1 generation pups are offspring of pups selected from the PR3 generation. Subsequent generations are bred completely within selected lines.

generation, the High USV line diverged from Random controls, and has displayed significantly higher USV rates in every generation since. This rise in USV rates was displayed as a significant increase in each succeeding generation until S_5, when High line USV rates leveled off. The data are convincing evidence that isolation-induced USV rate is a heritable trait that is highly responsive to selection.

CORRELATED PHENOTYPES: TWO-MINUTE SCREENING DATA AT 10 DAYS OF AGE. *Analyses in the fifth selected generation (S_5).* Since we were interested in learning whether any correlations existed between USV rates and other behavioral and physiological phenotypes, we analyzed other variables measured in isolation (see Tables 2 and 3 for description) in the S_5 generation. These revealed no line differences in other behaviors or physiological variables measured in isolation at that point (Brunelli *et al.*, 1997).

Analyses in the S_9–S_{12} generations. In contrast to the S_5 generation, by the S_9–S_{12} generations (litter mean data pooled across generations), the two selected lines had become separated on additional parameters. First, the selected lines differed in body weight at 10 days of age, $F(2, 268) = 6.276$, $p < .002$, in that Low line pups weighed significantly less than both the High, $p < .028$, and Random lines, $p < .003$, which did not differ (Figure 3).

Second, Low line pups lost more body heat during the 2 min of testing than High and Random line pups, $F(2, 264) = 5.254$, $p < .006$ (Figure 4). This effect was much weakened when weight was covaried with Line, *Line* $F(2, 263) = 2.722$, $p = .068$; *Weight* $F(1, 263) = 23.117$, $p = .000$, suggesting that increased heat loss in the Low line was primarily a function of their smaller body size and greater relative surface area rather than reflecting any significant difference in thermoregulatory functioning.

In addition, the lines differed significantly in their frequency of urination during the 2 minute test, $F(2, 268) = 7.626$, $p < .001$: High line pups urinated more than both Low and Random line pups, both $p < .001$, which did not differ (Figure 5).

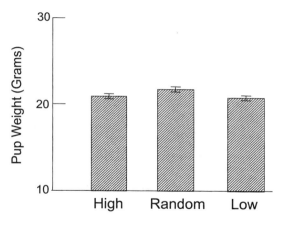

Figure 3. Body weight in grams of pups at postnatal day 10 (ordinate). Results of pooled data from the S_9–S_{12} generation litters ($N = 164$) in the High ($n = 59$), Random ($n = 47$), and Low ($n = 58$) USV lines (abscissa).

Although highly statistically significant, it is not clear whether, compared to massive differences in USV, small differences in urination between lines have biological significance. However, low doses of NMDA to 10-day-old pups both increase rates of isolation-induced USV and induce significantly more urination (Winslow, Insel, Trullas, & Skolnick, 1990). Both findings would be predicted in infant rats based on anxiogenic effects of the NMDA–glycine receptor system in adult rat stress models (e.g., Adamec, Burton, Shallow, & Budgell, 1999; Dunn, Corbett, & Fielding, 1989), and on the excitatory role of central NMDA in the control of micturation (e.g., Vera & Nadelhaft, 1991; Yoshiyama, Roppolo, & de Groat, 1994). A simple hypothesis arising from this suggestion is that High line pups should show greater

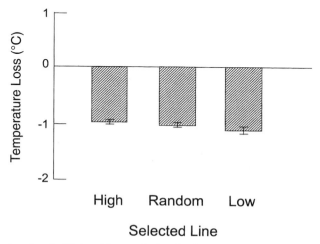

Figure 4. Temperature loss in 10-day-old pups during 2 min of isolation (axillary temperature before minus axillary temperature after 2-min test). Zero baseline represents pretest axillary temperature. Results of pooled data from the S_9–S_{12} generation litters ($N = 164$) in the High ($n = 59$), Random ($n = 47$), and Low ($n = 58$) USV lines (abscissa).

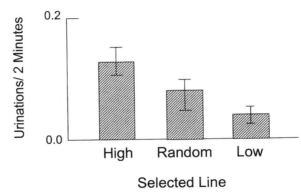

Figure 5. Urination in 10-day-old pups during 2 min of isolation. Results of pooled data from the S_9–S_{12} generation litters ($N = 164$) in the High ($n = 59$), Random ($n = 47$), and Low ($n = 58$) USV lines (abscissa).

central NMDA activation in isolation, which could be measured using a variety of neuropharmacologic methods.

Finally, a principal components analysis in the S_9–S_{12} generation modeled on the progenitor generations described above evaluated associations between variables measured in isolation in each of the three lines (Tables 3A–3C). The S_9–S_{12} Random line is randomly bred and therefore most comparable to the PR generation (with the inevitable loss of genetic heterogeneity that comes with breeding a small,

TABLE 3A. PRINCIPAL COMPONENTS ANALYSIS OF
ASSOCIATIONS AMONG ISOLATION-INDUCED
BEHAVIORS: COMBINED S_9, S_{10}, S_{11}, S_{12}
GENERATIONS—RANDOM LINE

Variable	Component[a]			
	1	2	3	4
Axillary T (°C)[b]	0.82	—	—	—
Axillary T (°C)[c]	0.91	—	—	—
Ambient T (°C)	0.38	−0.67	—	—
Weight (g)	0.34	0.43	0.59	—
Square cross	—	0.31	0.85	—
Pivot[d]	—	−0.72	—	—
Rear	—	0.74	—	—
Defecate/urinate	—	—	—	−0.66
Face wash	—	—	—	0.75
USV	—	0.45	—	−0.69
Percentage variance[e]	18.78	20.73	18.77	18.22

[a]Numbers in columns are coefficients called "loadings," which express the association of each variable to the component. Only loadings >0.30 are shown for each component.
[b]Pretest axillary temperature.
[c]Posttest axillary temperature.
[d]Pivot corresponded to a 360° rotation of the pup on its rear axis, a locomotor variable that was added after the selection program began.
[e]Percent of variance in the entire data set accounted for by each component.

TABLE 3B. PRINCIPAL COMPONENTS
ANALYSIS OF ASSOCIATIONS AMONG
ISOLATION-INDUCED BEHAVIORS: COMBINED
S_9, S_{10}, S_{11}, S_{12} GENERATIONS—HIGH LINE

Variable	Component[a]		
	1	2	3
Axillary T (°C)[b]	0.79	—	—
Axillary T (°C)[c]	0.87	—	—
Ambient T (°C)	0.60	−0.54	—
Weight (g)	—	0.77	—
Square cross	0.65	—	—
Pivot[d]	0.43	—	—
Rear	—	0.88	—
Defecate/urinate	—	—	−0.67
Face wash	—	—	0.82
USV	—	—	−0.77
Percentage variance[e]	26.23	18.59	18.22

[a]Numbers in columns are coefficients called "loadings,"
which express the association of each variable to the
component. Only loadings >0.30 are shown for each
component.
[b]Pretest axillary temperature.
[c]Posttest axillary temperature.
[d]Pivot corresponded to a 360° rotation of the pup on its
rear axis, a locomotor variable that was added after the
selection program began.
[e]Percent of variance in the entire data set accounted for
by each component.

TABLE 3C. PRINCIPAL COMPONENTS ANALYSIS OF
ASSOCIATIONS AMONG ISOLATION-INDUCED
BEHAVIORS: COMBINED S_9, S_{10}, S_{11}, S_{12}
GENERATIONS—LOW LINE

Variable	Component[a]			
	1	2	3	4
Axillary T (°C)[b]	0.87	—	—	—
Axillary T (°C)[c]	0.88	0.37	—	—
Ambient T (°C)	0.65	—	0.40	—
Weight (g)	—	0.66	—	—
Square cross	—	0.82	—	—
Pivot[d]	—	—	0.74	—
Rear	—	0.68	−0.44	—
Defecate/urinate	—	—	—	0.80
Face wash	—	—	−0.49	−0.64
USV	−0.54	—	—	0.48
Percentage variance[e]	24.84	17.27	13.46	13.18

[a]Numbers in columns are coefficients called "loadings," which express the association of each variable to the component. Only
loadings >0.30 are shown for each component.
[b]Pretest axillary temperature.
[c]Posttest axillary temperature.
[d]Pivot corresponded to a 360° rotation of the pup on its rear axis, a
locomotor variable that was added after the selection program
began.
[e]Percent of variance in the entire data set accounted for by each
component.

reproductively isolated group of animals; Crabbe *et al.*, 1990). After many generations, defecate/urinate behavior, face washing, and USV remained intercorrelated, suggesting stability in the relationships between these variables in randomly bred animals (Table 3A). This was also true of both the High and Low USV lines in the S_9–S_{12} generations (Tables 3B and 3C), suggesting that selection for differentially high and low USV rates had not altered that stability. If anything, that stability was strengthened in the High USV line (Table 3B), judging by the magnitude of the loadings on component 3. In the Low USV line (Table 3C), while loadings for these variables were still maintained on the same component, it appeared that selection for low rates of USV had reduced the extent of the relationship of USV to that component: loadings for this variable were split between the thermoregulatory component (component 1) and the anxiety/emotionality component (component 4). Other variables which had previously loaded on components 2 and 3 (Table 2) in the PR generations were variously loaded on components 2 and 3 in the S_9–S_{12} generations, and generally suggested associations that could be interpreted as a variable combination of maturity, locomotion, and thermoregulation in each of the lines. Such variability in the structure of principal components generally indicates multiply regulated systems (Maier *et al.*, 1988; Ramos & Mormede, 1998).

CORRELATED PHENOTYPES: OTHER CONTEXTS. The question of context specificity of the selected response becomes important when appraising the extent to which the response reflects central nervous system functioning. Thus, if this response were specific only to the first 2 min of isolation, we might find that pups from selected lines would not generalize their vocalization responses to other situations encountered in isolation. Alternatively, we might find that selective breeding produced extremes of vocalizing in High and Low line pups, irrespective of the context in which they found themselves in isolation.

Context specificity and context generality in the selected lines were tested based on responses of pups to unfamiliar adult males in isolation. Isolated 7- to 14-day-old pups suppress vocalization when exposed to adult males (Brunelli *et al.*, 1998; Shair *et al.*, 1997). They also exhibit immobility, a pup analogue of adult freezing (defined as the absence of movement, with the head stationary and elevated from the floor; Brunelli *et al.*, 1998; Shair *et al.*, 1997; Takahashi, 1992a), mobilization of analgesia (Wiedenmayer & Barr, 1998), and biologically significant elevations in plasma levels of ACTH (Takahashi, 1992a,b), suggesting a defensive response to a potential predator (Mennella & Moltz, 1988; Paul & Kuperschmidt, 1975). As shown in Figure 6, small samples of 10-day-old pups from all three lines showed complete suppression of USV during 10-min contact with an adult Wistar male, in contrast to control pups isolated for the same time period. Thus, the high rates of USV in High line pups is not simply a nonspecific recalibration of some central pattern generator, but is responsive to the context in which the pup is observed.

Suppression of USV was accompanied by immobility during exposure to the male (Figure 7). High line pups showed significantly greater immobility, *Line* $F(2, 153) = 3.984$, $p = .021$, compared to Low line pups ($p < .01$), with Random line pups intermediate between the two selected lines. There was also a tendency in this small sample for High line pups to urinate more than Low line pups, with Random line pups intermediate between the two, $F(2, 160) = 2.744$, $p = .067$.

This experiment confirms that High and Low selected line pups are showing species-typical responses in USV to adult males. High line pups suppressed USV

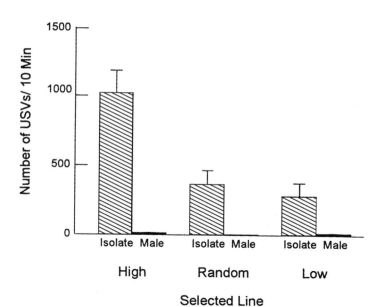

Figure 6. USV rates shown by samples of 10-day-old pups of the S_{10} generation litters (2 pups/litter) of the High ($n = 8$), Random ($n = 10$), and Low ($n = 6$) USV lines in response to 10-min periods of isolation versus 10-min exposure to a strange, anesthetized adult male.

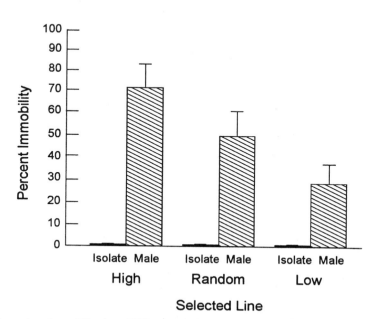

Figure 7. Percentage immobility (per 40 15-sec periods in 10 min) shown by pups from each line during 10-min isolation versus exposed to an anesthetized adult Wistar male.

responses to the adult male: in this they were not qualitatively different from either the unselected controls or from Wistar strain rat pups (Brunelli *et al.*, 1998; Shair *et al.*, 1997). On the other hand, the two lines appear to show differences in two quite different fear-related behaviors, immobility and urination, in directions that would be predicted by their USV responses in isolation: that is, High line pups exhibited more of the these fearful behaviors and Low line pups less.

SIGNIFICANCE OF CORRELATIONS FOUND IN SELECTED LINES. Not all of the predictions based on correlations found in the PR generations so far have been fulfilled. Notice that lines do not differ on face washing, which has been significantly correlated with USV in the principal component analyses in the PR and S_9–S_{12} generations.

What we do find is that after a number of selected generations, different profiles of associations are occurring in the High and Low USV lines. This supports a polygenic model of inheritance, that of multiple genes contributing to variation in USV. In a polygenic system, the longer a bidirectional selective breeding program is carried out and if care is taken to prevent inbreeding, the more likely it is that more correlated phenotypes will emerge (DeFries, 1981; Hyde, 1981). This is because early in selection, the genes which contribute most to variance in the phenotype will segregate first (Snustad *et al.*, 1997), whereas those that produce minor variation in the phenotype will take longer for their effects to be shown (Hyde, 1981). An instance of this was a program in which mice were selected for fast (high locomotor activation) and slow (depressed locomotor activity) responses to ethanol (Shen *et al.*, 1996). In early generations, these mice showed no other sensitivities to the effects of alcohol, whereas by generations S_{36} to S_{37}, selected lines were differentially sensitive to ethanol's central sedative effects ("slow") or excitatory effects ("fast"). Similarly, Colombo *et al.* (1995) reported differences in the percent of entries into the open arm of the plus maze (File, 1992), indicating greater anxiety in the S_{35} generation of two rat strains originally selectively bred for alcohol preferences.

Should it be the case that in future generations 10-day-old Low line pups continue to show smaller body size and High line pups continue to show more urination in isolation and more freezing and urinating in response to fear stimuli (with concomitant underlying neurochemical alterations), the implication would be that multiple genes controlling USV were manifesting different phenotypic associations. In that case alleles expressing "high" and those expressing "low" USV would likely show different genetic associations with behavioral and physiological processes, accounting for different phenotypic associations in the two lines. Classic experimental crosses such as those used in SHR-WKY or SHR-BB/OK lines described earlier to generate F_2 hybrids between High and Low lines will allow the detection of such associations. Combined with mapping strategies, locations of QTLs will provide insights into how genes controlling "high" and "low" USVs interact with other genes.

Alternatively, the current results may be mediated by other mechanisms entirely, including random genetic events, weak genetic correlations, sampling errors, or restricted ranges of variability (Crabbe *et al.*, 1999; DeFries, 1981; Henderson, 1989). To test the reliability of these associations, replicate selected lines will be necessary (Crabbe *et al.*, 1990; DeFries, 1981; Henderson, 1989). For instance, maternal effects (Myers, Brunelli, Squire, Shindledecker, & Hofer, 1989; Roubertoux *et al.*, 1990) may be mediating the decreased body size in Low line pups. It is possible that in selecting for low rates of USV in Low line pups, rates of some other behavioral correlate may be decreased or missing so that dams are not provided with sufficient stimulation to induce adequate maternal nursing behavior (Stern & Johnson, 1990). This would ultimately affect weight gain in the Low line. Illustrating this possibility, Myers, Brunelli, Squire, *et al.* (1989) characterized the postnatal maternal environments of SHR and WKY rats with similar results. Making careful observations of naturally occurring maternal activities, the authors found that although the qualitative structure of maternal behaviors was similar for SHR and WKY dams, highly significant differences were found in the frequency of occurrence of maternal behaviors between the strains. SHR dams assumed nursing postures associated with milk delivery more often than WKY mothers (Cierpial, Shasby, & McCarty, 1987). SHR pups were consistently more active than WKY pups. In a subsequent study (Myers, Brunelli, Shair, Squire, & Hofer, 1989), cross-fostering F_1 pups to SHR and WKY mothers eliminated differences in nursing behavior of SHR and WKY mothers, demonstrating that maternal behavior was sensitive to phenotypic differences in pups. In our selectively bred USV lines, studies are currently underway to test for such correlated environmental maternal effects through naturalistic behavioral observations of mothers and infants. A concurrent postnatal cross-fostering study between the lines is examining the effects of maternal care on USV rates and behavior in cross-fostered infants. Special attention will be paid to nursing interactions and weight gain, but also to interactions which are known to affect fear-mediated behaviors in adulthood (Caldji *et al.*, 1998; Liu *et al.*, 1997). Not only will these studies provide information on maternal–environmental–genetic interactions, but could delineate mechanisms whereby central pathways mediate such effects.

Development of USV in Selected Lines: Heterochrony in Infancy? Figure 8 illustrates how heterochrony (selection-induced change in the onset or offset of the development of USV) could be involved in the differentiation of USV rates in the two selected lines, and alternative scenarios. In Figure 8A, a change at 10 days would alter rates of USV at all other postnatal ages: this would mean that there is stability in postnatal development. Figures 8B and 8C show that selection could work through shifts in developmental timing of USV. Figure 8B shows a right shift in the developmental curve, suggesting a retention of high infantile rates of USV at later ages (neoteny); the reverse (Figure 8C) would be a left shift, suggesting accelerated development of reduced USV rates with age (progenesis; McKinney & McNamara, 1991). Or selection might be unique to USV rates at postnatal (PN) day 10 (Figure 8D).

A developmental study in a small subsample of S_5 generation pups at PN days 3, 10, 14, and 18 (Figure 9) revealed that as in PR generations, the three lines continued to exhibit their highest rates at PN day 3. At PN day 3, both High and Low line pups showed significantly higher USV rates than Random line pups, $F(2, 9) = 9.51, p < .01$. At PN day 10, USV rates in High and Low line pups were significantly different from Random line pups in predicted directions, $F(2, 24) = 3.19, p < .05$, which was also true at PN day 14, $F(2, 24) = 5.28, p < .05$. Overall, the pattern of development of USV in the High line was shifted to the right, consistent with the hypothesis that selection for high rates is occurring through neoteny (Figure 8B), whereas Low line pups exhibited a developmental pattern consistent with progenesis (Figure 8C). These data provide modest support for the hypothesis that heterochronic processes are acting to produce altered rates of USV in the two selected lines in opposite

SUSAN A.
BRUNELLI AND
MYRON A. HOFER

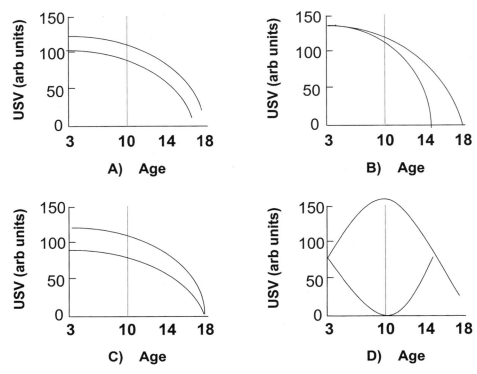

Figure 8. Theoretical development of postnatal USV rates based on predictions of heterochrony and alternative hypotheses.

directions during the postnatal period. But as Figure 9 also shows, for all three lines, USV rates went to virtually zero at 18 days, showing that selection had not extended development of USV beyond the postnatal period by five generations of selective breeding.

Studies are underway to revisit this question in a later generation to determine whether continuing selective pressure will clarify the picture. The most likely outcome is that USV will remain relatively fixed in a pattern of development and not extend beyond the postnatal period. If it is anchored to that particular period, the 40- to 50-kHz isolation-induced call could be thought of as an instance of a class of purely infantile behaviors, like postnatal suckling (Hall, 1975), fetal behavior (Lickliter, 1995; S. R. Robinson & Smotherman, 1995), play (Oppenheim, 1981), or early learning (Alberts, 1987). Such behaviors are called "ontogenetic adaptations" (Oppenheim, 1981) because they are thought to be specifically adapted to the particular ecological niche occupied by an organism at a particular stage of development (Oppenheim, 1981; West & King, 1987). In fact, it has been proposed that the process of development may be viewed as a sequence of adaptive reorganizations in response to a series of discrete adaptive niches encountered during the lifetime of an organism (Alberts, 1987). One of the implications of the concept of the ontogenetic niche is that features or processes adapted to ontogenetic niches and displayed early in life are not necessarily precursors or antecedents of adult characteristics (Oppenheim, 1980) and display no developmental continuity per se. In short, ontogenetic adapta-

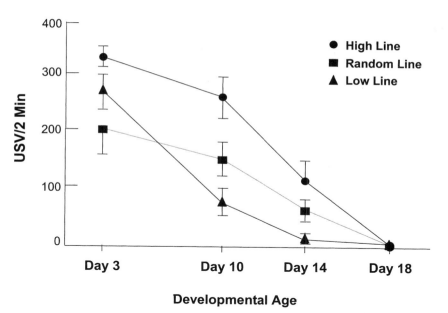

Figure 9. Samples of the S_5 generation observed at postnatal days 3, 10, 14, and 18 in the High, Random, and Low USV lines. (Reprinted from Brunell *et al.*, 1997. Copyright 1997, Wiley–Liss, Inc., a subsidiary of John Wiley & Sons, Inc.)

tions promote the survival of young organisms at the time they are seen, without reference to any future abilities or functions.

Nevertheless, the fact that a trait identified as an ontogenetic adaptation may not appear again later in development does not logically preclude a role underlying functional organization in development (Oppenheim, 1980, 1981; West & King, 1987). In this scheme, if USV may be regarded as a motor output that subserves different functions, then it may serve simultaneously or sequentially (or both) as a functional constituent of a social attachment system, a communicatory system, a thermoregulatory system, and a system mediating fear or anxiety, the disparate elements of which are developing at different rates relative to one another. In other words, USV may be one of a number of components or part-processes which appear across functional domains during juvenile development (Turkewitz & Devenny, 1993) and in adults. In that case, the disappearance of 40-kHz vocalizations at a certain point does not imply the disappearance of any functional system to which it was linked. Rather, it could mean that another motor output (either arising *de novo* or present all along) may now become primarily associated with that system, a situation which will give the appearance of discontinuity in development, but will actually be a reorganization of the underlying system. In the theoretical model we are proposing for USV, as we have for motor sequences in play (Brunelli & Hofer, 1990), these behaviors would both subserve functional systems that are organized and reorganized, using, sharing, and discarding different components to meet the needs of the organism throughout development. As a developmental–genetic model, it would suggest that selection pressure on the 10-day USV response is likely to be recruiting genes which are linked to the USV phenotype throughout development. Some of these may produce variations in central systems, others in peripheral systems at different times in the life of the organism. Examination of other measures in different contexts over development will allow us to test hypotheses about genetic correlations between central processes mediating anxiety and the selected USV phenotype as we continue to exert selection pressure on the two USV lines.

SUMMARY AND PREDICTIONS

In this final section, we will summarize and integrate the several components of this chapter by using the genetic and developmental principles described earlier, to predict possible outcomes of our selective breeding approach and to discuss how different results could help answer basic questions about the organization of early affect states, their genetic basis, and the processes underlying the development and evolution of affective responses from infant to adult.

In the first section, we outlined the evidence for isolation calling rate in young rats as an infantile trait expressing the intensity of an affective response and sharing a number of developmental, behavioral, and psychopharmacologic properties with human clinical anxiety states and with childhood separation anxiety in particular. There is only indirect evidence for heritability of this trait in either animals or humans, and the means by which vulnerability to early separation may be passed from one generation to the next is currently of considerable research interest (Klein, 1995; Suomi, 1997). In this regard, the present report demonstrates that a complex, multifaceted infantile behavior, isolation calling, can be selected for rather quickly and that the basis of this selection could be evaluated empirically.

When we began our project, a number of developmental researchers predicted that we would not succeed in obtaining any significant differences between lines selected for high and low USV rates within 15 or 20 generations. Variation in this trait, they thought, was primarily the result of individual differences in early experience. Others warned us that artificial selection would only produce artifactual changes such as structural defects in the organs for sound production, or gross and nonspecific changes such as altered sensory thresholds or motor inhibition. Selective breeding in the laboratory for calling rate, some asserted, cannot change an established communicative behavior that is important for survival. Evolutionary change in such a system can only take place in the animal's normal habitat, with natural selection acting on variation in overall fitness within the social group.

These diverse predictions emphasized to us how little we really knew about how selection can act on behavior, and particularly on an infantile trait. Indeed, we could find no previous studies on experimental selection during early development in mammals. Moreover, the biochemical and physiological pathways mediating USV production, as for most behaviors, are complex and involve many steps at which differential gene expression could control emission rates. If successful, at what point in this system might we predict that selection would act?

The infant's isolation call has been viewed within several different conceptual frameworks and selective breeding can be used to test alternative views. For example, a communication model would not predict line differences in heart rate or adrenocortical responses to isolation, whereas an affective or social separation model would. USV responses in rat pups can also be viewed as having evolved as a thermoregulatory adaption and therefore represent primarily a physiological response to thermal aspects of isolation rather than a behavioral response within a social/

communicative adaptation. In this case we would expect to see major differences in thermogenic responses to isolation in the three lines.

We can anticipate that as selection progresses, increasing numbers of individuals will be produced in the High line that fall outside the upper limit of the original range of USV observed in the progenitor stock and/or that pups produce their vocalizations by novel mechanisms, either centrally or peripherally, such as two short ultrasonic pulses in a single expiration in the High line instead of the normal single call. With the recruitment of alleles from several genes contributing to the trait, novel combinations will occur, producing novel traits (pleiotropic effects). These pups could provide us with an opportunity to identify novel mechanisms that would not be evident in the progenitor strain. Furthermore, we can predict that selective breeding will reveal some parallel changes in other systems that we had not expected to contribute to the selected trait, thus revealing novel physiological mechanisms for USV regulation. If coselected in this way, these traits need not be genetically linked to USV production and would segregate away from the USV trait in the second generation of interbreeding of the two lines (F_2). Thus we anticipate that selective breeding will reveal the presence of novel mechanisms for USV regulation in a readily recognizable form for future analysis.

One of the major strategies of this study has been to attempt to use selective breeding to create two lines of infants that differ selectively in the neural substrates underlying the USV response to isolation. For example, we hope to find differences between lines in c-*fos* (early-immediate gene) activation only in those areas of the brain that regulate the USV response, whereas other brain areas mediating other responses to isolation will show similar levels of c-*fos* activation. This "subtraction enhancement" should give us a map of the brain areas that are activated differentially in the two different lines. We predict that this map will involve not only the motor pathway for USV production, but also the neural substrates for an affective system responding differentially to isolation in the two strains. By integrating what is currently known about the neuroanatomy of distress vocalization in a variety of experimental animals (Jurgens, 1994) and the clinical evidence in patients with panic disorder (Rauch, Savage, Alpert, Fischman, & Jenike, 1997), we can consider that the cingulate cortex, periaqueductal gray, the medial hypothalamus, the septal area, and the bed nucleus of the stria terminalis are areas most likely to be highly divergent in the two lines. We will also be able to map the distribution of serotonin, opioid, benzodiazepine, NMDA, and other candidate neuromodulator systems in the two lines. The kinds of differences we might expect to see are illustrated in a more extreme form in two closely related species of voles that differ markedly in affiliative behaviors (Insel & Shapiro, 1992). In the highly parental prairie vole, oxytocin receptor density was highest in the prelimbic cortex, bed nucleus of the stria terminalis, nucleus accumbens, midline nuclei of the thalamus, and the lateral amygdala. In contrast, the montane vole, which has only low levels of affiliative behavior, showed little oxytocin binding in those areas; instead showing high levels in the lateral septum, hypothalamus, and the central amygdala. These chemo-anatomic mapping studies will help us answer the questions raised earlier: where in the pathway from gene to behavior is the selection process acting, and does it act simply on a distal motor output site such as the periaqueductal gray, or are higher integrative systems involved?

The results of our search for neuromodulator systems should give us clues to the specific genetic loci at which alterations have occurred. For example, genes for

5-HT$_{1A}$, 5-HT$_{1B}$, 5-HT$_{2A}$, and the serotonin transporter have been cloned and their loci identified. Both 5-HT$_{1A}$ and 5-HT$_{1B}$ are known to control USV (see Table 1). Specific screens would be used to look for alterations at those sites since these are candidate genetic sites at which our selection may have acted. Another approach would be to examine the genetic sites already identified by QTL analysis in rats that were previously bred for high and low levels of a trait consisting of behavioral inhibition and defecation/urination in a novel bare test arena (open field). In our study, pups in the progenitor stock, pups in the high USV line, and adults in the same line also showed increased urination and/or defecation. Possibly one of the three sites identified in those studies also may be altered in our selection for a different anxiety trait at a different age. This will be possible through the availability of dense marker maps for the rat (D. M. Brown *et al.*, 1998; Szpirer *et al.*, 1999) on which to genotype animals and look for polymorphisms between the two strains. Following other studies of selectively bred rat strains (e.g., Carr *et al.*, 1998), we would likely inbreed the High and Low USV lines (~20 generations), and an F$_2$ intercross generation would allow the detection of QTLs influencing USV and other, potentially correlated phenotypes. Mapping these to specific loci could provide clues as to what genes will be associated with USV. As suggested above, we might find that the 5-HT$_{1A}$ receptor maps to a particular region of a chromosome implicated in a QTL analysis, allowing the formulation of hypotheses with regard to the role of serotonin in the mediation of isolation-induced USV and the eventual location of a gene or set of genes in that region. Alternatively, once a candidate system was identified, knock-out or transgenic technology could reproduce alterations in mice. Combined with gene knockout technology, in which mice lack specific targeted genes (e.g., the 5-HT$_{1B}$ knockout Sandou *et al.*, 1994), putative genes identified by QTL can be tested for specificity in a behavioral response (Crabbe *et al.*, 1999).

In the foregoing, we have explored what we hope to learn about the infant's separation call through an experimental analysis of the pathways within the infant organism extending from genes to behavior. Now we turn to what we hope to learn about the role of the infants' external environment in the expression of the selected trait through development. Here we are not so concerned with direct effects, such as the influence of ambient temperature, as with the route from gene to behavioral phenotype through maternal behavior. One such anticipated effect is that the very high (or low) USV rates of the pups may well alter the behavior of their mothers. USV has been shown to increase maternal licking, inhibit rough maternal tactile contact, and elicit pup carrying, all interactions known to have long-lasting effects on pup behavior and brain development (Myers *et al.*, 1989; Plotsky & Meaney, 1993). These USV-induced changes in maternal behavior, in turn, have the potential to shape the development of the pup's USV regulation, a feedback or feed-forward effect. The second possible route is through a change in adult maternal behavior that is an adult expression of the altered infantile trait. For example, a lasting change in emotionality that was the result of selection for an infantile anxiety trait could also be expressed in the adult as altered maternal behavior toward her pups. The altered interaction that ensued could either support or suppress expression of the infantile trait in the next generation. We anticipate that there may be such secondary effects stemming from altered postnatal maternal behavior, and we will test for them by maternal behavior observation and cross-fostering. There are also possible prenatal and/or maternal cytoplasmic effects which could be detected by ova transplantation.

Perhaps the most important developmental issue that we may be able to study in our model involves the question of possible changes in the development of USV and other behaviors after the divergence of the two lines at 10 days of age. Infantile trait selection may even allow us a direct experimental approach to the long-debated relationship between the processes of development and those underlying evolution. Garstang, one of the most eloquent of those to first disagree with Haeckel's famous dictum, put it neatly: "Ontogeny does not recapitulate phylogeny, it creates it" (Garstang, 1929). But the processes by which this "creation" may take place remain speculative. Developmental processes (environmental as well as genetic influences) can be viewed as a major source of phenotypic variation on which natural selection can act (Bonner, 1974, 1993). The concept of heterochrony allows us to conceptualize more precisely how early developmental differences can have unforeseen long-term correlates both in development and evolution. The word "heterochrony" describes both an observation of certain patterns in evolution and an inferred process by which development creates evolution. Differences between ancestors and descendants often appear to be the result of changes in the rate, duration, or timing of developmental events in one organ or system in relation to others. In some instances, small changes in one system during development may cause extensive displacements of form and organization later in development, and have been proposed as a major mechanism for macroevolution. In our study, we could hypothesize, for example, that pups selected for high USV rate in midinfancy (10 days of age) may be those that tend to retain the high USV rates of the newborn period into mid-infancy, thus delaying the weaning age reduction in USV. This variation in the developmental trajectory of the USV control system may be progressively amplified by selective breeding to the point that the USV isolation response continues into adolescence or even adulthood. Strong interactions may also occur with schedules of development in other physiologically or genetically related systems. This is one possible way that widespread and essentially unpredictable downstream effects create complex differences in patterns of behavior in our two strains. We would predict that interaction effects of this sort would be most likely to take place in other systems mediating affect regulation in adulthood, but they would not follow any simple linear model. Paradoxically, *reductions* in affect response intensity might be found in adults of the High line as a result of the nonlinear interactions likely to proceed from displacement of one developmental schedule relative to another closely related one.

In the discussion above, we have from time to time related our selective breeding to natural selection in evolution. Clearly this is not an exact parallel, but in contrast to many laboratory studies using this approach, such as selection for the animals' extreme responses to substances of abuse, selection for isolation calling rate is actually likely to occur in nature. In some habitats in which predators abound and dams must travel far in foraging, silent pups will survive most often, whereas in other conditions, such as marginal nesting areas with frequent scattering of pups due to flood or winds, high-calling-rate pups will be easiest for the mother to find and to retrieve. The need for rapid evolutionary response to changes in the cost–benefit ratio of infant isolation calling in the rat's evolutionary past may account for the rapid selection we have observed thus far in our study. We are dealing with a response system that may have evolved to be evolutionarily flexible and in the two extreme lines we may be tapping the potential for different adaptive strategies rather than the extremes of a single trait. If true, this would have implications for the results we should anticipate. The Low line may show responses in other systems that differ in kind as well as in intensity from those in the High line. Consequently, we may find that different neuromodulators and different neural substrates will be found changed in the two lines, each differing from the random controls in their own particular patterns. The alternative, which we think less likely, is that High and Low line pups represent extreme levels of one feature of an affective display in an evolved communication system within which the pups gain advantage in their relations either with their mother or with predators. Thus, although speculative, it may be possible to find clues in the results of our selective breeding study as to the selection pressures that were present in the recent evolutionary history of the rat.

REFERENCES

Abel, E. (1991). Behavior and corticosteroid response of Maudsley reactive and nonreactive rats in the open field and forced swimming test. *Physiology and Behavior, 50,* 151–153.

Adamec, R. E., Burton, P., Shallow, T., & Budgell, J. (1999). NMDA receptors mediate lasting increases in anxiety-like behavior produced by the stress of predator exposure– implications for anxiety associated with posttraumatic stress disorder. *Physiology and Behavior, 65,* 723–737.

Alberts, J. R. (1987). Early learning and ontogenetic adaptation. In N. A. Krasnegor, E. M. Blass, M. A. Hofer, & W. P. Smotherman (Eds.), *Perinatal development: A psychobiological perspective* (pp. 11–37). Orlando, FL: Academic Press.

Albinsson, A., Bjork, A., Svartengren, J., Klint, T., & Anderson, G. (1994). Preclinical pharmacology of FG5893: A potential anxiolytic drug with high affinity for both 5-HT$_{1A}$ and 5-HT$_{2A}$ receptors. *European Journal of Pharmacology, 261,* 285–294.

Allin, J. T., & Banks, E. M. (1971). Functional aspects of ultrasound production by infant albino rats (*Rattus norvegicus*). *Animal Behaviour, 20,* 174–185.

Arborelius, L., Owens, M. J., Plotsky, P. M., & Nemeroff, C. B. (1999). The role of corticotropin-releasing factor in depression and anxiety disorders. *Journal of Endocrinology, 160,* 1–12.

Archer, J. (1973). Tests for emotionality in rats and mice: A review. *Animal Behaviour, 21,* 205–235.

Barr, G. A., & Wang, S. (1992). Tolerance and withdrawal to chronic morphine treatment in the week-old rat pup. *European Journal of Pharmacology, 215,* 35–42.

Barr, G. A., & Wang, S. (1993). Behavioral effects of chronic cocaine treatment in the week-old rat pup. *European Journal of Pharmacology, 233,* 143–149.

Barr, G. A., Wang, S., & Carden, S. E. (1994). Aversive properties of the κ opioid agonist U50,488 in the week-old rat pup. *Psychopharmacology, 113,* 422–428.

Beart, P. M., Summers, R. J., Stephenson, J. A., Cook, C. J., & Christie, M. J. (1990). Excitatory amino acid projections to the periaqueductal gray in the rat: A retrograde transport study utilizing D[3H] aspartate and [3H] GABA. *Neuroscience, 34,* 163–176.

Belknap, J. K., Metten, P., Helms, M. L., O'Toole, L. A., Angeli-Gade, S., Crabbe, J. C., & Phillips, T. J. (1993). Quantitative trait loci (QTL) applications to substances of abuse: Physical dependence studies with nitrous oxide and ethanol in BXD mice. *Behavior Genetics, 23,* 213–222.

Bell, R. W., Nitschke, W., Gorry, T. H., & Zachman, T. A. (1971). Infantile stimulation and ultrasonic signaling: A possible mediator of early handling phenomena. *Developmental Psychobiology, 4,* 181–191.

Belzung, C., & Le Pape, G. (1994). Comparison of different behavioral test situations used in psychopharmacology for measurement of anxiety. *Physiology and Behavior, 56,* 623–628.

Berton, O., Ramos, A., Chaouloff, F., & Mormède, P. (1997). Behavioral reactivity to social and nonsocial stimulations: A multivariate analysis of six inbred rat strains. *Behavioral Genetics, 27,* 155–166.

Blanchard, D. C., & Blanchard, R. J. (1988). Ethoexperimental approaches to the biology of emotion. *Annual Review of Psychology, 39,* 43–68.

Blanchard, D. C., Blanchard, R. J., Tom, P., & Rodgers, R. J. (1990). Diazepam changes risk assessment on an anxiety/defense test battery. *Psychopharmacology, 101,* 511–518.

Blanchard, D. C., Griebel, G., Rodgers, R. J., & Blanchard, R. J. (1998). Benzodiazepine and serotonergic modulation of antipredator and conspecific defense. *Neuroscience and Biobehavioral Reviews, 22,* 597–612.

Blanchard, R. J., Shepherd, J. K., Rodgers, R. J., Magee, L., & Blanchard, D. C. (1993). Attenuation of antipredator defensive behavior in rats following chronic imipramine. *Psychopharmacology, 110,* 245–253.

Blanchard, R. J., Griebel, G., Henrie, J. A., & Blanchard, D. C. (1997). Differentiation of anxiolytic and panicolytic drugs by effects on rat and mouse defense test batteries. *Neuroscience and Biobehavioral Reviews, 21,* 783–789.

Blass, E. M., & Brunson, L. (1990). Interference between opioid and nonopioid mechanisms of calming in 10-day-old rats. *Neuroscience Abstracts*, 16, 211.

Blass, E. M., & Fitzgerald, E. (1988). Milk-induced analgesia and comforting in 10-day-old rats: Opioid mediation. *Pharmacology, Biochemistry and Behavior*, 29, 9–13.

Blass, E. M., & Kehoe, P. (1987). Behavioral characteristics of emerging opioid systems in newborn rats. In N. A. Krasnegor, E. M. Blass, M. A. Hofer, & W. P., Smotherman, (Eds.), *Perinatal development: A psychobiological perspective* (pp. 61–82). New York: Academic Press.

Blass, E. M., & Shide, D. J. (1993). Endogenous cholecytokinin reduces vocalization in isolated 10-day-old rats. *Behavioral Neuroscience*, 107, 488–492.

Blizard, D. A. (1981). The Maudsley reactive and nonreactive strains: A North American perspective. *Behavior Genetics*, 11, 469–489.

Blumberg, M. S., Efimova, I. V., & Alberts, J. R. (1992). Thermogenesis during ultrasonic vocalization by rat pups isolated in a warm environment: A thermographic analysis. *Developmental Psychobiology*, 24, 497–510.

Bonner, J. T. (1974). *On development*. Cambridge, MA: Harvard University Press.

Bonner, J. T. (1993). *Life cycles: Reflections of an evolutionary biologist*. Princeton, NJ: Princeton University Press.

Bowlby, J. (1969). *Attachment*, Vol. 1. New York: Basic Books.

Brain, P. F., Kusumorini, N., & Benton, D. (1991). Anxiety in laboratory rodents: A brief review of some recent behavioral developments. *Behavioral Processes*, 25, 71–80.

Broadhurst, P. L. (1961). Analysis of maternal effects in the inheritance of behaviour. *Animal Behaviour*, 9, 129–141.

Broadhurst, P. L. (1975). The Maudsley reactive and non-reactive strains of rats: A survey. *Behavior Genetics*, 5, 299–319.

Brown, D. M., Matise, T. C., Koike, G., Simon, J. S., Winer, E. S., Zangen, S., McLaughlin, M. G., Shiozawa, M., Atkinson, O. S., Hudson, J. R., Jr., Chakravarti, A., Lander, E. S., & Jacob, H. J. (1998). An integrated genetic linkage map of the laboratory rat. *Mammalian Genome*, 9, 521–530.

Brown, R. E. (1986). Social and hormonal factors influencing infanticide and its suppression in adult male Long-Evans rats (*Rattus norvegicus*). *Journal of Comparative Psychology*, 100, 155–161.

Brunelli, S. A., & Hofer, M. A. (1990). Parental behavior in juvenile rats: Environmental and biological determinants. In N. A. Krasnegor, & R. S. Bridges (Eds.), *Mammalian parenting: Biochemical, neurobiological, and behavioral determinants* (pp. 372–399). New York: Oxford University Press.

Brunelli, S. A., & Hofer, M. A. (1996). Development of ultrasonic vocalization responses in genetically heterogeneous National Institute of Health (N:NIH) rats. II. Associations among variables and behaviors. *Developmental Psychobiology*, 29, 517–528.

Brunelli, S. A., Keating, C. C., Hamilton, N. A., & Hofer, M. A. (1996). Development of ultrasonic vocalization responses in genetically heterogeneous National Institute of Health (N:NIH) rats. I. Influence of age, testing experience and associated factors. *Developmental Psychobiology*, 29, 507–516.

Brunelli, S. A., Vinocur, D. D., Soo-Hoo, D., & Hofer, M. A. (1997). Five generations of selective breeding for ultrasonic vocalization (USV) responses in N:NIH strain rats. *Developmental Psychobiology*, 31, 255–265.

Brunelli, S. A., Masmela, J. R., Shair, H. N., & Hofer, M. A. (1998). Effects of biparental rearing on ultrasonic vocalization (USV) responses of rat pups (*Rattus norvegicus*). *Journal of Comparative Psychology*, 112, 1–13.

Cagiano, R., Sales, G. D., Renna, G., Racagni, G., & Cuomo, V. (1986). Ultrasonic vocalization in rat pups: Effects of early postnatal exposure to haloperidol. *Life Sciences*, 38, 1417–1423.

Cairns, R. B. (1993). Belated but bedazzling: Timing and genetic influence in social development. In G. Turkewitz & D. A. Devenny (Eds.), *Developmental time and timing* (pp. 61–84). Hillsdale, NJ: Erlbaum.

Cairns, R. B., MacCombie, D. J., & Hood, K. E. (1983). A developmental-genetic analysis of aggressive behavior in mice: I. Behavioral outcomes. *Journal of Comparative Psychology*, 97, 69–89.

Cairns, R. B., Hood, K. E., & Midlam, J. (1985). On fighting in mice: Is there a sensitive period for isolation effects? *Animal Behaviour*, 33, 49–65.

Cairns, R. B., Garièpy, J.-L., & Hood, K. E. (1990). Development, microevolution, and social behavior. *Psychological Review*, 97, 49–65.

Caldji, C., Tannenbaum, B., Sharma, S., Francis, D., Plotsky, P. M., & Meaney, M. J. (1998). Maternal care during infancy regulates the development of neural systems mediating the expression of fearfulness in the rat. *Proceedings of the National Academy of Sciences of the USA*, 95, 5335–5340.

Carden, S. E., & Hofer, M. A. (1990a). Independence of benzodiazepine and opiate action in the suppression of isolation distress in rat pups. *Behavioral Neuroscience*, 104, 457–463.

Carden, S. E., & Hofer, M. A. (1990b). Socially mediated reduction of isolation distress in rat pups is blocked by naltrexone but not by RO15–1788. *Behavioral Neuroscience*, 106, 421–426.

Carden, S. E., & Hofer, M. A. (1990c). The effects of opioid and benzodiazepine antagonists on dam-induced reductions in rat pup isolation distress. *Developmental Psychobiology, 23*, 797–808.

Carden, S. E., & Hofer, M. A. (1991). Isolation-induced vocalization in Wistar rat pups is not increased by naltrexone. *Physiology and Behavior, 49*, 1279–1282.

Carden, S. E., Barr, G. A., & Hofer, M. A. (1991). Differential effects of specific opioid receptor agonists on rat pup isolation calls. *Developmental Brain Research, 62*, 17–22.

Carden, S. E., Davachi, L., & Hofer, M. A. (1994). U50,488 increases ultrasonic vocalizations in 3-, 10-, and 18-day-old rat pups in isolation and the home cage. *Developmental Psychobiology, 27*, 65–83.

Carlier, M., Roubertoux, P. L., & Pastoret, C. (1991). The Y chromosome effect on intermale aggression in mice depends on the maternal environment. *Genetics, 129*, 231–236.

Carr, L. G., Foroud, T., Bice, P., Gobbett, T., Ivashina, J., Edenberg, H., Lumeng, L., & Li, T. K. (1998). A quantitative trait locus for alcohol consumption in selectively bred rat lines. *Alcoholism: Clinical and Experimental Research, 22*, 884–887.

Castanon, N., Perez-Diaz, F., & Mormède, P. (1995). Genetic analysis of the relationships between behavioral and neuroendocrine traits in Roman High and Low avoidance rat lines. *Behavior Genetics, 25*, 371–384.

Chua, S. C. Jr. (1997). Monogenetic models of obesity. *Behavior Genetics, 27*, 277–284.

Cierpial, M. A., Shasby, D. E., & McCarty, R. (1987). Patterns of maternal behavior in the spontaneously hypertensive rat. *Physiology and Behavior, 39*, 633–637.

Coe, C. L., & Erickson, C. M. (1997). Stress decreases lymphocyte cytolytic activity in the young monkey even after blockade of steroid and opiate hormone receptors. *Developmental Psychobiology, 30*, 1–10.

Coe, C. L., Rosenberg, L. T., & Levine, S. (1988). Effect of maternal separation on the complement system and antibody response in infant primates. *International Journal of Neuroscience, 40*, 289–302.

Colombo, G., Agabio, R., Lobina, C., Reali, R., Zocci, A., Fadda, F., & Gessa, G. L. (1995). Sardinian alcohol-preferring rats: A genetic animal model of anxiety. *Physiology and Behavior, 57*, 1181–1185.

Cook, M. N., Ramos, A., Courvoisier, H., & Moisan, M. P. (1998). Linkage mapping of alpha 3, alpha 5, and beta 4 neuronal nicotinic acetylcholine receptors to rat chromosome 8. *Mammalian Genome, 9*, 177–178.

Coplan, J. D., Gorman, J. M., & Klein, D. F. (1992). Serotonin related functions in panic-anxiety: A critical overview. *Neuropsychopharmacology, 6*, 189–200.

Crabbe, J. C. (1999). Animal models in neurobehavioral genetics: Methods for estimating genetic correlation. In B. C. Jones & P. Mormède (Eds.), *Neurobehavioral genetics: Methods and applications* (pp. 121–138). Boca Raton, FL: CRC Press.

Crabbe, J. C., & Belknap, J. K. (1992). Genetic approaches to drug dependence. *Trends in Pharmacological Sciences, 13*, 212–219.

Crabbe, J. C., & Phillips, T. J. (1993). Selective breeding for alcohol withdrawal severity. *Behavior Genetics, 23*, 171–177.

Crabbe, J. C., Phillips, T. J., Kosobud, A., & Belknap, J. K. (1990). Estimation of genetic correlation: Interpretation of experiments using selectively bred and inbred animals. *Alcoholism: Clinical and Experimental Research, 14*, 141–151.

Crabbe, J. C., Belknap, J. K., & Buck, K. J. (1994). Genetic animal models of alcohol and drug abuse. *Science, 264*, 1715–1723.

Crabbe, J. C., Phillips, T. J., Buck, K. J., Cunningham, C. L., & Belknap, J. K. (1999). Identifying genes for alcohol and drug sensitivity: Recent progress and future directions. *Trends in Neurosciences, 22*, 173–179.

Crusio, W. R. (1999). An introduction to quantitative genetics. In B. C. Jones & P. Mormède (Eds.), *Neurobehavioral genetics: Methods and applications* (pp. 13–42). Boca Raton, FL: CRC Press.

Cuomo, V., Cagiano, R., Renna, G., De Salvia, A., & Racagni, G. (1987). Ultrasonic vocalization in rat pups: Effects of early postnatal exposure to SCH 23390 (a DA_1-receptor antagonist) and sulpiride (a DA_2-receptor antagonist). *Neuropharmacology, 26*, 701–705.

Darwin, C. (1859/1979). *The origin of species.* New York: Avenel Books.

DeCasper, J., & Fifer, W. P. (1980). Of human bonding: Newborns prefer their mother's voices. *Science, 208*, 1174–1176.

DeFries, J. C. (1969). Pleiotropic effects of albinism on open field behavior in mice. *Nature, 221*, 65–66.

DeFries, J. C. (1981). Current perspectives on selective breeding: Example and theory. In G. E. McClearn, R. A. Dietrich, and V. G. Erwin (Eds.), *Development of animal models as pharmacogenetic tools* (pp. 11–35). Rockville, MD: U.S. Department of Health and Human Services, Public Health Service: Alcohol, Drug Abuse & Mental Health Administration.

De Fries, J. C., Hyde, J. S., Lynch, C., Petersen, D., & Roberts, R. C. (1981). The design of selection experiments. In G. E. McClearn, R. A. Dietrich, & V. G. Erwin (Eds.), *Development of animal models as pharmacogenetic tools* (pp. 269–275). Rockville, MD: U.S. Department of Health and Human Services, Public Health Service: Alcohol, Drug Abuse & Mental Health Administration.

Deitrich, R. A. (1993). Selective breeding for initial sensitivity to ethanol. *Behavior Genetics, 23*, 153–162.

Denenberg, V. H. (1969). Open-field behavior in the rat: What does it mean? *Annals of the New York Academy of Sciences, 159*, 852–859.

Devinsky, O., Morrell, M. J., & Voigt, B. A. (1995). Contributions of anterior cingulate cortex to behaviour. *Brain, 118*, 279–306.

Devor, E. J. (1993). Why there is no gene for alcoholism. *Behavior Genetics, 23*, 145–151.

Dunn, R. W., Corbett, R., & Fielding, S. (1989). Effects of 5-HT$_{1A}$ receptor agonists and NMDA receptor antagonists in the social interaction test and the elevated plus maze. *European Journal of Pharmacology, 169*, 1–10.

File, S. E. (1992). Behavioral detection of anxiolytic action. In J. M. Elliott, D. J. Heal, & C. A. Marsden (Eds.), *Experimental approaches to anxiety and depression* (pp. 25–44). London: Wiley.

File, S. E. (1995). Animal models of different anxiety states. In G. Biggio, E. Sanna, & E. Costa (Eds.), *GABA-A receptors and anxiety: From neurobiology to treatment* (pp. 93–113). New York: Raven Press.

File, S. E., Ouagazzal, A. M., Gonzalez, L. E., & Overstreet, D. H. (1999). Chronic fluoxetine in tests of anxiety in rat lines selectively bred for differential 5-HT1A receptor function. *Pharmacology, Biochemistry and Behavior, 62*, 695–701.

Flaherty, C. F., Greenwood, A., Martin, J., & Leszczuk, M. (1998). Relationship of negative contrast to animal models of fear and anxiety. *Animal Learning and Behavior, 26*, 397–407.

Flint, J., Corley, R., DeFries, J. C., Fulker, D. W., Gray, J. A., Miller, S., & Collins, A. C. (1995). A simple genetic basis for a complex psychological trait in laboratory mice. *Science, 269*, 1432–1435.

Frankel, W. N., & Schork, N. J. (1996). Who's afraid of epistasis? *Nature Genetics, 14*, 371–373.

Gardner, C. R. (1985). Inhibition of ultrasonic distress vocalizations in rat pups by chlordiazepoxide and diazepam. *Drug Development Research, 5*, 185–193.

Gardner, C. R., & Budhram, P. (1987). Effects of agents which interact with central benzodiazepine sites on stress-induced ultrasounds in rat pups. *European Journal of Pharmacology, 134*, 275–283.

Garièpy, J.-L. (1994). The mediation of aggressive behavior in mice: A discussion of approach-withdrawal processes in social adaptations. In G. Greenberg, K. E. Hood, & E. Tobach (Eds.), *Behavioral development in comparative perspective: The approach–withdrawal theory of T. C. Schneirla* (pp. 231–284). New York: Garland Press.

Garièpy, J.-L., Gendreau, P. L., & Lewis, M. H. (1995). Rearing conditions alter social reactivity and D$_1$ dopamine receptors. *Psychopharmacology, Biochemistry and Behavior, 51*, 767–773.

Garièpy, J.-L., Lewis, M. H., & Cairns, R. B. (1996). Genes, neurobiology, and aggression: Time frames and functions of social behaviors in adaptation. In D. M. Stoff & R. B. Cairns (Eds.), *Aggression and violence: Genetic, neurobiological, and biosocial perspectives* (pp. 41–63). Mahwah, NJ: Erlbaum.

Garstang, W. (1929). The origin and evolution of larval forms. *British Association for the Advancement of Science Reports, 1928*, 77–98 [cited in McKinney & McNamamara, 1991, p. 10].

Gershenfeld, H. K., Neumann, P. E., Mathis, C., Crawley, J. N., Li, X., & Paul, S. M. (1997). Mapping quantitative trait loci for open-field behavior in mice. *Behavior Genetics, 27*, 201–210.

Gonzalez, L. E., File, S. E., & Overstreet, D. H. (1998). Selectively bred lines of rats differ in social interaction and hippocampal 5-HT1A receptor function: A link between anxiety and depression? *Pharmacology, Biochemistry and Behavior, 59*, 787–792.

Gottlieb, G. (1987). The developmental basis of evolutionary change. *Journal of Comparative Psychology, 101*, 262–271.

Gould, S. J. (1977). *Ontogeny and phylogeny*. Cambridge, MA: Harvard University Press.

Graeff, F. G. (1994). Neuroanatomy and neurotransmitter regulation of defensive behaviors and related emotions in animals. *Brazilian Journal of Medical and Biological Research, 27*, 811–829.

Graham, M., & Letz, R. (1979). Within-species variation in the development of ultrasonic signaling of preweanling rats. *Developmental Psychobiology, 12*, 129–136.

Grant, P. R. (1986). *Ecology and evolution of Darwin's finches*. Princeton, NJ: Princeton University Press.

Gunnar, M. R., Larson, M. C., Hertsgaard, L., Harris, M. L., & Brodersen, L. (1992). The stressfulness of separation among nine-month-old infants: Effects of social context variables and infant temperament. *Child Development, 63*, 290–303.

Hahn, M. E., Hewitt, J. K., Adams, M., & Tully, T. (1987). Genetic influences on ultrasonic vocalizations in young mice. *Behavior Genetics, 17*, 155–166.

Hahn, M. E., Hewitt, J. K., Schanz, N., Weinreb, L., & Henry, A. (1997). Genetic and developmental influences on infant mouse ultrasonic calling. I. A diallel analysis of the calls of 3-day-olds. *Behavior Genetics, 27*, 133–143.

Hahn, M. E., Karkowski, L., Weinreb, L., Henry, A., Schanz, N., & Hahn, E. M. (1998). Genetic and developmental influences on infant mouse ultrasonic calling: II. Developmental patterns in the calls of mice 2–12 days of age. *Behavior Genetics, 28,* 313–325.

Hall, W. G. (1975). Weaning and growth in artificially reared rats. *Science, 190,* 1313–1315.

Hansen, C., & Spuhler, K. (1984). Development of the National Institutes of Health genetically heterogeneous rat stock. *Alcoholism: Clinical and Experimental Research, 8,* 477–479.

Härd, E., & Engel, J. (1988). Effects of 8-OH-DPAT on ultrasonic vocalization and audiogenic immobility reaction in pre-weanling rats. *Neuropharmacology, 27,* 981–986.

Härd, E., & Engel, J. (1991). Ontogeny of ultrasonic vocalization in the rat: The influence of neurochemical transmission systems. In T. Archer & S. Hansen (Eds.), *Behavioral biology: Neuroendocrine axis* (pp. 37–52). Hillsdale, NJ: Erlbaum.

Härd, E., Engel, J., & Musi, B. (1982). The ontogeny of defensive reactions in the rat: Influence of monoamine transmission systems. *Scandinavian Journal of Psychology (Supplement I), 1982,* 90–96.

Härd, E., Engel, J., & Lindh, A. S. (1988). Effect of clonidine on ultrasonic vocalization in preweaning rats. *Journal of Neurotransmission, 73,* 217–237.

Hart, F. H., & King, J. A. (1966). Distress vocalizations of young in two subspecies of *Peromyscus maniculatus. Journal of Mammology, 47,* 287–293.

Harvey, A. T., & Hennessy, M. B. (1995). Corticotropin-releasing factor modulation of the ultrasonic rate of isolated pups. *Developmental Brain Research, 87,* 125–134.

Harvey, A. T., Moore, H., Lucot, J. B., & Hennessy, M. B. (1994). Monoamine activity in anterior hypothalamus of guinea pig pups separated from their mothers. *Behavioral Neuroscience, 108,* 171–176.

Henderson, N. D. (1989). Interpreting studies that compare high- and low-selected lines on new characters. *Behavior Genetics, 19,* 473–502.

Hendley, E. D., & Ohlsson, W. G. (1991). Two new inbred rat strains derived from SHR: WKHA, hyperactive, and WKHT, hypertensive, rats. *American Journal of Physiology, 261,* H583–H589.

Hendley, E. D., Atwater, D. G., Myers, M. M., & Whitehorn, D. (1983). Dissociation of genetic hyperactivity and hypertension in SHR. *Hypertension, 5,* 211–217.

Hendley, E. D., Wessel, D. J., & Van Houten, J. (1986). Inbreeding of Wistar-Kyoto strain with hyperactivity but without hypertension. *Behavioral and Neural Biology, 45,* 1–16.

Hennessy, M. B. (1986). Multiple, brief maternal separations in the squirrel monkey: Changes in hormonal and behavioral responsiveness. *Physiology and Behavior, 36,* 245–250.

Hennessy, M. B. (1997). Hypothalamic–pituitary–adrenal responses to brief social separation. *Neuroscience and Biobehavioral Reviews, 21,* 11–29.

Hennessy, M. B., & Moorman, L. (1989). Factors influencing cortisol and behavioral responses to maternal separation in guinea pigs. *Physiology and Behavior, 36,* 245–250.

Hennessy, M. B., & Weinberg, J. (1990). Adrenocortical activity during conditions of brief maternal separation in preweaning rats. *Behavioral and Neural Biology, 54,* 42–55.

Hennessy, M. B., O'Neil, D. R., Becker, L. A., Jenkins, R., Williams, M. T., & Davis, H. N. (1992). Effects of centrally administered corticotropin-releasing factor (CRF) and alpha-helical CRF on the vocalizations of isolated guinea pig pups. *Pharmacology, Biochemistry and Behavior, 43,* 37–43.

Hinde, R. A. (1970). *Animal behaviour: A synthesis of ethology and comparative psychology,* 2nd ed. New York: McGraw-Hill.

Hinrichsen, C. F. L., & Ryan, A. T. (1981). Localization of laryngeal motoneurons in the rat: Morphological evidence for dual innervation? *Experimental Neurology, 74,* 341–355.

Hofer, M. A. (1970). Cardiac and respiratory function during sudden prolonged immobility in wild rodents. *Psychosomatic Medicine, 32,* 633–647.

Hofer, M. A. (1996). Multiple regulators of ultrasonic vocalization in the infant rat. *Psychoneuroendocrinology, 21,* 203–217.

Hofer, M. A., & Shair, H. N. (1978). Ultrasonic vocalization during social interaction and isolation in 2-week-old rats. *Developmental Psychobiology, 11,* 495–504.

Hofer, M. A., & Shair, H. N. (1991). Independence of ultrasonic vocalization and thermogenic responses in infant rats. *Behavioral Neuroscience, 105,* 41–48.

Hofer, M. A., & Shair, H. N. (1992). Ultrasonic vocalization by rat pups during recovery from hypothermia. *Developmental Psychobiology, 25,* 511–528.

Hofer, M. A., Brunelli, S. A., & Shair, H. N. (1993). Ultrasonic vocalization responses of rat pups to acute separation and contact comfort do not depend on maternal thermal cues. *Developmental Psychobiology, 26,* 81–95.

Hofer, M. A., Brunelli, S. A., & Shair, H. N. (1994). Potentiation of isolation-induced vocalization by brief exposure of rat pups to maternal cues. *Developmental Psychobiology, 27,* 503–517.

Hofer, M. A., Brunelli, S. A., Masmela, J., & Shair, H. N. (1996). Maternal interactions prior to separation potentiate isolation-induced calling in rat pups. *Behavioral Neuroscience, 110,* 1158–1167.

Hood, K. E., & Cairns, R. B. (1988). A developmental–genetic analysis of aggressive behavior in mice: II. Cross-sex inheritance. *Behavior Genetics, 18*, 605–619.

Hyde, J. S. (1981). A review of selective breeding programs. In G. E. McClearn, R. A. Deitrich, & V. G. Erwin (Eds.), *Development of animal models as pharmacogenetic tools* (pp. 59–77). Rockville, MD: U.S. Department of Health and Human Services, Public Health Service, Alcohol, Drug Abuse and Mental Health Administration.

Insel, T. R., & Harbaugh, C. R. (1989). Central administration of corticotrophin releasing factor alters rat pup isolation calls. *Pharmacology, Biochemistry and Behavior, 32*, 197–201.

Insel, T. R., & Hill, J. L. (1987). Infant separation distress in genetically fearful rats. *Biological Psychiatry, 22*, 783–786.

Insel, T. R., & Shapiro, L. E. (1992). Oxytocin receptor distribution reflects social organization in monogamous and polygamous voles. *Proceedings of the National Academy of Sciences of the USA, 89*, 5981–5985.

Insel, T. R., & Winslow, J. T. (1991a). Central administration of oxytocin modulates the infant rat's response to social isolation. *European Journal of Pharmacology, 203*, 149–152.

Insel, T. R. & Winslow, J. T. (1991b). Rat pup ultrasonic vocalizations: An ethologically relevant behavior responsive to anxiolytics. In *Animal models in pharmacology: Advances in pharmacological sciences* (pp. 15–36). Basel: Birkhauser.

Insel, T. R., Gelhard, R. E., & Miller, L. P. (1989). Rat pup isolation distress and the brain benzodiazepine receptor. *Developmental Psychobiology, 22*, 509–525.

Insel T. R., Hill, J. L., & Mayor, R. B. (1986). Rat pup ultrasonic vocalization isolation calls: Possible mediation by the benzodiazepine receptor complex. *Pharmacology, Biochemistry and Behavior, 24*, 1263–1267.

Insel, T. R., Miller, R., Gerhard, R., & Hill, J. (1988). Rat pup ultrasonic isolation cells and the benzodiazepine receptor. In J. D. Newman (Ed.), *The physiological control of mammalian vocalization* (pp. 331–342). New York: Plenum Press.

Jones, R. B., Mills, A. D., Faure, J.-M., & Williams, J. B. (1994). Restraint, fear, and distress in Japanese Quail genetically selected for long or short tonic immobility reactions. *Physiology and Behavior, 56*, 529–534.

Joyce, M. P., & Carden S. E. (1999). The effects of 8-OH-DPAT and (±)-pindole on isolation-induced ultrasonic vocalizations in 3-, 10-, and 14-day-old rats. *Developmental Psychobiology, 34*, 109–117.

Jürgens, U. (1994). The role of the periaqueductal grey in vocal behaviour. *Behavioral Brain Research, 62*, 107–117.

Jürgens, U. (1998). Neuronal control of mammalian vocalization, with special reference to the squirrel monkey. *Naturwissenschaften, 85*, 376–388.

Jürgens, U., & Lu, C.-L. (1993). The effects of periaqueductally injected transmitter antagonists on forebrain-elicited vocalization in the squirrel monkey. *European Journal of Neuroscience, 5*, 735–741.

Jürgens, U., & Pratt, R. (1979). Role of the periaqueductal grey in vocal expression of emotion. *Brain Research, 167*, 367–378.

Kehne, J. H., McCloskey, T. C., Baron, B. M., Chi, E. M., Harrison, B. L., Whitten, J. P., & Palfreyman, M. G. (1991). NMDA receptor complex antagonists have potential anxiolytic effects as measured with separation-induced ultrasonic vocalizations. *European Journal of Pharmacology, 193*, 283–292.

Kehoe, P., & Blass, E. M. (1986a). Opioid-dependent mediation of separation distress in 10 day old rats: Reversal of stress and maternal stimuli. *Developmental Psychobiology, 19*, 385–398.

Kehoe, P., & Blass, E. M. (1986b). Behaviorally functional opioid systems in infant rats: II. Evidence for pharmacological, physiological, and psychological mediation of pain and stress. *Behavioral Neuroscience, 100*, 624–630.

Kehoe, P., & Boylan, C. B. (1992). Cocaine-induced effects on isolation stress in neonatal rats. *Behavioral Neuroscience, 106*, 374–379.

Kehoe, P., & Boylan, C. B. (1994). Behavioral effects of kappa-opioid-receptor stimulation on neonatal rats. *Behavioral Neuroscience, 108*, 418–423.

Kehoe, P., & Harris, J. C. (1989). Ontogeny of noradrenergic effects on ultrasonic vocalizations in rat pups. *Behavioral Neuroscience, 103*, 1099–1107.

Kehoe, P., Clash, K., Skipsey, K., & Shoemaker, W. J. (1996). Brain dopamine response in isolated 10-day-old rats: Assessment using D2 binding and dopamine turnover. *Pharmacology, Biochemistry and Behavior, 53*, 41–49.

Klein, R. G. (1995). Is panic disorder associated with childhood separation anxiety disorder? *Clinical Neuropharmacology, 18*, S7–S114.

Klöting, I., Berg, S., Kovacs, P., Voigt, B., & Schmidt, S. (1997). Diabetes and hypertension in rodent models. *Annals of the New York Academy of Sciences, 20*, 64–84.

Knapp, D. J., Overstreet, D. H., & Crews, F. T. (1998). Brain 5-HT1A receptor autoradiography and hypothermic responses in rats bred for differences in 8-OH-DPAT sensitivity. *Brain Research, 782,* 1–10.

Knardahl, S., & Hendley, E. G. (1990). Association between cardiovascular reactivity to stress and hypertension or behavior. *American Journal of Physiology, 259,* 248–257.

Kurtz, M. M., & Campbell, B. A. (1994). Paradoxical autonomic responses to aversive stimuli in the developing rat. *Behavioral Neuroscience, 108,* 962–971.

Lagerspetz, K. M. J. (1961). Genetic and social causes of aggressive behavior in mice. *Scandinavian Journal of Psychology, 2,* 167–173.

Lagerspetz, K. M. J., & Wuorinen, K. (1965). A cross-fostering experiment with mice selectively bred for aggressiveness and non-aggressiveness. *Reports of the Institute of Psychology of the University of Turku, 17,* 1–6.

Lander, E. S., & Schork, N. J. (1994). Genetic dissection of complex traits. *Science, 265,* 2037–2048.

Larson, C. R., & Kistler, M. K. (1986). The relationship of periacqueductal gray neurons to vocalization and laryngeal EMG in the behaving monkey. *Experimental Brain Research, 63,* 596–606.

Larson, M. C., Gunnar, M. R., & Hertsgaard, L. (1991). The effects of morning naps, car trips, and maternal separation on adrenocortical activity in human infants. *Child Development, 62,* 362–372.

Lecanuet, J.-P., Fifer, W. P., Krasnegor, N. A., & Smotherman, W. P. (1995). *Fetal development: A psychobiological perspective.* Hillsdale: Erlbaum.

LeDoux, J. (1996). *The emotional brain: The mysterious underpinnings of emotional life.* New York: Simon & Schuster.

LeDoux, J. E., Sakaguchi, A., & Reis, D. J. (1983). Strain differences in fear between spontaneously hypertensive and normotensive rats. *Brain Research, 277,* 137–143.

LeRoy, I. (1999). Quantitative trait loci (QTL) mapping. In B. C. Jones & P. Mormède (Eds.), *Neurobehavioral genetics: Methods and applications* (pp. 69–76). Boca Raton, FL: CRC Press.

LeRoy, I., Perez-Diaz, F., Cherfouh, A., & Roubertoux, P. L. (1999). Preweanling sensorial and motor development in laboratory mice: Quantitative trait mapping. *Developmental Psychobiology, 34,* 139–158.

Lewis, M. H., Gariépy, J.-L., Gendreau, P. J., Nichols, D. E., & Mailman, R. B. (1994). Social reactivity and D1 dopamine receptors: Studies in mice selectively bred for high and low levels of aggression. *Neuropsychopharmacology, 10,* 115–122.

Li, T. K., Lumeng, L., & Dolittle, D. P. (1993). Selective breeding for alcohol preference and associated responses. *Behavior Genetics, 23,* 163–170.

Lickliter, R. (1995). Embryonic sensory experience and intersensory development in precocial birds. In J.-P. Lecanuet, W. P. Fifer, N. A. Krasnegor, & W. P. Smotherman (Eds.), *Fetal development: A psychobiological perspective* (pp. 281–294). Hillsdale, NJ: Erlbaum.

Liebsch, G., Montkowski, A., Holsboer, F., & Landgraf, R. (1998). Behavioral profiles of two Wistar rat lines selectively bred for high or low anxiety-related behavior. *Behavioral Brain Research, 94,* 301–310.

Lipp, H.-P., & Wolfer, D. P. (1999). Natural genetic variation of hippocampal structures and behavior. In B. C. Jones & P. Mormède (Eds.), *Neurobehavioral genetics: Methods and applications* (pp. 217–235). Boca Raton, FL: CRC Press.

Liu, D., Diorio, J., Tannenbaum, B., Caldji, C., Francis, D., Freedman, A., Sharma, S., Pearson, D., Plotsky, P. M., & Meaney, M. J. (1997). Maternal care, hippocampal glucocorticoid receptors and pituitary-adrenal responses to stress. *Science, 277,* 1659–1662.

MacLean, P. D. (1985). Brain evolution relating to family, play and the separation call. *Archives of General Psychiatry, 42,* 405–417.

MacLean, P. D., & Newman, J. D. (1988). Role of the midline frontolimbic cortex in production of the isolation call of squirrel monkeys. *Brain Research, 450,* 111–123.

Maier, S. E., Vandenhoff, P., & Crowne, D. P. (1988). Multivariate analysis of putative measures of activity, exploration, emotionality, and spatial behavior in the hooded rat (*Rattus norvegicus*). *Journal of Comparative Psychology, 102,* 378–387.

Mathis, C., Neumann, P. E., Gershenfeld, H., Paul, S. M., & Crawley, J. N. (1995). Genetic analysis of anxiety-related behaviors and responses to benzodiazepine-related drugs in AXB and BXA recombinant inbred mouse strains. *Behavior Genetics, 25,* 557–568.

McClearn, G. E. (1993). Genetics, systems, and alcohol. *Behavior Genetics, 23,* 223–230.

McClearn, G. E., Deitrich, R. A, & Erwin, V. G. (Eds.). (1981). *Development of animal models as pharmacogenetic tools.* Rockville, MD: U.S. Department of Health and Human Services, Public Health Service, Alcohol, Drug Abuse and Mental Health Administration.

McCormick, C. M., Kehoe, P., & Kovacs, S. (1998). Corticosterone release in response to repeated, short episodes of neonatal isolation: Evidence of sensitization. *International Journal of Developmental Neuroscience, 16,* 175–185.

McGregor, I. S., Dastur, F. N., McLellan, R. A., & Brown, R. E. (1996). Cannabinoid modulation of rat pup ultrasonic vocalizations. *European Journal of Pharmacology, 313,* 43–49.

McKinney, M. L., & McNamara, K. J. (1991). *Heterochrony: The evolution of ontogeny*. New York: Plenum Press.

Mennella, J. A. & Moltz, H. (1988). Infanticide in rats: Male strategy and female counter-strategy. *Physiology and Behavior, 42,* 19–28.

Michel, G. F., & Moore, C. L. (1995). *Developmental psychobiology: An interdisciplinary science*. Cambridge, MA: MIT Press.

Miczek, K. A., Tornatzky, W., & Vivian, J. (1991). Ethology and neuropharmacology: Rodent ultrasounds. In *Animal models in psychopharmacology: Advances in pharmacological sciences*. (pp. 410–427). Basel: Birkhäuser.

Moisan, M. P., Courvoisier, H., Bihoreau, M. T., Gaugier, D., Hendley, E. D., Lathrop, M., James, M. R., & Mormède, P. (1996). A major quantitative trait locus influences hyperactivity in the WKHA rat. *Nature Genetics, 14,* 471–473.

Myers, M. M. (1992). Behavioral and cardiovascular traits: Broken links and new associations. *Annals of the New York Academy of Sciences, 662,* 84–101.

Myers, M. M., Brunelli, S. A., Squire, J. M., Shindledecker, R. D, & Hofer, M. A. (1989). Maternal behavior of SHR rats and its relationship to offspring blood pressures. *Developmental Psychobiology, 22,* 29–53.

Myers, M. M., Brunelli, S. A., Shair, H. N., Squire, J. M., & Hofer, M. A. (1989). Relationships between maternal behavior of SHR and WKY dams and adult blood pressures of cross-fostered F_1 pups. *Developmental Psychobiology, 22,* 55–67.

Naito, H., & Tonoue, T. (1987). Sex differences in ultrasound distress call by rat pups. *Behavioral Brain Research, 25,* 13–21.

Naito, H., Nakamura, A., & Inoue, M. (1998). Ontogenic changes in responsiveness to benzodiazepine receptor ligands on ultrasonic vocalizations in rat pups. *Experimental Animal, 47,* 89–96.

Naumenko, E. V., Popova, N. K., Nikulina, E. M., Dygalo, N. N., Shishkina, G. T., Borodin, P. M., & Markel, A. L. (1989). Behavior, adrenocortical activity, and brain monoamines in Norway rats selected for reduced aggressiveness towards man. *Pharmacology, Biochemistry, and Behavior, 18,* 85–91.

Newman, J. D. (1991). Vocal manifestations of anxiety and their pharmacological control. In S. E. File (Ed.), *Psychopharmacology of anxiolytics and antidepressants* (pp. 251–260). New York: Pergamon Press.

Newman, J. D. (1992). The primate isolation call and the evolution and physiological control of human speech. In J. Wind *et al.* (Eds.), *Language origin: A multidisciplinary approach* (pp. 301–321). Dordrecht, The Netherlands: Kluwer.

Nitschke, W., Bell, R. W., Bell, N. J., & Zachman, T. (1972). Distress vocalizations of young in three inbred strains of mice. *Developmental Psychobiology, 5,* 363–370.

Nitschke, W., Bell, R. W., Bell, N. J., & Zachman, T. (1975). The ontogeny of ultrasounds in two strains of *Rattus norvegicus*. *Experimental Aging Research, 1,* 229–242.

Noirot, E. (1968). Ultrasounds in young rodents: II. Changes with age in albino rats. *Animal Behavior, 16,* 129–134.

Noirot, E. (1972). Ultrasounds and maternal behavior in small rodents. *Developmental Psychobiology, 5,* 371–387.

Noirot, E., & Pye, D. (1969). Sound analysis of ultrasonic distress calls of mouse pups as a function of their age. *Journal of Zoology, 162,* 71–83.

Nosten, M., & Roubertoux, P. L. (1988). Uterine and cytoplasmic effects on pups' eyelid opening in two inbred strains of mice. *Physiology and Behavior, 43,* 161–171.

Nutt, D. J. (1988). Benzodiazepine receptor ligands. *Neurotransmissions, 4,* 1–5.

Okon, E. E. (1970a). The effect of environmental temperature on the production of ultrasounds by isolated nonhandled albino mouse pups. *Journal of Zoology, 162,* 71–83.

Okon, E. E. (1970b). The ultrasonic responses of albino mouse pups to tactile stimuli. *Journal of Zoology, 162,* 485–492.

Okon, E. E. (1971). The temperature relations of vocalization in infant golden hamsters and Wistar rats. *Journal of Zoology (London), 164,* 227–237.

Okamoto, K., & Aoki, K. (1963). Development of a strain of spontaneously hypertensive rats. *Japanese Circulation Journal, 27,* 282–293.

Olivier, B., Molewijk, H. E., van Oorschot, R., van der Heyden, J. A. M., Ronken, E., & Mos, J. (1998). Rat pup ultrasonic vocalization: Effects of benzodiazepine receptor ligands. *European Journal of Pharmacology, 358,* 117–128.

Olivier, B., Molewijk, H. E., van der Heyden, J. A. M., van Oorschot, R., Ronken, E., Mos, J., & Miczek, K. A. (1998). Ultrasonic vocalizations in rat pups: Effects of serotonergic ligands. *Neuroscience and Biobehavioral Reviews, 23,* 215–227.

Oppenheim, R. W. (1980). Metamorphosis and adaptation in the behavior of developing organisms. *Developmental Psychobiology, 13,* 353–356.

Oppenheim, R. W. (1981). Ontogenetic adaptations and retrogressive processes in the development of the nervous system and behavior: A neuroembryological perspective. In R. J. Connolly, & H. F. R.

Prechtl (Eds.), *Maturation and development: Biological and psychobiological perspectives* (pp. 73–109). Philadelphia: Lippincott.

Osborne, K. A., Robichon, A., Burgess, E., Butland, S., Shaw, R. A., Coulthard, A., Pereira, H. S., Greenspan, R. J., Sokolowski, M. B. (1997). Natural behavior polymorphism due to a cGMP-dependent protein kinase of *Drosophila. Science, 277*, 834–836.

Oswalt, G. L., & Meier, G. W. (1975). Olfactory, thermal, and tactual influences on infantile ultrasonic vocalization in rats. *Developmental Psychobiology, 8*, 129–135.

Overstreet, D. H., Rezvani, A. H., Pucilowski, O., Gause, L., & Janowsky, D. S. (1994). Rapid selection for serotonin-1A sensitivity in rats. *Psychiatric Genetics, 4*, 57–62.

Overstreet, D. H., Rezvani, A. H., Knapp, D. J., Crews, F. T., & Janowsky, D. S. (1996). Further selection of rat lines differing in 5-HT-1A receptor sensitivity: Behavioral and functional correlates. *Psychiatric Genetics, 6*, 107–117.

Panksepp, J. (1998). *Affective neuroscience: The foundations of human and animal emotions.* New York: Oxford Press.

Panksepp, J., Meeker, R., & Bean, D. H. (1980). The neurochemical control of crying. *Pharmacology, Biochemistry and Behavior, 12*, 437–443.

Paterson, A. H. (1995). Molecular dissection of quantitative traits: Progress and prospects. *Genome Research, 5*, 321–333.

Paul, L., & Kuperschmidt, J. (1975). Killing of conspecific and mouse young by male rats. *Journal of Comparative and Physiological Psychology, 88*, 755–763.

Pedersen, P. E., & Blass, E. M. (1981). Prenatal and postnatal determinants of the 1st suckling episode in albino rats. *Developmental Psychobiology, 15*, 349–355.

Pine, D. S., & Grun, J. (1999). Childhood anxiety: Integrating developmental psychopathology and affective neuroscience. *Journal of Adolescent Psychopharmacology, 9*, 1–12.

Plomin, R., DeFries, J. C., & McClearn, G. E. (1991). *Behavioral genetics: A primer.* New York: W. H. Freeman.

Plotkin, H. C. (Ed.) (1988). *The role of behavior in evolution.* Cambridge: MIT Press.

Plotsky, P. M., & Meaney, M. J. (1993). Early postnatal experience alters hypothalamic corticotropin-releasing factor (CRF) mRNA, median eminence CRF content and stress-induced release in rats. *Molecular Brain Research, 18*, 195–200.

Ramboz, S., Sandou, F., Amara, D. A., Belzung, C., Segu, L., Misslin, R., Baht, M. C., & Hen, R. (1996). 5-HT1B receptor knockout—Behavioral consequences. *Behavioral Brain Research, 73*, 305–312.

Ramos, A., & Mormede, P. (1998). Stress and emotionality: A multidimensional and genetic approach. *Neuroscience and Biobehavioral Reviews, 22*, 33–57.

Rauch, S. L., Savage, C. R., Alpert, N. M., Fischman, A. J., & Jenike, M. A. (1997). The functional neuroanatomy of anxiety: A study of three disorders using positron emission tomography and symptom provocation. *Biological Psychiatry, 42*, 446–452.

Rex, A., Barth, T., Vagged, J.-P., Daemon, A. M., & Fink, H. (1994). Effects of cholecystokinin tetrapeptide and sulfated cholecystokinin octapeptide in rat models of anxiety. *Neuroscience Letters, 172*, 139–142.

Reznick, D. N., Shaw, F. H., Rodd, F. H., & Shaw, R. G. (1997). Evaluations of the rate of evolution in natural populations of guppies (*Poecilia reticulata*). *Science, 275*, 1934–1935.

Roberts, L. H. (1975). Evidence for the laryngeal source of ultrasonic and audible cries of rodents. *Journal of Zoology (London), 175*, 243–257.

Robinson, D. J., & Udine, B. (1982). Ultrasonic calls produced by three laboratory strains of *Mus musculus. Journal of Zoology (London), 197*, 383–389.

Robinson, S. R., & Smotherman, W. P. (1995). Habituation and classical conditioning in the rat fetus: Opioid involvements. In J.-P. Lecanuet, W. P. Fifer, N. A. Krasnegor, & W. P. Smotherman (Eds.), *Fetal development: A psychobiological perspective* (pp. 295–314). Hillsdale, NJ: Erlbaum.

Roubertoux, P. L., & Carlier, M. (1988). Differences between CBA/H and NZB mice on intermale aggression. II. Maternal effects. *Behavior Genetics, 18*, 175–184.

Roubertoux, P. L., Nosten-Bertrand, M., & Carlier, M. (1990). Additive and interactive effects of genotype and maternal environment. *Advances in the Study of Behavior, 19*, 205–247.

Roubertoux, P. L., Carlier, M., Degrelle, H., Haas-Dupertuis, M.-C., Phillips, J., & Moutier, R. (1994). Co-segregation of intermale aggression with pseudoautosomal region of the Y chromosome in mice. *Genetics, 135*, 225–230.

Roubertoux, P. L., Martin, B., LeRoy, I., Beau, J., Marchaland, C., Perez-Diaz, F., Cohen-Salmon, C., & Carlier, M. (1996). Vocalizations in newborn mice: Genetic analysis. *Behavior Genetics, 26*, 427–437.

Roubertoux, P. L., Mortaud, S., Tordjman, S., LeRoy, I., & Degrelle, H. (1986). Behavior-genetic analysis and aggression: The mouse as a prototype. In M. Sabourin, F. Craik, & M. Robert (Eds.), *Advances in psychological science*, Volume 2: *Biological and cognitive aspects* (pp. 3–29). Montreal: Psychology Press.

Roubertoux, P. L., Nosten-Bertrand, M., Cohen-Salmon, & l'Hotellier, L. (1992). Behavioral development: A tool for genetic analysis in mice. In D. Goldowitz, D. Wahlsten, & R. E. Wimer (Ed.),
Techniques for the genetic analysis of brain and behavior (pp. 423–441). New York: Elsevier.

Rutherford, S. L., & Lindquist, S. (1998). Hsp90 as a capacitor for morphological evolution. *Nature, 396,*
336–342.

Rutter, M., Silberg, J., O'Connor, T., & Simonoff, E. (1999a). Genetics and child psychiatry: I Advances in
quantitative and molecular genetics. *Journal of Child Psychology and Psychiatry, 40,* 3–18.

Rutter, M., Silberg, J., O'Connor, T., & Simonoff, E. (1999b). Genetics and child psychiatry: II. Empirical
research findings. *Journal of Child Psychology and Psychiatry, 40,* 19–55.

Sales, G. S. (1979). Strain differences in the ultrasonic behavior of rats (Rattus norvegicus). *American
Zoologist, 19,* 513–527.

Sales, G. S., & Smith, J. C. (1978). Comparative studies of the ultrasonic calls of infant murid rodents.
Developmental Psychobiology, 11, 595–619.

Sandnabba, N. K. (1993). Female aggression during gestation and lactation in two strains of mice selected
for isolation-induced intermale aggression. *Behavioral Processes, 30,* 157–164.

Sandnabba, N. K. (1996). Selective breeding for isolation-induced intermale aggression in mice: Associated responses and environmental influences. *Behavior Genetics, 26,* 477–488.

Saudou, F., Amara, D. A., Dierich, A., LeMeur, M., Ramboz, S., Segu, L., Baht, M. C., & Hen, R. (1994).
Science, 265, 1875–1878.

Scott, P. A., Cierpial, M. A., Kilts, C. D., & Weiss, J. M. (1996). Susceptibility and resistance of rats to stress-
induced decreases in swim-test activity: A selective breeding study. *Brain Research, 725,* 217–230.

Scearce-Levie, K., Viswanathan, S. S., & Hen, R. (1999). Locomotor response to MDMA is attenuated in
knockout mice lacking the 5-HT1B receptor. *Psychopharmacology (Berlin), 141,* 154–161.

Scherer, K. R. (1985). Vocal affect signaling: A comparative approach. *Advances in the Study of Behavior,
15,* 189–244.

Schouten, W. G. P. (1988). Development of ultrasonic vocalization in the rat (*Rattus norvegicus*). In J.
Unshelm, G. van Putten, K. Zeeb, & I. Ekesbo (Eds.), *Proceedings of the International Congress on
Applied Ethology in Farm Animals* (pp. 1–5). Darmstadt, Germany: Kuratorium für Technik und
Bauwesen in der Landwirtschaft (KTBL).

Shair, H. N., Masmela, J. R., Brunelli, S. A., & Hofer, M. A. (1997). Potentiation and inhibition of
ultrasonic vocalization of rat pups: Regulation by social cues. *Developmental Psychobiology, 30,* 195–200.

Shen, E. H., Dorow, J. D., Huson, M., & Phillips, T. I. (1996). Correlated responses to selection in FAST
and SLOW mice: Effects of ethanol on ataxia, temperature, sedation and withdrawal. *Alcoholism,
Clinical and Experimental Research, 20,* 688–696.

Shoemaker, W. J., & Kehoe, P. (1995). Effect of isolation conditions on brain regional enkephalin and
beta-endorphin levels and vocalizations in 10-day-old rat pups. *Behavioral Neuroscience, 109,* 117–122.

Smotherman, W. P., Bell, R. W., Starzec, I., Elias, J., & Zachman, T. A. (1974). Maternal responses to infant
vocalizations and olfactory cues in rats and mice. *Behavioral Biology, 12,* 55–66.

Snustad, D. P., Simmons, M. J., & Jenkins, J. B. (1997). *Principles of genetics.* New York: Wiley.

Sokoloff, G., & Blumberg, M. S. (1997). Thermogenic, respiratory, and ultrasonic responses of week-old
rats across the transition from moderate to extreme cold exposure. *Developmental Psychobiology, 30,*
181–194.

Stanton, M. E., Wallstrom, J., & Levine, S. (1986). Maternal contact inhibits pituitary–adrenal stress
responses in preweanling rats. *Developmental Psychobiology, 20,* 1311–1145.

Stanton, M. E., Gutierrez, Y. R., & Levine, S. (1988). Maternal deprivation potentiates pituitary–adrenal
stress responses in infant rats. *Behavioral Neuroscience, 102,* 692–700.

Stern, J. M., & Johnson, S. K. (1990). Ventral somatosensory determinants of nursing behavior in Norway rats. I. Effects of variations in the quality and quantity of pup stimuli. *Physiology and Behavior, 47,*
933–1011.

Suomi, S. J. (1997). Early determinants of behaviour: Evidence from primate studies. *British Medical
Bulletin, 53,* 170–184.

Szpirer, C., Szpirer, J., Van Vooren, P., Tissir, F., Simon, J. S., Koike, G., Jacob, H. J., Lander, E. S., Helou, K.,
Klinga-Levan, K., & Levan, G. (1999). Gene-based anchoring of the rat genetic linkage and cyto-
genetic maps. *Transplant Proceedings, 31,* 1541–1543.

Takahashi, L. K. (1992a). Developmental expression of defensive responses during exposure to conspecific adults in preweanling rats (*Rattus norvegicus*). *Journal of Comparative Psychology, 106,* 69–77.

Takahashi, L. K. (1992b). Ontogeny of behavioral inhibition induced by unfamiliar adult male conspecifics in preweanling rats. *Physiology and Behavior, 52,* 493–498.

Thom, S. (1988). *The effects of benzodiazepines and beta-carbolines on rat pup ultrasonic isolation calls.* Ph.D.
dissertation, City University of New York, New York.

Turkewitz, G., & Devenny, D. A. (1993). Timing and the shape of development. In G. Turkewitz & D. A. Devenny (Eds.), *Developmental time and timing* (pp. 1–11). Hillsdale, NJ: Erlbaum.

Uni, Z., Gutman, M., Leitner, G., Landesman, E., Heller, D., & Cahaner, A. (1993). Major histocompatibility complex class IV restriction fragment length polymorphism markers in replicated meat-type chicken lines divergently selected for high or low early immune response. *Poultry Science, 72,* 1823–1831.

Vera, P. L., & Nadelhaft, I. (1991). MK-801, a non-competitive NMDA receptor antagonist, produces facilitation of the micturition reflex in awake, freely-moving rats. *Neuroscience Letters, 134,* 135–138.

Vivian, J. A., Barros, H. M., Manitiu, A., & Miczek, K. A. (1997). Ultrasonic vocalizations in rat pups: Modulation at the gamma-aminobutyric acid(A) receptor complex and the neurosteroid recognition site. *Journal of Pharmacology and Experimental Therapeutics, 282,* 318–325.

Weinstock, M. (1997). Does prenatal stress impair coping and regulation of the hypothalamic–pituitary–adrenal axis? *Neuroscience and Biobehavioral Reviews, 21,* 1–10.

Weller, A., & Blass, E. M. (1988). Behavioral evidence for cholecystokinin-opiate interactions in neonatal rats. *American Journal of Physiology, 255,* R901–R907.

Weller, A., & Gispan, D. (2000) A cholecystokinin receptor antagonist blocks milk-induced but not maternal-contact-induced decrease of ultrasonic vocalization in rat pups. *Developmental Psychobiology, 37,* 35–43.

West, M. J., & King, A. P. (1987). Settling nature and nurture into an ontogenetic niche. *Developmental Psychobiology, 20,* 549–562.

Wetzel, D. M., Kelley, D. B., & Campbell, B. A. (1980). Central control of ultrasonic vocalizations in neonatal rats: I. Brain stem motor nuclei. *Journal of Comparative and Physiological Psychology, 94,* 596–605.

Whitney, G. D., Nyby, J., Cable, J. R., & Dizinno, G. A. (1978). Genetic influences on 70kHz ultrasound production of mice (*Mus musculus*). *Behavior Genetics, 8,* 574–578.

Wiedenmayer, C. P., & Barr, G. A. (1998). Ontogeny of defensive behavior and analgesia in rat pups exposed to an adult male rat. *Physiology and Behavior, 63,* 261–269.

Winslow, J. T., & Insel, T. R. (1990). Serotonergic and catecholaminergic reuptake inhibitors have opposite effects on the ultrasonic isolation calls of rat pups. *Neuropsychopharmacology, 3,* 51–59.

Winslow, J. T., & Insel, T. R. (1991a). The infant rat separation paradigm: A novel test for novel anxiolytics. *Trends in Pharmacological Sciences, 12,* 402–404.

Winslow, J. T., & Insel, T. R. (1991b). Serotonergic modulation of the rat pup ultrasonic isolation call: Studies with 5HT$_1$ and 5HT$_2$ subtype-selective agonists and antagonists. *Psychopharmacology, 105,* 513–520.

Winslow, J. T., & Insel, T. R. (1993). Effects of central vasopressin administration to infant rats. *European Journal of Pharmacology, 223,* 101–107.

Winslow, J. T., Insel, T. R., Trullas, R., Skolnick, P. (1990). Rat pup isolation calls are reduced by functional antagonists of the NMDA receptor complex. *European Journal of Pharmacology, 190,* 11–21.

Xu, L., Anwyl, R., De Vry, J., & Rowan, M. J. (1997). Effect of repeated ipsapirone treatment on hippocampal excitatory synaptic transmission in the freely behaving rat: Role of 5-HT1A receptors and relationship to anxiolytic effect. *European Journal of Pharmacology, 323,* 59–68.

Yajima, Y., & Hayashi, Y. (1983). Ambiguous motoneurons discharging synchronously with ultrasonic vocalization in rats. *Experimental Brain Research, 50,* 359–366.

Yongue, B. G., & Myers, M. M. (1988). Cosegregation analysis of salt appetite and blood pressure in genetically hypertensive and normotensive rats. *Clinical and Experimental Hypertension: Theory and Practice A, 10,* 323–343.

Yongue, B. G., & Myers, M. M. (1989). Further evidence for genetic independence of blood pressure and salt appetite in spontaneously hypertensive and Wistar-Kyoto rats. *Clinical and Experimental Hypertension: Theory and Practice A, 11,* 25–33.

Yoshiyama, M., Roppolo, J. R., & de Groat, W. C. (1994). Interactions between glutamergic and monoaminergic systems controlling the micturition reflex in the urethane-anesthetized rat. *Brain Research, 639,* 300–308.

Zákány, J., Gérard, M., Favier, B., & Duboule, D. (1997). Deletion of a *HoxD* enhancer induces transcriptional heterochrony leading to transposition of the sacrum. *EMBO Journal, 16,* 4393–4402.

Zhang, S. P., Davis, P. J., Bandler, R., & Carrive, P. (1994). Brain stem integration of vocalization: Role of the midbrain periaqueductal gray. *Journal of Neurophysiology, 72,* 1337–1356.

Zhang, S. P., Bandler, R., & Davis, P. J. (1995). Brain stem integration of vocalization: Role of the nucleus retroambigualis. *Journal of Neurophysiology, 74,* 2500–2512.

Zimmerberg, B., Brunelli, S. A., & Hofer, M. A. (1994). Reduction of rat pup ultrasonic vocalizations by the neuroactive steroid allopregnanalone. *Pharmacology, Biochemistry and Behavior, 47,* 735–738.

12

The Ontogeny of Motivation

Hedonic Preferences and Their Biological Basis in Developing Rats

Aron Weller

Introduction

This chapter examines the ontogeny of motivated behavior, particularly the origins of early attractions. It focuses on the biological foundations of preferences in an animal model, the laboratory rat. A developmental approach is utilized to examine the appearance over ontogeny of different behavioral tendencies and how they change. This approach can reveal components of motivation as they appear (at different ontogenetic times) and allow the examination of the physiological substrate (hormonal, neural, etc.) of these developing components separately. This approach can also specify how experience contributes to motivational change by building on states that are, by definition, rewarding to newborns. These states, concerned with energy conservation and with stimulation of the central nervous system (CNS) for brain growth and development, will be discussed and the position advanced here of intrinsically rewarding systems will be supported.

I will focus on early development, mainly on the postnatal and weaning period that spans the rat's first month of life. The chapter examines the classes of attractors, inherent and acquired, that are provided by the maternal nest, siblings, and dam. I then discuss how these attractors support the foundation of acquired preferences and then focus on the role of one specific peptide system (cholecystokinin, CCK) in mediating some preferences and choices of rat pups. The contribution of this system will be discussed because it is one of the few relevant physiological systems examined in this mostly unexplored field and is an exemplar of a system gaining force through

Aron Weller   Developmental Psychobiology Laboratory, Department of Psychology, Bar Ilan University, Ramat-Gan 52900, Israel.

Developmental Psychobiology, Volume 13 of *Handbook of Behavioral Neurobiology*, edited by Elliott Blass, Kluwer Academic / Plenum Publishers, New York, 2001.

its primary function of energy conservation. The role of other neurochemical systems that may contribute to choice and to infant–mother attachment, including opioids, oxytocin, norepinephrine and vasopressin, among others, have recently been reviewed (Blass, 1996; Leon, 1992a; Nelson & Panksepp, 1998). Another possibly relevant neurochemical link may be the dopamine D2 receptor, which has been recently shown to mediate the formation of partner preferences in adult female prairie voles (Wang *et al.*, 1999).

A number of themes are emphasized in this chapter. First, remarkably, rats are capable of goal-directed behavior from birth. The initial potential of reward signals is wide and narrows with time in a number of systems ranging from ingestion to sibling preference and maternal recognition. The reports of Johanson, Polefrone, and Hall (1984) and of Moran, Schwartz, and Blass (1983a, b), showing that newborn rats will push a lever to obtain milk or brain stimulation and the increased integration of the motor patterns elicited by brain stimulation as the infant matures are now classic studies of the availability of motor and affect systems from birth on.

Second, the classes of attractors that are available to all mammals at birth flow from two themes: energy conservation and brain neurotransmission. Thus, newborns are attracted to soft, warm textures, thereby decreasing their surface:mass ratio, will suckle interminably (even nonnutritive suckling), and are apparently rewarded by substances that are either directly involved in food processing such as CCK or ones (such as opioids) that are released by high-energy tastes and flavors such as sweets and fats. They will also work for brief periods of excitation ("behavioral activation"), which appear to cause release of norepinephrine (Sullivan & Wilson, 1994).

Third, early experience in which an arbitrary stimulus is paired with one of the changes just described determines the selection of objects toward which the already integrated motor patterns and affective responses will be directed during early ontogeny, the juvenile stage, and adulthood. Because the mother is the source of the arbitrary stimuli encountered naturally, her role in shaping positive preferences will be highlighted in this chapter.

Fourth, neurochemical systems that modulate the degree to which specific hedonic preferences are expressed in infants are extremely primitive relative to their adult status. This represents an important uncharted domain in developmental biology because we do not know how maturation of these systems will influence the representation of early experiences and the affective changes to which they have given rise.

Fifth, it is important to state at the outset that infant motivational and choice systems differ from adult systems. Very young (day 6) rats appear to find electric shock rewarding, at least at relatively low levels (Camp & Rudy, 1988). This may reflect the excitation of central circuits (based on behavioral patterns) generated by the shock. There is therefore an asymmetry in infant rats between peripheral stimulation and central change. In addition, the same may also be said about peripheral change. Thus, rats prefer the taste of a substance that had caused them mild illness (Kehoe & Blass, 1986a). In this instance it would appear that release of opioids in response to slight malaise was disproportionately high. Pretreatment with naloxone caused an aversion to the stimulus that precipitated low levels of malaise.

The ontogeny of motivation in general, and as partially depicted in this chapter (see also the chapter by Holmes in this volume for a discussion of experiential determinants of kin recognition), has implications in domains as far-reaching as behavioral ecology and clinical psychology because understanding the proximal events that lead to choice in the laboratory may also shed light on the events and mechanisms that do so in the field. Moreover, understanding these affiliative mechanisms as they may naturally unfold can provide insights into the decomposition of these systems or how "wrong" targets can be chosen. To the extent that we identify these situations and their underlying neurochemical and molecular bases, the possibility of effective treatment is improved.

DEFINITIONS

In this chapter the term hedonic will be used to reflect a domain of approach/avoidance, pleasure/displeasure. The term "initial preference " will be used to refer to the newborn infant's initial set of choices (irrespective of the prenatal source of these choices). These initial preferences (often called "natural preferences") and the patterns of changes that occur in them during the postnatal, preweanling, and weaning period are the major issues to be reviewed in detail in this chapter.

THEORETICAL APPROACHES

TWO DIMENSIONS: DISCRIMINATIVE AND AFFECTIVE

Theories of hedonics and motivation differentiate between two relatively independent dimensions: discriminative and affective (e.g., Cabanac, 1971, 1979; Young, 1966). The discriminative dimension reflects the ability to detect and identify a stimulus from a stimulus layout (by sensing and assessing the quality, quantity, and intensity of the stimulus). In contrast, the affective dimension reflects the amount of "pleasure" or "displeasure" that the animal experiences, as inferred from the willingness of an animal to approach or work for a stimulus or event (see Figure 1). Some behaviors, feeding, for example, lend themselves well to this dichotomy. According to Berridge and Grill (1983), ingestive responses (e.g., including mouth movements and tongue protrusion) differ from aversive responses (e.g., head shak-

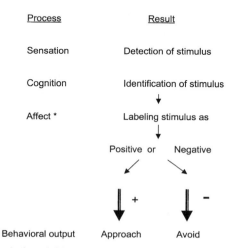

Figure 1. A simplified theoretical model for the processes involved in hedonic preference. *Note*: In some cases affect may be processed simultaneously with cognition.

ing and paw treading), presumably reflecting the animal's state and recent and past histories with the ingesta. Thus, a sugar solution would produce an ingestive reaction and a bitter quinine solution would produce an aversive response, but this can change with degree of satiety or hunger (or thirst) and the animal's history with the solution. Affective decisions are rarely unidimensional (Berridge, Flynn, Schulkin, & Grill, 1984). Appetitive and aversive components may be processed separately, in parallel, or sequentially with the behavioral outcome, depending upon competition between the relative strengths of these two components (Toates, 1986).

DEVELOPMENTAL ELABORATION: FROM APPROACH/WITHDRAWAL TO SEEKING AND AVOIDANCE

Newborn infant rats orient to and move toward particular sources ("topotaxis"; Fraenkel & Gunn, 1961) upon contact, especially to other animals present in the nest, i.e., the mother and siblings. This results in aggregation into huddles of pups, a behavior that contributes to thermoregulation (Schank & Alberts, 1997) and infant–mother proximity, especially when infants are extremely dependent on their mothers as sources of heat, food, and fluid. Contact with the dam and siblings provides one means of learning about mother, siblings, and available food, distinguishing these animals and objects from others of their kind, and responding accordingly. Particular states (e.g., excitation, thermoneutrality) become linked with particular stimuli (e.g., siblings) and, at least in humans, elaboration of the cognitive aspects of the specific emotion (Barrett & Campos, 1987). In animals, sibling preferences during infancy later are elaborated into more complex actions such as sibling defense and alliances.

Rosenblatt (1983, 1987) proposed a developmental theory of approach/ withdrawal processes based on Schneirla's (1939, 1959, 1965) writings. According to this view, newborns first approach low-intensity tactile and thermal stimuli and withdraw from high-intensity stimulation. These simple reactions gradually develop into more elaborate seeking and avoidance/escape behaviors, respectively. According to this view, there is an important developmental transition regarding the sensory stimuli: from initially responding only to stimulus intensity, the infant later responds to other qualitative and quantitative characteristics of the stimuli. Schneirla (1965) described the neonate's motivation as "pushed" by its state of arousal when responding to the stimulus. Later in development, the particular sensory aspects of the stimuli are established as incentives, and the infant now can avoid or seek them. Rosenblatt (1983) further suggested that while the touch and thermal senses are utilized for both the simple and more elaborate stages, olfaction is mainly used for the second process, in particular to (1) identify the source of stimulation, the incentive, (2) provide orientation information to it, and (3) specify the particular response to be made (through conditioning).

This approach, useful in its time, requires updating. As indicated, low intensity stimulation can be best understood from the perspective of the infant being attracted to aspects of its mother that conserve heat, such as warm temperatures and soft skin. Moreover, studies during the past 15 years have demonstrated that infants also find rewarding brief, intense stimulation, whether delivered naturally in the nest through the anogenital licking by the mother or experimentally through brain stimulation or food infusion, or even shock. The basis for excitation being rewarding has not yet been established. Presumably it facilitates oxygenation of the brain and activation of the reward circuitry that has been studied in adults.

Identifying stimulation as attractive or rewarding within the context of energy conservation leads to specific predictions as to what classes of stimulation, internal or external (to the animal), will be rewarding. Thus, neuropeptides such as endorphins that are released by sweet taste and fat flavors should contribute to energy conservation and be a natural reward to suckling infants. Likewise, CCK, released by the act of sucking and during the passage of milk through the gut (see review in the following sections of this chapter), should also prove rewarding, within the context of energetics. In agreement with this, research findings indicate a positive, rewarding role for opioids and CCK, at least in the sense that pairing a novel odor with these peptides (administered exogenously) results in conditioned preferences in infant rats (Kehoe and Blass, 1986b; Weller & Blass, 1990). What has not been established are the limits of this view in terms of how many categories within the energetics context will be rewarding during development.

To the point of the current review, there is considerable evidence for olfactory plasticity in rodents even before birth. For example, the chemical properties (the taste/odor) of the amniotic fluid contribute prenatally to postnatal conditioned preferences and aversions (e.g., Pedersen & Blass, 1982; Stickrod, Kimble, & Smotherman, 1982). It is now clear (see chapter by Johnson and Leon in this volume) that olfactory receptors are distributed topographically on the olfactory membrane (Buck & Axel, 1991) and that this topographic representation appears to continue in rats to the first central relay stations in the glomerular layer of the olfactory bulb. In turn, through more central olfactory projections, the characteristics of odors become catalogued according to the events with which they are associated or which they predict. The linkage of sensory, affective, and (already available) motor mechanisms then determines future affect and guides future behaviors. This shaping of affective systems may have long-lasting impact, even into adulthood (Blass, 1992; Fillion & Blass, 1986).

THE PATTERN OF THE YOUNG RAT'S PREFERENCES: SENSORY MODALITIES

TACTILE PREFERENCES

There are clear tactile preferences in newborns, although it is difficult to rule out the involvement of olfaction. Fresh wood shavings as bedding material were reported to reduce ultrasonic vocalizations (USV) in nest-deprived rat pups compared to a situation without wood shavings (Oswalt & Meier, 1975). Artificial fur was more efficient in reducing USV than a smooth, warm, artificial surrogate model (Hofer & Shair, 1980). In addition, the Harlows' classic experiments in which infant rhesus monkeys preferred to huddle with a terry cloth surrogate mother over a metal-wire model that provided milk are extremely relevant here (Harlow & Harlow, 1962).

In our studies (Rakover-Atar & Weller, 1997) on the hierarchy of early tactile preference, rat pups 5 and 10 days of age given preference tests among four unfamiliar textures preferred a sponge texture over all three other textures, a rug over plywood and metal-wire, and plywood over the metal-wire net texture. This initial tactile scale may both constrain future learning and provide a basis for building more elaborate profiles of conditioned preferences as the growing infant is exposed to new environmental experiences.

Infants appear to have clear gustatory preferences and aversions. For example, intraoral infusions of saccharin or sodium chloride reduced USV in rat pups, whereas water or quinine infusions resulted in USV rates comparable to the no-infusion control group (Kehoe & Sakurai, 1991; although see Graillon, Barr, Young, Wright, & Hendricks, 1997, for human newborns in whom quinine briefly arrested spontaneous crying during administration). Sweet taste eliminates crying of infant rats and humans and also produces analgesia that endures beyond the point of infusion (Blass, 1996). These effects of sweet tastes have been connected to opioid release and the resulting calm state may induce energy conservation (Rao, Blass, Brignol, Marino, & Glass, 1997). Human infants also appear to have differential preferences, or at least reactions to tastes as exhibited by their facial expressions. Steiner (1974, 1977) showed that newborn infants lick and suck in response to sucrose (sweet) and show clearly different expressions in response to sour and bitter tastants. This has since been investigated and replicated in more detail (Ganchrow, Steiner, & Munif, 1983; Rosenstein & Oster, 1990), with investigators using additional physiological and behavioral measures to assess infants' "hedonic responsiveness" (Soussignan, Schaal, Marlier, & Jiang, 1997). A similar approach to Steiner's analysis of facial expressions, named "taste reactivity," has been applied to rats. Grill and Norgren (1978) and Grill and Berridge (1985) have in fact described two clearly distinct profiles of behaviors, ingestive and aversive, in rats receiving oral infusions of taste solutions. The results of applying this approach to infant rats will be summarized later in this section.

The complexity of infants' hedonic preferences can be appreciated through studies of taste preference reported by Moe (1986). Previous reports had shown that at 2–3 days of age, rat pups reject hypertonic saline relative to water (Wirth & Epstein, 1976; Bruno, 1981), drink less of a 5% saline solution than a 2.5% solution, and drink more water than either solution (Hall & Bryan, 1978), and that weanling rats show a greater saline preference than adults (Midkiff & Bernstein, 1983). Moe (1986) investigated the development of salt preferences by infusing saline, quinine, or ammonium chloride solutions into the mouth via a catheter and measuring intake. She reported concentration-dependent changes in intake of NaCl solutions that were qualitatively the same at all ages tested, P6, P12, and P18, as in adults. The lower concentrations were preferred over water, whereas pups were either indifferent to or rejected higher concentrations. However, the preference–aversion curve was shifted from the adult curve in an age-dependent manner; the curve was relatively flat and broadly extended in the young ages and more convex and restricted in range on P18 and in adults. The pattern of results with ammonium chloride was similar. The main developmental difference here was, however, that whereas adult rats do not prefer this taste at any concentration tested, the pups showed preference–aversion curves. Moe (1986) attributed this to the "inability" of the infants to detect the bitterness of ammonium chloride, because they did not show a strong (rejection) response to quinine until P18. However, one should be aware that several studies have shown that the electrophysiologic responses recorded from peripheral taste nerves are relatively large in preweanling rat pups, and the magnitude of these responses does not change significantly with age from infancy to adulthood (see Mistretta & Bradley, 1988, for a review).

Ganchrow, Steiner, & Munif (1986) examined taste reactivity (motor displays in the face and head regions) in newborn rats during the first 4 postnatal days. They reported that licking and rhythmic mouth movements were most saliently elicited by sweet stimuli, head movements and gaping characterized the reaction to quinine, and that the reaction to sour often contained components of both sweet and bitter. Schwartz and Grill (1985) reviewed the earlier literature on early taste reactivity:

> Rat pups and adults share some general responses to ... gustatory stimuli ... sucrose is ingested. NaCl stimuli elicit concentration-dependent acceptance and rejection in both pup and adult. Acid and quinine stimuli are avoided and actively rejected by both pups older than 9 days and adults. Although there are general similarities, many differences exist as well. For instance, quinine infusions elicit aversive behavior only in pups older than day 9. Furthermore, pups younger than day 12 prefer hypertonic saline to water, while adults avoid hypertonic saline. In addition, procedures differ or data are unavailable, making comparisons impossible. (Schwartz & Grill, 1985, pp. 379–380)

Although adult and infant systems are similar qualitatively, quantitative differences abound that may reveal how sensory and affective systems change over time (for a review see the chapter by Hill in this volume).

TEMPERATURE

Infant mammals seek heat around the time of birth and even prefer warm metallic coils to cold fur; in thermally graded alleys, infant rats (and other mammals) orient and move along the thermal gradient from cool to warm if they are not immobilized by the cold (Kleitman & Satinoff, 1982; Leon, 1986; Satinoff, 1991). It is possible that these approach responses are elaborated into more complex, affective preferences, expressed, for example, toward warm, comforting social companions and lasting beyond the early age in which the infant is unable to thermoregulate. Alberts has shown that neonatal rats thermoregulate behaviorally, moving in and out of the huddle, and a simplified version of this behavior, based on the assumption that pups have a initial preference for approaching a sibling when active, has been recently modeled by a computer simulation (Schank & Alberts, 1997). Research from my laboratory, discussed below, suggests that the infant rat's preference for a warmer versus a colder temperature is mediated by the peptide CCK. Thus, it is possible that early thermal preferences share a common physiological mediation with preferences toward other aspects of the mother, and that the basic attraction toward a warm object could be a building block to more selective social attractions. Guenaire, Costa, and Delacour (1979, 1982a, b) showed that rat pups can respond to the "incentive value" of a thermal stimulus apart from their natural tendency to orient toward heat. Research by Alberts and May (1984) further suggested that this ability is used for learning in the nest. Briefly, a series of studies showed that thermotactile stimuli such as maternal care from a lactating or a nonlactating foster mother or contact with a warm inanimate object were "rewarding," inducing conditioned olfactory filial preferences. Interestingly, filial preferences formed to an odor associated with a heated tube were of similar magnitude to those formed to an odor paired with a lactating foster mother, suggesting that the thermotactile stimulus was the crucial "unconditioned" aspect in the associative process.

OLFACTION

To my knowledge, there have not been any studies on pure olfactory attractants in newborn mammals. It is not clear what might be learned from this effort because of concentration issues and also possible prenatal experiential biasing through sampling of amniotic fluid, which changes composition in accordance with the pregnant mother's diet. This is of considerable biological interest because the rats' initial nipple attachment is determined by its prenatal olfactory experiences. Newborn rats are attracted toward odors emanating from the dam or littermates (Alberts & May, 1984; Brown, 1982; Galef & Kaner, 1980; Pedersen & Blass, 1982). Later, they are particularly attracted toward the mother's anal excreta (which contains the pheromone "cecotrophe") during the third and fourth weeks of life (Leon, 1974). Moltz argued that the basis for preference for the pheromone carried in the feces of lactating females is its rich bacteria content that protects the pup's gut against acute enteritis, facilitates lipid absorption from the gut, and enhances the deposition of central myelin (Lee & Moltz, 1984a, b). In contrast, rat pups usually avoid novel odors, and thus demonstrate, within the olfactory modality, a categorical "affective scale."

Rat pups prefer maternal and nest smells to those of other rats and odors of other rats to relatively neutral odors such as clean woodshavings (Brown, 1982). Infant rats avoid lemon odor, peppermint odor, and the odor of fresh cedar shavings (Barnett & Spenser, 1953; Cornwell-Jones, 1979; Cornwell-Jones & Sobrian, 1977; Leon, Galef, & Behse, 1977). This may reflect the fact that these odors are mediated through the trigeminal system and may be irritating. This aversion can be overcome experientially. Rearing for 3 days in cedar shavings abolished the cedar aversion (Cornwell et al., 1996). Eighteen-day-old rats preferred maternal odors most often, then pine shavings, and peppermint was preferred the least (Brown & Willner, 1983). Recently, using two-odorant choice tests, 7- and 25-day-old rat pups showed relative aversions to peppermint, orange, and lemon odors compared to fresh home bedding, with no significant difference among the three former odors (Amiri, Dark, Noce, & Kirstein, 1998). In a choice between banana and peppermint odors, weanlings preferred banana, whereas infants showed no preference (Amiri et al., 1998). A major interesting issue remains—how can these initial aversions be overcome during the natural course of events and, more generally, how is the individual's profile of hedonic preferences and aversions molded?

Engaging reviews of the ontogeny of olfactory preferences in humans by Doty (1981) and Schaal (1988) described both initial and later-emerging preferences. I will only illustrate with an example from recent research on initial preferences toward breast odors. Newborn infants prefer a breast treated with amniotic fluid over an untreated breast; however, the number of 2- to 5-day-old babies that selected one over the other was not statistically significant (Varendi, Porter, & Winberg, 1997). This preference may reflect the newborns' prenatal experiences or may be due to odor intensity causing a heightened state of arousal (and presumably reward) in the newborns. The faded odor preference for amniotic fluid after birth was influenced by postnatal experience (maternal contact and breastfeeding). Interestingly, 2- and 4-day-old bottle-feeding infants showed longer head orientations (a measure of preference) toward the odor of amniotic fluid than toward the odor of formula milk, showing the resilience of this preference during 2–4 days of milk exposure (Marlier, Schaal, & Soussignan, 1998). In extending these findings to autonomic and other behavioral measures in 3-day-old infants, Soussignan et al. (1997) found good agreement between facial and preference measures, but the autonomic measure was discrepant. This is in keeping with other studies on affect in human infants (Blass, 1999).

Further research suggests that early olfactory experiences of human infants with a flavor in the mother's milk (e.g., garlic) modifies how the infants respond to that flavor in subsequent feedings (Mennella & Beauchamp, 1993). Furthermore, exposure to the flavor of carrots in the mother's milk has recently been shown to alter the amount of carrot-flavored cereal ingested at weaning (Mennella & Beauchamp, 1999). As Mennella and Beauchamp (1998) conclude in their review article, "Because the chemical senses are not only functioning during infancy, but change during development, breast-fed infants may be afforded an opportunity to learn about the flavor of the foods of their people long before solids are introduced" (p. 209).

AUDITION AND VISION

In rat pups, audition and vision become available at about the start of weaning (circa 2 weeks). It would be of great interest to understand how the new (auditory and visual) sensory information regarding social objects such as the mother and siblings is integrated in the pup's previous representations, which were based upon other sensory modalities, and if this changed the pup's "attitude" or preference toward them. In humans, there is a small literature on intermodal congruity (see Pineau & Streri, 1990, and Streri & Molina, 1993, for haptic–visual agreement; see Spelke, 1981, for visual–auditory agreement). To my knowledge, however, studies have not been performed in any species in which one natural attribute of the mother (her voice, for example) was linked experimentally with another to assess the formation of multidimensional representations of the dam or siblings.

EARLY PREFERENCES AND EXPERIENCE: LONG-TERM EFFECTS

Newborn pups locate the nipple for the first time by orienting to an odor, that of the amniotic fluid, which the dam licks on her ventrum during the birth process (Teicher & Blass, 1977). This vital postnatal olfactory attraction is apparently based on the fetus' experience *in utero* with the chemosensory (taste/odor) properties of the amniotic fluid: adding an artificial odor prenatally to the amniotic fluid resulted in preferential postnatal attachment of the newborn rat to the nipples of a dam coated with the same odor stimulus (Pedersen & Blass, 1982). Thus, the fetus' prenatal experience is shaping or focusing the individual's future (postnatal) specific hedonic preferences (see Coopersmith & Leon, 1988, for discussion of the role of excitation in the formation of early olfactory preferences).

Early postnatal olfactory experiences can similarly strongly influence a rat's future hedonic preferences. When dams were scented with an artificial odor during the lactation period the odor later affected the pups' olfactory preference (as weanlings; Galef & Kaner, 1980) and ejaculation latency during copulation (in adults; Fillion & Blass, 1986). These findings suggest that subtle olfactory (and other sensory) influences within the pup's micro-environment during the sensitive pre-weanling period may have formative, long-term effects upon the individual's future hedonic and social choices. The rats in the study by Fillion and Blass (1986), for example, were isolated for months from the odor that they had experienced neonatally before encountering it again as adults during their initial reproductive experience. Thus, the adult ejaculatory pattern of preference or at least sexual enthusiasm in this case was determined by a sensory "impression" originating in the early postnatal period. The functional significance of shortened latencies is not known.

Another example of early impressions with potentially long-term implications comes from the work of Galef on social transmission of diet preferences. Here, a brief (30 min) exposure of pups 21 days of age and older to an anesthetized conspecific whose face had been dusted with a novel diet produced a preference for that diet (Galef & Kennett, 1987). Even earlier in development, exposure to the mother's diet via her milk and smelling her mouth during the preweaning period appear to bias the dietary self-selection at weaning (Galef, 1992; Galef & Sherry, 1973).

Finally, there is extensive research (reviewed by Holmes, 1988; and this volume) documenting the development of social preferences for genetically related conspecifics. Together with "phenotype matching," these preferences often are based upon experience: familiarity with one's rearing associates, who are typically related conspecifics. As rats have strong olfactory preferences that are either natural/initial or acquired at an early age, it is of much interest to know how these play a role in the development of social, dietary, and other preferences under natural conditions. Unfortunately, very few studies have examined this issue outside of the laboratory.

PREFERENCES

COMPONENTS OF "AFFECTIVE SYSTEMS": SENSORY MECHANISMS, AFFECTIVE LABELS, AND MEMORY

In order for an animal to respond "affectively" to impacting stimuli, it has to be equipped with transducers to act on the stimuli and a central neurological system to produce sensations of "pleasure" and "displeasure" in response to particular relevant stimuli and to remember the properties of the transduction. The cognitive component of this "central affective system" labels the incoming sensory stimuli, thus sorting them into hedonically positive and negative categories. This is based upon previous experience recalled from memory and inborn tendencies. The "affective" component of this central system links sensations and emotional responses to the now labeled stimuli. In many cases, the affective process may operate simultaneously, in parallel with the cognitive-labeling process.

The neural representation of the sensory event, now labeled and experienced as pleasant or not at some given level of intensity, will then be stored in memory for retrieval and use in the future. Such a "central affective system" can be regarded as the neurobiological basis for many of the individual's choices and preferences, providing direction and motivational impact to the behavioral course throughout ontogeny.

How would this work? After an infant has detected a specific stimulus, as judged by its reactions to it, the affective system would apply a particular valence to the stimulus, reflecting the peripheral and central properties of the stimulus, the infant's state, its history with the stimulus, and the specific motor patterns that the stimulus elicits. This complex is stored and is available for association with conjoint stimulation that predicted the occurrence of the stimulus. Future encounter with this stimulus context would trigger the affective system again, this time including the impact of memory, which could potentiate or habituate the sensation and/or the response. Future analyses must also take into account the fact that affective systems themselves change over time so that what had previously been classified as positive may on occasion take on negative affective tone. An elaboration of the examples provided above (Camp & Rudy, 1988; Moran et al., 1983a, b) is helpful here.

Consider the case in which infant rats experience either the flavor of milk or an electric shock. In both instances the animal rolls around, makes gaping movements, hyperextends its body as when withdrawing milk from a nipple, and dorsoflexes, as do female rats during copulation. Note that these are the same patterns elicited in 3- to 6-day-old rats by electrical stimulation of the medial forebrain bundle (MFB). Interestingly, even when an initial preference or aversion exists, the hedonic preferences themselves can change over time. Thus, 6-day-old pups express a preference for an odor paired with shock! By day 12, a differential "affective label" or category is attached to these two stimuli: milk = positive/appetitive/attractive; i.e., pups will work specifically for milk when hungry. When brain stimulation is paired with milk delivered to the mouth, day 9 rats eliminate nonfeeding behaviors and focus exclusively on ingestive responses. This stands in sharp contrast to their behaviors of only 3 days earlier when conjoint stimulation of food and MFB stimulation increased the likelihood of all behaviors. Twelve-day-old pups respond differentially, mouthing and probing continues in response to milk, but no longer to shock; flinching and escaping from the shock become the elicited actions (Moran *et al.*, 1983a, b). Appropriately, they also acquire an odor aversion after odor–shock pairing (Camp & Rudy, 1988). These ontogenetic changes in affective labeling would appear to reflect the dramatic changes in the underlying neural (neurochemical and neuroanatomic) systems which occur in the postnatal period in the rat (Leon, 1992a, b). However, we do not know which aspects of neural maturation underlie the above-described patterns of hedonic reversal and improved integration. One of the many candidates may be the noradrenergic system, in which dramatic changes occur in responsivity over the first 2 weeks of the rat's life, changes that affect the patterns of learning new olfactory preferences (Leon, 1992a; Sullivan & Wilson, 1994).

ONTOGENETIC TRANSITIONS IN HEDONIC VALUE: FROM SUCKLING TO WEANING

Changes in hedonic value become manifest around weaning. For instance, neonatal rat pups totally ignore dry food, but as weaning approaches they show progressively greater preferences for food and water (Almli, 1971; Thiels, Alberts, & Cramer, 1990), ingesting free foods for the first time on postnatal day 14 (Redman & Sweeney, 1976). Importantly, changes in infant motivation track those of spontaneous ingestion.

By evaluating changes in individual pup preferences for nutritive suckling, nonnutritive suckling, feeding, and drinking in a Y-maze over the first 4 weeks of the rat's life, i.e., through the weaning period, Blass and his colleagues characterized the changes in appetitive behavior from suckling to feeding. The ontogenetic pattern reflects changes in control of intake and in hedonic priorities. Pups 7–23 days of age preferred nonnutritive suckling over rooting into the gauze-covered ventrum of their mother on the other side of the maze (Kenny & Blass, 1977). Nutritive suckling is preferred over nonnutritive suckling at 17 and 21 days of age, but not at 10 and 12 days (Kenny, Stoloff, Bruno, & Blass, 1979). The appearance of this later preference on day 17, only a few days after the pup starts to taste food, was not dependent on suckling experience from day 11 to 17 (Kenny *et al.*, 1979).

In rats 17–32 days old, the decision of whether to eat a familiar food or to suckle depended upon the rats' age, availability of the dam at the feeding site, nutritive versus nonnutritive suckling opportunities, and the quality of the food offered (Stoloff & Blass, 1983). More older than younger pups chose the food arm of the maze. At all ages, the opportunity of maternal contact at the feeding arm increased the choice to feed by about 20%. In a choice of foods, milk was preferred over dry chow by a majority of the pups 24 days of age and older. Suckling was the dominant choice over feeding on day 17; on days 21 and 24 the opportunity of a lactating versus a nonnutritive nipple increased the preference for suckling over feeding. On days 28 and 32 rats did not suckle in the maze and choice was affected mainly by the opportunity to contact the dam.

When given a choice between nonnutritive suckling and drinking water, the opportunity to contact the dam strongly influenced the pups' choice behavior (Bruno, Blass, & Amin, 1983). In this study of pups 15, 20, and 25 days of age, drinking behavior was more readily expressed in the presence of the dam; the occurrence increased with age. In addition, cellular dehydration (induced by an injection of hypertonic saline) increased drinking, but only when the dam was accessible. When she was not, dehydrated 15-day-old pups continued to choose the arm that allowed maternal contact, avoiding the water. Dehydrated 20- and 25-day-old pups did not express a preference.

These studies have clearly shown through the two-choice test methodology that the hedonic value of nutritive and nonnutritive suckling, food, and contact with the dam undergo dramatic shifts during the period in which the rat changes from a suckling to a weanling. In the beginning of this period, (nonnutritive) suckling of the dam appears to be of very strong value. The nutritive product of this suckling only affects the pups' choice in the third week of life, as feeding and drinking independently emerge as increasingly attractive options. Contact with the dam, a combined thermal, olfactory, tactual, and social stimulus, remains of great value to the pups throughout this period of transitions. Social reinforcement remains a reward throughout life. Whereas the infant's major rewarding object is its mother, an animal's companions and mate have this role in later developmental phases of life.

Benefits of Initial Preferences

As a rule, studies of responses to stimuli from various modalities suggest that basic sensory pleasure appears to be adapted to the defense of homeostasis (Cabanac, 1990), in accord with the general notions set forth toward the beginning of the century by Cannon (1918, 1932; Cannon & Washburn, 1912) and especially Richter (see Blass, 1976). In general, the hedonic rating of a stimulus often (although not always; Toates, 1986) reflects the value of the stimulus as a promoter of survival needs (Cabanac, 1979). To illustrate, when a furless altricial rat pup moves from a warm to a warmer area or prefers to huddle with its littermates over being alone (Alberts, 1978; Kleitman & Satinoff, 1982; Leon, 1986), it is displaying a preference with a metabolic/energetic advantage: maintaining body temperature at a much lower physiological cost than through thermogenesis (see the chapter by Blumberg in this volume). Similarly, the preference infant rats exhibit toward their mother over home nest shavings (2–10 days old; Polan & Hofer, 1998) and even over another (nonlactating) female (16-days old: Leon & Moltz, 1971; 10-days old: Nyakas & Endroczi, 1970) results in gaining or conserving energy, a crucial advantage for altricial infants that are totally dependent upon the mother at this early stage of life (Blass, 1996,1999). In an interesting contrast, rat pups (12 and 20 days old) prefer an odor associated with their littermates over another odor associated with their dam (Hepper, 1987a). While the advantage of such a preference is unclear, it may be in the establishment of kin recognition based on familiarity and preference of sibling odors (Hepper, 1987a, 1991; see the chapter by Holmes in this volume for a discussion of this complex issue).

Although many initial preferences seem advantageous, it must be clearly stated that the basis by which natural selection is acting over evolutionary time to produce a particular pattern of infantile preferences is not demonstrated by ontogenetic studies. Indeed they can not inform us if it is adaptive, has been selected, or represents some other manifestation. Therefore, energetic/metabolic proximal benefits provided by particular choices of the infant, although suggestive, should be regarded as highly speculative in relation to distal advantages, i.e., adaptive significance.

The Shaping of Young Rats' Affective Preferences

By definition, altricial mammals are initially totally dependent upon their mother as the source of nutrition, water, minerals, and antibodies. To attain maternal care they must maintain contact with her, stay in the nest in her absence, or signal their separation from her outside of the nest. Infant rats remain in the nest during their mother's periods of absence, are attracted to her when she is present, orienting toward her, approaching, attaining, and maintaining contact. Young rats, and the young of other altricial mammals, are equipped with an initial inclination or attraction toward various stimulus aspects of the dam and nest (e.g., their warmth and odor). The basic inclination toward the mother would appear to be important for the infant's survival and has been proposed to be the basis for infant–mother attachment (Bowlby, 1980). As a rule of thumb, contact and suckling with their positive thermal and energetic consequences allow the newborn of many species, including humans, to learn the details of the specific preferences to *this* dam, *these* siblings, *this* nest (Hogan, 1988; and this volume) and to be attracted to them. This information is acquired both in the womb and during the postnatal period (Blass, 1996, 1999; Leon, 1992a, b; Porter & Winberg, 1999; Terry & Johanson, 1996).

Many studies, expanding upon the embryological concept of sensitive periods (see the chapter by Oppenheim in this volume) for the establishment of social behavior, have found multiple pathways through which newborn mammals are oriented to, learn about, and "attach" to a parent/caretaker. This system is available prenatally in newborn rats, which prefer the odor of their own amniotic fluid compared to that of other rats (Hepper, 1987b). Thus, learning about properties of the mother and kin, such as relevant odors and sounds, starts *in utero* (in humans; DeCasper & Fifer, 1980; Hepper, 1990). This experience may be important in the formation of social preferences after birth (Hepper, 1991). In an important observation, Blass (1996) noted that all mammals learn about their mothers through a mother-induced change in state (to either marked excitation or calm) during the interactions surrounding nursing and suckling. For example, infant rats are activated by extreme aspects of the nursing interaction, such as vigorous licking by the dam before nursing, and during milk withdrawal. These and other, more calming aspects of nursing, such as the act of suckling itself and the taste of milk (partially mediated through opioid and CCK pathways), have been shown to support learning about specific maternal features (for details see Blass, 1995, 1996; Sullivan & Wilson, 1994). Human newborns can also establish representations of their mothers, by particular sensory dimensions, also through state changes similar to those found in rats, and probably through the same mechanisms. These, too, are elaborated and enriched over time. This topic is covered in detail by Blass (1999).

As the infant develops it utilizes associative and nonassociative (e.g., habituation to exposure alone) capacities to expand its knowledge, developing positive and negative reactions to novel stimuli and modifying previous attractions and aversions with further experience (Galef, 1981; Leon, 1992a, b; Rudy, 1991; Terry & Johanson, 1996). For infant rats it seems that exposure to a stimulus in the context of the dam provides a strong basis for developing a preference toward it. In a recent study in rats (Terry & Johanson, 1996), dams were fed either regular diet or diet with a unique odor (eucalyptol) during gestation and postpartum until testing. Infant rats (3–9 days old) clearly preferred eucalyptol-scented bedding over unscented bedding, spending 80% of the test time over the eucalyptol-scented bedding compared with 14–22% of the time for control rats. Thus, sensory aspects of the mother and the nest that are available to the infant while it is being rewarded by positive, preferred aspects of the mother and nest situation (including mother's warmth, milk, contact, licking, etc., and their underlying physiology, e.g., opioid, CCK, and norepinephrine release) can strongly shape the infant rat's profile of hedonic preferences.

Another study of olfactory learning highlights the interactions of mechanisms contributing to the shaping of affective preferences. Galef and Kaner (1980), following work by Leon *et al.* (1977), reported that simple exposure of pups to a (naturally/ initially) mildly aversive odorant (peppermint) from birth to day 21 produced a substantial peppermint preference. If the odor exposure continued until day 33, however, the preference was not maintained. Yet, if the peppermint extract was spread on the dam for the first 33 postnatal days, preference was maintained to day 33 (Galef & Kaner, 1980). Thus, experiencing an odor within the context of excitation and calm provided by the dam, and probably other aspects of the nest and siblings, provide a basis for learning and shaping the infant rat's preferences. Because this finding of altered behavior has been reported for different behaviors in a number of contexts, two follow-up questions are posed: First, what particular aspects of the dam–pup interaction are the rewarding (preferred) "unconditioned stimuli" that, when paired with the novel stimulus, produce a conditioned preference? Second, what are the underlying mechanisms?

Shaping Preferences by Learning from the Mother and the Nest

Neurobiological Mechanisms.   Rat pups can learn to respond positively to various aspects of the mother. For example, the flavor of milk (or sugar, or fat) reduces isolation-induced ultrasonic vocalization (USV) and elevates thermal pain response thresholds (analgesia) in infant rats (Blass & Fitzgerald, 1988; Blass & Shide, 1993; Blass, Fitzgerald & Kehoe, 1987; Shide & Blass, 1989), although most likely through different neural systems (Ren, Blass, Zhou, & Dubner, 1997). Studies with pharmacologic antagonists have implicated the opioid system in these effects and the CCK system in the effect of milk and fat (but not sugar) on USV, but not analgesia (Blass & Shide, 1993; Weller & Gispan, 2000). These studies will be discussed again later in this chapter. In addition, rats prefer an odor that was paired with oral infusions of fat or sucrose, and this was reversed by the opioid antagonist naltrexone (Shide & Blass, 1991). Thus, these taste stimuli appear to be analgesic, calming, and rewarding, supporting the shaping of new preferences through specific neurochemical systems. In human and rat infants, sucrose and nonnutritive suckling have been shown to be analgesic, calming (Blass & Watt, 1999), and rewarding (Shide & Blass, 1991). The similarity suggests the possibility of opioid mediation in the human phenomena as well.

An important aspect of the care provided by the rat mother, her licking of the pups, has been mimicked by stroking with a fine artist's brush or a cotton swab. When this brushing was paired with exposure to a novel odor on postnatal days 1–18, a conditioned olfactory preference resulted on day 19 (Coopersmith & Leon, 1984). Further studies showed an apparent sensitive period for this phenomenon. For the preference to be expressed on day 19, training had to take place on days 1–8; training on days 1–4 or starting after day 7 did not result in a preference on day 19 (Leon, 1992a). The neural basis of this learning has been studied in detail. Briefly, the tactile stimulation used in this paradigm evokes a clear response by noradrenergic fibers in the locus coerulus that project to the olfactory bulb. This projection is evident from postnatal day 1 (Nakamura, Kimura, & Sakaguchi, 1987) and is functional from at least the first postnatal week (Wilson & Leon, 1988). Olfactory learning modifies olfactory bulb responses to subsequent presentations of the conditioned stimulus odor, and the noradrenergic input is necessary for associative learning to occur in the infant rats (Leon, 1992a; Sullivan & Wilson, 1994).

Investigators have also shown that initial aversion of infant rats toward one type of bedding, cedar, could be altered by rearing in the cedar odor for a few days (Cornwell *et al.*, 1996; Cornwell-Jones, 1979). Because studies in adult and weaned rats suggested that central norepinephrine modulates the learning of olfactory preferences, noradrenergic mediation of bedding-odor preferences was recently examined. Permanent forebrain depletion of norepinephrine by injection of a neurotoxin on postnatal day 0 or on days 5–6 did not alter initial pine versus cedar preferences nor impair the learned preference to cedar on day 12 or 16 following exposure to cedar in the nest (Cornwell *et al.*, 1996). The norepinephrine system may be more relevant for adaptation to new home odors at older, postweaning ages, as shown by the same team of investigators (Cornwell-Jones, Decker, Gianulli, Wright, & McGaugh, 1990). Note that the ability of the preweanling pups to habituate to the novel cedar odor while experiencing severe noradrenergic depletions is surprising now that we realize the dependence of the pups' olfactory associative learning on central norepinephrine in the research by Leon, Sullivan, and co-workers. One of many possible explanations is that in preweanling rats, norepinephrine mediates the explicit associative paired learning, as in Leon's studies, but not the more general habituation of aversion occurring over 3 days of odor exposure, as in Cornwell's experiments.

THE IMPACT OF PREFERENCES ON LEARNING Before moving on, another point should be emphasized. The young pup's preference may not be modified solely by learning. The pup's preferences themselves may affect the pup's learning ability. Shelly Rakover-Atar and I discovered in 6- and 9-day-old pups differential patterns of learning and association between an unconditioned stimulus (US; contact with a sibling pup) and a tactile conditioned stimulus (CS), depending upon the relative preference of the CS (Rakover-Atar & Weller, 1997). Specifically, a rug floor texture, which was preferred by all pups, produced only a moderate pattern of acquisition, whereas pairing the US with a less preferred plywood texture produced a robust increase in CS preference and a different pattern of acquisition (see Figure 2). This is in accordance with research in adult rats showing that the form of the Pavlovian appetitive conditioned response is influenced by various characteristics of the conditioned stimulus (Holland, 1977, 1980).

It is also of interest that while control (unconditioned, rug-exposed) pups tended to habituate to the rug texture, experimental pups exposed to rug–US pairing exhibited similar rug-preference levels at baseline and after conditioning, i.e., no habituation (Rakover-Atar & Weller, 1997). Note that a similar phenomenon was reported by Holland (1977) in adults. Here, pairing a visual CS (light) with food

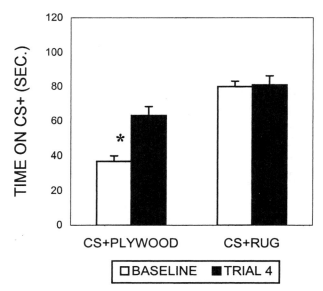

Figure 2. Mean (± SE) preference for the conditioned stimulus (CS) during baseline trials (mean baseline, white bars) and after conditioning (dark bars) in two experimental groups: rats conditioned with a plywood CS and rats conditioned with a rug CS. (n = 32/texture, age = 6 or 9 days). *p < .05 compared to mean baseline level. (Reprinted from Rakover-Atar & Weller, 1997, with permission.).

delivery prevented habituation of the orienting response (rearing) normally seen to the visual CS. Experimental (CS–US pairing) animals exhibited the rearing response throughout the training procedure, whereas control (unpaired) animals showed habituation of the rearing response over trials.

The pups conditioned in our study with the relatively preferred rug texture were not at "ceiling" response level; that is, they could have increased their preference levels even more than they did. As indicated in Figure 2, this group's mean preference scores were around 80 sec, about two-thirds of the maximum level possible, which was 120 sec. We also reported that 6- and 7-day old rats preferred cotton balls as bedding material (mean = 107.3 sec) over plywood (mean = 15.7 sec), showing that conditioning could have further increased the preference level to the rug texture beyond the levels found (Rakover-Atar & Weller, 1997). Furthermore, the rug texture was selected after pretests had shown that it was relatively less preferred when compared with a sponge texture (as described above in the section on the young rat's preferences). Thus, same-age pups are capable of expressing greater preference than was expressed after conditioning to the rug texture in our study.

Increased preference toward the floor texture could be direct, i.e., reflecting a differential level of "conditionability" or ability to be conditioned to, according to the CS's relative level of initial preference. Alternatively, the effect could be happening by changing the CS–US complex, or the reward value of the US; thus, contact with a sibling on a wood floor may cause a different affective change than contact with a sibling on a carpet. This is a crucial developmental issue because we do not know the rules of change caused by various forms of stimulation. That is, the gain in positive affect may be set by the context of the change that is age dependent, among other things (Kehoe & Blass, 1986a).

At this point I will examine the role of a physiological system with widespread effects on physiology and behavior that has rarely been discussed in a developmental, noningestive, and non-drug-related hedonic preference context. This section will provide information on cholecystokinin (CCK) and evidence for its role in behavioral development. The subsequent section will focus on CCK's contribution toward infant rats' interactions with the dam and preferences and attractions toward aspects of mother and nest.

CCK

Most research on the polypeptide gut hormones has focused on how they affect digestion, nutrient transport, and metabolism. Of the many (over 20) gut hormones identified, behavioral research has focused mainly on CCK, a peptide released from mucosal cells in the proximal small intestine, and in particular on its C-terminal octapeptide (CCK-8). CCK is also a neuropeptide, abundantly distributed in the central nervous system, but peripheral and central pools of CCK are separated by the blood–brain barrier in adults (Passaro, Debas, Oldenorf, & Yamada, 1982). To my knowledge, data on the efficacy of this barrier in infants is not available.

Peripheral CCK is released from cells in the mucosa of the small intestine into the surrounding intercellular space (paracrine secretion) and into the blood (hormonal secretion). CCK suppresses feeding in many species, and other behavioral effects have been documented as well, e.g., on analgesia, sleep, memory, anxiety, and sexual behavior (Crawley & Corwin, 1994). Among its diverse physiological effects, CCK promotes growth of the gastrointestinal tract and the pancreas, induces the release of nutrient-digesting enzymes, regulates gastric emptying, and increases the efficiency of insulin-mediated glucose utilization (Dockray, 1981; Liddle, 1997). CCK binds to two receptors, which differ in their binding potencies in regard to CCK-related peptides (Moran, Robinson, Goldrich, & McHugh, 1986). CCK_A receptors are localized mainly in the gastrointestinal tract, but are also found on the vagus and in discrete brain areas, specifically the nucleus of the solitary tract, the area postrema, interpeduncular nucleus, median raphe, dorsal medial hypothalamus, and nucleus accumbens. CCK_B receptors, which show the same binding profile as peripheral gastrin receptors in the stomach, are widely distributed throughout the CNS (Moran *et al.*, 1986). Studies with selective pharmacologic receptor antagonists have shown that some behavioral effects of CCK are mediated selectively by one receptor (e.g., feeding inhibition is mediated by CCK_A and not by CCK_B receptors; Moran, Ameglio, Schwartz, & McHugh, 1991; Smith, Tyrka, & Gibbs, 1991), whereas other effects are facilitated by one receptor subtype and attenuated by the other (e.g., memory on a social recognition test: Lemaire, Piot, Roques, Bohme, & Blanchard, 1992; Lemaire, Bohme, Piot, Roques, & Blanchard, 1994).

PERIPHERAL CCK IN INFANCY

Plasma CCK levels are lower in exclusively breast-fed 9-month-old human infants than in formula-fed controls. Infants in both groups however increased plasma CCK levels postprandially (Salmenpera *et al.*, 1988). CCK levels increase immediately after nursing (breast feeding) and also 30 and 60 min later (Uvnas-Moberg, Marchini, & Winberg, 1993). The initial rise is probably due to the sucking act itself, the latter increase to milk in the intestine. The initial, suckling-induced CCK increase would appear to be "nonnutritive" by definition, since the milk would not have had time to reach the intestine and release CCK directly. It remains to be seen what component of the nursing situation (maternal stimuli such as mother's odor, touch, holding, etc., and the tactual aspects of the nipple in the infant's mouth and/or the motor act of suckling on the infant's part) stimulates CCK release, and whether this secretion is reflexive or conditioned, i.e., learned during the first few days of life.

In rats the peripheral CCK system is functional in the early postnatal period. CCK receptors are abundant and widely distributed in the gut (Robinson, Moran, Goldrich, & McHugh, 1987). Furthermore, as shown in Figure 3, compared to littermates taken from the nest (and receiving a gastric load of isotonic saline), pups separated in a group from their mother overnight had lower plasma CCK levels and pups reunited with a dam for 1 hr had higher CCK levels (Weller *et al.*, 1992). The particular aspects of deprivation (e.g., nutritional deficit, dehydration, lack of contact, etc.) and of reunion (e.g., milk intake, suckling, maternal odor, warmth and contact, pup-licking, etc.) that are responsible for these changes have yet to be elucidated. Preliminary findings in dog puppies further suggest increases in plasma CCK levels during sucking and also when the puppies were returned to their littermates after a period of separation (Uvnas-Moberg, Widstrom, Marchin, & Winberg, 1987). This suggests that the warmth, touch, or odor of the other puppies elicits CCK release in the subjects (even without sucking or ingestive stimulation).

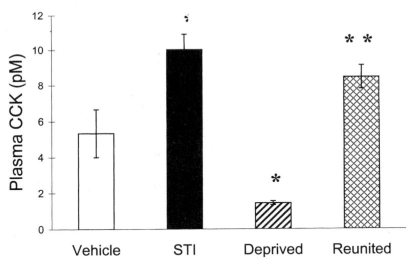

Figure 3. Mean (+ SE) plasma concentrations of CCK (pM) were determined by the dissociated pancreatic acinar cell bioassay in four groups of 9- to 12-day-old rat pups. Nondeprived pups that received intragastric soybean trypsin inhibitor (STI) had significantly higher plasma concentration of CCK than pups receiving isovolumetric, intragastric saline (Vehicle). In overnight deprived pups, reunion with the dam (Reunited) produced plasma CCK levels that were significantly higher than the plasma concentration of CCK in pups that were deprived, but were not permitted access to the dam (Deprived). Plasma concentration of CCK in the deprived pups (Deprived) not given access to the dam was also significantly lower than the plasma concentration of CCK in nondeprived pups that were infused intragastrically with saline (Vehicle). *Significantly different from deprived; **Significantly different from vehicle, determined by Duncan's multiple range test ($p < .05$). (Reprinted from Weller *et al.*, 1992, with permission. Copyright 1991 Elsevier Sciences.)

According to Uvnas-Moberg (1989), CCK helps coordinate the developing infant's digestion, metabolism, and growth. CCK, along with other gut hormones, may also contribute to behavioral development and help shape the pattern of the infant's interaction with its mother (this point is supported by data reviewed by Uvnas-Moberg, 1989, and in the next sections). These behavioral effects could be limited to the realm of ingestive behavior: suckling and feeding. CCK effects extend beyond ingestion, however; in rats CCK also mediates other important nonnutritive aspects of infant behavioral development such as reacting to separation distress. CCK effects may also extend to nonnutritive aspects of developing infant–mother and infant–sibling relations.

Over the last 10 years my colleagues and I have been investigating the role of CCK in mediating ingestive and other behaviors of infant rats. Endogenous CCK, bound by CCK_A receptors, reduces independent feeding in infant rats when they eat their first meal away from the dam (Weller, Smith, & Gibbs, 1990). Furthermore, we have recently shown that CCK_A receptors mediate a portion of the intake reduction produced by a preload of corn oil (but not glucose, 2-deoxy-D-glucose, or maltose), in infant rats (Weller, Gispan, & Smith, 1997; Weller, Gispan, Armony-Sivan, Ritter, & Smith, 1997). These findings suggest that as in adults the CCK system can function selectively in preweanling rats, mediating the emergence of a specific regulatory system. Interestingly, CCK, like other inhibitors of feeding such as gastric distention, does not affect the preweanling infant rat's intake during suckling from the (anesthetized) dam. It is only after the first 2 weeks of life, when the weaning process has started and rats eat directly from the environment, with suckling coming under the same motivational and physiological controls as feeding (Hall, 1990), that CCK becomes effective and reduces intake (Blass, Beardsley & Hall, 1979; D. Lorenz, 1994).

Cholecystokinin and Infant–Mother Attachment in Infant Rats

Other research has shown that CCK may be involved in affective responses and in infant–mother attachment. Specifically, we have studied separation-induced ultrasonic vocalization (USV) in rat pups. As shown in Figure 4, administration of CCK to 11-day-old rats significantly and selectively reduced USV; pain threshold or activity levels were not affected (Weller & Blass, 1988a). Blass and Shide (1993) further showed that corn oil or milk infused into the pup's mouth reduced USV and that the selective CCK_A-receptor antagonist devazepide (Chang & Lotti, 1986; Lotti *et al.*, 1987) blocked this effect. Sucrose infusion worked through a different pathway: sucrose-reduced USV was not blocked by devazepide (Blass & Shide, 1993). Recently, we have shown that while this antagonist blocks the USV-quieting effect of a milk meal, it is ineffective against USV-calming produced by returning the pup to its (anesthetized) dam (Weller & Gispan, 2000). Yet, devazepide increased USV in pups returned (alone) to their active dam (Weller & Dubson, 1998). The contrast between potentiation with the active dam and the inefficacy with the anesthetized dam, which brings to mind the different mechanisms underlying "passive potentiation" (following separation from an anesthetized dam) and "active potentiation" (from an active dam) of USV (Hofer, Masmela, Brunelli, & Shair, 1998, 1999), requires further

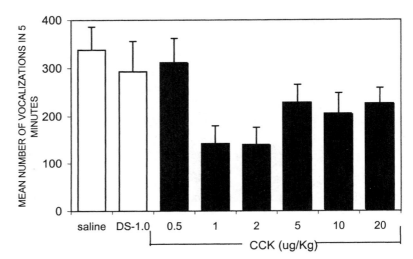

Figure 4. Mean number ($\pm$ SE) of isolation-induced ultrasonic vocalizations (USV) produced by 11-day-old rats after one of six doses of CCK-8, desulfated CCK (DS-1.0 μg/kg, white bar) or isotonic saline (black bar). N = 10–12/bar. Statistical analyses showed that treatment with 1 and 2 μg/kg CCK significantly reduced USV compared to the saline, DS, and 0.5 μg/kg CCK. (Reprinted from Weller & Blass, 1988a, with permission. Copyright 1988 American Physiological Society.)

examination: clearly the mechanisms differ. Nevertheless, the general conclusion is that there are multiple pathways for quieting of infant USV. Sucrose is apparently working through opiate receptors (Blass *et al.*, 1987). Milk and fat may be activating both CCK and opiate receptors (Blass & Fitzgerald, 1988; Blass & Shide, 1993; Shide & Blass, 1989; Weller & Gispan, 2000). The suckling act, e.g., of a nonnutritive pacifier is likely employing these and/or other pathways, etc.

Just to add a note of additional complexity to the link between CCK and USV, it seems likely that while CCK_A receptors are implicated in USV reduction, CCK_B receptors in the brain may mediate an opposite effect. Rex, Barth, Voight, Domeney, and Fink (1994) reported USV increase (!) following intraperitoneal administration of butyloxycarbonyl-CCK-4 (BOC-CCK-4), a form of CCK with greater affinity for B than A receptor types (Harhammer, Schafter, Henklein, Ott, & Repke, 1991). In addition, these researchers reported similarly "anxiogenic" responses to BOC-CCK-4 administration in three adult models of anxiety: the elevated plus maze, a two-compartment black and white box, and a conflict paradigm based on novelty suppressed feeding in food-deprived rats (Rex *et al.*, 1994). These findings imply a more subtle, balancing, type of regulation by CCK, where the behavioral output, USV, and possibly other aspects of affective states/traits like anxiety are the function of the relative degree of activation of the two receptor types in unidentified anatomic sites. (For an overview of all neurochemical systems implicated in the mediation of USV, see Table 1 in the chapter by Brunelli and Hofer in this volume; cf. Hofer, 1996; Winslow & Insel, 1991).

Note that this is not the first report of opposite effects of CCK_A and CCK_B receptors: Lemaire *et al.* (1992, 1994) discussed evidence for "endogenous cholecystokininergic balance" in social memory in adult rats. CCK modulates pain signals in adult rats through CCK_B (but not CCK_A) receptors (Wiertelak, Maier, & Watkins,

1992), whereas it is neutral in infant rats (Weller & Blass, 1988). The effect in adults is particularly interesting. Opiate-induced analgesia at the spinal level has been shown to be blocked by a stimulus that predicts safety; this effect is mediated by the "safety stimulus" releasing endogenous CCK in the spinal cord and operating via CCK_B (but not CCK_A) receptors (Wiertelak *et al.*, 1992). This process may contribute to the regulation of pain and to development of opiate tolerance.

Coping with the distress of maternal separation is just one of the many aspects of attachment and the infant–mother interaction. The involvement of CCK as one of multiple, possibly redundant, modulating neurochemical systems, in most of the other portions of the infant's early formative social relationships has yet to be evaluated. We have recently shown (Weller & Dubson, 1998) that the rat dam was found more frequently in proximity with her 6- to 9-day-old pup just returned to her if the pup had previously received devazepide (compared to vehicle-treated controls). In addition, as mentioned above, devazepide-treated pups emitted more USV than controls. In accordance, pups treated with 1 μg/kg CCK received less body licking than vehicle controls. In addition, dams hovered and crouched over devazepide-treated pups more than over pups treated with 1 μg/kg CCK (Weller & Dubson, 1998). These findings show that blockade of the pups' CCK_A receptors by devazepide resulted in pup-directed maternal behaviors.

But what stimuli were devazepide-treated pups presenting to their dams that resulted in this altered maternal treatment? Vision, audition, and olfaction are the sensory modalities most likely implicated. Thus, if pups were more active in the presence of the dam (pup activity was, unfortunately, only examined away from the dam) this may have elicited increased maternal attention and care via visual (or tactile) stimuli. Olfaction may be involved because administration of other hormones (e.g., testosterone, dihydrotestosterone, or progestins) to rat pups (Birke & Sadler, 1985; Moore, 1982) has been shown to alter their odor, thus affecting levels of pup (anogenital) licking by their dams (Birke & Sadler, 1985; Moore & Power, 1992; Richmond & Sachs, 1983). Regarding audition, lactating rats orient toward pups emitting USV and tend to retrieve the pups when found (Allin & Banks, 1972; Noirot, 1973), apparently responding differentially to males and females according to vocalization levels (Naito & Tonoue, 1987). Therefore, the devazepide-treated pups' elevated USV levels could have accounted for their increased levels of maternal care. An alternative, more integrative possibility is that the devazepide-treated pups may have had a lower threshold to respond with USV to maternal licking/grooming, initiated by detection through another sensory modality (maternal licking and handling appear to increase USV; S. A. Brunelli & A. Weller, unpublished observations, 1999). Although this possibility would suggest that the USV increase was secondary, it could still serve a feedback role within the mother–infant interaction, further increasing maternal care and proximity.

Thus, endogenous CCK in rat pups, together with other hormones, e.g., steroids (Birke & Sadler, 1985; Moore, 1982) and neurochemicals, e.g., opioids (Kehoe & Blass, 1986b, c; Shide & Blass, 1989), affect infant distress and infant–mother interactions. These results suggest that endogenous CCK pacifies, calms, and quiets preweanling pups and this in turn modulates infant–mother interaction and maternal care. We do not know what constitutes ideal care. One may speculate, therefore, that CCK may play a role in shaping infant behavior and contribute to individual differences, for example, in the processes of weaning and maturing toward independence.

In a separate line of investigation, we have addressed the ability of CCK, as an unconditioned stimulus, to influence early learning. A novel, unfamiliar odor was paired with intraperitoneally administered CCK (controls received isotonic saline) in a noningestive context. As shown in Figure 5, this produced a conditioned preference for that conditioned odor on a subsequent test day (Weller & Blass, 1990). This effect was unique to the preweaning period and was blocked by pretreatment with devazepide (Weller & Blass, 1988b). Separate studies showed that exposure to the conditioned odor could reduce USV and feeding (Weller & Blass, 1989; Weller, Blass, Smith, & Gibbs, 1995). These results suggest, indirectly, that CCK increase is a positive/rewarding physiological event in the preweaning rat (if the reverse were the case, we would have expected to find conditioned aversion). As such, CCK release may mediate a portion of the infant's early attraction to the dam and of the pup's attraction toward stimulus aspects of the dam and nest, such as their odor, texture, and warm temperature. Furthermore, the findings suggest that CCK may provide a positive, unconditioned cue that can be associated with other "conditioned" stimuli in the nest situation. This could be a possible physiological mediating mechanism for learning by association about important (positive) aspects of the nest situation, such as olfactory, tactile, and thermal cues emitted by the mother and the siblings. We are currently examining this possibility. Other physio-

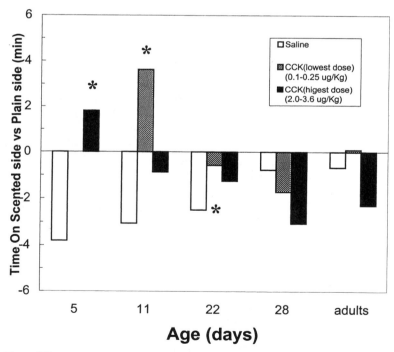

Figure 5. Mean difference (relative preference) scores in rats of various ages after odor–CCK or odor–control pairings (*p < .05). Only highest and lowest doses used in this series of studies are presented in this summary figure. Group means at all doses used were compared separately at each age by ANOVA. (Reprinted from Weller & Blass, 1990, with permission. Copyright 1990 by the American Psychological Association.)

logical systems such as opioids, oxytocin, and norepinephrine may play similar, overlapping and interacting roles with CCK (Blass, 1996; Nelson & Panksepp, 1996; Sullivan & Wilson, 1994). As their influence increases they may gain greater affective control, reducing the frequency and duration of nest area vocalizations when the mother egresses the nest as weaning termination approaches.

DOES CHOLECYSTOKININ MEDIATE AFFECTIVE PREFERENCE IN THE INFANT RAT?

Newborn mammals tend to be attracted to and prefer the dominant stimulus aspects of the mother and nest, such as their odor, texture, and warm temperature (Rosenblatt, 1983). As elaborated above, this is essential; pups that fail to be attracted to the dam and nest will most likely fail to survive. Recently, Michal Shayit and I have set up a series of test conditions in which infant rats were presented with a choice between two levels of one modality, for example, a warmer versus a cooler floor area. The difference between the stimuli on each side of the test arena was purposely moderate; preliminary studies allowed the development of testing conditions in which pups showed, on the average, a 66% preference for one condition. This low level of preference was needed in order to examine whether devazepide pretreatment could increase relative preference. The three tests that we employed examine relative preference toward (1) maternal odor (Leon, 1974), (2) a furry rug texture versus a plywood texture (as in Rakover-Atar & Weller, 1997), or (3) a warm (31–32°C) versus a cooler (about 25°C) floor temperature (Satinoff, 1991). The first two tests were employed on 11- to 12-day-old pups; the last, on 9-day-old pups, aiming at a sensitive time in thermoregulatory development (Satinoff, 1991). Pups received IP doses of 600 μg/kg or 1000 μg/kg devazepide or vehicle.

Previously, we showed that an initial positive reinforcer, contact with a sibling pup, changed patterns of texture preference when a specific texture is paired with the filial/social contact (Rakover-Atar & Weller, 1997). It now appears that patterns of initial preferences toward stimulus aspects of the nest and dam may be mediated by pathways utilizing CCKA receptors (Shayit & Weller, 1997). Figure 6 shows the overall pattern of results when combining the three separate tests. The figure shows the percentage of time spent by the pups on the side of the test arena nearest to the stimuli representing the dam and nest (the rest of the time was spent on the other side of the arena, away from the stimulus). Statistical analysis showed that the group receiving 600 μg/kg devazepide spent significantly more time overall on the side near the nest stimulus compared to the control group. (We are not sure why the higher dose was less effective; it could be less specific.)

Following these promising, preliminary results (Shayit & Weller, 1998), the findings regarding odor and texture preferences have been recently advanced by a more formal study, using the dose of 600 μg/kg devazepide (Shayit & Weller, 2001). I will discuss here in some detail the results of the temperature-preference portion of the results, including a follow-up study examining two possible alternative explanations (Shayit & Weller, 1998).

In 9-day-old rats, 600 and 1000 μg/kg devazepide significantly ($p <. 05$) increased preference rates for a warmer versus a cooler temperature (time on the warmer side) by 39% and 24%, respectively, compared to vehicle (Shayit & Weller, 1997). Control studies have reduced the likelihood of explaining these results by two important variables. First, the results of the temperature preference test could have

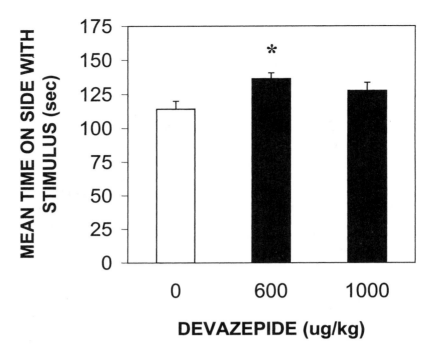

Figure 6. Mean time spent on the side of the test arena containing a nest-related stimulus (maternal odor, a rug, or warm temperature) in a 180-sec preference test. Before the test, pups received an IP injection of the CCK$_A$ receptor antagonist devazepide (0, 600, or 1000 μg/kg). ($n = 12$–17/stimulus/dose; age = 9 or 11–12 days).

been affected by devazepide changing the pups' body temperature. Measurements, however, show that the (axillary and rectal) body temperature of the devazepide-treated pups was not different from that of controls under our testing conditions (Shayit & Weller, 1998).

Second, it was possible that the increased preference levels found in the devazepide-treated pups could be accounted for by an unselective increase in activity levels in these pups. Specifically, if devazepide increased locomotor, horizontal activity levels, pups would be more likely to encounter the side of the arena with the preferred stimulus in their hyperactive state, and thus to know its location, allowing the expression of their initial preference. However, detailed analysis of videotapes of pups in the preference test arena did not support this possibility (Shayit & Weller, 1998). No significant difference was found between vehicle-treated and devazepide (600 μg/kg)-treated pups on any of the activity measures employed: number and duration of immobile episodes, body movements without locomotion, locomotion; number of strides, twitches, head-raising, wall climbing, and grooming episodes. Thus we conclude that devazepide-induced increase in the degree of temperature preference (and by extension, if replicated in the other modalities, in preference toward specific sensory aspects of the nest and dam) appears to be relatively specific to the behavioral choice itself; the effect apparently cannot be accounted for by changes in the pups' body temperature or general activity level.

It is of interest to note that in adult rats, devazepide did not change the preference shown for the odor of juvenile rats versus their own odor (Lemaire *et al.*,

1992). Thus, it is possible that the effects that we found may be unique to preweanling rats. The findings reported here with devazepide are preliminary, awaiting further replication. Furthermore, attribution of these effects uniquely to CCK_A receptors will be possible only after a dose–response study (with devazepide) and comparison of devazepide's efficacy with that of CCK_B receptor antagonists.

In summary, the pattern of results obtained with a CCK_A antagonist suggests that a functional CCK system may be needed for the infant to *maintain* normal levels of maternal/nest attraction. (Interestingly, *establishment* of the infant–mother bond in sheep may be differently modulated by CCK receptors; Nowak *et al.*, 1997). *When the regular action of the CCK system is disrupted (by CCK_A blockade) it appears that pups show increased motivation to approach maternal stimuli,* as evident by the increased preference levels in our studies. *This suggests that endogenous CCK is involved in mediating the reward provided by maternal stimuli and that CCK_A-blocked pups are "unsatiated" by maternal stimuli and therefore they persist in pursuing them.* An additional perspective (similar to Panksepp's ideas on the opiates; see Nelson & Panksepp, 1988) is that if CCK is serving as an anxyolitic agent, then infants that had received devazepide should seek calming from an external source.

Endogenous CCK may thus be playing a role in mediating the pup's attraction toward and positive reaction to olfactory, tactual, and thermal aspects of the dam. This is consistent with reduction of separation-induced USV by exogenous and endogenous CCK (Blass & Shide, 1993; Weller & Blass, 1988a; Weller and Gispan, 2000) as CCK may produce a state of lesser distress, alertness, and pursuit of the dam. The findings of conditioned odor preferences and conditioned odor USV quieting resulting from CCK–odor pairings in preweanling rats (Weller & Blass, 1989, 1990) further suggest that CCK is producing a positive/rewarding signal.

By analogy, a similar type of mechanism has been attributed to CCK (and devazepide) in modulating the incentive value of foods after a period of feeding in adults. Specifically, Balleine and Dickinson (1994) and Balleine, Davies, and Dickinson (1995) suggest that CCK acts as a signal within the feeding system that reduces the incentive value of foods after a period of feeding. CCK attenuated and devazepide increased the incentive value (roughly equivalent to the preference level) of foods to which adult rats were reexposed when undeprived (this reexposure reduces preference rate in their paradigm). Endogenous CCK thus appears to mediate the incentive value of food such that when CCK levels are high, incentive is low, and when CCK is not readily available, feeding incentive increases. It would be of interest if devazepide-treated rats showed other motivated increases in hunger. In any event the present data are analogous in the domains of social reactivity. The patterns reported here are consistent with the idea that CCK is anxiolytic in preweanling rats (at least at the A-type receptor) so that an increase in its availability at this receptor decreases an interest in nest objects, thereby supporting weaning.

CONCLUSIONS

Interesting questions arise when conceptualizing CCK as a mechanism or signal for incentive value or as an antianxiety agent. For example, what will be the role of CCK in mediating early learning in the nest about new positive and negative stimuli? It is possible that CCK could provide baseline levels of motivation (or incentive) that would modulate the strength of the reward and the infant's state to influence learning and performance. Is CCK important only for stimuli that are socially potent at a specific period of development?

Regarding the modulation of affective states, CCK may participate through CCK_A receptors, along with opioids and other neuropeptides, in calming rats distressed by the mother's absence or during the process of leaving the nest. This particular signal may help balance the anxiogenic/panicogenic message provided by CCK's action at CCK_B receptors (Bradwejn & Koszycki, 1994; Rex et al., 1994) and other anxiogenic signals such as corticotropin-releasing factor (CRF) (Coplan & Lydiard, 1998). An individual's level of trait or state anxiety is most surely the result of many complex sources of influence, both endogenous and exogenous (situational, social, experiential). The studies described above suggest that CCK may play a mixed "positive" and "negative" role as part of the endogenous, physiological basis of affective state and reactivity. CCK is ideally situated in the digestive chain to be involved in this process as its release is triggered by milk fat, and it appears to be easily conditioned. Conceivably, CCK might be released in advance of milk actually arriving at the intestine, probably as a conditioned response to the preingestive and perioral cues surrounding the initiation of a nursing bout.

SUMMARY AND CONCLUSIONS

I have described patterns of preferences of newborn and infant rats, with occasional examples from human infants, to argue for widespread conservation of traits and mechanisms surrounding suckling. Newborns discern and discriminate among stimuli within different sensory modalities, and, as a rule, are attracted to classes of stimuli that conserve energy. This is also expressed in changes in facial expressions in human infants that are interpreted as pleasing. With development, these primitive attractors serve as unconditioned stimuli for learning about particular features that allow them to identify particular members of the species: odors for rats, faces for human infants. Thus, this general process gives rise to particular identification, changes in affect, and increased tendencies to approach or avoid certain classes of stimulation.

We can safely say that only the surface has been touched during our journey toward understanding infant motivation and the basis of preference. We are reminded of the contribution of excitation and the paradox of being attracted to certain stimuli that are shortly thereafter avoided; shock and low levels of malaise, for example. The gains in infancy seem to differ from those in adults; indeed, we do not know the importance of specific versus general changes in response to particular stimuli such as shock. Another domain that we have only identified concerns packaging of behaviors well in advance of their appearance, e.g., feeding and reproduction.

Despite this we have progressed since Moran wrote his influential review on motivation for the 1988 edition of this *Handbook*. Multiple physiological pathways appear to support initial and learned affective preferences. Some of these are of importance in infancy, but seem to disappear (i.e., lose their modulatory influence on this behavior) around weaning. One example of this is CCK's ability to serve as a positive unconditioned stimulus in infant rats until day 21, but not on day 28 or in adults (Weller & Blass, 1990). Other pathways, such as opioids affecting analgesia, may play a continuous role from infancy through adulthood, but the behavioral events that engage these systems change radically (Zeifman, Delaney, & Blass, 1996). And then there are pathways that only become relevant for the juvenile or adult because the behavior in question only appears at these developmental stages (e.g., mediation of social play and sexual or maternal behavior, respectively).

The bases for transitional (days/weeks) controls during the nursing period has yet to be resolved. We do not know if organizing principles are involved or if each system has been organized on a case by case basis so that similarities are more apparent than real. It is possible that the system is shifted to modulate other behaviors in adulthood. Alternatively, it may still be relevant for the same behavioral system (e.g., positive affect or stress reactivity) in adulthood as in infancy; the response pattern may just have changed and the researchers may just not have found adequate ways to assess it. The possibility also exists that the "disappearing" system plays a role in mediating an ontogenetic adaptation that is useful for the vulnerable young animal and then becomes irrelevant.

Physiological mediation of affect can take place at many levels (see Figure 1). A neurochemical may play a permissive role (like that attributed by Sullivan & Wilson, 1994, to norepinephrine in the development of learned olfactory preferences). The case of CCK, appearing to mediate the incentive value of a stimulus, and the possible opposite roles of CCK_A and CCK_B receptors in mediating reward and memory, both discussed above, highlight the complexity of the neural mediation of affect. A further level of complexity is in the interaction of various neurochemical systems. These may be producing similar outputs (e.g., CCK and opioids, reducing USV), or different, even opposing outputs. Acting together, all possible combinations, from mutual inhibition to mutual potentiation, are possible, indeed likely, as more families of receptor subtypes are identified.

Future research should provide the opportunity to understand the complex patterns of neural mediation of affect and preferences in infants. This can provide a basic science to advance the understanding of choices, likes and dislikes, major components of the individual's personal style, and the milieu in which he or she is raised.

Acknowledgments

Research in the author's laboratory is funded by the Israel Science Foundation and the US–Israel Binational Science Foundation.

A portion of this chapter was written while the author was a Fellow at the Department of Psychiatry, Columbia University College of Physicians and Surgeons.

The author is grateful to his colleagues and students who participated in the research reported and in discussions of its implications, and to Elliott M. Blass for careful editing of the manuscript.

REFERENCES

Alberts, J. R. (1978). Huddling by rat pups: Group behavioral mechanisms to temperature regulation and energy conservation. *Journal of Comparative and Physiological Psychology, 92,* 231–245.

Alberts, J. R., & May, B. (1984). Nonnutritive thermotactile induction of filial huddling in rat pups. *Developmental Psychobiology, 17,* 161–181.

Allin, J. T., & Banks, E. M. (1972). Functional aspects of ultrasound production by infant albino rats. *Animal Behaviour, 20,* 175–185.

Almli, C. R. (1971). The ontogeny of the onset of drinking and plasma osmotic pressure regulation. *Developmental Psychobiology, 6,* 147–158.

Amiri, L., Dark, T., Noce, K. M., & Kirstein, C. L. (1998). Odor preferences in neonatal and weanling rats. *Developmental Psychobiology, 33,* 157–162.

Balleine, B. W. & Dickinson, A. (1994). The role of cholecystokinin in the motivational control of instrumental action. *Behavioral Neuroscience, 108,* 590–605.

Balleine, B., Davies, A., & Dickinson, A. (1995). Cholecystokinin attenuates incentive learning in rats. *Behavioral Neuroscience, 109,* 312–319.

Barnett, S. A., & Spenser, M. M. (1953). Responses of wild rats to offensive smells and tastes. *British Journal of Animal Behaviour, 1,* 32–37.

Barrett, K. C., & Campos, J. J. (1987). Perspectives on emotional development II: A functionalist approach to emotions. In J. D. Osofsky (Ed.), *Handbook of infant development* (2nd ed.; pp. 555–578). New York: John Wiley & Sons.

Berridge, K. C., & Grill, H. J. (1983). Alternating ingestive and consummatory responses suggest a two-dimensional analysis of palatability in rats. *Behavioral Neuroscience, 97,* 563–573.

Berridge, K. C., Flynn, F. W., Schulkin, J., & Grill, H. J. (1984). Sodium depletion enhances salt palatability in rats. *Behavioral Neuroscience, 98,* 652–660.

Birke, L. I. A., & Sadler, D. (1985). Maternal behavior of rats and effects of neonatal progestins given to the pups. *Developmental Psychobiology, 18,* 85–99.

Blass, E. M. (Ed.) (1976). *The psychobiology of Curt Richter.* Baltimore, MD: York Press.

Blass, E. M. (1992). The ontogeny of motivation: Opioid bases of energy conservation and lasting affective change in rat and human infants. *Current Directions in Psychological Science, 1,* 116–120.

Blass, E. M. (1995). The ontogeny of ingestive behavior. In A. Morrison & S. Fluharty (Eds.), *Progress in psychobiology and physiological psychology* (Vol. 16, pp. 1–51). New York: Academic Press.

Blass, E. M. (1996). Mothers and their infants: Peptide-mediated physiological, behavioral and affective changes during suckling. *Regulatory Peptides, 66,* 109–112.

Blass, E. M. (1999). The ontogeny of human infant face recognition: Orogustatory, visual, and social influences. In P. Rochat (Ed.), *Early social cognition: Understanding others in the first months of life* (pp. 35–65). Mahwah, NJ: Erlbaum.

Blass, E. M., & Fitzgerald, E. (1988). Milk-induced analgesia and comforting in 10-day-old rats: Opioid mediation. *Pharmacology, Biochemistry and Behavior, 29,* 9–13.

Blass, E. M., & Shide, D. J. (1993). Endogenous cholecystokinin reduces vocalization in isolated 10-day-old rats. *Behavioral Neuroscience, 107,* 488–492.

Blass, E. M., & Watt, L. B. (1999). Suckling- and sucrose-induced analgesia in human newborns. *Pain, 83,* 611–623.

Blass, E. M., Beardsley, W., & Hall, W. G. (1979). Age-dependent inhibition of suckling by cholecystokinin. *American Journal of Physiology, 236,* E567–E570.

Blass, E. M., Fitzgerald, E., & Kehoe, P. (1987). Interactions between sucrose, pain and isolation distress. *Pharmacology, Biochemistry and Behavior, 26,* 483–489.

Bowlby, J. (1980). *Attachment and loss, Vol. 1, Attachment.* New York: Basic Books.

Bradwejn, J., & Koszycki, D. (1994). The cholecystokinin hypothesis of anxiety and panic disorder. *Annals of the New York Academy of Sciences, 713,* 273–282.

Brown, R. E. (1982). Preferences of pre- and post-weaning Long-Evans rats for nest odors. *Physiology and Behavior, 29,* 865–874.

Brown, R. E., & Willner, J. A. (1983). Establishing an "affective scale" for odor preference of infant rats. *Behavioral and Neural Biology, 38,* 251–260.

Bruno, J. P. (1981). Development of drinking behavior in preweanling rats. *Journal of Comparative and Physiological Psychology, 95,* 1016–1027.

Bruno, J. P., Blass, E. M., & Amin, F. (1983). Determinants of suckling versus drinking in weanling albino rats: Influence of hydrational state and maternal contact. *Developmental Psychobiology, 16,* 177–184.

Buck, L., & Axel, R. (1991). A novel multigene family may encode odorant receptors: A molecular basis for odor recognition. *Cell, 65,* 175–187.

Cabanac, M. (1971). Physiological role of pleasure. *Science, 173,* 1103–1107.

Cabanac, M. (1979). Sensory pleasure. *Quarterly Review of Biology, 54,* 1–29.

Cabanac, M. (1990). Taste: The maximization of multidimensional pleasure. In E. D. Capaldi & T. L. Powley (Eds.), *Taste, experience and feeding* (pp. 28–42). Washington, DC: American Psychological Association.

Camp, L. L. & Rudy, J. W. (1988). Changes in the categorization of appetitive and aversive events during postnatal development of the rat. *Developmental Psychobiology, 21,* 25–42.

Cannon, W. B. (1918). The physiological basis of thirst. *Proceedings of the Royal Society of London, 90B,* 283–301.

Cannon, W. B. (1932). *The wisdom of the body.* New York: Norton.

Cannon, W. B., & Washburn, A. L. (1912). An explanation of hunger. *American Journal of Physiology, 29,* 441–454.

Chang, R. S. L., & Lotti, V. J. (1986). Biochemical and pharmacological characterization of an extremely potent and selective nonpeptide cholecystokinin antagonist. *Proceedings of the National Academy of Sciences of the USA, 83,* 4923–4926.

Coopersmith, R., & Leon, M. (1984). Enhanced neural response following postnatal olfactory experience in Norway rats. *Science, 225,* 849–851.

Coopersmith, R., & Leon, M. (1988). The neurobiology of olfactory learning. In E. M. Blass (Ed.), *Handbook of behavioral neurobiology,* Volume 9, *Developmental psychobiology and behavioral ecology* (pp. 283–308). New York, Plenum Press.

Coplan, J. D., & Lydiard, R. B. (1998). Brain circuits in panic disorder. *Biological Psychiatry, 44,* 1264–1276.

Cornwell, C. A., Chang, J. W., Cole, B., Fukada, Y., Gianulli, T., Rathbone, E. A., McFarlane, H., & McGaugh, J. L. (1996). DSP-4 treatment influences olfactory preferences of developing rats. *Brain Research, 711,* 26–33.

Cornwell-Jones, C. A. (1979). Olfactory sensitive periods in albino rats and golden hamsters. *Journal of Comparative and Physiological Psychology, 93,* 668–676.

Cornwell-Jones, C. A. & Sobrian, S. K. (1977). Development of odor-guided behavior in Wistar and Sprague-Dawley rat pups. *Physiology and Behavior, 19,* 685–688.

Crawley, J. N. & Corwin, R. L. (1994). Biological actions of cholecystokinin. Peptides, 15, 731–755.

Cornwell-Jones, C. A., Decker, M. W., Gianulli, T., Wright, E. L., & McGaugh, J. L. (1990). Norepinephrine reduces the effects of social and olfactory experience. *Brain Research Bulletin, 25,* 643–649.

DeCasper, A. J. & Fifer, W. P. (1980). Of human bonding; newborns prefer their mothers voices. *Science, 208,* 1174–1176.

Dockray, G. J. (1981). Cholecystokinin. In S. R. Bloom, & J. M. Polack (Eds.), *Gut hormones* (pp. 228–239). Edinburgh: Churchill Livingstone.

Doty, R. L. (1981). Olfactory communication in humans. *Chemical Senses, 6,* 351–376.

Fillion, T. J., & Blass, E. M. (1986). Infantile experience with suckling odors determines adult sexual behavior in male rats. *Science, 231,* 729–731.

Fraenkel, G. S., & Gunn, D. L. (1961). *The orientation of animals.* New York: Dover.

Galef, B. G., Jr. (1981). Development of flavor preference in man and animals: The role of social and nonsocial factors. In *Development of perception* (Vol. 1, pp. 411–431). New York: Academic Press.

Galef, B. G., Jr. (1992). Weaning from mother's milk to solid foods. The developmental psychobiology of self-selection of foods by rats. *Annals of the New York Academy of Sciences, 662,* 37–52.

Galef, B. G., Jr., & Kaner, H. C. (1980). Establishment and maintenance of preference for natural and artificial olfactory stimuli in juvenile rats. *Journal of Comparative and Physiological Psychology, 94,* 588–595.

Galef, B. G., Jr., & Kennett, D. J. (1987). Different mechanisms for social transmission of diet preference in rat pups of different ages. *Developmental Psychobiology, 20,* 209–215.

Galef, B. G., Jr., & Sherry, D. F. (1973). Mother's milk: A medium for transmission of cues reflecting the flavor of mother's diet. *Journal of Comparative and Physiological Psychology, 83,* 374–378.

Ganchrow, J. R., Steiner, J. E., & Munif, D. (1983). Neonatal facial expressions in response to different quality and intensities of gustatory stimuli. *Infant Behavior and Development, 6,* 473–484.

Ganchrow, J. R., Steiner, J. E., & Canetto, S. (1986). Behavioral display to gustatory stimuli in newborn rat pups. *Developmental Psychobiology, 19,* 163–174.

Graillon A, Barr R G, Young S. N., Wright J. H., & Hendricks, L. A. (1997). Differential response to intra-oral sucrose, quinine and corn oil in crying human newborns. *Physiology and Behavior, 62,* 317–325.

Grill, H. J. & Berridge, K. C. (1985). Taste reactivity as a measure of the neural control of palatability. *Progress in Psychobiology and Physiological Psychology, 11,* 2–61.

Grill, H. J., & Norgren, R. (1978). The taste reactivity test. I. Mimetic responses to gustatory stimuli in neurologically normal rats. *Brain Research, 143,* 263–279.

Guenaire, C., Costa, J. C., & Delacour, J. (1979). Thermosensibilitè et conditionnement instrumental chez le rat nouveau-nge. *Physiology and Behavior, 22,* 837–840.

Guenaire, C., Costa, J. C., & Delacour, J. (1982a). Discrimination spatiale avec reinforcement thermique chez le jeune rat. *Physiology and Behavior, 28,* 725–731.

Guenaire, C., Costa, J. C., & Delacour, J. (1982b). Conditionement operant avec reinforcement thermique chez le rat nouveau-nè. *Physiology and Behavior, 29,* 419–424.

Hall, W. G. (1990). The ontogeny of ingestive behavior: Changing control of components in the feeding sequence. In E. M. Stricker (Ed.), *Handbook of behavioral neurobiology,* Volume 10: *Neurobiology of feed and fluid intake* (pp. 77–123). New York: Plenum.

Hall, W. G., & Bryan, T. E. (1981). The ontogeny of feeding in rats: IV. Taste development as measured by intake and behavioral responses to oral infusions of sucrose and quinine. *Journal of Comparative and Physiological Psychology, 95,* 240–251.

Harhammer, R., Schafer, U., Henklein, P., Ott, T., & Repke, H. (1991). CCK-8-related C-terminal tetrapeptides: Affinities for central CCKB and peripheral CCKA receptors. *European Journal of Pharmacology, 209,* 263–266.

Harlow, H. F., & Harlow, M. K. (1962). Social deprivation in monkeys. *Scientific American, 207*, 136–146.

Hepper, P. G. (1987a). Rat pups prefer their siblings to their mothers: Possible implications for the development of kin recognition. *Quarterly Journal of Experimental Psychology, 39B*, 265–271.

Hepper, P. G. (1987b). The amniotic fluid: An important priming role in kin recognition. *Animal Behaviour, 35*, 1343–1346.

Hepper, P. G. (1990). Fetal olfaction. In D. W. Macdonald & S. Natynzcuk (Eds.), *Chemical signals in vertebrates* V (pp. 282–288). Oxford: Oxford University Press.

Hepper, P. G. (1991). Recognizing kin: Ontogeny and classification. In P. G. Hepper (Ed.), *Kin recognition* (pp. 259–288). Cambridge: Cambridge University Press.

Hofer, M. A. (1996). Multiple regulators of ultrasonic vocalization in the infant rat. *Psychoneuroendocrinology, 21*, 203–217.

Hofer, M. A. & Shair, H. N. (1980). Sensory processes in the control of isolation-induced ultrasonic vocalization by 2-week-old rats. *Journal of Comparative and Physiological Psychology, 94*, 271–279.

Hofer, M. A., Masmela, J. R., Brunelli, S. A., & Shair, H. N. (1998). The ontogeny of maternal potentiation of the infant rat's isolation call. *Developmental Psychobiology, 33*, 189–202.

Hofer, M. A., Masmela, J. R., Brunelli, S. A., & Shair, H. N. (1999). Behavioral mechanisms for active maternal potentiation of isolation calling in rat pups. *Behavioral Neuroscience, 113*, 51–61.

Hogan, J. A. (1988). Cause and function in the development of behavior systems. In E. M. Blass (Ed.), *Handbook of behavioral neurobiology*, Volume 9, *Developmental psychobiology and behavioral ecology* (pp. 63–106). New York, Plenum Press.

Holland, P. C. (1977). Conditioned stimulus as a determinant of the form of the Pavlovian conditioned response. *Journal of Experimental Psychology: Animal Behavior Processes, 3*, 77–104.

Holland, P. C. (1980). Influence of visual conditioned stimulus characteristics on the form of the Pavlovian appetitive conditioned responding in rats. *Journal of Experimental Psychology: Animal Behavior Processes, 6*, 81–97.

Holmes, W. G. (1988). Kinship and the development of social preferences. In E. M. Blass (Ed.), *Handbook of behavioral neurobiology*, Volume 9, *Developmental psychobiology and behavioral ecology* (pp. 389–413). New York, Plenum Press.

Johanson, I. B., Polefrone, J. M., & Hall, W. G. (1984). Appetitive conditioning in neonatal rats: Conditioned ingestive responding to stimuli paired with oral infusion of milk. *Developmental Psychobiology, 17*, 357–381.

Kehoe, P., & Blass, E. M. (1986a). Conditioned aversions and their memories in 5-day-old rats during suckling. *Journal of Experimental Psychology: Animal Behavior Processes, 12*, 40–47.

Kehoe, P., & Blass, E. M. (1986b). Behaviorally functional opioid systems in infant rats: II. Evidence for pharmacological, physiological and psychological mediation of pain and stress. *Behavioral Neuroscience, 100*, 624–630.

Kehoe, P., & Blass, E. M. (1986c). Opioid-mediation of separation distress in 10-day-old rats: Reversal of stress with maternal stimuli. *Developmental Psychobiology, 19*, 385–398.

Kehoe, P., & Sakurai, S. (1991). Preferred tastes and opioid-modulated behaviors in neonatal rats. *Developmental Psychobiology, 24*, 135–148.

Kenny, J. T., & Blass, E. M. (1977). Suckling as an incentive to instrumental learning in preweanling rats. *Science, 196*, 898–899.

Kenny, J. T., Stoloff, M. L., Bruno, J. P., & Blass, E. M. (1979). Ontogeny of preference for nutritive over nonnutritive suckling in albino rats. *Journal of Comparative and Physiological Psychology, 93*, 752–759.

Kleitman, N., & Satinoff, E. (1982). Thermoregulatory behavior in rat pups from birth to weaning. *Physiology and Behavior, 29*, 537–541.

Lee, T. M., & Moltz, H. (1984a). The maternal pheromone and brain development in the preweanling rat. *Physiology and Behavior, 33*, 385–390.

Lee, T. M., & Moltz, H. (1984b). The maternal pheromone and deoxycholic acid in the survival of preweanling rats. *Physiology and Behavior, 33*, 931–935.

Lemaire, M., Piot, O., Roques, B., Bohme, G. A. & Blanchard, J.-C. (1992). Evidence for an endogenous cholecystokininergic balance in social memory. *NeuroReport, 3*, 929–932

Lemaire, M., Bohme, G. A., Piot, O., Roques, B., & Blanchard, J.-C. (1994). CCK-A and CCK-B selective receptor agonists and antagonists modulate olfactory social recognition in male rats. *Psychopharmacology, 115*, 435–440.

Leon, M. (1974). Maternal pheromone. *Physiology and Behavior, 13*, 441–453.

Leon, M. (1986). Development of thermoregulation. In E. M. Blass (Ed.), *Handbook of behavioral neurobiology*, Volume 8, *Developmental psychobiology and developmental neurobiology* (pp. 297–322). New York, Plenum Press.

Leon, M. (1992a). Neuroethology of olfactory preference development. *Journal of Neurobiology, 23*, 1557–1573.

Leon, M. (1992b). The neurobiology of filial learning. *Annual Review of Psychology, 43*, 377–398.

Leon, M., & Moltz, H. (1971). Maternal pheromone: Discrimination by preweanling albino rats. *Physiology and Behavior, 7*, 265–267.

Leon, M., Galef, B. G., Jr., & Behse, J. H. (1977). Establishment of pheromonal bonds and diet choice in young rats by odor pre-exposure. *Physiology and Behavior, 18*, 387–391.

Liddle, R. A. (1997). Cholecystokinin cells. *Annual Review of Physiology, 59*, 221–242.

Lorenz, D. (1994). Effects of CCK-8 on ingestive behaviors of suckling and weanling rats. *Developmental Psychobiology, 27*, 39–52.

Lotti, V. J., Pendleton, R. G., Gould, R. J., Hanson, H. M., Chang, R. S. L., & Clineschmidt, B. V. (1987). *In vivo* pharmacology of L-364,718, a new potent nonpeptide peripheral cholecystokinin antagonist. *Journal of Pharmacology and Experimental Therapeutics, 241*, 103–109.

Marlier, L., Schaal, B., & Soussignan, R. (1988). Bottle-fed neonates prefer an odor experienced *in utero* to an odor experienced postnatally in the feeding context. *Developmental Psychobiology, 33*, 133–145.

Mennella, J. A., & Beauchamp, G. K. (1993). The effects of repeated exposure to garlic-flavored milk on the nursling's behavior. *Pediatric Research, 34*, 805–808.

Mennella, J. A., & Beauchamp, G. K. (1998). Early flavor experiences: Research update. *Nutrition Reviews, 56*, 205–211.

Mennella, J. A., & Beauchamp, G. K. (1999). Experience with a flavor in mother's milk modifies the infant's acceptance of flavored cereal. *Developmental Psychobiology, 35*, 197–203.

Midkiff, E. E., & Bernstein, I. L. (1983). The influence of age and experience on salt preference in the rat. *Developmental Psychobiology, 16*, 385–394.

Mistretta, C. M., & Bradley, R. M. (1986). Development of the sense of taste. In E. M. Blass (Ed.), *Handbook of behavioral neurobiology*, Volume 8, *Developmental psychobiology and developmental neurobiology* (pp. 205–236). New York, Plenum Press.

Moe, K. E. (1986). The ontogeny of salt preference in rats. *Developmental Psychobiology, 19*, 185–196.

Moore, C. L. (1982). Maternal behavior of rats is affected by hormonal condition of pups. *Journal of Comparative and Physiological Psychology, 96*, 123–129.

Moore, C. L., & Power, K. L. (1992). Variation in maternal care and individual differences in play, exploration, and grooming of juvenile Norway rat offspring. *Developmental Psychobiology, 25*, 165–182.

Moran, T. H, Schwartz, G. J., & Blass, E. M. (1983a). Organized behavioral responses elicited by lateral hypothalamic electrical stimulation in neonatal rats. *Journal of Neuroscience, 3*, 10–19.

Moran, T. H, Schwartz, G. J., & Blass, E. M. (1983b). Stimulation induced ingestion in neonatal rats. *Developmental Brain Research, 7*, 197–204.

Moran, T. H., Robinson, P. H., Goldrich, M. S., & McHugh, P. R. (1986). Two brain cholecystokinin receptors: Implications for behavioral actions. *Brain Research, 362*, 175–179.

Moran, T. H., Ameglio, P. J., Schwartz, G. J., & McHugh, P. R. (1991). Blockade of type A, not type B, CCK receptors attenuates satiety actions of exogenous and endogenous CCK. *American Journal of Physiology, 262*, R46–R50.

Naito, H., & Tonoue, T. (1987). Sex difference in ultrasound distress calls by rat pups. *Behavioral Brain Research, 25*, 13–21.

Nakamura, S., Kimura, F., & Sakaguchi, F. (1987). Postnatal development of electric activity in the locus coerulus. *Journal of Neurophysiology, 58*, 510–524.

Nelson, E. E., & Panksepp, J. (1998). Brain substrates of infant–mother attachment: Contributions of opioids, oxytocin, and norepinephrine. *Neuroscience and Biobehavioral Reviews, 22*, 437–452.

Noirot, E. (1973). Ultrasounds and maternal behavior in small rodents. *Developmental Psychobiology, 5*, 371–387.

Nowak, R., Goursaud, A. P., Levy, F., Orgeur, P., Schaal, B., Belzung, C., Picard, M., Meunier–Salaven, M. C., Alster, P., & Urnaes–Moberg, K. (1997). Cholecystokinin receptors mediate the development of a preference for the mother by newly born lambs. *Behavioral Neuroscience, 111*, 1375–1382.

Nyakas, C., & Endroczi, E. (1970). Olfaction guided approaching behaviour of infantile rats to the mother in maze box. *Acta Physiologica Academiae Scientiarum Hungaricae, 38*, 59–65.

Oswalt, G. L., & Meier, G. W. (1975). Olfactory, thermal, and tactual influences on infantile ultrasonic vocalization in rats. *Developmental Psychobiology, 8*, 129–135.

Passaro, E., Debas, H., Oldenorf, W., & Yamada, T. (1982). Rapid appearance of intraventricularly administered neuropeptides in the peripheral circulation. *Brain Research, 241*, 335–340.

Pedersen, P. E., & Blass, E. M. (1982). Prenatal and postnatal determinants of the first suckling episode in albino rats. *Developmental Psychobiology, 15*, 349–355.

Pineau, A., & Streri, A. (1990). Intermodal transfer of spatial arrangement of the component parts of an object in infants aged 4–5 months. *Perception, 19,* 795–804.

Polan, H. J., & Hofer, M. A. (1998). Olfactory preference for mother over home nest shavings by newborn rats. *Developmental Psychobiology, 33,* 5–20.

Porter, R. H., & Winberg, J. (1999). Unique salience of maternal breast odors for newborn infants. *Neuroscience and Biobehavioral Reviews, 23,* 439–449.

Rakover-Atar, S., & Weller, A. (1997). The influence of natural preference for tactile stimuli on appetitive learning in rat pups. *Developmental Psychobiology, 30,* 29–39.

Rao, M., Blass, E. M., Brignol, M. M., Marino, L., & Glass, L. (1997). Reduced heat loss following sucrose ingestion in premature and normal human newborns. *Early Human Development, 48,* 109–116.

Redman, R. S., & Sweeney, L. R. (1976). Changes in diet and patterns of feeding activity in developing rats. *Journal of Nutrition, 106,* 615–626.

Ren, K., Blass, E. M., Zhou, Q., & Dubner, R. (1997). Suckling and sucrose ingestion suppress persistent hyperalgesia and spinal Fos expression after forepaw inflammation in infant rats. *Proceedings of the National Academy of Sciences of the USA, 94,* 1471–1475.

Rex, A., Barth, T., Voigt, J.-P., Domeney, A. M., & Fink, H. (1994). Effects of cholecystokinin tetrapeptide and sulfated cholecystokinin octapeptide in rat models of anxiety. *Neuroscience Letters, 172,* 139–142.

Richmond, G., & Sachs, B. D. (1983). Maternal discrimination of pup sex by rats. *Developmental Psychobiology, 17,* 347–356.

Robinson, H., Moran, T. H., Goldrich, M., & McHugh, P. R. (1987). Development of cholecystokinin binding sites in rat upper gastrointestinal tract. *American Journal of Physiology, 252,* G529–G534.

Rosenblatt, J. S. (1983). Olfaction mediates developmental transition in the altricial newborn of selected species of mammals. *Developmental Psychobiology, 16,* 347–375.

Rosenblatt, J. S. (1987). Biological and behavioral factors underlying the onset and maintenance of maternal behavior in the rat. In N. A. Krasnegor, E. M. Blass, M. A. Hofer & W. P. Smotherman (Eds.), *Perinatal development* (pp. 323–341). Orlando, FL: Academic Press.

Rosenstein, D., & Oster, H. (1990). Differential facial responses to four basic tastes in newborns. *Child Development, 59,* 1555–1568.

Rudy, J. W. (1991). Development of learning: From elemental to configural associative networks. In C. Rovee-Collier & L. Lipsitt (Eds.), *Advances in infancy research* (Vol. 7, pp. 247–289). Norwood, NJ: Ablex.

Salmenpera, L., Perheentupa, J., Siimes, M. A., Adrian, T. E., Bloom, S. R., & Aynsley-Green, A. (1988). Effects of feeding regimen on blood glucose levels and plasma concentrations of pancreatic hormones and gut regulatory peptides at 9 months of age: Comparison between infants fed with milk formula and infants exclusively breast-fed from birth. *Journal of Pediatric Gastroenterology and Nutrition, 7,* 651–656.

Satinoff, E. (1991). Developmental aspects of behavioral and reflexive thermoregulation. In H. N. Shair, G. A. Barr, & M. A. Hofer (Eds.), *Developmental psychobiology: New methods and changing concepts* (pp. 169–188). New York: Oxford University Press.

Schaal, B. (1988). Olfaction in infants and children: Developmental and functional perspectives. *Chemical Senses, 13,* 145–190.

Schank, J. C., & Alberts, J. R. (1997). Self-organized huddles of rat pups modeled by simple rules of individual behavior. *Journal of Theoretical Biology, 189,* 11–25.

Schneirla, T. C. (1939). A theoretical consideration of the basis for approach–withdrawal adjustments in behavior. *Psychological Bulletin, 37,* 501–502.

Schneirla, T. C. (1959). An evolutionary developmental theory of biphasic processes underlying approach and withdrawal. *Nebraska Symposium on Motivation, 1,* 1–42.

Schneirla, T. C. (1965). Aspects of stimulation and organization in approach/withdrawal processes underlying vertebrate behavioral development. In D. S. Lehrman, R. A. Hinde, and E. Shaw (Eds.), *Advances in the study of behavior* (Vol. 1, pp. 1–74). New York: Academic Press.

Schwartz, G. J., & Grill, H. J. (1985). Comparing taste-elicited behaviors in adult and neonatal rats. *Appetite, 6,* 373–386.

Shayit, M., & Weller, A. (1998). *Relative preference of neonatal rats towards stimuli representing the nest is increased by a CCK-A receptor antagonist.* Paper presented at the 31st Annual Meeting of the International Society for Developmental Psychobiology. Orleans, France.

Shayit, M., & Weller, A. (2001). Cholecystokinin receptor antagonists increase the rat pup's preference towards maternal-odor and rug texture. *Developmental Psychobiology, 38,* 164–173.

Shide, D. J., & Blass, E. M. (1989). Opioid-like effects of intraoral infusions of corn oil and polycose on stress reactions in 10-day-old rats. *Behavioral Neuroscience, 103,* 1168–1175.

Shide, D. J., & Blass, E. M. (1991). Opioid mediation of odor preferences induced by sugar and fat in 6-day-old rats. *Physiology and Behavior, 50,* 961–966.

Smith, G. P., Tyrka, A., & Gibbs, J. (1991). Type-A CCK receptors mediate the inhibition of food intake and activity by CCK-8 in 9- to 12-day-old rat pups. *Pharmacology, Biochemistry and Behavior, 38,* 207–210.

Soussignan, R., Schaal, B., Marlier, L., & Jiang, T. (1997). Facial and autonomic responses to biological and artificial olfactory stimuli in human neonates: Re-examining early hedonic discrimination of odors. *Physiology and Behavior, 62,* 745–758.

Spelke, E. S. (1981). The infant's acquisition of knowledge of bimodally specified events. *Journal of Experimental Child Psychology, 31,* 279–299.

Steiner, J. E. (1974). The human gustofacial response. In J. F. Bosma (Ed.), *Oral sensation and perception: Development in the fetus and infant.* Washington, DC: U.S. Government Printing Office.

Steiner, J. E. (1977). Facial expressions of the neonate infant indicating the hedonics of food-related chemical stimuli. In J. M. Weiffenback (Ed.), *Taste and development: The genesis of sweet preference.* Washington, DC: U.S. Government Printing Office.

Stickrod, G., Kimble, D. P., & Smotherman, W. P. (1982). *In utero* taste odor aversion conditioning in the rat. *Physiology and Behavior, 28,* 5–7.

Stoloff, M. L., & Blass, E. M. (1983). Changes in appetitive behavior in weanling-age rats: Transition from sucking to feeding behavior. *Developmental Psychobiology, 16,* 439–453.

Streri, A., & Molina, M. (1993). Visual–tactual transfer between objects and pictures in 2-month-old infants. *Perception, 22,* 1299–1318.

Sullivan, R. M., & Wilson, D. A. (1994). The locus coerulus, norepinephrine, and memory in newborns. *Brain Research Bulletin, 35,* 467–472.

Teicher, M. H., & Blass, E. M. (1977). First suckling response of the newborn albino rat: The roles of olfaction and amniotic fluid. *Science, 198,* 635–636.

Terry, L. M., & Johanson, I. B. (1996). Effects of altered olfactory experiences on the development of infant rats' responses to odors. *Developmental Psychobiology, 29,* 353–377.

Thiels, E., Alberts, J. R., & Cramer, C. P. (1990). Weaning in rats: II. Pup behavior. *Developmental Psychobiology, 23,* 495–510.

Toates, F. (1986). *Motivational systems.* Cambridge, UK: Cambridge University Press.

Uvnas-Moberg, K. (1989). The gastrointestinal tract in growth and reproduction. *Scientific American, 261* (July), 60–65.

Uvnas-Moberg, K., Widstrom, A. M., Marchin, G., & Winberg, J. (1987). Release of GI hormones in mother and infant by sensory stimulation. *Acta Paediatrica Scandinavica, 76,* 851–860.

Uvnas-Moberg, K., Marchini, G., & Winberg, J. (1993). Plasma cholecystokinin concentrations after breast feeding in healthy 4 day old infants. *Archives of Disease in Childhood, 68,* 46–48.

Varendi, H., Porter, R. H., & Winberg, J. (1997). Natural odour preferences of newborn infants change over time. *Acta Paediatrica, 86,* 985–990.

Wang, Z., Yu, G., Cascio, C., Liu, Y., Gingrich, B., & Insel, T. R. (1999). Dopamine D2 receptor-mediated regulation of partner preferences in female prairie voles (*Microtus ochrogaster*): A mechanism for pair bonding? *Behavioral Neuroscience, 113,* 602–611.

Weller, A. & Blass, E. M. (1988a). Behavioral evidence for cholecystokinin–opiate interactions in neonatal rats. *American Journal of Physiology, 255,* R901–R907.

Weller, A., & Blass, E. M. (1988b). Cholecystokinin-induced conditioned odor-preference is blocked by the selective antagonist L-364,718. *Society for Neuroscience Abstracts, 14,* 199.

Weller, A., & Blass, E. M. (1989). 'Conditioned olfactory calming': Further evidence for positive effects of cholecystokinin peptide in infant rats. *Eastern Psychological Association Abstracts, 60,* 25.

Weller, A., & Blass, E. M. (1990). Cholecystokinin conditioning in rats: Ontogenetic determinants. *Behavioral Neuroscience, 104,* 199–206.

Weller, A. & Dubson, L. (1998). A CCK_A-receptor antagonist administered to the neonate alters mother–infant interactions in the rat. *Pharmacology, Biochemistry and Behavior, 59,* 843–851.

Weller, A., & Gispan, I. H. (2000). A cholecystokinin receptor antagonist blocks milk-induced but not maternal-contact-induced decrease in ultrasonic vocalization in rat pups. *Developmental Psychobiology, 37,* 35–43.

Weller, A., Smith, G. P., & Gibbs, J. (1990). Endogenous cholecystokinin reduces feeding in young rats. *Science, 247,* 1589–1591.

Weller, A., Corp, E. C., Tyrka, A., Ritter, R. C., Brenner, L., Gibbs, J., & Smith, G. P. (1992). Trypsin inhibitor and maternal reunion increase plasma cholecystokinin in neonatal rats. *Peptides, 13,* 939–941.

Weller, A., Blass, E. M., Smith, G. P. & Gibbs, J. (1995). Odor-induced inhibition of intake after pairing of odor and CCK-8 in neonatal rats. *Physiology and Behavior, 57,* 181–183.

Weller, A., Gispan, I. H. & Smith, G. P. (1997a). Characteristics of glucose and maltose preloads that inhibit feeding in 12-day-old rats. *Physiology and Behavior, 61,* 819–822.

Weller, A., Gispan, I. H., Armony-Sivan, R., Ritter, R. C., & Smith, G. P. (1997b). Preloads of corn oil inhibit independent ingestion on postnatal day 15 in rats. *Physiology and Behavior, 62,* 871–874.

Wiertelak, E. P., Maier, S. F., & Watkins, L. R. (1992). Cholecystokinin antianalgesia: Safety cues abolish morphine analgesia. *Science, 256,* 830–833.

Wilson, D. A., & Leon, M. (1988). Noradrenergic modulation of olfactory bulb excitability in the postnatal rat. *Developmental Brain Research, 42,* 69–75.

Winslow, J. T., & Insel, T. R. (1991). The infant rat separation paradigm: A novel test for novel anxiolytics. *Trends in Pharmacological Science, 12,* 402–404.

Wirth, J. B., & Epstein, A. N. (1976). Ontogeny of thirst in the infant rat. *American Journal of Physiology, 320,* 188–198.

Young, P. T. (1966). Hedonic organization and regulation of behavior. *Psychological Review, 73,* 59–86.

Zeifman, D., Delaney, S., & Blass, E. M. (1996). Sweet taste, looking, and calm in 2- and 4-week-old infants: The eyes have it. *Developmental Psychology, 32,* 1090–1099.

13

Taste Development

DAVID L. HILL

INTRODUCTION

The developing gustatory system has the complex task of processing and organizing an ever-increasing array of sensory stimuli. During ontogeny animals must be able to recognize food and to appropriately reject toxic foods that induce adverse or lethal consequences at the age when they begin to sample substances from their environment. In response to toxic stimuli, the neural taste message should be accurate, reliable, and probably not change significantly with age. In contrast, the ability to have an alterable neural message to other food classes during development is adaptive. There are multiple examples of both "static" and "plastic" processing in the developing gustatory system, and much of this chapter is devoted to expanding upon these ideas and their implications. I also consider the gustatory system as a major component in energy homeostasis; it must meet changes in nutritive demands and developmentally related gastrointestinal processes with an accurate, complex processing of relevant sensory stimuli. It is this relatively unexplored aspect of the gustatory system that may have the greatest relevance during development.

Since the oropharyngeal cavity is the anterior outpost of the digestive system, gustatory status may reflect and complement that of more distal physiological systems. For example, electrolyte balance is regulated in large part by gut and renal function, which changes markedly during preweaning and early postweaning periods (e.g., Spitzer, 1980). Since alterations in circulating electrolyte balance has corresponding physiological effects on the gustatory system (Bradley, 1973), it is possible that age-related changes in electrolyte levels determined by gut and renal development may influence age-related changes in gustatory function. While no experimental data exist that directly confirm this hypothesis, there is a high correlation between changes in electrolytes, such as increased plasma sodium levels, and

DAVID L. HILL Department of Psychology, P.O. Box 400400, University of Virginia, Charlottesville, Virginia 22904.

Developmental Psychobiology, Volume 13 of *Handbook of Behavioral Neurobiology*, edited by Elliott Blass, Kluwer Academic / Plenum Publishers, New York, 2001.

increased taste responses during the first three postnatal weeks in rats (Ferrell, Mistretta, & Bradley, 1981; Hill & Almli, 1980; Jelinek, 1961, Yamada, 1980). Reciprocally, the gustatory system impacts visceral function by way of reflexes originating in the mouth. For example, the taste of sugar triggers the preabsorptive pancreatic release of insulin in adult animals. The reflex presumably occurs in anticipation of nutritive load, i.e., before the meal reaches the stomach (Berridge, Grill, & Norgren, 1981). During development these reflexes may direct early events in digestion and absorption and may be associated with certain tastes that induce specific digestive cascades, as proposed to occur in adults (Berridge *et al.*, 1981). Therefore, the developing taste system may interact reciprocally with developmental processes of the gut. These topics have received little attention in developmental studies, but would yield important insights into the emergence of gut/gustatory interactions.

In addition to "driving" taste-elicited ingestive behaviors, the developing gustatory system also participates in forming associations between taste stimuli and internal physiological states. The most widely studied example of this association is conditioned taste aversion. Strong, one-trial associations are made between the taste of a solution and illness (Garcia & Koelling, 1966). This requires an intact and functional taste system for the acquisition and the maintenance of the conditioned response (Kehoe & Blass, 1985; Martin & Alberts, 1979). The functional maturation of the taste system would be responsible for the maturation of one of the afferent limbs of conditioned responses (Bernstein & Courtney, 1987; Midkiff & Bernstein, 1983; Moe, 1986); proper associations may be made only if the proper neural signal from the gustatory system is integrated accurately with physiological events (Formaker & Hill, 1990). More recent developmental studies provide other powerful examples of how the gustatory system functions beyond merely discriminating taste stimuli. Specifically, it is also involved in modulating emotional responses. Blass's (1996, 1999) work shows that the taste of sucrose allays pain perception and enhances contact motivation in young animals and humans via opioid pathways at both spinal and supraspinal levels (Ren, Blass, Zhou, & Dubner, 1997).

Therefore, an understanding of the developing gustatory system at the behavioral, neurophysiologic, anatomic, and molecular levels will provide not only insights into sensory processing, but also a richer understanding of homeostasis and affect. It is my goal in this chapter to describe the neurobiological substrates of taste with particular reference to neurophysiologic function and to morphology. The interested reader should also examine other recent reviews with differing emphases (Hill & Mistretta, 1990; Mistretta, 1991, 1998; Mistretta & Hill, 1995; Stewart & Hill, 1993; Stewart, DeSimone, & Hill, 1997).

Past and Current Status of Developmental Gustatory Biology

In the edition of this handbook published in 1986, Charlotte Mistretta and Robert Bradley had the task of presenting findings and their thoughts on gustatory ontogeny (Mistretta & Bradley, 1986). They wrote during an especially important and exciting time in the life of an emerging research area. Their scientific approach, which reflected the field as a whole, was to examine systems-level phenomena that characterized peripheral and central gustatory development. As with any other system, the direction of future research resides with outcomes of these initial descriptive studies. The earlier findings remain influential because they demonstrate the wide-ranging functional and morphological changes in both peripheral and central taste structures. Through this framework, contemporary cellular and molecular biology has identified some of the mechanisms through which gustation becomes assembled functionally and anatomically.

Although the synthesis of systems and molecular approaches must continue to advance our understanding of gustatory development, there is a danger of the field being swept into successively more molecular realms with a loss of focus on the larger issues of how developmental processes affect both the organism and behavior. Loss of this focus will lead to a failure to exploit the unique and most interesting properties of taste as a model experimental system and its relationship with other developing systems so that molecular findings would occur without reference to a larger frame of reference. Fortunately, the field of chemical senses research has a tradition of successfully incorporating new techniques into the larger conceptual framework provided by work on behavior and neurophysiology.

DEVELOPMENT OF GUSTATORY PAPILLAE AND TASTE BUDS

Collections of sensory receptor cells responsible for taste transduction form taste buds that are dispersed throughout the oropharyngeal cavity (Figure 1). In mammals, the majority of taste buds reside within epithelial structures termed papillae (Figure 1A). Access of stimuli to taste bud cells occurs via taste pores, which are simply an absence of epithelia in the papillae (Figure 1B). Taste bud cells in all papillae have a characteristic columnar orientation; surrounding epithelial tissue does not (Figure 1). This columnar orientation has significant functional implications. The apical domain of a taste receptor cell has microvilli (Figure 1B), specialized processes where most functional membrane receptor elements for taste transduction are located. The physical barrier between microvilli and the rest of the cell is provided by tight junctions (Figure 1B). The two cellular domains defined by the location of tight junctions are the apical domain (above the tight junctions) and basolateral domain (below the tight junctions). These cellular barriers make it unlikely that the majority of taste stimuli contact membrane components in the basolateral domain. Although recent findings have modified this dogma somewhat (Ye, Heck, & DeSimone, 1993), most transduction processes occur apically. Finally, nerve profiles that innervate taste bud structures are located throughout the extent of the taste bud. (Note: The synapses shown in Figure 1B are simplified; they ramify extensively throughout the taste bud; Beidler, 1969.)

Although taste buds have characteristic morphologies, they are located in morphologically distinct papillae throughout the oropharyngeal cavity (see Figure 2). Taste buds on the anterior tongue are located in fungiform papillae, which are eminences in the dorsal epithelium (see also Figure 1A). Taste buds in the posterior tongue are contained in foliate or circumvallate papillae, which, unlike fungiform papillae, have trenches that invaginate into the surface of the tongue (Figure 2). Taste buds in the rodent palate are either located anteriorly in the nasoincisor duct or posteriorly in the soft palate. Taste buds on the soft palate are not encased within papillae, but are embedded within the tissue. Finally, collections of taste buds without papillae are located in the epiglottis, larynx, and pharynx (see Figure 2). Much of peripheral gustatory development is shaped by papillae development (see below). More importantly, however, the distinct receptor populations have their own innervation and have different functional characteristics.

Primarily because of accessibility, most of the information about taste bud mor-

DAVID L. HILL

A

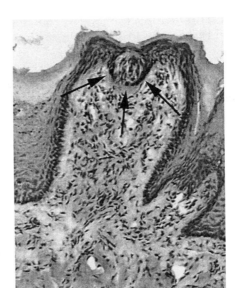

B

oral cavity

taste pore and microvilli

tight junctions

taste bud cells

synapse

basement membrane

afferent taste fibers

epithelium

nerve profile

basal cell

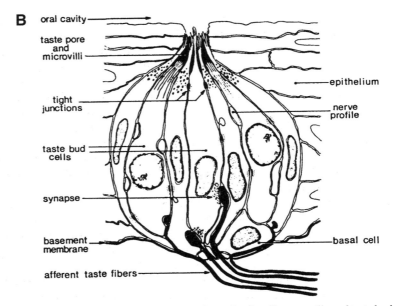

Figure 1. (A) Photomicrograph of a coronal section through a fungiform papilla and taste bud (denoted by arrows). (B) Schematic of a taste bud and associated structures. (Reprinted from Mistretta, 1989, with permission. Copyright 1989 New York Academy of Sciences.)

phology and function has come from studies of lingual gustatory structures. Indeed, much of the developmental anatomy of the peripheral gustatory system has focused on how taste bud development on the tongue is coordinated with the developmental morphology of associated papillae. This reflects the view that taste bud development can be understood best within the context of papillae development. For example, processes related to papillae formation are important for the developing taste system because it is where many of the taste receptors take up residence, and at least

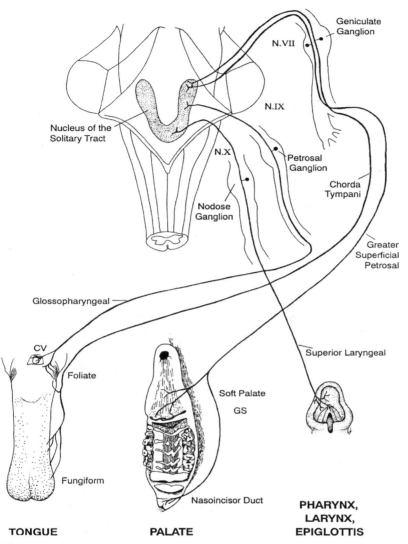

TONGUE　　　　**PALATE**　　　　**PHARYNX,
LARYNX,
EPIGLOTTIS**

Figure 2. Diagram of taste receptor areas on the rat tongue, palate, pharynx, larynx, and epiglottis. Taste buds are found in fungiform, foliate, and circumvallate (CV) papillae on the tongue, in the nasoincisor duct, soft palate and geschmachstreiffen (GS) on the palate, and on the surface of the pharynx, larynx, and epiglottis. The chorda tympani nerve carries afferent information from taste buds in fungiform papillae and from taste buds located in anterior foliate papillae. The glossopharyngeal nerve carries information from taste buds in the CV and in posterior foliate papillae. The greater superficial petrosal nerve innervates palatal taste buds and the superior laryngeal nerve carries information from taste buds in the pharynx, larynx, and epiglottis. The nodose, petrosal, and geniculate ganglia contain cell bodies of the superior laryngeal, glossopharyngeal, and the chorda tympani and greater superficial petrosal nerves, respectively. Gustatory nerves project to the nucleus of the solitary tract in the medulla as shown.

indirectly determine taste bud densities. The following sections, therefore, will focus on fungiform papillae and the taste buds contained within these papillae (see Figure 1A).

PAPILLAE FORMATION

Papillae are composed of epithelia and mesenchyme, with taste buds located in the epithelia and supporting tissue in the mesenchyme (Figure 1). Fungiform papillae receive their primary innervation via the trigeminal nerve, whereas the nervous supply to taste buds occurs via the chorda tympani nerve, a branch of the facial nerve (Figure 2). Therefore, there is clear segregation of tissue types and nervous supply in the sensory organ of taste (i.e., taste buds) from the supportive tissue (i.e., fungiform papillae). As will be presented in the following sections, there are significant interactions between these structures during development that determine their final morphologies.

INDUCTION AND DIFFERENTIATION

Gustatory papillae form independently of innervation. The first hint of papillae formation in rats appears at embryonic day (E) 14 (gestation = 21 days). By E15, the emerging papillae are apparent in light and electron microscopic speci-

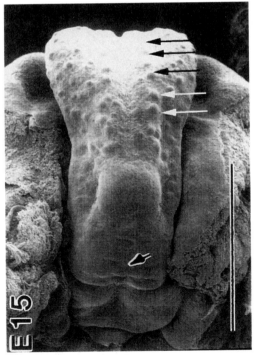

Figure 3. Scanning electron micrograph of a rat tongue at embryonic day 15. The short arrows points to the single circumvallate papilla on the back of the tongue and the series of longer arrows point to a row of fungiform papillae. Scale bar = 1.0 mm. (Reprinted from Mbiene *et al.*, 1997, with permission. Copyright 1997 Wiley–Liss, Inc. a Subsidiary of John Wiley & Sons, Inc.)

mens (Figure 3). By E16, gustatory papillae are highly visible on the tongue, occupying distinct rostral to caudally located rows (Mbiene, MacCallum, & Mistretta, 1997; Mistretta, 1972). Histologically, papillae formation precedes tissue invasion by trigeminal neurons (Mbiene & Mistretta, 1997). Thus, based on timing alone, papillae emergence on the dorsal surface of the tongue is not induced by innervation.

Organ culture studies have verified that innervation is not necessary for papillae formation (Mbiene *et al.*, 1997). Entire rat tongues cultured beginning at E13 or E14 grow and differentiate sufficiently to form fungiform papillae. In fact, the patterned rows characteristic of tongues obtained from older fetuses are also seen on cultured tongues (Figure 3). Since the cultured tongue is dissected without the trigeminal nerve, it is clear that papillae can form initially without this nerve. However, the processes that induce papillae differentiation in culture are time limited. Tongues cultured at E19 and older do not form normal papillae. Instead, they degenerate after they are formed initially (Mbiene *et al.*, 1997). Thus, innervation seems necessary for structure maintenance.

PAPILLAE MAINTENANCE

More direct evidence for papillae maintenance being dependent on innervation comes from nerve cut studies in early postnatal rats. Sectioning the chorda tympani nerve and the lingual branch of the trigeminal nerve on postnatal day (PN) 1 results in papillae loss through degeneration over the following 21 days (Nagato, Matsumoto, Tanioka, Kodama, & Toh, 1995). Sectioning only the chorda tympani in PN10 rats also causes 80% of fungiform papillae to degenerate within 30 days, even though the lingual nerve is left intact (Sollars & Bernstein, 1996; Sollars, Shuler, & Hill, 1996).

TASTE BUD FORMATION

Controversy abounds concerning processes involved in early taste bud formation. One long-standing view held that taste buds required innervation for formation because nerves reached the dorsal epithelia just before the appearance of rudimentary taste buds (e.g., Farbman, 1965). Therefore, simply based on timing, taste nerves were believed to induce taste bud development.

Recent studies on both nonmammalian and mammalian taste bud development have reassessed the mechanisms underlying taste bud induction. Classic embryological experiments show that taste buds differentiate independently from nerve fibers and are derived from local tissues. Using axolotls as the experimental model, grafts of presumptive oropharyngeal tissue were placed ectopically on the trunk of host embryos (i.e., away from their normal sites) prior to innervation and prior to formation of taste buds. The grafts develop well-differentiated taste buds, even in the absence of neurites (Barlow, Chien, & Northcutt, 1996). Accompanying tissue culture studies in which the presumptive axolotl oropharyngeal region was cultured prior to innervation verified taste bud development independent of innervation; cultured oropharyngeal tissue forms many taste buds even in the absence of neurons (Barlow *et al.*, 1996). Experiments with *trkB* −/− knockout mice, which lack the tyrosine kinase receptor for the neurotrophin brain-derived neurotrophic factor (BDNF), provide additional support for the idea that taste bud induction does not require innervation. In the absence of *trkB*, knockout mice lose the neurons that usually innervate taste buds, yet taste buds continue to develop even though many are small (Fritzsch, Sarai, Barbacid, & Silos-Santiago, 1997). These two lines of study challenge the dogma that taste bud induction is dependent upon innervation.

The issue is not resolved, however, because other recent studies support the neural dependence view. For example, taste bud numbers in knockout mice missing genes for BDNF are severely reduced, and this is highly correlated with reduced numbers of innervating neurons (Jones, Farinas, Backus, & Reichardt, 1994; Liu, Ernfors, Wu, & Jaenisch, 1995). This result is consistent with a neural dependence view of taste bud induction because decreased numbers of innervating neurons should have a corresponding effect on decreased numbers of taste buds. Since mice were examined only after taste buds were induced, it is not clear whether the reduced number of taste buds reflects an initial failure of structure formation or whether taste buds develop in a rudimentary form and then quickly degenerate.

These collective findings hold implications for later morphological and functional development and for underlying taste-related behaviors. It is during late gestational and the first one to two postnatal weeks in rat that (1) the total number of innervating neurons is determined, (2) the size and type of sensory organ is determined, and (3) the refinement of nerve to target matchings occurs. All of these structure–function relationships can be related to molecules produced by target tissue (attractant and repellent) that act on innervating neurons and, reciprocally, to molecules that are produced by neurons that target taste receptor cells. In short, these processes reflect proper matches between nerve and targets. It is nearly certain that an ensemble of molecules have varying functions during well-defined periods of development to determine the three processes noted above. One family of molecules that likely plays some role(s) is neurotrophic factors.

Neurotrophic Influences

Although it is beyond the scope of this chapter fully to examine the role of neurotrophins in gustatory development, reflections on some key findings will be useful. Neurotrophins are likely involved in a number of developmental processes, ranging from neuronal survival to refinement of how single gustatory neurons innervate proper taste bud cells (i.e., neuron/target matching). It is clear from other sensory systems that neurotrophic factors help determine the proper number of neurons that innervate targets (e.g., Davies & Lumsden, 1984). In the case of the gustatory system, recent data suggest that the neurotrophins are probably also responsible for this function. Specifically, BDNF has been strongly implicated as being a neurotrophin in taste buds located on the anterior tongue (e.g., Nosrat & Olson, 1995). Presumably, neurons that successfully compete for a limited supply of BDNF live, whereas those that do not receive sufficient amounts are eliminated (e.g., Fritzsch *et al.*, 1997). Accordingly, receptors for these neurotrophins (e.g., trkB) should be found on neurons that innervate anterior tongue taste buds (i.e., on the chorda tympani neurons). Knockout mice with a disruption of the neurotrophin receptor trkB have only 5% of geniculate ganglion cells that live past birth (Fritzsch *et al.*, 1997), which is consistent with the idea that gustatory neuron numbers are determined in part by amount of neurotrophins produced by taste buds. It is unlikely, however, that neurotrophins also function as guidance cues that direct pioneering gustatory neurons to the rudimentary taste buds. Emerging data indicate that the developing gustatory system uses a different strategy for the initial innervation of presumptive taste buds, at least on the anterior tongue. Instead of attractant molecules, the chorda tympani nerve is directed at fungiform papillae by being repelled instead of attracted by specific molecules (Rochlin *et al.*, 1999).

Once the overall number of neurons that comprise taste nerves is determined early in development, precise matches between target (taste buds) and individual neurons must then be accomplished. Thus, even though the total number of peripheral gustatory neurons is determined, the subsequent process of matching individual taste buds with the appropriate innervation by single neurons is needed. This process is essential for establishing the number of taste buds innervated by a single peripheral gustatory neuron (i.e., receptive field organization of single taste neurons). This important process will be discussed below.

RECEPTIVE FIELDS

Receptive fields in the peripheral gustatory system are defined by the number and location of taste buds innervated by a single gustatory fiber. The earliest work delineating receptive field development and organization was in fetal and early postnatal sheep (Mistretta, Gurkan, & Bradley, 1988). An inverted "U" function most accurately describes the relationship between developmental time and the number of taste buds innervated by single chorda tympani neurons. About 50 taste buds are innervated by a single gustatory neuron very early in development. By birth, receptive fields become larger, with an average of 100 taste buds. Then the number of taste buds innervated by individual taste neurons decreases postnatally, thereby resulting in mature terminal fields. The change from small, to high, to small receptive fields again illustrates a highly dynamic system of neuron/target interactions with age. One of the implications of this morphological remodeling with age is that the taste responses transmitted to central gustatory structures change dramatically with age. As an aside to these data illustrating how morphological remodeling has functional consequences, the decrease in size in receptive fields is accompanied by increase in sodium taste responses in sheep chorda tympani neurons (Nagai, Mistretta & Bradley, 1988) (see section, Peripheral Development of the Gustatory System—Physiology, for discussion).

Complementary developmental data on how many neurons innervate a single taste bud are also available. This is somewhat of a different question than that related to receptive fields because it examines the number of neurons that innervate single taste buds instead of the number of taste buds innervated by single neurons. By labeling individual fungiform taste buds in rats with a retrograde tracer (Figure 4), we determined how many geniculate ganglion cells containing the cell bodies of chorda tympani fibers (Figure 2) innervated taste buds at different ages (Krimm & Hill, 1998). There appears to be a random arrangement of innervating neurons per taste bud at all ages. In PN10–30 rats, there is no apparent relationship between the size of the taste bud and the number of innervating neurons. However, beginning at PN40 and continuing through adulthood, a highly ordered relationship emerges when the number of ganglion cells matches the corresponding taste bud volume. The number of neurons that innervate each taste bud is related to taste bud size; the larger the taste bud, the greater the number of innervating ganglion cells.

Recent studies demonstrate that the final number of ganglion cells innervating a taste bud is established early in development (Krimm & Hill, 1998). The mean number of ganglion cells that innervate single taste buds is similar from PN10 to PN40 in rat; only taste bud size increases with age (Krimm & Hill, 1998). Moreover, single taste buds labeled at PN10 and again at PN40 have the majority of geniculate ganglion cells double-labeled (Krimm & Hill, 2000). That is, the majority of neurons that innervate a taste bud at PN10 continue to innervate it through 40 days. It is the growth in taste buds that changes with age; the size of the taste bud in 40-day-old rats

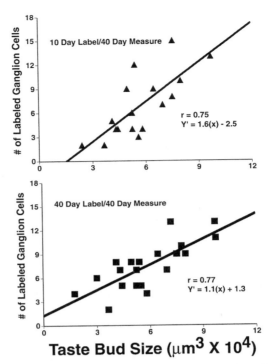

Figure 4. (Above) The function of taste bud size versus number of labeled ganglion cells when taste buds are labeled at PN10 and then allowed to develop to PN40. The number of neurons that innervate taste buds at PN10 predicts the size of the taste bud 30 days later. (Below) The same function as shown in panel A, except both the label and measurements occur at PN40.

is predicted by the number of geniculate ganglion cells labeled at 10 days (Figure 4). These findings indicate that the neural "template" for the receptive field is determined early and becomes matched with taste bud size later in development. Determining the manner in which the "template" becomes formed early in development (perhaps prenatally) would be a major breakthrough in understanding general principles in nerve/target interactions. For example, the emerging taste bud may dictate the number of surviving neurons by way of the amount of neurotrophic factors produced by the bud. The greater the amount of neurotrophic factor produced (e.g., BDNF), the greater the number of innervating neurons. Once the complement of neurons is determined (i.e., the template), the total number of neurons that innervate a given taste bud may subsequently determine the taste bud's size.

Peripheral Development of the Gustatory System—Physiology

Response Development

The functional development of the peripheral gustatory system is characterized by large and dramatic increases in response magnitudes to select stimuli. For example, the chorda tympani nerve in sheep (Mistretta & Bradley, 1983) and rat (Ferrell et al., 1981; Hill & Almli, 1980; Yamada, 1980), which innervates taste receptors on the anterior two thirds of the tongue (Figure 2), is poorly responsive to sodium salts during early development. In sheep, where there is a significant period of taste development prenatally (Bradley & Mistretta, 1973), low responses to NaCl occur from the last trimester and extend until weaning (Mistretta & Bradley, 1983). In rats, development of the peripheral taste system is almost entirely postnatal (Farbman, 1965; Mistretta, 1972); low responses occur soon after birth and then increase in magnitude through weaning (Ferrell *et al.*, 1981; Hill & Almli, 1980; Yamada, 1980). In both species, sodium salts elicit the largest magnitude response of all stimulus classes at adulthood. In contrast, the developing gustatory system is responsive as soon as it begins functioning to some other, non-sodium salt, stimuli. For example, citric acid and ammonium chloride produce vigorous responses in PN2 rats (Hill & Almli, 1980). Therefore, unlike other sensory systems where peripheral neurons and receptor cells are "sluggish" to all stimuli (Coleman, 1990), the taste system responds with large magnitudes to some stimuli early in development (Hill, Mistretta, & Bradley, 1982).

Amiloride and Salt Taste Development

Our understanding of gustatory physiology has been advanced significantly during the past 15 years with the advent of a pharmacologic agent, amiloride, that blocks epithelial sodium ion channels and prevents sodium salt taste transduction. In adult rats, lingual application of amiloride reduces NaCl taste responses by 70–80% (Figure 5) and responses to sodium acetate completely (e.g., Formaker & Hill, 1988). The difference in the amount of suppression relates to multiple or single transduction pathways for NaCl and sodium acetate, respectively, and have significant effects on the perception of halogenated verses nonhalogenated sodium salts (see Stewart *et al.*, 1997). The following summarizes the developmental studies that have used the epithelial blocker.

Increased sensitivity to sodium occurs concomitantly with sensitivity to amiloride (Figure 5). Small sodium taste responses from the chorda tympani nerve in rats less than 13 days old are affected little by amiloride (Figure 5). However, the sodium taste response becomes larger through adulthood, and so, too, does the absolute amount of response suppression due to amiloride (Hill & Bour, 1985; Sollars & Bernstein, 1994) (Figure 5). Thus, increased sensitivity to sodium appears to reflect an increase in functional amiloride-sensitive sodium channels (Hill & Bour, 1985).

These findings provide critical information regarding sodium response ontogeny. First, they point to the cellular locus in the development of sodium responses. Second, they establish a developing transduction pathway that is relatively well characterized in gustatory and nongustatory tissue (e.g., Avenet & Lindemann, 1988; DeSimone & Ferrell, 1985; Garty & Benos, 1988). Finally, specific predictions concerning amiloride-sensitive sodium channels on taste-related behaviors, such as sodium appetite, can be tested experimentally.

The behavioral consequences of the suppressed sodium salt responses in adult rats treated with amiloride can now be compared with that of young rats whose sodium responses are sluggish and in which amiloride is relatively less effective (Figure 5). One means of achieving this behaviorally is through conditioned taste aversion paradigms in which a solution paired with an illness is subsequently avoided, as are solutions of a similar taste. Indeed, generalization gradients reflect similarity among solutions. The conditioned aversion paradigm also reveals which component of a compound determines avoidance behavior. Thus, pairing 0.1 M

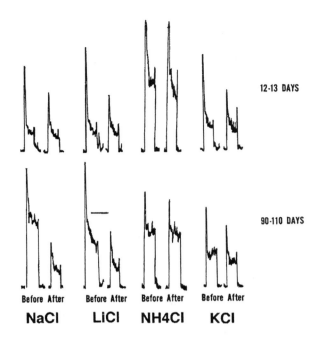

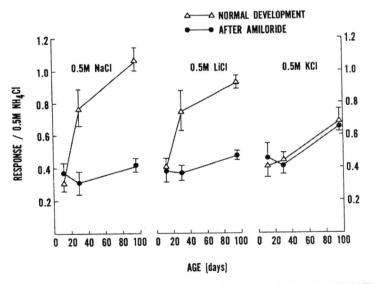

Figure 5. (Above) Integrated chorda tympani responses to 0.5 M NaCl, 0.5 M LiCl, 0.5 M NH$_4$Cl, and 0.5 M KCl before and after application of amiloride in a 12-day postnatal rat and in an adult rat. (Below) Mean response rates of 0.5 M NaCl, 0.5 M LiCl, and 0.5 M KCl compared to responses to 0.5 M NH$_4$Cl in rates ages 12–13 days, 29–31 days, and 90–110 days, before application of amiloride (normal development) and after amiloride (Reprinted from Hill and Bour, 1985, with permission. Copyright 1985 Elsevier Science).

NaCl with illness produces a profound taste aversion to 0.1 M NaCl and to all stimuli that taste similar to it, such as 0.1 M sodium acetate (Hill, Formaker, & White, 1990) (Figure 6). Interestingly, solutions that share the anion, chloride, but do not have the sodium cation (e.g., 0.1 M KCl) will also be avoided, but not with the same magnitude as for sodium salts (Figure 6). Therefore, rats use at least two perceptual components to NaCl taste, the one to sodium being the more dominant. Solutions that do not taste like the conditioned stimulus, sucrose, for example, are not avoided (Hill *et al.*, 1990). With these standards for NaCl taste perception in adult rats, the effects of amiloride mixed with the conditioned stimulus of NaCl could be determined (Hill *et al.*, 1990). Rats conditioned to the taste of NaCl mixed in amiloride, thereby blocking sodium channels, do not use the taste of sodium in the conditioned stimulus when avoiding other solutions. Rather, they appear to use the taste of chloride to avoid salts that seem to taste "bitter" or "sour" (Hill *et al.*, 1990) (Figure 6). These findings parallel the behavior of immature rats not treated with amiloride. Rats of age PN25–30 exhibit widespread taste aversions to all monochloride salts if the conditioned stimulus is NaCl (Hill *et al.*, 1990) (Figure 7). Like adults that had exposure to amiloride, weanling rats do not discriminate the taste of NaCl from other salts (Figures 6 and 7). This suggests that sodium discrimination reflects the development of the amiloride-sensitive sodium channel. The boundaries of the phenomenon have not been identified behaviorally.

The specific "amiloride deficits" have made these rats interesting targets for immunohistochemical and biophysical investigations. With the use of a polyclonal

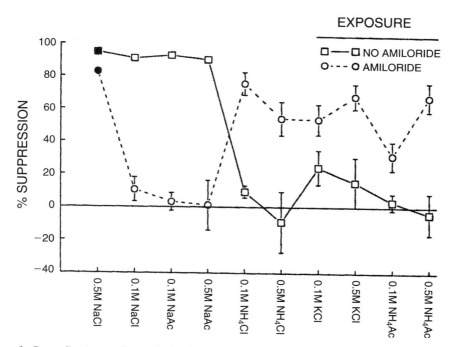

Figure 6. Generalization gradients obtained from rats exposed to amiloride and from rats not exposed to amiloride when the conditioned stimulus was 0.5 M NaCl. (Reprinted from Hill *et al.*, 1990, with permission. Copyright 1990 American Psychological Associations.) Note: The data for each group are connected by lines for ease of comparison; however, the functions are not continuous.

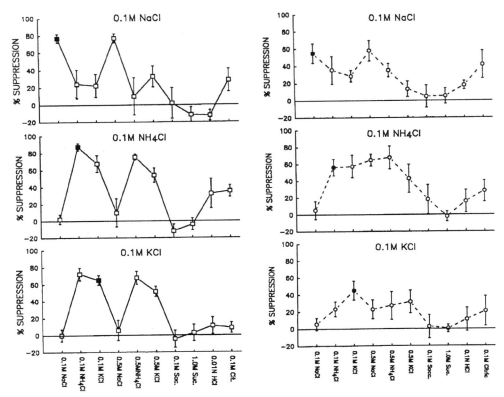

Figure 7. Generalization gradients obtained from adult rats (left) and rats 25–35 days old (right) when the conditioned stimulus was 0.1 M NaCl (top panels), 0.1 M NH$_4$Cl (middle panel), or 0.1 M KCl (bottom panel). Chemical abbreviations are shown along the abscissa: Sac., saccharin; Suc., sucrose; Cit., citric acid. (Reprinted from Formaker & Hill, 1990, with permission. Copyright 1990 American Psychological Association.) Note: The data for each group are connected by lines for ease of comparison; however, the functions are not continuous.

antibody directed at the amiloride channel, immunopositive taste buds are unexpectedly seen on the tongue of PN1 rats (Stewart, Lasiter, Benos, & Hill, 1995). This is surprising since chorda tympani responses from early postnatal rats have small (or no) responses to sodium salts and little sensitivity to amiloride (Hill & Bour, 1985; Sollars & Bernstein, 1994). It is suggested, therefore, that at least some form of the channel is present in early developing taste buds and that the channels become functional later in development (Stewart *et al.*, 1995). Alternatively, the channels may be functional in taste receptor cells of young rats, but not be available to sodium ions that stimulate the apical domain (Stewart *et al.*, 1995). Kossel, McPheeters, Lin, and Kinnamon (1997) supported the latter alternative. After removing taste buds from tongues of PN0–30 rats, they used whole-cell patch-clamp recordings of stimulus-induced conductances. Therefore, the functions of the apical and basolateral portions of the cell are not assessed separately (see Figure 1). They found that receptor cells from fungiform taste buds in PN2 rats are as sensitive to amiloride as are cells from PN30 rats; sodium-related conductances are suppressed in taste bud cells similarly throughout development (Kossel *et al.*, 1997). The relative lack of amiloride sensitivity in neural recordings (Hill & Bour, 1983; Sollars & Bernstein, 1995) suggests that whole-cell recordings detect functional amiloride-sensitive channels that are not normally reached by chemical stimuli in intact tissue because of their location in the basolateral domain of taste receptor cells (Kossel *et al.*, 1997).

Further support of the initial availability of functional channels on the taste receptor membrane in PN14 rats comes from biophysical analyses. In the first study, stimulus-induced ionic currents were recorded from dorsal lingual epithelia that was stripped from the tongue in PN14 to adult rats, but with intact taste buds (Settles & Mierson, 1993). Therefore, stimuli only have contact with the portion of taste receptor cells normally stimulated (i.e., the apical portion) and not to the entire cell (see Figure 1). Receptor cell polarity was maintained. Sodium conductances increase during development along with a corresponding increase in amiloride sensitivity (Settles & Mierson, 1993). In a second biophysical study on the ontogeny of fungiform taste bud sodium and amiloride sensitivity, an *in vivo* voltage clamp procedure in which voltage was applied across lingual epithelia in anesthetized rats was combined with neurophysiologic recordings from the chorda tympani nerve (Hendricks, Stewart, Heck, DeSimone, & Hill, 2000). Thus, ions of a stimulus are driven across lingual epithelia in live (albeit anesthetized) rats while recording the neurophysiologic taste response. This approach generates parameters such as channel densities (i.e., numbers/unit volume) and channel affinities (i.e., function) (Ye *et al.*, 1993) to reveal the underlying mechanisms of increased sodium taste responses during development. The number of functional amiloride-sensitive channels on the anterior tongue increases dramatically with age (Hendricks *et al.*, 2000) (Figure 8). Furthermore, the channels also become more efficient (i.e., have higher affinities for sodium salts) with age (Figure 8). Thus, two age-related processes occur: more channels are added to taste receptor cells and the functional channels become more efficient in transducing sodium. The net effect is a significantly larger response to sodium salts with age.

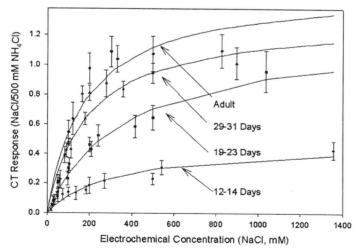

Figure 8. Curves of best fit of the NaCl response in the chorda tympani of rats 12–14, 19–23, and 29–31 days old and adults expressed relative to the electrochemical concentration of NaCl obtained from *in vivo* voltage clamp experiments. The increase in the maximum response denotes an increase in the functional sodium transduction elements, while a shift in the half-maximal response to the left related to increased efficiency of the transduction elements. These two measures are obtained from biophysical models used by Ye, Heck, and DeSimone, (1993). (Reprinted from Hendricks *et al.*, 2000, with permission. Copyright 2000 The American Physiological Society.)

These apparent divergent sets of findings may be explained by translocation of the channel from the basolateral to the apical domain. The translocation of functional amiloride channels in polarized epithelial cells has been demonstrated in a variety of tissues (Garty & Benos, 1988; Garty & Palmer, 1997) and is not unique to gustatory tissue. This results from a number of factors responsible for movement of functional channels and/or structural alterations of channels *in situ* (Garty & Benos, 1988; Garty & Palmer, 1997). Nonetheless, this leaves open the issue as to what factor(s) makes the channels functional. Evidence suggests that hormonal (e.g., vasopressin) and/or growth factors (e.g., insulin-like growth factors 1 and 2) that develop well before the manifestation of the sodium taste system are involved. These diffusible agents may determine how the amiloride-sensitive sodium channel is formed and then regulated (Hill & Przekop, 1988; Przekop, Mook, & Hill, 1990; Stewart & Hill, 1996; also see below, section Early Dietary Manipulations).

Gustatory Ganglia

Although much has been learned about the peripheral gustatory system by studies of taste receptor cells and of nerves that innervate these cells, virtually nothing is known about the development of cell bodies of the taste nerves and the structures in which they reside. However, the recent attention to induction and the maintenance of taste buds (see section Taste Bud Formation) has focused on developmental processes resident to the ganglia of taste nerves. The interaction between target (e.g., taste buds) and nerves (e.g., the chorda tympani) directs a number of developmental processes. For example, the survival of neurons that innervate emerging taste buds likely depends on proper amounts of neurotrophic factors produced by the taste buds (see section Taste Bud Formation). As noted earlier, an abundance of neurotrophins would likely lead to an excess of peripheral gustatory neurons (i.e., geniculate ganglion neurons), and diminished amounts of neurotrophins would likely lead to a loss of peripheral gustatory neurons (e.g., Fritzsch *et al.*, 1997). It also follows that alterations in ganglion cell number influence the maintenance of taste buds and/or taste cells, the receptive fields of peripheral neurons, and the central development of gustatory structures.

Central Development of the Gustatory System—Anatomy

Structures along the gustatory pathway undergo dramatic morphological changes during development. For example, chorda tympani fibers begin synapse formation in the rostral pole of the nucleus of the solitary tract (NTS; see Figure 2) as early as PN1 in rat (Lasiter, Wong, & Kachele, 1989), even though some neurons may arrive centrally much earlier (Scott & Atkinson, 1998). However, the area in which the nerves synapse in the NTS, the terminal field, does not reach its full size until PN25 (Lasiter, 1992). Chorda tympani axons migrate caudally in the NTS and envelop local neurons and neurons that project to the second-order central gustatory relay in the parabrachial nucleus (PBN). Neurons postsynaptic to chorda tympani fibers in the NTS also show dramatic changes in morphology, with dendritic lengths and dendritic branching increasing approximately threefold between PN8 and PN25 (Lasiter, 1992; Lasiter & Kachele, 1990). The collective increase in pre- and postsynaptic neurites in the NTS with age indicates that an increasingly large afferent message (see section Peripheral Development of the Gustatory System—

Physiology) may influence circuit formation (e.g., Lasiter, 1992; Mistretta & Hill, 1995).

The NTS to PBN projection does not parallel that of the chorda tympani nerve to the NTS: it starts much later (Lasiter & Kachele, 1988, 1989). Second-order neurons that project from the NTS to the PBN do not begin until PN7 and their development is slower, attaining completion at about PN60 (Lasiter, 1992). It is not clear whether this asymmetry in projection has functional implications; however, this prolonged postnatal development may contribute to both the pre- and post-synaptic plasticity.

In contrast to gustatory terminations in the rostral NTS and neurons resident in this area, fibers of the glossopharyngeal nerve, which innervates taste receptors on the posterior tongue (Figure 2), do not enter the intermediate zone of the NTS until PN9–10 in rat (Lasiter, 1992). Moreover, the rostral–caudal maturation of this field does not occur until approximately PN45 (Lasiter, 1992). Compared to the long delay in pre- and postsynaptic maturation in the rostral NTS, only 1 week separates the morphological maturation of pre- and postsynaptic elements in the intermediate NTS. The maturation progression is not even: significant maturation of the NTS occurs prior to the beginning of PBN development, suggesting that signals coming from more distal structures guide the morphological development of the next gustatory relay (Lasiter, 1992). These factors alone suggest that there should be profound functional and behavioral changes with age that correspond to the morphological changes.

Rather little is available concerning the morphological development of the PBN and its more rostral relays. The delayed development of presynaptic elements in the PBN is continued in postsynaptic cells. Dendrites of PBN neurons arborize significantly between PN16 and PN35 in rat (Lasiter & Kachele, 1988), the period during which projections to the next relay in the thalamic taste area occur and metabolic activity of PBN neurons increases (Lasiter & Kachele, 1988). These later developing morphological changes suggest that functional response maturation of the PBN may differ from those in the NTS. Indeed, taste-guided behaviors in which the PBN participates should differ from those determined by the NTS. For example, gustatory coding in the NTS may predominate for early taste-guided behaviors (e.g., early food-seeking behaviors), whereas the PBN may be involved in later developing behaviors (e.g., behaviors integrating taste and visceral inputs). Unfortunately, there is no direct evidence on this point.

CENTRAL DEVELOPMENT OF THE GUSTATORY SYSTEM—PHYSIOLOGY

The initial studies examining taste responses from NTS neurons in rat (Hill, Bradley, & Mistretta, 1983) and sheep (Bradley & Mistretta, 1980; Mistretta & Bradley, 1978a) demonstrated that central response development mirrors peripheral development. Response magnitudes to sodium salts increase profoundly in NTS neurons during development. In contrast, NTS neuron responses to other stimuli, such as to NH_4Cl, do not change with age. There is a complete lack of sodium sensitivity in NTS neurons to sodium salts in fetal sheep, whereas the chorda tympani nerve is responsive, albeit modestly (Bradley & Mistretta, 1980; Mistretta & Bradley, 1978a). This complete lack of responses is not found in rats as young as PN14, which is the youngest age examined (Hill *et al.*, 1983); it may be found in younger rats. Furthermore, in both rat (Hill *et al.*, 1983) and sheep (Bradley & Mistretta, 1980;

Mistretta & Bradley, 1978a), mature NTS responses are achieved later in development compared to the chorda tympani (Bradley & Mistretta, 1973; Ferrell *et al.*, 1981; Hill & Almli, 1980; Mistretta & Bradley, 1983; Yamada, 1980). For example, in rats, mature taste responses to sodium salts in the chorda tympani occur at postnatal weeks 3–4. In the NTS, mature responses appear about 1 week later. Therefore, central taste response development is not tightly linked to age-related peripheral changes (Bradley & Mistretta, 1980; Hill *et al.*, 1983; Mistretta & Bradley, 1978a). Synaptic organization or reorganization may occur during development in the NTS, which delays the response development beyond that of the chorda tympani nerve. This probably results in a degraded neural signal across the first gustatory relay to some stimuli during the immature NTS period.

The delay cannot be attributed to poorly developed NTS neurons; functional membrane characteristics mature before the taste-elicited responses. In fact, the membrane parameters of resting membrane potential, action potential, and discharge properties change the most between PN5 and PN15 in rats, with mature values reached by postnatal day 20 (Bao, Bradley, & Mistretta, 1995). This provides further evidence that changes in the synapses and not the NTS neurons per se are what account for the prolonged functional changes postnatally.

Similar developmental patterns are observed in the next synaptic relay, the PBN. Responses to taste stimuli that change developmentally at more peripheral levels also occur in the PBN (Hill, 1987a). However, PBN neurons increase in response magnitudes to all stimuli from weaning to early adulthood, including responses to stimuli that are very effective in eliciting NTS responses during early development. Therefore, PBN neurons appear to be "sluggish" to all stimuli early in development and then sensitivity gradually increases with age. The delayed functional maturation at successively higher synaptic levels suggests that central anatomic and/or neurochemical events are responsible for the unique developmental patterns at each neural relay. This also holds open the possibility that gustatory experience affects the higher order neural ontogeny. Again, the behavioral consequences either in terms of threshold, preference, or motivation for a taste solution are not known.

EXPERIENCE-INDUCED ALTERATIONS IN THE DEVELOPING TASTE SYSTEM

As the basis of normal gustatory development becomes better understood, it is possible to assess how environmental factors affect morphological and physiological development (Mistretta & Bradley, 1978b). That is, normal functional and morphological developmental patterns provide the standards to which experimentally related effects can be compared. Receptor cell destruction, altered taste experience, and early dietary manipulations have been used to examine the consequence of these histories on a number of morphological, physiological, and behavioral measures.

RECEPTOR CELL DESTRUCTION

An experimental strategy borrowed from analyses of other sensory systems to determine the role of neuronal activity on development has been to eliminate or markedly reduce afferent activity (Coleman, 1990). In the taste system, Lasiter and colleagues destroyed taste receptors on the anterior tongue by cautery at various ages postnatally and then examined the effects on terminal field organization in the NTS (Lasiter & Kachele, 1990). Destruction of taste buds at PN2–7 permanently reduces the chorda tympani terminal field in the NTS by 30%, a reduction that persists even though taste buds regenerate on the anterior tongue (Lasiter & Kachele, 1990). The smaller terminal field cannot be attributed to fewer neurons projecting to the NTS because normal numbers of geniculate ganglion neurons are maintained (Lasiter & Kachele, 1990). Decreased terminal field size probably reflects receptor damage-induced changes in neuronal activity during a sensitive period (Lasiter & Kachele, 1990). However, neither peripheral nor central functional responses were recorded in these rats. Therefore, it is not clear whether the early destruction may have altered activity-dependent mechanisms responsible for central terminal field expansion. Regardless of the actual mechanisms (i.e., activity versus non-activity dependent), subsequent work indicates the site of cellular change responsible for the central field changes resides in the geniculate ganglion and not in the terminal field (Lasiter & Bulcourf, 1995). Thus, at least peripheral gustatory neurons are affected by receptor destruction.

Altered Taste Experience

Unlike the strategy of destruction which influences perception of all gustatory stimuli, a more specific approach is to alter experience through dietary manipulation. Early efforts met with limited success and often provided conflicting results (Capretta & Rawls, 1974; Rozin, Gruss, & Berk, 1979; Warren & Pfaffmann, 1959; Wurtman & Wurtman, 1979). Indeed, one conclusion from the earliest studies was that the gustatory system is remarkably resistant to environmental manipulations (Wurtman & Wurtman, 1979). However, more recent work using selective postnatal exposure to taste stimuli in conjunction with work using early prenatal dietary manipulations have begun to etch the limits of the environmental parameters on the functional and morphological development of the gustatory system. More importantly, however, it shows us important features of gustatory development by perturbing normal morphological and functional development.

Selective Exposure to Taste Stimuli. Through artificial rearing procedures developed by Hall (1979), rat pups were isolated from their mothers by PN4 and received diet intragastrically (Lasiter, 1995; Lasiter & Diaz, 1992); they were selectively exposed to taste stimuli for specific periods. Limited orochemical stimulation is sufficient to produce normal axonal and terminal field development in the NTS, i.e., like that in rats raised with their mothers (Lasiter, 1995; Lasiter & Diaz, 1992) (Figure 9). Water infusions alone from postnatal days 4 to 10 do not produce normal NTS terminal fields (Lasiter, 1995). Like rats receiving early receptor damage (Lasiter & Kachele, 1990), rats exposed only to water early in development fail to show the caudal migration of the chorda tympani terminal field and proper development of the other terminal fields (Figure 9A). However, stimulation with either NaCl, lactose, whole rat milk, or dialyzed rat milk also produces normal fields when exposure occurs between PN4 and PN10 (Lasiter & Diaz, 1992) (Figure 9A). At least 3 days of exposure to 150 mM NaCl begun at PN4 also produces normal terminal fields (Lasiter, 1995) (Figure 9B). Therefore, results from early receptor damage and from selective taste exposure studies demonstrate experimentally that the normal caudal migration of the chorda tympani field and the development of the other fields in the NTS are dependent on intact taste receptors early in development and are dependent upon certain taste stimulation that occurs soon after birth.

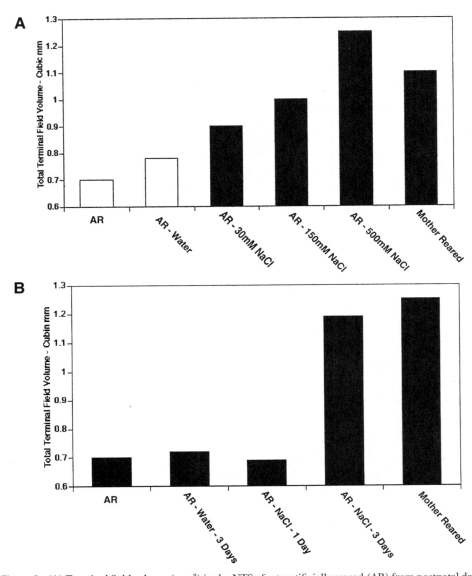

Figure 9. (A) Terminal field volume (mm³) in the NTS of rats artificially reared (AR) from postnatal days 4–10 with no taste stimulation (AR), with water as the stimulus (AR–water), with either 40 mM NaCl, 150 mM NaCl, or 500 mM NaCl (AR–30mM NaCl, AR–150mM NaCl, AR–500mM NaCl, respectively), or reared by their mother (Mother Reared). (B) Results of limited stimulation studies. Dependent measures are identical to those shown in panel A. Stimulation with 150 mM NaCl was only effective in inducing normal terminal field development when delivered for 3 days between postnatal days 4 and 7. (Adapted from Lasiter, 1995.)

EARLY DIETARY MANIPULATIONS. Restriction of maternal dietary sodium by reducing sodium content in food from 1% to 0.03% beginning on or before 8 days postconception and continued in offspring through at least PN28 results in dramatically reduced neurophysiologic responses to sodium salts in the chorda tympani (Figure 10) (Hill, 1987b; Hill & Przekop, 1988; Hill, Mistretta, & Bradley, 1986).

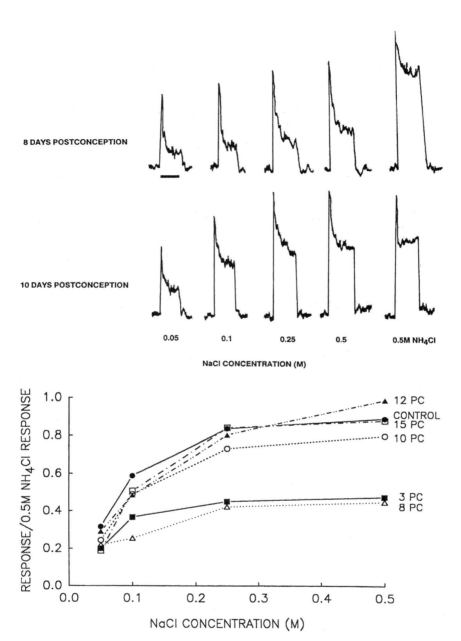

Figure 10. (Above) Integrated responses from the chorda tympani nerve to a concentration series of NaCl and 0.5 M NH$_4$Cl in a rat that had a 0.03% NaCl diet imposed at embryonic day 8 or embryonic day 10. Scale bar = 15 sec. (Below) Mean response rates from the chorda tympani nerve to a concentration series of NaCl in rats sodium-restricted at postconception (PC) day 3, 8, 10, 12, and 15 and from rats fed the control diet. (Reprinted from Hill and Przekop, 1988, with permission. Copyright 1988 American Association for the Advancement of Science.)

Responses to NaCl are reduced by as much as 60% in sodium-restricted rats (Figure 10). In contrast, taste responses to NH_4Cl and nonsalt stimuli are unaffected (Hill, 1987; Hill & Przekop, 1988; Hill *et al.*, 1986), showing that severe sodium restriction causes selective deficits in chorda tympani nerve function. In addition to showing the specificity of the response, these results also eliminate the concern that drastic sodium restriction induces nonspecific alterations in gustatory function. The selective decrease in sodium salt-elicited responses reflects the absence of functional amiloride-sensitive sodium channels (Hill, 1987b; see above, section Peripheral Development of Gustatory System—Physiology). Moreover, corroborative data on the mechanism responsible for the lack of sodium sensitivity has been provided recently through *in vivo* voltage clamp procedures. Developmentally restricted rats have 90% fewer amiloride channels in the apical domain of taste receptor cells (Ye, Stewart, Heck, Hill, & DeSimone, 1993).

The period of vulnerability to sodium deficiency (on or before E8) occurs before taste buds first appear on the anterior tongue, suggesting that stimulus–receptor interactions are not crucial for the formation of amiloride-sensitive sodium channels (Hill & Przekop, 1988) (Figure 11). Interestingly, mother's milk and the offspring's blood electrolyte levels are protected from the dietary manipulation (Hill, 1987b; Stewart, Tong, McCarty, & Hill, 1993). Thus, the consequences of early dietary manipulations are probably not induced by altering the fluids that circulate about the tongue. Moreover, it is unlikely that the effect is directly due to altered sodium levels because of the integrity of the circulating fetal plasma (Kirksey, Pike, & Callahan, 1962). It is possible, however, that the altered dietary sodium levels affect other physiological systems in mother or fetus, which in turn may impact the expression of amiloride channels almost three weeks later (Figure 11). The chal-

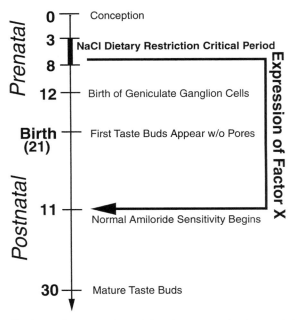

Figure 11. Time line showing major events in peripheral gustatory development. It is proposed that a factor(s) (i.e., Factor X) is affected by early dietary restriction, but it does not have a functional consequence until postnatal day 11, when normal amiloride sensitivity begins.

lenge is to reconstruct the series of events leading up to the failed functional channel. Candidate factors (denoted as "Factor X" in Figure 11) that may be influenced by the dietary manipulation include hormones and growth factors (Stewart *et al.*, 1997). Based on alterations in the amiloride-sensitive sodium channel in toad urinary bladder (Blazer-Yost, Cox, & Furlanetto, 1989), one would predict a similar fate in the gustatory system following reduction of insulin-like growth factor (IGF1). Reduction of IGF1 would likely lead to an alteration in the amiloride-sensitive sodium channel, as documented in other tissue (e.g., Blazer-Yost *et al.*, 1989).

According to immunohistochemical studies, some components of the channel present in the taste receptor cells of both sodium-restricted and normal rats by PN1 are not functional. That is, taste buds in developmentally sodium restricted rats are positively labeled for the amiloride-sensitive channel antibody (Stewart *et al.*, 1995). The deficits, therefore, are not in the early formation of the channel. They may reflect a lower circulating level of IGF1 or the inability of the nascent channel to respond normally to IGF1. Attempts to alter sodium levels after the critical E3–8 period or use less severe restriction procedures (e.g., 0.05% NaCl instead of 0.03% NaCl) fail to produce changes in chorda tympani function (Bird & Contreras, 1987; Hill & Przekop, 1988).

Although the initial effects on taste receptor cell function are dramatic, the dietary effects on the peripheral gustatory system are reversible. Chorda tympani responses in restricted rats fed a NaCl-replete diet after weaning recover to control levels within 15 days, with a corresponding increase in functional amiloride-sensitive sodium channels (Hill, 1987b). Remarkably, function can be restored during adulthood by a single, 30-ml ingestive bout of physiological saline (Przekop *et al.*, 1990; Stewart & Hill, 1996). Importantly, this recovery is not dependent on direct stimulation of taste receptor cells with sodium. Sodium-restricted rats allowed to drink physiological saline sufficient to produce recovery but not allowed to absorb sodium due to a preinjection of the diuretic drug furosemide do not recover normal chorda tympani function (Przekop *et al.*, 1990). Therefore, the sodium transducer is not under control of stimulation of the receptor with sodium. The 10–12 days needed for recovery suggests that postingestive effects are targeted toward newly proliferating cells because taste receptor cell turnover probably occurs at about the time that functional recovery is seen (i.e., about 10 days) (Stewart & Hill, 1996). Thus, it is the newly forming taste receptor cells that appear to have the functional sodium channels and not ones present when "recovery" begins. Recovery is dependent on age-related processes that begin at or beyond the age of weaning because some of these factors present in normal milk are not used by early restricted rats for recovery of the response. Rats born to sodium-restricted rats and then cross-fostered to control mothers for the first 2 weeks postnatally before being returned to a restricted dietary regimen fail to recover normal chorda tympani function (Phillips, Stewart, & Hill, 1995). If normal milk promotes formation of functional amiloride-sensitive channels, then normal sodium taste function should have occurred in these cross-fostered pups.

As indicated by specific response deficits in the chorda tympani, not all populations of taste receptors are susceptible to these early dietary manipulations. In this regard, receptors on the palate which physically oppose those on the anterior tongue (see Figure 2) are not affected by early sodium restriction. Greater superficial petrosal (GSP) nerve recordings reveal that early sodium-restricted rats respond normally to palatal stimulation with sodium and nonsodium salts (Sollars &

Hill, 1998, 2000) (Figure 12). Thus, only a restricted population of taste receptors is vulnerable to early NaCl restriction. The bases of the differential susceptibilities are not known.

The consequences of early dietary sodium restriction also extend to taste bud morphology in the anterior tongue. Taste bud volume in adult rats sodium-restricted throughout development is that of approximately 3-week-old normal rats (Krimm & Hill, 1999). There is approximately a 30% reduction in average taste bud size in sodium-restricted compared to control rats (Figure 13A). Although taste bud size is reduced, the number of ganglion cells that innervate individual taste buds is not

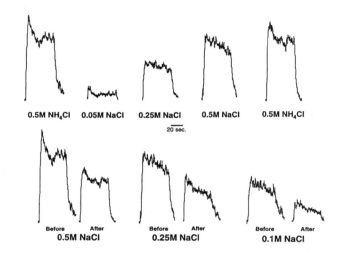

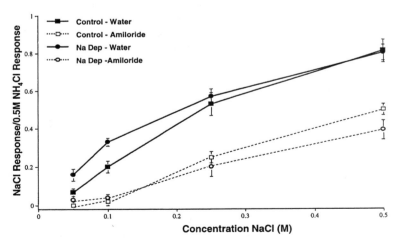

Figure 12. (Above) Integrated responses from the greater superficial petrosal nerve (GSP) to a concentration series of NaCl and to 0.5 M NH$_4$Cl (top), and to 0.5, 0.25, and 0.1 M NaCl before and after application to amiloride (bottom). (From Sollars & Hill, 2000.) (Below) Mean relative responses of the GSP in control of rats and developmentally sodium-restricted rats to a concentration series of NaCl before and after amiloride.

reduced; the mean number of geniculate ganglion cells that innervate taste buds on the anterior tongue in restricted rats is similar to those in controls (Krimm & Hill, 1999). Thus, there is a mismatch between ganglion cell number and taste bud volume because taste buds are smaller than controls but have the appropriate number of neurons innervating them (Figure 13B). Upon repletion with dietary sodium, taste buds grow so that taste bud volume is eventually matched with the appropriate number of ganglion cells that innervate them (Krimm & Hill, 1999) (Figure 13). Therefore, the adjustment in the ratio of taste bud volume to number of innervating neurons is due to changes in taste bud size and not to neuronal reorganization during sodium repletion (Krimm & Hill, 1999). Although it is tempting to speculate about the functional and behavioral consequences of this altered pattern of nerve/target development, there are no data available to make such comparisons. Only through single-fiber recordings of the chorda tympani in sodium-restricted rats during the recovery period can absolute statements be made about strength and quality of the afferent signal.

In light of these multiple morphological and functional effects on peripheral gustatory development, central gustatory deficits, at least for chorda tympani-driven

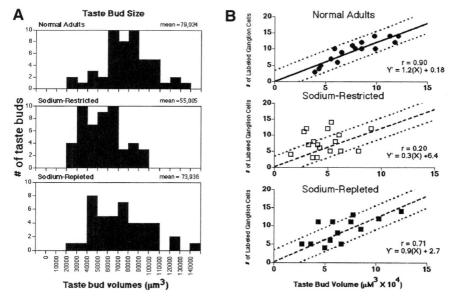

Figure 13. (A) Histograms showing the number of taste buds of different volumes for normal adult rats (top), sodium-restricted rats (middle), and sodium-repleted rats (bottom). Taste buds of sodium-restricted rats are smaller than those of normal adult rats and of sodium-repleted rats. (B) Top: Number of labeled geniculate ganglion cells plotted against taste bud volumes for taste buds sampled from the tongue midregion in normal adults. The 95% confidence intervals for the regression line are displayed by the dotted lines. The number of labeled geniculate ganglion cells is also plotted against taste bud volumes for taste bud volumes for taste buds sampled from the tongue midregion in sodium-restricted (middle) and in sodium-repleted rats (bottom). The regression lines in the top and bottom panels are replotted from panel A and are shown as a dashed line. There is no relationship between taste bud volume and number of labeled ganglion cells in sodium-restricted rats. However, there is a positive correlation between taste bud volume and number of labeled ganglion cells in sodium-repleted rats. The correlation coefficients and equations for regression lines for each group are shown in the bottom right of the respective panel. (Reprinted from Krimm and Hill, 1999, with permission. Copyright Wiley–Liss, Inc., a subsidiary of John Wiley & Sons, Inc.)

processes, are predicted. Dietary sodium restriction during pre- and postnatal development produces both abnormally distributed and irregularly shaped chorda tympani terminal fields (King & Hill, 1991). In fact, a 9-day period of sodium restriction from postconception day 3 to 12 produces a permanent alteration in the chorda tympani terminal field (Krimm & Hill, 1997). Thus, a presynaptic morphological alteration at the first central gustatory relay is caused by dietary restriction initiated during the brief interval in which chorda tympani neurons are born (approximately E12). Although severe sodium restriction must occur during early embryonic development, expression of the altered effects is delayed until the terminal field normally expands (i.e., during the first three postnatal weeks) (Walker & Hill, 1995). Such a delayed effect of the early dietary manipulation may provide insights into chorda tympani terminal field development. The dietary-induced sodium deficits may be determined during ganglion cell birth (at about E12; Altman & Bayer, 1982), even though the deficits do not become manifest until stimulus-induced activity is most pronounced (i.e., beginning at postnatal day 11). Although the underlying mechanism is not understood, recent evidence from labeling single chorda tympani fibers shows that chorda tympani arbors are maintained normally in sodium-restricted rats (Thaw, Walker, & Hill, 1996), yet some neuronal processes project beyond the normal boundaries of the rostral NTS before they terminate.

It is critical to note that the central anatomic effects are limited to the chorda tympani field; the size and topography of the projections from another gustatory nerve, the lingual–tonsilar branch of the glossopharyngeal, are unaffected by dietary manipulations (King & Hill, 1991). The terminal field of the GSP appears normal also, even though located in the rostral pole of the NTS continuous with the terminal field of the chorda tympani (Sollars & Hill, 2000). This is not surprising given normal function of the GSP in developmentally restricted rats (Sollars & Hill, 2000). That is, there is a clear dichotomy between the chorda tympani and GSP in the physiological and anatomic susceptibility to early sodium restriction. Such a dichotomy may relate to the developmental time course of the two nerves and the associated molecular and cellular events that occur within each system.

In contrast, receptor cell damage (Lasiter & Kachele, 1990) in PN2 rats and restricted taste experience in PN4–10 rats (Lasiter, 1995) results in a reduced, instead of an enlarged, chorda terminal field in the NTS. The basis of these differences from early NaCl restriction is not clear. Early dietary sodium restriction may influence different cellular mechanisms or influence the same mechanism but in opposite ways (e.g., increased or decreased growth factor availability at the target). Interestingly, the influences resulting from receptor cell damage and dietary manipulations are confined to the chorda tympani nerve and associated receptors and do not extend to all afferent taste systems (King & Hill, 1991; Lasiter & Kachele, 1990; Sollars & Hill, 2000).

There are also striking postsynaptic morphological alterations. Specific cell types in the rostral pole of the adult rat NTS restricted of dietary NaCl since E3 show pronounced increases in dendritic length and number, whereas others are spared (King & Hill, 1993) (Figure 14). The affected cells remain permanently enlarged after rats are fed the NaCl-replete diet (Figure 14). Neurons affected by dietary restriction are putative relay neurons (Lasiter & Kachele, 1988). These data strongly suggest that the morphological effects of early dietary manipulations are specific to certain cell types and may relate to their function (King & Hill, 1993). It is unclear, however, whether changes in terminal field organization affect the dendritic organization of NTS or vice versa.

The consequences of the early environmental manipulations are also expressed functionally. NTS neurons in sodium-restricted rats respond with lower response frequencies to sodium salts, while activity to nonsodium salts and to nonsalt stimuli are normal (Vogt & Hill, 1993) (Figure 4). Sodium-restricted rats fed a NaCl-replete diet for at least 5 weeks at adulthood have NTS neurons that are more responsive to sodium salts than in controls (Vogt & Hill, 1993) (Figure 14). Accordingly, these neurons have an apparent shift in the stimulus to which they respond best. The central functional changes, like the morphological changes, are permanent and specific. Finally, the terminal fields of NTS neurons in the PBN are similar among control, sodium-restricted, and "recovered" rats (Walker & Hill, 1998). Thus, the PBN is resistant to the dietary-induced changes seen at the lower neural levels. This resistance may reflect different developmental processes and convergence of vis-

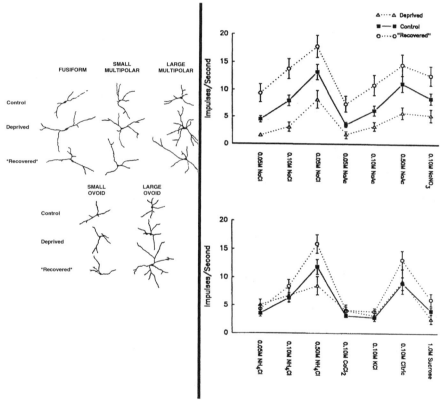

Figure 14. Left: Camera lucida drawings of morphological cell types in the NST of control, sodium-deprived, and "recovered" rats. In fusiform and large multipolar cells there is a significant increase in the length of dendrites between control and NaCl-deprived animals. Following recovery, the fusiform cells show a similar total dendritic length compared to controls, whereas dendrites on the large ovoid cells are significantly longer compared to those in either control or deprived animals. (Reprinted from King & Hill, 1993, with permission. Copyright 1993 Wiley–Liss, Inc., a subsidiary of John Wiley & Sons, Inc.) Right: Means and standard errors for response frequencies for NST neurons in control, sodium-deprived, and "recovered" rats for sodium taste stimuli (top) and nonsodium stimuli (bottom). Responses to sodium salts are lower than those of controls in deprived rats but are greater than those of controls and recovered rats. The responses to nonsodium stimuli are generally the same for controls as for deprived and recovered rats. (From Vogt & Hill, 1993.)

ceral and multiple gustatory inputs in the PBN (chorda tympani, GSP, and glosso-pharyngeal nerves) (Herbert, Moga, & Saper, 1990; Ogawa, Hayama, & Ito, 1982). Behavioral studies are needed here also.

Findings from experiments in which receptors were damaged during early development (Lasiter & Kachele, 1988, 1991), in which very young rat pups were given limited experience to specific taste stimuli (Lasiter, 1995), and in which dietary sodium was manipulated prenatally (Hill, 1987b) all have shown that the gustatory system is "plastic" both in the periphery and in the brain stem. Identifying the underlying mechanisms of the deficit induction and repair will provide understanding of normal developmental processes.

Epilogue

There is a tendency, as shown here, to become increasingly specialized. While this provides a richer understanding of the mechanisms involved in development, it courts the danger of losing a systems-level approach. This is especially acute in an area in which few scientists are active, as in gustatory development research. Consequently, questions related to how the underlying neurobiological substrates determine taste-related behaviors become more and more distant, as conveyed in my final discussion. The danger of studying the biology of the taste system without maintaining a healthy dialogue among all levels of analysis is that the real function of the system is never understood. The gustatory system's "special" function may be homeostatic as well as sensory–perceptual. As such, it may be more fruitful to explore the commonalties with the development of gut physiology (i.e., a regulatory system) instead of commonalities with exteroceptive systems (i.e., a "pure" sensory system).

From a more developmental perspective, it is important to identify the interactions between the developing gustatory system and other physiological systems. For example, do factors that are generally associated with gut physiology (e.g., insulin and insulin-like growth factors) impact taste system organization during early development? Cellular proliferation rates are clearly controlled by growth factors that may originate from the gut (e.g., Cohick & Clemmons, 1993). Therefore, taste bud ontogeny may depend upon proper proliferative factors originating in the gut. Reciprocally, the development of central gustatory structures (e.g., the rostral pole of the NTS) could influence the development of neural pathways that control nongustatory processes. The chorda tympani nerve appears to determine the development of the rostral NTS (e.g., Lasiter & Kachele, 1990). In turn, this could influence the development of other brainstem structures. Indeed, the intermediate NTS is affected by structural changes in more rostral areas (Lasiter & Kachele, 1990). Therefore, it is possible that the rostral NTS may organize processes involved in mastication, swallowing, and digestion throughout development.

I think it important that we explore these relationships and do not relegate the developing taste system to only categorizing stimuli as "sweet," "sour," "bitter," and "salty" as the animal is exposed to increasingly more foodstuffs. We must start to analyze gustation as a broadly diverse system that interacts extensively with other developing physiological and cognitive systems. For example, omnivores are exposed to increasingly more foodstuffs at weaning onset. These foods must pass the portal of the oropharynx to cause diverse gastric, hormonal, caloric, and systemic changes. One may hope that future reviews of gustatory ontogeny will be able to draw on research evaluating how these factors and their underlying mechanisms affect the development of gustatory morphology and function.

The field has traditionally ignored more cognitive factors in gustatory development. With the exception of Hogan's famous studies in the establishment of food versus nonfood categories in chickens (see Hogan, 1986, for review), I know of no other efforts that have addressed this fundamental issue of food versus nonfood categorization. This also essentially holds for complex discriminations between edible and inedible foods and for the transmission of food traditions across generations. Work has started in these areas among cognitive and developmental psychologists, but with rare exception have any of these issues received the attention of the gustatory community that is uniquely qualified to identify their underlying mechanisms. As in other domains of psychobiology and behavioral ecology (see chapter by Holmes & Weller in this volume), the mechanisms underlying gustatory development should now be placed in a broader perspective.

Acknowledgments

Preparation of this chapter and unpublished research reported herein were supported by National Institutes of Health Grants R01 DC-00407 and P01 DC-03576.

References

Altman, J., & Bayer, S. (1982). Development of the cranial nerve ganglia and related nuclei in the rat. *Advances in Anatomical Embryology and Cell Biology, 74,* 1–90.

Avenet, P., & Lindemann, B. (1988). Amiloride-blockable sodium current sin isolated taste receptor cells. *Journal of Membrane Biology, 105,* 245–255.

Bao, H., Bradley, R. M., & Mistretta, C. M. (1995). Development of intrinsic electrophysiological properties in neurons from the gustatory region of rat nucleus of solitary tract. *Developmental Brain Research, 86,* 143–154.

Barlow, L. A., Chien, C.-B., & Northcutt, R. G. (1996). Embryonic taste buds develop in the absence of innervation. *Development, 122,* 1103–1111.

Beidler, L. M. (1969). Innervation of rat fungiform papilla. In C. Pfaffmann (Ed.), *Olfaction and taste III* (pp. 352–369). New York: Rockefeller.

Bernstein, I. L., & Courtney, L. (1987). Salt preference in the preweanling rat. *Developmental Psychobiology, 20,* 443–453.

Berridge, K., Grill, H. J., & Norgren, R. (1981). Relation of consummatory responses and preabsorptive insulin release to palatability and learned taste aversions. *Journal of Comparative and Physiological Psychology, 95,* 363–382.

Bird, E., & Contreras, R. J. (1987). Maternal dietary NaCl intake influences weaning rats' salt preferences without affecting taste nerve responsiveness. *Developmental Psychobiology, 20,* 111–130.

Blass, E. M. (1996). Mothers and their infants: Peptide-mediated physiological, behavioral and affective changes during suckling. *Regulatory Peptides, 66,* 109–112.

Blass, E. M. (1999). The ontogeny of human infant face recognition: Orogustatory, visual and social influences. In P. Rochat (Ed.) *Early social cognition: Understanding others in the first months of life* (pp. 35–66). Mahwah, NJ: Erlbaum.

Blazer-Yost, B. L., Cox, M., & Furlanetto, R. (1989). Insulin and IGF I receptor-mediated Na$^+$ transport in toad urinary bladders. *American Journal of Physiology, 357,* C612–C620.

Bradley, R. M. (1973). Investigations of intravascular taste using perfused rat tongue. *American Journal of Physiology, 225,* 300–304.

Bradley, R. M., & Mistretta, C. M. (1973). The gustatory sense in foetal sheep during the last third of gestation. *Journal of Physiology (London), 231,* 271–282.

Bradley, R. M., & Mistretta, C. M. (1980). Developmental changes in neurophysiological taste response from the medulla in sheep. *Brain Research, 191,* 21–34.

Capretta, P. J., & Rawls, L. H. (1974). Establishment of a flavor preference in rats: Importance of nursing and weaning experience. *Journal of Comparative and Physiological Psychology, 86,* 670–673.

Cohick, W. S., & Clemmons, D. R. (1993). The insulin-like growth factors. *Annual Review of Physiology, 55,* 131–153.

Coleman, J. R. (1990). Development of sensory systems in mammals, New York: Wiley.

Davies, A., & Lumsden, A. (1984). Relation of target encounter and neuronal death to nerve growth factor responsiveness in the developing mouse trigeminal ganglion. *Journal of Comparative Neurology, 223*, 124–137.

DeSimone, J. A., & Ferrell, F. (1985). Analysis of amiloride inhibition of chorda tympani taste response of rat to NaCl. *American Journal of Physiology, 249*, R52–R61.

Farbman, A. I. (1965). Electron microscope study of the developing taste bud in the rat fungiform papilla. *Developmental Biology, 11*, 110–135.

Ferrell, M. F., Mistretta, C. M., & Bradley, R. M. (1981). Development of chorda tympani taste responses in rat. *Journal of Comparative Neurology, 198*, 37–44.

Formaker, B. K., & Hill, D. L. (1988). An analysis of residual NaCl taste response after amiloride. *American Journal of Physiology, 255*, R1002–R1007.

Formaker, B. K., & Hill, D. L. (1990). Alterations of salt taste perception in the developing rat. *Behavioral Neuroscience, 4*, 356–364.

Fritzsch, B., Sarai, P. A., Barbacid, M., & Silos-Santiago, I. (1997). Mice with a targeted disruption of the neurotrophin receptor *trk*B lose their gustatory ganglion cells early but do develop taste buds. *International Journal of Developmental Neuroscience, 15*, 563–576.

Garcia, J., & Koelling, R. A. (1966). Relation of cue to consequence in avoidance learning. *Psychonomic Science, 4*, 123–124.

Garty, H., & Benos, D. J. (1988). Characteristics and regulatory mechanisms of the amiloride-blockable Na⁺ channel. *Physiological Review, 68*, 309–337.

Garty, H., & Palmer, L. G. (1997). Epithelial sodium channels: Function, structure, and regulation. *Physiological Reviews, 77*, 359–396.

Hall, W. G. (1979). The ontogeny of feeding in rats: I. Ingestive and behavioral responses to oral infusions. *Journal of Comparative and Physiological Psychology, 93*, 977–1000.

Hendricks, S. J., Stewart, R. E., Heck, G. L., DeSimone, J. A., & Hill, D. L. (2000). Development of rat chorda tympani sodium responses: Evidence for age-dependent changes in global amiloride-sensitive Na⁺ channel kinetics. *Journal of Neurophysiology, 84*, 1531–1544.

Herbert, H., Moga, M. M., & Saper, C. B. (1990). Connections of the parabrachial nucleus with the nucleus of the solitary tract and the medullary reticular formation in the rat. *Journal of Comparative Neurology, 293*, 540–580.

Hill, D. L. (1987a). Development of taste responses in the rat parabrachial nucleus. *Journal of Neurophysiology, 2*, 481–495.

Hill, D. L. (1987b) Susceptibility of the developing rat gustatory system to the physiological effects of dietary sodium deprivation. *Journal of Physiology (London), 393*, 423–434.

Hill, D. L., & Almli, C. R. (1980). Ontogeny of chorda tympani nerve responses to gustatory stimuli in the rat. *Brain Research, 20*, 310–313.

Hill, D. L., & Bour, T. C. (1985). Addition of functional amiloride-sensitive components to the receptor membrane: A possible mechanism for altered taste responses during development. *Developmental Brain Research, 20*, 310–313.

Hill, D. L., & Mistretta, C. M. (1990). Developmental neurobiology of salt taste sensation. *Trends in Neuroscience, 13*, 188–195.

Hill, D. L., & Przekop, P. R. (1988). Influences of dietary sodium on functional taste receptor development: A sensitive period. *Science, 241*, 1826–1828.

Hill, D. L., Mistretta, C. M., & Bradley, R. M. (1982). Developmental changes in taste response characteristics of rat single chorda tympani fibers. *Journal of Neuroscience, 2*, 782–790.

Hill, D. L., Bradley, R. M., & Mistretta, C. M. (1983). Development of taste responses in rat nucleus of solitary tract. *Journal of Neurophysiology, 50*, 879–895.

Hill, D. L., Mistretta, C. M., & Bradley, R. M. (1986). Effects of dietary NaCl deprivation during early development on behavioral and neurophysiological taste responses. *Behavioral Neuroscience, 100*, 390–398.

Hill, D. L., Formaker, B. K., & White, K. S. (1990). Perceptual characteristics of the amiloride-suppressed NaCl taste response in the rat. *Behavioral Neuroscience, 104*, 734–741.

Hogan, J. A. (1986). Cause and function in the development of behavior systems. In E. Blass (Ed.), *Handbook of behavioral neurobiology*, Volume 8, *Developmental psychobiology and developmental neurobiology* (pp. 63–106). New York: Plenum Press.

Jelinek, J. (1961). The development of the regulation of water metabolism. V. Changes in the content of water, potassium, sodium, and chloride in the body and body fluids of rats during development. *Physiologica Bohemoslovia, 10*, 249–258.

Jones, K. R., Farinas, I., Backus, C., & Reichardt, L. F. (1994). Targeted disruption of the BDNF gene perturbs brain and sensory neuron development but not motor neuron development. *Cell, 76*, 989–999.

Kehoe, P., & Blass, E. M. (1985). Gustatory determinants of suckling in albino rats 5–20 days of age. *Developmental Psychobiology, 18*, 67–82.

King, C. T., & Hill, D. L. (1991). Dietary sodium chloride deprivation throughout development selectively influences the terminal field organization of gustatory afferent fibers projecting to the rat nucleus of the solitary tract. *Journal of Comparative Neurology, 303*, 159–169.

King, C. T., & Hill, D. L. (1993). Neuroanatomical alterations in the rat nucleus of the solitary tract following early maternal NaCl deprivation and subsequent NaCl repletion. *Journal of Comparative Neurology, 333*, 531–542.

Kirksey, A., Pike, R. L., & Callahan, J. A. (1962). Some effects of high and low sodium intakes during pregnancy in a rat. *Journal of Nutrition, 77*, 43–51.

Kossel, A. H., McPheeters, M., Lin, W., & Kinnamon, S. C. (1997). Development of membrane properties in taste cells of fungiform papillae: Functional evidence for early presence of amiloride-sensitive sodium channels. *Journal of Neuroscience, 17*, 9634–9641.

Krimm, R. F., & Hill, D. L. (1997). Early prenatal critical period for chorda tympani nerve terminal field development. *Journal of Comparative Neurology, 378*, 254–264.

Krimm, R. F., & Hill, D. L. (1998). Innervation of single fungiform taste buds during development in rat. *Journal of Comparative Neurology, 398*, 13–24.

Krimm, R. F., & Hill, D. L. (1999). Early dietary sodium restriction disrupts the peripheral anatomical development of the gustatory system. *Journal of Neurobiology, 39*, 218–226.

Krimm, R. R., & Hill, D. L. (2000). Neuron/taret matching between chorda tympani neurons and taste buds during postnatal rat development. *Journal of Neurobiology, 48*, 98–106.

Lasiter, P. S. (1992). Postnatal development of gustatory recipient zones within the nucleus of the solitary tract. *Brain Research Bulletin, 28*, 667–677.

Lasiter, P. S. (1995). Effects of orochemical stimulation on postnatal development of gustatory recipient zones within the nucleus of the solitary tract. *Brain Research Bulletin, 38*, 1–9.

Lasiter, P. S., & Bulcourf, B. B. (1995). Alterations in geniculate ganglion proteins following fungiform receptor damage. *Developmental Brain Research, 89*, 289–306.

Lasiter, P. S., & Diaz, J. (1992). Artificial rearing alters development of the nucleus of the solitary tract. *Brain Research Bulletin, 29*, 407–410.

Lasiter, P. S., & Kachele, D. L. (1988). Postnatal development of the parabrachial gustatory zone in rat: Dendritic morphology and mitochondrial enzyme activity. *Brain Research Bulletin, 21*, 79–94.

Lasiter, P. S., & Kachele, D. L. (1989). Postnatal development of protein P-38 ('synaptophysin') immunoreactivity in pontine and medullary gustatory zones of rat. *Developmental Brain Research, 48*, 27–33.

Lasiter, P. S., & Kachele, D. L. (1990). Effects of early postnatal receptor damage on development of gustatory recipient zones within the nucleus of the solitary tract. *Brain Research, 55*, 57–71.

Lasiter, P. S., & Kachele, D. L. (1991). Effects of early postnatal receptor damage on dendritic development in gustatory recipient zones of the rostral nucleus of the solitary tract. *Developmental Brain Research, 61*, 197–206.

Lasiter, P. S., Wong, D. M., & Kachele, D. L. (1989). Postnatal development of the rostral solitary nucleus in rat: Dendritic morphology and mitochondrial enzyme activity. *Brain Research Bulletin, 22*, 313–321.

Liu, X., Ernfors, H., Wu, H., & Jaenisch, R. (1995). Sensory but not motor deficits in mice lacking NT4 and BDNF. *Nature, 375*, 238–241.

Martin, L. T., & Alberts, J. R. (1979). Taste aversions to mother's milk: The age-related role of nursing in acquisition and expression of a learned association. *Journal of Comparative and Physiological Psychology, 93*, 430–455.

Mbiene, J.-P., & Mistretta, C. M. (1997). Initial innervation of embryonic rat tongue and developing taste papillae: Nerves follow distinctive and spatially restricted pathways. *Acta Anatomica, 160*, 139–158.

Mbiene, J.-P., MacCallum, D. K., & Mistretta, C. M. (1997). Organ cultures of embryonic rat tongue support tongue and gustatory papilla morphogenesis *in vitro* without intact sensory ganglia. *Journal of Comparative Neurology, 377*, 324–340.

Midkiff, E. E., & Bernstein, I. L. (1983). The influence of age and experience on salt preference of the rat. *Developmental Psychobiology, 16*, 385–394.

Mistretta, C. M. (1972). Topographical and histological study of the developing rat tongue, palate and taste buds. In J. F. Bosma (Ed.), *Third symposium on oral sensation and perception* (pp. 163–187). Springfield, IL: Thomas.

Mistretta, C. M. (1989). Anatomy and neurophysiology of the taste system in aged animals. In C. Murphy, W. S. Cain, & D. M. Hegsted (Eds.), *Nutrition and the chemical senses in aging: Recent advances and current research needs* (pp. 277–290). New York: New York Academy of Sciences.

Mistretta, C. M. (1991). Developmental neurobiology of the taste system. In T. V. Getchell, R. L. Doty, L. M. Bartoshuk, & J. B. Snow (Eds.), *Smell and taste in health and disease* (pp. 35–64). New York: Raven Press.

Mistretta, C. M. (1998). The role of innervation in induction and differentiation of taste organs: Introduction and background. In C. Murphy (Ed.), *Olfaction and taste XII: An international symposium* (pp. 1–13). New York: New York Academy of Sciences.

Mistretta, C. M., & Bradley, R. M. (1978a). Taste responses in sheep medulla: Changes during development. *Science, 202*, 535–537.

Mistretta, C. M., & Bradley, R. M. (1987b). Effects of early sensory experience on brain and behavioral development. In G. Gottlieb (Ed.), *Studies on the development of behavior and the nervous system* (Vol. 4, pp. 215–247). New York: Academic Press.

Mistretta, C. M., & Bradley, R. M. (1983). Neural basis of developing salt taste sensation: Response changes in fetal, postnatal, and adult sheep. *Journal of Comparative Neurology, 215*, 199–210.

Mistretta, C. M., & Bradley, R. M. (1986). Development of the sense of taste. In E. Blass (Ed.), *Handbook of behavioral neurobiology*, Volume 8, *Developmental psychobiology and developmental neurobiology* (pp. 205–236). New York: Plenum Press.

Mistretta, C. M., Gurkan, S., & Bradley, R. M. (1988). Morphology of chorda tympani fiber receptive fields and proposed neural rearrangements during development. *Journal of Neuroscience, 8*, 73–78.

Mistretta, C. M., & Hill, D. L. (1995). Development of the taste system: Basic neurobiology. In R. L. Doty (Ed.), *Handbook of clinical olfaction and gustation* (pp. 635–668). New York: Marcel Dekker.

Moe, K. (1986). The ontogeny of salt preference in rats. *Developmental Psychobiology, 19*, 185–196.

Nagai, T., Mistretta, C. M., & Bradley, R. M. (1988). Developmental decrease in size of peripheral receptive fields of single chorda tympani nerve fibers and relation to increasing NaCl taste sensitivity. *Journal of Neuroscience, 8*, 64–72.

Nagato, T., Matsumoto, K., Tanioka, H., Kodama, J., & Toh, H. (1995). Effect of denervation on morphogenesis of the rat fungiform papilla. *Acta Anatomica, 153*, 301–309.

Nosrat, C. A., & Olson, L. (1995). Brain-derived neurotrophic factor mRNA is expressed in the developing taste bud-bearing tongue papillae of rat. *Journal of Comparative Neurology, 360*, 698–704.

Ogawa, H., Hayama, T., & Ito, S. (1982). Convergence of input from tongue and palate to the parabrachial nucleus neurons of rats. *Neuroscience Letters, 28*, 9–14.

Philips, L. M., Stewart, R. E., & Hill, D. L. (1995). Cross fostering between normal and sodium-restricted rats: Effects on peripheral gustatory function. *American Journal of Physiology, 269*, R603–R607.

Przekop, P., Mook, D. G., & Hill, D. L. (1990). Functional recovery of the gustatory system after sodium deprivation during development: How much sodium and where. *American Journal of Physiology, 259*, R786–R791.

Ren, K., Blass, E. M., Zhou, Q., & Dubner, R. (1997). Suckling and sucrose ingestion suppress persistent hyperalgesia and spinal Fos expression after forepaw inflammation in infant rats. *Proceedings of the National Academy of the USA, 94*, 1471–1475.

Rochlin, M. W., O'Conner, R., Giger, R. J., Verhaagen, J., Tessier-Lavigne, M., & Farbman, A. I. (1999). Neuropilin-1 dependent repulsion guides geniculate axons destined for lingual taste buds. *Chemical Senses, 24*, 552.

Rozin, P., Gruss, L., & Berk, G. (1979). Reversal of innate aversions: Attempts to induce a preference for chili peppers in rats. *Journal of Comparative and Physiological Psychology, 93*, 1001–1014.

Scott, L., & Atkinson, M. E. (1998). Target pioneering and early morphology of the murine chorda tympani. *Journal of Anatomy, 192*, 91–98.

Settles, A. M., & Mierson, S. (1993). Ion transport in rat tongue epithelium *in vitro*: A developmental study. *Pharmacology, Biochemistry, and Behavior, 46*, 83–88.

Sollars, S. E., & Bernstein, I. L. (1994). Amiloride sensitivity in the neonatal rat. *Behavioral Neuroscience, 108*, 981–987.

Sollars, S. I., & Bernstein, I. L. (1996). Neonatal chorda tympani transection alters adult preference for ammonium chloride in the rat. *Behavioral Neuroscience, 110*, 551–558.

Sollars, S. I., & Hill, D. L. (1998). Taste responses in the greater superficial petrosal nerve: Substantial sodium salt and amiloride sensitivities demonstrated in two rat strains. *Behavioral Neuroscience, 112*, 991–1000.

Sollars, S. I., & Hill, D. L. (2000). Lack of functional and morphological susceptibility of the greater superficial petrosal nerve to developmental dietary sodium restriction. *Chemical Senses, 25*, 719–727.

Sollars, S. I., Shuler, M. G., & Hill, D. L. (1996). Disappearance of fungiform papillae and taste pores in rats with unilateral chorda tympani section at 10 days postnatal. *Chemical Senses, 21*, 673.

Spitzer, A. (1980). *The kidney during development: Morphology and function.* New York: Masson.

Stewart, R. E., & Hill, D. L. (1993). The developing gustatory system: Functional, morphological and behavioral perspectives. In S. Simon & S. Roper (Eds.), *Mechanisms of taste perception* (pp. 127–158). Boca Raton, FL: CRC Press.

Stewart, R. E., & Hill, D. L. (1996). Time course of saline-induced recovery of the gustatory system in sodium-restricted rats. *American Journal of Physiology, 270,* R704–R712.

Stewart, R. E., Lasiter, P. S., Benos, D. J., & Hill, D. L. (1995). Immunohistochemical correlates of peripheral gustatory sensitivity to sodium and amiloride. *Acta Anatomica, 153,* 310–319.

Stewart, R. E., Tong, H., McCarty, R., & Hill, D. L. (1993). Altered gustatory development in Na$^+$-restricted rats is not explained by low Na$^+$ in mothers' milk. *Physiology and Behavior, 53,* 823–826.

Stewart, R. E., DeSimone, J. A., & Hill, D. L. (1997). New perspectives in gustatory physiology: Transduction, development, and plasticity. *American Journal of Physiology, 272,* C1–C26.

Thaw, A. K., Walker, B. R., & Hill, D. L. (1996). NTS neurons of developmentally sodium restricted rats show abnormal dendritic lengths before adulthood. *Chemical Senses, 21,* 679.

Vogt, M. D., & Hill, D. L. (1993). Enduring alterations in neurophysiological taste responses after early dietary sodium deprivation. *Journal of Neurophysiology, 69,* 832–841.

Walker, B. R., & Hill, D. L. (1995). Emergence of afferent fiber terminal field alterations in sodium restricted rats. *Society for Neuroscience Abstracts, 21,* 1657.

Walker, B. R., & Hill, D. L. (1998). Developmental sodium restriction and gustatory afferent terminal field organization in the parabrachial nucleus. *Physiology and Behavior, 64,* 173–178.

Warren, R. P., & Pfaffmann, C. (1959). Early experience and taste aversion. *Journal of Comparative and Physiological Psychology, 52,* 263–266.

Wurtman, J. J., & Wurtman, R. J. (1979). Sucrose consumption early in life fails to modify the appetite of adult rats for sweet foods. *Science, 205,* 321–322.

Yamada, T. (1980). Chorda tympani responses to gustatory stimuli in developing rats. *Japanese Journal of Physiology, 30,* 631–643.

Ye, Q., Heck, G. L., & DeSimone, J. A. (1993). Voltage dependence of the rat chorda tympani response to Na$^+$ salts: Implications for the functional organization of taste receptor cells. *Journal of Neurophysiology, 70,* 167–178.

Ye, Q., Stewart, R. E., Heck, G. L., Hill, D. L., & DeSimone, J. A. (1993). Na$^+$-restricted rats lack functional Na$^+$ channels in taste cell apical membranes: Proof by membrane voltage perturbation. *Journal of Neurophysiology, 70,* 1713–1716.

14

Infant Stress, Neuroplasticity, and Behavior

PRISCILLA KEHOE AND WILLIAM SHOEMAKER

INTRODUCTION

The fulfillment of an individual's genome of the central nervous system ontogenetically involves a series of processes collectively termed *developmental plasticity* (Perry & Pollard, 1998). These processes govern cellular migration, synapse formation, and other aspects of the orderly development of the nervous system (Z. Hall, 1992). The term *plasticity* is used because of the presence and absence of transmitters, growth factors, and hormones that influence the appearance of cells and synaptic connections defining the species. Perturbations of these processes result in abnormal development (Perry & Pollard, 1998). Developing systems are not "fixed and immutable" but are susceptible to disturbances that reflect severity and point in developmental time (Perry, 1997). *Experiential plasticity* defines the neural changes that may occur following exogenous stimulation or social restriction. For fetuses and newborns *experiential plasticity* is superimposed upon the genetically synchronized developmental plasticity and both processes proceed simultaneously during early development. *Experiential plasticity* is understandable within the context of early stages of neural development; rate and asymptote of brain growth and maturity are very much dependent on environmental conditions and how those conditions impact the infant.

Newborns depend primarily on their mothers to provide the appropriate environment and substrates (e.g., food, water, warmth) for proper growth and sensory–motor development (Hofer 1983; Hofer & Shair, 1982). Although there is a wide

PRISCILLA KEHOE   Department of Psychology, Trinity College, Hartford, Connecticut 06106.   WILLIAM SHOEMAKER   Department of Psychiatry, University of Connecticut Health Center, Farmington, Connecticut 06030.

Developmental Psychobiology, Volume 13 of *Handbook of Behavioral Neurobiology*, edited by Elliott Blass, Kluwer Academic / Plenum Publishers, New York, 2001.

range of sensory stimulation provided by the mother and which is idiosyncratic among species, it is generally agreed that stimulation outside that range is harmful. Because of the range of individual differences in both behavior and underlying structure and because changes at one level are not necessarily manifest at a different level, research efforts have often taken extreme forms of experimental manipulation. The present chapter identifies less severe environmental challenges that affect both behavior and structure, thereby producing a model through which the consequences of early stimulation or restriction can be empirically determined. Many of the studies to be discussed involve maternal neglect (human), isolation from the dam and siblings (rodents), or being raised either entirely by an inanimate surrogate or only with peers (monkeys). These "stress" paradigms all produce levels of stimulation above or below the boundaries that animals normally encounter and result in behavioral and physiological abnormalities. It is perhaps advantageous to consider the level of stimulation as the critical variable rather than level of stress (or the notion of "good stress" or "bad stress," or in Selye's terms, eustress and distress). The term "stress" carries many connotations, whereas "stimulation" is always present and seems more objective. What constitutes good, "safe nurturing" is that the stimulation the neonate receives is predictable, gradual, and attuned to the organism's developmental stage (Perry & Pollard, 1998).

The idea of stress may be advantageously integrated within the general concept of homeostasis, maintaining the constancy of the internal milieu, first identified by the French physiologist Claude Bernard (1813–1878) and later modified by the American physiologist Walter Cannon (1871–1945), who coined the term "homeostasis" to describe the process and endpoints in constancy (Cannon, 1932). The American psychobiologist Curt Richter devoted his long and distinguished career to behavioral homeostasis and its underlying mechanisms (Richter, 1976). More current thinking hypothesizes that noxious, threatening, or unpredictable elements in the environment or the absence of a necessary component of the normal environment trigger a complex of behavioral, physiological, and biochemical responses to either neutralize or escape from noxious stimulation. Unlike the characteristics of internal stability, which can easily be measured and manipulated, emotional stability is relatively inaccessible. We must therefore be satisfied with homeostasis as a metaphor for a system that subjectively seeks equilibrium. Failure to remove stressful events or the removal process itself can profoundly affect emotional, motivational, or cognitive processes and their underlying brain mechanisms (Bremner et al., 1993; Maier & Seligman, 1976). It is the purpose of this chapter to discuss our work using an apparently benign perturbation that does not compromise basal growth and behavior. At the same time, however, this regimen leaves the rat more vulnerable to pharmacologic and experiential stressors presented during neonatal, juvenile, and adult periods and results in changes in brain mechanisms that affect emotional and motivational processes.

We see this as one answer to the challenge of identifying the normal experiential range for infants of a species and of defining stress at the boundary of this range. This is complicated by the fact that even very altricial newborns are endowed with behavioral and physiological means of adjusting to short-term environmental disturbances such as modest perturbations in ambient temperature and isolation from dam, nest, and siblings as might occur during nest transport. Isolated rats emit ultrasonic vocalizations that both attract the mother and improve their own thermal status (Blumberg & Alberts, 1990, 1991; Hofer & Shair, 1980; Kehoe & Blass, 1986b, c; Smotherman, Bell, Starzec, Elias, & Zachman, 1974).

Thus, mothers provide the structured nest environment through which neonates may adjust to a certain range of perturbations. Infant capacities increase during development as their motor, sensory, and cognitive ranges expand. In addition, there are specialized capacities, *ontogenetic adaptations* (Alberts & Cramer, 1988; Cramer & Blass, 1983; Oppenheim, 1981), such as suckling, that are manifest during development and then disappear. The principle behind ontogenetic adaptation states that early stages of development evolve "specific morphological, biochemical, physiological and behavioral mechanisms which are different from the adult and which may require modification, suppression or even destruction before the adult stage can be obtained" (Oppenheim, 1981). For example, as rat pups become less reliant on the mother, they produce fewer isolation-induced vocalizations that are in a narrower ultrasonic frequency range, perhaps attracting less attention from predators (Brudzynski, Kehoe, & Callahan, 1999). With development, both the boundaries that distinguish "normal" from "stressful" stimulation and the classes of response used to meet external challenges change in the increasingly mobile, better insulated, and more independent infants/weanlings.

The time span during which environmental stimulation impacts ontogeny to produce permanent functional or structural change has been referred to as a "sensitive" period. Sensitive periods can be short, e.g., only a few days of blindfolding during week 4 or 5 of life in a kitten can result in a substantial loss of connections in visual cortical cells (Hubel & Wiesel, 1962, 1970; Wiesel & Hubel, 1965), or a longer time frame encompassing weeks or months of postnatal development. In the domain of stress research, cocaine exposure has dramatically different outcomes depending on life stage. Cocaine acts as an antagonist at the presynaptic membrane transporter for those neurons that utilize the monoamine transmitters, norepinephrine, dopamine, and serotonin (Cooper, Bloom, & Roth, 1996). In the presence of cocaine, transmitter levels increase in the synaptic cleft, resulting in specific physiological changes, including stereotyped motor activity and enhanced reward-related behaviors. This, in turn, causes additional neurotrophic effects in brain regions that receive strong monoamine innervation, namely cerebral cortex and limbic structures. Levitt and his colleagues described abnormalities in dendritic development (Jones, Fischer, & Levitt, 1996), enhanced γ-aminobutyric acid (GABA)-immunoreactive neurons in the cingulate gyrus (Wang, Levitt, Grayson, & Murphy, 1996), and reduction in dopamine receptor-mediated activation of striatal G-proteins following prenatal cocaine that persisted up to 100 days (Wang, Yeung, & Friedman, 1995). Because monoamine transmission has a neurotrophic effect on target sites during development (Lauder, 1993; Levitt, Harvey, Freidman, Simansky, & Murphy, 1997), the altered levels of transmitter or functionally altered receptors during development result in changes in brain structure and function that do not occur when these amines are increased in adults. Interestingly, the most pronounced anatomic changes were in the anterior cingulate and prefrontal cortices, both limbic targets of dense dopaminergic input (Levitt, Reinoso, & Jones, 1998).

Regulation of neurotransmitter receptors also differs between developing and mature individuals. In the classic example, in adults, receptor dynamics after a decrease in acetylcholine at the peripheral neuromuscular junction results in muscle hypersensitivity to acetylcholine, i.e., denervation supersensitivity (Cooper *et al.*, 1996). Also in adults, the number of brain dopamine receptors increases following chronic treatment with dopamine antagonists (Seeman & VanTol, 1994). In contrast, during central nervous system (CNS) development, both transmitter and receptor levels increase at the same time (Schlumf, Lichtensteiger, Shoemaker, &

553

INFANT STRESS, NEUROPLASTICITY, AND BEHAVIOR

Bloom, 1980), indicating that the mechanisms that govern receptor up- and down-regulation during development are suspended, uncoupled, or not yet in place. Importantly, receptor responses to alterations in transmitter tonus during these early phases would not be predictable from knowledge of the adult mechanism for that transmitter system. Prenatal drug exposure in rats to the potent dopamine receptor antagonist haloperidol eliminated dopamine receptor regulation in response to neurotransmitter level changes later in life, although some aspects of adult dopamine receptor functions were intact (Rosengarten & Friedhoff, 1979). In this regard, peripheral adrenergic neuron depletion of transmitter content by neonatal 6-hydroxydopamine treatment resulted in decreased density and responsivity of peripheral beta-adrenergic receptors. The same depletion in adults actually increases receptor numbers. Thus, although there is plasticity in both neonatal and adult stages, reactions can occur in opposite directions to the same drug treatment, emphasizing CNS vulnerability during its assembly (Burns, 1980).

A frequently expressed notion concerning stress during development is that the young of a species are resilient and recover from stresses and trauma better than do adults (Perry & Pollard, 1998). In fact, the opposite is true in certain domains. A stressful event such as isolation from the mother and siblings can easily overwhelm neonates, causing an immediate reaction particularly in the hypothalamic–pituitary–adrenal (HPA) axis (McCormick, Kehoe, & Kovacs, 1998). Neural signals are transduced into endocrine responses at the level of the mediobasal hypothalamus, stimulating the release of corticotropin-releasing factor (CRF), then pituitary adrenocorticotropic hormone (ACTH), and in turn, an increase in adrenal glucocorticoids (Gibbs, 1986; Plotsky, 1987; Rivier & Plotsky, 1986). Glucocorticoids assist in increased energy availability and immune suppression, necessary responses to "acute" stress. It is then advantageous for the organism to return to the prestress basal functioning of the HPA axis. Meaney *et al.* (1993) studied HPA responsiveness to prolonged postnatal stress (3–6 hr of maternal separation) to show that early experiences have long-term effects on hypothalamic CRF neurons influencing the glucocorticoid negative-feedback system. Additionally, the hippocampus, a late-maturing part of the brain and one that contains many glucocorticoid receptors, is sensitive to, and affected by, early environmental stress. In the animal model, at least, hypothalamic and hippocampal changes persisted throughout life, affecting the organism's later response to novelty or immobilization stress with increased plasma ACTH and corticosterone levels (Plotsky & Meaney, 1993) as well as increased glucocorticoid receptor binding (Meaney *et al.*, 1996). These findings are consistent with the idea that neonatal stress experiences result in permanent alterations of the HPA axis and are evident when the animal is challenged.

A neurobiological model for learning and memory through which the effects of early stress can be evaluated is Hebb's theory that prolonged stimulation produces structural changes in neural circuits (Hebb, 1949). The cellular mechanisms of this model, especially in the hippocampus, have been a major undertaking of neurobiologists (Squire & Zola-Morgan, 1991). Hippocampal cells demonstrate long-term potentiation when stimulated with high-frequency pulses, which not only has long lasting effects (hours to weeks), but occurs quickly and is reinforced by repetition (Kim & Yoon, 1998).

It is of interest in this context that the hippocampus has numerous glucocorticoid receptors involved in the negative feedback of stress hormones (Kim & Yoon, 1998). Hippocampal stress involvement can be demonstrated by stress-induced changes in the structure and function of various neuronal components (Kim &

Stress-induced plasticity may alter the basal functioning of the CNS and/or the response to later environmental challenges depending on the nature and intensity of the original traumatic event and the age at which it occurs (Perry & Pollard, 1998). In the developing organism, adaptation to a variety of ecological niches reflects a naturally occurring plasticity, whereas traumatic events superimposed on these changes may then disrupt ongoing function, particularly in areas of the brain recruited in response to threat such as the limbic–HPA axis (Meaney *et al.*, 1996; Perry & Pollard, 1998; Plotsky & Meaney, 1993).

Depending on the type and level of perinatal stress, detectable changes in physical appearance, growth rate, resting-state physiology, or spontaneous behaviors will provide direction for structural causality. This has been seen in embryological manipulations (Galler, Shumsky, & Morgane, 1996). Prenatally restricting necessary cellular substrates, protein, for example, results in smaller body and brain weights, but essentially normal adult brain structure and basic physiological and behavioral responses within normal limits (Gressens *et al.*, 1997; Morgane *et al.*, 1993; Morgane, Bronzino, Austin-LaFrance, & Galler, 1996; Tonkiss & Galler, 1990). However, these prenatally protein-malnourished animals have some hippocampal abnormalities and abnormal responses to stress. Specifically, prenatal protein malnutrition caused a reduction in dendritic spines and dendritic branching in the granule cells of the dentate gyrus (Diaz-Cintra *et al.*, 1991). Almeida, Tonkiss, and Galler (1996) suggest that higher levels of exploration in aversive conditions, such as the elevated plus maze, can be interpreted as either a greater impulsiveness or lower anxiety in rats that were gestationally malnourished.

Earlier Studies of Neonatal Stress

Within the neurobiological frame of reference, we now establish the historical trail that has led to current studies of neonatal stress. In a landmark series of studies on neonatal stress in rhesus monkeys, Harry Harlow and his students reported that although infant rhesus monkeys raised by surrogate mothers performed normally on cognitive tests (Gluck, Harlow, & Schiltz, 1973; Harlow, Schiltz, Harlow, & Mohr, 1971) such as two-choice discrimination tests and delayed response tests, they were extremely dysfunctional socially (Harlow, 1958, 1959; Harlow & Harlow, 1962; Harlow & Suomi, 1971; Novak & Harlow, 1974). This seminal work of Harlow spawned a variety of designs related to more subtle forms of infant isolation and rearing conditions (Suomi, 1991).

In one design, newborn rhesus monkeys raised with peers for the first 6 postnatal months and compared with mother-reared monkeys showed relatively normal social behavioral repertoires in familiar and stable social settings (Higley, Suomi, & Linnoila, 1991; Higley *et al.*, 1992). However, they were hyperreactive both physiologically and behaviorally to later challenges such as brief social separation. The response of free-ranging rhesus monkeys to natural forms of social separation, such as occurs with the first postpartum estrus, is marked by disturbances in mother–infant relationships (Berman, Rasmussen, & Suomi, 1994). Once the mothers began mating again, the infants demonstrated stress-related behaviors and changes in play activity, that is, males played more and females played less.

Other primates isolated from their mothers early in life, such as squirrel monkeys and chimpanzees, do not exhibit the extreme social abnormalities found in the isolated rhesus monkeys (Rogers & Davenport, 1970; Rosenblum 1968). Rhesus monkeys may be exceptional; they are high-strung, socially vigilant primates that routinely engage in agonistic interactions. Some species tend to spend less time in contact with the mother and may perhaps better tolerate separation episodes. In general, though, infants of all species experiencing isolation have shown at least some abnormal social behavior, albeit sometimes subtle. Rosenblum has developed a laboratory model with bonnet macaques that keeps intact the mother–infant dyad but places different foraging demands on the mother that force her to leave the infant for variable lengths of time (Rosenblum & Paully, 1984; Smith, Coplan, Trost, Scharf, & Rosenblum, 1997). Infants of mothers that were placed in a variable foraging situation that was unpredictable from week to week were found constantly clinging to the mother, showed reduced play and exploration, seemed emotionally disturbed, and showed signs of depression. Thus, when mother monkeys, especially rhesus, are unavailable to their infants for unpredictable periods of time, the mother–infant relationship is less secure, aspects of normal development are interrupted, and deviant behaviors are more likely to appear.

Whether using a peer-reared design or a variable foraging demand design in primate research, the outcomes demonstrate the enduring behavioral alterations in adults exposed to stressful early rearing. As will be discussed later, we have incorporated a similar subtle separation design in our rodent studies, that is, the separation from the mother is only for 1 hr/day for 8 days. This design permits normal nutrition and growth rates in the isolated offspring. In fact, when looking at normative baseline behaviors and physiology, these isolated rodents appear normal. It is primarily when they are challenged that differences appear, the same phenomena displayed by the rhesus monkeys stressed as infants. Suomi (1991, p. 111) summarizes this best:

> More recent studies have examined the long term consequences of more subtle variation in early rearing environments. Monkeys reared from birth without mothers but with extensive peer contact develop relatively normal social behavioral repertoires and function well in the familiar and stable social settings. However, peer-reared monkeys display extreme behavioral and physiological reactions to environmental challenges such as brief social separation later in life. In contrast, monkeys reared by unusually nurturant foster mothers appear to develop effective strategies for coping with subsequent environmental challenges.

Variations in early rearing environments have profound and enduring effects. Peer-reared rhesus monkeys or those with unpredictable maternal presence display abnormalities when challenged later in life. As juveniles and adults, following stressful situations, they show increased levels of stress hormones as well as increased serotonin and norepinephrine metabolism which correlated with established anxiety and aggression differences between peer-only and mother-reared monkeys (Higley, Suomi, & Linnoila, 1991).

In the 1960s, a number of researchers described the effects of extreme handling and maternal separation in rats during the early postnatal period on consequences during adulthood (Ader, 1969; Hess, Denenberg, Zarrow, & Pfeifer, 1969; Levine, 1962). The reasoning was that corticosterone may help organize emotional behavior during development, similar to the effect of sex hormones on sexual behavior (Denenberg & Zarrow, 1970). These studies, which incorporated a variety of manipulations, some severe, suggested that experiences of the organism during the neonatal period modified later endocrine and behavioral response to stress. It was found that "handling" attenuated later HPA responsiveness to placement in a novel environment and response to the stress of electric shock (Levine, Chevalier, & Korchin, 1956). In addition, the neonatally handled rats displayed an increase in open field activity (Denenberg & Smith, 1963), suggestive of lower anxiety and/or higher impulsiveness.

In recent studies of infant stress in rodents, variants of maternal separation are utilized. The paradigms utilized to study maternal separation vary dramatically from 15 min to 6 hr daily some time during the first 3 weeks postnatally with predictably variable outcomes (F. S. Hall, Wilkinson, Humby, & Robbins, 1999; Mathews, Robbins, Everitt, & Caine, 1999; Meaney *et al.*, 1996; Zimmerberg & Shartrand, 1992). Because some 6-hr maternal separation methods resulted in smaller body and brain weights (Zimmerberg & Shartrand, 1992), we have employed a short-term (1 hr/day) neonatal isolation that does not cause changes in growth rate or body weight nor influence brain and behavior measures in rats tested in basal conditions. When these rats are challenged, however, marked differences appear between neonatally stressed and control rats. While the specific underlying mechanisms are unknown, brief periods of maternal separation and isolation produce profound changes in response to later challenges such as drug administration or stress. The trauma-induced neuroplasticity superimposed on "normal" ontogenetic plasticity is enduring and potentially pathological.

More recent series of handling/stress studies by Meaney *et al.* (1996) evaluated maternal separation and handling, assessing HPA responses to stress as adults. In these studies, litters of rat pups are removed from the nest and placed together in small containers for various time periods (Francis *et al.*, 1996; Heim, Owens, Plotsky, & Nemeroff, 1997; Meaney *et al.*, 1996; Plotsky & Meaney, 1993). In general, these studies (see Meaney *et al.*, 1996, for a review) show that as adults (a) male rat pups removed from the dam daily and huddled for 15 min (defined as "handling") for the first 21 days show attenuated fearfulness in novel environments and decreases in HPA activation when challenged with a wide variety of stressors, (b) male rat pups removed from the dam and huddled for 3 or 6 hr (defined as "stressed") for the first 2 weeks show an increase in HPA activation to novelty or restraint stress and decreased glucocorticoid receptor binding in both hippocampus and hypothalamus, whereas nonhandled controls had hormonal levels intermediate between short- and long-separated pups, and (3) "handling" alone reduced the often neurotoxic effects of glucocorticoids associated with stress, whereas the longer maternal separation ("stress") produced animals vulnerable to the exaggerated negative effects of later stress. The authors state that "in animals carrying such a vulnerability, its consequences would emerge only under conditions of stress" (Meaney *et al.*, 1996).

Although we share a similar interest with Meaney and his colleagues in the consequences of neonatal stress and the role of handling, one major focus of our work has been on *neurochemical consequences* at various life stages as well as during the time of the stressful experience. Moreover, we were interested in examining the effects of early stress without the possible complications of nutritional deficits that occur with prolonged maternal separation. To accomplish this, we found that a shorter period of maternal separation prevented any possible weight difference. However, it was then necessary to isolate the animal individually to obtain long-term changes instead of in a huddle, where there are a number of familiar stimuli (tactile, olfactory, thermal, and auditory).

In our initial experiments, we used a within/heterogenous-litter design, that is, one in which half the litter is handled and isolated and the other half only handled, which means being sexed, weighed, and marked, then returned to the dam. The within-litter design seeks to control for the historical maternal influence of that litter and uniqueness of that family, and, when used in these experiments, maximizes the differences between the treatment groups rather than litter effects. However, there are both advantages and disadvantages to using a within-litter design. On the positive side, it controls for a particular dam's caretaking and fewer litters are needed. The disadvantage is the possible differential mothering that occurs when the isolated pups are removed and then returned to the nest. Thus, an important methodological consideration is the quality of the care each pup receives in the nest during the reunion following the separation. As revealed by the study examining maternal licking and later HPA responsiveness, maternal care given during the reunion of the separated pup with the mother is a vital factor in terms of effects of stress that the infant has just experienced (Liu *et al.*, 1997).

It has been found that rat dams exhibit differential caretaking to male and female pups with greater retrieval and licking given to the male pups in the litter (Moore & Chadwick-Dias, 1986; Moore & Morelli, 1979). When the nest is disturbed or the pups removed from the nest, the dam often exhibits stereotypic behaviors, such as burrowing, grooming, and circling, which can be taken as signs of stress (Wilkins, Logan, & Kehoe, 1998). It is possible that the stressful state of the dam could lead to unusual or overreactive maternal behavior accompanied by differential gender alteration. The interaction between handling and sex that we and others have observed (Weiner, Feldon, & Ziv-Harris, 1987) demonstrates prominent differences between handled and nonhandled animals. In fact, in our studies using a within-litter design, handled males and females showed some differential effects to psychostimulant and restraint challenges. The handled males resembled their isolated male siblings, whereas the handled females resembled nonhandled females from separate control litters (Kehoe, Shoemaker, Triano, Callahan, & Rappolt, 1998).

In later studies we used a between/homogeneous-litter design, which consists in treating an entire litter as either handled, isolated, or nonhandled (Shoemaker *et al.*, 1998). We found that such litters may receive less-variable caretaking and thus this procedure provides less variability in the individual pup measures. What is important here is to underscore the influence of the maternal–infant interaction on the aspects of the pup's brain and behavioral functioning that influence the animal's response to challenges later in life.

Our initial hypothesis was that the rapidly developing neonatal nervous system would be particularly susceptible to shifts in neurochemical balance produced by stress. Since many neurotransmitters and peptides play important neurotrophic roles in brain development, the question is whether prolonged or repeated activation of particular systems would upset the delicate sequence that results in normal brain development (Emerit, Riad, & Hamon, 1992; Lauder, 1993). A group of neurotransmitters that is known to play a role in brain development and differentiation are the monoamines and, in particular, dopamine (Feigenbaum & Yanai, 1984) and serotonin (Lauder, 1993; Whitaker-Azmitia & Azmitia, 1986). The ontogeny of dopamine innervation in rat (Bruinink, Lichtensteiger, & Schlumpf, 1983) and human (Olsen, Boreus, & Seiger, 1973) and the appearance of dopamine receptors in the rat (Murrin & Zeng, 1989; Neal & Joyce, 1992) and human (Unis, Roberson, Robinette, Ha, & Dorsa, 1998) have been studied extensively. In fact, similarities in dopamine ontogeny between rat and human are strong enough for the rat to be used as a model of dopamine-based early-onset psychiatric disorders (Unis, 1995).

Several neuropeptides have also been shown to affect development in the nervous system (see review by Lauder, 1993) but the endogenous opioids have received the greatest attention. Since the recognition that endorphin and enkephalins develop very early in fetal life (Bayon, Shoemaker, Bloom, Mauss, & Guillemin, 1979), studies have shown the growth effects that ensue when these substances are blocked from interacting with their normal targets (Zagon & McLaughlin, 1985, 1986).

Endogeneous opioids modulate vocalizations of infants separated from their mothers in several vertebrate species including rodents (Carden & Hofer, 1990, 1991; Kehoe, 1988), birds (Panksepp, Meeker, & Bean, 1980), dogs (Panksepp, Herman, Conner, Bishop, & Scott, 1978), humans (Blass & Ciaramitaro, 1994), and non-human primates (Harris & Newman, 1988; Kalin, Shelton, & Barksdale, 1988). The evidence for endogenous opioid involvement in distressed responses is chiefly indirect, inferred from studies with opiate agonists (morphine) (Kehoe & Blass, 1986a) and antagonists (naloxone and naltrexone) (Carden & Hofer, 1990, 1991; Kehoe & Blass, 1986a–c). Moreover, prenatal opioid receptor blockade with 50 mg/kg naltrexone daily through gestation had neurobehavioral consequences as well as growth effects (McLaughlin, Tobias, Max Lang, & Zagon, 1997). Therefore, when we began our studies of brain systems involved in stress during the neonatal period, we focused on the dopamine and endogenous opioid systems.

EFFECTS OF ACUTE ISOLATION ON POSTNATAL DAY 10

PRECURSOR STUDIES

Kehoe and Blass (1986b, c, 1989) demonstrated a graded relationship between the severity of stress and ultrasonic vocalization (USV) and analgesia in 10-day-old rats, with the former decreasing and the latter increasing with stress. Kehoe and Blass suggested that isolation stress triggered a release of endogenous opioids that induced analgesia as demonstrated by increased paw lift latency to heat and reduced vocalization during isolation, presumably through similar mechanisms, as both were naloxone-reversible (Kehoe & Blass, 1986a, b, d).

BRAIN OPIOID PEPTIDE LEVELS IN ISOLATED 10-DAY-OLD RAT PUPS

Since the release of the opioids in these earlier studies was inferred pharmacologically and not measured directly (Bayon *et al.*, 1979), several brain regions were analyzed for β-endorphin and enkephalins via radioimmunassay in 10-day-old rats separated from mother and siblings or taken directly from the nest (Shoemaker & Kehoe, 1995). Because there is no uptake mechanism for peptides released at the synapse and because released peptides are rapidly degraded enzymatically, decreases in peptide levels over time can be interpreted as increased release from terminals. Significant decreases of β-endorphin were seen in the midbrain (containing the periaquaductal gray) after isolation, but only if the isolated rat was housed in familiar bedding (Figure 1).

Enkephalin levels in midbrain were decreased with any form of isolation, but enkephalin in the brain stem revealed the same specificity (bedding present during isolation) as β-endorphin. No change was observed in either peptide in the hypothalamus, septum, or amygdala after isolation compared with controls, indicating regional specificity. Thus, opioid peptide responsiveness is sensitive to small changes in environmental conditions and when isolation conditions are changed so that "familiar objects" are no longer in the environment, opioid systems become less, rather than more, involved. It is of interest that the familiar condition most closely

PRISCILLA KEHOE
AND WILLIAM
SHOEMAKER

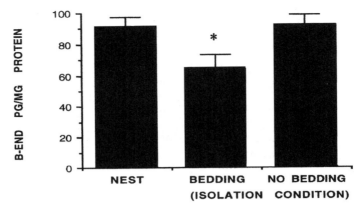

Figure 1. β-Endorphin levels in midbrain regions from 10-day-old rat pups either taken directly from the nest or after 5 min of isolation under two different conditions. The bedding condition was familiar, but the no-bedding condition (in a bare cup) was unfamiliar or novel. There were 10–20 rats in each treatment group. Ambient temperature was 30°C. *Significance = $p < .02$. (Reprinted from Shoemaker & Kehoe, 1995, with permission. Copyright 1995 by the American Psychological Association.)

resembles the isolation condition in the Kehoe and Blass studies in which naltrexone markedly increased vocalizations. The condition in which naltrexone only marginally affected USV was isolation without any familiar stimulus (Carden & Hofer, 1990).

Shoemaker and Kehoe (1995) examined vocalization of isolated 10-day-old rat pups in familiar (bedding) and unfamiliar environments. Naltrexone (1.0 mg/kg) or saline was given prior to isolation. Naltrexone increased vocalizations in pups tested in familiar bedding (Figure 2). The level of vocalizing in the bedding plus naltrexone condition is almost equivalent to the no-bedding saline condition, suggesting that the familiar bedding suppresses calling through opioid pathways that would

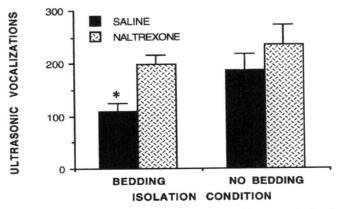

Figure 2. Total number of ultrasonic vocalizations made by 10-day-old rat pups during 5 min of isolation from the nest. Pups received either saline or naltrexone IP and then were placed in either a cup with bedding (familiar condition) or no bedding (novel condition). There were 22 pups in each treatment condition. *Significance = $p < .05$. (Reprinted from Shoemaker & Kehoe, 1995, with permission. Copyright 1995 by the American Psychological Association.)

otherwise occur if there were no familiar objects in the environment. This suppression is unmasked by the opiate antagonist and raises the level of vocalizations to the no-bedding condition. Naltrexone modestly increased vocalization in the no-bedding group, but the variability in this group of rats was almost twice that of the familiar group, resulting in failure to reach statistical significance.

Carden and Hofer (1991) also reported increased variability in vocalizations of pups isolated in unfamiliar conditions, a reversal of the suppression of vocalization by familiar cues (e.g., an anesthetized dam), and a return to original isolation levels by naltrexone. D'Amato *et al.* (1992) showed that mouse pups isolated in unfamiliar bedding produced more USV and showed higher plasma corticosterone levels than littermates isolated in nest bedding, indicating that isolation in an unfamiliar environment is more stressful. Thus, these results reinforce the notion that the endogenous opioid system, along with its well-known role in pain modulation, is responsive to subtle environmental cues designating familiarity. The lower endogenous opioid secretion found in the more stressful condition seems paradoxical, especially in light of opioid release to combat pain. The demands of the system differ, however; familiar odors suppress vocalizations which under natural conditions could attract predators to the nest, the source of the familiar odor. For pain, endogenous opioids facilitate both coping and recovery.

Brain Dopamine Response in Isolated 10-Day-Old Rat Pups

In addition to the endogenous opioids, the dopamine neurotransmitter system develops early and is activated by isolation in neonates (Cabib, Puglisi-Allegra, & D'Amato, 1993; Tamborski, Lucot, & Hennessy, 1990). We have measured dopamine release in the striatum and septum after a single 5-min isolation at 10 days of age. Dopamine release was measured by *in vivo* receptor binding using $_3$H-raclopride, *ex vivo* binding, also using $_3$H-raclopride, and dopamine turnover measures (Kehoe, Clash, Skipsey, & Shoemaker, 1996). Homogenate binding of dopamine receptors assured that changes in brain, using *in vivo* binding, or changes in tissue slices, using *ex vivo* binding, were due to competition of the labeled ligand with released dopamine and not to a change in the absolute number (B_{max}) or affinity (K_d) of dopamine receptors caused by the isolation. All three measures demonstrated that dopamine was released in the septum and striatum during the 5-min isolation period and that varying the environment of the isolation to more unfamiliar conditions produced greater dopamine release (Table 1).

Both dopamine and endogenous opioids participate in the response to stress in the neonate, but the particulars of each system's involvement differ. The maximum response in endogenous opioid release occurs in the periaquaductal grey and in the brain stem in pups that are isolated in a familiar environment, whereas the maximum response of the mesolimbic dopamine system occurs in its targets when pups are isolated in a novel environment. Thus, we have clearly identified two major neurochemical response systems and some of their distribution properties activated during infant stress of isolation: opioid and dopamine systems.

Dopamine contributions may be better understood from the perspective of future coping to stress in rats and humans as seen in behavioral sensitization. Basically, behavior sensitization was originally described as the enhanced behavioral response to psychomotor stimulants (amphetamine, cocaine, etc.) following earlier repeated exposure to these drugs (Kuczenski & Segal, 1988). In both animal models and humans who use these drugs, repeated exposure results in greater responsive-

TABLE 1. CHANGE FROM THE NEST CONDITION DUE TO ISOLATION
TREATMENTS FOR EACH OF THREE METHODS OF ESTIMATING
DOPAMINE RELEASE[a]

Method	Site of dopamine release	Percentage change for given environment	
		Familiar	Novel
In vivo binding	Striatum	12	34
	Septum	9	38
Ex vivo binding	Striatum	28	47
	Septum	23	31
Dopamine turnover	Striatum	6	38
	Septum	20	41

[a]Variations in percentage change may be due to the individual techniques utilized.

ness (i.e., sensitization) even when a long time has elapsed from the initial exposure to current exposure. In humans, the sensitization takes the form of heightened pleasure, giving the drug increased salience in the individual's value system (Robinson, 1988). In animals, the sensitization takes the form of increased activity levels and enhanced mesolimbic dopamine release following the recent drug exposure compared to animals receiving the drug for the first time (Kalivas & Alesdatter, 1993; Kalivas & Duffy, 1989; Kalivas, Duffy, Abhold, & Dilts, 1988; Kuczenski & Segal, 1987, 1988; Robinson, Jursons, Bennett, & Bentgen, 1988; Segal & Kuczenski, 1992). What is significant to our own work were the observations that (1) the mesolimbic dopamine system was intimately involved (Kalivas *et al.*, 1988; Kuczenski & Segal, 1988) and (2) that stress could substitute for the initial exposure to produce a similar sensitization when exposed to a psychomotor stimulant for the first time (Antelman, 1988).

Sensitization is a form of learning that has many features of long-term-potentiation (LTP) (Sarvey, Burgard, & Decker, 1989). Hypersensitive responses, whether induced by drugs or stress, lasts for months, suggesting lasting changes in circuit strength after exposure. Although the cellular and subcellular mechanisms of sensitization have proven difficult to explore in mammalian brains, the similarity to LTP allows us to draw inferences concerning underlying mechanisms, particularly because excitatory amino acid release from presynaptic terminals have been implicated in the mechanism of LTP (Bliss, Douglas, Errington, & Lynch, 1986; Dolphin, Errington, & Bliss, 1982; Overton, Richards, Berry, & Clark, 1999; Shors & Servatius, 1995; Svendsen, Tjolsen, Rygh, & Hole, 1999). Sensitization by repeated stress or repeated drug administration and LTP has the characteristic that the sensitizing event must be repetitious in order to obtain the change in responsiveness.

There is additional evidence for long-lasting central changes following repeated stress. First, the noncompetitive N-methyl-D-aspartate (NMDA) antagonist MK-801 blocks psychostimulant-induced sensitization (DeMontis *et al.*, 1992; Karler, Calder, Chaudhry, & Turkanis, 1989; Marek, Ben-Eliyahu, Vaccarino, *et al.*, 1991; Marek, Ben-Eliyaha, Gold, *et al.*, 1991) and response to stress (Shors & Mathew, 1998). This strengthens the analogy to LTP since NMDA antagonists also prevent the initiation of LTP (Collingridge, Kehl, & McLennan, 1983; Harris, Ganong, & Cotman, 1984; Harris, Ganong, Monaghan, Watkins, & Cotman, 1986; Wigstrom & Gustafson,

1984). These observations also point to an aspect of excitatory amino acid release as a critical component of both LTP and sensitization.

Second, morphological alterations following sensitization have been demonstrated in the form of increased dendritic spines and branches in the nucleus accumbens and prefrontal cortex following amphetamine and cocaine sensitization (Robinson & Kolb, 1997, 1999). Sunanda Rao and Raju (1995) described increased dendritic width and spines or excrescences of hippocampal pyramidal neurons following stress sensitization in adult rats. These anatomic changes therefore appear to endure for at least several weeks and perhaps much longer.

These observations raised the question of whether the *repeated* isolation of rat pups that results in release of substances such as dopamine and opioid peptides could produce changes in the postsynaptic neuron that initiates or participates in the cascade of events culminating in altered responsiveness, i.e., sensitization. Is peptide and dopamine release increased in mesolimbic structures through daily repeated isolation? Does repeated neurotransmitter release alter their postsynaptic targets? Is there evidence for tolerance, or sensitization? Could repeated isolations be detected in anatomic, physiological, or behavioral changes? Are these changes lasting or short-lived? Framed in neurobiological terms, one may ask, does repeated infant stress alter brain structure or function, and, if so, do these changes endure?

Brain and Behavioral Effects of Repeated Isolation

Short-Term Effects of Repeated Neonatal Stress

Given the known characteristics of HPA responsiveness to stress in neonates and adults, it was important to assess the hormonal response of the pups in our isolation paradigm to see whether the repeated daily isolations produced habituation or sensitization. Using a between-litter design in which entire litters were handled, nonhandled, or isolated, we found that corticosterone release in 9-day-old pups that had been isolated from their mother, siblings, and nest daily for 1 hr on postnatal days (PN) 2–8 was markedly elevated on their PN9 isolation relative to pups that were isolated for the first time at PN9 (McCormick *et al.*, 1998). Both control groups (handled litters and nonhandled litters) showed similar baseline levels and an increase in plasma corticosterone after isolation on PN 9 (Figure 3). Repeated isolation, however, greatly enhanced the isolation-induced corticosterone release. Thus, repeated daily stress of the neonate potentiates the HPA response to a later episode of stress while not changing basal levels. It also provides a mechanism for further neuroplasticity since elevated corticosterone levels can produce neuroplastic changes especially in the hippocampus (Sapolsky, 1992).

Moreover, neonatal isolation probably influenced the central dopaminergic system. Ten-day-old rats repeatedly isolated for 1 hr daily in a between-litter design comparing only isolated and nonhandled groups from PN2 to PN9 were extremely hyperactive compared to the control pups following amphetamine (0.5 mg/kg) on PN10 (Kehoe, Shoemaker, Arons, Triano, & Suresh, 1998) (Figure 4), although the activity of isolates did not differ from nonhandled control levels during 15 min of isolation when saline-treated. This supports the idea that previous stress amplifies responsiveness to future psychostimulant administration.

The neurochemical effects of isolation as assessed by measuring dopamine turnover rates in the dorsal striatum, septum, and hypothalamus after the activity

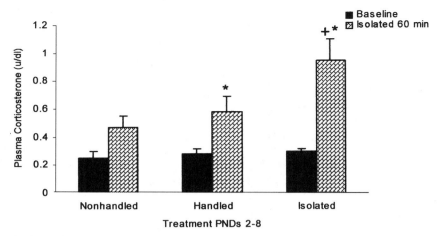

Figure 3. Mean plasma corticosterone levels in the three treatment groups after 60 min of isolation on PN9 (open histograms). Previously isolated pups had significantly higher levels following 60 min of isolation than nonhandled and handled control pups ($n = 10–15$ in each condition). Previous isolation does not cause an increase in baseline corticosterone as shown in filled histograms. *Significantly different from the respective nonhandled baseline ($p < .05$); [+]significantly different from handled and nonhandled treatment groups during isolation on PN9 ($p < 0.05$).

testing yielded unexpected findings. "Isolates" that received injections of saline did not differ in their activity levels relative to control pups (Figure 4). Yet dopamine turnover in the septum and hypothalamus was about twice that of nonhandled controls, whereas striatal dopamine turnover was at normal levels (Figure 5). Thus, it is the mesolimbic dopamine system that appears to be differentially affected by the repeated isolation, not the nigrostriatal system.

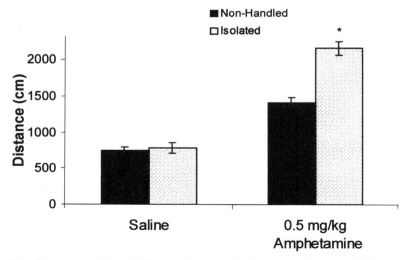

Figure 4. The distance traveled on PN10 over a 15-min period for previously nonhandled or isolated rats following saline or amphetamine 0.5 mg/kg IP ($n = 25$ per treatment). *Significant at $p < .0001$. This represents a statistically reliable interaction of isolation and treatment. (Reprinted from Kehoe, Shoemaker, Arons, *et al.*, with permission. Copyright 1998 by the American Psychological Association.)

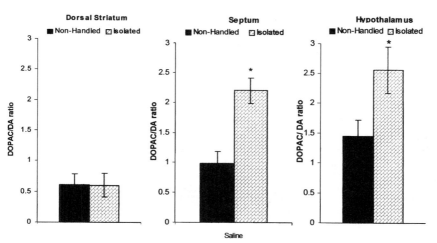

Figure 5. Dihydroxyphenylacetic acid (DOPAC)/dopamine (DA) ratio of the striatum, septum, and hypothalamus of 10-day-old pups representing the two treatment groups (isolated days 2–9 or non-handled) given IP saline and tested in a novel chamber for 15 min. For septum, $n = 19$ for each group; for striatum, $n = 15$; for hypothalamus, nonhandled, $n = 14$; for hypothalamus, isolated, $n = 17$. *Significant at $p < .001$. (Reprinted from Kehoe, Shoemaker, Arons, *et al.*, with permission. Copyright 1998 by the American Psychological Association.)

In vivo microdialysis provides more direct evidence that repeated isolation produces changes in brain dopamine. Using a between-litter design, we found that a single dose of amphetamine (0.5 mg/kg) in isolated pups caused an almost twofold increase in dopamine release in the nucleus accumbens compared to nonhandled controls given amphetamine, whereas nonhandled and isolates given saline did not differ (Figure 6). These findings suggest that isolation has affected mesolimbic dopamine function either through enhanced release, diminished uptake, or a hypertrophy of dopamine terminals. We now examine whether these changes endure.

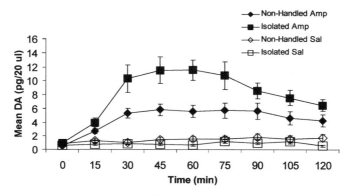

Figure 6. Mean dopamine (DA) levels (pg per 20 μl) obtained by *in vivo* microdialysis every 15 min on day 10 of the two treatment groups (nonhandled or isolated) prior to and after an IP injection of saline (Sal) or amphetamine (Amp) 0.5 mg/kg. For nonhandled amphetamine, $n = 9$; for nonhandled saline, $n = 10$; for isolated amphetamine, $n = 12$; for isolated saline, $n = 11$. Main effect of treatment, $p < .001$. (Reprinted from Kehoe, Shoemaker, Arons, *et al.*, with permission. Copyright 1998 by the American Psychological Association.)

PRISCILLA KEHOE
AND WILLIAM
SHOEMAKER

Long-term isolation-induced changes in stimulant-induced locomotion and mesolimbic dopamine release were determined in weanlings (PN27) that had been isolated from PN2 to PN9 (Kehoe, Shoemaker, Triano, Hoffman, & Arons, 1996). This study used a within-litter design in which half of the animals in each litter were isolated and the others weighed and marked and returned to their mother without isolation, labeled as handled. Entire litters of nonhandled controls were also assessed. Juvenile rats isolated as neonates traveled a greater distance than either handled or nonhandled controls following amphetamine injection. Differential activity was only obtained with a fairly high dose of intraperitoneal (IP) amphetamine, 7.5 mg/kg, as compared to 2.0 mg/kg (Kehoe, Shoemaker, *et al.*, 1996). By comparison, the effective doses for neonates are 0.5 mg/kg (Kehoe, Shoemaker, Arons *et al.*, 1998) and 1.0 mg/kg (Spear, Shalaby & Brick, 1980) and for adults is 2.0 mg/kg (Kalivas, Sorg, & Hooks, 1993; Segal & Kuczenski, 1992). The higher dose was predicted because juvenile rats are extremely hyposensitive to amphetamine (Anderson, Lyss, & Teicher, 1998).

Changes in the mesolimbic dopamine system were confirmed by *in vivo* microdialysis of the nucleus accumbens in a separate group of animals given 7.5 mg/kg IP amphetamine as a challenge. Dopamine release in the previously isolated juveniles following this dose of amphetamine was elevated (Figure 7). In fact, peak dopamine response in isolates was threefold greater than that of control nonhandled and handled rats. Thus, repeated isolation for 1 hr daily during the first week of life changed both brain and behavioral responsiveness to a pharmacologic challenge in weanlings that endured for at least 2 weeks.

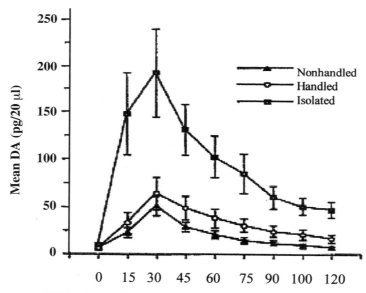

Figure 7. Dopamine (DA) levels in dialysates of ventral striatum of the three treatment groups of juvenile rats. Mean dopamine (pg/20 μl) at baseline and eight post-amphetamine time points. Nonhandled, *n* = 10; nonisolated, *n* = 15; isolated, *n* = 14. Main effect of treatment, *p* < .002. (Reprinted from Kehoe, Shoemaker, Arons, *et al.*, with permission. Copyright 1998 by the American Psychological Association.)

This further suggests that the mesolimbic dopamine system is chiefly affected by isolation. The nucleus accumbens has received a great deal of attention for its role in reward/reinforcement behavior. Repeated psychostimulant exposure in adult animals produced hypersensitivity of dopamine responsiveness to later psychostimulant challenges in the nucleus accumbens, similar to that produced by amphetamine in previously isolated animals. The convergence of findings from repeated isolation stress and repeated psychostimulant exposure reinforces the notion that the mesolimbic dopamine system plays a central role in setting the "tone" for the brain's responsiveness to stimulation.

However, more recent discussions about the role of the mesolimbic dopamine system garnered from neurophysiological data (Redgrave, Prescott, & Gurney, 1999) point out that responses in this system are not restricted to reinforcing stimuli. Redgrave *et al.* (1999) suggest that the role of dopamine is to focus attention of the animal to salient stimuli without regard to their potential reward (food, sex, novelty) or potential for harm (predators, aggressive conspecifics, strange loud noises). Given their newer perspective, it would be important to study the neuroplastic changes that occur in the dopamine system by expanding the testing of responses from psychomotor stimulants to testing in response to stressful stimuli. In fact, several studies have demonstrated cross-sensitization between stress and psychostimulants (Ahmed, Stinus, LeMoal, & Cador, 1995; Cabib *et al.*, 1993; Prasad, Sorg, Ulibarri, & Kalivas, 1995; Sorg & Kalivas, 1991, 1993). A frequently employed stressor that can be used in rats from juvenile to adult stages is restraint. Mills, Bruckert, and Smith (1990) and Kellogg, Awatramani, and Piekut (1998) used restraint stress in juveniles to study a variety of brain and peripheral measures.

In our study of isolates at PN27 using a within-litter design (Kehoe, Triano, Glennon, & Daigle, 1997), locomotor activity was measured upon release from 1 hr of restraint (Figure 8). Compared to nonhandled and handled controls, isolated rats had significantly lower activity scores. The isolated rats demonstrate suppression of activity compared to the control group. Thus, in the presence of amphetamine, isolated animals show exaggerated behavioral activity and exaggerated nucleus

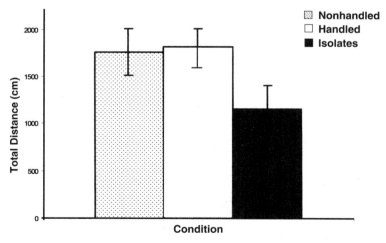

Figure 8. Distance traveled of juvenile rats (PN27–30) for three treatment groups (nonhandled, $n = 9$; handled, $n = 13$; and isolates, $n = 10$) upon release from a 1-h restraint. Significant at $p < .02$.

accumbens dopamine responsiveness; however, during the application of restraint, opposite and divergent results are seen.

In vivo microdialysis of the nucleus accumbens of weanlings from a within-litter design revealed no differences in baseline dopamine levels, but dopamine levels in isolates decreased during restraint and remained low for the ensuing hour (Figure 9). However, nonhandled rats' dopamine levels increased during restraint and then tapered off, whereas those of handled rats remained relatively flat, and thus was intermediate between those of nonhandled rats and isolates. The pattern of dopamine release in nonhandled restrained controls resembles that described for restrained adult mice (Imperato, Angelucci, Casolini, Zacchi, & Puglisi-Allegra, 1992). The basis for the increase in dopamine levels of control rats produced by the restraint is not known and has been the subject of considerable speculation.

The dopamine system activity was suppressed in the isolated rats. They did not show a rise and fall of dopamine levels during the onset and offset of the imposed restraint. Additionally, previously isolated juveniles express an attenuated corticosterone response 20 min after the initiation of the restraint compared to nonhandled controls (McCormick, Rood, & Kehoe, 1998). An explanation of such an inhibition of a neurochemical and glucocorticoid response has remained elusive. To summarize, previously isolated rats show significantly reduced levels of activity following restraint as well as reduced nucleus accumbens dopamine levels. Whereas it appears that there is a correlation between the behavior and nucleus accumbens dopamine, they may not be related as cause and effect since nucleus accumbens dopamine levels are not necessarily correlated with activity levels.

Our studies at the juvenile stage demonstrated that neonatal isolation produced lasting (at least 2 weeks) changes in the mesolimbic dopamine system and behaviors that are differentially responsive to psychostimulants and to stress. However, there would be considerably more confidence in assigning these isolation-induced alterations to a lasting neuroplastic modification if it could be demonstrated that these alterations endure into adulthood.

To summarize, it is clear that the effects of isolation from PN2 to PN9 can be seen not only at PN10, but also at PN27, more than 2 weeks after the last isolation

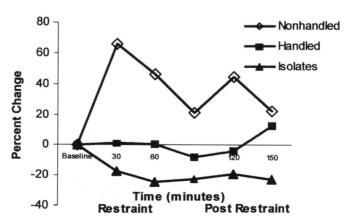

Figure 9. The percentage change from baseline of dopamine levels in juvenile rats from three treatment groups (nonhandled, $n = 14$; handled, $n = 15$; isolates, $n = 13$) following 1 hr of restraint. Main effect of treatment, $p < .001$.

session. It is also clear that not only is the age at testing a factor in determining outcome, but that the nature of the challenge (i.e., whether pharmacological or environmental) also interacts with age to produce differential outcomes. Thus, while challenge with amphetamine at both ages produces increases in locomotor activity and mesolimbic dopamine release in isolates compared to controls, the environmental challenge (placement in a novel environment for PN10 and restraint for 1 hr for PN27) produced complex results. Whether the complexities are a result of the challenge applied (i.e., restraint) or of the age of the animal will be examined in the next section for isolated animals tested as adults.

Studies on Adults Isolated as Neonates

Locomotion and stereotyped behavior in isolated and control adult rats, PN70–90, using a within-litter design were evaluated after pharmacological (1.0 or 2.0 mg/kg amphetamine) or environmental (restraint) challenges (Kehoe, Shoemaker, Triano, *et al.*, 1998). Controls consisted of handled littermates as well as separate litters that had never been handled. Previously isolated animals were hyperactive and showed greater stereotypy following either dose of amphetamine (Kehoe, Shoemaker, Triano, *et al.*, 1998). Figure 10A displays the activity level and Figure 10B the stereotypy scores following the 2.0 mg/kg dose.

As seen in Figures 10A and 10B, in both measures of amphetamine responsiveness, isolates are more active than controls. In addition, Figure 10B (stereotypy) reveals a main effect of time as well as an interaction of the treatment and time, that is, during the second half of the session isolates maintained the highest stereotypy scores. Interestingly, previously handled adults show a reduced level of stereotypy to amphetamine, a finding at present unexplained, but in agreement with those of Meaney *et al.* (1996) for handled adult rats and their subdued response to chal-

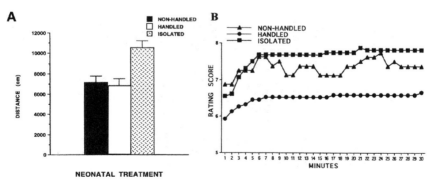

Figure 10. (A) Activity level (distance) of adults after 2.0 mg/kg amphetamine (*n* = 10–12 of each treatment). Main effect of isolation, $p < .0006$. (B) Stereotypy scores for each minute of the 30-min session for the amphetamine-injected (2.0 mg/kg) animals. Stereotypy was scored according to the Ellinwood and Balster (1974) rating scale in which 0–4 is normal activity and 5–8 various types of stereotypy. All animals receiving amphetamine scored above 5 and all animals receiving saline scored below 5. There was a significant effect of treatment ($p < .035$), time ($p < .0001$), and interaction of time and treatment ($p < .007$). In the second and third 10-min epochs, isolates are significantly elevated in their stereotypy scores compared to nonhandled controls, $p < .04$, and handled controls, $p < .03$. (Reprinted from Kehoe, Shoemaker, Arons, *et al.*, with permission. Copyright 1998 by the American Psychological Association.)

lenges. At this dose of amphetamine (2.0 mg/kg), no sex difference was present. However, at the 1.0 mg/kg amphetamine dose, a sex difference was present, and it was clear that the females had higher distance scores than males. The isolated animals had significantly higher stereotypy scores and a more rapid achievement of stereotypy. In particular, female isolates reached the highest level of stereotypy. Interestingly, the previously isolated males did not demonstrate a statistically different amount of activity when measured by distance traveled. However, they clearly showed a greater level of stereotypy than the nonhandled or handled controls, demonstrating sensitization in adults isolated as neonates (Kehoe, Shoemaker, Triano, *et al.*, 1998).

As in the case with juvenile animals, adult locomotor activity was measured after 1 hr of restraint. There was general agreement with the juvenile data (Figure 11). Previously isolated adult rats showed significantly suppressed locomotion following the restraint compared to nonhandled controls. Following 1 hr of restraint stress, neonatally isolated females exhibited a significant reduction in locomotor activity compared with nonhandled and handled female controls. Furthermore, the two female control groups did not differ from each other. In contrast, handled adult males as well as previously isolated males traveled significantly less compared with nonhandled males following restraint.

Change in nucleus accumbens dopamine levels during 1 hr of restraint in adults (Shoemaker *et al.*, 1998) produced sex-specific results (Kehoe *et al.*, 1997). Using a between-litter design, we found that control rats and previously isolated females when placed in restraint show a rise in dopamine levels during the initial 15- or 20-min period (Figure 12). Dopamine levels in previously isolated males, however, decrease upon initiation of restraint. This decreased level remains fairly constant

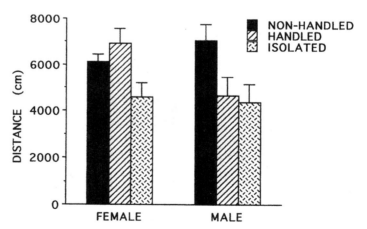

Figure 11. Locomotor activity of both nonhandled and isolated adult rats after release form 1 hr of restraint stress (*n* = 8–10 in each treatment group). There is a significant treatment-by-sex effect, *p* = .05. A one-way analysis of variance (ANOVA) on females revealed a significant treatment effect, *p* = .01. Handled females did not differ from nonhandled female controls, *p* = .33. Both nonhandled and handled females traveled significantly farther than isolated females, *p* = .010 and *p* = .04, respectively. The one-way ANOVA on males indicated a significant treatment effect, *p* = .034. The locomotor response of handled males was similar to that of previously isolated males, and both of these groups were significantly decreased compared with nonhandled males, *p* = .032 and *p* = .05, respectively. (Reprinted from Kehoe, Shoemaker, Arons, *et al.*, with permission. Copyright 1998 by the American Psychological Association.)

throughout the session including following release from restraint. In contrast, non-handled males and all females display dopamine levels above baseline post-restraint. It is interesting that juvenile male and female isolates respond with suppression of dopamine during and after restraint, whereas only postpubertal male isolates and not females continue with this stress-induced dopamine inhibition. It is difficult to understand why this dopamine suppression occurs only in isolated adult males and not isolated adult females, especially when both isolated males and isolated females respond similarly on activity measures following restraint stress (Figure 11). However, notice that handled males in Figure 11 are similar to isolated males; this could indicate that these sex differences are due more to being male than a specific inter-action between isolation and restraint. It does signify, however, that nucleus accumbens dopamine release does not necessarily correlate with locomotor activity.

The increase in nucleus accumbens dopamine during a stress such as restraint is difficult to interpret when one invokes a reinforcement–positive reward hypothesis for dopamine function in this part of the brain. However, a new perspective by Redgrave *et al.* (1999) states that the mesolimbic dopamine system is activated when the animal is in the process of switching attentional and behavioral selections to unexpected, important stimuli. This would explain the response to both rewarding stimuli (e.g., reinforcing psychomotor stimulants) and aversive stimuli (e.g., re-straint). What it does not explain is why isolated juveniles and isolated adult males do not respond with increased dopamine, but demonstrate lower levels upon restraint, suggesting that these animals have nonresponsive attentional mechanisms.

Similar to dopamine, corticosterone is also attenuated during stress in isolated animals (Ogawa *et al.*, 1994). It was found that isolated animals had a smaller increase in corticosterone in response to restraint stress than did nonhandled animals as juveniles (McCormick *et al.*, 1998), a finding also seen in adults, both males and females, in our laboratory (C. M. McCormick & P. Kehoe, unpublished, 1999). This finding of attenuated stress-induced corticosterone response was similar to results of Ogawa *et al.* (1994) in adults that had been maternally deprived as infants for 4.5 hr daily for the first 3 weeks of life.

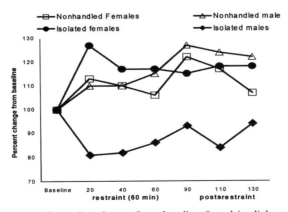

Figure 12. Mean percentage dopamine change from baseline found in dialysates obtained from the nucleus accumbens during the 1-h restraint period and 1-h postrestraint periods of nonhandled adult males and females, and adult males and females that had been isolated as neonates ($n = 8$ or 9 in each treatment group). Main effect of treatment, $p < .05$.

The following points can be made from these observations: (1) Daily, 1-hr isolation from PN2 to PN9 produces changes in adult behavior and neurochemical responses at least in the nucleus accumbens as well as HPA alterations. (2) There are sexually dimorphic physiological responses in adults. This suggests either that these differences are latent, i.e., present from early age but not evident, or that sex of the individual has a strong influence on the expression of these sensitized responses (Camp & Robinson, 1988).

Given that the effects of infant stress endured to adult stages and that there are strong parallels between stress-related behavioral sensitization and long-term potentiation (LTP), it was plausible to suggest hippocampal involvement. We tested the effects of neonatal isolation on hippocampal LTP using a widely practiced paradigm. Rats isolated from their mothers for 1 hr/day during postnatal days 2–9 were surgically prepared at 70–90 days of age with stimulating and recording electrodes placed in the medial perforant pathway of the hippocampus and the hippocampal dentate gyrus, respectively (Kehoe & Bronzino, 1999). No significant effect of sex or neonatal isolation treatment was obtained for baseline electrophysiological measures such as excitatory postsynaptic potential (EPSP) slope, a measure of synaptic activation, or population spike amplitude (PSA), a measure of the population cellular response. In order to rule out baseline differences in hippocampal cell excitability in female adult rats, we measured the response of dentate granule cells for one estrus cycle and showed no pretetanization enhancement in the evoked response in either controls or previously stressed rats. Following 10-Hz, 5-sec stimulation, there was a significant treatment and sex effect (Figure 12). Long-term potentiation gets its name from the lasting effect that the intense stimulation produces. In nonhandled controls, the potentiation remains for about 72 hr (3 days) after the stimulation has occurred. During the induction of LTP, PSA values were significantly enhanced in both isolated males and females and were significantly longer in duration when compared to the unhandled control group. Additionally, we observed that females, particularly female "isolates" (Figure 13A), took longer to reach baseline levels than males (Figure 13B). Taken together, these results indicate that repeated infant isolation stress enhances LTP induction and duration in both males and females.

Infant isolation enhances LTP in adults as well as juveniles (Bronzino, Kehoe, Austin-LaFrance, Rushmore, & Kurdian, 1996; Kehoe & Bronzino, 1999; Kehoe, Hoffman, Austin-LaFrance, & Bronzino, 1995). The functional importance of enhanced and prolonged LTP found in previously stressed juveniles and adults is unknown. Complete profiles of cognitive functioning of previously isolated adults remain to be obtained. The stress-induced neural modifications in LTP obtained from previously stressed adults may be linked with sensitization to psychostimulants that endures through adulthood (Kehoe, Shoemaker, *et al.*, 1996; Kehoe, Shoemaker, Arens, *et al.*, 1998; Kehoe, Shoemaker, Triano, *et al.*, 1998; Kolta, Scalzo, Ali, & Holson, 1990; Robinson & Becker, 1986; Segal & Kuczenski, 1992). Baseline electrophysiological measures of cells in the hippocampus were similar in isolates and nonhandled adults, as was spontaneous locomotion and baseline dopamine levels. Thus far, effects of neonatal isolation have been revealed only through pharmacological challenge, demands of environmental stress, or hippocampal high-frequency stimulation.

The importance of these observations involving hippocampal LTP is that LTP serves as the paragon of neuroplasticity and that the hippocampal formation is intimately involved with learning and memory functions. It is also possible that what is encoded by the changes in the hippocampus are the altered cognitive and perceptual differences producing abnormal responses of limbic circuits when the animal is challenged at a later time. Thus, the early isolation paradigm has a profound effect on the plasticity of the brain region that is a pivotal structure for cognitive and emotional circuitry.

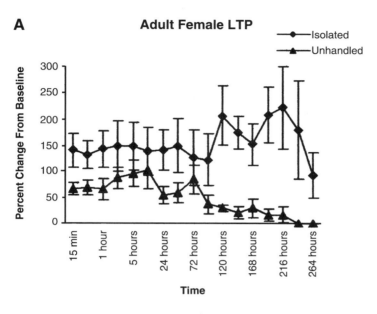

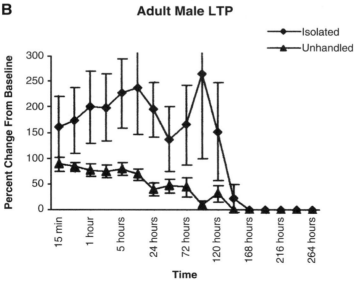

Figure 13. Mean (±SEM) percentage change in population spike amplitude measures from isolated and nonhandled (A) females (16) and (B) males ($n = 15$). An ANOVA test performed on the data indicated no sex differences for the first 24 hr and a significant effect of treatment between the isolates and nonhandled animals at a number of time points including both the 1-hr and the 24-hr posttetanization points (significant at $p < .05$).

We have presented data on a number of our published studies as well as historical and background material that is helpful in placing these studies in context. In all of our studies, animals were isolated individually from PN2 to PN9. We have assessed the effects of this 1-hr isolation experience at three different ages, neonatal, juvenile, and adult stages. We have measured locomotor activity, mesolimbic dopamine release, and, in some studies, plasma corticosterone response. In an attempt to glean overall trends in these various studies, Table 2 summarizes all results.

Because little or no effects of the isolation treatment were seen in baseline behaviors, measurements were gathered after a challenge was given to the animal. Two types of challenges were used: a pharmacological challenge (dose of amphetamine) or an environmental challenge (a stress). The stresses must be age-specific, so the stress for the neonate was placement in a novel environment, whereas for the juvenile and adult, the stress was restraint for 1 hr. The arrows in Table 2 indicate the direction of the response of the isolated animals compared to the appropriate controls.

The response to the psychomotor stimulant is consistent across all ages: increased locomotion and increased mesolimbic dopaminergic tone. (We have not measured corticosterone in animals receiving amphetamine.) The responses to stress are more variable. For the neonates, placement in a novel environment did not produce differences in locomotor behavior, but isolates had increases in dopamine. The restraint stress at the juvenile and adult ages produced decreases in behavioral activity, decreases in mesolimbic dopamine tone, and a decreased corticosterone response (C. M. McCormick & P. Kehoe, unpublished, 1999). Thus, the effect of isolation is not a generalized supersensitivity reaction, but an exaggeration of the appropriate response direction as compared with controls. This supports the hypothesis that a neuroplastic change occurred in response to the repeated isolation and that the changes endured through adult stages, and in general, limbic system activation appeared to be exaggerated in the direction that is normally caused by the stimulation. This position is supported by the studies looking at hippocampal LTP in these animals, where isolates had greatly exaggerated amplitude and duration of responses. This suggests, broadly, that early stress, as used in our laboratories, does not reorganize brain or behavioral systems so much as it exaggerates the dominant process caused by particular challenges.

TABLE 2. THE RESPONSE OF ISOLATED ANIMALS COMPARED TO THAT OF CONTROLS[a]

Age at testing	Type of challenge	Activity response to challenge	Mesolimbic dopamine	Plasma corticosterone response
Neonate	Stress (locomotion in novel environment)	↔	↑	↑
	Amphetamine	↑	↑	ND
Juvenile	Stress (1-hr restraint)	↓	↓	↓
	Amphetamine	↑	↑	ND
Adult	Stress (1-hr restraint)	↓	♂↓ ♀↑	↓[c]
	Amphetamine	↑	↑[b]	ND

[a]ND, Not done.
[b]F. S. Hall *et al.* (1999).
[c]Unpublished.

DISCUSSION

575

INFANT STRESS,
NEURO-
PLASTICITY, AND
BEHAVIOR

We have shown that removing infant rats from the nest, dam, and siblings for 1 hr daily from postnatal days 2–9 causes enduring neurobiological and behavioral changes. The effects are often subtle, however, and must be revealed through environmental or pharmacological challenges. Activity of isolated animals is exaggerated by a psychostimulant but is actually dampened following periods of immobilization, possibly for conditioning reasons. Neurochemical changes are parallel: mesolimbic dopamine release is enhanced following amphetamine administration but is suppressed during and after immobilization stress. Infant isolation caused greater and longer lasting hippocampal potentials in adults to high-frequency stimulation. Thus, repeated infant isolation induces neurobiological and behavioral changes that may be functionally linked through specific changes in limbic circuits that require either substantial activation for their exacerbation, as in pharmacological treatment, or a repeat of the inducing condition, meaning a stressful experience, to show exaggeration compared to control rats.

We do not know what the critical environmental and endogenous events are that cause these changes. Moreover, we do not know the limits of this model for identifying parallels with human infant stress. To begin this possible linkage, we present the general factors discussed in this chapter in Figure 14. Although difficult to quantify, we see the level of stimulation encountered during infancy as a predictor of immediate and long-term changes. We do not specify class of stimulation because we do not know which domains influence future morphology and behavior. Nonetheless, there are intensity-related changes in brain morphology and behavior associated with level of stimulation. Together these three domains and their interactions may provide direction for clinical research and the further use of these animal models as clinical understanding develops.

THEORETICAL CONSIDERATIONS

One theoretical consideration that remains open is whether different levels of stimulation lie on a continuum, anchored by lack of stimulation or neglect representing one end and severe stress or trauma representing the other end, with optimal or beneficial levels of stimulation occupying the middle ground. In this regard the Yerkes–Dodson function seems to be a reasonable first approximation; either extreme stimulation causes potential negative effects (Figure 14). Rather little is known about the effects of social restriction on later behavior.

It is undoubtedly true that these early manipulations are linked to learning and memory systems. These considerations impact our understanding of central neuroplasticity. As mentioned earlier, neuroplastic responses have been increasingly identified in studies related to changes underlying learning and memory. In cerebral cortex, plasticity has been shown through changes in representational maps (Kaas, 1991). Although change can be induced in juvenile or adult cortices, they are most pronounced in neonates and may involve different interneuronal circuitry (Kaczmarek, Kossut, & Skangiel-Kramska, 1997). Dendritic change in the hippocampus after restraint stress in adults (Sunanda Rao & Raju, 1995) and increased dendritic length and dendritic spines in the nucleus accumbens and medial frontal cortex after chronic psychostimulant exposure (Robinson & Kolb, 1997) are examples of *experiential-induced, structural plasticity*. Our studies have shown the enduring nature of rather modest but early interventions. The discovery of plastic changes in struc-

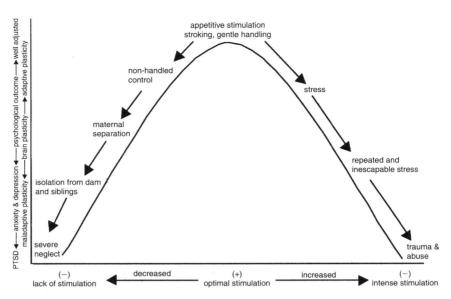

Figure 14. Representation of the animal model and its relation to clinical outcomes.

tures of the limbic system correlates with behavioral changes seen after stresses and psychomotor stimulant exposure (Robinson, 1988; Segal & Kuczenski, 1992; Sorg & Kalivas, 1991, 1993). In fact, such animal studies are used as models for psychopathology including posttraumatic stress disorder (PTSD), anxiety, depression, and attention deficit hyperactivity disorder (ADHD) (Perry & Pollard, 1998). The psychostimulation studies, with their more readily available mechanisms for experimental analyses, provide a potential salient link of early behavioral stress with later pathology.

In this regard, developmental models demonstrating altered responsiveness to stress, abnormalities in the HPA axis, increased activity of the mesolimbic dopamine system, and increased anxiety and fear may map on to clinical entities of children exposed to neglect or abuse. This line of thinking has been advanced therapeutically by clinical and animal studies in which tactile stimulation helps reverse certain conditions stemming from early trauma. Thus, gently stroking isolated infant rat pups with a soft brush, mimicking maternal stimulation, reverses trauma-induced changes in growth hormone and corticosterone production (Evoniuk, Kuhn, & Schanberg, 1979). Additionally, prenatally stressed rat pups treated with postnatal handling or adoption by foster dams providing increased licking and carrying behavior exhibited less anxiety as adults and a reversal of the HPA deficits (Takahashi, 1998). Likewise, premature infants, including those exposed to cocaine, gain weight faster, score higher on mental and motor development, and are discharged earlier from the hospital when given daily programs of tactile stimulation (Cornell & Gottfried, 1976; Field, 1990, 1995; Field, Scafidi, & Schanberg, 1987; Scarr-Salapatek & Williams, 1973). In general, stimulated infants were more active and appeared healthier. Furthermore, tactile stimulation increased levels of serum norepinephrine and epinephrine, but not dopamine, cortisol, or growth hormone (Kuhn *et al.*, 1991), suggesting that the sympathetic nervous system was primarily affected. Thus, stabilizing premature infant–adult interactions, including predictable tactile stimulation, reverses the clinical isolation experienced in neonatal intensive care units. It is of interest that mothers in Third World countries are equally successful in raising their premature infants through continuous ventral–ventral contact.

Clinical Implications

With the advent of the industrial revolution, many children became displaced from their families and were placed in foundling homes. Incredibly, the survival rate for such children was less than 1% (Kesson, 1965). As better nutrition and hygiene were introduced, mortality rate declined, but the phenomena of "cot death" continued and many of these children were found dead in their beds without an explainable cause. In several landmark studies, Spitz (1945, 1946) determined that the sensory and cognitive neglect of the child in the orphanages was a major factor in their physical and psychological pathology (Spitz & Wolf, 1946). Even when conditions improved that allowed increased longevity in such institutionalized children, Ribble (1944) found a high incidence of psychological disturbance. Although current practices in treating infants isolated for medical purposes and in orphanages feature frequent handling and stimulation, individual cases of extreme neglect continue (Chisolm, Carter, Ames, & Morison, 1995). Using modern neurobiological techniques, Perry and Pollard (1997) reported decreased frontal–occipital circumference as well as cortical atrophy in 28 cases of severe global neglect using magnetic resonance imaging and computerized tomography images. The consequences of neglect are extremely acute today, especially in the current political and martial climate in which a high percentage of children in war-torn areas are separated from their parents and housed in orphanages or displaced person camps for months or longer (Wolff & Fesseha, 1998).

Persistent threatening or unpredictable environments for infants or children result in abnormal behavior, e.g., uncontrollable aggression (Perry, 1997). Moreover, exposure to repeated stressors generally elevates HPA axis function. This continuous activation, as seen in rat studies, could cause hippocampal damage and impaired glucose utilization and metabolism. In fact, abused children do show hippocampal and limbic abnormalities (Glod & Teicher, 1996; Ito, Teicher, Glod, & Ackerman, 1998; Teicher *et al.*, 1997). In one study, 115 consecutive admissions to a child and adolescent psychiatric hospital were reviewed for evidence of early abuse, characterized as psychological abuse, neglect, or definite physical or sexual abuse, or no abuse. No differences were noted between abused and nonabused patients in the prevalence of abnormal neurological examinations or abnormal neuropsychologic tests. Abnormal electroencephalogram (EEG) studies were seen in 27% of the psychologically abused patients, but twice that amount in children who had been abused physically or sexually. Nonabused patients showed EEG abnormalities in only 19.2% of cases. It is important to note that the nonabused group in this study are individuals admitted with a mental or psychiatric diagnosis and were not "normal controls."

The major distinction between the abused and nonabused patients in this study was in left-sided frontotemporal abnormalities (Ito *et al.*, 1993). This finding coincides with many other studies that describe patients with a variety of mental disorders having abnormalities in the prefrontal cortex, a part of the limbic system. While these studies must be replicated and expanded, they indicate that young people have suffered both behavioral and brain abnormalities as a result of stress early in life. These abnormal EEG patterns, however, cannot pinpoint a locus of pathology or even determine how the EEG abnormality is related to the mental condition. This is an area where our animal models can be most useful.

FUTURE CONSIDERATIONS

The animal model and supporting data described here provide certain parallels with the human condition. Neonatal isolation stress clearly affects developmental plasticity in multiple areas of the brain. These changes in neuroplasticity in the animal model may provide insights into alterations occurring in the human CNS. For example, since traumatic experiences or neglect affect the developing brain, particularly during sensitive periods, human infants are likely to be more vulnerable to decreased levels of stimulation during periods of rapid synaptogenesis, which, in turn, could lead to specific neuropathologies such as PTSD, ADHD, behavioral impulsivity, anxiety, and depressive disorders as well as vulnerability to drug abuse (Perry & Pollard, 1998). Thus, identifying the levels of stimulation and the critical time period that produce changes in developmental plasticity represents an important research area for developmental neuroscientists. Likewise, identifying classes of developmental stress in various species and the later range of experience that precipitates aberrant reactions remains a major goal of animal models.

Finally, it is perhaps not premature to propose that animal experimenters begin to search for mechanisms that underlie these neuroplastic changes. Even though much work remains to explore the various parameters of isolation and traumatic stresses, much could be accomplished using the models already at hand to begin the painstaking process of mechanistic understanding. This approach will be the start of possible intervention into the process and guidance toward improved directions for rehabilitation.

REFERENCES

Ader, R. (1969). Early experiences accelerate maturation of the 24-hour adrenocortical rhythm. *Science, 163*, 233–238.

Ahmed, S. H., Stinus, L., Le Moal, M., & Cador, M. (1995). Social deprivation enhances the vulnerability of male Wistar rats to stressor- and amphetamine-induced behavioral sensitization. *Psychopharmacology, 117*, 116–124.

Alberts, J. R., & Cramer, C. P. (1988). Ecology and experience. Sources of means and meaning of developmental change. In E. M. Blass (Ed.), *Handbook of behavioral neurobiology*, Volume 9, *Developmental psychobiology and behavioral ecology* (pp. 1–39). New York: Plenum Press.

Almeida, S. S., Tonkiss, J., & Galler, J. R. (1996). Prenatal protein malnutrition affects avoidance but not escape behavior in the elevated T-maze test. *Physiology and Behavior, 60*, 191–195.

Anderson, S. L., Lyss, P. J., & Teicher, M. H. (1998). Maturational changes in the dopamine system do not explain amphetamine subsensitivity during adolescence. Abstract. Presented at the International Society for Developmental Psychobiology, Orleans, France.

Antelman, S. M. (1988). Stressor-induced sensitization to subsequent stress: Implications for the development and treatment of clinical disorders. In P. W. Kalivas & C. D. Barnes (Eds.), *Sensitization in the nervous system* (p. 227–256). Caldwell, NJ: Telford Press.

Bayon, A., Shoemaker, W. J., Bloom, F. E., Mauss, A., & Guillemin, R. (1979). Perinatal development of the endorphin and enkephalin-containing systems in the rat brain. *Brain Research, 179*, 93–101.

Berman, C. M., Rasmussen, K. L. R., & Suomi, S. J. (1994). Responses of free-ranging rhesus monkeys to a natural form of social separation. I. Parallels with mother–infant separation in captivity. *Child Development, 65*, 1028–1041.

Blass, E. M., & Ciarmitaro, V. (1994). Oral determinants of state, affect, and action in newborn infants. *Monographs of the Society for Research in Child Development, 59*, 1–96.

Bliss, T. V. P., Douglas, R. M., Errington, M. L., & Lynch, M. A. (1986). Correlation between long-term potentiation and release of endogenous amino acids from dentate gyrus of anaesthetized rats. *Journal of Physiology, 377,* 391–408.

Blumberg, M. S., & Alberts, J. R. (1990). Ultrasonic vocalizations by rat pups in the cold: An acoustic by-product of laryngeal barking? *Behavioral Neuroscience, 104,* 808–817.

Blumberg, M. S., & Alberts, J. R. (1991) Both hypoxia and milk derivation diminish metabolic heat production and ultrasound emission by rat pups during cold exposure. *Behavioral Neuroscience, 105,* 1030–1037.

Bremner, J. D., Scott, T. M., Delaney, R. C., Southwick. S. M., Mason, J. W., Johnson, D. R., Innis, R. B., McCarthy, G., & Charney, D. S. (1993). Deficits in short-term memory in posttraumatic stress disorder. *American Journal of Psychiatry, 150,* 1015–1019.

Bronzino, J. D., Kehoe, P., Austin-La France, R. J., Rushmore, R. J., & Kurdian, J. (1996). Neonatal isolation alters LTP in freely moving rats: Sex differences. *Brain Research Bulletin, 41,* 175–183.

Brudzynski, S. M., Kehoe, P., & Callahan, M. (1999). Sonographic structure of isolation-induced ultrasonic calls of rat pups. *Developmental Psychobiology, 34,* 195–204.

Bruinink, A., Lichtensteiger, W., & Schlumpf, M. (1983). Pre- and postnatal ontogeny and characterization of dopaminergic D_2, serotonergic S_2 and spirodecanone binding sites in rat forebrain. *Journal of Neurochemistry, 40,* 1227–1236.

Burns, E. M. (1980). Depressed endogenous norepinephrine during beta-adrenergic receptor ontogeny. In H. Parvez and S. Parvez (Eds.), *Biogenic amines in development* (pp. 663–682). New York: Elsevier/North-Holland Biomedical Press.

Cabib, S., Puglisi-Allegra, S., & D'Amato, F. (1993). Effects of postnatal stress on dopamine mesolimbic system responses to aversive experiences in adult life. *Brain Research, 604,* 232–239.

Camp, D. M., & Robinson, T. E. (1988). Susceptibility to sensitization. II. The influence of gonadal hormones on enduring changes in brain monoamines and behavior produced by repeated administration of D-amphetamine or restraint stress. *Behavioral Brain Research, 30,* 69–88.

Cannon, W. B. (1932). *The wisdom of the body.* New York: Norton.

Carden, S. E., & Hofer, M. A. (1990). Independence of benzodiazepine and opiate action in the suppression of isolation distress in rat pups. *Behavioral Neuroscience, 104,* 160–166.

Carden, S. E., & Hofer, M. A. (1991). Isolation-induced vocalizations in Wistar rat pups is not increased by naltrexone. *Physiology and Behavior, 49,* 1279–1282.

Chisholm, K., Carter, M. C., Ames, E. W., & Morison, S. J. (1995). Attachment security and indiscriminately friendly behavior in children adopted from Romanian orphanages. *Development and Psychopathology, 7,* 283–294.

Collingridge, G. L., Kehl, S. J., & McLennan, H. (1983). Excitatory amino acids in synaptic transmission in the Schaffer collateral-commissural pathway of the rat hippocampus. *Journal of Physiology, 334,* 33–46.

Cooper, J. R., Bloom, F. E., & Roth, R. H. (1996). *The biochemical basis of neuropharmacology.* New York: Oxford University Press.

Cornell, E. H., & Gottfried, A. W. (1976). Intervention with premature human infants. *Child Development, 47,* 32–39.

Cramer, C. P., & Blass, E. M. (1983). Mechanisms of control of milk intake in suckling rats. *American Journal Physiology, 245,* R154–R159.

D'Amato, F. R., Cabib, S., Puglisi-Allegra, S., Pataechioli, F. R., Cigliana, G., Maccari, S., & Angelucci, L. (1992). Effects of acute and repeated exposure to stress on the hypothlamo-pituitary–adrenocortical activity in mice during postnatal development. *Hormones and Behavior, 26,* 474–485.

DeMontis, M. G., DeVoto, P., Meloni, D., Gambarana, C., Giorgi, G., & Tagliamonte, A. (1992). NMDA receptor inhibition prevents tolerance to cocaine. *Pharmacology, Biochemistry and Behavior, 42,* 179–182.

Deneberg, V. H., & Smith, S. A. (1963). Effects of infantile stimulation and age upon behavior. *Journal of Comparative and Physiological Psychology, 66,* 533–535.

Denenberg, V. H., & Zarrow, M. X. (1970). Infantile stimulation, adult behaviour and adrenocortical activity. In S. Kazda and V. H. Denenberg (Eds.), *The postnatal development of phenotype* (pp. 123–132). Prague: Academia.

Diaz-Cintra, S., Cintra, L., Galvan, A., Aguilar, A., Kemper, T., & Morgane, P. J. (1991). Effects of prenatal protein deprivation on postnatal development of granule cells in the fascia dentata. *Journal of Comparative Neurology, 310,* 356–364.

Dolphin, A. D., Errington, M. L., & Bliss, T. V. P. (1982). Long-term potentiation of the perforant path *in vivo* is associated with increased glutamate release. *Nature, 297,* 496–498.

Ellinwood, E. H., & Balster, R. J. (1974). Rating the behavioral effects of amphetamine. *European Journal of Pharmacology, 28,* 35–41.

Emerit, M. B., Riad, M., & Hamon, M. (1992). Tropic effects of neurotransmitters during brain maturation. *Biology of the Neonate, 62,* 193–201.

Evoniuk, G. E., Kuhn, C. M., & Schanberg, S. M. (1979). The effect of tactile stimulation on serum growth hormone and tissue ornithine decarboxylase activity during maternal deprivation in rat pups. *Communications in Psychopharmacology, 3*, 363–370.

Feigenbaum, J. J., & Yanai, J. (1984). Normal and abnormal determinants of dopamine receptor ontogeny in the central nervous system. *Progress in Neurobiology, 23*, 191–225.

Field, T. (1990). Alleviating stress in newborn infants in the intensive care unit. In B. M. Lester & E. Z. Tronick (Eds.), *Stimulation and the preterm infant: The limits of plasticity* (pp. 1–9). Philadelphia, PA: Saunders.

Field, T. (1995). Cocaine exposure and intervention in early development. In M. Lewis & M. Bendersky (Eds.), *Mothers, babies, and cocaine* (pp. 355–368). Hillsdale, NJ: Erlbaum.

Field, T., Scafidi, F., & Schanberg, S. (1987). Massage of preterm newborns to improve growth and development. *Pediatric Nursing, 13*, 385–387.

Francis, D., Diorio, J., LaPlante, P., Weaver, S., Seckl, J. R., & Meaney, M. J. (1996). The role of early environmental events in regulating neuroendocrine development: Moms, pups, stress, and glucocorticoid receptors. In C. F. Ferris & T. Grisso (Eds.), *Understanding aggressive behavior in children* (pp. 136–152). New York: New York Academy of Sciences.

Galler, J. R., Shumsky, J. S., & Morgane, P. J. (1996). Malnutrition and brain development. In A. Walker (Ed.), *Nutrition in pediatrics* (pp. 196–212). New York: Dekker.

Gibbs, D. M. (1986). Vasopressin and oxytocin:hypothalamic modulators of the stress response: A review. *Psychoneuroendocrinology, 11*, 131–139.

Glod, C. A., & Teicher, M. H. (1996). Relationship between early abuse, post-traumatic stress disorder, and activity levels in prepubertal children. *Journal of the American Academy of Child and Adolescent Psychiatry, 35*, 1384.

Gluck, J. P., Harlow, H. F., & Schiltz, K. A. (1973). Differential effect of early enrichment and deprivation on learning in the rhesus monkey (*Macaca mulatta*). *Journal of Comparative and Physiological Psychology, 84*, 598–604.

Gressens, P., Muaku, S. M., Besse, L., Nsegbe, E., Gallego, J., Delpech, B., Gaultier, C., Evrard, P., Ketelslegers, J. M., & Maiter, D. (1997). Maternal protein restriction early in rat pregnancy alters brain development in the progeny. *Developmental Brain Research, 103*, 21–35.

Hall, F. S., Wilkinson, L. S., Humby, T., & Robbins, T. W. (1999). Maternal deprivation of neonatal rats produces enduring changes in dopamine function. *Synapse, 32*, 37–43.

Hall, Z. (1992). *An introduction to molecular neurobiology.* Sunderland, MA: Sinauer.

Harlow, H. F. (1958). The nature of love. *American Psychologist, 12*, 673–685.

Harlow, H. F. (1959). Love in infant monkeys. *Scientific American, 200*, 68.

Harlow, H. F., & Harlow, M. K. (1962). Social deprivation in monkeys. *Scientific American, 207*, 137–146.

Harlow, H. F., & Suomi, S. J. (1971). Social recovery by isolation-reared monkeys. *Proceedings of the National Academy of Sciences of the USA, 68*, 1534–1538.

Harlow, H. F., Schiltz, K. A., Harlow, M. K., & Mohr, D. J. (1971). The effects of early adverse and enriched environments on the learning ability of rhesus monkeys. In L. E. Jarrard (Ed.), *Cognitive processes of nonhuman primates* (pp. 121–148). New York: Academic Press.

Harris, E. W., & Newman, J. D. (1988). Primate models for the management of separation anxiety. In J. D. Newman (Ed.), *The physiological control of mammalian vocalization* (pp. 321–330) New York: Plenum Press.

Harris, E. W., Ganong, A. H., & Cotman, C. W. (1984). Long-term potentiation in the hippocampus involves activation of N-methyl-D-aspartate receptors. *Brain Research, 323*, 132–137.

Harris, E. W., Ganong, A. H., Monaghan, D. T., Watkins, J. C., & Cotman, C. W. (1986). Action of 3-((±)-2-carboxypiperazin-4-yl)-propyl-1-phosphonic acid (CPP): A new and highly potent antagonist of N-methyl-D-aspartate receptors in the hippocampus. *Brain Research, 382*, 174–177.

Hebb, D. O. (1949). *The organization of behavior: A neuropsychological theory.* New York: Wiley.

Heim, C., Owens, M. J., Plotsky, P. M., & Nemeroff, C. B. (1997). Persistent changes in corticotropin-releasing factor systems due to early life stress: Relationship to the pathophysiology of major depression and post-traumatic stress disorder. *Psychopharmacology Bulletin, 33*, 185–192.

Hess, J. L., Denenberg, V. H., Zarrow, M. X., & Pfeifer, W. D. (1969). Modification of the corticosterone response curve as a function of handling in infancy. *Physiology and Behavior, 4*, 109–112.

Higley, J. D., Suomi, S. J., & Linnoila, M. (1991). CSF monoamine metabolite concentrations vary according to age, rearing, and sex, and are influenced by the stressor of social separation in rhesus monkeys. *Psychopharmacology, 103*, 551–556.

Higley, J. D., Hopkins, W. D., Thompson, W. W., Byrne, E. A., Hirsch, R. M., & Suomi, S. J. (1992). Peers as primary attachment sources in yearling rhesus monkeys (*Macaca mulatta*). *Developmental Psychology, 28*, 1163–1171.

Hofer, M. A. (1983). The mother–infant interaction as a regular of infant physiology and behavior. In

L. A. Rosenblum & H. Moltz (Eds.), *Symbiosis in parent–offsring interactions* (pp. 61–75). New York: Plenum Press.

Hofer, M. A., & Shair, H. (1980). Sensory processes in the control of isolation-induced ultrasonic vocalization by 2-week-old rats. *Journal of Comparative and Physiological Psychology, 94,* 271–279.

Hofer, M. A., & Shair, H. (1982). Control of sleep–wake states in the infant rat by features of the mother–infant relationship. *Developmental Psychobiology, 15,* 229–243.

Hubel, D. H., & Wiesel, T. N. (1962). Receptive fields, binocular interaction and functional architecture in the cat's visual cortex. *Journal of Physiology, 160,* 106–154.

Hubel, D. H., & Wiesel, T. N. (1970). The period of susceptibility to the physiological effects of unilateral eye closure in kittens. *Journal of Physiology (London), 206,* 419–436.

Imperato, A., Angelucci, L., Casolini, P., Zocchi, A., & Puglisi-Allegra, S. (1992). Repeated stressful experiences differently affect limbic dopamine release during and following stress. *Brain Research, 577,* 194–199.

Ito, Y., Teicher, M. N., Glod, C. A., Harper, D., Magnus, E., & Gelbard, H. A. (1993). Increased prevalence of electrophysiological abnormalities in children with psychological, physical, and sexual abuse. *Journal of Neuropsychiatry, 5,* 401–408.

Ito, Y., Teicher, M. N., Glod, C. A., & Ackerman, E. (1998). Preliminary evidence for aberrant cortical development to abused children: A quantitative EEG study. *Journal of Neuropsychiatry and Clinical Neuroscience, 298,* 298.

Jones, L., Fischer, I., & Levitt, P. (1996). Region-specific alteration of dendritic development in the cerebral cortex following prenatal cocaine exposure. *Cerebral Cortex, 6,* 431–435.

Kaas, J. H. (1991). Plasticity of sensory and motor maps in adult mammals. *Annual Review of Neuroscience, 14,* 137–167.

Kaczmarek, L., Kossut, M., & Skangiel-Kramska, J. (1997). Glutamate receptors in cortical plasticity: Molecular and cellular biology. *Physiological Reviews, 77,* 217–255.

Kalin, N. H., Shelton, S. E., & Barksdale, C. M. (1988). Opiate modulation of separation-induced distress in non-human primates. *Brain Research, 440,* 285–292.

Kalivas, P. W., & Alesdatter, J. E. (1993). Involvement of *N*-methyl-D-aspartate receptor stimulation in the centra tegmental area and amygdala in behavioral sensitization to cocaine. *Journal of Pharmacology and Experimental Therapeutics, 267,* 486–495.

Kalivas, P. W., & Duffy, P. (1989). Similar effects of daily cocaine and stress on mesocortico–limbic dopamine neurotransmission in the rat. *Biological Psychiatry, 25,* 913–928.

Kalivas, P. W., Duffy, P., Abhold, R., & Dilts, R. P. (1988). Sensitization of mesolimbic dopamine neurons by neuropeptides and stress. In P. W. Kalivas & C. D. Barnes (Eds.), *Sensitization in the nervous system* (pp. 119–144), Caldwell, NJ: Telford Press.

Kalivas, P. W., Sorg, B. A., & Hooks, M. S. (1993). The pharmacology and neural circuitry of sensitization to psychostimulants. *Behavioral Pharmacology, 4,* 315–334.

Karler, R., Calder, L. D., Chaudhry, I. A., & Turkanis, S. A. (1989). Blockade of "reverse tolerance" to cocaine and amphetamine by MK-801. *Life Sciences, 45,* 599–606.

Kehoe, P. (1988). Opioids, behavior, and learning in mammalian development. In N. Adler & E. M. Blass (Eds.), *Handbook of behavioral neurobiology*, Volume 9, *Developmental psychobiology and behavioral ecology* (pp. 309–347). New York: Plenum Press.

Kehoe, P., & Blass, E. M. (1986a). Behaviorally functional opioid systems in infant rats: I. Evidence for olfactory and gustatory classical conditioning. *Behavioral Neuroscience, 100,* 359–367.

Kehoe, P., & Blass, E. M. (1986b). Central nervous system mediation of positive and negative reinforcement in neonatal albino rats. *Developmental Brain Research, 27,* 69–75.

Kehoe, P., & Blass, E. M. (1986c). Opioid mediation of separation distress in 10-day-old rats: Reversal of stress with maternal stimuli. *Developmental Psychobiology, 19,* 385–398.

Kehoe, P., & Blass, E. M. (1986d). Behaviorally functional opioid systems in infant rats: II. Evidence for pharmacological, physiological and psychological mediation of pain and stress. *Behavioral Neuroscience, 100,* 624–630.

Kehoe, P., & Blass, E. M. (1989). Conditioned opioid release in ten-day-old rats. *Behavioral Neuroscience, 103,* 423–428.

Kehoe, P., & Bronzino, J. D. (1999). Neonatal stress alters LTP in freely moving male and female adult rats. *Hippocampus, 9,* 651–658.

Kehoe, P., Hoffman, J., Austin-LaFrance, R., & Bronzino, J. (1995). Neonatal isolation enhances hippocampal LTP in freely-moving juvenile rats. *Experimental Neurology, 136,* 89–97.

Kehoe, P., Clash, K., Skipsey, K., & Shoemaker, W. J. (1996). Brain dopamine response in isolated 10-day-old rat pups: Assessment using D_2 binding and dopamine turnover. *Pharmacology, Biochemistry and Behavior, 53,* 41–49.

Kehoe, P., Shoemaker, W. J., Triano, L., Hoffman, J., & Arons, C. (1996). Repeated isolation in the neonatal rat produces alterations in behavior and ventral striatal dopamine release in the juvenile following amphetamine challenge. *Behavioral Neuroscience, 110,* 1435–1444.

Kehoe, P., Triano, L., Glennon, C., & Daigle, A. (1997). Juvenile rats stressed as infants exhibit differential dopamine and activity levels following restraint. *Society for Neuroscience Abstracts, 23,* 227.

Kehoe, P., Shoemaker, W. J., Arons, C., Triano, L., & Suresh, G. (1998). Repeated isolation stress in the neonatal rat: Relation to brain dopamine systems in the 10-day-old rat. *Behavioral Neuroscience, 112,* 1466–1474.

Kehoe, P., Shoemaker, W. J., Triano, L., Callahan, M., & Rappolt, G. (1998). Adults stressed as neonates show exaggerated behavioral responses to both pharmacological and environmental challenges. *Behavioral Neuroscience, 112,* 116–125.

Kellogg, C., K., Awatramani, G. B., & Piekut, D. T. (1998). Adolescent development alters stressor-induced Fos immunoreactivity in rat brain. *Neuroscience, 83,* 681–689.

Kesson, W. (1965). *The child.* New York: Wiley.

Kim, J. J., & Yoon, K. S. (1998). Stress: Metaplastic effects in the hippocampus. *TINS, 21,* 505–509.

Kolta, M. G., Scalzo, F. M., Ali, S. F., & Holson, R. R. (1990). Ontogeny of the enhanced behavioral response to amphetamine-pretreated rats. *Psychopharmacology, 100,* 377–382.

Kuczenski, R., & Segal, D. S. (1988). Psychomotor stimulant-induced sensitization: Behavioral and neurochemical correlates. In P. W. Kalivas & C. D. Barnes (Eds.), *Sensitization in the nervous system* (pp. 175–206). Caldwell, NJ: Telford Press.

Kuhn, C. M., Schanberg, S. M., Field, T., Symanski, R., Zimmerman, E., Scafidi, F., & Roberts, J. (1991). Tactile/kinesthetic stimulation effects on sympathetic and adrenocortical function in preterm infants. *Journal of Pediatrics, 119,* 434–440.

Lauder, J. M. (1993). Neurotransmitters as growth regulatory signals: Role of receptors and second messenger systems. *Trends in Neuroscience, 16,* 223–239.

Levine, S. (1962). Plasma-free corticoid response to electric shock in rats stimulated in infancy. *Science, 135,* 795–796.

Levine, S., Chevalier, J. A., & Korchin, S. J. (1956). The effects of early shock and handling on later avoidance learning. *Journal of Personality, 24,* 475–493.

Levitt, P., Harvey, J. A., Freidman, E., Simansky, K., & Murphy, E. H. (1997). New evidence for neurotransmitter influences on brain development. *Trends in Neuroscience, 20,* 269–274.

Levitt, P., Reinoso, B., & Jones, I. (1998). The critical impact of early cellular environment on neuronal development. *Preventive Medicine, 27,* 180–183.

Liu, D., Diorio, J., Tannenbaum, B., Caldji, C., Francis, D., Freeman, A., Sharma, S., Pearson, D., Plotsky, P. M., & Meaney, M. J. (1997). Maternal care, hippocampal glucocorticoid receptors, and hypothalamic–pituitary–adrenal responses to stress. *Science, 277,* 1659–1662.

Maier, S. F., & Seligman, M. E. P. (1976). Learned helplessness: Theory and evidence. *Journal of Experimental Psychology: General, 103,* 3–46.

Marek, P., Ben-Eliyahu, S., Vaccarino, A., & Liebeskind, J. C. (1991). Delayed application of MK-801 attenuates development of morphine tolerance in rats. *Brain Research, 558,* 163–165.

Marek, P., Ben-Eliyahu, S., Gold, M., & Liebeskind, J. C. (1991). Excitatory amino acid antagonists (kynurenic acid and MK-801) attenuate the development of morphine tolerance in the rat. *Brain Research, 547,* 77–81.

Matthews, K., Robbins, T. W., Everitt, B. J., & Caine, S. B. (1999). Repeated neonatal maternal separation alters intravenous cocaine self-administration in adult rats. *Psychopharmaoclogy* (Berlin), *141,* 123–134.

McCormick, C. M., Kehoe, P., & Kovacs, S. (1998). Corticosterone release in response to repeated, short episodes of neonatal isolation: Evidence of sensitization. *International Journal of Developmental Neuroscience, 16,* 175–185.

McCormick, C. M., Rood, B., & Kehoe, P. (1998). Neonatal isolation alters corticosterone response to restraint stress in juvenile but not adult rats. *Society for Neuroscience Abstracts, 24,* 117.

McEwen, B. S., & Sapolsky, R. M. (1995). Stress and cognitive function. *Current Opinion in Neurobiology, 5,* 205–216.

McLaughlin, P. J., Tobias, S. W., Max Lang, C., & Zagon, E. S. (1997). Opioid receptor blockade during prenatal life modifies postnatal behavioral development. *Pharmacology, Biochemistry and Behavior, 58,* 1075–1082.

Meaney, M. J., Bhatnagar, S., Larocque, S., McCormick, C., Shnaks, N., Sharma, S., Smythe, J., Viau, V., & Plotsky, P. M. (1993). Individual differences in the hypothalamic–pituitary–adrenal stress response and the hypothalamic CRF system. *Annals of the New York Academy of Sciences, 297,* 70–85.

Meaney, M. J., Diorio, J., Francis, D., Widdowson, J., LaPlante, P., Caldji, C., Seckl, J. R., & Plotsky, P. M. (1996). Early environmental regulation of forebrain glucocorticoid receptor gene expression: Implications for adrenocortical responses to stress. *Developmental Neuroscience, 18,* 49–72.

Mills, E., Bruckert, J. W., & Smith, P. G. (1990). Influence of postnatal maternal stress on blood pressure and heart rate of juvenile and adult rat offspring. *Departmental Psychobiology, 23,* 839–847.

Moore, C. L., & Chadwick-Dias, A. M. (1986). Behavioral responses of infant rats to maternal licking: Variations with age and sex. *Developmental Psychobiology, 19,* 427–438.

Moore, C. L., & Morelli, G. A. (1979). Mother rats interact differently with male and female offspring. *Journal of Comparative and Physiological Psychology, 93,* 677–684.

Morgane, P. J., Austin-LaFrance, R., Bronzino, J., Tonkiss, J., Diaz-Cintra, S., Cintra, L., Kemper, T., & Galler, J. R. (1993). Prenatal malnutrition and development of the brain. *Neuroscience Behavioral Review, 17,* 91–128.

Morgane, P. J., Bronzino, J. D., Austin-LaFrance, R. J., & Galler, J. R. (1996). Malnutrition, central nervous system effects. *Encyclopedia of Neuroscience, 717,* 1–9.

Murrin, L. C., & Zeng, W. (1989). Dopamine D_1 receptor development in the rat striatum: Early localization in striosomes. *Brain Research, 480,* 170–177.

Neal, B. S., & Joyce, J. N. (1992). Neonatal 6-OHDA lesions differentially affect striatal D_1 and D_2 receptors. *Synapse, 11,* 35–46.

Novak, M. A., & Harlow, H. F. (1974). Social recovery for the first year of life: 1. Rehabilitation and therapy. *Developmental Psychology, 11,* 453–465.

Ogawa, T., Mikuni, M., Kuroda, Y., Muneoka, K., Mori, K. J., & Takahashi, K. (1994). Periodic maternal deprivation alters stress response in adult offspring, potentiates the negative feedback regulation of restraint stress-induced adrenocortical response and reduces the frequencies of open field-induced behaviors. *Pharmacology, Biochemistry and Behavior, 49,* 961–967.

Olsen, L., Boreus, O., & Seiger, A. (1973). Histochemical demonstration and mapping of 5-hydroxytryptamine and catecholamine-containing neuron systems in the human fetal brain. *Zeitschrift für Anatomie und Entwicklungsgeschichte, 139,* 259–282.

Oppenheim, R. W. (1981). Ontogenetic adaptations and retrogressive processes in the development of the nervous system and behavior: A neuroembryological perspective. In K. Connelly & H. Prechtl (Eds.), *Maturation and development: Biological and psychological perspectives.* London: Spastica Society.

Overton, P. G., Richard, C. D., Berry, M. S., & Clark, D. (1999). Long-term potentiation at excitatory amino acid synapses on midbrain dopamine neurons. *Neuroreport, 10, 2,* 221–226.

Panksepp, J., Herman, B., Conner, R., Bishop, P., & Scott, J. P. (1978). The biology of social attachments: Opiates alleviate separation distress. *Biological Psychiatry, 13,* 607–618.

Panksepp, J., Meeker, R., & Bean, N. J. (1980). The neurochemical of crying. *Pharmacology, Biochemistry and Behavior, 12,* 437–443.

Perry, B. D. (1997). Incubated in terror: Neurodevelopmental factors in the "cycle of violence." In D. Osofsky (Ed.), *Children in a violent society* (pp. 124–149). New York: Guilford Press.

Perry, B. D., & Pollard, R. (1997). Altered brain development following global neglect in early childhood. *Society for Neuroscience Abstracts, 23,* 1625.

Perry, B. D., & Pollard, R. (1998). Homeostasis, stress, trauma, and adaptation: A neurodevelopmental view of childhood trauma. *Child and Adolescent Psychiatric Clinics of North America, 7,* 8-1–8-19.

Plotsky, P. M. (1987). Regulation of hypophysiotropic factors mediating ACTH secretion. *Annals of the New York Academy of Sciences, 512,* 205–217.

Plotsky, P. M., & Meaney, M. J. (1993). Early, postnatal experience alters hypothalamic factor (CRF) mRNA, median eminence CRF content and stress-induced release in adult rats. *Brain Research, 18,* 195–200.

Prasad, B. M., Sorg, B. A., Ulibarri, C., & Kalivas, P. W. (1995). Sensitization to stress and psychostimulants. Involvement of dopamine transmission versus the HPA axis. *Annals of the New York Academy of Sciences, 771,* 617–625.

Redgrave, P., Prescott, T. J., & Gurney, K. (1999). Is the short-latency dopamine response too short to signal reward error? *TINS, 22,* 146–151.

Ribble, M. A. (1944). Infantile experience in relation to personality development. In J. McV. Hunt (Ed.), *Personality and behavior disorders* (pp. 621–651). New York: Ronald Press.

Richter, C. P. (1976). *The psychobiology of Curt Richter.* Baltimore, MD: York Press.

Rivier, C. L., & Plotsky, P. M. (1986). Mediation by corticotropin releasing factor (CRF) of adenohypophysical hormone secretion. *Annual Review of Physiology, 48,* 475–494.

Robinson, T. E. (1988). Stimulant drugs and stress; factors influencing the susceptibility to sensitization. In C. Barnes & P. Kalivas (Eds.), *Sensitization of the central nervous system* (pp. 145–173). Caldwell, NJ: Telford Press.

Robinson, T. E., & Becker, J. B. (1986). Enduring changes in brain and behavior produced by chronic amphetamine administration: A review and evaluation of animal models of amphetamine psychosis. *Brain Research Review, 11,* 157–198.

Robinson, T. E., & Kolb, B. (1997). Persistent structural modifications in nucleus accumbens and prefrontal cortex neurons produced by previous experience with amphetamine. *Journal of Neuroscience, 17,* 8491.

Robinson, T. E., & Kolb, B. (1999). Alterations in the morphology of dendrites and dendritic spines in the nucleus accumbens and prefrontal cortex following repeated treatment with amphetamine or cocaine. *European Journal of Neuroscience, 11,* 1598–1604.

Robinson, T. E., Jursons, P. A., Bennett, J. A., & Bentgen, K. M. (1988). Persistent sensitization of dopamine neurotransmission in ventral striatum (nucleus accumbens) produced by prior experience with amphetamine: A microdialysis study in freely moving rats. *Brain Research, 462,* 211–222.

Rogers, C. M., & Davenport, R. K. (1970). Chimpanzee maternal behavior. In G. H. Bourne (Ed.), *The chimpanzee* (pp. 361–368). Basel: Karger.

Rosenblum, L. A. (1968). Mother–infant relations and early behavioral development in the squirrel monkey. In L. Rosenblum & R. W. Cooper (Eds.), *The squirrel monkey* (pp. 207–234). New York: Academic Press.

Rosenblum, L. A., & Paully, G. S. (1984). The effects of varying environmental demands on maternal infant behavior. *Child Development, 55,* 305–314.

Rosengarten, H., & Friedhoff, A. J. (1979). Enduring changes in dopamine receptor cells of pups from drug administration to pregnant and nursing rats. *Science, 203,* 1133–1135.

Sapolsky, R. M. (1992). How do glucocorticoids endanger the hippocampal neuron? In *Stress, the aging brain, and the mechanisms of neuron death* (pp. 223–258). Cambridge, MA: MIT Press.

Sarvey, J. M., Burgard, E. C., & Decker, G. (1989). Long-term potentiation: Studies in the hippocampal slice. *Journal of Neuroscience Methods, 28,* 109–124.

Scarr-Salapatek, S., & Williams, M. L. (1973). The effects of early stimulation on low-birth-weights infants. *Child Development, 44,* 94–101.

Schlumpf, M., Lichtensteiger, W., Shoemaker, W. J., & Bloom, F. E. (1980). Fetal monoamine systems: Early stages and cortical projections. In H. Parvez & S. Parvez (Eds.), *Biogenic amines in development* (pp. 567–590). New York: Elsevier/North-Holland Biomedical Press.

Seeman, P., & VanTol., H. H. M. (1994). Dopamine receptor pharmacology. *TIPS, 15,* 264–270.

Segal, D. S., & Kuczenski, R. (1987). Individual differences in responsiveness to single and repeated amphetamine administration: Behavioral characteristics and neurochemical correlates. *Journal of Pharmacology and Experimental Therapeutics, 242,* 917–926.

Segal, D. S., & Kuczenski, R. (1992). *In vivo* microdialysis reveals a diminished amphetamine-induced DA response corresponding to behavioral sensitization produced by repeated amphetamine pretreatment. *Brain Research, 571,* 330–337.

Shoemaker, W. J., & Kehoe, P. (1995). Effect of isolation conditions on brain regional enkephalin and β-endorphin and vocalizations in 10-day-old rat pups. *Behavioral Neuroscience, 109,* 117–122.

Shoemaker, W. J., Kehoe, P., Antolik, C., Norrholm, S., Geary, M., & Fong, D. (1998). Altered stress-induced dopamine release in adult rats previously isolated as neonates. *Society for Neuroscience Abstracts, 24,* 451.

Shors, T. J., & Mathew, P. R. (1998). NMDA receptor antagonism in the lateral/basolateral but not central nucleus of the amygdala prevents the induction of facilitated learning in response to stress. *Learning and Memory, 5,* 220–230.

Shors, T. J., & Servatius, R. J. (1995). Stress-induced sensitization and facilitated learning require NMDA receptor activation. *Neuroreport, 6,* 677–680.

Smith, E. L., Coplan, J. D., Trost, R. C., Scharf, B. A., & Rosenblum, L. A. (1997). Neurobiological alterations in adult nonhuman primates exposed to unpredictable early rearing. Relevance to posttraumatic stress disorder. *Annals of the New York Academy of Sciences, 821,* 545–548.

Smotherman, W. P., Bell, R. W., Starzec, J., Elias, J., & Zachman, T. A. (1974). Maternal responses to infant vocalizations and olfactory cues in rats and mice. *Behavioral Biology, 12,* 55.

Sorg, B. A., & Kalivas, P. W. (1991). Effects of cocaine and footshock stress on extracellular dopamine levels in the ventral striatum. *Brain Research, 559,* 29–36.

Sorg, B. A., & Kalivas, P. W. (1993). Effects of cocaine and footshock stress on extracellular levels in the medial prefrontal cortex. *Neuroscience, 53,* 695–703.

Spear, L. P., Shalaby, I. A., & Brick, J. (1980). Chronic administration of haloperidol during development: Behavioral and psychopharmacological effects. *Psychopharmacology, 70,* 47–58.

Spitz, R. A. (1945). Hospitalism. *Psychoanalytic study of the child, 1,* 53–74.

Spitz, R. A. (1946). Hospitalism: A follow-up report on investigation described in Volume 1, 1945. *Psychoanalytic Study of the Child, 2,* 113–117.

Spitz, R. A., & Wolf, K. A. (1946). Anaclitic depression: An inquiry into the genesis of psychiatric conditions in early childhood. *Psychoanalytic Study of the Child, 2,* 313–342.

Squire, L. R., & Zola-Morgan, S. (1991). The medial temporal lobe memory system. *Science, 253,* 1380–1386.

Sunanda Rao, M. S., & Raju, T. R. (1995). Effect of chronic restraint stress on dendritic spines and excrescences of hippocampal CA3 pyramidal neurons—A quantitative study. *Brain Research, 694,* 312–317.

Suomi, S. J. (1991). Early stress and adult emotional reactivity in rhesus monkeys. *Ciba Foundation Symposium; New Series, 156,* 171–188.

Svedsen, F., Tjolsen, A., Rygh, L. J., & Hole, K. (1999). Expression of long-term potentiation in single wide dynamic range neurons in the rat is sensitive to blockade of glutamate receptors. *Neuroscience Letters, 259,* 25–28.

Takahashi, L. L. (1998). Prenatal stress: Consequences of glucocorticoid on hippocampal development and function. *International Journal of Developmental Neuroscience, 16,* 199–207.

Tamborski, A., Lucot, J. B., & Hennessy, M. B. (1990). Central dopamine turnover in guinea pigs during separation from their mothers in a novel environment. *Behavioral Neuroscience, 104,* 607–611.

Teicher, M. H., Ito, Y., Glod, C. A., Anderson, S. L., Dumont, N., & Ackerman, E. (1997). Preliminary evidence for abnormal cortical development in physically and sexually abused children using EEG coherence and MRI. *Annals of the New York Academy of Sciences, 821,* 160–175.

Tonkiss, J., & Galler, J. R. (1990). Prenatal protein malnutrition and working memory performance in adult rats. *Behavioral Brain Research, 40,* 95–107.

Unis, A. S. (1995). Developmental molecular psychopharmacology in early-onset psychiatric disorder: From models to mechanisms. *Child and Adolescent Psychiatric Clinics of North America, 4,* 41–57.

Unis, A. S., Roberson, M. D., Robinette, R., Ha, J., & Dorsa, D. M. (1998). Ontogeny of human brain dopamine receptors I. Differential expression of [^{3}H]-SCH23390 and [^{3}H]-YM09151-2 specific binding. *Developmental Brain Research, 106,* 109–117.

Wang, H. Y., Yeung, J. M., & Friedman, E. (1995). Prenatal cocaine exposure selectively reduces mesocortical dopamine release. *Journal of Pharmacology and Experimental Therapeutics, 273,* 492–498.

Wang, X.-H., Levitt, P., Grayson, D. R., & Murphy, E. H. (1996). Intrauterine cocaine exposure of rabbits: Persistent elevation of GABA-immunoreactive neurons in anterior cingulate cortex but not in visual cortex. *Brain Research, 689,* 32–46.

Weiner, I., Feldon, J., & Ziv-Harris, D. (1987). Early handling and latent inhibition in the conditioned suppression paradigm. *Developmental Psychobiology, 20,* 233–240.

Wiesel, T. N., & Hubel, D. H. (1965). Comparison of the effects of unilateral and bilateral eye closure on cortical unit responses in kittens. *Journal of Neurophysiology, 28,* 1029–1040.

Wilkins, A. S., Logan, M., & Kehoe, P. (1998). Postnatal pup brain dopamine depletion inhibits maternal behavior. *Pharmacology, Biochemistry, and Behavior, 58,* 867–873.

Whitaker-Azmitia, P. M., & Azmitia, E. C. (1986). Autoregulation of fetal serotonergic neuronal development: Role of high affinity serotonin receptors. *Neuroscience Letters, 67,* 307–312.

Wigstrom, H., & Gustafsson, B. (1984). A possible correlate of the postsynaptic condition for long-lasting potentiation in the guinea pig hippocampus in vitro. *Neuroscience Letters, 44,* 327–332.

Wolff, P. H., & Fesseha, G. (1998). The orphans of Eritrea: Are orphanages part of the problem or part of the solution? *American Journal of Psychiatry, 155,* 1319–1324.

Zagon, I. S., & McLaughlin, P. J. (1985). Opioid antagonist-induced regulation of organ development. *Physiology and Behavior, 34,* 507–511.

Zagon, I. S., & McLaughlin, P. J. (1986). Opioid antagonist (naltrexone) modulation of cerebellar development: Histological and morphometric studies. *Journal of Neuroscience, 6,* 1424–1432.

Zimmerberg, B., & Shartrand, A. M. (1992). Temperature-dependent effects of maternal separation on growth, activity, and amphetamine sensitivity in the rat. *Developmental Psychobiology, 25,* 213–226.

15

Science Lies Its Way to the Truth ... Really

MEREDITH J. WEST AND ANDREW P. KING

PROLOGUE

In science today, the study of behavior seems to be something someone does to get to somewhere else. Behavior affords a gateway to physiology, neuroscience, and genetics. The quality of such integrative work thus begins and ends with the quality of knowledge about behavior. How do we assay behavioral quality and its integrative potential? We ask this question because we believe many beliefs about the nature of behavior are wrong, albeit wrong in a right way, a scientific way. In this chapter, we illustrate some of the issues in the study of behavior to be considered before its transport into new domains. We call for renewed emphasis on the tasks actually confronting organisms as they develop and learn and on the social context in which behavior typically occurs, a context too often excised and ignored in current work. Without a focus on task and context, we cannot connect knowledge about biology to knowledge about behavior. Without such a focus, we cannot prevent a cascade of changes in the meaning, relevance, and utility of behavioral knowledge as it is subjected to the process of integration.

> In Africa a thing is true at first light and a lie by noon and you have no more respect for it than the lovely weed-fringed lake you see across the sun-baked plain. You have walked across that plain in the morning and you know no such lake is there. But now it is there absolutely true, beautiful and unbelievable. (Hemingway, 1998, p. iv.)

MEREDITH J. WEST AND ANDREW P. KING   Department of Psychology, Indiana University, Bloomington, Indiana 47405.

Developmental Psychobiology, Volume 13 of *Handbook of Behavioral Neurobiology*, edited by Elliott Blass, Kluwer Academic / Plenum Publishers, New York, 2001.

MEREDITH J.
WEST AND
ANDREW P. KING

In 1988, for the prior edition of this volume in the *Handbook*, we wrote about the nature and nurture of behavior (West, King, & Arberg, 1988). We described what we saw as a neglected element in discussions of ontogeny, the inherited niche, i.e., the physical and social contexts that accompany genes. It was a promising theme, we thought, outlining ways to weave ecology firmly into ontogeny. Since then, we have seen our "nature–nurture–niche" notion surface here and there (e.g., Drickamer, Vessey, & Meikle, 1996; Michel & Moore, 1995), but the converted already believed it, and the unconverted remained unaware of it. It was a message in a bottle that did not travel far. But, unbeknown to us, we were floating our modest idea at a time that a supertanker of an idea, the Human Genome Project, was about to be launched. As we write this in 2000, the HHS *Genome* has already reached an important port of entry, human chromosome 22, and will soon reach others. All the maps from these journeys are to be published in a book projected to be as tall as the Washington Monument (550 feet). One of the project's captains has dubbed the monumental work, "the book of life" (Collins, 1999). Our bottle hugs the shore.

The arrival of genomics deserves grand words and even grandiose opinions because the revolution in our knowledge about molecular biology is astounding. But, as students of behavior and development, we believe that interest in the molecular has come at the cost of interest in the nonmolecular. Such neglect would not matter if the nonmolecular level could now be discarded as one might dispatch with outer packaging. The multidisciplinary mergers do, however, include behavior, either as a probe to explore another level or as a phenomenon to be explained by finding previously unseen physiological or genetic roots (Wilson, 1998). Moreover, integration of molecular and nonmolecular systems includes animal models and humans. The stakes are high in getting the payoff matrix right between the benefits of using behaviors simple enough to study at many levels and the costs of compromising the integrity of the behaviors we seek to understand. The stakes are also high because the science of the future forces us to revisit some of the most entrenched and darkest ideas about behavior at a molar level. What theory will guide bioengineering of organisms, biocleansing of "flawed" genotypes, or biological determinism made simple?

Despite these perils, common sense tells that we may be taking the state of progress of science too seriously. The gaps in knowledge within areas of science are still great enough to make talk about domesticating genetics or decommissioning evolution just talk. Moreover, although science involves big successes, it also involves big mistakes, hence, our title, "Science lies its way to the truth." Scientists take concepts and methods, ones known to be only partly right, and create a narrative to frame their theories and their work. Thus, the trial and error that defines the scientific process leads us to regard the hoopla about the human genome project as enjoyable, but not intoxicating. It is early in the day.

As we look ahead, however, our wariness stems from the real possibility that the Human Genome Project or other grand multidisciplinary enterprises may end up pursuing mirages. If, as we propose, the solidity of many behavioral phenomena are being assumed without sufficient questioning, linkages between different levels of integration become troublesome. We start with the premise that it is hard not to link a behavioral layer with a genetic layer. Progress to this point has depended heavily on reductionism. Molecular biology is not only the dominant field in biology, but the predominant metaphor for science. When bringing together knowledge from different levels, we will have to be selective and sensitive to which phenomena can survive disconnection and reconnection. What metrics should we use? And what is really being connected? Although genes are now visible and amenable to dissection and manipulation, the kinds of behavior–gene connections proposed are of two kinds. First, there is the search for the concrete, i.e., "biological" genes, specific patterns of proteins that match up to specific changes in behavior. Second, there are "abstract" or "evolutionary" genes, of which memes are most famous, that fend for themselves in a theoretical struggle for survival (Griffiths & Neumann-Held, 1999). The specific locus of their functions is not the issue for the latter, but is for the former. The distinctions made by Griffiths and Neumann-Held between biological and abstract genes is much like that of D. O. Hebb, who made an analogous distinction between the conceptual nervous system and the central nervous system. He did so to distinguish thinking about what we assume the nervous system must do and what we actually know it to do (Hebb, 1949).

A conceptual frame for abstract genes and behavior can overlap with biological genes and behavior, but the excitement at present is about biological linkages, how G, C, A, and T, in all their combinations, link to higher systems. Our focus here is on the nature of that pursuit. How do biological genes map to real behaviors? The use of genes in a more metaphorical way is of less concern to us here. But the potential for confusion between the two is great. The never-ending debate about the genetics of IQ caricatures the possible downside of inattention to the nature of measures and the kinds of linkages being made. At times, both geneticists and psychologists have had to remind the other of the frailty or outright falseness of one another's measures. Even more dubious connections now fill the popular media, in the syntactic frame of "genes for X." There have been reports of genes linked to things like the wearing of flannel shirts all the way to the ability to form the plurals of nouns (Gopnik & Goad, 1997). In the case of genes for plural nouns, what was remarkable was how easily people overlooked the far more pervasive language deficiencies of the families; plural nouns were far from the root of these families' biological problems.

Errors or exaggerations in science are of course routine, that is what science is all about—today's fiction is yesterday's truth. But there is no mechanism in science to evaluate the currency and meaning of data for those outside of the specialization that created them. Given the number of specializations, the problem becomes one of monumental scale. Moreover, people and textbook or popular science writers tend to fossilize good stories, either because they do not know they are wrong or because good, simple stories sell. A finding true at first light may get lots of attention, but not the finding making it a lie by noon. Less spectacular data may not make it into print, let alone into a textbook's second edition! But, by then, the now-outdated original information from the first edition may have skipped across integrative borders into other texts and other research programs. Such error-prone transfers of knowledge are even more likely given the particulate nature of knowledge: who can possibly embrace so many parts?

From Textbook Biology to a Social Task Biology

The path to integrative progress means changes in basic behavioral approaches, especially laboratory-based models. We need an ecologically based method of choos-

MEREDITH J.
WEST AND
ANDREW P. KING

ing and defining robust scientific problems on which to use integrative methods. We need such a method because the organisms and genomes we seek to understand developed in response to ecologies filled with contextual demands. Most laboratory work does not, however, take those ecologies or demands seriously (but for a model paradigm see Timberlake, 1993; Timberlake & Silva, 1994).

Of even more importance is that many of the species we study also evolved in worlds where stimulation, information, and knowledge are socially constrained. We contend that many individuals do not have direct access to potential stimulation (thermal, tactile, visual, olfactory, taste, or any cross-modal connections among these). Nor do organisms often have direct access to proximate reinforcement systems to consolidate and shape perceptual experiences. Instead, for many species, these basic resources are delivered via a "social gateway" in which older organisms or peers facilitate or constrain sensory exposure, timing, and duration. Of even more consequence is that some stimulation is not even available to a single conspecific, but must occur in an aggregation of conspecifics—thermal regulation in a rat huddle is an elegant example (Alberts, 1978). The social gateway, however, has been bypassed in most laboratory research in perception, cognition, and neuroscience. We shuttle the animal through the social gateway even before the work begins! Consider a very simple example. Golden hamsters (*Meseocricetus auratus*) are not considered particularly social animals. But adult females can be taught a novel means of obtaining food, a fishing technique in which they use paws and teeth to reel in a chain with food attached at the end. They can learn this by experimenters' shaping the response through 10 days of extensive training (Previde & Poli, 1996). The pups of trained mothers show the entire response in only 1 day of observation if a trained mother is present and using the response. Pups of untrained mothers do not show the behavior at all; they try whatever technique their untrained mother tries. In the life of a small rodent, a difference of 9 days in learning efficiency is huge. But, by and large, studies of observational or social learning are rare in comparison to the number of studies of individually shaped or conditioned animals. Because of the bias toward the latter approach, the search for the genetic roots of learning ability is aligned to the model of individual learning. While such data are important in that the capacity to learn individually is abundantly apparent, it unfortunately bypasses the question of what might be the evolutionarily preferred or selected route for learning. This is hardly a new point in comparative psychology: as Harlow (1965) noted, "the Skinner box enables one to demonstrate discrimination learning in a greater number of trials than any other method" (p. 98).

Thus, as we discuss the idea of a task biology, we highlight two features. We argue for a change in the basic unit of analysis from "behavior" to "task." By task, we mean a functionally rooted set of acts performed by one *or more* organisms in an actual ecology, tasks ranging from foraging to reproduction to communication. Second, we assume that many functional outcomes, such as successful foraging, cannot be genetically dissected by looking only at animals as individuals. The very existence of certain phenomena depend on the social composition of a species and the pattern of interaction among individuals (Giraldeau, Caraco, & Valone, 1994).

As a way to illustrate some features of tasks and social dependences, we provide several examples below of "textbook knowledge" based on familiar laboratory practices. We do so to expose the inherent frailty of many behavioral explanations when the behaviors have not been grounded in an ecological context. We do so as well to show how the light of day can change lovely truths into misleading mirages.

Imprinting refers to the tendency of newly hatched precocial young to form a rapid social attachment, an attachment often thought to be irreversible. Who can forget the pictures of ducklings paddling behind Konrad Lorenz? In part it is the recognition factor of the behavior that made it endure and in part it was the finding itself. At that time in history, the dominant theory of learning consisted of the laws of classical or operant conditioning. Imprinting did not fit with such general laws of learning and thus broke important new ground (Gottlieb, 1971; Sluckin, 1965).

Although few still believe in the irreversibility of imprinting, now even the basic phenomenon is under scrutiny. When the door was opened to let in some of the behavior's natural complexity, the simple story cracked under social pressure. As it turns out, imprinting unfolds quite differently when the young hatchlings have access to their siblings, with whom they have had auditory contact as eggmates. The ages at which ducklings learn and remember the experimenter-designated stimulus of model hens, for example, are not the same if ducklings receive sibling stimulation (Gottlieb, 1993; Johnston & Gottlieb, 1985; Lickliter & Gottlieb, 1987; Lickliter, Dyer, & McBride, 1993). Exposure and time with siblings can also promote or delay maternal imprinting. Indeed, if the mother is not seen frequently, siblings imprint on one another. Thus, despite first appearances, reliable imprinting on the mother is dependent on other social contingencies, a more fragile process than often portrayed. Interactions with siblings occur at the same point in developmental times as interactions with the mother. It is the dynamics of the co-occurrence of these two tasks that needs to be emphasized to understand imprinting as it actually happens.

SCIENCE LIES ITS WAY TO THE TRUTH ... REALLY

LOOK AGAIN

Stories about the evolution of a behavior are especially vulnerable to textbook ossification if the original story is intuitively logical. The tendency to fall for "just-so-ism" or, more formally, the "adaptationist programme" (Gould & Lewontin, 1979), is widespread. When things make sense, we are lulled into complacency. The moth that is becoming a myth serves as an elegant example of the dangers (Majerus, 1998; Millar & Lambert, 1999). The peppered moth, *Biston betularia*, showed changes in wing coloration during and after the industrial revolution from light to dark to light again. These changes are cited in many textbooks as an example of natural selection in modern times, as the tree trunks the moths were presumed to rest against also turned from light to dark to light as a result of changes in sources of pollution. The correlated changes in color were thought to have been favored as a mimicry response to prevent predation by birds. But questions about the phenomenon have come from many directions, including problems with replication, knowledge about vision in birds, and about the actual resting places of moths. The story assumes moths rest against the lichen on the bark of the aforementioned tree trunks. But, in a triumph for context, scientists cannot find peppered moths on tree trunks, and no one is actually sure where the moths rest (the undersides of branches seem more likely). But how then would mimicry work to deter predation? The true story may be more complicated, involving fluctuations in moth and bird populations, or what the pollutants do directly to the moths. In any case, the tasks for the moth and for the predator no longer can be captured in the bold strokes of black and white theorizing.

MEREDITH J.
WEST AND
ANDREW P. KING

At a certain age, human infants make what is called the "A not B" error. Infants are shown two locations, A and B, and then are shown or allowed to reach for a toy at point A for several trials. The object is then placed at point B. Under many conditions, infants then reach at A, not B. It is a very reliable phenomenon for infants of a certain age, good enough to bank on for class demonstrations, etc. It has also been the subject of countless experiments and permutations to provide a coherent explanation. Some have said it shows the absence of certain mental structure, glitches in an emerging theory of mind. But simple tests can change the error rate, manipulations that focus on the idea that the baby's mistake is born out of time-linked patterns of perception and action, not absent mental modules (Smith, Thelen, Titzer, & Mclin, 1999). Babies repeat the arm movement to A when it "should" go to B because their arm keeps going where it has been going. If procedures are used to make the baby more aware of arm movements or if the arm movements are made less familiar, perseveration greatly decreases. Within the same session, it is hard to argue for mental structures that simply come and go within the same child, but changes in the proprioceptive feedback from a limb are highly plausible. Thus, the "A not B" test can have vastly different developmental explanations. A baby's reach in space is not independent of context. The issue is the experimenter's theoretical reach. If we do not first analyze the tasks we create for our subjects, we cannot hope to explain bigger things such as mind or the structure of knowledge, or at least not seriously (Thelen & Smith, 1994).

PSYCHOLOGY'S BLACK RAT

Some textbook illusions come from the tendency to persist not with the same task or setting, but the same animal. Few would dispute that the "white rat" has mascot status in psychology. Most associate the white rat with how individual animals learn in a Skinner box or maze. But rats as a phyletic group are highly social and highly successful, generally outwitting humans' attempts to contain them because of their ability to learn socially (see Galef's, 1982, 1996, work on social learning of food preferences). The social part of rats' normal task biology is much less widely communicated in our basic texts, if mentioned at all, it is in a special section on observational learning, which is treated as the exception, not the rule. Studies of social learning in black rats are probably even less often cited, as textbooks tend to jump from phylum to phylum, not rat to rat.

In the 1980s, Terkel and his associates discovered that black rats (*Rattus rattus*) could strip the bark off the cones of Jerusalem pine trees, trees planted during the 20th century (Terkel, 1996) That the rats were in such a habitat at all was also a surprise and represented an opportunity to watch animals exploit a previously unavailable niche. Terkel discovered that adults from another population never exposed to cones did not learn the skill of pine cone stripping by handling the cones or by observation, and that young rats from the population had to learn to strip cones. In a series of laboratory experiments, the role of the mother became clear as she could demonstrate the behavior as well as provide materials for trial and error. Although learning from the mother was important, it also turned out that access to partially stripped cones was enough for some adults to acquire the habit and to facilitate the learning in the young. Thus, pine cone stripping is a social behavior available for the "taking" if one has access to adult demonstrators and the materials.

What prevents its spread to new species such as wood mice or squirrels, species also found in the new niche of the pine forest? The spread is constrained because the behavior is enmeshed in a social safety net. In all their years of work, the investigators have never found partially stripped cones lying about in the wild. Access to these learning tools comes only through interactions with one's mother from whom cones can be snatched. Thus, a social task one would think to be feasible, learning by finding half-eaten cones, is not available. Another species or population would have to steal the social system, mother, pups, and all, to get the behavior. That is the power of a social system of distributed intelligence, a system not visible at all in the laboratory world of the "model" species, the white rat, or in the tasks we typically ask it to perform.

Integration or Cubism?

What holds these examples together is that each demonstrates that need to view behavior within a system and to understand the task from the organisms' point of view. In each case, when only part of the behavioral whole is accentuated, or if others parts are too highly discounted, we are led astray. Connecting information made up of excised parts often leads to frustrating examples of integration, producing cubist images, an eye here, a leg there, half a body somewhere else.

In a recent *Science* report, Iacoboni *et al.* (1999) ask about cortical mechanisms of imitation. Using functional magnetic resonance imaging (fMRI) techniques, they present data suggesting enhanced signal intensity in three areas of the brain (area 44 and parietal cortex) when human adults are asked to imitate a finger movement as opposed to watch a finger movement or perform one as a result of specific instructions. The authors take these data as evidence of the presence of a cortical "imitation mechanism," an internal module tuned or resonant to the perception of "matching" the behavior of another. The conclusion requires evaluation, however, in light of the tortuous course taken. First, the original example of imitation given by the authors, the premise for a neural resonance mechanism, is neonatal imitation of human facial and manual gestures. Newborns have been reported to imitate tongue protrusion in response to seeing an adult perform it. It is a textbook favorite and those outside the field seem wholeheartedly to embrace it (Wilson, 1998). The problem is that it may be a mirage. Jones (1996) has shown that one can also elicit similar rates of tongue protrusion when babies see a set of randomly blinking lights. Thus, it is not clear that infants really do imitate at very young ages. As a result, the rationale stated by Iacoboni *et al.* for imitation is really a leap of faith, not a step forward based on solid facts: imitation must be a very simple neural mechanism because newborns can imitate. But Iacaboni *et al.* do not study young babies. They ask adult subjects to do something quite different from the babies—to follow instructions about when to imitate. But, in keeping with Jones' finding, the conditions for imitation, as opposed to observation, were perceptually more dynamic. Thus, with more controls, it is not clear what the task really is from the subject's point of view.

The authors also bring to bear data from nonhuman primates (where the kind of motor imitation is not described) to argue for the particular location of the neural substrate they believe to be implicated and invoke another kind of imitation, children learning language, as another behavioral premise to help delineate the neural search. And so, we have three kinds of imitation, all of which are very different tasks. But all are now connected in cubist fashion to create a picture of the underlying substrate for imitation and to support the idea of a "direct matching mechanism" to effect imitative learning.

It is important to follow the trail of substitutions used to produce an integrative picture. Babies are hard to study and even harder to use in fMRI procedures, and so adults and a new kind of imitation were substituted, a common bypass technique in integrative work. There is no problem as long as the reader *knows* that to keep the conceptual patient alive, there was the need for such surgery. But if the investigators themselves do not know that the starting premise, neonatal imitation, has another explanation, just what is being integrated? This study is not unusual in the number of biological proxies employed nor the number of generalizations from different levels of analysis.

Thus, to summarize, as we look over the time since the last edition of this volume in the *Handbook*, we have serious concerns about the possibility of *veridical* vertical integration through levels of analysis. Perhaps the image is beautiful, but the pathway to its realization is tortuous. But, as the unrelenting accumulation of facts goes on, new reasons emerge making the need for some new kind of organization necessary. We now face an obesity of bite-sized facts, unaccommodated by current theoretical structures. Perhaps it is the sheer volume of scientific stuff that is at the root of calls for multidisciplinary work. It is a call for help.

Our proposal to focus on tasks suggests ways to change the kind of informational units as they are generated to allow them to cohere and form robust, but manageable, units. Task biology is a way to embed behaviors within an ecological framework and thus exploit natural structure (E. J. Gibson, 1969; J. J. Gibson, 1966; Lehrman, 1970; Oyama, 1985; Thelen & Smith, 1994). In the preceding paragraphs, we talked, at the general levels of the nature of science and nature of behavior, about the problems of ignoring context. In the next section, we explore specific experiences, in particular, a scientific "lie" of our own making. We want to recount how hard it can be to edit a mirage. Our plan is to describe old and new studies and show how the more recent social approaches altered previous conclusions about the nature of productive units of analysis in the study of communication. These experiences should help to explain our growing interest in a task-oriented approach and our skepticism as to how such an approach fits with the prevailing images of integrative work.

Early in the Day

When we began to explore vocal development in the 1970s, we had two goals. First, we wanted to focus on naturally occurring early experiences. Second, we wanted to track how a developmental system worked when things went right, not when things went wrong. Both of these goals were reactions to then popular developmental paradigms in which the early experiences imposed on animals were quite alien for the species involved. Moreover, the more frequent practice was to remove, rather than add, social experiences so as to disable development and then try to rehabilitate it (e.g., Harlow's studies of attachment or lack thereof; Harlow & Harlow, 1962). In contrast, we sought a system where the constructive properties of development were visible, but in which developmental change occurred at a slow enough pace (not the 1-day window of imprinting) to allow us to see the process.

We chose to focus on song learning. Singing is something a bird naturally does, but it depends on certain kinds of experiences. Singing is conspicuous, but the nature of its ontogeny is not (Kroodsma, 1978, 1996; Slater, 1983). Learning to sing is compounded out of the acts of vocalizing, socializing, discriminating, attending, imitating, improvising, rehearsing, and memorizing, to name only the most obvious. The study of birdsong is also natural for a task-oriented biology, as there is a clear outcome, how song affects the organism's ability to reproduce, an outcome with implications for understanding incipient processes of speciation. Singing is also an example of social learning, as species-typical songs pass from one generation to the next.

The last decade of study of birdsong has been productive, but as overwhelming as the rest of science. Although Tinbergen's questions of cause, development, function, and adaptive value remain the mantra, the field has followed the trend of science in general of increased specialization. Thus, to get answers to all four of Tinbergen's questions for the same species is difficult. Part of the problem is that a major outcome of the last several decades has also been one that does not easily fit into Tinbergen's schema, the finding of tremendous inter- and intraspecific diversity. Even among the less than 5–10% of all 4000 or so songbird species that have been well studied, each new fact seems to halve or quarter the scope of a previous generalization. Some birds sing many different melodies, others just one; some birds share songs with neighbors, some do not; some species imitate, others improvise; some learn only when young, others for much longer, some birds that learn song are technically not songbirds, but parrots or parakeets. Thus, even this single sea of facts has risen to daunting size.

One rule appears to hold: young learners (typically males) need to hear adults sing to develop species-typical song (Kroodsma, 1982; Kroodsma & Baylis, 1982). This statement is also true for the species we have studied: the brown-headed cowbird, *Molothrus ater*. To read our own early work and others' reports of our work up to and including the present, however, the cowbird would seem to be the exception to the rule. The textbook story of the cowbird, which we have helped to create, often reads as if cowbirds are the exception to the rule, i.e., when reared without adult males, young males produce a species-acceptable song, *unlike* the ontogeny of other songbirds. What is the truth?

Because cowbirds are brood parasites, it has always been easy to suppose that their song development differed from that of other passerines (Lehrman, 1970, 1974; Mayr, 1974). The species had originally appealed to us because its upbringing suggested that it might display exceptions to common developmental rules, as the young are not raised by their own species, but by a variety of host species (Friedmann, 1929). We thought that viewing exceptions would more clearly demarcate general rules and the conditions supporting them; others had expressed similar opinions (e.g., Lehrman, 1974; Mayr, 1974).

In 1977, we published data suggesting that we had found such a difference (King & West, 1977). In that study, we had hand-reared naive male cowbirds which had never heard adult male song and recorded their songs when they reached sexual maturity. The songs the naive males produced differed on several acoustic dimensions from a sample of over 400 songs of wild males from the same geographic area. Thus, the naive males appeared to be like other songbirds. This was not what the idiosyncratic cowbird was "supposed" to do. But we also played the songs back to female cowbirds, along with the songs of other species and control songs from a wild male. As expected, females responded more to cowbird songs than those of other species. The surprise was that the females responded most to the atypical songs of the naive males. Thus, although naive male songs were abnormal, to females, they seemed supernormal.

Now here was a good story: cowbirds naturally deprived of early experience have

MEREDITH J.
WEST AND
ANDREW P. KING

just the fail-safe mechanism a brood parasite should have! As we wrote, here was "an independent system designed to insure identification during the most important context, the breeding season" (King & West, 1977, p. 1004). Maybe their songs are atypical, but they work. The mechanism seemed quite simple: nature had innately endowed young males with sexy song elements. The "just-so-ish" gist of the story undoubtedly acted as a magnet to hold the parts together and worked against us when we tried to pull it apart.

As it turns out, we were wrong on three counts. First, across playback experiments, on average, female cowbirds do not find atypical songs of acoustically naive males more stimulating than the songs of wild, successfully breeding males (West, King, & Freeberg, 1994). We had unwittingly created conditions favoring the naive males' songs, as is discussed below. Second, we now know that socially and acoustically naive males cannot obtain actual copulations with their atypical songs, even if the songs are acoustically potent (Freeberg, King, & West, 1995; West, King, & Freeberg, 1996). Third, the social experience of females affects playback responsiveness and mate choice and thus may change which songs they find attractive (Freeberg, 1996, 1997).

Concerning the first point, years of playback tests of females housed without males have revealed that females' copulatory responsiveness changes with the range of song stimulation presented. Thus, if females only hear the songs of acoustically naive males, they may respond on roughly 50% of the trials, as, given their male-deprived state, these songs are the "best" available. If species-typical songs from wild males are introduced into the same playback set, however, responding to the songs of the naive males quickly drops relative to the wild males' songs, with only some overlap between the least effective wild songs and the most effective naive songs (West *et al.*, 1994). The relativity of female responsiveness could have affected the original 1977 experiment. The 1977 females probably responded more to the atypical songs of the naive males because we did not test the full range of variation in wild cowbird song. The songs of the naive males may not have been *super*normal, the songs of the wild males we played back may have been *sub*normal. Hand-reared acoustically naive males produce a much wider range of variation than do wild males, thus some naive males' songs can be among the very best, others among the very worst. We will explain more about the second and third counts subsequently, but in general, as we learned more about the actual task of courtship and mating, we saw ever increasing limitations to our proposition about an innate safety net.

We have written many times about these facts-turned-fictions, but we still find the old facts in new places (Ball & Hulse, 1998; Michel & Moore, 1995). Part of the reason is our own doing: we often discussed the seemingly anomalous nature of the original finding, supernormal song in naive males, in an attempt to focus attention on the need to understand more current work. Our strategy did not work: what was remembered was the possibility of supernormal behavior, the mirage, and not the possibility of contextual relativity. Even after a paper entitled, "Innate is not enough," we found our colleagues dubbing cowbirds as exceptions to songbirds' developmental rules (West, King, & Duff, 1990). In that paper, we recounted the many studies after 1977 showing that such males could not succeed because females used characteristics other than song to choose mates, such as dominance and courtship persistence. We also showed that adult males attended to very effective songs from young males and would attack naive males singing such songs. Thus, supernormality, even if it did develop, could be detrimental. These males were like Kuo's rat-loving kittens or other *in vitro* preparations (Kuo, 1967). Like others, we had created animals with behaviors that had no context outside the laboratory. But, still, the message was not clear.

The Play's the Thing ...

During the last few years, we have taken a more direct approach to understanding the task of song development by focusing on its goal, successful courtship. At the same time, we have used contexts in which the animals lived in groups or were tested in groups to begin to incorporate the social composition of learning. To see what social and vocal skills males must have to succeed, we have looked at the parts of the story in a more connected manner. Thus, we carried out a series of studies in which the major players were naive male cowbirds. We followed them from 75 days of age until 2 or 3 years of age. We exploited social surroundings to look directly for any fail-safe system based on vocal behavior. The findings summarized below come from three longitudinal studies measuring the effects of social experience on vocal outcomes (Freeberg, 1996; Freeberg *et al.*, 1995; King, West, & Freeberg, 1996; West *et al.*, 1996; West, King, & Freeberg, 1997). We have omitted many details in favor of synopses of the studies. All the details are published. In writing up the studies, however, we found that the conventional elements of style for journal writing did not capture the dramas we had watched unfold. Although each year's study was a true experiment with independent and dependent variables, it seemed more like a summer theater production, with us as producers, not experimenters. That sense came from the considerable backstage efforts we needed in order to introduce more elements of social settings into each scene and because the script was organic and dynamic. We did not know what the major players would do: in this case, young male cowbirds being asked to use their song outside the confines of a sound chamber.

To create male cowbirds with different social and song histories, we housed young male cowbirds individually in sound-attenuating chambers. Half of the males were provided with several female cowbirds from their local area (Fmales). The other half we housed with canaries (Cmales). By the time the action took place in the spring, the young males had had extensive social exposure to either their female cowbird or canary companions, but had not been exposed to the sights or sounds of adult male cowbirds since they were fledglings.

The first act took place in early May after we had removed the young males from their female or canary companions and put them in two flight cages by group. To look at the males' behavior when faced with new cage mates, we watched each male, in two separate sessions, while in a test flight cage with two unfamiliar female cowbirds and two canaries. Thus, we retained much of the simplicity of their original housing, but allowed the males more choice. Would the males discriminate their species and spend more time with cowbirds than canaries? Would they court the females? In this context, we began to see unfold what our playback methods could not reveal. The Fmales sang and courted the female cowbirds, albeit only haphazardly compared to what normally reared males would do. But the Cmales appeared to court the canaries. Even when females approached and solicited courtship, the Cmales' attention was on the new canaries. We saw no safety net for species identification during the breeding season, the time period we had targeted in our 1977 explanation. Moreover, the core behavior, the male's song, began to differentiate before our eyes: singing to a female appeared very different from singing to a male or singing directed to no individual.

The subsequent acts occurred in large aviaries, where life was very different for the previously sheltered C and Fmales. In the aviaries were various classes of birds: canaries, starlings, female cowbirds from two geographic areas, and eventually adult male cowbirds from the same respective populations. There were, of course, many other differences including rich opportunities for foraging to encounters with resident frogs in the aviaries' ponds. Thus, the C and Fmales could do many things, many of which had nothing to do with courtship. In these socially rich conditions, we looked to see if the Cmales would give up their unreciprocated pursuit of the canaries and to see whether the Fmales could court female cowbirds selectively. Would the Fmales show more attention to unfamiliar females from their own population as opposed to females from a distant area? And, what would the males do with one another? We gave the birds time to show us what they could or could not do. These were not 20-min tests, but 2 or 3 hr of observation 7 days/week for 6 weeks.

The Cmales showed no measurable changes in their social orientation, Indeed, it seemed all the more amazing to watch in the large expanse of the aviaries. The Cmales would fly by female cowbirds with no more attention than they gave to the starlings. But the Fmales, who had attended to females when in the confines of the flight cage, now focused most of their attention on each other. They reminded the observers of teenagers at a dance where the boys talk together across the room from the girls. Some of the Fmales did sing to females but were rarely successful in obtaining a copulation or maintaining the persistent following and accompaniment of the female as seen in nature or in our aviaries with wild-caught males. The Fmales also showed no selectivity toward their own population. Were the females making the yearling males look bad? Unlikely. The females were responsive when sung to and did copulate with the two males who sang the most to them. They also attended to wild males who were attracted to the goings-on in the aviary and sang to the birds through the wire. Thus, the F and Cmales' lack of interest in females did not appear to originate in actions by the females.

We wondered if role models would help and so midway through the breeding season, we brought in adult males. The setting reorganized almost immediately, but not to the young males' advantage. Females that had been rarely seen were now highly visible in front of the adult males. We saw typical courtship and strong patterns of mate assortment by geographic population, but only among the newly added adults. When we repeated the entire production over the next 2 years with a new cohort from a different population, we saw essentially the same thing, naive male cowbirds appeared largely uninterested and highly nonselective when attending to females (although we saw less overt attention to canaries). Even when females did adopt copulatory postures, males did not mount. About 40% of the time, males sang and only watched as females solicited in front of them. In today's vernacular, the males didn't get it.

We wanted to publish the data from the first cohort in bold print as "COWBIRD COURTS CANARY" to end any argument about the flaws in the 1977 study. The new studies were sweeping dismissals of our original proposition of the existence of a built-in identification program geared especially to mate recognition. Under the social conditions present in the more complex settings, we saw that behavioral pieces did not suddenly or gradually snap into place, an image fostered in many ethological works by connotations of terms such as innate releasing mechanisms. Prior to this work in the aviaries, we had focused on whether the birds could develop the appropriate behavioral forms in the first place. It had not occurred to us to ask if the animals would know how to connect the behaviors to appropriate goals. These data also put the female playback procedure in a new light: we realized the importance of the seemingly innocuous tasks that we carried out on behalf of the males, i.e., choosing a song and getting the song close to a female before playing it. The importance of proximity is paramount. Indeed, we had learned by using the playback bioassay that even very small changes increasing the distance at which songs were recorded greatly depressed female responding. The song is a close-distance communication signal: the male is often less than a foot away singing at about 85–90 dB ... hardly a whispered sweet nothing.

In sum, what we did by creating a more social setting was to join together many potential parts of a young male cowbird's ecological world and put him in its midst. By parts, we mean the social tasks possible with different classes of individuals Thus, in the aviaries, we had different species, ages, and sexes of cowbirds, individuals varying in experience and geographic background. We also changed the social dynamics periodically by adding or removing birds as we tested courtship competence. Thus, when asking the males to do more than one thing at a time, we were specifically requiring that song be placed in a realistic contextual state, allowing us to see how much power it did or did not have as a single behavior, a different kind of bioassay than playbacks to females. After using that setting, it is even difficult to call song a single behavior because whether it is directed to males or females or no one produces different immediate consequences.

What was missing among our subjects was the ability to connect behaviors to outcomes. Lorenz (1957) called the presence of such automatic connectivity "the hereditary teaching machine," and considered it essential to understanding how innate behaviors worked, i.e., how behavioral pieces were shaped to fit together and work as a unit. He assumed the connectivity was also innate: in cowbirds, it was the connectivity, not the behaviors, that was missing. What the birds did not seem to know was what to do with their behaviors. In fact, the arrow-like progression does not always work in stickleback (Bolyard & Rowland, 1996; Rowland, 1994). Like cowbirds, social learning and perception of context affects recognition and response to cues so that behaviors are directed in other ways. Even the famous red belly of the female stickleback does not always elicit courtship from a male.

THE FEMALE BIOASSAY OF SONG: BUYER BEWARE

In telling this story, it is at about this point that someone usually says, "OK, so males can learn, but isn't the safety net in the female? Maybe you were closer to the truth than you thought." The answer is no, the female does not have a genetically constrained safety net for mate recognition. There are three parts to the female's story. First, we must say more about the functional cornerstone of so much of our work, the female bioassay. Then, we discuss her actual role in song development. As it turns out, looking at males and females as a unit does suggest a safety net of sorts, but one woven experientially out of social interaction between the sexes. Third, we will show how the system might work in a study of cultural transmission.

Thorndike noted that scientists, being human, are quick to see the marvelous rather than the mundane (Thorndike, 1911/1965). The result of this tendency may be to read more into what is actually there in a single measure if it is at the marvelous end of the continuum. Certainly to us, the bioassay of song function we were using resided at the marvelous end. The response is rapid and unambiguous: the sound goes in the female's ear and down her spine, sometimes before the song is even over.

At the time we began using the bioassay, the only alternative was to judge the normality or typicality of males' songs by looking visually at sonograms of song structure. While this measure has many uses, it is functionally constrained because our eyes are not tuned to female cowbirds' ears. Indeed, the parts of the song that first struck us as most atypical turned out to be ones that did not influence female responsiveness. We judged the terminal whistle to be the most differentiating feature, but females' copulatory postures happened before the whistle.

Playback tests have now been carried out for a number of songbird species, but we suspect that we hold the record for continuous use of the song bioassay, having done at least one playback study of female cowbirds every year from 1973 to 1997 (Searcy, 1992). We have gained an extraordinary appreciation for the experimental context and task demands. We now know that female responsiveness to song in the context of playback testing has limitations that must be well understood in order to interpret outcomes. For example, responsiveness is relative to the kind and amount of stimulation received, as noted earlier. For example, we have done playbacks to learn about specific acoustic dimensions of song where all we have played to females are songs from another cowbird subspecies; they respond frequently in this context (King & West, 1983a). They would not, however, respond as frequently or at all if they also heard songs of their own subspecies. Thus, the bioassay only gives the "relative appeal" of a song in that single context. Thus, one must identify what aspects of playback responding carry across contexts.

We have identified one parameter of female responding that may tap a stable form of individual variation, a female's pattern of choosiness. In the 24 years of testing, we find stable patterns of individual variation among females within and between experiments. Some females are highly responsive, some less so, and each year 1 or 2 never respond at all out of a sample of 20 or so. To calibrate the females across yearly cohorts, i.e., to see if the range of responsiveness is similar, we have often used the same set of playback songs across some years. To do so, we created a special playback set composed of six songs we knew were consistent winners and six that were consistent losers. In the lab, it became known as the good–bad test tape. We found extraordinary stability across sets of females in their reactions to the songs, each agreeing with previous years' females with correlations above $rs = +0.80$ (if all the females were from the same geographic population) (West & King, 1986).

As we studied the problem, we realized, however, that females expressed their preferences differently: some responded very frequently, and thus, although they responded more often to good songs, they also responded on a number of trials to bad songs. Other females responded less often in general, but when they did respond, it was to the same good songs as the other females, and rarely to bad songs. The latter females were thus more "choosy" or more selective, conserving copulatory energy for the best songs. It took several years to realize that every year we saw these two "types" of females. We were also led to this measure of choosiness because we had watched some females in the playback context and in breeding aviaries. The females that seemed most reactive to song, the least choosy, were often the ones not systematically courted by the males, whereas the females that looked as if they were made of granite when males came near were vigorously pursued.

These individual differences were then used in a study of the anterior forebrain pathway within the song control system (Hamilton, King, Sengelaub, & West, 1997). We were especially interested in one nucleus, lMAN, as it has been implicated in initial song acquisition in male songbirds and was thus a logical candidate for receptive song acquisition in nonsinging females (Bottjer & Arnold, 1986; Bottjer,

Halsema, Brown, & Miesner, 1989; Nixdorf-Bergweiler, Lips, & Heinemann, 1995). It also was monomorphic between male and female cowbirds, an unusual finding relative to other songbirds where the female does not sing. We measured the volume of lMAN in a set of females following playback of the good–bad test tape. Selectivity or choosiness was highly positively correlated with volume of lMAN. Subsequently, the effect has been replicated in a second set of female cowbirds from another geographic area, using different "good–bad" songs.

No other significant correlates of choosiness were found with other song-related structures. Thus, nonsinging female cowbirds possess a "song" nucleus typically described as necessary for initial acquisition of song in singing males and its size is comparable across sexes. Moreover, consistent individual differences among females mapped extraordinarily well onto neural variation in lMAN, providing an anatomic validation of the playback assay as tapping a reliable source of variation among females. Moreover, it taps variation in perceptual selectivity at the level of individual differences in songs from males within a population, the level at which the discrimination occurs in nature.

Despite these elegant biological correlates, the female song bioassay simply cannot bear the burden of explanation that we originally assumed. It is a beautiful behavior and seductively believable, but true only in some light. It can be used profitably and reliably to learn much about song structure, the measure we wanted it to replace or supplement. Thus, we can use it to identify the kinds of acoustic properties that are necessary (but not sufficient) for a stimulating song to lead to copulation. No matter how much we learn about its neural structure or variation across females, however, it alone cannot predict copulatory success of males in the actual social task nature has set. Indeed, when one watches females in such a social setting or in the wild, one would wonder if females ever respond so positively or animatedly to a male. Thus, the power and concomitant danger of the female bioassay is its deceptive simplicity. It has fine features but fine print.

Mum's the Word: Breaking the Female Wall of Silence

These limitations of the use of the bioassay still leave unanswered questions about the development of female preferences, whether measured by playback or by watching courtship. In 1983, we concluded that females possessed an experientially "closed" system, a conclusion that was premature, and constitutes the second reason that the safety net cannot be assumed to be genetically provided (King & West, 1983b). We made the statement about lack of modifiability on the basis of a traditional laboratory design in which we housed acoustically naive, hand-reared female cowbirds with males from their own area or from another subspecies: 25 of 26 females preferred playbacks of their own populations' songs. The songs of the two subspecies differed acoustically in several ways, including the presence or absence of certain song elements altogether. We did find one population in Oklahoma, which is near a subspecies border, where similar housing did broaden females' preferences (King, West, & Eastzer, 1986). In general, however, under these conditions, social experience most often had no effect on female discrimination in playback tests of geographically distant song variants, a level of discrimination we admit may be ecologically irrelevant. Our reason for now believing in an open system for females does not rest on having studied only populational preferences. It rests on having misconstrued the nature of the task of the female as primarily a listener and social bystander during song development.

The first major clue to the active nature of forming perceptual preferences came when we realized that the procedures we had used to modify females had had the opposite effect. We discovered that the males we had been using as potential modifiers of female preference were in fact being modified by the females: their song repertoires were changing during the period of time the males were employed to affect female perception (King *et al.*, 1986; West & King, 1985). The females were not hearing stable renditions of the songs of the other subspecies of males, but were gradually exposed to changing sounds as males began to include the song elements found in the female's home population. When male cowbirds were housed with canaries during the same time period, such changes did not occur in the males' repertoires (King & West, 1988). What were males and females doing? What was the task from their point of view?

This finding led us to study social interactions during development, well before the time of mate choice. We found that female cowbirds have the capacity to respond socially to emergent acoustic patterns from males (West & King, 1988a). We have found the effect in three populations representing the three subspecies of cowbirds. The most conspicuous evidence of social shaping was the discovery of the female's use of brief wing movements coincident with the male's singing of certain songs. Songs receiving wing strokes were more likely to be retained by males and to be more potent releasers of copulatory postures relative to other songs from the same male not receiving wing strokes.

Wing strokes and other social actions of adult females are associated with different rates of progress through stages of vocal development of the young males. We have found that greater social interaction between the sexes leads to faster acquisition of stereotyped songs, earlier cessation of practice, and more potent song types. A major component of the social interactions is whether females stay when males sing or fly away, sometimes even before the song is finished (Smith, King, & West, 2000). Female proximity gives males an opportunity to receive more social feedback to individual songs and, conversely, gives females more chances to shape content by more subtle cues such as wing stroking or beak or body movements.

Discovering the wing stroke and female shaping of song was a major break-through because it demonstrated beyond all doubt that female birds that do not sing can still affect the song learning process. Thus, imitation could not be the only route to song learning as had been proposed. The data showed that operant learning or shaping now had to be brought back into the picture, having been banished from the birdsong field at about the same time it was ruled out for imprinting (see also Marler & Nelson, 1993; Nelson & Marler, 1994). The discovery of female signaling is also a neat story because it also reveals a communication system embedded within the vocal learning system, a gestural system in which the female is the sender and the male is the receiver. But we hoped we have learned something in two decades and do not want to isolate wing stroking out of the entire system of yearlong social interaction. For our purposes in this chapter, the wing stroke is most appropriately viewed as a symbol of the kinds of ongoing, organizing processes that have often been missed in developmental studies because the actual nature of social interactions has been inferred, not investigated.

But the fact of female shaping still begs the question of possible female modifiability. Wing stroking and other responses could reflect innate or socially modified female perceptual biases. They probably reflect both. Female cowbirds appear to need no postnatal experience to show preferential responding to cowbird song as opposed to the song of heterospecifics, a broad level of discrimination that may not come into play at all when birds breed, having been together in flocks for almost 1 year (West & King, 1988b). Although females appear to discriminate cowbird song with no postnatal experience, however, the prior work does not rule out that social experience naturally changes or fine tunes their preference; females show wide variation in choosiness. Is selectivity affected by social experience?

We are accumulating evidence suggesting that social experience does affect female song discrimination. For example, we compared the preferences of juvenile and adult females for atypical versus typical song from males of their local population. Adult females made finer distinctions among songs than did yearling females. These data suggest that juvenile females may not give as specific or fine-tuned a feedback to males, a proposition we still need to test (A. P. King, unpublished data). Ongoing work with adult and juvenile females in aviaries shows some effects: young males housed in an aviary with adult females were more successful at obtaining copulations with unfamiliar females than were males housed with yearling females (S. Schlossberg, unpublished data). Thus, there are differences in perceptual skills that seem related to experience.

BIASING FEMALE MATE PREFERENCES: OPENING THE CLOSED PROGRAM

Powerful evidence of the implications of such plasticity comes from studies of mate preferences, not simply song preferences, carried out in our lab. In a 2-year-long study, Freeberg (1996, 1997) studied the mate preferences of females given typical and atypical song stimulation. In a longitudinal design, he showed that social preferences of juvenile females for potential mates differed when the females were given different social experiences during their first year. The contrasting social experience consisted of residence in large aviaries with other young females and males from their own local population and with male and female adults either from their natal area or from a distant area. But the young females were not forced by physical constraints to be near the adult males or females, or the other juveniles, for that matter. Moreover, they could hear and see wild birds outside of the aviary. In the original work, the sound attenuating chambers offered none of these liberties.

Under these conditions, Freeberg found female mate preferences were predicted by social experience, not by geographic/genetic background. His design ruled out mate copying and another experiment reinforced the direction of effects: it is the female, not the male, that controls whether copulations occur. But another test of the durability of the social effect followed. More remarkably, the learned differences in preferences were transmitted to a new generation of young males and females whose models were the former pupils from the original study, the experiential F1's.

Freeberg's work has several important implications. First, it shows the power of a social context. We had assumed in the 1983 study that housing females for almost an entire year with males from another cowbird population in triads in sound attenuating chambers would increase the chances of finding malleability, perhaps even malleability not normally seen because of the unnatural context. Our assumption was in part intuition and part tradition. The conventional procedure for testing innate preferences in many animals is housing of the sort we used. Instead, we found malleability when females were faced with a seemingly more complex setting. Thus, failures in other species to find malleability may reflect hitherto unrecognized inhibitory effects of standard isolate housing (but see Payne, Payne, Woods, & Sorenson, 2000).

Second, Freeberg's data show that female copulatory responsiveness toward males is malleable and male responsiveness to females is malleable. If that is the case, where is the net? What holds the system together? To understand this question again requires we step back and consider function. What are the advantages of (1) female shaping of male song and (2) social shaping of female preferences to reflect local experience?

Necessity Is the Mother of Vocal Invention

A critical task for all females, but especially in parasitic species or in species where males offer no resources, is mate assessment. What can a female cowbird use? Is it to her advantage to have an epigenetic stake in song ontogeny? We suggest that her involvement in song ontogeny allows her to implant quality detectors to be used by her local population later on.

We begin with the repeated finding from playback and mating studies that female cowbirds from local populations share preferences for specific song types. The work on choosiness was based on the finding that females show extremely high concordance within a set of songs. They "agree" as to the best and the worst and even in between show only moderate variation. The level of concordance among local females in responsiveness to song suggests that they are listening to or for the same acoustic features across different males. As it turns out, recent work looking at the specific acoustic features affected by female social interactions with males during song development show that across males, the same acoustic properties are affected even when males are with different females. Moreover, females that are most choosy (reserve all of their responding for only the best songs) have a different effect on males than do females that are less choosy. The males with the choosier females ultimately develop smaller repertoires of high playback quality than do the males with the seemingly more responsive but less selective females (King & West, 1989). We suspect that this effect is a simple form of partial reinforcement: males learn to pay more attention to feedback from choosy females because it occurs less often and thus has a higher "signal to noise" ratio. Thus, given social shaping by females, high homogeneity in preference should foster high vocal homogeneity among local males, at least for the parts of the song that elicit copulatory responding (the first half of the song).

If the acoustic features that females use in song or mate discrimination are ones that females can influence during song ontogeny, females would then be choosing males that display the most past attentiveness to female signals. But, and this is a critical feature, the attentiveness need not be specific to a given male, i.e., we in no way mean that females prefer or even interact with the same individuals they may have interacted with earlier in the year. Indeed, in playback testing, we find no evidence at all that females respond more or less to the specific males with which they have been housed than they do to other males in the same experimental condition, but they do discriminate by condition (West & King, 1980). This is an effect we have seen in over 20 cohorts. Thus, the female's effect is generic.

As local females share a percept, however, the males most attentive to females in the spring should show evidence of a widely shared percept in their songs. Thus, song becomes a potential index of male sensitivity to females. In the case of the cowbird, for a male to be affected by females' ontogenetic signals, he needs to do more than simply sing, as we have shown in several studies. He needs to be alert to changes in a female's behavior in order to connect his immediate behavior to her immediate response. Males that, for whatever reason, do not direct songs to females from close proximity do not develop potent songs.

Our studies of captive males' courtship show that some males, even with potent songs, fare poorly in courtship if they do not also display active attention to females (chasing, approaching, coming back for more after being pecked or ignored) and active attention to males (dominance, aggressiveness, and defense of females) (West, King, & Eastzer, 1981; West et al., 1996). The latter form of interaction is equally important, although we have not emphasized it in this chapter. As noted earlier, males also discriminate potent songs and attend differently to males according to their social and vocal behavior. Thus, the female-implanted song structure is a potential handicap and, as such, only serves the females' interest more completely as she assesses how males handle their vocal legacy.

Thus, when a female is courted by a male whose vocalizations contain female-shaped features (e.g., potency) and male-shaped courtship skills (e.g., dominance), she is using his compounded history of successful learning. This history can be operationally defined by measuring how effective males are at courting and copulating with the same females over time (consort persistence level, CPL). CPL is measure of the number of days in which a male meets a criterion of directing a certain number of songs at close proximity to the same female during a sampling period, divided by the number of days in the aviary (Eastzer, King, & West, 1985). CPL is thus implicitly assembled out of song structure, song use, and social vigilance to males and females, and all of these behaviors are assembled out of the nested parts we have been discussing. Not all males in our aviaries, even wild-caught ones, seem to bring all of the parts together. For example, there are males that do not court but whose songs turn out to be highly potent releasers of copulatory postures. There are other males who sing and chase females frequently but also sing and chase males. In both cases, these males' CPLs are low. Thus, the female's part in stimulating social learning is one means to assay the degree of social learning.

How does social malleability in the female seen in Freeberg's work fit into this scenario? We suspect that Freeberg's effect rested on females' opportunity to see adult males and females interact with one another and then with adults and juveniles. These experiences may have narrowed the female's preferences because the narrower vocal output correlated with social output of the males preferred by older females. In Terkel's studies of black rats, the only access to partially stripped pine cones (to practice stripping) came through the mother–pup bond. In cowbirds, the only access for young females to see the full range of song and its effects may come from life in a flock, in the company of other females, something we could not simulate in the prior work using small enclosures.

These data also suggest that CPL is the kind of unit on which females may assess males. The data also suggest the importance of visual attentiveness to that unit. The latter might seem obvious and not that noteworthy, but most theories of song learning and its neural substrates have not looked for visual involvement because they were focused on a vocal part, not the social whole. If we are right that individual components of CPL such as potency or repertoire size (number of song types) or directing of songs to males or females must be assimilated into larger units, perhaps we should be able to see evidence of such integration in the song control region of the avian brain in the same way we saw that choosiness mapped onto female perception. If so, it might add credibility to the practice used here for the CPL of "lumping" rather than "splitting," something many fear to do because it often smacks of a cover-up rather than a discovery.

The avian forebrain in songbirds contains two connected pathways formed by a series of discrete nuclei that appear to have differentiated roles in the motor production and sensory perception and learning of song. It is a distributed system with complex connectivity, perhaps in keeping with the complexity of the compound nature of the function of vocal behavior. To be brief, we studied the brains of adult males whose song and courtship histories were known in detail, with data from several years of each male (Hamilton, King, Sengelaub, & West, 1998). We looked at the traditional measure of song used in many such studies, song repertoire size, and found no correlations with any nucleus. We did find a strong but negative correlation between song potency and Area X, a nucleus thought to play a role in assimilation of developmental experience, perhaps in the winnowing of song to fewer patterns. But we also looked at visual areas in light of our knowledge of male's visual attention to females. Nucleus rotundus (Rt) is the largest nucleus in the avian visual system. Rt has been implicated in visual processing of texture, motion, shape, and color in pigeons (*Columbia livia*; Shimizu & Karten, 1993; Shimizu, Katz, & Cook, 1997; Wang, Jiang, & Frost, 1993). These qualities are clearly key components of social attention to conspecifics and thus could affect attentive singing behavior. Volume of Rt was most positively correlated with CPL (with less strong, but reliable correlations with potency and directed singing, but none with repertoire size). The data are the first to implicate a visual nucleus in song use and to show the utility of compound measures such as CPL. It is further anatomic validation of the role of social learning. Thus, male's visual attentiveness, the skill we believe unlocks his ability to learn from females and males, is important throughout ontogeny and through the period of actual reproduction. And a female's assessment of a male's visual attentiveness may be facilitated when he uses vocal patterns that males and females mutually shape. Thus, the safety net is not prebuilt into either sex, but socially built up between them.

These inferences lead to certain predictions about social life in cowbirds. Our data suggest that birds pay attention to age and sex of individuals and can learn different things from such attention. Do the birds show any signs of attention to these dimensions? The data also suggest that social access to different age or sex classes may be important at certain points in the year. In the laboratory experiments, we provided the access, but what if the birds are left to themselves? We are now beginning to answer these questions using more complex social settings. We are completing an investigation of social context in a setting even more unstructured than Freeberg's. We spent 1 year watching a group of 74 cowbirds, juvenile and adult males and females, in a very large aviary (the aviary is as long as a football field, with multiple roosts and perching and foraging areas). We measured social assortment by gathering data on "nearest neighbor" patterns, tallying which birds were within 1 foot of one another during sampling sessions. We also measured singing patterns and social context during singing. Would we see order at group and individual levels? Would we see the possibility of social transmission in patterns of proximity (Smith, King, & West, 2001).

Briefly, we saw strong social assortment by age and by sex: young birds perched near each other, adults with each other, and the sexes segregated as well. Thus, we saw order at the level of attention to age and sex. There were some departures from these patterns, however, and these departures indicate windows of access for learning. First, in the late winter, young and adult males were found together much more often than in the fall when the ages remained apart. Second, throughout the year, adult females showed the least assortment by class. Said another way, they had more neighbors from all the classes, meaning juvenile males had access to them from the beginning. The adult females' pattern of sociality suggests that the social niche exists in more socially complex settings for the kinds of learning we saw in lab settings. Third, the young birds were more social, and were found in the company of some other bird more often than were adults. Among the juveniles, however, the frequency of social contact still varied greatly. We followed the young males through their first breeding season looking at courtship competence. Even though all the young males had the same general social access of adult males and females, their CPLs ranged from 0 to 1. Thus, the opportunity for social interaction does not always translate into action or consequences. Moreover, as hoped, we found correlations between social assortment and breeding success: juvenile males who associated more with adult males and juvenile males who showed more frequent social contacts had higher CPLs.

These data thus begin to lay out an epigenetic landscape created by the birds themselves. Given multiple access to food, shelter, perches, and roosts, the design of the aviary did not force the social order that appeared. The kinds of social dynamics we saw are also found in wild avian populations. The observations also suggested that learning is available only through negotiating a social structure set by age and gender biases. In follow-up work, we are looking at aviaries where not all ages and classes are present. One of the current aviaries is composed only of adult and juvenile females. With no males around, the ambient social atmosphere does not involve reacting to song. Throughout the fall and early winter, we have not seen segregation by sex; adult and juvenile females intermix freely. Thus, attention to age may depend on attention to gender. It is hard to resist seeing the females as living a relaxed life, as they experience a context with no competition for males or by males. But will the young females have learned to be discriminating listeners? That is the next question on the agenda.

A CALL FOR TASK BIOLOGY: AN INTEREST IN REALITY

> Consider the fruit fly, *Drosophila melanogaster*, which thousands of geneticists have studied. For over 80 years, it has been at the center of genetic research ... That effort has paid off ... we can do astounding things: we can clone fly cells and remove and insert genes at will ... But after all that investment and knowledge we still do not know how the fruit fly survives through the winter. (Suzuki & Knudston, 1989, pp. 21–22, cited in Michel & Moore, 1995.)

By reviewing our work in detail, we hope to have demonstrated that the question of the reliability or quality of behaviors used as focal points in research must be guided by ecological principles. Once behaviors are taken too far afield of their natural settings, the nature of the knowledge acquired loses functional value. What unites the cowbird work to the earlier textbook examples is the danger of studying behavior without attention and respect for the contexts present in the environments of species. Hence, the challenge before us is for an ecological task biology to trace connections between the tasks on which we base our knowledge and the contexts in which organisms actually display them. Establishing a task biology, i.e., paying closer attention to what animals really do, is, however, not enough. We have to look at more

MEREDITH J.
WEST AND
ANDREW P. KING

animals doing more things. We need to expand horizons on the tasks and on the organisms we study. When Charles Darwin took his voyage around the world, he was shaken as much, if not more, by the variation he saw in human form as by the varieties of animal and plant forms (Rozzi, 1999). How could he reconcile the mannerisms of his English colleagues with those of South American tribesmen? To Darwin, the latter acted like animals, sexually exuberant and morally vacant primates. Here was the literal "descent of man" in the flesh. Had Darwin not left England, he would not have witnessed variation that struck so important a chord. What Darwin witnessed was an assembling of the parts of human behavior in a manner and to a degree he thought not possible for one and the same species. It was culture shock writ large.

When conceiving of a task biology, Darwin's experience looms large in our thinking. We believe that some of the inertia against adopting contextual approaches comes from fear of having to reconcile the variations in behavior one sees when contexts shift. It is not easy to introduce change into paradigms. Just try to shake researchers loose from college sophomores or white rats, let alone direct them to South America! In 1989, Kroodsma pointed out possible problems with playback designs used to test songbirds' responses to the songs and suggested changes; changes requiring wider sampling of behavior across different populations of songs and birds (Kroodsma, 1989). These changes were important because of the ongoing debates about the existence and function of song dialects, acoustic variation among songs within neighboring populations of the same species. Many of the concerns he expressed eventually led to needed changes in methodology and less reliance on studies of single populations. But, not, we should note without a fuss. The original article was published in *Animal Behaviour*, as were the many short communications from those who felt wounded by the article or who simply disagreed on other methodological grounds. Several editors of the journal began to call it the "Kroodsma" section of the journal. We recommend readers take a look because it shows how hard it can be to introduce change, especially when the changes contain the implication that some past work could not now be considered true or definitive. And keep in mind that the journal published only the most mannerly end of the continuum (Kroodsma, 1992).

We have made suggestions comparable to Kroodsma's about the asocial conditions of the tradition of work termed "animal learning," work that now forms the basis for many conceptualizations of the workings of the central nervous system. If Kroodsma temporarily hit a stone wall, we continue to hit the rock of Gibraltar. We do not know what argument to use to persuade researchers that a wider sampling of the conditions in which animals learn is essential. Perhaps the decline in the numbers of researchers interested in such paradigms reveals the ultimate answer; if we are patient, their students will see it themselves. But knowledge about social learning exists now and already shows the need to consider how the narrow use of tasks employed in animal models could serve to defeat integrative goals.

The study of learning is thus at a decision point. We can continue to learn about parts, in even more exquisite technological detail, or we can study animals in social contexts where animals must do more than one thing at a time. A contextual or task biology has to begin with actual tasks and guard against oversimplifying the animal's options. More of the behavioral parts, the separated abilities, must be examined at the same time in the same settings, and then in different settings at different times. One of the great surprises we have found in putting animals in larger groups and giving them more social freedom is that it has not deprived us of the ability to see or measure developmental change. Behavior and its ontogeny is a complex system, as the term is formally used in the lexicon of adaptive systems or complexity theory. A fundamental finding of such research is the imbalance between effects on the behavior of an excised part and the system from which it was excised. The greatest effect is on that which is removed and not on the system from which it was removed, e.g., think of cutting a leaf from a plant or taking a single worker ant from a colony (Bar, 1997). Isolating individuals from the social system in which they normally live and from the environment to which they are adapted can produce the same imbalance and lead to distortion in theories. The new approaches vastly increase the burden of description to the point that new technological tools are needed to increase feasibility. Fortunately, video and audio technology, biological telemetry, behavior-based robotics, voice recognition data collection, computational models, and oceanic amounts of raw computing power are now here to assist in handling what may seem like added complexity, but which in the end, may be a restoration of order. Thus, we may be able to use the newly available computational power to find patterns in social settings hard to see with the naked eye and then to vary these patterns in simulations to guide the empirical steps (Schank & Alberts, 1998).

END OF THE DAY

> It is easy to look, but learning to see is a more gradual business, and it sneaks up on you by stealth, the sign that it is happening is the fact that you are not bored by the unspectacular. (Hughes, 1999, p. 12.)

We stated at the outset that the study of behavior had become a transfer station for many scientists, not a place to take up permanent residence. In a "drive-thru" age, simple, particulate behaviors, with no social or ecological strings, are the easiest to package and sell. We hope to have made some progress in these pages in stimulating interest in arranging longer visits and creating bigger bundles. Such changes are needed to address what is disquieting about current integrative science. It is not the idea itself, but its present instantiation. That instantiation includes a remarkable confidence in downwardly spiraling reductionism, with only a nod toward the possibility that behavior in context is poorly studied and poorly understood.

And it is on this very point that the mission of the HHS *Genome* causes concern. We cannot expect molecular geneticists to know the fragile state of behavioral knowledge about concepts such as intelligence, aggression, imitation, or risk taking. These constructs may be especially attractive to molecular geneticists because they seem structurally similar to diseases, constructs with which the field is already familiar. Although diseases are softer categories than many might believe, they are clearly robust entities compared to behavioral phenomena. If those using genomics do not understand the fragility of behaviors, they may be not only wasting time seeking things that are not there, but they may also place society in a perilous state. The call for more behaviorists to work directly with geneticists will help, but the behaviorists will need a real backbone to resist restricting the range of phenomena and even the organisms (use mice, not rats, we are told) to those that geneticists see as tractable (Azar, 2000). If we use mice, we should use mice (plural) with work to do, with the tasks tapping the workings of ecology and evolution. Honeybee genomics is setting a fascinating example, as the natural busyness of bees has not been separated from the business of science (Page & Robinson, 1991).

We have argued strongly for creating methods and measures that capture socially emergent phenomena, behaviors co-defined by settings and by individuals. The reality of emergent outcomes is not hard to illustrate. Consider the matter of producing RBIs (runs batted in) by a hitter in baseball. RBIs require hitting ability, but also a pitcher, fielders, catcher, umpires, AND the performance of the previous batters. The matter may sound trivial when couched in terms of baseball (but see Gould, 1996, for evolutionary analyses). An RBI, however, is a direct measure of social dynamics and many of the measures of human behavior we seek to understand bear analogous properties. Now consider whether there could be a gene or genes for producing RBIs. Specifying the task dynamics for such a search swiftly exposes the asocial and nondynamic assumptions of the HHS *Genome*. The never-ending discussion among true sports fans of a player's or team's "statistics" reveals a respect for natural history we need to recover.

An emergent measure we discussed earlier in connection with cowbirds, persistence in courtship (CPL), requires singing ability but also requires receivers and rivals whose roles and abilities are as differentiated and as necessary as baseball teammates are to achieving an RBI. Ecological, temporal, social, and individual probabilities must combine to afford the possibility of the outcome. Thus, another way to state our reservations about the mapping of genes to behaviors, even in a metaphorical sense, is to suggest that it perpetuates the very dangerous idea that functional behaviors are *literally* self-contained. We believe that this is the mirage luring some would-be mappers onto the HHS *Genome*–the illusory view that the behavior we see before us is a property of the individual organism. We may be focusing at one moment in time on only this organism, but the contributions of those we are simultaneously ignoring are no less real.

And so we return to our original message in a bottle. Genes inherit a nested set of environments that lead to a succession of biological products, some of which are mappable to familiar categories such as cells, limbs, organs, reflexes, movements, etc. But, at all of these stages, effects from other organisms and environments co-occur—hidden correlates with stealth-like properties easily missed if we are not looking. But the effects are real and measurable. To abide by contextual rules will require an active effort to restructure our science. Part of that process will be looking at past work to see how the original need to simplify, reduce, or take out of context has affected knowledge as a whole. Part of the process will be recognizing that the lives of real organisms have somehow slipped out of the big picture. Part of the process will be the willingness to collide with present paradigms. At times, the forward momentum of big science makes these goals seem as improbable as being able to hold back an ocean with one's hands (Seaton, 1992). But the history of science is also a story of highly improbable successes and extraordinary amounts of effort. Confronting an ocean, however hard, is what science is about … not maintaining a mirage… however lovely, however grand.

Acknowledgments

The title of the chapter is from a poem by an unknown poet. MJW heard the line spoken in an NPR broadcast of "All Things Considered" and scribbled it down while driving to the airport to pick up Elliott Blass sometime around 1986. Any leads on the author would be greatly appreciated. The NSF funded the research. We thank D. J. White for comments and M. H. Goldstein for important discussions early in the process. But it was Elliott Blass' Brooklyn Dodgers-honed perseverance, cajoling, and comments that made the chapter a reality.

REFERENCES

Alberts, J. R. (1978). Huddling by rat pups: Group behavioral mechanisms of temperature regulation and energy conservation. *Journal of Comparative and Physiological Psychology, 92*, 231–240.

Azar, B. (2000). Wanted: behavioral researchers with a penchant for genetics. *Monitor on Psychology, 31*, 36–39.

Bar. (1997). *Dynamics of complex systems*. Reading, MA: Addison–Wesley.

Ball, G. F., & Hulse, S. H. (1998). Birdsong. *American Psychologist, 33*, 37–58.

Bolyard, K. J., & Rowland, W. J. (1996). Context-dependent response to red coloration in stickleback. *Animal Behaviour, 52*, 923–927.

Bottjer, S. W., & Arnold, A. P. (1986). The ontogeny of vocal learning in songbirds. In E. M. Blass (Ed.), *Handbook of behavioral neurobiology*, Volume 8, *Developmental psychobiology and developmental neurobiology*. (pp. 129–161). New York: Plenum Press.

Bottjer, S. W., Halsema, K. A., Brown, S. A., & Miesner, E. A. (1989). Axonal connections of a forebrain nucleus involved with vocal learning in zebra finches. *Journal of Comparative Neurology, 279*, 312–326.

Collins, F. (June 27, 2000). In N. Angier (Au.), A pearl and a hodgepodge: human DNA. *The New York Times*, p. 1.

Drickamer, L. C., Vessey, S. H., & Meikle, D. (1996). *Animal Behavior*. Dubuque, IA: Brown.

Eastzer, D. H., King, A. P., & West, M. J. (1985). Patterns of courtship between cowbird subspecies: Evidence for positive assortment. *Animal Behaviour, 33*, 30–39.

Freeberg, T. M. (1996). Assortative mating in captive cowbirds is predicted by social experience. *Animal Behaviour, 52*, 1129–1142.

Freeberg, T. M. (1997) *Cultural transmission of behaviors facilitating assortative courtship and mating in cowbirds (Molothrus ater)*. Ph.D. dissertation, Indiana University, Bloomington, Indiana.

Freeberg, T. M., King, A. P., & West, M. J. (1995). Social malleability in cowbirds (*Molothrus ater artemisiae*): Species and mate recognition in the first 2 years of life. *Journal of Comparative Psychology, 109*, 357–367.

Friedmann, H. (1929). *The Cowbirds: A study in the biology of social parasitism*. Springfield, IL: Thomas.

Galef, B. G. (1982). Studies of social learning in Norway rats: A brief review. *Developmental Psychobiology, 15*, 279–296.

Galef, B. G. (1996). Social enhancement of food preferences in Norway rats: A brief review. In C. M. Heyes & B. G. Galef (Eds.), *Social learning in animals: The roots of culture* (pp. 49–64). San Diego, CA: Academic Press.

Gibson, E. J. (1969). *Principles of perceptual learning and development*. New York: Appleton-Century-Crofts.

Gibson, J. J. (1966). *The senses considered as perceptual systems*. Boston: Houghton-Mifflin.

Giraldeau, L.-A., Caraco, T., & Valone, T. J. (1994). Social foraging: individual learning and cultural transmission of innovations. *Behavioral Ecology, 5*, 35–43.

Gopnik, M., & Goad, H. (1997). What underlies inflectional error patterns in genetic dysphasia? *Journal of Neurolinguistics, 19*, 109–137.

Gottlieb, G. (1971). *Development of species identification in birds*. Chicago: University of Chicago Press.

Gottlieb, G. (1993). Social induction of malleability in ducklings: Sensory basis and psychological mechanism. *Animal Behaviour, 45*, 707–719.

Gould, S. J. (1996). *Full house*. New York: Three Rivers Press.

Gould, S. J., & Lewontin, R. C. (1979). The spandrels of San Marco and the Panglossian paradigm: A critique of the adaptationist programme. *Proceedings of the Royal Society of London, Series B, 205*, 581–598.

Griffiths, P. E., & Neumann-Held, E. M. (1999). The many faces of genes. *BioScience, 49*, 656–662.

Hamilton, K. S., King, A. P., Sengelaub, D. R., & West, M. J. (1997). A brain of her own: A neural correlate of song assessment in a female songbird. *Neurobiology of Learning and Memory, 68*, 325–332.

Hamilton, K. S., King, A. P., Sengelaub, D. R., & West, M. J. (1998). Visual and song nuclei correlate with courtship skills in brown-headed cowbirds. *Animal Behaviour, 56*, 973–982.

Harlow, H. F. (1965). Mice, monkeys, men, and motives. In H. Fowler (Eds.), *Curiosity and exploratory behavior* (pp. 91–103). New York: Macmillan.

Harlow, H. F., & Harlow, M. K. (1962). Social deprivation in monkeys. *Scientific American, 207*, 136–146.

Hebb, D. O. (1949). *The organization of behavior*. New York: Wiley.

Hemingway, E. (1998). *True at first light*. New York: Scribner.

Hughes, R. (1999). *A jerk on one end: Memoirs of a mediocre fisherman.* New York: Ballantine.

Iacoboni, M., Woods, R., Brass, M., Bekkering, H., Mazziotta, J. C., & Rizzolatti, G. (1999). Cortical mechanisms of imitation. *Science, 286,* 2526–2528.

Johnston, T. D., & Gottlieb, G. (1985). Effects of social experience on visually imprinted maternal preferences in Peking ducklings. *Developmental Psychobiology, 18,* 261–271.

Jones, S. S. (1996). Imitation or exploration? Young infants' matching of adults' oral gestures. *Child Development, 67,* 1952–1969.

King, A. P., & West, M. J. (1977). Species identification in the North American cowbird: Appropriate responses to abnormal song. *Science, 195,* 1002–1004.

King, A. P., & West, M. J. (1983a). Dissecting cowbird song potency: Assaying a song's geographic identity and relative appeal. *Ethology, 63,* 37–50.

King, A. P., & West, M. J. (1983b). Female perception of cowbird song: A closed developmental program. *Developmental Psychobiology, 16,* 335–342.

King, A. P., & West, M. J. (1988). Searching for the functional origins of cowbird song in eastern brown-headed cowbirds (*Molothrus ater ater*). *Animal Behaviour, 36,* 1575–1588.

King, A. P., & West, M. J. (1989). Presence of female cowbirds (*Molothrus ater ater*) affects vocal improvisation in males. *Journal of Comparative Psychology, 103,* 39–44.

King, A. P., West, M. J., & Eastzer, D. H. (1986). Female cowbird song perception: Evidence for different developmental programs within the same subspecies. *Ethology, 72,* 89–98.

King, A. P., West, M. J., & Freeberg, T. M. (1996). Social experience affects the process and outcome of vocal ontogeny in two populations of cowbirds. *Journal of Comparative Psychology, 110,* 276–285.

Kroodsma, D. E. (1978). Aspects of learning in the ontogeny of bird song: Where, from whom, when, how many, which, and how accurately? In G. M. Burghardt & M. Bekoff (Eds.), *The development of behavior* (pp. 215–230). New York: Garland Press.

Kroodsma, D. E. (1982). Learning and the ontogeny of sound signals in birds. In D. E. Kroodsma & E. H. Miller (Eds.), *Acoustic communication in birds* (pp. 1–24). New York: Academic Press.

Kroodsma, D. E. (1989). Suggested experimental designs for song playbacks. *Animal Behaviour, 37,* 600–609.

Kroodsma, D. E. (1992). Much ado creates flaws. *Animal Behaviour, 44,* 580–582.

Kroodsma, D. E. (1996). Ecology of passerine song development. In D. E. Kroodsma & E. H. Miller (Eds.), *Ecology and evolution of acoustic communication in birds* (pp. 3–19). Ithaca, NY: Cornell University Press.

Kroodsma, D. E., & Baylis, J. R. (1982). Appendix: A world survey of evidence for vocal learning in birds. In D. E. Kroodsma & E. H. Miller (Eds.), *Acoustic communication in birds* (pp. 311–337). New York: Academic Press.

Kuo, Z. Y. (1967). *The dynamics of behavioral development: An epigenetic view.* New York: Random House.

Lehrman, D. S. (1970). Semantic and conceptual issues in the nature–nuture problem. In L. R. Aronson, E. Tobach, D. S. Lehrman, & J. S. Rosenblatt (Eds.), *Development and evolution of behavior: Essays in memory of T. C. Schneirla* (pp. 17–52). San Francisco: Freeman.

Lehrman, D. S. (1974). Can psychiatrists use ethology? In N. F. White (Eds.), *Ethology and psychiatry* (pp. 187–196). Toronto: University of Toronto Press.

Lickliter, R., & Gottlieb, G. (1987). Social specificity: Interaction with own species is necessary to foster species species-specific maternal preferences in ducklings. *Developmental Psychobiology, 21,* 311–321.

Lickliter, R., Dyer, A. B., & McBride, T. (1993). Perceptual consequences of early social experience in precocial birds. *Behavioural Processes, XX,* 1–16.

Lorenz, K. (1957). Companionship in bird life. In C. H. Schiller (Eds.), *Instinctive behavior: The development of a modern concept* (pp. 82–128). New York: International Universities Press.

Majerus, M. E. N. (1998). *Melanism: Evolution in action.* Oxford: Oxford University Press.

Marler, P., & Nelson, D. A. (1993). Action-based learning: A new form of developmental plasticity in bird song. *Netherlands Journal of Zoology, 43,* 91–103.

Mayr, E. (1974). Behavior programs and evolutionary strategies. *American Scientist, 62,* 650–659.

Michel, G. F., & Moore, C. L. (1995). *Developmental Psychobiology.* Cambridge, MA: MIT Press.

Millar, C., & Lambert, D. (1999). *BioScience, 49,* 1021–1023.

Nelson, D. A., & Marler, P. (1994). Selection-based learning in bird song development. *Proceedings of the National Academy of Sciences of the USA, 91,* 10498–10501.

Nixdorf-Bergweiler, B. E., Lips, M. B., & Heinemann, U. (1995). Electrophysiological and morphological evidence for a new projection of LMAN-neurons towards area X. *Neuroreport, 6,* 1729–1732.

Oyama, S. (1985). *The ontogeny of information: Developmental systems and evolution.* Cambridge: Cambridge University Press.

Page, J., R. E., & Robinson, G. E. (1991). The genetics of division of labor in honey bee colonies. *Advances in Insect Physiology, 23,* 118–169.

Payne, R. B., & Payne, L. L. (1997). Social learning of bird song: Field studies of indigo buntings and village indogobirds. In C. T. Snowdon & M. Hausberger (Eds.), *Social influences on vocal development* (pp. 57–84). Cambridge: Cambridge University Press.

Payne, R. B., Payne, L. L., Woods, J. L., & Sorenson, M. D. (2000). Imprinting and the origin of parasite–host species associations in brood parasitic indigobirds, *Vidua chalybeata. Animal Behaviour, 59,* 69–81.

Previde, P. E., & Poli, M. D. (1996). Social learning in the golden hamster (*Mesocricetus auratus*). *Journal of Comparative Psychology, 110,* 203–208.

Rowland, W. J. (1994). Proximate determinants of stickleback behavior. In M. A. Bell & S. A. Foster (Eds.), *The evolutionary biology the threespine stickleback* (pp. 297–344). Oxford: Oxford University Press.

Rozzi, R. (1999). The reciprocal links between evolutionary–ecological sciences and environmental ethics. *BioScience, 49,* 911–921.

Schank, J. C., & Alberts, J. R. (1997). Self-organized huddles of rat pups modeled by simple rules of individual behavior. *Journal of Theoretical Biology, 189,* 11–25.

Searcy, W. A. (1992). Measuring responses of female birds to male song. In P. K. McGregor (Eds.), *Playback and studies of animal communication* (pp. 175–189). New York: Plenum Press.

Seaton, M. (1992). *The sea among the cupboards.* Minneapolis, MN: New Rivers Press.

Shimizu, T., & Karten, H. J. (1993). The avian visual system and the evolution of the neocortex. In H. P. Ziegler & H.-J. Bishof (Eds.), *Vision, brain, and behavior in birds* (pp. 103–114). Cambridge MA: MIT Press.

Shimizu, T., Katz, J. S., & Cook, R. G. (1997). Effects of thalamic and telencephalic lesions on visual texture discrimination in birds. *Society for Neuroscience Abstracts, 23,* 453.

Slater, P. J. B. (1983). Bird song learning: Theme and variations. In A. H. Brush & G. A. Clark Jr. (Eds.), *Perspective in ornithology* (pp. 475–499). Cambridge: Cambridge University Press.

Sluckin, W. (1965). *Imprinting and early earning.* Chicago: Aldine.

Smith, L. B., Thelen, E., Titzer, R., & Mclin, D. (1999). Knowing in the context of acting: The task dynamics of the A-not-B error. *Psychological Review, 106,* 235–260.

Smith, V. I., King, A. P., & West, M. J. (2000). A role of her own: Female cowbirds influence male song development. *Animal Behavior, 60,* 599–609.

Smith, V. I., King, A. P., & West, M. J. (2001). The context of social learning: Association patterns in a captive flock of brown headed cowbirds. *Animal Behavior,* in press.

Suzuki, D., & Knudston, P. (1989). *Genetics: The ethics of engineering life.* London: Unwin & Hyman.

Terkel, J. (1996). Cultural transmission of feeding behaviors in the Black rat (*Rattus rattus*). In C. M. Heyes & B. G. Galef (Eds.), *Social learning in animals: The roots of culture* (pp. 17–47). San Diego, CA: Academic Press.

Thelen, E., & Smith, L. B. (1994). *A dynamic systems approach to the development of cognition and action.* Cambridge, MA: MIT Press.

Thorndike, E. L. (1911/1965). *Animal intelligence.* New York: Hafner.

Timberlake, W. (1993). Behavior systems and reinforcement: An integrative approach. *Journal of the Experimental Analysis of Behavior, 60,* 105–128.

Timberlake, W., & Silva, F. J. (1994). Observation of behavior, inference of function, and the study of learning. *Psychonomic Bulletin and Review, 1,* 73–88.

Wang, Y. C., Jiang, S., & Frost, B. J. (1993). Visual processing in pigeon nucleus rotundus: Luminance, color, motion, and looming subdivisions. *Visual Neuroscience, 10,* 239–254.

West, M. J., & King, A. P. (1980). Enriching cowbird song by social deprivation. *Journal of Comparative and Physiological Psychology, 94,* 263–270.

West, M. J., & King, A. P. (1985). Social guidance of vocal learning by female cowbirds: Validating its functional significance. *Ethology, 70,* 225–235.

West, M. J., & King, A. P. (1986). Song repertoire development in male cowbirds (*Molothrus ater*): Its relation to female assessment of song. *Journal of Comparative Psychology, 100,* 296–303.

West, M. J., & King, A. P. (1988a). Female visual displays affect the development of male song in the cowbird. *Nature, 334,* 244–246.

West, M. J., & King, A. P. (1988b). Vocalizations of juvenile cowbirds (*Molothrus ater ater*) evoke copulatory responses from females. *Developmental Psychobiology, 21,* 543–552.

West, M. J., King, A. P., & Eastzer, D. H. (1981). Validating the female bioassay of cowbird song: Relating differences in song potency to mating success. *Animal Behaviour, 29,* 490–501.

West, M. J., King, A. P., & Arberg, A. A. (1988). An inheritance of niches: The role of ecological legacies in ontogeny. In E. M. Blass (Ed.), *Handbook of Behavioral Neurobiology*, Volume 9, *Developmental psychobiology and behavioral ecology* (pp. 41–62). New York: Plenum Press.

West, M. J., King, A. P., & Duff, M. A. (1990). Communicating about communicating: When innate is not enough. *Developmental Psychobiology, 23,* 585–598.

West, M. J., King, A. P., & Freeberg, T. M. (1994). The nature and nurture of neophenotypes. In L. A. Real (Eds.), *Behavioral mechanisms in evolutionary ecology* (pp. 238–257). Chicago: University of Chicago Press.

West, M. J., King, A. P., & Freeberg, T. M. (1996). Social malleability in cowbirds: New measures reveal new evidence of plasticity in the eastern subspecies (*Molothrus ater ater*). *Journal of Comparative Psychology, 110,* 15–26.

West, M. J., King, A. P., & Freeberg, T. M. (1997). Building a social agenda for birdsong. In C. T. Snowdon & M. Hausberger (Eds.), *Social influences on vocal development* (pp. 41–56). Cambridge: Cambridge University Press.

Wilson, E. O. (1998). *Consilience: The unity of knowledge.* New York: Knopf.

Index

615